Air Pollution Modeling and Its Application II

NATO • Challenges of Modern Society

A series of edited volumes comprising multifaceted studies of contemporary problems facing our society, assembled in cooperation with NATO Committee on the Challenges of Modern Society.

Volume 1 AIR POLLUTION MODELING AND ITS APPLICATION I
Edited by C. De Wispelaere

Volume 2 AIR POLLUTION: Assessment Methodology and Modeling
Edited by Erich Weber

Volume 3 AIR POLLUTION MODELING AND ITS APPLICATION II
Edited by C. De Wispelaere

Volume 4 HAZARDOUS WASTE DISPOSAL
Edited by John P. Lehman

Air Pollution Modeling and Its Application II

Edited by

C. De Wispelaere

Prime Minister's Office for Science Policy
National Research and Development Program on Environment
Brussels, Belgium

Published in cooperation with
NATO Committee on the Challenges of Modern Society

PLENUM PRESS • NEW YORK AND LONDON

Library of Congress Cataloging in Publication Data

International Technical Meeting on Air Pollution Modeling and Its Application (12th: 1981: SRI International)
Air pollution modeling and its application II.

(NATO challenges of modern society; v. 3)
"Proceedings of the Twelfth International Technical Meeting on Air Pollution Modeling and Its Application, held August 25–28, 1981, at SRI International, Menlo Park, California"—Verso t.p.
Includes bibliographical references and index.
1. Air—Pollution—Mathematical models—Congresses. I. Wispelaere, C. de. II. North Atlantic Treaty Organization. Committee on the Challenges of Modern Society. III. Title. IV. Series: NATO challenges of modern society series; v. 3.
TD881.I59 1981 628.5′32′0724 82-13119
ISBN 0-306-41115-6

Proceedings of the Twelfth International Technical Meeting on
Air Pollution Modeling and Its Application, held August 25-28, 1981,
at SRI International, Menlo Park, California

Printed in the United States of America

PREFACE

In 1969 the North Atlantic Treaty Organization established the Committee on the Challenges of Modern Society. Air Pollution was from the start one of the priority problems under study within the framework of the pilot studies undertaken by this Committee. The organization of a yearly symposium dealing with air pollution modeling and its application is one of the main activities within the pilot study in relation to air pollution.

After being organized for five years by the United States and for five years by the Federal Republic of Germany, Belgium, represented by the Prime Minister's Office for Science Policy Programming, became responsible in 1980 for the organization of this symposium.

This volume contains the papers presented at the 12th International Technical Meeting on Air Pollution Modeling and its Application held at SRI International, Menlo Park, California in the USA from 25th to 28th August 1981. The meeting was jointly organized by the Prime Minister's Office for Science Policy Programming, Belgium and SRI International, USA. The conference was attended by 109 participants and 51 papers have been presented. The members of the selection committee of the 12th I.T.M. were A. Berger (Chairman, Belgium), W. Klug (Federal Republic of Germany), L.E. Niemeyer (United States of America), L. Santomauro (Italy), J. Tikvart (United States of America), M.L. Williams (United Kingdom), H. Van Dop (The Netherlands), C. De Wispelaere (Coordinator, Belgium).

The main topic of this 12th I.T.M. was Physical-chemical reactions in plumes. On this topic two review papers were presented : one paper dealing with "Chemical transformation in plumes" by C.S. Burton, M.K. Liu, P.M. Roth, C. Seigneur and G.Z. Whitten, Systems Applications, Inc.,U.S.A. and another one "A comparison between chemically reacting plume models and windtunnel experiments by P.J.H. Builtjes, T.N.O., The Netherlands.

Other topics of the conference were: air trajectory models for air pollution transport, advanced mathematical techniques in air pollution modeling, evaluation of model performance in practical applications and finally particular studies in the field of air pollution modeling.

On behalf of the selection committee and as organizer and editor I should like to record my gratitude to all participants who made the meeting so stimulating and the book possible. Among them I particulary mention the chairmen and rapporteurs of the different sessions. Thanks also to the local organizing committee, especially Dr. W.B. Johnson and Miss Valerie Ramsay who was the Conference Secretary. Finally it is a pleasure to record my thanks to Miss A. Vandeputte for preparing the papers, and Misses C. Bonnewijn, M.L. Koekelbergh and L. Mongaré for typing the papers.

C. De Wispelaere
Operational Director
of the National R-D
Program on Environment

CONTENTS

1: PHYSICAL-CHEMICAL REACTIONS IN PLUMES

Chairmen: W. Klug
P. Roth

Rapporteurs: R. Stern
P. J. H. Builtjes

CHEMICAL TRANSFORMATION IN PLUMES

C. Shepherd Burton, Mei-Kao Liu, Philip M. Roth,
Christian Seigneur, and Gary Z. Whitten

Systems Applications Inc.
101 Lucas Valley Road
San Rafael, California 94903

INTRODUCTION

It is now recognized that a variety of chemical transformations occurs in plumes. The types of transformations, their rates of occurrence, and the magnitudes of the resultant pollutant concentrations depend on the chemical composition of a plume and of the ambient air entrained in a plume, the elevation and temperature of a plume, and the meteorological conditions (including solar intensity) experienced by a plume. It is postulated that these transformations contribute to atmospheric haziness and discoloration, atmospheric acidity, and extended sulfate and ozone concentrations in rural areas. In addition, it is postulated that transformations in plumes contribute to brief, elevated ground-level concentrations of nitrogen dioxide.

The topic of chemical transformations in plumes, taken literally, is broadly encompassing; we cannot hope to cover all aspects of it--not even all the principal aspects. Thus,we have adopted some guidelines as to the areas that we will cover.

> Attention is primarily given to a review of chemistry and chemical transformations. The physical processes, such as transport and dispersion, are discussed only in terms of their influence on the chemistry. Deposition and scavenging processes are not considered.

> Emphasis is given to the review and synthesis of topics. Reporting of specific research studies is minimized.

> Topical areas are segmented as follows :

- Knowledge gained through observational (field) programs--section 2.
- Knowledge of atmospheric chemical reactions--section 3.
- The encoding of known chemical and physical processes in plume models--section 4.
- Research and development needs--section 5.

> The examination of atmospheric chemistry is organized according to the following groupings :

- Atmospheric SO_2 oxidation.
- The chemistry of nitrogen oxides in plumes.

> The focus, in terms of air quality impacts and thus reaction products, includes

- Ozone, NO_2, and photochemical oxidation products.
- Species potentially involved in visibility reduction.
- Precursors to the formation of acid precipitation.

Following the list of references at the end of the paper we have included a list of recommended readings that, taken together, constitute a comprehensive discussion of the topics treated in the paper.

OBSERVATIONAL PROGRAMS

This section presents an overview of observational programs that have as their focus the study of chemical transformations in plumes and the environmental effects of pollutants found in plumes. The section does not present an exhaustive review of plume field programs. Rather, it summarizes the findings of some of the primary programs that have provided the observational bases for improving the knowledge of plume chemistry and for developing and evaluating mathematical models. Table 1 lists some of these programs. Each program and the corresponding findings are briefly described in the remainder of this section.

Project MISTT (Midwest Interstate Sulfur Transformation and Transport)

The main objective of Project MISTT was to study the transformation of SO_2 to sulfate in plumes emitted from a power plant--the Labadie power plant, from refineries--the Wood River complex, and from an urban area--the St. Louis area. This study, sponsored by the U.S. Environmental Protection Agency (EPA), involved a large number of participants including academic institutions, private firms, and government agencies. The program took place during July and August of 1973 and 1974, February, July, and August, 1975, and July and August, 1976. It involved the use of two instrumented aircraft, an instrumented van, and three mobile, single-theodolite pilot balloon units.

The main results of Project MISTT can be summarized as follows:

1. Secondary pollutants, such as sulfates and ozone formed in plumes, may be transported over distances of 50 to 200 kilometers.

2. Ground-level concentrations of SO_2 emitted from tall stacks are less than those emitted from short stacks. However, SO_2 emissions from tall stacks lead to higher sulfate formation than do SO_2 emissions from short stacks because of diminished SO_2 deposition on the ground.

3. Visibility impairment is mainly due to secondary aerosol formation.

4. Because of adverse weather conditions that occurred during field studies, the processes involved in heterogeneous SO_2 oxidation could not be quantitatively assessed. However, homogeneous SO_2 oxidation processes could be studied with satisfactory accuracy.

Project CARB (California Air Resources Board)

In a study sponsored by the State of California Air Resources Board, ground-level measurements of plume concentrations of nitrogen oxides, ozone, sulfur dioxide, sulfates, nitrates, and SF_6 tracer were conducted within 19 km downwind of the Haynes steam plant and the Alamitos generating station. This field study differs from most of the others described here, in that it is concerned with power plant plumes in an urban environment. The principal conclusions of the study are :

1. The photostationary-state relationship among NO, NO_2 and O_3 was in agreement with the measurements.

2. Secondary sulfate formation was negligible during the first 90 minutes of plume transport.

3. Sulfuric acid formed in the plume probably displaced the ammonia/nitric acid/ammonium nitrate equilibrium, since lower ammonium nitrate concentrations were observed in the plume than in the background.*

Project EPRI (Electric Power Research Institute)

The Electric Power Research Institute contracted with Meteorology Research, Inc. (MRI), Systems Applications, Inc. (SAI), and the University of Washington Cloud and Aerosol Research Group to study (from 1975 to 1977), by means of airborne measurements, the chemical transformations that occur in coal- and gas-fired power plants. The power plants studied were the Four Corners near Farmington, New Mexico; Cunningham, at Hobbs, New Mexico; Wilkes near Longview, Texas; Centralia in Washington; and a power plant near Jefferson, Texas.

The conclusions drawn from the results of these field studies are discussed in the references given in table 1. They are summarized here :

> Net formation of ozone in power plant plumes is not common. The entrainment of ambient ozone and ozone precursors into the plume accounts for the observed plume ozone levels.

> The photostationary-state relationship among NO, NO_2 and O_3 was verified within the degree of uncertainty of the measurements.

> Conversion of SO_2 to aerosol sulfate was on the order of a few tenths of one percent for the Centralia power plant and a few percent for the Four Corners power plant. No conclusions were drawn as to the mechanisms that govern gas-to-particle conversion.

Project TREATS (Tennessee Regional Atmospheric Transport Study)

Project TREATS was sponsored by the EPA as part of the Federal Interagency Energy/Environmental Research and Development program and was initiated by the Tennessee Valley Authority. The main objectives of this project were to study the phenomena that affect regional pollutant levels and interregional transport of sulfur dioxide and sulfates. The first studies were conducted during the spring of 1976 and the summer of 1977. Seven coal-fired power plants are located in the region studied, i.e., the Tennessee valley region.

---

* Sulfuric acid formed in the plume reacts with the gas-phase ammonia to form ammonium sulfate. Then, the product of the HNO_3 and NH_3 partial pressures is less than the saturation value, and particulate ammonium nitrate decomposes.

Table 1. Some Observational Programs Focusing on Plume Chemistry, Physics, and Optics

Program	Date of Field Program	Plumes Studied	Principal Processes Studied	Sponsor	References
MISTT	1973–1976	Labadie power plant Wood River refinery complex St. Louis urban plume	Sulfate and ozone formation Secondary aerosol formation	EPA	Wilson (1978) White et al. (1976)
CARB	September 1974			CARB	Richards, Avol, and Harker (1976)
EPRI	1975–1977	Four Corners power plant Cunningham power plant Jefferson power plant Centralia power plant Wilkes power plant	Ozone-nitrogen oxides chemistry SO_2-sulfate chemistry	EPRI	Ogren et al. (1976) Hegg, Hobbs, and Radke (1976) Hegg et al. (1977) Hobbs et al. (1979)
TREATS	Spring 1976, Summer 1977	Seven power plants in the Tennessee valley region	Regional data--sulfur and sulfates	EPA	Crawford and Reisinger (1980)
STATE	August 1978	Cumberland steam plant	Sulfate formation	EPA	Schiermeier et al. (1979)
VISTTA	1978–1981	Four Corners power plant San Manuel smelter Navajo power plant Kincaid power plant LaCygne power plant Labadie power plant Douglas smelter	Plume chemistry Aerosol dynamics Plume visual effect	EPA	Zwicker et al. (1981) Blumenthal et al. (1981) Richards et al. (1981) Seigneur et al. (1980)
SEAPC	August 1980	Navajo power plant	Plume chemistry Aerosol dynamics Plume dispersion Plume visual effects	SRP	
MAP3S	1976–1981	Oak Creek power plant St. Louis urban plume	Sulfate and nitrate formation	DOE	Miller et al. (1978) MAP3S (1979)

Several conclusions regarding the long-range transport of sulfur compounds were drawn from these field programs. The principal findings can be summarized as follows :

1. Measurements of SO_2 concentrations generally showed lower values in the afternoon than in the morning. No such trend was observed for sulfate, nitrate, or ammonium ions.

2. The emissions from power plants located in the Tennessee valley region did not account for the total sulfur budget. Evidence strongly suggests that upwind anthropogenic sources and, possibly, biogenic sources contribute to the sulfur concentrations measured.

3. Sulfur dioxide, sulfate, nitrate and ammonium concentrations depend on mean direction of wind flow, time of day, and atmospheric stability.

Project STATE (Sulfur Transport And Transformation in the Environment)

The goal of the EPA-sponsored Project STATE was to investigate the effect on air quality of pollutants released far upwind and to relate the emissions of sulfur dioxides to the observed ambient levels of sulfates. Field measurements conducted during the program included sulfur dioxide, sulfates, ammonia, ammonium ion, total acidity, hydrocarbon species, nitrogen oxides, ozone, primary aerosols, and meteorological parameters. The first phase of the program--the Tennessee Plume Study (TPS)--which took place in August 1978, was designed to study the air quality impact of the Cumberland steam plant located at Cumberland City, Tennessee. Data analysis has not yet been completed. Plume transport, dispersion, transformation and removal processes will be analyzed and the data will be used for air quality model evaluation.

The second phase of the STATE program was concerned with the analysis of a Prolonged Elevated Pollution Episode (PEPE) caused by the stagnation of a high-pressure center over the northeastern United States. Although it was mainly a regional field study program, it involved the study of urban and large point source plumes for one to two days of transport.

Project VISTTA (Visibility Impairment due to Sulfur Transport and Transformation in the Atmosphere)

The EPA-sponsored VISTTA field programs were designed to study the relative contributions of natural and anthopogenic sources to visibility impairment and to provide a detailed data base that could be used for the evaluation of visibility models. The June-July and December 1979, field programs were conducted in the vicinity of a coal-fired power plant--the Navajo Generating Station near Page,

Arizona. The activities comprised studies of plume chemistry, plume visibility impairment, and regional haze. The data obtained included airborne and ground-level measurements of trace-gas chemistry, aerosol size distribution and chemical composition, meteorological parameters, emissions rates and characteristics, and telephotometer measurements of the plume visual effects.

The principal findings related to plume chemistry can be summarized as follows :

1. Primary nitrogen dioxide emissions are less than 6 percent of the total NO_x emissions.

2. Plume nitric oxide is transformed into nitrogen dioxide as ambient ozone is entrained into the plume.

3. Nitric acid is formed during the day mainly by the reaction of NO_2 with OH, at night by the reaction of NO_2 with O_3.

4. No aerosol nitrate formation was observed.

5. Sulfate formation was observed, mainly in the 0.01 to 0.1 μm aerosol diameter range at downwind distances of 25 to 100 km. Reaction of SO_2 with OH appeared to be the primary source of sulfate,and ammonium sulfate was probably the nucleating species.

The February 1981 VISTTA field program took place at the Kincaid power plant in Illinois and at the LaCygne power plant in Kansas. The data obtained during this program are currently being analyzed. The August 1981 VISTTA field program will be oriented toward the study of the long-range transport of the Los Angeles urban plume, plumes of smelters in Arizona, and the Labadie power plant plume in Missouri. In addition, the analysis of the 1978 VISTTA data obtained at the Four Corners power plant and at the San Manuel smelter is being completed (Zwicker et al., 1981).

Project SEAPC (Source Emissions and Plume Characterization)

The August 1980 SEAPC field program was sponsored by the Salt River Project (SRP) to study the chemistry, physics, and optics of the plume of the coal-fired Navajo Generating Station near Page, Arizona. The program consisted of making a coordinated set of measurements aimed at identifying the physics and chemistry of primary and secondary pollutants. Data were gathered concerning trace-gas concentrations (NO, NO_2, O_3, SO_2, HNO_3, NH_3), secondary sulfate and nitrate aerosol, aerosol size distributions, emissions rates and meteorological conditions. In addition, airborne measurements of plume geometry were obtained from lidar data, and ground-level telephotometer measurements of the plume visual effects were conducted.

The SEAPC program was designed to complete the data base obtained for the Navajo plume during the 1979 VISTTA field programs. The data are currently being analyzed; they will be used to evaluate reactive plume models.

Project MAP3S (Multistate Atmospheric Power Production Pollution Study)

The U.S. Department of Energy is currently commissioning this study to determine the distances over which air quality may be affected by power production sources, the relative importance of power plant emissions and other source emissions with respect to air pollution, the relative roles of nitrogen oxides and sulfur oxides in the formation of secondary pollutants, the presence of chemical species causing adverse health effects, the role of fly ash in neutralizing acidic particles, and the frequency of occurence of pollution episode conditions.

Although this study is regional, it includes the study of power plant and urban plumes as well. Ground-level and aircraft sampling were carried out in plumes and at a regional scale. The St. Louis urban plume and the Oak Creek power plant plume were studied in 1976. The plume of the Oak Creek power plant was sampled as far as 120 km downwind from the source. Under these long-range transport conditions, which included the urban plume of the Milwaukee area as the background, Miller et al., (1978) found that ozone levels exceeding those of the background were observed in the plume. The measured nonmethane hydrocarbon/NO_x ratio in the background was on the order of 30, a level sufficient for the formation of ozone when NO_x from the plume is added to the background air under high solar irradiation. It was also suggested that the conversion rates of SO_2 to sulfate may be higher under urban plume conditions than in relatively clean areas. Although transformation of SO_2 and NO_x to sulfates and nitrates was shown to occur in concentrated plumes, it was concluded that most of the oxidation of SO_2 and NO_x occurs at low concentrations in dilute plumes or at the regional scale. This suggests that the major oxidation processes for SO_2 and NO_x involve chemical species from the background air, such as OH radicals and O_3. The determining factor for high sulfate and nitrate levels at the regional scale is therefore the oxidation of SO_2 and NO_x in dilute plumes rather than in concentrated plumes.

Principal Findings

The observational programs reported here led to some important findings regarding the chemical transformations occuring in plumes and their interactions with plume dynamics. In this section, we summarize the program results that made significant contributions to our knowledge of plume chemistry.

The results of the MISST program provided evidence of the transformation of SO_2 to sulfates during the long-range transport of plumes and, in addition, it showed that aerosol sulfate formation occurs in the nuclei and accumulation modes, i.e., at aerosol sizes of less than about 3 μm in diameter. Furthermore, it appears that sulfate formation is slow near the source and that its rate of formation increases as the plume is transported downwind and becomes mixed with ambient air. It was also concluded that elevated emissions of SO_2, such as those from tall stacks, lead to greater sulfate formation than ground-level emissions since the residence time of SO_2 is longer in an elevated plume as a result of reduced deposition on the ground (Wilson, 1978). These results were later confirmed by other plume studies.

The VISTTA program showed that slow rates of sulfate formation may be observed if the plume is stable and if it is released in a clean and dry environment. The dependence of the sulfate formation rate on the total amount of solar radiation received by the plume (as found in other important studies) was fairly evident in the 1979 VISTTA results (Wilson and McMurry, 1981). Although this finding suggests the importance of homogeneous oxidation of SO_2 in sulfate formation, uncertainty in the measurements precludes ruling out other mechanisms.

The interactions of plume nitrogen oxides and ambient ozone have been of substantial interest because of the possibility of photochemical smog formation in plumes. In the EPRI studies conducted in the southwestern United States, no ozone formation in excess of background levels was observed (Ogren et al., 1976; Hegg, Hobbs, and Radke, 1976). However, in the midwestern and eastern United States, ozone formation in plumes in excess of background levels has been measured (Davis, Ravishankara and Fischer, 1979; Miller et al., 1978; and Meagher et al., 1981). These studies found that when the plume is transported and diluted far downwind, i.e., at distances on the order of 500 km, ambient hydrocarbons are mixed with the nitrogen oxides in the plume, thereby allowing photochemical smog chemistry to take place.

The formation of NO_2 in a plume is limited by the mixing of the plume NO with the background O_3. Plume studies, such as the VISSTA and EPRI programs, have shown that the NO, NO_2 and O_3 concentrations in the plume are mainly governed by the following reactions :

$$NO + O_3 \rightarrow NO_2 + O_2$$

$$O + O_2 + M \rightarrow O_3$$

$$NO_2 \xrightarrow{h\nu} NO + O$$

These reactions occur fast enough to establish a steady state--the so-called photostationary state--between NO, NO_2, and O_3. Such a state has been verified in plume measurements (White 1977; Richards et al., 1981).

When ozone exceeding background levels are observed in the plume (under conditions in which the plume mixes with polluted background air), NO_2 formation is no longer limited by O_3 diffusion into the plume, but rather by photochemical smog chemical kinetics. However, the photostationary-state relationship among NO, NO_2 and O_3 still applies.

The formation of HNO_3 in plumes has been a subject of growing concern because of its possible effect on the acidity of precipitation. The VISTTA studies have shown that HNO_3 is formed during the day mainly by the reaction of plume NO_2 with background OH radicals. The formation rate of nitric acid is about 3 to 10 times faster than that of sulfate (Richards et al., 1981). Recent studies suggest that HNO_3 is also formed at night at rates on the order of a few percent per hour when the plume NO_2 reacts with the background O_3 to form NO_3 radicals (which ultimately lead to HNO_3 formation).

Finally, current research is being carried out to investigate the different pathways involved in the oxidation of SO_2 to sulfate, the processes leading to nitric acid formation at night, and NO, NO_2, and O_3 chemistry under varying background concentrations of hydrocarbons, solar irradiation, temperature and humidity.

The plume study projects discussed here do not represent all the observational programs that have been conducted in the past few years to investigate chemical transformations occuring in atmospheric plumes. Furthermore, field studies are still being performed to improve our understanding of chemical processes as they occur in plumes. Some of those that are currently in progress or that are being analyzed include VISTTA, MAP3S, the Cold Weather Plume Study (which was conducted in February 1981 at the Kincaid power plant in Illinois to study plume chemistry under cold weather conditions), and the Tennessee Valley plume studies.

ATMOSPHERIC CHEMISTRY

This section summarizes the current knowledge of the atmospheric chemical reactions of SO_2 and NO_x (and their transformation to sulfate, NO_2, nitrates and O_3) as determined from field programs and laboratory experiments. The following sections discuss the regional aspects of chemistry where "clean tropospheric" chemistry and "polluted-atmosphere" chemistry must both be taken into account. Finally, the chemical models used to describe those atmospheric transformations are discussed.

Atmospheric SO_2 Oxidation

The recent review by Newman (1981), as well as most other reviews that were previously carried out, divides the oxidation of SO_2 into homogeneous and heteorogeneous pathways. In this context, heterogeneous implies any process not exclusively occuring in the gas phase. Historically, primary concern has focused on sulfate. However, sulfite has received some attention because of studies that suggest the occurence of adverse health effects resulting from this species (Gunnison and Palmes, 1973) and because some atmospheric measurement methods cannot distinguish between sulfate and sulfite. Even though significant quantities of SO_2 may be removed from the atmosphere by heteorogeneous means, key compounds may be involved in the removal processes that were generated in the atmosphere by homogeneous chemistry. Specifically, ozone and hydrogen peroxide have recently been shown to play significant roles in the SO_2 oxidation that can occur in water droplets (Penkett et al., 1979), and the presence of high NO_2 concentrations has been postulated to influence the oxidation of SO_2 as well (Cofer, Schryer, and Rogowski, 1980). Yet, these oxidants are generated by the smog-type reactions that occur in the gas phase throughout the troposphere (Bufalini, Gay, and Brubaker, 1972).

Homogeneous SO_2 Chemistry

A recent review of homogeneous SO_2 chemistry by Atkinson and Lloyd (1981) extended the review by Calvert et al. (1978). The review by Calvert et al. (1978) explored a wide range of reactions and identified the most significant reactions. The brief review by Atkinson and Lloyd (1981) used information presented in recent studies to aid in reducing the number of reactions that appear to significantly affect SO_2 oxidation. However, the current literature indicates that one pathway ($RO_2 + SO_2$) suggested for consideration by Calvert et al. (1978), but considered to be insignificant in the atmosphere by Atkinson and Lloyd (1981), is still of uncertain importance (Kan, Calvert, and Shaw, 1981). Atmospheric observations are discussed in the aforementioned review by Newman (1981), who examined the mechanisms that can be supported by recent data. These three sources form the basis for the discussion that follows. Only three pathways appear to be in any way significant as gas-phase sinks for SO_2.

Reaction with the hydroxyl radical (OH). Many studies seem to indicate that this reaction may be the most significant pathway to sulfate in the atmosphere. However, both the rate constant for the reaction of atmospheric conditions and the series of steps necessary to produce sulfate in the atmosphere through this reaction are difficult to determine in laboratory studies.

The currently recommended atmospheric rate constant for the SO_2-OH reaction is 1500 ppm^{-1} min^{-1} at 298K, with a temperature dependence factor of $(T/298)^{-2.7}$ and an uncertainty factor of 1.5 (Atkinson and Lloyd, 1981). Newman (1981) has estimated the fraction of sulfate formed via homogeneous (gas phase) chemistry in several power plant plumes; the uncertainty factor in these fractions was also estimated to be 1.5. The uncertainty in the rate constant is mainly a consequence of the need to extrapolate laboratory findings to atmospheric conditions.

The sequence of steps involved in converting SO_2 to H_2SO_4 involve H_2O and presumably NO_x and HO_2 (Davis, Ravishankara and Fischer, 1979). For the practical purpose of atmospheric modeling, accounting for the role of H_2O is not important since the only product observed to date is H_2SO_4, which rapidly appears when even trace quantities of H_2O present (Niki et al., 1980). However, the nature of the involvement of NO_x and HO_2 are still uncertain; nevertheless, these species and OH are apparently important to the formation of both sulfate and ozone. How these two "formation chemistries" might affect each other is an area requiring further study.

The reaction of oxygen atoms $O(^3P)$ with SO_2. As pointed out by Calvert et al. (1978), under rare conditions such as those involving power plant plumes containing high NO_2 concentrations and low O_3 concentrations near the stack, this reaction may be important. Under typical atmospheric conditions, however, the reaction is of minimal importance. The near-stack conditions require daylight and high NO_2 concentrations (20 ppm). These conditions can support production of concentrations of $O(^3P)$ atoms sufficient to generate an oxidation rate of 1.5 percent per hour. This temporary near-stack rate is actually comparable to that of the OH + SO_2 pathway at a time when the plume has proceeded downwind.

H_2SO_4 is formed via the reaction sequence

$$O(^3P) + SO_2 + M \ (M = \text{air}) \rightarrow SO_3 + M$$

$$SO_3 + H_2O + M \rightarrow H_2SO_4 + M$$

Atkinson and Lloyd (1981) recommend a rate constant of

$$K^T = 49 \times \exp\left[1009(1/298 - 1/T)\right] \ ppm^{-1} \ min^{-1}$$

for the first reaction, which represents the rate-limiting step.

The reaction of SO_2 with intermediates in O_3-alkene chemistry
This reaction pathway can occur only in the presence of both O_3 and

alkenes. Calvert et al. (1978) and Atkinson and Lloyd (1981) both suggest that even under such conditions, this pathway is of minor importance, though the authors acknowledge that considerable uncertainty exists. Subsequent to the report of Niki et al. (1977), several studies have theorized that SO_2 oxidation in the presence of O_3 and alkenes occurs via biradicals that form from the products of O_3-alkene reactions :

$$RR'C = CR''R''' + O_3 \rightarrow RCR'OO\cdot + \text{products}$$

$$RCR'OO\cdot + SO_2 \rightarrow RR'CO + SO_3$$

However, the biradicals are expected to react faster with NO and NO_2 than with SO_2, as discussed by Atkinson and Lloyd (1981), who also propose several other possible competing pathways for reactions of the biradicals. Unfortunately, the relative rate constants for all of the biradical possibilities are not known at this time.

Heterogeneous SO_2 chemistry

Heterogeneous SO_2 is more difficult to characterize than homogeneous chemistry, since the steps leading to sulfate involve microphysical processes that interact with the chemistry. The state of knowledge, therefore, is less well developed than that for gas-phase chemistry. Recent reviews concerning the atmospheric oxidation of SO_2 via heterogeneous pathways were carried out by Peterson and Seinfeld (1980); Middleton, Kiang, and Mohnen (1980); Kaplan, Himmelblau and Kanaoka (1981); and Newman (1981). The approach taken in each was quite different. The review of Peterson and Seinfeld (1980) is organized according to type of pathway, whereas the review of Newman (1981) is organized according to the type of plume occuring where observations were made. The latter also included discussions related to homogeneous oxidation in the same plume.

In some studies, the observed SO_2 oxidation could be explained on the basis of the OH homogeneous pathway discussed earlier. For example, Wilson and McMurry (1981) showed a linear relationship between the formation of sulfate and the total ultra-violet insolation. The observations of Hegg and Hobbs (1978) and those of Roberts and Williams (1979) could also be completely explained on the basis of hydroxyl radical chemistry. Recent results reported by Blumenthal et al. (1981) involving clean, dry conditions and those reviewed by Gillani, Kohli, and Wilson (1981) involving "polluted", humid conditions are also consistent with the premise of a homogeneous SO_2 oxidation pathway dominated by the SO_2 and OH reaction. Nevertheless, many observations seem to require consideration of heterogeneous chemistry. These observations usually involve aerosols near the source and water droplets such as fog or clouds elsewhere.

Heterogeneous oxidation is believed to occur on particle surfaces and in the liquid phase. Oxidation on aerosol surfaces has been reported to occur on carbon (Brodzinsky et al., 1980) and is enhanced by acidity and the presence of NO_2 (Cofer, Schryer, and Rogowski, 1980). Oxidation of fly ash (Mamane and Pueschel, 1979) as well as freshly emitted urban aerosols has been studied (Judeikis, Steward, and Wren, 1978). However, all the reactions may involve adsorbed water films (Newman, 1981).

The key mechanisms identified in aqueous oxidation involve hydrogen peroxide, ozone, and other materials such as metals. Central to all aqueous mechanisms are the rate and amount of SO_2 that enter the liquid phase from the gas phase. On the one hand, SO_2 solubility decreases with acidity and on the other hand, oxidation to form sulfuric acid increases the acidity. A complicating factor exists for wet surfaces, since sulfuric acid draws more water onto the surface, thereby decreasing the acidity. Yet another complexity stems from the recently discovered phenomenon of enhanced solubility observed in the surface of growing water droplets (Matteson, 1978).

Liquid-phase oxidation by H_2O_2. Penkett et al. (1979) showed that the rate of oxidation of SO_2 actually increased with acidity. The reaction was found to be first order in dissolved sulfite and hydrogen peroxide. Under simulated atmospheric conditions, Penkett et al. (1979) also demonstrated that the H_2O_2 mechanism could be important in the formation of cloud droplet acidity. Other known mechanisms tend to decrease sharply in effectiveness with increased acidity.

Reaction with ozone. Penkett et al. (1979) showed that ozone could be more effective than H_2O_2 above pH 6 in oxidizing SO_2. Larson, Horike, and Harrison (1978) concluded that in clouds (H_2O about 0.6 g/m^3), the ozone mechanism could produce SO_2 oxidation rates of 1 to 4 percent per hour.

Reactions with other species. As discussed by Peterson and Seinfeld (1980), mechanisms involving other species are often lengthy and the derived rate expressions are largely empirical. The reactions attributed to the presence of metal ions are presented in Table 2, which was taken from the review of Peterson and Seinfeld (1980). According to the review of Newman (1981), rigorous models for heterogeneous chemistry have not yet been constructed that compare in scientific quality to the rigorous models developed for homogeneous chemistry. Much of the observed data now available can be simulated using homogeneous models, though for the highest reported SO_2 oxidation rates, heterogeneous mechanisms must still be postulated. Historically, lack of consideration of heterogeneous chemistry has been suggested as an explanation for the differences

Table 2. Metal Ion-Catalyzed. Liquid-Phase Oxidation of SO_2

Type of Mechanism	Rate Coefficient and/or Expression	Comments	Ref.
Cu^{2+} catalyst; mannitol inhibitor	$k_s = 0.013 + 2.5[Cu^{2+}]$	25 °C	56
Metal salts		Formation of complexes such as $[O_2\ Mn(SO_3)_2]^{2-}$ and rapid oxidation	
Fe^{2+} catalyst with and without NH_3		Conversion rate = 1.8×10^{-4} % min^{-1}	62
Metal salts; $2SO_2 + 2H_2O + O_2 \rightarrow 2H_2SO_4$	SO_2 conversion rate = 0.09% min^{-1} for Mn, 0.15-1.5% min^{1} for Fe	Theoretical study; rates for Mn and Fe depend on many factors; rate for Fe-catlyzed oxidation is pH dependent	63
SO_2 oxidation catalyzed by metal salts; $Mn^{2+} + SO_2 \rightleftarrows Mn \cdot SO_2^{2+}$ $2MN \cdot SO_2^{2+} + O_2 \rightleftarrows [(Mn \cdot SO_2^{2+})_2 \cdot O_2]$ $\rightleftarrows 2MN \cdot SO_3^{2+}$ $HSO_4^- + H^+ \rightleftarrows H_2SO_4$	$-\frac{d[SO_2]}{dt} = k_1\ [Mn^{2+}]_0^2$ $k_1 = 2.4 \times 10^5\ M^{-1}\ sec^{-1}$	Neligible SO_4^{2-} formation for RH < 95%; similar mechanism may be responsible for catalysis by other metal salts	64
SO_2 oxidation catalyzed by NH_3; $2SO_2 + 2H_2O + O_2 \xrightarrow{catalyst} 2H_2SO_4$	SO_2 conversion rate ~ 0.03% min^{-1} with Mn^{2+} levels typical of urban industrial atmosphere; ~ 0.33% min^{-1} with levels typical of plume from coal-powered plant	Oxidation rate estimated by extrapolation to atmospheric conditions	35
Sulfite oxidation catalyzed by cobalt ions; free radical mechanism; CO(III) reduced	$-\frac{d[SO_2]_g}{dt} = k[CO(H_2O)_6^{3+}]^{1/2}[SO_3^{2-}]^{3/2}$	Could not determine specific value for k	65
SO_2 oxidation by O_2 with trace Fe catalyst	$-\frac{d[S(IV)]}{dt} = k[Fe(III)][S(IV)]$ $k = 100\ M^{-1}\ sec^{-1}$; SO_2 conversion rate ~ 3.2% day^{-1} in fog assuming 28 $\mu g/m^{-3}$ SO_2 and 10^{-6}M Fe(III)	Possibility of Fe(III) contamination discussed	52
SO_2 oxidation catalyzed by Fe	$\frac{d[SO_4^{2-}]}{dt} = K_0K_S^2[H_2SO_3]^2[Fe^{3+}]/[H^+]^3$ K_s = 1st dissociation constant of H_2SO_3	Rate increases rapidly with RH and decreases by about one order of magnitude with 5°C increase in temperature	66
SO_2 oxidation catalyzed by Fe	Same as above, except K_0 a complex function of $[Fe^{3+}]$	Rate dependence changes from $[SO_2]^2/[H^+]^3$ to $[SO_2]/[H^+]$ as pH or $[SO_2]$ increases	67
Mn and Fe catalysts	$\frac{d[SO_4^{2-}]}{dt} = k[SO_2]_g$	8°C and 25°C; 2.1-mm-diam droplets; 10^{-6}-10^{-4} M for Mn and Fe; SO_2 concentrations 0.01 - 1.0 ppm; in pH range 2 - 4.5 the catalytic effectiveness was $Mn^{2+} > Fe^{2+} > Fe^{3+}$; increase in T from 8°C to 25°C caused an increase in MN^{2+} catalyzed oxidation rate of 5 - 10 in pH range 2 - 4.5	68
Oxidation in homogeneous aqueous phase of rain water; metal concentrations 10^{-7}-10^{-6} M for Mn and 10^{-6}-10^{-5} M for Fe	$\frac{d[SO_4^{2-}]}{dt} = k_0[SO_2]_g$ $k_0 = 1.95 \times 10^{16} \exp - \frac{23,000}{RT}$	Rainwater pH 3.2 - 5.2	

Source: Peterson and Seinfeld (1980).

in results between observed data and models containing only homogeneous chemistry. The extent of "necessary" heterogeneous chemistry needed to adequately simulate observations tends to vary with the choice of the homogeneous mechanism and the rate selected for the OH and SO_2 reaction.

The Chemistry of Nitrogen Oxides in Plumes

In this section we discuss NO_x chemistry, starting with the emitted species--NO and NO_2-- and examine the possible chemical pathways to nitrates (HNO_3, NH_4NO_3, etc.) and ozone. Most of the NO_x in plumes is emitted as NO and is then converted to NO_2 by chemical reactions. For ease of conceptualization, we trace the chemical processes as the plume proceeds downwind, since the relative importance of different reactions is governed, to a large extent, by the age of the plume. Eventually, the NO_x is converted to nitrates, in some instances to peroxyacetylnitrate (PAN); except for PAN, the nitrates are inert and represent the species eventually removed from the atmosphere. In our discussions, we follow the progress of the plume as it is transported downwind: first, NO-to-NO_2 conversions are considered, with the principal reactions involving NO and O_2; then reactions involving NO, NO_2 and O_3; finally, reactions involving hydrocarbons. Second, inorganic and organic nitrate formation are considered.

The NO-NO_2-O_2 Chemistry near the Source*

High concentrations of NO react with molecular oxygen to produce NO_2 by means of the molecular reaction

$$2NO + O_2 \rightarrow 2NO_2 \qquad (1)$$

This reaction rate is only slightly temperature-dependent, increasing as the temperature decreases. However, the rate depends on the square of the NO concentration, so the importance of Reaction 1 decreases rapidly as the NO concentration decreases. Consequently, this reaction is important near the stack during the early stages of plume dilution. The amount of NO converted to NO_2 by this process can be on the order of 5 to 10 percent.

The NO-NO_2-O_3 Reactions

As a plume moves downwind, ambient ozone is entrained into the plume. The reaction of NO_x in the plume with ozone at night is different from that which occurs during the day. The reaction among NO, NO_2 and O_3 is simplest at night because photolysis does not occur. NO reacts rapidly with O_3 to form NO_2 :

* Sources that emit both NO_x and hydrocarbons are not considered in this section

$$NO + O_3 \rightarrow NO_2 + O_2 \quad (2)$$

Because of this reaction and others, including the reaction of O_3 with surfaces, the O_3 concentration near the ground decreases during the night. However, the higher O_3 concentrations formed in urban atmospheres during daytime conditions can persist through the night in a stable layer of air above the surface, where the O_3 is not exposed to either the ground or ground-based NO emissions.

When the sun rises, the reactions occuring at night continue, but NO_x chemistry becomes more complex because photolysis reactions begin. Among the most important reactions is

$$NO_2 + h\nu \rightarrow NO + O \quad (3)$$

Most of the oxygen atoms (O) produced in Reaction 3 react with oxygen molecules to produce O_3 :

$$O + O_2 + M \rightarrow O_3 + M \quad (4)$$

Reactions 2, 3, and 4 constitute a cycle involving NO, NO_2, and O_3. This cycle is presented schematically in the upper portion of Figure 1 (the effects of hydrocarbons are discussed later). Most of the time during daylight hours when sufficient NO_x is present, the concentrations of NO, NO_2, and O_3 are related, to within perhaps 10 percent, according to a steady-state equation given by

$$k_3 [NO_2] = k_2 [NO] [O_3]$$

or

$$[O_3] = \frac{k_3 [NO_2]}{k_2 [NO]} \quad (5)$$

where k_3 and k_2 are the rate constants for Reactions 3 and 2, respectively. Since k_3 depends on the intensity of sunlight, which varies diurnally, seasonally, and spatially, the steady-state concentrations of NO, NO_2 and O_3 depend on solar intensity.

Since NO represents the majority of the nitrogen oxides emissions in plumes, O_3 levels in plumes will be depleted below background levels, since NO reacts rapidly with the entrained O_3. As NO is converted to NO_2 with increasing distance downwind, O_3 levels will increase in accordance with Equation 5. When the plume becomes dispersed, plume ozone concentrations will tend to be the same as the background concentration, and the ratio of NO_2 and NO concentrations in the dilute plume is then given by Equation 5. The validity of the photostationary-state relationship during the daytime has been demonstrated in plumes within the level of uncertainty of the measurements.

At night, NO_2 photolysis does not provide the O_3 production pathway. Consequently, the O_3 entrained in the plume is depleted by reaction with NO (reaction 2) and with NO_2 :

$$NO_2 + O_3 \rightarrow NO_3 + O_2 \qquad (6)$$

Ozone Generation

As mentioned earlier, ozone concentrations above background levels have been observed in plumes at distances far downwind of the source. Generation of ozone in the troposphere is a photochemical process resulting from the chemical interaction between nitrogen oxides and hydrocarbons. Since hydrocarbon concentrations must be sufficiently high to induce smog formation, ozone formation in plumes occurs when the plume mixes with background air rich in hydrocarbons. The chemistry of photolytically induced ozone formation is summarized next.

The oxidation of hydrocarbons in smog is essentially a photochemically driven, combustion-like process that culminates in the production of CO_2 and H_2O. After an initial oxidation step, an aliphatic hydrocarbon (one that does not contain double bonds or aromatic rings) disintegrates by a stepwise process like the highly simplified process shown in Figure 2. In this figure, R represents an alipathic hydrocarbon group such as methyl (CH_3-) or ethyl (CH_3CH_2-). Note that the reaction sequence in this figure is nearly cyclical, but the final aldehyde, (R-1)CHO, has one less $-CH_2$-group than the initial aldehyde, RCHO. Different types of hydrocarbons enter this stepwise oxidation process at different points : aliphatic hydrocarbons (called paraffins) react with O or OH· to produce $RCO_3^{\cdot}$ and $RO_2^{\cdot}$ in various amounts; aldehydes react with OH·, as shown in Figure 2. This sequence omits many reactions, but it shows that for each $-CH_2$ group oxidized to CO_2, three molecules of NO can be converted to NO_2. Conversion of NO to NO_2 increases the O_3 concentration. Numerous research studies discuss ozone formation in plumes (see, for example, Niki, Daby, and Weinstock, 1972 ; Demerjian, Kerr and Calvert, 1974; Hecht, Seinfeld and Dodge, 1974; Nicolet, 1975; Calvert and McQuigg, 1975; Chameides and Walker, 1976 ; and Whitten, Hogo and Killus, 1980).

By the process shown in Figure 2, nonmethane hydrocarbons (NMHC) convert NO to NO_2 and generate a host of reactive intermediates. NMHC thus plays two important roles in NO_x chemistry :

> By converting NO to NO_2, NMHC increases O_3 formation from a given amount of NO_x.

> By generating intermediates that react with NO_x, NMHC increases the rate at which NO_x is consumed.

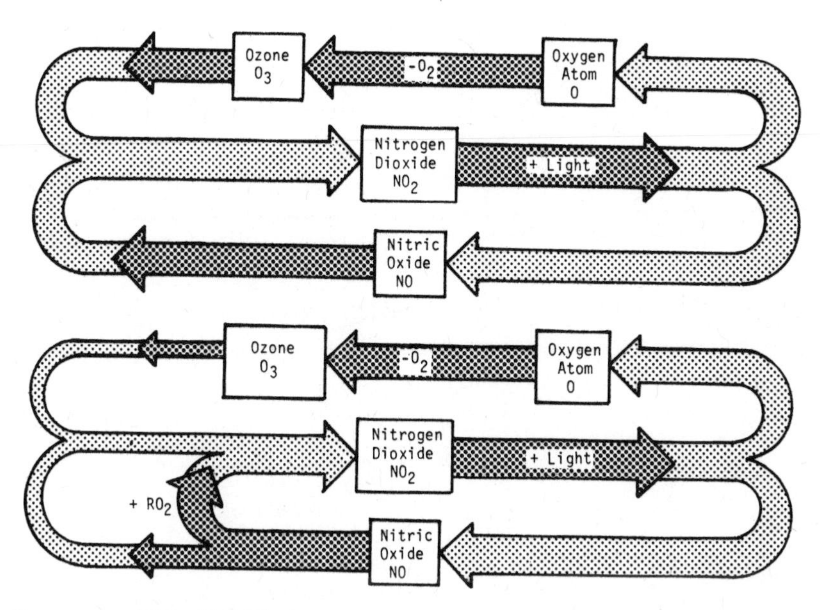

Source : NAS (1977).

Figure 1. The $NO-NO_2-O_3$ Cycle in Air Contaminated with NO_x only (Above) and with NO_x and Hydrocarbons (Below). Above, the dissociation of nitrogen dioxide by sunlight forms equal numbers of nitric oxide molecules and oxygen atoms. The latter are rapidly converted to ozone molecules. The ozone then reacts with the nitric oxide, again on a 1:1 basis, to reform nitrogen dioxide. Only a small steady-state concentration of ozone results from this cycle. Below, when hydrocarbons, aldehydes, or other reactive contaminants are present, they can form peroxy radicals that oxidize the nitric oxide, pumping it directly to nitrogen dioxide. This leaves very little of the nitric oxide to react with the ozone, so the ozone builds up to large concentrations.

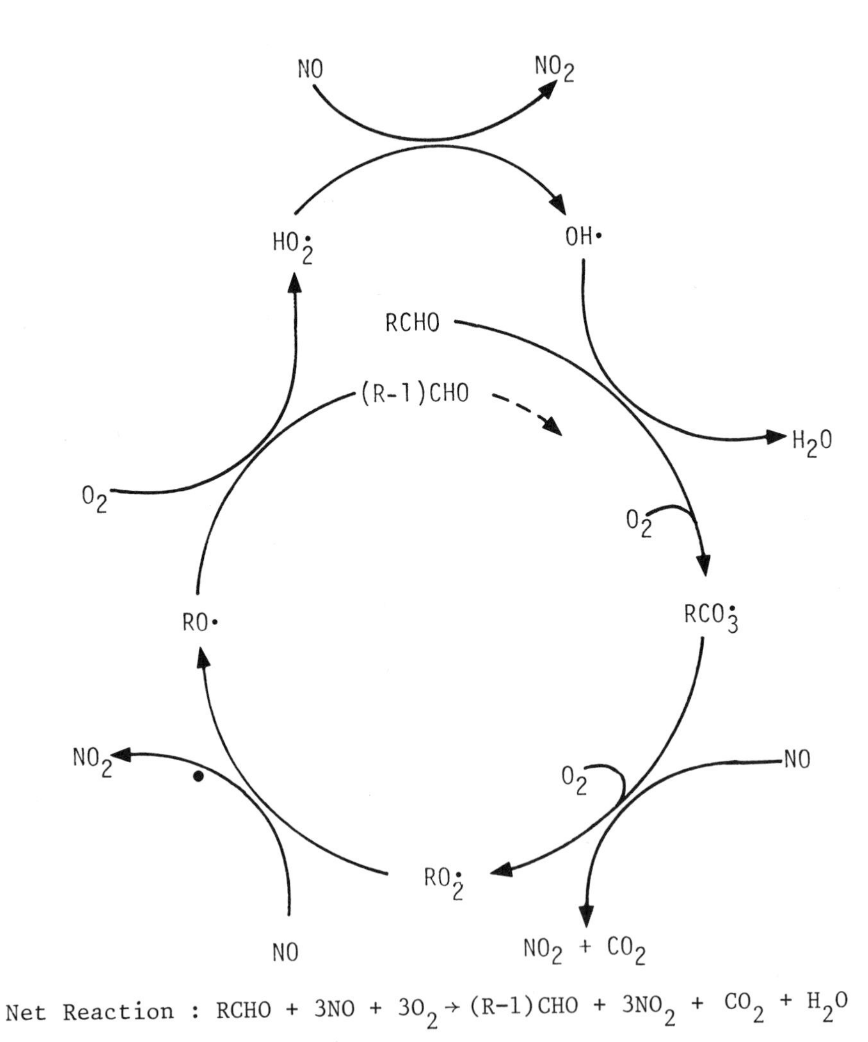

Net Reaction : $RCHO + 3NO + 3O_2 \rightarrow (R-1)CHO + 3NO_2 + CO_2 + H_2O$

Figure 2. Simplified Description of Hydrocarbon Oxidation in Smog

The Chemistry of HNO_3 and Nitrates

A primary removal mechanism, or sink, for NO_2 in plumes in daylight periods appears to be its reaction with OH· (hydroxyl radical), which leads to the formation of nitric acid :

$$OH\cdot + NO_2 \rightarrow HNO_3 \qquad (7)$$

During the day, when OH· is present, NO_2 may be consumed at a rate of 10 percent per hour by this reaction. Another possibly important sink for NO_2 is the sequence of reactions

$$NO_2 + O_3 \rightarrow NO_3 + O_2 \qquad (6)$$

$$NO_3 + NO_2 \rightarrow N_2O_5 \qquad (8)$$

$$N_2O_5 + H_2O \rightarrow 2HNO_3 \qquad (9)$$

This reaction sequence occurs even at night, and during the daylight period, the following reactions also occur :

$$NO_3 + h\nu \rightarrow NO + O_2 \qquad (10)$$

$$NO_3 + h\nu \rightarrow NO_2 + O \qquad (11)$$

$$N_2O_5 \rightarrow NO_3 + NO_2 \qquad (12)$$

The rate constant of Reaction 6 is a factor of 40 lower than the rate of Reaction 2 :

$$NO + O_3 \rightarrow NO_2 + O_2 \qquad (2)$$

Therefore, little NO_3 is formed when the NO_2-NO ratio is low. The importance of Reaction 9 in the atmosphere is uncertain because it occurs both in the gas phase and on surfaces. Surface reactions may be important under some circumstances, but this type of chemistry is difficult to verify or model. An upper limit for the rate constant of Reaction 9 has recently been estimated for the total of gas and heterogeneous conversion in the nighttime Los Angeles atmosphere by Platt et al. (1980a). Their value of 2×10^{-6} ppm^{-1} min^{-1} corresponded to a loss rate of NO_2 to nitric acid of about 30 percent per hour.

Formation of nitric acid in plumes has also been observed at night in several instances (Richards et al., 1981). A possibly important consequence of Reactions 6, 8, and 9 is that in urban areas, although nitric acid formation near the surface essentially ceases, it can continue aloft. Nitric acid formation ceases at the surface during periods of darkness for two reasons : the rate of Reaction 7 becomes zero in the absence of sunlight; the rate of

Reaction 6 becomes extremely slow or zero because during the evening period Reaction 2 consumes all of the available O_3. However, above the ground, Reaction 6 can continue during the evening hours because there is a continuous source of ozone from the air outside the plume that can be entrained into the expanding plume.

Although the chemistry of atmospheric NO_3^- aerosols is discussed in more detail in the next section, it is necessary to consider it here as well, because gas-phase reactions that lead to nitrates are important to overall gas-phase chemistry. The reactions by which NO_x is transformed to aerosols or to gas-phase nitrates usually involve the loss of some radical species in addition; thus, such reactions can have a moderating effect on the rate of smog formation.

Figure 3, taken from Orel and Seinfeld (1977), outlines the paths of HNO_3 and NO_3^- formation in aerosols. We have just discussed path 1--the gas-phase photochemical formation of HNO_3 by Reaction 7 and Reaction 9. Paths 2 and 3 describe the reaction of HNO_3 with NH_3 and the direct absorption of HNO_3 into particles. Path 3 is important only in the presence of a liquid aerosol; i.e., at humidities above about 62 percent (the deliquescence point of NH_4NO_3). Note that the presence of sulfuric acid in an aerosol droplet might lower the pH sufficiently to convert nonvolatile inorganic nitrates into volatile HNO_3, thus reducing the total NO_3^- concentration (Harker, Richards, and Clark, 1977). Path 4 involves formation of NO_3^- within the droplet after absorption of NO and NO_2; the kinetics is uncertain. Path 5 relates to direct absorption of NH_3. Organic nitrates, formed by path 6, are discussed below.

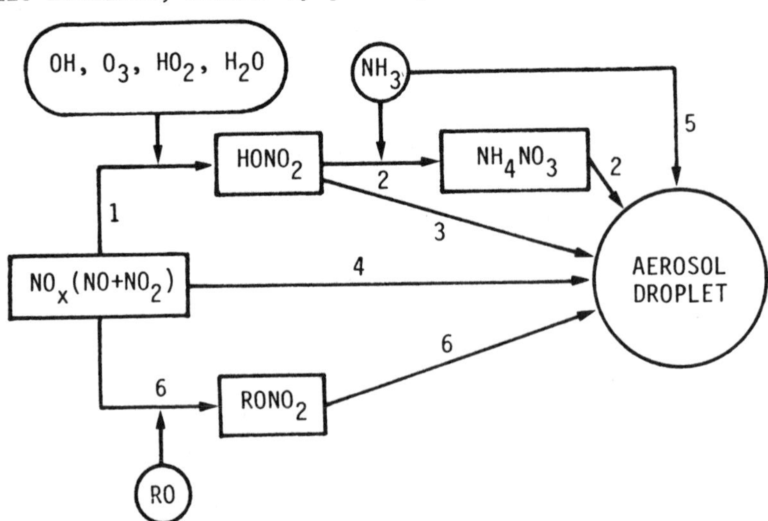

Source : After Orel and Seinfeld (1977)

Figure 3. Nitrate Formation Mechanisms in Atmospheric Aerosols

Reactions of NO_x with reactive organic intermediates include the following :

$$RO_2 + NO \rightarrow RONO_2 \quad (R \geqslant C_4) \tag{10}$$

$$RCO_3^{\cdot} + NO_2 \rightleftarrows RCO_3NO_2 \rightarrow \text{Sink} \tag{11}$$

$$RO_2^{\cdot} + NO_2 \rightleftarrows RO_2NO_2 \tag{12}$$

$$RO\cdot + NO \rightarrow RONO \tag{13}$$

$$RONO + h\nu \rightarrow RO\cdot + NO \tag{14}$$

$$R{=}R' + NO_3 \rightarrow NO_x, \text{ dinitrates} \tag{15}$$

$$\text{Cresols} + NO_3 \rightarrow \text{Nitrocresols}, HNO_3 \tag{16}$$

Reaction 10, which produces organic nitrates, is known to be important for aliphatic R groups equal to, or larger than, C_4 (Darnal et al., 1976). Although this type of reaction has not been reported for R groups other than pure aliphatics, Atkinson et al. (1980) expect that reactions similar to Reaction 10 are important to the chemistry of aromatics. Reaction 11 produces peroxyacylnitrates, of which peroxyacetylnitrate is the most common. Many of these compounds, which are powerful eye irritants, react with vegetation. The peroxynitrates that form as a result of Reaction 12 are apparently only temporary sinks for NO_x, since the rapid back-reactions merely reproduce the reactants; however, little is known of other possible reactions of the peroxynitrates, whose competition with the back-reaction could increase the importance of peroxynitrates as a sink for NO_x.

Reactions 12 and 13 both act as temporary radical sinks or dampers during the daytime. For instance, a sudden increase in the number of OH radicals will be damped by the formation of HONO, which will then photolyze to release radicals as shown in Reaction 14. HONO is not expected to be a primary sink of NO_x because during the day, most of the HONO formed photolyzes to produce NO. At night little HONO is formed because little OH· is present. Another HONO formation reaction,

$$NO + NO_2 + H_2O \rightarrow 2HONO \tag{17}$$

is slow in the gas phase. This reaction occurs more rapidly on particles; its importance as an NO_x sink is uncertain. However, Platt et al., (1980b) have observed the buildup of HONO during the night. At dawn, HONO can serve as an important source of OH radicals, though the amount of NO_x involved ($\lesssim$ 5 ppb) may be minor.

Reaction 15 eventually leads to dinitrate formation as shown by Akimoto et al. (1978), but the actual mechanism involved is still uncertain (Atkinson and Lloyd, 1981). Cresols and phenols, which are intermediates in the oxidation of aromatics, can react via Reaction 16. Such reactions are fast (Carter, Winer, and Pitts, 1981), and they constitute a significant NO_x sink when aromatics are present (Atkinson et al., 1980).

NO_2 can also react with alkoxyl radicals, RO·, to form alkyl nitrates :

$$RO\cdot + NO_2 \rightarrow RONO_2 \tag{18}$$

Unlike Reaction 10, which increases in rate with increased size of the R group, Reaction 18 seems to be independent of R (Batt, McCulloch and Milne, 1975). Alkyl nitrates apparently are not formed by Reaction 18 in concentrations in photochemical smog that are high compared with concentrations formed through the pathway involving Reaction 10. For example, Altshuller et al. (1967) found a maximum of 0.05 moles of methyl nitrate formed per mole of propylene consumed in a series of smog chamber experiments. Because simple alkyl nitrates exhibit high vapor pressures, they are expected to remain in the gas phase rather than in aerosols. However, O'Brien, Holmes and Bockian (1975) suggested that formation of such alkyl nitrates would account for the observed inhibition of aerosol formation in the presence of high NO_2 concentrations.

Conclusions

At this writing, it can be concluded that the photochemical gas-phase sulfur dioxide oxidation pathway involving OH-radical attack has been identified as being important, perhaps explaining the greater portion of the observed SO_2 oxidation rates in humid conditions and all of the SO_2 oxidation in dry, clean conditions. When some form of liquid water is present, the heterogeneous pathway involving hydrogen peroxide appears to be important as well.

Since both the key homogeneous pathway and the key heterogeneous pathway to sulfate involve species influenced by the ozone chemistry (OH and hydrogen peroxide), atmospheric sulfate concentrations must necessarily be a function of the level of control imposed on the emissions of ozone precursors. However, atmospheric modeling studies of typical scenarios are needed before generalizations can be made as to the expected effectiveness of controlling hydrocarbon and nitrogen oxides emissions (relative to the control of SO_2 emissions) in reducing sulfate concentrations.

Oxidation of NO to NO_2 is reasonably well understood both near and far downwind of the source. Away from the source, NO_2 formation is dominated by the reaction of NO and O_3 and by the resultant steady-state relationship among NO, NO_2, O_3 and sunlight. At farther downwind distances, the treatment of NO_2 and O_3 formation requires consideration of the influence of hydrocarbons whenever these species are present.

The principal chemical sinks of NO_2 are (inorganic and organic) nitrates. Inorganic nitrates consist primarily of nitric acid, and, in the presence of sufficient quantities of NH_3, it appears that inorganic nitrates may be formed under both daytime and nighttime conditions. Organic nitrate formation appears important only in the presence of hydrocarbons and under daylight conditions.

CODIFICATION OF CURRENT KNOWLEDGE : A REVIEW OF REACTIVE PLUME MODELS

A variety of mathematical models has been developed and utilized over the past decade to simulate the physical and chemical processes that take place in a plume. These models, adopting either an Eulerian fixed coordinate or a Lagrangian moving system, are designed to address one or more observed characteristics of the plume with varying degrees of emphasis. They range from simple analytic solutions to complex numerical models. To provide a framework for sorting out the large number of plume models discussed in the literature, a brief description of the evolution of a plume seems to be in order.

As shown in Figure 4a, effluents emanating from an industrial stack generally undergo five stages of evolution related to physical aspects of the plume, which according to Hilst (1978) can be identified as follows.

> First stage--plume rise primarily due to the momentum flux of the exhaust stream.

> Second stage--plume rise and entrainment due to the buoyancy of the exhaust stream.

> Third stage--a transitional zone between the buoyancy-dominated plume and the ambient-turbulence-dominated plume.

> Fourth stage--plume dispersion primarily due to ambient turbulence.

> Fifth stage--horizontal transport and dispersion due to ambient wind and turbulence.

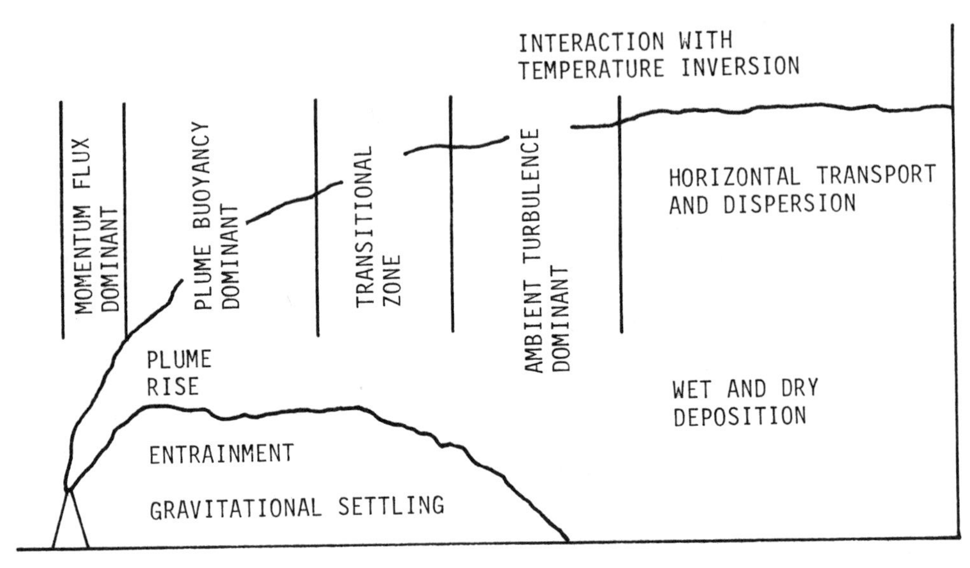

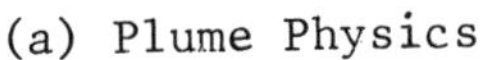
(a) Plume Physics

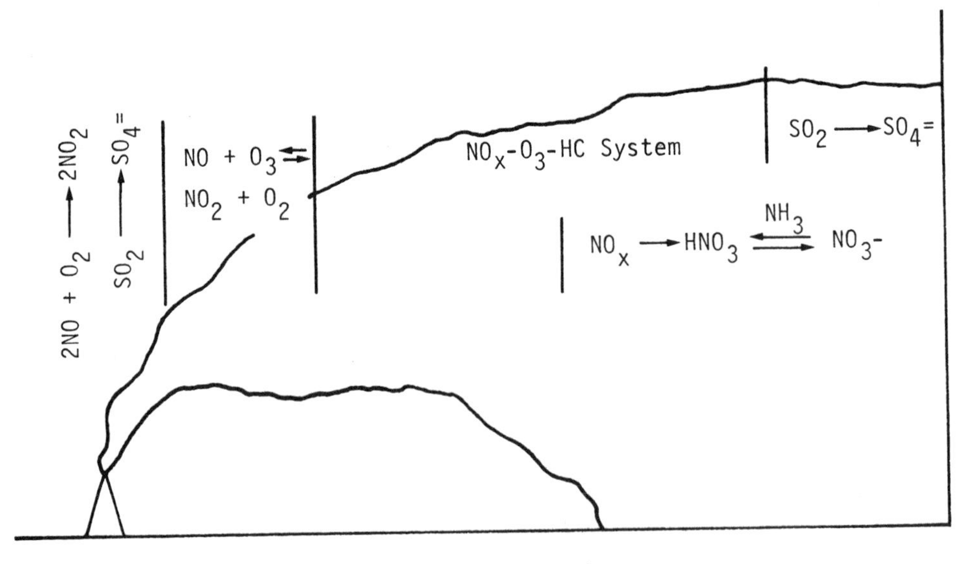

(b) Plume Chemistry

Figure 4. Evolution of Plume Physics and Plume Chemistry

Attendant gravitational settling of heavy particles and wet and dry deposition of gaseous and particulate pollutants will also occur at appropriate stages of plume development.

The corresponding chemical evolution of the plume is shown in Figure 4b. These processes were discussed in detail in the previous section.

Catalog of Plume Models

The various physical and chemical processes operative in the atmosphere play differing roles at each stage of plume evolution. The ultimate goal of the plume model is to simulate the significant processes and their interactions at each stage. This section presents a general overview of the development of atmospheric plume models. This overview is restricted to physically based models. Furthermore, only models employing short averaging times--on the order of one hour-- are considered.

For the purpose of stratifying the numerous plume models that have been developed, a triangulum has been prepared (see Figure 5). Each of the three apexes of this triangulum denotes one of the three principal processes :

- Dynamics
- Chemistry
- Turbulence.

Included in the first category are plume rise, plume transport, and in a macroscopic sense, dispersion. In this regard, the Gaussian plume model assembled by Turner (1970)--indicated as A at the lower left apex of the triangulum--has now become a standard plume model for treating chemically inert species. Considering only a few fast and reversible chemical reactions (e.g., those involving NO, NO_2 and O_3), Peters and Richards (1977), Forney and Giz (1980a and b), Forney and Heffner (1981) and deBower (1976) have developed reactive plume models based on the Gaussian framework, the entrainment theory and the numerical approach, respectively.

At the lower right apex of the triangulum, gas-phase chemical kinetic mechanisms and heterogeneous aerosol models applicable to both plume and urban situations are exemplified by models developed by Whitten, Hogo and Killus (1980) and Basset, Gelbard and Seinfeld (1981). Slightly to the left of this apex, reactive plume models are indicated that can accomodate a full set of chemical kinetics.

To simulate plume dynamics, these models adopt a simple box approach (Levine, 1980; Gelinas and Walton, 1974); retain the Gaussian modeling framework (Stewart and Liu, 1981; Seigneur, 1982; Hov and Isaksen, 1981; and Lusis, 1976); or use K-theory in a gridded framework (Eltgroth and Hobbs, 1979).

The interaction of chemical reactions with turbulence has raised considerable interest since the study of Donaldson and Hilst conducted in 1972. Concentration inhomogeneities caused by turbulent fluctuations may yield a slower apparent rate for the fast reaction between NO and O_3 as it becomes diffusion-limited (Eschenroeder and Martinez, 1970).

At the top of the triangulum is the representation of turbulent dynamics as exemplified by models developed by Lumley (1978). Application to buoyant plume problems of the gradient-transport approach and a turbulent closure scheme has been carried out, for example, by Liu et al. (1976) and Yamada (1979), respectively. A more general model based on the invariant second-order closure method was proposed by Donaldson (1969) and has been applied to buoyant plume problems by Teske, Lewellen and Segun (1978). Finally, Hilst et al. (1973) extended this approach to include chemical reactions and to effectively join all three processes. Although the potential of their proposed modeling technique has yet to be fully explored, it does represent a well-balanced approach to treat both the physics and chemistry of plumes. As a final note in the treatment of turbulence and chemistry, all of the aforementioned models adopt a momentum formulation based on Reynolds decomposition and ensemble averages. It is interesting to note that O'Brien, Meyers and Benkovitz (1976) have developed a reactive plume model employing an approach based on the probability density function (p.d.f.). Although the p.d.f. would contain spectral information equivalent to an infinite set of moments, practically it can handle only single bimolecular chemical reactions.

A summary of the plume models discussed here is presented in Table 3. The table identifies the treatment of chemistry in each model, i.e., NO_x and SO_2 chemistry and secondary aerosol formation; the physical treatment of the plume, i.e., the technique used to treat plume dispersion (Gaussian model, K-theory, simple box model, or a more detailed treatment of the interaction between turbulence and chemical reactions); and the spatial resolution offered by the model (1-D corresponds to crosswind plume average, 2-D to vertically well-mixed plumes, and 3-D to a full spatial resolution). Some model may be more appropriate for the treatment of near-source processes (e.g., vertical resolution and the interactions between turbulence and chemical reactions), whereas others will apply better at large downwind distances (e.g., assuming vertically well-mixed plumes).

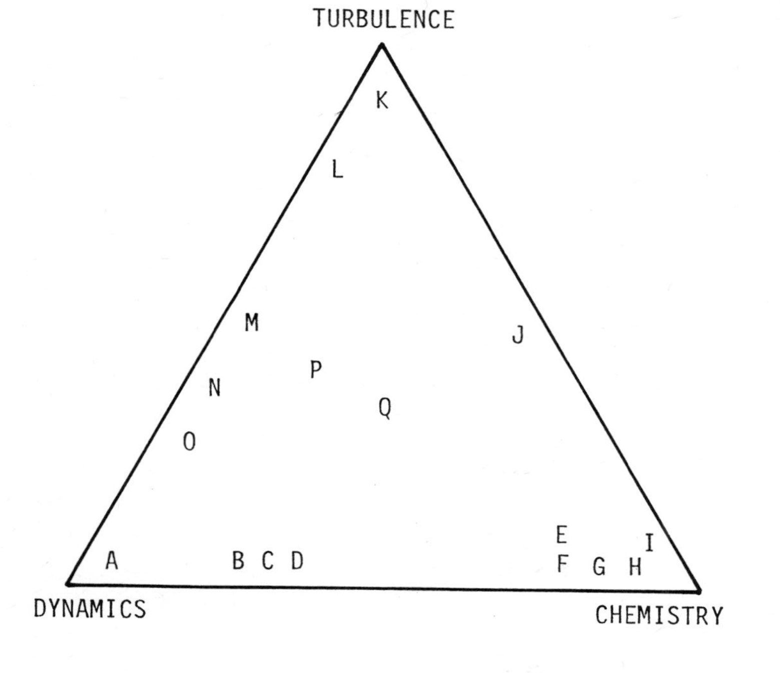

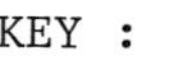
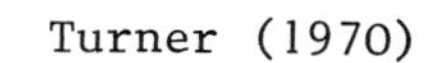

KEY :

A Turner (1970)

B Peters and Richards (1977)

C Forney and Giz (1980a, b); Forney and Heffner (1981)

D deBower (1976)

E Stewart and Liu (1981)

F Gelinas and Walton (1974)

G Levine (1980)

H Whitten, Hogo and Killus (1980)

I Basset, Gelbard and Seinfeld (1981)

J Donaldson and Hilst (1972)

K Lumley (1978)

L Mathieu and Gence (1978)

M Donaldson (1973); Teske et al.(1978)

N Yamada (1979)

O Liu et al. (1976)

P O'Brien, Meyers and Benkovitz (1976)

Q Hilst et al. (1973)

Figure 5. A Triangular Diagram for Plume Models

TABLE 3. Summary of Plume Models

Model	Chemistry*			Physics	
	NO_x Chemistry	SO_2 Chemistry	Aerosol Formation	Plume Dispersion	Model Dimensionality
Turner (1970)	-	First-order oxidation	-	Gaussian	3-D
Peters and Richards (1977)	3 reactions steady-state	-	-	Gaussian	3-D
Forney and Giz (1980b)	3 reactions steady-state	-	-	Box model	1-D
Forney and Heffner (1981)	1 hypothetical bimolecular reaction		-	Entrainment theory	3-D
deBower (1976)	Seinfeld (1971) 11 reactions	-	-	K-theory	3-D
Stewart and Liu (1981), Seigneur (1982)	Carbon-Bond Mechanism 70 reactions		Sulfate aerosol	K-theory	2-D
Lusis (1976)	General		-	Gaussian	2-D
Hov and Isaksen (1981)	General		-	Gaussian	2-D
Eltgroth and Hobbs (1979)	37 reactions		Sulfate aerosol	K-diffusion	3-D
Gelinas and Walton (1974)	41 reactions (Strato-spheric)	-	-	Box model	1-D
Levine (1980)	45 reactions		-	Box model	1-D
Whitten, Hogo, and Killus (1980)	Carbon-Bond Mechanism 68 reactions		-	Box model	1-D

TABLE 3 (Concluded)

	Chemistry*			Physics	
Model	NO_x Chemistry	SO_2 Chemistry	Aerosol Formation	Plume Dispersion	Model Dimensionality
Basset, Gelbard, and Seinfeld (1981)	-	First-order oxidation	Multicomponent sulfate aerosol	Box model	1-D
Donaldson and Hilst (1972)	1 hypothetical bimolecular reaction		-	Turbulence-reaction interaction	1-D
Shu, Lamb, and Seinfeld (1978)	3 reactions steady-state	-	-	Turbulence-reaction interaction	1-D
Carmichael and Peters (1981)	3 reactions steady-state	-	-	Turbulence-reaction interaction	1-D
Lumley (1978)	-	-	-	Turbulence closure method	3-D
Mathieu and Gence (1978)	-	-	-	Turbulence closure method	3-D
Donaldson (1969)	-	-	-	Turbulence closure method	3-D
Teske, Lewellen, and Segun (1978)	-	-	-	Turbulence closure method	3-D
Yamada (1979)	-	-	-	Turbulence closure method	3-D
Liu et al. (1976)	-	-	-	K-theory	3-D
O'Brien, Meyers, and Benkovitz (1976)	1 hypothetical bimolecular reaction		-	Spectral method	3-D
Hilst et al. (1973)	General			Turbulence closure method	3-D

Treatment of Chemistry

Gas-Phase Chemistry Models

Mathematical modeling of atmospheric chemistry (described in the prior section) has been the subject of considerable effort in the past decade. We consider here a brief overview of models of gas-phase and aerosol chemistry that have been proposed in recent years. These models can be cast into three broad categories :

1. Simple kinetic models, which offer an analytical solution for the chemical species concentrations.

2. Detailed kinetic models, which describe in detail the chemical mechanisms of specific species that can be used as surrogates to represent the complex mixtures in the atmosphere.

3. Lumped kinetic models, which describe atmospheric chemistry by lumping classes of hydrocarbons according to either their chemical characteristics (e.g., paraffins, olefins, aromatics, aldehydes, and others) or their chemical structures (e.g. types of chemical bonds).

Next, we briefly discuss these approaches, giving particular attention to their applications in the study of plume chemistry.

Simple kinetic models use pseudo-first-order rates and quasi-steady-state approximations to represent a limited set of chemical reactions. Examples of this approach in plume modeling are the representation of SO_2 oxidation to sulfate by a pseudo-first-order reaction (e.g., Alam and Seinfeld, 1981) and the photostationary-state approximation for the NO-NO_2-O_3 triad (e.g. Peters and Richards, 1977; Forney and Giz, 1980b).

Detailed models have been developed to study the kinetic behavior of specific compounds such as propylene, n-butane, acetaldehyde, and formaldehyde, as well as to treat the inorganic chemistry. Examples of such mechanisms are those developed by Demerjian, Kerr and Calvert (1974), Carter et al. (1979) and Whitten, Killus and Hogo (1980). These mechanisms have been evaluated by means of smog chamber data and have performed satisfactorily. However, the number of reactions involved can be on the order of several hundred. The mechanisms simulate actual atmospheric phenomena. Thus, their size often precludes their use in air quality models that must also take into account transport and dispersion processes.

An example of the use of a detailed mechanism is the Empirical Kinetics Modeling Approach (EKMA) developed by the EPA. The chemistry of complex urban hydrocarbon mixtures is treated through inclusion of the detailed chemistry of a simple propylene-butane mixture.

This type of approach has certain advantages : the assurance of mass conservation of carbon and the elimination of uncertainty and confusion introduced by the need to speficy parameters (such as the fraction of a reactant following reaction path 1 versus path 2) that cannot be estimated or derived easily through implementation calculations. The disadvantage of the approach stems from the uncertainty and confusion introduced in relating various complex mixtures of hydrocarbons to some combination of propylene and butane.

More concise chemical mechanisms, which still provide a fairly accurate description of atmospheric chemistry, have been developed for inclusion in air quality models. These models (generally referred to as lumped mechanisms) use some procedure for aggregating species, and in certain instances, reactions to reduce the total number of reactions considered. These broad classes of lumped mechanisms can be defined as follows.

Lumped mechanisms based on a single surrogate species for all hydrocarbons (e.g. Hecht and Seinfeld, 1972; Martinez and Eschenroeder, 1972). These mechanisms have not performed well because they have not been able to describe the wide range of reactivities among hydrocarbons.

Lumped mechanisms based on several surrogate species in which each species represents a class of hydrocarbon--e.g., paraffins, olefins, aromatics, aldehydes, or biogenic hydrocarbons. Hydrocarbon representation through this species class approach was first adopted by Hecht, Seinfeld, and Dodge (1974) and has been used more recently by Falls and Seinfeld (1978) and by K. L. Demerjian (private communication, 1979). This approach can give satisfactory results when compared with smog chamber and atmospheric data. The disadvantages of this approach stem primarily from the difficulty associated with proper use of such mechanisms. Gelinas and Skewes-Cox (1977) addressed this usage problem for a revision of the mechanism by Hecht, Seinfeld, and Dodge (1974).

Source categories of a wide range of hydrocarbons have to be related to the three surrogate species used in such mechanisms. The parameters originally used to test such mechanisms may simultaneously provide good simulations of some smog chamber data and good carbon or mass balance. However, the necessary use of average molecular weights or surrogate equivalents to convert source categories into the lumped species also requires adjustment of the parameters to achieve and maintain carbon balance. On the other hand, if the parameters are adjusted to provide optimum performance of ozone generation for some smog chamber or atmospheric data, then carbon conservation may be achieved. For regional modeling over large distances and long times, attention to carbon conservation is important.

Lumped mechanisms based on the chemical structure of the species. This concept, well known in chemical kinetics, has been used by Whitten, Hogo, and Killus (1980) in the development of the Carbon-Bond Mechanism. In this approach, the hydrocarbons are represented through their chemical bonding structure (single paraffinic bonds, double olefinic bonds, double aromatic bonds, carbonyl groups). This approach offers several advantages : carbon mass is conserved and a reasonable representation of the range of reactivities of hydrocarbons is provided. Carbon balance is achieved by explicitly specifying all carbon atoms when converting source categories into the lumped species (unreactive hydrocarbons are not treated in the chemical mechanism but can be carried along in the overall model). The mechanism then maintains carbon balance explicitly by assigning a specific number of carbon atoms to each lumped species used in the mechanism.

Specification of source categories is more straightforward than in the lumped surrogate mechanisms, as shown by the example of octene-1. Six of the eight carbon atoms are single bonded, and yet the molecule is technically an olefin. The CBM treats this molecule as six paraffinic carbon atoms and two olefinic carbon atoms. Directions for use of the lumped surrogate mechanisms differ, but this molecule might be treated as a pure olefin or as a mixture of olefin and paraffin. Either treatment leads to individual, arbitrary, and sometimes confusing approaches to handling the reactivity, intermediate products, and carbon balance.

The Carbon-Bond Mechanism-- as well as other atmospheric chemistry mechanisms--is being updated continuously on the basis of new information and the findings of evaluations that utilize smog chamber data (Killus and Whitten, 1981).

Aerosol Models

Models of secondary aerosol formation can be cast into three broad categories :

1. Simple lumped aerosol models involving only a few parameters.

2. Detailed aerosol models based on the general dynamic equation (GDE).

3. Lumped aerosol models based on the GDE.

Simple lumped models can be derived by considering that aerosol dynamics can be represented by a few lumped processes. An example of a lumped process is gas-to-particle conversion in the accumulation mode. Whitby, Vijayakumar, and Anderson (1980) have developed such a model involving five lumped processes. A disadvantage of this approach is that the parameters defining the lumped processes may vary according to the application of the model.

Detailed aerosol models (based on the numerical solution of the general dynamic equation) describe nucleation, condensation, and aerosol coagulation, as well as removal and emission processes (e.g., Middleton and Brock, 1976; Gelbard and Seinfeld, 1979). However, these models are too detailed to be included in air quality models.

Therefore, an alternative representation of aerosol dynamics must be adopted if the aerosol model is to be incorporated in an air quality model that involves atmospheric transport and dispersion processes as well as gas-phase chemistry. To this end, lumped aerosol models have been developed. Two main approaches have been considered for the development of such models.

1. The lognormal representation can be used since the aerosol distribution can be represented as three distinct modes (nuclei, accumulation, and coarse) that can be approximated by three lognormal distributions (e.g., Eltgroth and Hobbs, 1979).

2. The sectional representation, in which the aerosol distribution is approximated by a step function, can be used (e.g., see Basset, Gelbard and Seinfeld, 1981). In comparison with the lognormal representation, this approach offers the advantage of being mathematically exact, and it can include the treatment of aerosol chemistry by coupling each aerosol section with an aerosol chemistry model such as that developed by Peterson and Seinfeld (1979).

Comparisons of lumped aerosol models with detailed aerosol models have shown that the lumped approach based on a sectional representation of the aerosol distribution offers the best compromise between model accuracy and computational costs.

Mechanism Evaluation and Use

The construction of a mechanism and the evaluation of its performance are inextricably related. Mechanisms are not merely assembled and then evaluated and used. The construction process involves many iterative steps before achieving the specification of a mechanism that maintains acceptable balance among the current knowledge of atmospheric chemistry, the sophistication of the overall model, the details of the data, and the quality of simulation using appropriate test data sets.

The approach taken in the development of the Carbon-Bond Mechanism (CBM) is somewhat different from that used for many of the other mechanisms. Along with the original formulation and its various updates, the CBM was based on explicit mechanisms for several specific species, such as formaldehyde, acetaldehyde, ethylene propylene, butene-2, butane, methylglyoxyl, and toluene. Smog chamber experiments with NO_x were used to test the explicit chemistry of each species. Hence, the CBM represents a condensation of these explicit mechanisms. Once finalized, all mechanisms are usually then tested further, often by other scientists, using atmospheric data as well as other smog chamber data employing complex mixtures of hydrocarbons and NO_x.

One final problem to be solved before using a chemical mechanism in an atmospheric study involves the numerical method to be employed. Based on many years of successful use, the integration scheme originally developed by Gear (1971) is known to provide reliable results. However, use of this type of numerical scheme can lead to computational difficulties (such as excessive storage requirements and computing time requirements and, thus, high costs) when coupled with large atmospheric models. Hence, the original chemical mechanism may need to be approximated by the use of steady-state species in order to reduce the size and numerical stability problems associated with the differential equations that mathematically embody the mechanism.

A problem related to the choice and use of steady-state species is the difference between nighttime and daytime chemistry. The photochemically generated intermediates that exist during daylight hours are often good choices for steady-state species, but such species are essentially nonexistent during the nighttime. Hence, two distinct chemistry packages are often used--one for daytime and the other for nighttime. However, careful attention must be given to the dawn and twilight transition periods to ensure that the model used performs satisfactorily. Performance is usually tested by running a "Gear-based" model using simplified meteorology through similar transitions to check the reliability of the larger model. The steady-state approximations are typically tested in a similar fashion. The meteorologically related parts of the larger model can usually be simplified to approximate a smog chamber experiment that has been simulated by means of a "Gear-based" scheme.

Conclusions Regarding Chemical Mechanisms

From our survey of plume models that in one way or another incorporate a representation of atmospheric chemistry, the following observations can be made :

1. Plume models vary widely as to the extent of representation incorporated. The number of reactions treated (for homogeneous chemistry) ranges for 1 to 3 (low) to 35 to 70 (high).

2. Many models treat NO_x-O_3 chemistry, a number treat homogeneous SO_2 chemistry, and a few treat secondary aerosol formation.

3. In general, homogeneous gas-phase processes are better understood--and thus more satisfactorily represented--than are heterogeneous processes.

4. Plume models displaying a detailed treatment of dynamics and turbulence typically incorporate a simplified, or a highly restrictive, chemical representation only. Conversely, those models that attempt to deal fully with sophisticated atmospheric chemistry are usually based on a limited treatment of turbulence and dynamics. Exceptions include models proposed by Hilst et al. (1973) and O'Brien, Meyers and Benkovitz (1976).

5. Among the models displaying a sophisticated treatment of chemistry, there is not yet agreement as to one currently "most acceptable" representation, whether it be for NO_x, SO_2 or secondary aerosol. However, the principal treatments within a category share much in common and, consequently, may well produce similar predictions in many circumstances. (No formal comparison of chemical representations among the plume models exists. However, Jeffries (1981) has compared mechanisms using outdoor smog chamber data.)

The treatment of atmospheric transformations can be handled through the use of several different kinetic schemes currently available. However, the reliability and ease of use can differ among them. In addition to known performance characteristics, the choice of kinetic mechanism should take into consideration : the basis for conversion of atmospheric species into the species used in the mechanism; the extent to which mass or carbon balance is likely to be maintained; possible inclusion of species in steady state; and the sensitivity to adjustment of "arbitrary" parameters.

Given these observations, the status of knowledge concerning chemistry discussed in the previous section of this paper, and the experiences of investigators examining the comparative performance of chemical mechanisms, we currently conclude that :

1. NO_x-O_3-hydrocarbon chemistry is reasonably well understood and fairly well represented in the better mechanisms and, thus, in the plume models incorporating them. However, there is not now agreement on a "best" representation in plumes. Current differences seem to center on the advantages and disadvantages of the two types of "lumped" mechanisms discussed earlier in this section, i.e., mechanisms using surrogate species to represent a category of hydrocarbons, and those based on the chemical structure of the hydrocarbons--the carbon-bond mechanisms.

2. SO_2 chemistry is not completely understood. There is some general agreement that homogeneous oxidation of SO_2 is primarily due to reaction with OH radicals. Some pathways involved in the heterogeneous oxidation of SO_2, either on the aerosol surface or in the aerosol liquid phase, have been identified. However, the detailed mechamisms are not well established, and there is still considerable uncertainty regarding the kinetics of the heterogeneous oxidation reactions.

3. The basic processes that lead to gas-to-particle conversion of H_2SO_4 and HNO_3 are well understood and can be modeled successfully. Most uncertainties in the modeling of secondary aerosol formation are due to our lack of knowledge concerning the oxidation reactions of SO_2 and NO_2 that take place in liquid aerosols. Mathematical techniques for modeling the formation and evolution of secondary aerosols have been developed. The most promising approach appears to be the sectional method (e.g., that of Basset, Gelbard, and Seinfeld, 1981).

Evaluation of Predictive Performance

The predictive performance of plume models, like that of most other types of air quality models, typically is first carried out by the model developers. Moreover, the efforts of the developer tend to be the primary evaluative undertaking for any one model. These initial evaluative efforts, because of their developmental nature, generally involve an evolutionary process of preliminary evaluation, diagnostic analysis to identify problems restricting the model's performance, model modification, and further evaluation. The cycle continues until "satisfactory performance" is achieved for the case(s) under study.

Generally, this type of scrutiny does not constitute a full range of testing. Moreover, one or more of the following deficiencies may characterize the efforts.

> Availability or use of only a limited or restrictive data base. Potential inadequacies in the data base include :

- Types or classes of data that are missing or that have not been collected, such as hydrocarbon speciation, vertical soundings, concentrations aloft, and others.

- Too few cases or too limited a range of cases.

- Uncertain or inaccurate data.

> Evaluation conditions that do not sufficiently stress the model, so that if it is inadequate, the model is forced to display, through its predictions the consequences of the inadequacies (i.e., poor prediction).

> Potential, albeit unintentional, evaluative and interpretative bias on the part of the developers ("It is my 'baby'; I must display it in the best possible light").

To be fair, few observational data bases collected are suitably designed for model performance evaluation. Those that may qualify include :

1. The EPRI plume model validation study (Bowne and Hilst, 1981)

2. The TVA plume study (Meagher et al., 1981)

3. The Lake Michigan plume study (Miller et al., 1978).

The first study focused on nonreactive pollutant species only, i.e., SO_2 and an inert tracer. The second and third studies primarily relied on a few cross-plume measurements of chemical species made by aircraft. Clearly, the model developer has been limited heavily by the availability of suitable data bases.

From our experiences in model evaluation and our review of the literature,we conclude that :

1. Adequate, reliable, and comprehensive programs of evaluation for reactive plume models have yet to be carried out.

2. Proper evaluation of predictive performance of plume models is inherently difficult to achieve. A suitable and complete data base is expensive and difficult to collect,some of the measurements are often rather uncertain (such as cross-plume transects), and the opportunity to plan an observational program to match the model evaluation objectives rarely arises.

3. It is possible, perhaps likely, that many models have not been subjected to the type of evaluative stress that leads to confidence in the model's future use. Again, by "stress" we mean evaluation under conditions in which the model is likely to fail if it is deficient.

4. It would be desirable to establish the performance of existing models--both absolutely and relative to each other--through a series of carefully designed observational programs. If the models prove to be deficient, a subsequent program of diagnosis and model development should follow. The recent activities of

EPRI-- in nonreactive plume modeling and regional-scale modeling--provide examples of soundly planned evaluative and developmental programs.

Currently, we must rely on the existing body of model evaluation literature. Prior to adopting any model, a user should acquaint himself fully with :

> Its formulation.

> The assumptions underlying the formulation, and the extent to which they are likely to apply.

> The input requirements, and the expected increase in predictive uncertainty if certain of the inputs are unavailable.

> Past performance evaluation activities, including the meteorological conditions for which they were carried out.

> The likelihood of inadequacies in model prediction becoming evident under the conditions studied (if the model is assumed to be deficient).

Overall Model Characterization

The codifying of plume models through use of the triangulum, as described earlier, provides but a first step in characterization. This approach is limited in that it does not account for certain other attributes of interest in model selection. We would like to suggest ways in which this mode of characterization might be extended :

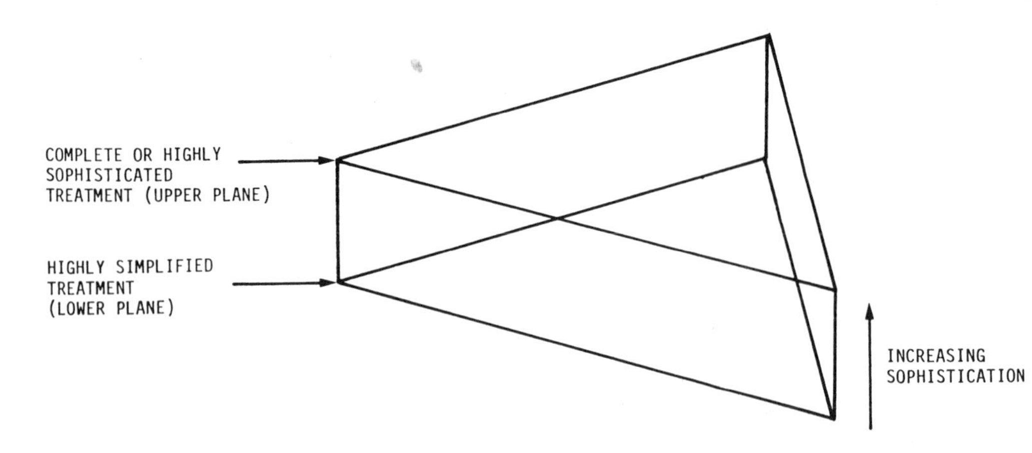

Figure 5.

> Extend the triangulum upward to include a third dimension that indicates the degree of sophistication of the treatment.

The two-dimensional triangulum of Figure 5 is the projection of all points in the above three-dimensional figure, ignoring degree of sophistication.

> Demark, in the three-dimensional diagram, subvolumes that constitute regimes of "acceptability of formulation" of a model for a given use (categories of uses are discussed later in this section).

> For a specified problem and thus a designated use, select a model that falls in the appropriate region of acceptability.

Although this characterization scheme is conceptually straightforward, it may be difficult to (1) unambiguously locate "points" in the three-dimensional space, and (2) unambiguously define the regions of acceptability. Practuality, then remains to be explored.

Finally, evaluation of performance must be a part of the characterization process. We might enter a "point" into the three-dimensional diagram as an asterisk (*) if performance had been suitably evaluated and the model was deemed to be acceptable.*

Plume Models and Their Uses

The question inevitably arises as to which of the available models we recommend for use. To address this question, one must be rather specific :

> What is the issue of concern?

> What specific technical-scientific questions derive?

> For these questions, what models fall into the to-be-defined "region of acceptability" for use?

> Of these models, which have been adequately evaluated, including being satisfactorily stressed?

Clearly, several of the requisites are not currently met--codification of performance evaluations, defined "regions of acceptability" (or, equivalently, agreed upon standards of performance based on sound measures of performance), and specified categories of use. For all of these, further work is needed and, therefore, we cannot at this time make recommendations for model selection.

* We have not attempted to deal with measures of, or standards for, model acceptability (see AMS, 1980; Fox, 1981).

Model Selection under the Present State of Knowledge

Because the type of use to which a model is put is an essential determinant in its selection, this factor enters into the construction of the three-dimensional diagram. We further consider the type of use, focusing first on the nature and sophistication of formulation.

We noted earlier that models vary greatly in the sophistication of treatment of chemistry, dynamics, and turbulence. Some models treat one of these classes of phenomena rather fully and the other two more simply; other models display reasonably thorough descriptions of two classes of phenomena. However, sophisticated treatment of all three classes is currently impractical; problems in formulation and in solution are difficult to surmount. Therefore, one inevitably gives up sophistication in the treatment of at least one class of these phenomena-- and often of two.

Model selection is principally influenced by the nature of the issue under study, and, in turn, by the technical components of the problem. Consider, for example, the study of an NO_x plume at night, with concern about the conversion to nitric acid. If nitric acid formation is dependent on the presence of ozone and water vapor--as we now believe it to be--and if the rate of formation is dependent on the rate at which ambient air mixes with plume materials (as it is likely to be under typically quiescent night-time conditions), it may be important to codify in some detail both the chemical rate processes and the mass transfer (or mixing) processes. Thus, emphasis will be placed on selecting models of some sophistication in their treatment of homogeneous and heterogeneous NO_x chemistry and turbulent dispersion. The user may then be required to select a model displaying only a simplified treatment of plume dynamics, such as standard parameterizations.

Consider, as a second example, the problem of interstate transport of SO_2 from large point sources and the determination of culpability of sources in their contribution to elevated sulfate levels in the "downwind" state. Here, plume dynamics (transport) and SO_2-sulfate chemistry should be treated in a complete manner. Treatment of turbulence, however, might be ignored or highly simplified. If the general flow patterns are well defined or understood, then their treatment might also be somewhat simplified.

These examples indicate the importance of defining overall objectives, translating them into scientific-technical issues, examining the technical attributes of the physical situation at the outset, and using this information judiciously and thoughtfully in model selection.

Model Uses

How, in fact, are reactive plume models likely to be used? Following are some types of questions that they might be "asked" to address :

1. Is an existing plant, through the contribution of its emissions to ground-level concentrations of NO_2, significantly impeding progress toward attainment of the annual or, in some states, the short-term NO_2 standard? If so, what emissions reductions are needed to ensure attainment of the standard?

2. Where should a planned new plant be located to minimize its contributions to SO_2 concentration increases and, thus to prevent exceeding the Class I PSD increment?

3. What is the potential air quality impact of emissions from two distinct sources commingling and reacting along the transport path?

4. Is an existing source or group of sources contributing to, or will a proposed source or group of sources cause, visibility impairment (i.e., reduction in visual range or atmospheric discoloration) in the vicinity of recognized scenic vistas?

5. Are any sources contributing to elevated secondary particulate concentrations, and if so, what is the magnitude of the contribution?

6. What sources are contributing to atmospheric acid burdens, and what is the magnitude of their contributions?

In thinking about these examples, as well as other questions, it may be helpful to categorize uses according to three general groupings --permit (or impact) evaluations, general planning, and research and development activities.

Permit Evaluations. Regulations in place in the United States at this time suggest that the substances of most concern involving chemical transformations are secondary particulates (sulfate and nitrates), nitrogen dioxide, atmospheric acidity (sulfuric and nitric acid), and ozone. The specific concern may entail "point" estimates of ground-level concentrations for averaging periods of an hour to a year or aerial concentration estimates that are spatially and temporally averaged, as in the case of visibility impairment. The relationships between these reactive pollutants and the pertinent regulations in the United States are shown in Table 4.

Table 4. Relationships between reactive pollutants and regulation(s) pertinent to plumes in the United States

Pollutant	Regulatory Relevance in United States	Remarks
Nitrogen dioxide	Health-based ambient standard; annual averaging period	One-hour averaging time in California; short-term standard also being considered nationally
	Visibility impairment--atmospheric discoloration (short-term effect)	
Secondary particulates (sulfate and nitrate)	Health-based ambient standard for total suspended particulates (TSP); 24-hour average and annual average	Revision of TSP standard under consideration that focuses on particulates < 10 μm in diameter
	Visibility impairment--reduction in visual range (short-term effect)	
Ozone	Health- and welfare-based standard; one-hour averaging time	
Atmospheric acidity (sulfuric and nitric acid)	Air-quality-related values of potential importance in new source review action	Other regulatory and legislative actions under consideration

New source review rules, including nonattainment and PSD programs, are of primary concern. At present, no definitive guidelines for selecting reactive models for such uses have been issued by the U.S. EPA. The demand for sophistication in this class of modeling activities is highly variable.

Planning. The state implementation planning process requires the development of plans for progress toward attainment of the NAAQS. For reactive pollutants, the focus is on secondary particulates, annual-average NO_2, and hourly ozone. Adequate treatment of hydrocarbon-NO_x-O_3 chemistry is obviously essential for short-term estimation. Annual-average NO_2 is estimated by a variety of techniques, most of which do not depend on the simulation of dynamics. Urban areas contain the majority of the nonattainment areas; thus, treatment of plumes is usually inextricably tied to treatment of area and line source impacts. Such modeling procedures are complex (see, for example, Seigneur, Roth and Wyzga, 1981), and they have not been addressed in this paper.

As the regulatory framework is modified or tailored to changing needs, novel concepts emerge, often placing new demands on "the bag of tools". The bubble concept, for example, may require thoughtful treatment of groups of sources in modeling. For example, if a net reduction in emissions from a group of sources is contemplated, will pollutant concentrations remain constant or will they decrease if control is preferentially applied to the more elevated sources?

Research and Development. In this area, all pollutants are of interest; sophistication in representation of physical and chemical processes is generally both appropriate and necessary; and modification, innovation, and extension of knowlegde or modeling capability is encouraged. Frontier, or cutting-edge, issues--transboundary transport, regional visibility degradation, acidic precipitation--are frequently of interest.

In addition to contributing to the advancement of atmospheric sciences, plume models possessing state-of-the-art treatment of chemical and physical phenomena can :

1. Facilitate planning and design of field and laboratory measurement programs.

2. Provide guidance in establishing and modifying research policies.

3. Facilitate identification, evaluation, and selection of regulatory policy options, including less sophisticated alternative modeling approaches.

4. Augment, in lieu of observations, the evaluation of less sophisticated models, provided that a sound basis exists for using the more sophisticated models as a standard of performance.

RESEARCH AND DEVELOPMENT NEEDS

In carrying out a review of this type, one inevitably confronts the reality that much of that which is of interest is not known or well understood. We conclude this review with a summary of topics dealing with chemical transformations in plumes that merit further investigation.

This paper has discussed the following topics in sequence : observational studies, current understanding of atmospheric chemistry encodement of existing knowledge of plume models; and the performance selection and use of models. Our first impulse was to categorize our recommendations similarly. However, on reflection, it is apparent that many aspects of defining and carrying out studies are interrelated. Model performance is evaluated using observational data. Modeling is used, in part, to aid in planning observational studies. Thus, the activities of model development, performance evaluation, and execution of field programs are linked. We therefore offer our recommendations without categorization.

In general, there is a need to mount definitive observational studies aimed at evaluating model performance. When model performance is found to be adequate, the data collected should serve as a basis for diagnostic analysis and model development. Currently, the paucity of data bases suitable for use in performance evaluation probably poses a greater barrier to progress than do deficiencies in model formulations.

Field programs to date have typically focused on observations aloft, in contrast to observations at the surface. Although observations aloft are essential, the demand for predictive accuracy is centered primarily at the earth's surface. Future planning of field studies should reflect this need.

The rate of mixing of reactants and intermediates is often the limiting step in a reaction sequence. In practical situations in whi this is the case and in which the results of prediction are sensitiv to variations in assumed rates, it is essential to model this phenomenon properly and to apply the model properly. The nature of chemical versus mixing (or diffusion) rate limitations is well understood; it must be encoded in models if they are to be of value in practical circumstances.

Acceptable treatment of nitrate and sulfate formation must be developed and encoded in models for evaluative testing. Although the homogeneous reactions governing sulfur dioxide formations are known, these reactions may account for only 25 to 50 percent of the observed formation rates in humid, "dirty" atmospheres; only in the clean, dry West do such reactions adequately explain the observations. Considerably more effort is needed to advance descriptions of surface reactions and the reactions occuring in liquid particles.

Suitable nitrate mechanisms also await development. Of particular interest are the mechanisms of nighttime and daytime formation of nitric acid and ammonium nitrate. In urban areas, interest is extended to organic nitrates. Field studies of plumes are needed that are directed at obtaining nitrogen balances in urban and rural atmospheres under both moist and dry conditions and during both daytime and nighttime. Such programs should pursue obtaining observations of NH_3, NO_3, HNO_3, N_2O_5 and (if at all possible) HONO, in addition to NO_2, O_3 and hydrocarbons. Particulate nitrate measurements are also necessary. Plume (or at least surface) measurements of OH and H_2O_2 should be made as well. Tracer techniques should be employed and observations should be carried out to 100 to 150 km.

Models embodying the treatment of chemical and physical dynamics are typically suitable for prediction only over short periods of time. Models incorporating dynamics that can be used for the estimation of seasonally averaged concentrations of NO_2, nitric acid, and nitrates are needed.

Treatment of the vertical stratification of the atmosphere, the storage of pollutants aloft, and the dynamics of plume/strata interactions are all inadequately understood. Because of the importance of these phenomena in some circumstances, they merit detailed investigation, with the goal of improving simulation models.

Treatment of the formation, dispersion, growth, and deposition of fine (diameters < 15 μm) particles, composed of secondary aerosols, is a topic of concern. Increasing attention will be given to the health effects of fine particles, particularly in light of the growing use of coal as a fuel.

Knowledge of wet and dry deposition processes is still inadequate. Generally, deposition losses are not significant in elevated plumes; they can be quite significant, however, in plumes that "touch down" reasonably near the source. In cases in which the impact of concern is cumulative--such as vegetation damage--deposition phenomena can be of considerable interest. The greatest need here is probably for observational studies; lack of data currently limits the development of suitable model representations.

Plume interactions with clouds can materially influence chemical transformation processes and the physical loss of material from the mixed layer. Fundamental investigations are required in this area.

Because this paper has focused on chemical transformations, comments concerning research and development needs are directed primarily to that area of study. However, several topics dealing with dynamics (i.e., transport) and turbulence (i.e., dispersion and mixing) merit investigation as well. Moreover, for the spatial and temporal scales of interest, certain of these research needs supersede those involving chemical transformations per se. Topics, meriting investigation include the explicit treatment of dispersion dynamics at 10 to 100 kilometers (or more) distance from the source, non-steady-state plume behavior, bifurcation phenomena, and interaction of a plume with stable layer(s) aloft (including penetration of an inversion layer).

REFERENCES

Akimoto, H., et al. (1978), "Formation of Propylene Glycol 1,2 Dinitrate in the Photooxidation of a Propylene-Nitrogen Oxides-Air System", J. Environ. Sci. Health, Vol. A13, pp. 677-686.

Alam, M. K., and Seinfeld, J. H. (1981), "Solution of the Steady-State, Three Dimensional Atmospheric Diffusion Equation for Sulfur Dioxide and Sulfate Dispersion from Point Sources", Atmos. Environ., Vol. 15, pp. 1221-1225.

Altshuller, A. P. et al. (1967), "Chemical Aspects of the Photooxidation of the Propylene-Nitrogen Oxide System", Environ. Sci. Technol., Vol. 1, pp. 899-914.

AMS (1980), Proc. American Meteorological Society Workshop on Dispersion Model Performance, 8-11 September 1980, Woods Hole, Massachusetts.

Atkinson, R., and Lloyd, A. C. (1981) "Evaluation of Kinetic and Mechanistic Data for Modeling of Photochemical Smog", J. Phys. Chem. Ref. Data, to appear in Vol. 10. Also available as Atkinson R., and Lloyd, A. C. (1980), ERT Document No. P-A040, Environmental Research and Technology, Inc., Westlake Village, California.

Atkinson, R., et al. (1980), "A Smog Chamber and Modeling Study of the Gas Phase NO--Air Photooxidation of Toluene and the Cresols", Int. J. Chem. Kinet., Vol. 12, pp. 779-837.

Basset, M., Gelbard, F. and Seinfeld, J. H. (1981), "Mathematical Model for Multicomponent Aerosol Formation and Growth in Plumes", Atmos. Environ., Vol. 15, pp. 2395-2406.

Batt, L., McCulloch, R. D. and Milne, R. T. (1975), "Thermochemical and Kinetic Studies of Alkyl Nitrites (RONO)-D(RONO), the Reactions between RO· and NO, and the Decomposition of RO·", Int. J. Chem. Kinet., Symposium No. 1, pp; 441-461.

Blumenthal, D. L., et al. (1981), "Effects of a Coal-Fired Power Plant and Other Sources on Southwestern Visibility (Interim Summary of EPA's Project VISTTA)", Atmos. Environ., Vol. 15, pp. 1955-1969.

Bowne, N. E., and Hilst, G. R. (1981), " Overview of the EPRI Plume Model Validation Project", Fifth Symposium on Turbulence, Diffusion and Air Pollution, American Meteorological Society, Atlanta, Georgia.

Brodzinsky, R., et al. (1980), "Kinetics and Mechanism for the Catalytic Oxidation of Sulfur Dioxide on Carbon in Aqueous Suspensions", J. Phys. Chem., Vol. 84, pp. 3354-3358.

Bufalini, J. J., Gay, B. W. and Brubaker, K. L. (1972), "Hydrogen Peroxide Formation from Formaldehyde Photooxidation and Its Presence in Urban Atmospheres", Environ. Sci. Technol., Vol. 6 pp. 816-821.

Calvert, J. G., et al. (1978), "Mechanism of the Homogeneous Oxidation of Sulfur Dioxide in the Troposphere", Atmos. Environ., Vol. 12, pp. 197-226.

Calvert, J. G., and McQuigg, R. D. (1975), "The Computer Simulation of the Rates and Mechanisms of Photochemical Smog Formation", Int. J. Chem. Kinet., pp. 113-154.

Carmichael, G. R., and Peters, L. K. (1981), "Application of the Mixing-Reaction in Series Model to NO_x-O_3 Plume Chemistry", Atmos. Environ., Vol. 15, No. 6, pp. 1069-1074.

Carter, W. P. L., Winer, A. M., and Pitts, J. N. Jr. (1980) "Kinetics of the Gas Phase Reactions of the Nitrate Radical with Phenol and the Cresols", J. Phys. Chem., in press.

Carter, W. P. L., et al. (1979), "Computer Modeling of Smog Chamber Data : Progress in Validation of a Detailed Mechanism for the Photooxidation of Progress and n-Butane in Photo-chemical Smog", Int. J. Chem. Kinet., Vol. 11, pp. 45-101.

Chameides, W. L., and Walker, J. C. G. (1976), "A Time-Dependent Photochemical Model for Ozone near the Ground", J. Geophys. Res., Vol. 81, pp. 413-420

Cofer, W. R., Schryer, D. R. and Rogowski, R. S. (1980), "The Enhanced Oxidation of SO_2 by NO_2 on Carbon Particulates", Atmos. Environ., Vol. 14, pp. 571-575.

Crawford, T. L., and Reisinger, L. M. (1980), "Transport and Transformation of Sulfur Oxides through the Tennessee Valley Region", EPA-600/7-80-126, U.S. Environmental Protection Agency, Research Triangle Park, North Carolina.

Darnal et al., (1976), "Importance of NO_2 + NO in Alkyl Nitrate Formation from C_4-C_6 Alkane Photooxidations under Simulated Atmospheric Conditions", J. Phys. Chem., Vol. 80, p. 1948.

Davis, D. D., Ravishankara, A. R. and Fischer, S.(1979), "SO_2 Oxidation via the Hydroxyl Radical : Atmospheric Fate of HSO_x Radicals", Geophys. Res. Lett., Vol. 6, pp. 113-116.

deBower, K. (1976), "A Method of Modeling Chemically Reactive Plumes", Report n° 45, Atmospheric Science Group, the University of Texas College of Engineering, Austin, Texas.

Demerjian, K. L., Kerr, J. A., and Calvert, J. G. (1974), "The Mechanism of Photochemical Smog Formation", in Advances in Environmental Science and Technology, J. N. Pitts, Jr. and R. L. Metcalf, eds. (John Wiley & Sons, New York, New York).

Donaldson, C. DuP. (1973), "Atmospheric Turbulence and the Dispersal of Atmospheric Pollutants", EPA R4-73-016a, prepared by Aeronautical Research Associates of Princeton, Inc., Princeton, New Jersey, for U.S. Environmental Protection Agency, Office of Research and Monitoring, Washington, D.C.

Donaldson, C. DuP., and Hilst, G. R. (1972), "Effect of Inhomogeneous Mixing on Atmospheric Photochemical Reactions", Environ. Res. Technol., Vol. 6, No. 9, pp. 812-816.

Donaldson, C. DuP. (1969), "A Computer Study of an Analytical Model of Boundary Layer Transition", AIAA J., Vol. 7, pp. 272-278.

Eltgroth, M. W., and Hobbs, P. V. (1979), "Evolution of Particles in the Plumes of Coal-Fired Power Plants. II. A Numerical Model and Comparisons with Field Measurements", Atmos. Environ. Vol. 13, pp. 953-975.

Eschenroeder, A. Q., and Martinez, J. R. (1972), "Photochemical Smog and Ozone Reactions", Adv. Chem. Series, Vol. 113, p. 101.

Eschenroeder, A. Q., and Martinez, J. R. (1970), "Analysis of Los Angeles Atmospheric Reaction Data from 1968 and 1969", General Research Corporation, Santa Barbara, California.

Falls, A. H., and Seinfeld, J. H. (1978), "Continued Development of a Kinetic Mechanism for Photochemical Smog", Environ. Sci. Technol., Vol. 12, pp. 1398-1406.

Forney, L. J., and Heffner, D. A. (1981), "Slow Second Order Reactio[n] in Power Plant Plumes", Georgia Institute of Technology, Atlant[a] Georgia, Paper 81-60.3 presented at APCA 74th Annual Meeting, June 1981, Philadelphia, Pennsylvania.

Forney, L. J., and Giz, Z. G. (1980a), "Slow Chemical Reactions in Power Plant Plumes : Application to Sulfates", Atmos. Environ., Vol. 14, pp. 553-541.

Forney, L. J., and Giz, Z. G. (1980b), "Fast Reversible Reactions in Power Plant Plumes : Application to the Nitrogen Dioxide Photylytic Cycle", Atmos. Environ., Vol. 15, pp. 345-352.

Fox, D. G. (1981), "Judging Air Quality Model Performance", Bull. Am Meteorol. Soc., Vol. 62, pp. 599-609.

Gear, C. W. (1971), "The Automatic Integration of Ordinary Differential Equations", Comm. ACM, Vol. 14, pp. 176-179.

Gelbard, F., and Seinfeld, J. H. (1979), "The General Dynamic Equation for Aerosols--Theory and Application to Aerosol Formation and Growth", J. Colloid Interface Sci., Vol. 68, pp. 363-381.

Gelinas, R. J., and Skewes-Cox, P. D., "Tropospheric Photochemical Mechanisms", J. Phys. Chem., Vol. 81, pp. 2468-2479.

Gelinas, R. J., and Walton, J. J. (1974), "Dynamic-Kinetic Evolution of a Single Plume of Interacting Species", J. Atmos. Sci., Vol. 31, pp. 1807-1813.

Gillani, N. V., Kholi, S., and Wilson, W. E. (1981), "Gas-to-Particle Conversion of Sulfur in Power Plant Plumes : I. Parameterization of the Gas Phase Conversion Rate", Atmos. Environ., Vol. 15, pp. 2293-2314.

Gunnison, A. F., and Palmes, E. D. (1973), "Persistence of Plasma S-Sulfonates Following Exposure of Rabbits to Sulfite and Sulfur Dioxide", Toxicol. Appl. Pharmacol., Vol. 24, p. 255

Harker, A. B., Richards, L. W., and Clark, W. E. (1977), "The Effect of Atmospheric SO_2 Photochemistry upon Observed Nitrate Concentrations in Aerosols", Atmos. Environ., Vol. 11, pp. 87-91.

Hecht, T. A., Seinfeld, J. H., and Dodge, M. C. (1974), "Further Development of a Generalized Kinetic Mechanism for Photochemical Smog", Environ. Sci. Technol., Vol. 8, pp. 327-339.

Hecht, T. A., and Seinfeld, J. H. (1972), "Development and Validation of a Generalized Mechanism for Photochemical Smog", Environ. Sci. Technol., Vol. 6, pp. 47-57.

Hegg, D. A., and Hobbs, P. V. (1978), "Oxidation of Sulfur Dioxide in Aqueous Systems with Particular Reference to the Atmosphere", Atmos. Environ., Vol. 12, pp. 241-253.

Hegg, D. A., Hobbs, P. V., and Radke, L. F. (1976), "Reactions of Nitrogen Oxides, Ozone and Sulfur in Power Plant Plumes", EPRI EA-270, Electric Power Research Institute, Palo Alto, California.

Hegg, D. A., et al. (1977), "Reactions of Ozone and Nitrogen Oxides in Power Plant Plumes", Atmos. Environ., Vol. 11, pp. 521-526.

Hilst, G. R. (1978), "Plume Model Validation", EPRI EA-917-SY, Electric Power Research Institute, Palo Alto, California.

Hilst, G. R., et al. (1973), "The Development and Preliminary Applications of an Invariant Coupled Diffusion and Chemistry Model", NASA CR-2295, National Aeronautical and Space Administration, Washington, D.C.

Hobbs, P. V., et al. (1979), "Evolution of Particles in the Plumes of Coal-Fired Power Plants. I. Deductions from Field Measurements", Atmos. Environ., Vol. 12, pp. 935-951.

Hov, O., and Isaksen, I. S. A. (1981), "Generation of Secondary Pollutants in a Power Plant Plume : A Model Study", Atmos. Environ., Vol. 15, Vol. 15, pp. 2367-2376.

Jeffries, H. E. (1981), "Effect of Substituting Chemistry in EKMA Level II Analysis", Proc. EKMA Workshop, 15,16 December 1981, U.S. EPA, Research Triangle Park, North Carolina.

Judeikis, H. S., Steward, T. B., and Wren, A. G. (1978), "Laboratory Studies of Heterogeneous Reactions of SO_2", Atmos. Environ., Vol. 12, pp. 1633-1642.

Kan, C. S., Calvert, J. G., and Shaw, J. H. (1981), "Oxidation of Sulfur Dioxide by Methylperoxy Radicals", J. Phys. Chem., Vol. 85, pp. 1126-1132.

Kaplan, D. J., Himmelblau, D. M., and Kanaoka, C. (1981), "Oxidation of Sulfur Dioxide in Aqueous Ammonium Sulfate Aerosols Containing Manganese as a Catalyst", Atmos. Environ., Vol. 15, pp. 763-773.

Killus, J. P., and Whitten,G. Z., (1981), "A User's Guide to the Carbon-Bond Mechanism", SAI No. 75-81-EF81-90, Systems Applications, Inc., San Rafael, California.

Larson, T. V., Horike, N. R., and Harrison, H. (1978), "Oxidation of Sulfur Dioxide by Oxygen and Ozone in Aqueous Solution : A Kinetic Study with Significance to Atmospheric Rate Processes", Atmos. Environ., Vol. 12, pp. 1597-1611.

Levine, S. Z. (1980), "A Model for Stack Plume Reactions with Atmospheric Dilution (Spread)", Environmental Chemistry Division, Department of Energy and Environment, Brookhaven National Laboratory, Upton, New York, presented at the Symposium on Plumes and Visibility : Measurements and Model Components, 10-14 November 1980, Grand Canyon, Arizona.

Liu, M. K., et al. (1976), "The Chemistry, Dispersion and Transport of Air Pollutants Emitted from Fossil Fuel Power Plants in California : Data Analysis and Emission Impact Model", SAI EF76-18R, Systems Applications, Inc., San Rafael, California

Lumley, J. (1978), "Turbulent Transport of Passive Contaminants and Particles : Fundamentals and Advanced Methods of Numerical Modeling", von Karman Institute for Fluid Dynamics Lecture Series on Pollutant Dispersal, 8-12 May 1978, Rhode-Saint-Genese Belgium.

Lusis, M. (1976), "Mathematical Modeling of Chemical Reactions in a Plume", Proc. Seventh International Technical Meeting on Air Pollution Modeling and Its Application, p. 831, 7-10, September 1976, Airlie House, Virginia.

Mamane, Y., and Pueschel, R. F. (1979), "Oxidation of SO_2 on the Surface of Fly Ash Particles under Low Relative Humidity Conditions", Geophys. Res. Lett., Vol. 6, pp. 109-112.

MAP3S (1979), "Multistate Atmospheric Power Production Pollution Study--MAP3S", DOE/EV-0400, progress report FY 1977 and FY 1978 U.S. Department of Energy, Office of Health and Environmental Research, Washington, D.C.

Martinez, J. R., and Eschenroeder, A. Q., (1972), "Evaluation of a Photochemical Pollution Simulation Model", General Research Corporation, Santa Barbara, California.

Mathieu, J., and Gence, J. N. (1978), "Remarks on Predictive Methods in Atmospheric Turbulence", von Karman Institute for Fluid Dynamics Lecture Series on Pollutant Dispersal, 8-12 May 1978, Rhode-Saint-Genese, Belgium.

Matteson, M. J. (1978), "Capture of Atmospheric Gases by Water Vapor Condensation on Carbonaceous Particles", Proc. Carbonaceous Particles in the Atmosphere, 20-22 March 1978, Lawrence Berkeley Laboratory Publication LBL-9037, published June 1979.

Meagher, J. F., et al. (1981), "Atmospheric Oxiation of Flue Gases from Coal-Fired Power Plants--A Comparison between Conventional and Scrubbed Plumes", Atmos. Environ., Vol. 15, pp. 749-762.

Middleton, P., and Brock, J. (1976), "Simulation of Aerosol Kinetics", J. Colloid Interface Sci., Vol. 54, pp. 249-264.

Middleton, P., Kiang, C. S., and Mohnen, V. A. (1980), "Theoretical Estimates of the Relative Importance of Various Urban Sulfate Aerosol Production Mechanisms", Atmos. Environ., Vol. 14, pp. 463-472.

Miller, D. F., et al., (1978), "Ozone Formation Related to Power Plant Emissions", Science, Vol. 202, p. 1186.

NAS (1977), "Medical and Biologic Effects of Environmental Pollutants: Ozone and Other Photochemical Oxidants", Committee on Medical and Biologic Effects of Environmental Pollutants, Division of Medical Sciences, Assembly of Life Sciences, National Research Council, National Academy of Sciences, Washington, D.C.

Newman, L. (1981), "Atmospheric Oxidation of Sulfur Dioxide as Viewed from Power Plant and Smelter Plume Studies", Atmos. Environ., Vol. 15, pp. 2231-2240.

Nicolet, M. (1975), "On the Production of Nitric Oxide by Cosmic Rays in the Mesosphere and Stratosphere", Proc. of the Fourth Conference on the Climatic Impact Assessment Program, DOT-TSC-OST-75-38, U.S. Department of Transportation, Washington, D.C., pp. 292-302.

Niki, H., et al. (1980), "Fourier Transform Infrared Study of the HO Radical Initiated Oxidation of SO_2", J. Phys. Chem., Vol. 84, pp. 14-16.

Niki, H., et al. (1977), "Fourier Transform IR Spectroscopic Observations of Propylene Dozonide in the Gas Phase Reaction of Ozone-cis-butene-formaldehyde", Chem. Phys. Lett., Vol. 46, pp. 327-330.

Niki, H., Daby, E. E., and Weinstock, B. (1972), Photochemical Smog and Ozone Reactions, Advances in Chemistry Series 113 (American Chemical Society, Washington, D.C.).

O'Brien, E. E., Meyers, R. E., and Benkovitz, C. M. (1976), "Chemically Reactive Turbulent Plumes", Third Symposium on Atmospheric Turbulence,Diffusivity, and Air Quality, American Meteorological Society, 19-22 October 1976, Raleigh, North Carolina.

O'Brien, R. J., Holmes, J. R., and Bockian, A. H. (1975), "Formation of Photochemical Aerosol from Hydrocarbons : Chemical Reactivity and Products", Environ. Sci. Technol., Vol. 9, pp. 568-576.

Ogren, J. A., et al. (1976), "Determination of the Feasibility of Ozone Formation in Power Plant Plumes", EPRI EA-307, Electric Power Research Institute, Palo-Alto, California.

Orel, A. E., and Seinfeld, J. H., (1977), "Nitrate Formation in Atmospheric Aerosols", Environ. Sci. Technol., Vol. 11, pp. 1000-1007.

Penkett, S. A., et al. (1979), "The Importance of Atmospheric Ozone and Hydrogen Peroxide in Oxidizing Sulphur Dioxide in Cloud and Rainwater", Atmos. Environ., Vol. 13, pp. 123-138.

Peters, L. K., and Richards, L. W. (1977), "Extension of Atmospheric Dispersion Models to Incorporate Fast Reversible Reactions", Atmos. Environ., Vol. 11, pp. 101-108.

Peterson, T. W., and Seinfeld, J. H. (1980), "Heterogeneous Condensation and Chemical Reaction in Droplets--Application to the Heterogeneous Atmospheric Oxidation of Sulfur Dioxide", Adv. Environ. Sci. Technol., Vol. 10, pp. 125-180.

Peterson, T. W., and Seinfeld, J. H. (1979), "Calculation of Sulfate and Nitrate Levels in a Growing, Reacting Aerosol", AIChE J., Vol. 25, pp. 831-838.

Platt, U., et al. (1980a), "Detection of NO_3 in the Polluted Troposphere by Differential Optical Absorption", Geophys. Res. Lett., Vol. 7, pp. 89-92.

Platt, U., et al. (1980b), "Observations of Nitrous Acid in an Urban Atmosphere by Differential Optical Absorption", Nature, Vol. 285 pp. 312-314.

Richards, L. W., et al. (1981), "The Chemistry, Aerosol Physics and Optical Properties of a Western Coal-Fired Power Plant Plume", Atmos. Environ., Vol. 15, pp. 2111-2134.

Richards, L. W., Avol, E. L., and Harker, A. B. (1976), "The Chemistry, Dispersion and Transport of Air Pollutants Emitted from Fossil Fuel Power Plants in California", State of California Air Resources Board, Sacramento, California.

Roberts, D. B., and Williams, D. J. (1979), "The Kinetics of Oxidation of Sulfur Dioxide within the Plume from a Sulphide Smelter in a Remote Region", Atmos. Environ., Vol. 13, pp. 1485-1499.

Schiermeier, F. A., et al. (1979), "Sulfur Transport and Transformation in the Environment (STATE) : A Major EPA Research Program Bull. Am. Meteorol. Soc., Vol. 60, p. 1303.

Seigneur, C. (1982), "A Model of Sulfate Aerosol Dynamics in Atmospheric Plumes", Atmos. Environ., Vol. 16, pp. 2207-2228.

Seigneur, C., Roth, P. M., and Wyzga, R. E., (1981), "Mathematical Modeling of Chemically Reactive Plumes in an Urban Environment", Proc. 12th International Technical Meeting on Air Pollution Modeling and Its Application, 25-28 August 1981, Palo Alto, California.

Seigneur, C., et al. (1980), "The Data Base of Plume Visibility Model Evaluation from the June, July, and December 1979 VISTTA Field Programs", Symposium on Plumes and Visibility : Measurements and Model Components, 10-14 November 1980, Grand Canyon, Arizona.

Shu, W. R., Lamb, R. G. and Seinfeld, J. H. (1978), "A Model of Second-Order Chemical Reactions in Turbulent Fluid--Part II. Application to Atmospheric Plumes", Atmos. Environ., Vol. 12, pp. 1695-1704.

Stewart, D. A., and Liu, M. K. (1981), "Development and Application of a Reactive Plume Model", Atmos. Environ., Vol. 15, pp. 2377-2393.

Teske, M. E., Lewellen, W. S. and Segun, H. S. (1978), "Turbulence Modeling Applied to Buoyant Plumes", EPA-600/4-78-050, U.S. Environmental Protection Agency, Research Triangle Park, North Carolina.

Turner, D. B. (1970), "Workbook of Atmospheric Dispersion Estimates", Publication No. 999-AP-26, National Air Pollution Control Administration, U. S. Public Health Service.

Whitby, K. T., Vijayakumar, R., and Anderson, G. (1980), "New Particle and Volume Formation Rates in Five Coal-Fired Power Plant Plumes", preprints, Symposium on Plumes and Visibility : Measurements and Model Components, 10-14, November 1980, Grand Canyon, Arizona.

White, W. H. (1977), "NO_x-O_3 Photochemistry in Power Plant Plumes : Comparison of Theory with Observation", Environ. Sci. Technol., Vol. 11, pp. 995-1000.

White, W. H., et al. (1976), "Midwest Interstate Sulfur Transformation and Transport Project : Aerial Measurements of Urban and Power Plant Plumes, Summer 1974", EPA-600/3-76-110, U.S. Environmental Protection Agency, Research Triangle Park, North Carolina.

Whitten, G. Z., Hogo, H., and Killus, J. P., (1980), "The Carbon-Bond Mechanism : A Condensed Kinetic Mechanism for Photochemical Smog", Environ. Sci. Technol., Vol. 14, pp. 690-700

Whitten, G. Z., Killus, J. P., and Hogo, H. (1980), "Modeling of Simulated Photochemical Smog with Kinetic Mechanisms", EF79-129, Systems Applications, Inc., San Rafael, California.

Wilson, J. C., and McMurry, P. H., (1981), "Secondary Aerosol Formation in the Navajo Power Plant Plume", Atmos. Environ., Vol. 15, pp. 2329-2339.

Wilson, W. E., (1978), "Sulfate in the Atmosphere : A Progress Report on Project MISTT", Atmos. Environ., Vol. 12. p. 537.

Yamada, T. (1979), "An Application of a Three-Dimensional Simplified Second-Moment Closure Numerical Model to Study Atmospheric Effects of a Large Cooling-Pond", Atmos. Environ., Vol. 13, pp. 693-704.

Zwicker, J. O., et al., (1981), "Chemistry and Visual Impact of the Plumes from the Four Corners Power Plant and San Manuel Smelter", final report to the Environmental Protection Agency, 1978 VISTTA measurements.

Recommended Reading

General Chemistry

Atkinson, R., and Lloyd, A. C. (1980), "Evaluation of Kinetic and Mechanistic Data for Modeling of Photochemical Smog", ERT P-A040, Environmental Research and Technology, Inc., Westlake Village, California.

Codata Task Group on Chemical Kinetics (1980), "Evaluation Rate Constants for Atmospheric Chemistry", J. Phys. Chem. Ref. Data.

Seinfeld, J. H. (1980), "Lectures in Atmospheric Chemistry", AIChE Monograph Series, Vol. 76, No. 12, American Institute of Chemical Engineers, New York, New York.

Killus, J. P., and Whitten, G. Z., (1981), "A New Carbon-Bond Mechanism for Air Quality Simulation Modeling", SAI Report No. 81245.

Aerosol Chemistry

Kaplan, D. J., Himmelblau, D. M., and Kanaoka, C. (1981), "Oxidation of Sulfur Dioxide in Aqueous Ammonium Sulfate Aerosols Containing Manganese as a Catalyst", Atmos. Environ., Vol. 15, pp. 763-773.
Peterson, T. W. and Seinfeld, J. H. (1980), "Heterogeneous Condensation and Chemical Reaction in Droplets--Application to the Heterogeneous Atmospheric Oxidation of SO_2," Adv. Environ. Sci. Technol., Vol. 10, pp. 125-180.
Proceedings of the International Symposium on Sulfur in the Atmospher (Dubrovnik, 7-14 September 1977), in Atmos. Environ., Vol. 12, Nos. 1-3, pp. 1-796, 1978.
Stelson, A. W., Friedlander, S. K., and Seinfeld, J. H. (1979),"A Note on the Equilibrium Relationship between Ammonia and Nitric Acid and Particulate Ammonium Nitrate", Atmos. Environ., Vol. 13, pp. 369-371.
Tang, I. N. (1980), "On the Equilibrium Partial Pressures of Nitric Acid and Ammonia in the Atmosphere", Atmos. Environ., Vol. 14, pp. 819-828.

Plume Measurements

Gillani, N. V., and Wilson, W. E. (1980), "Formation and Transport of Ozone and Aerosols in Power Plant Plumes", Annals of the N.Y. Academy of Science, New York, New York.
Hegg, D. A., and Hobbs, P. V. (1980), "Measurements of Gas-to-Particle Conversion in the Plume from Five Coal-Fired Electric Power Plants", Atmos. Environ., Vol. 14, pp. 99-116.
Newman, L. (1981), "Atmospheric Oxidation of Sulfur Dioxide as Viewed from Power Plant and Smelter Plume Studies", Atmos. Environ., Vol. 15, pp. 2231-2240.
White, W. H., (1977), "NO_x-O_3 Photochemistry in Power Plant Plumes: Comparison of Theory with Observation", Environ. Sci. Technol., Vol. 11, pp. 995-1000.
White, W. H., editor (1981), special issue "Plumes and Visiblity: Measurements and Model Components", Atmos. Environ., Vol. 15, No. 10/11, pp. 1785-2406.

A COMPARISON BETWEEN CHEMICALLY REACTING PLUME MODELS AND WINDTUNNEL EXPERIMENTS

P.J.H. Builtjes

MT/TNO
P.O. Box 342, 7300 AH Apeldoorn
The Netherlands

ABSTRACT

Experiments have been carried out in a windtunnel of a chemically reacting plume. A NO-plume in an O_3-environment has been studied and the conversion from NO to NO_2 has been determined.

Based on the results from these experiments and existing field studies, a critical review of chemically reacting plume models has been carried out.

It is shown that in principle these models should take into account the effect of inhomogeneous mixing and the correlation between concentration fluctuations.

INTRODUCTION

During the last decade there has been a growing concern about chemically reactive pollutants, and the impact of secondary pollutants on air quality. Attention has been given to photochemical episodes, the transformation of SO_2 to sulfate in a plume, and the conversion of NO to NO_2 emitted from a stack or a highway. Field experiments have been carried out, smog-chamber studies have been performed and theoretical considerations have been given. One of the unsolved problems in this area is the contribution to the groundlevel concentration of NO and NO_2, on an annual basis as well as for short time intervals, caused by the emission of NO_x from a stack (see for example Cole[4]). It is commonly accepted that the NO_x-emission of a burning process consists for over 95% of NO. During the following dispersion process chemical transformation takes place, in which O_3 plays a major role, which leads to a partial conversion from NO to NO_2. The practical problem to be solved is to determine this conversion rate for a given stack and situation.

A number of field experiments have been carried out which show a large range of time scales at which the 50%-conversion rate is reached. Time scales from several minutes to over one hour have been found, depending on the O_3-concentration and the meteorological con- found, depending on the O_3-concentration and the meterological conditions (6,11,22,32). Theoretical studies have been carried out, tion of chemical transformation of the emittant, so-called reactive plume models. Models with particular emphasis on chemistry or dispersion; and of different levels of sophistication have been constructed It has been tried to validate the different models against existing field data. However, due to the inevitable complexity of the field situations, for the chemistry as well as for the meteorological conditions it seems to be impossible at this moment to check some basic assumptions made in the different models, and to judge which model should be used in practice to determine the ground-level concentration of NO and NO_2 for a given situation.

In view of this, it was decided to try to simulate a chemically reacting plume in an atmospheric boundary layer wind tunnel to study such a plume in a controlled environment. The experimental data obtained in the wind tunnel can be used to validate chemically reactive plume models.

THE EXPERIMENTAL SET-UP

The experiments have been carried out in the atmospheric boundary layer tunnel at TNO-Apeldoorn. This wind tunnel has a test section with a length of 10 m, and a cross-section of 2.65 x 1.2 m2. A neutral atmospheric boundary layer is simulated on a scale of 1 : 500. For a detailed description the reader is referred to Builtjes (1). In the fully developed turbulent boundary layer with a thickness of about 0.8 m, a stack is placed with a height of h = 0.14 m. The surface consists of a carpet, simulating a grass-land roughness. Because a reasonable amount of chemical transformation has to be achieved within the test section it was decided to study a plume of NO emitted from the stack in an O_3-environment, and to run the wind tunnel at a very low wind speed. Using a maximum velocity of 0.5 m/s, the flow in the wind tunnel is still stable and the normal velocity profile for a neutral boundary layer over grass-land exists. In this situation the dispersion of the emission Q of an iner gas, SF_6, as well as the dispersion of NO without O_3, and of NO_x with O_3 has been investigated. The stack exit velocity is 0.4 m/s, equal to the mean wind velocity U at that height. The concentration c determined at the plume axis and at the heigt of the stack is given in Fig. 1 against the downwind distance in the wind tunnel x, together with the calculated concentration using the Gaussian plume model with the Pasquill dispersion coefficients of stability class D. The dispersion shows the normal Gaussian plume behaviour up to about 8 m from the stack, with a sligth tendency to a more stable atmosphere. Consequently, in principle chemically reacting plume experiments can be carried out at this low wind velocity in the wind tunnel, with the normal Gaussian plume behaviour for an inert species as a basis.

Six vertical pipes with about 50 emission holes each are used to bring O_3 into the wind tunnel. These pipes are situated just after the contraction, and just before the vortex generators and castelled barrier, based on the method of Counihan to create the turbulent boundary layer. The O_3 is created using an ozone generator with a capacity of 6 g/hour. In case no absorption takes place, the maximum O_3 -concentration in the wind tunnel would be around 600 ppb. However, due to losses, the O_3-concentration at the test-section is on the average 350 ppb. The inhomogenity in transverse direction is about 10%, from the stack downwind there is a slight decrease of O3 of about 8 ppb/m, due to wall absorption. The emission from the stack has a NO-concentration of 3900 ppm, which is delivered by a bottle of NO embedded in an N_2-atmosphere.

To determine the concentration, iso-kinetic air samples are led to an O_3- and NO_x-monitor. A Dashibi 1003. AH O_3-monitor has been used, based on the UV-absorption method. Measurements of NOx and NO have been performed by a Bendix 8101 CX-analyser, based on chemoluminescence with O3. NO_2-data have been found by subtracting NO- from NO_x-concentrations.

A principal difficulty is created by the delay time caused by the transport of the air sample taken through the tubes to the monitors. During this transport a continuous reaction of NO and O_3 could take place. To investigate these experiments have been performed at the same place in the flow using different delay times, created by using different lengths and thicknesses of sampling tubes. Delay times from 1.75 to 2.5. sec for O_3, and 1.3 to 2.2 for NO have been used. No systematic differences between measured concentrations for different delay times were found. Consequently, no reaction taking place in the sampling tubes could be detected. This can be explained by the fact that in the plume relatively large slabs containing either NO or O_3 are present, with a reaction taking place only at the interfaces of the slabs. In the air sample taken these slabs will still be present, also in the sampling tubes because the flow in the tubes is laminar and no mixing will occur. So, for a first order accuracy, the situation in the tubes can be viewed as chemically "frozen".

This assumption is supported by an experiment carried out with the same monitors and sampling tubes, but the air samples taken from a bottle containing known levels of NO and O_3, injected in the bottle in such a way that a well-mixed situation was present. In this experiment different delay times were leading to different concentration levels measured, which could be directly related to a continuous reaction between NO and O_3. A careful comparison between measurements of the plume and of the well-mixed bottle could probably lead to the determination of a mixing factor for the plume with a rather un-mixed situation close to the stack changing into a rather well-mixed situation more downstream.

To determine the concentration of NO, NO_2 and O_3 at a certain place the following procedure is carried out. Two sampling tubes, with iso-kinetic sampling devices, are placed close together in the flow. Five 1-minute averaged NO_x-values are determined, together with ten 30-second averaged O_3-values. Subsequently five 1-minute averaged NO-values are determined, also with ten 30-second averaged O_3-values. These data again are averaged, leading to one NO_x, NO, O_3 and $NO_2 = NO_x - NO$ concentration at that place. Afterwards the NO-plume is stopped and five 30-second averaged O_3-values are determined, leading to the O_3-environment level. The parameters for the experiment in the wind tunnel are given in Table 1.

SCALING LAWS AND TIME SCALES

A correct simulation of the atmospheric turbulent boundary layer in a wind tunnel, in combination with the dispersion from a stack of an inert pollutant, can only be achieved when the similarity criteria which govern the boundary layer and the dispersion are satisfied. That these criteria are satisfied for the disperion of an inert pollutant in the wind tunnel used has been shown previously (Cermak[2], Snijder[33], see also[1]). The question arises whether this is still the case for a chemically reacting plume. The reaction between NO and O_3 takes place on a molecular scale, and is governed, in the wind tunnel as well as in the field, by the molecular diffusivity.The length scales which cause the turbulent mixing are 500 times smaller in the wind tunnel than in the field. This leads to the fact that the smallest turbulence scale in the wind tunnel which causes mixing of about 0.5 mm, corresponds to a scale of about 0.25 m in the field. In other words, the turbulent spectrum in the wind tunnel is "smaller" than in the field. The reaction will take place at the interface between slabs of O_3 and NO, the extent of the reaction will be determined by the size of the surface of the interface. It seems logical to assume that this size is determined by relatively large eddies in the field, mainly of a scale larger than 0.25 m (see also Hegg[10]). When this is the case, the similarity criteria can be satisfied

Table 1. Parameters for the wind tunnel experiment in a neutral boundary layer

$(NO)_{x=o}$	= 3900 ppm
$(O_3)_{env.}$	= 0.350 ppm
stackheight h	= 0.14 m
integral length scale Λ_f (z=h)	= 0.3 m
turbulence intensity u'/U (z=h)	= 0.1
mean velocity U (z=h)	= 0.4 m/s
roughness length z_o	= 10^{-4} m
scale factor	= 500
boundary layer thickness	= 0.8 m

in the wind tunnel. In case eddies of a size smaller than 0.25 m in the field have a large influence on the extent of the reaction, the wind tunnel experiment will be a more imperfect model of the field situation.

For the situation of a NO-plume in an O_3-environment the ratio between reaction time scales and diffusion time scales plays an important role. In the wind tunnel, because no hydrocarbons are present, and light intensity plays a minor role, the only reaction taking place is

$$NO + O_3 \rightarrow NO_2 + O_2 \qquad (1)$$

For the conversion of NO to NO_2 the following reaction time scale can be given :

$$\tau_{NO} = \frac{1}{k < O_3 >} \qquad (2)$$

with k the rate coefficient given by reaction (1) and $< O_3 >$ the concentration of O_3.
Another time scale is given by

$$\tau_{O_3} = \frac{1}{k < NO >_{x=o}} \qquad (3)$$

with $< NO >_{x=o}$ the NO-concentration at the moment of emission. Several diffusion time scales can be defined.

The Lagrangian integral time scale I_L is given by

$$I_L = 0.4 \frac{\Lambda_f}{u'} \qquad (4)$$

in which Λ_f is the Eulerian integral length scale (macro scale) of the turbulence intensity in the mean flow direction u' (see for example Hinze [13]). A time scale T defined by

$$T = z_i/w \qquad (5)$$

is used by Shu[32], with z_i the mixing heigth and w the convective velocity scale. Pasquill [28] defines the following time scale for a plume :

$$\tau_D = R/2V \qquad (6)$$

in which R is the radius of the plume and V the mixing eddy velocity, with $V \cong 0.4\ U^{(10)}$. In the field, with for example values of $U \cong 7$ m/s, $u'/U \cong 0.1$, $\Lambda_f \cong 150$ m, $z_i \cong 800$ m, $w \cong 2$ m/s, $R \cong 125$ m, these time scales are approximately $I_L \cong 100$ sec, $T \cong 200$ sec, $\tau_D \cong 25$ sec. The choice of time scale to be used is not essential, in the further analysis, I_L will be used.

The following scaling law should hold :

$$\frac{T_p \quad U_p}{L_p} = \frac{T_m \quad U_m}{L_p} \qquad (7)$$

in which the subscript m is related to the model situation in the wind tunnel, p is the prototype situation in the field. With values of U_m = 0.5 m/s, U_p = 7 m/s (in accordance with U = 5 m/s at stack height), and L_p = 500 L_m, it follows T_p = 35 T_m ; so in the wind tunnel all time scales are a factor 35 smaller than in the field.

The reaction time scale in the windtunnel can be found from formula (2) and (3), using a value for the reaction rate k of 25 ppm $^{-1}$ min^{-1}, with $< O_3 >$ = 350 ppb and $<NO>_{x=o}$ = 3900 ppm. This leads to τ_{NO} = 6.9 sec, τ_{O_3} = 6.2 . 10^{-4} sec. In the field situation these time scales will be a factor 35 larger, leading to τ_{NO} = 240 sec, τ_{O_3} = 2,2 . 10^{-2} sec. Consequently, using again formula (2) and (3) the situation in the wind tunnel with $< O_3 >$ = 350 ppb and $< NO >_{x=o}$ = 3900 ppm has to be compared to a field situation with for only the reaction of NO with O_3, $< O_3 >$ = 10 ppb and $< NO >_{x=o}$ = 110 ppm, which means a low concentration of O_3, leading to a relatively slow reaction from NO to NO_2. Essential for the further analysis is the ratio between the reaction time scale which determines the conversion of NO to NO_2 and the diffusion time scale. For the wind tunnel, as well as for the comparable situation in the field the ratio is :

$$\left(\frac{I_L}{\tau_{chem}}\right)_m = \left(\frac{I_L}{\tau_{chem}}\right)_p = 0.4 \qquad (8)$$

with for the wind tunnel τ_{chem} = τ_{NO}. The experimental results obtained in the wind tunnel should be compared with a field situation of a second-order irreversible reaction determined by τ_{chem} with the above given ratio between diffusion and chemical time scale

THE EXPERIMENTAL RESULTS

The experiments have been carried out with an O_3-concentration in the environment of 0.350 ppm and the time scale ratio given in formula (8). Transverse concentration profiles of NO, NO_x and O_3 have been measured at the height of the stack at distances of x = 0.5, 1, 1.5, 2, 3, 4, and 5 m behind the stack. In this study no ground-level concentrations have been determined. However, with chemically reactive plume models validated, among others, against these experiments, ground-level concentra-

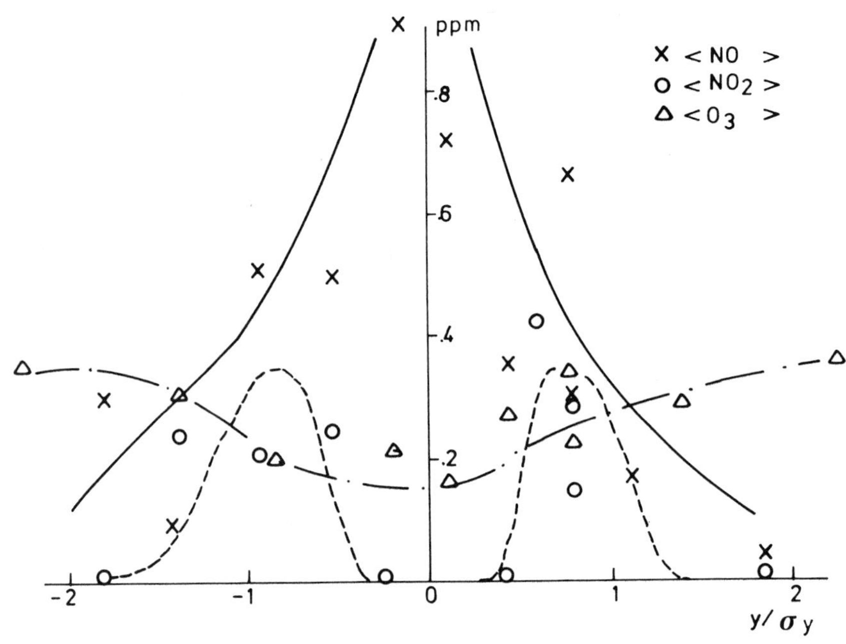

Fig. 2 Concentrations at x = 1 m

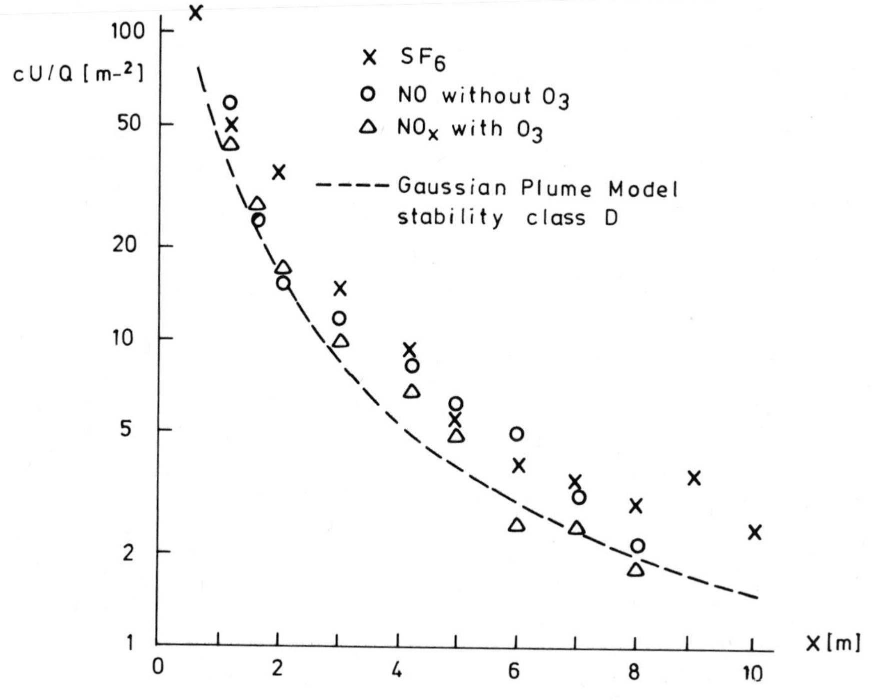

Fig. 1 Concentration at plume axis

trations could easily be calculated. Fig. 2 shows the concentration profile at x = 1 m. All the data are given in values determined in the wind tunnel. Field-distances can be found by multiplying by a factor 500, field-time scales by multiplying by a factor 35. The NO_2-data are determined by subtracting the measured NO_x and NO-concentration at that place. The transverse direction y is made dimensionless with σ_y according to the Pasquill-dispersion coefficient for stability class D :

$$\sigma_y = 0.128\ x^{0.907} \tag{9}$$

Fig. 3 shows the same profiles at x = 2 m, Fig. 4 at x = 3 m. The reproducibility of the measurements is about 15% for O_3, 25% for NO_x and 30% for NO, which explains the large scatter in the NO_2-data.

The O_3-profiles show a distinct decrease of O_3 in the plume. This is in agreement with all the fieldmeasurements (ref.6, 9, 10, 11, 14, 22, 23, 24, 27, 32, 35, 36, 38). Close behind the stack the NO_2-concentration at the plume-axis is zero, at the edge of the plume two separated peaks of NO_2 occur. This behaviour can be explained by the fact that O_3 is entrained in the plume first at the edges of the plume, and consequently a reaction zone will develop. This behaviour is typical of a diffusion controlled chemical reaction. The expected occurrence of NO_2-peaks has been mentioned by Shu(32) and Hegg(10), (see also the discussion in ref. 18) ; but field data hardly show this behaviour because in general only results rather far from the stack are given (an exception is shown in ref. 27). The results show that the NO_2-peaks disappear quickly, and can not be detected for distances in the wind tunnel further downstream than 2 m. Caused by the occurrence of the NO_2-peaks, the NO-profile will be smaller than the NO_x-profile and consequently cannot be described by a normal Gaussian plume behaviour. The maximum NO_2-concentration in the peaks is always smaller than the NO-concentration at that place, for this situation.

Next to the transverse profiles given in Fig. 2 - 4 more general and practical results can be obtained by regarding the situation on the plume axis downwind of the stack. Fig. 5 gives the NO_2-concentration in downwind direction. The maximum NO_2-concentrations found in the peaks are also given, which show a maximum NO_2-level at about x = 1 m. At x = 3 m, the NO_2-side peaks have disappeared. Fig. 6 gives the NO/NO_x-ratio at the plume axis, fig. 7 the ratio $O_3/(O_3)$ environment, both against the travelling time in the wind tunnel, t = x/U, with U = 0.4 m/s.

A 50%-conversion from NO to NO_2 is found in the wind tunnel at about 8 sec. Transferred to the field situation this conversion time would be around 5 minutes, or a distance of about 2km down-

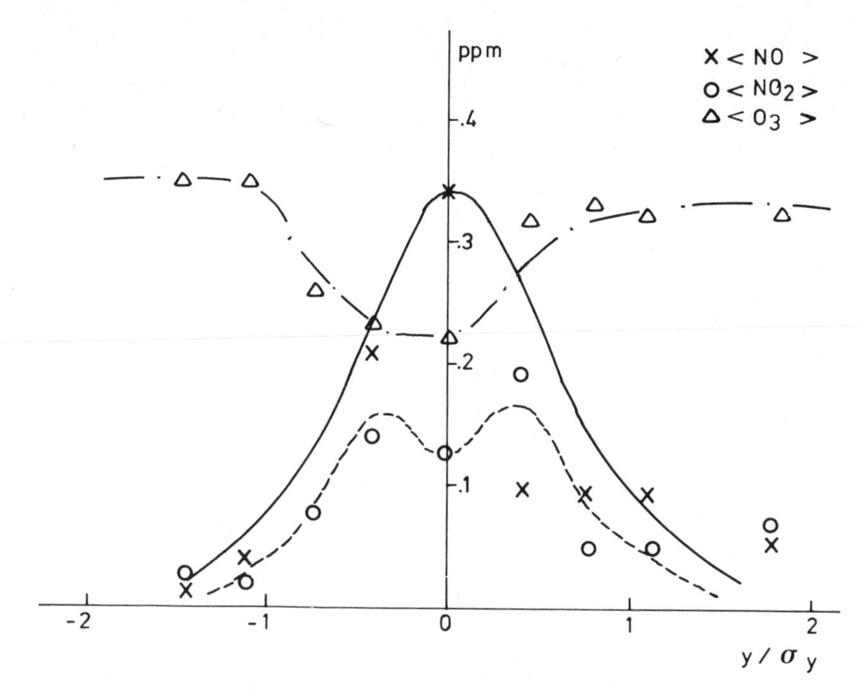

Fig. 3 Concentration at x = 2 m

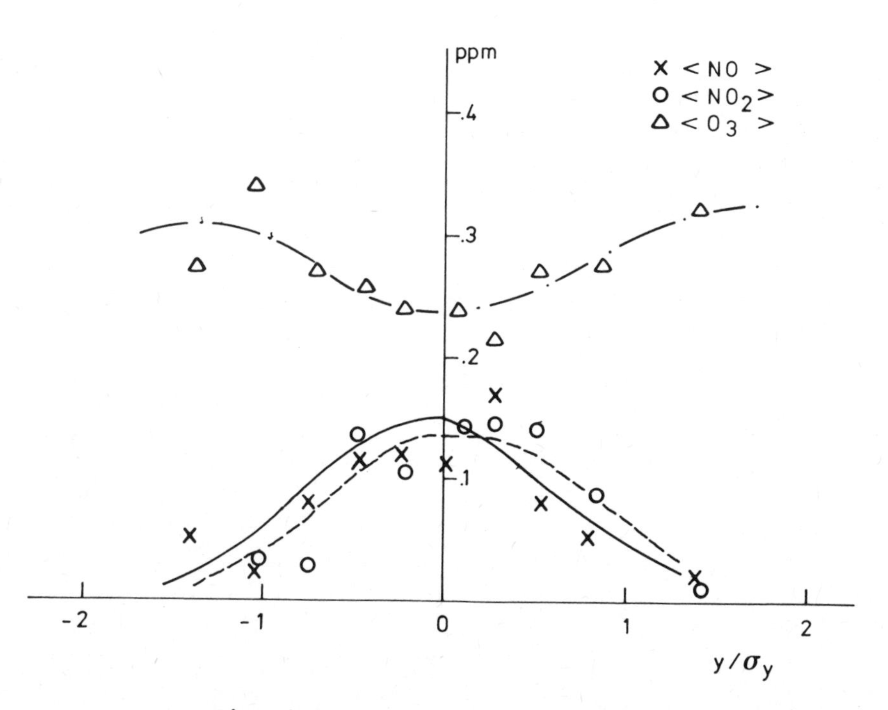

Fig. 4 Concentration at x = 3 m

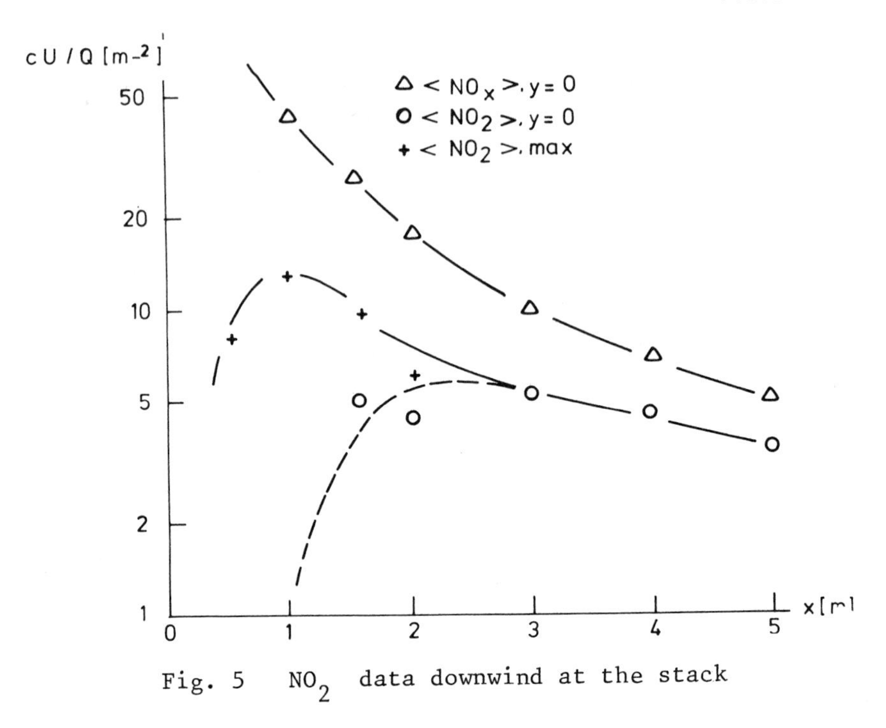

Fig. 5 NO_2 data downwind at the stack

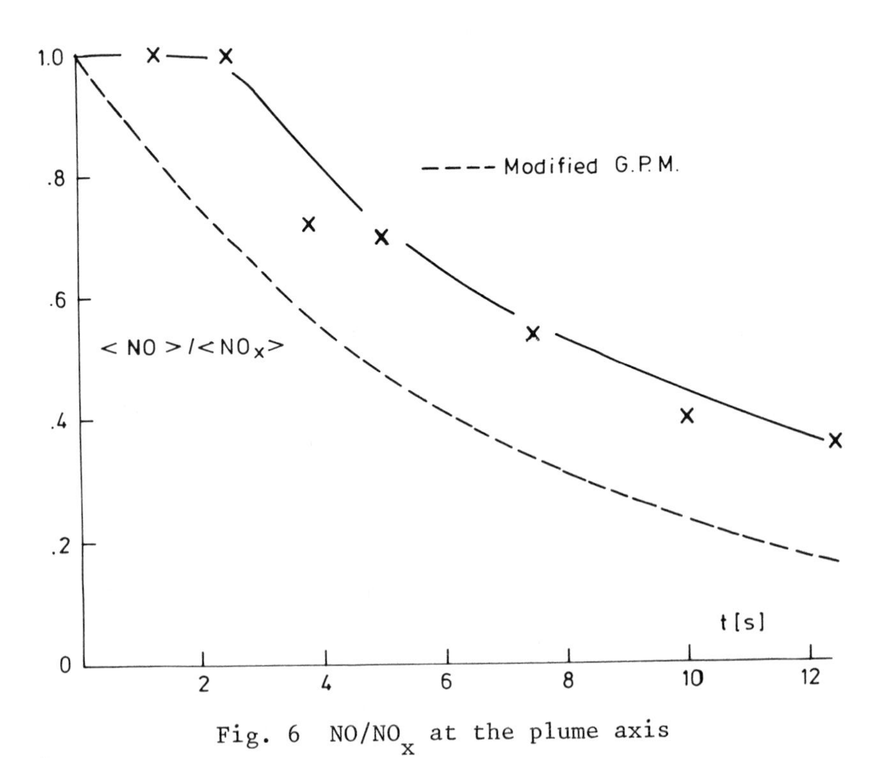

Fig. 6 NO/NO_x at the plume axis

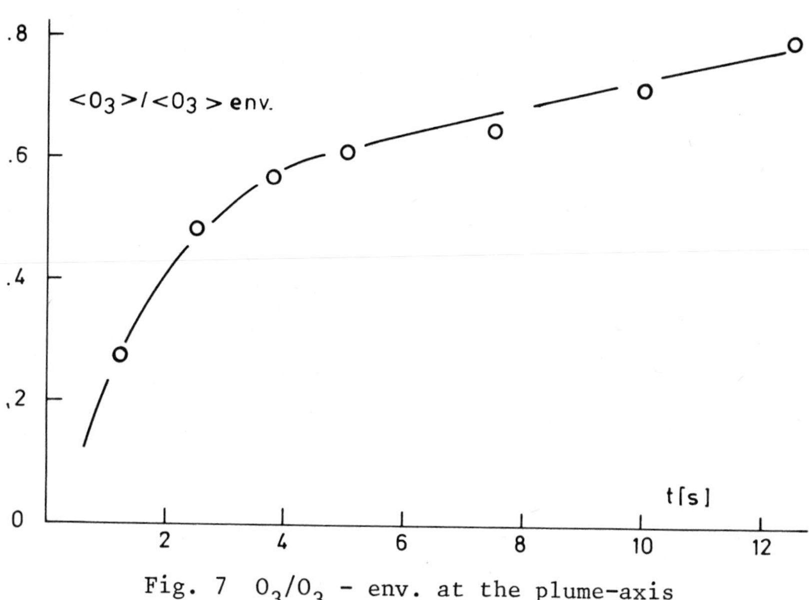

Fig. 7 O_3/O_3 - env. at the plume-axis

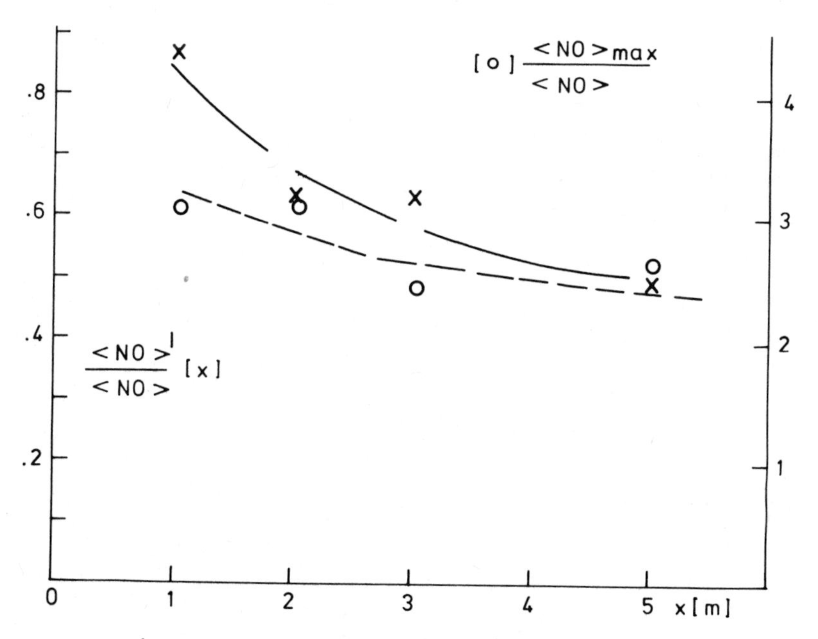

Fig. 8 NO-fluctuations at the plume-axis

wind of the stack. For a practical application, the results of the conversion rate as given in Fig. 6 are most important. In case this conversion rate should be known for every situation, different O_3-environment concentrations and different meteorological conditions, NO and NO_2-groundlevel concentrations could be determined. It is obvious that field studies and wind tunnel experiments alone will not be sufficient to obtain the desired information. Only with the aid of a validated chemically reactive plume model can the conversion rate be determined for every situation desired. The validation of such a theoretical model should be made against field studies and wind tunnel experiments.
Finally, some remarks will be made about measured NO-concentration fluctuations. The response time of the NO-analyser is about 1 sec, the time lag caused by the sampling tubes is also around 1 sec. Because the flow is laminar in the tube, a broadening of a stepwise change in NO-concentration will be less than 0.2 sec. At the plume axis, 10-minute recordings of NO-concentrations have been performed, after which the results are analysed for instantaneous values at intervals of 3 sec. Root-mean-square values of NO, $\langle NO \rangle'$, have been obtained, together with the second maximum value obtained during the recording. The $\langle NO \rangle'$-data have to be compared with concentration fluctuations in the field of every 100 sec, consequently only low-frequency fluctuations have been detected. The results are given in Fig.8. The RMS-value of NO decreases going downwind, as would be expected, also the maximum value seems to decrease ; these results, however, should be viewed to be tentative only.

The above given results are an illustration of the total number of existing data. In case detailed comparisons between these data and calculations with reactive plume models are being made and more information is needed the full data can be obtained from the author.

THE MODIFIED GAUSSIAN PLUME MODEL

Insight in the relevant processes which govern plume behaviour can be obtained by trying to adapt the simpliest possible model to the results, the modified Gaussian plume model. The diffusion equation for the concentration of NO_x in a plume can be written as :

$$U\frac{\partial}{\partial x}\langle NO_x\rangle = K_y\frac{\partial^2}{\partial y^2}\langle NO_x\rangle + K_z\frac{\partial^2}{\partial z^2}\langle NO_x\rangle \qquad (10)$$

in which K_y and K_z are the turbulent diffusivities in the y- and z- direction. With the appropriate values for K_y and K_z, equation (10) leads to the normal Gaussian plume model. The diffusion equation for NO, however, reads :

$$U\frac{\partial}{\partial x} \langle NO \rangle = K_y \frac{\partial^2}{\partial y^2} \langle NO\rangle + K_z \frac{\partial^2}{\partial z^2} \langle NO\rangle - k \langle NO\rangle\langle O_3\rangle - k \overline{\langle NO\rangle'\langle O_3\rangle'} \quad (11)$$

in which $\langle NO\rangle'$ is the turbulent fluctuation of the NO-concentration. The same equation holds for $\langle O_3\rangle$ Lamb[17], Shu[32]). In case the last term on the right-hand side is zero (or proportional to k $\langle NO\rangle \langle O_3\rangle$!), and $\langle O_3\rangle$ is constant, and K_y and K_z have the Gaussian plume values, the concentration of NO can be calculated by the Gaussian plume model, with the emission Q modified by

$$Q = Q_o \exp(-x/U\, \tau_{NO}) \quad (12)$$

in which Q_o is the emission of an inert species as NO_x and τ_{NO} is given by formula (2). In this situation the second-order reaction of NO is approached by a pseudo first-order reaction. However, this can only be done in case the chemical reaction is slow, relative to the diffusion process, which means $\tau_{chem} \gtrsim 10\ I_L$ which is in general the case for the oxidation of SO_2 (see Lusis, 22). However, in case $\tau_{chem} \lesssim I_L$, diffusion will play an important role in the transformation process (see Donaldson (5), Hegg (10), Kewley (16), Lamb (17)). Consequently, diffusion could be of importance for the wind tunnel experiment, and even more so for field situations with relatively high O_3- values in the surrounding air.

In this situation the O_3 will diffuse from the edges to the centre of the plume, and will be largely consumed by NO before the centre is reached, NO_2-peaks will be formed at the edges of the plume (see the description given by Hegg (16)). In combination with this process, the concentration fluctuation correlation $\overline{\langle NO\rangle'\langle O_3\rangle'}$ will be of importance. This correlation will be negative because in case $\langle NO\rangle$ is large, $\langle O_3\rangle$ will be small. This means that the change of NO with distance, or time, given by formula (11) will be retarded relative to a modified Gaussian plume model approach by the correlation term, as has been stated by Cole(4), Donalson (5), Hegg(10), Kewley(16), Thesche(26), Shu(32) and others. Also the inhomogeneous mixing of O_3 described before will create a retardation because the $\langle O_3\rangle$ -value in the term $\langle NO\rangle\langle O_3\rangle$ will be smaller. Consequently, both related processes will cause a retardation of the conversion of NO.

To investigate the importance of the possible retardation of the NO-conversion measured in the windtunnel the modified Gaussian plume model will be applied to the measured resulty. By assuming a constant concentration in the environment of 350 ppb O_3 the conversion of NO to NO_x can be directly obtained by

$$\langle NO\rangle = \langle NO_x\rangle \exp(-t/\tau_{NO}) \quad (13)$$

in which τ_{NO} = 6.9 sec. The calculated results are given in Fig.6. These results clearly show that the measured conversion rate is retarded relative to the calculated conversion rate which is an agreement with the results shown by Shu [32] for example. In case the retardation can be explained solely by the correlation term, this correlation term would be about equal to the term $\langle NO \rangle \langle O_3 \rangle$ at the 50%-conversion rate. It can be expected that for chemical time scales smaller than the value of 2.5 I_L used in this experiment, the correlation term would be even larger. To investigate this, experiments in the wind tunnel at higher O_3-concentrations should be carried out.

CHEMICALLY REACTING PLUME MODELS

As has been stated, for a practical situation, it is sufficient for determination of groundlevel concentrations of NO and NO_2 caused by an emission of NO from a stack, to know the conversion rates as given in Fig. 6, for every situation. The NO_x-concentration can be calculated with the normal Gaussian plume model. As has been shown this conversion rate cannot be calculated by a modified Gaussian plume model. Consequently, chemically reacting plume models have to be applied to determine conversion rates. In general, the conversion rate will primarily be a function of the ratio between the chemical time scale and the diffusion time scale, which means that the conversion rate is a function of the O_3-concentration in the surrounding air, and of the meteorological conditions which influence the dispersion process. In case $\tau_{chem} \lesssim I_L$, the chemical and diffusion process are connected and cannot be handled separately. The conversion rate will of course also be a function of light intensity, the amount of hydrocarbons present, etc. However, the reaction of NO with O_3 is of primary importance, and illustrates the interrelationship between chemistry and diffusion.

For practical applications it is not necessary to use a model which is capable of calculating the observed NO_2-peaks as shown in Figs. 2 and 3 ; these peaks are rather of scientific interest. On the other hand a model to be used should be more sophisticated than a modified Gaussian plume model, which means that a model must predict the retardation shown.

Four classes of reactive plume models exist :

I The single well-mixed box approach
II A 'divided' Gaussian plume approach
III Models which incorporate the correlation term
IV Higher-order closure schemes.

In principle, higher-order closure schemes could be used to

calculate the desired conversion rates. The NO-plume in an O_3-environment is an example of the class of bimolecular reactions with initially unmixed reactants in a turbulent fluid. Based on additional differential equations which describe the behaviour of the correlation term, and higher order correlations, calculations could be made. Some of these approaches have been described by Hill[12], Libby[19], Spalding[34], Murthy[25], O'Brien[26], and others [15,29, 30]. However, at this moment no real reactive plume calculations have been performed using this approach. This is mainly caused by the fact that the computations required would be extensive, and the fact that data to validate the models and their assumptions, are lacking. It seems unlikely that in the near future models of this type would be used for practical applications.

Some reactive plume models exist which use the concept of one single expanding well-mixed box. The box grows in size going downwind due to a dilution factor for example given by a Gaussian plume-model approach. Surrounding pollutants will be entrained into the box. This type of models is typical of models which incorporate a relatively large chemical reaction scheme. Main emphasis in these models is on chemistry, not on dispersion Examples are the models developed by Liu[20], Cocks[31], Varey [37], Isaksen[14], Forney[7]. However, these types of models will be inappropriate for describing the behaviour of a NO-plume in an O_3-environment. The models do not take into account inhomogeneous mixing, nor the correlation term, and will consequently not be able to predict the above-mentioned retardation. A modification of these types of models is given by Peters[31] and White [38] (see also[16]). They define an inert, conserved species which is for example the sum of two reactive species. This hypothetical inert species will be dispersed according to the Gaussian plume model behaviour.Afterwards a chemical calculation is carried out based on local chemical equilibrium. Also in this approach there is a complete separation of chemistry and dispersion, and consequently these models are inadequate to describe the conversion of NO in a plume.

More promising is the concept of the 'divided' Gaussian plume. The basis is the dispersion of an inert species according to the Gaussian plume model. To calculate chemical transformations, the Gaussian plume profiles are divided, either in boxes, see Liu[21], or in elliptical rings, see Lusis[22,23]. (see also[8]). In this way these models do take into account the inhomogeneous mixing, O_3 will be entrained from the edges of the plume going inwards. The few comparisons made between these models and field studies show a reasonable agreement, the models seem to describe the retardation (see Lusis[23]). It should be noticed that those models do not take into account the correlation term.

This correlation term is incorporated in the reactive plume model described by Lamb[17] and Shu[32] (see also[16]). The correlation term is expressed in terms related to either the chemistry or the mixing process. Calculations made with this model clearly show the retardation[32], the results are in agreement with field experiments. However, at this moment the model only calculates concentrations averaged in the transverse direction. This means that the model does not take into account the effect of inhomogeneous mixing. A remark has to be made regarding the suggestion made by Cole[4] and White[39]. They argue that for the calculation of a reactive plume, instead of the time-averaged dispersion coefficients of the Gaussian plume model, instantaneous dispersion coefficients have to be used. In terms of equation (11) this means that instead of, or next to, taking into account the correlation term and changing the $< O_3 >$ -value in the term $k < NO > < O_3 >$, the terms describing the diffusion process should be changed. This procedure would lead to the fact that in principle the diffusion terms would become chemistry-dependent, because the dispersion of an inert species can be described by the normal Gaussian plume model. From a physical point of view this proposed dependency of the dispersion terms on chemistry is unrealistic.

As a conclusion the following can be stated. The retardation relative to the modified Gaussian plume model calculation for the conversion rate of NO to NO_2 is caused by two related processes, inhomogeneous mixing and the correlation term. Reactive plume models to be used for practical applications must be capable of calculating this retardation, and consequently must take into account inhomogeneous mixing and the correlation term. The Liu-Lusis model takes into account only the inhomogeneous mixing, the Lamb-Shu model takes into account only the correlation term. Which of these effets is most relevant cannot be decided at the moment, especially because the two effects are highly interrelated. A little confusing is the fact that both the Liu-Lusis and Lamb-Shu models are in agreement with field experiments, which shoul mean that by incorporating in these models the "missing" term, an over-retardation would be found. Still, the incorporation of the missing term into the models will be of importance. Of course, testing of the models to field experiments and wind tunnel data will be of vital importance.

CONCLUSION

The following conclusions can be drawn :

- The conversion of NO to NO_2 in a NO-plume in an O_3-environment can be investigated in an atmospheric boundary-layer wind tunnel
- Experiments in the wind tunnel at a ratio of the chemical time scale to the diffusion time scale of about 2.5 clearly show a retardation of the conversion from NO to NO_2 relative

to a modified Gaussian plume-model approach
- In the wind tunnel two distinct NO_2-peaks at the edges of the plume have been found close to the stack.
- The retardation effect is caused by two interrelated processes, inhomogeneous mixing of O_3 and the turbulent concentration fluctuation correlation .
- The retardation effect can only be taken into account in chemically reacting plume models when at least one of the abovementioned processes is incorporated in the model, in principle both terms should be incorporated.
- The chemically reacting plume model approaches as carried out by Liu-Lusis and Lamb-Shu seems to be the most promising, although at this moment these models do not seem to be directly suited for practical applications.

ACKNOWLEDGEMENT

The author is indebted to mr. J. Duyzer for his advice in the set-up of the chemical measurements, and to mr. F. Balster and mr. P. Merrelaar for carrying out the measurements.

REFERENCES

1. Builtjes, P. J. H. and Vermeulen, P. E. J.
"Atmospheric boundary layer simulation in the PIA and MIA wind tunnels of TNO-Apeldoorn" - TNO - Rep. 80 : 0290 (1980).
2. Cermak, S.E.
"Laboratory simulation of the atmospheric boundary layer" AIAA - Journal Vol. 9,9, 1746 (1971).
3. Cocks, A. T. and Fletcher, I. S.
"A model of the gas-phase chemical reactions of power station plume constituents" - CERL - Rep. RD/L/R 1999 (1979).
4. Cole, H. S. and Summerhays, J. E.
"A review of techniques available for estimating short-term NO_2-concentrations" - J. of the Air Poll. Conts. Ass. 29,8, 812-817 (1979).
5. Donaldson, C. du P. and Hilst, G. R.
"Effect of inhomogeneous mixing on atmospheric photochemical reactions" - Environ. Sci. Techn. 6, 812 (1972).
6. Elshout, A. J. e.a.
"The oxidation of NO in plumes" - Elektrotechniek 56, 429-437 (1978) in Dutch.
7. Forney, L. T. and Giz, Z. G.
"Fast reversible reactions in power plant plumes : application to the nitrogen dioxide photolytic cycle" - Atm. Env. 15, 345 (1981).

8. Freiberg J.
"Conversion limit and characteristic time of SO_2 oxidation in plumes" - Atm. Env. 12, 339 (1978).
9. Guicherit, R. e.a.
"Conversion rate of nitrogen oxides in a polluted atmosphere" 11-th Int. Tech. Meeting on Air Pollution Modelling and its Application, pg. 378, Amsterdam (nov. 1980).
10. Hegg, D. e.a.
"Reactions of ozone and nitrogen oxides in power plant plumes" Atm. Env. 11, 521-526 (1977).
11. Hegg, D. A. and Hobbs, P. V.
"Measurements of gas-to-particle conversion in the plumes from five coalfired electrical power plants" - Atm. Env. 14, 99 (198(
12. Hill, J. C.
"Homogeneous turbulent mixing with chemical reaction" - Chapter from "Turbulent Chemical Reactions", 1976.
13. Hinze, J. O.
"Turbulence" - 2nd Ed. Mc Grawhill (1975)
14. Isaksen, I. S. A. e.a.
"A chemical model for urban plumes : test for ozone and particulate sulfur formation in St. Louis urban plume" - Atm. Env. 12. 599 (1978).
15. Janicke, J. and Kollman, W.
"Ein Rechenmodell für reagierende turbulente Schertströmungen in chemischen Nichtgleichgewicht" - Wärme- und Stoffübertragung 11 157-174 (1978).
16. Kewley, D. J.
"Atmospheric dispersion of a chemically reacting plume" Atm. Env. 12, 1895-1900 (1978).
17. Lamb, R. G. and Shu, W. R.
"A model for second-order chemical reactions in turbulent fluid Part 1 "Formulation and validation" - Atm. Env. 12, 1685-1694 (1978).
18. Lamb, R. G. and Richards, L. W.
Discussion in Atm. Env. 13, 1473 (1979).
19. Libby, P. A. and Williams, F. A.
"Turbulent reacting flows" - Topics in Applied Physics Vol. 44, Springer-Verlag Berlin (1980).
20. Liu, M. K.
"Development of a mathematical model for simulating power plant plumes" - 4th Int. Clean Air Congress, Tokyo, Japan (197
21. Liu, M. K. e.a.
"An improved version of the reactive plume model (RPM-II)" 9th Int. Technical Meeting on Air Pollution and its Applicatio Toronto, Canada (Aug. 1978).
22. Lusis, M.
"Mathematical modelling of chemical reactions in a plume" 7th Int. Technical Meeting on Air Pollution and its Applicatio Airlie, Virgina, USA (Sept. 1976).

23. Lusis, M.
"Mathematical modelling of dispersion and chemical reactions in a plume-oxidation of NO to NO_2 in the plume of a power plant" - Atm. Env. 12, 1231-1234 (1978).
24. Lusis, M. e.a.
"Plume chemistry studies at a Northern Alberta power plant" Atm. Env. 12, 2429-2437 (1978).
25. Murthy, S. N. B.
"Turbulent mixing in nonreactive and reactive flows" Plenum Press (1975).
26. O'Brien, F. E.
"Closure for stochastically distributed second-order reactants" Physics of Fluids 11, 9, 1883-1888 (1968).
27. Parungo, F. P. e.a.
"Chemical characteristics of oil refinery plumes in Los Angeles" Atm. Env. 14, 509 (1980).
28. Pasquill, F.
"Atmospheric Diffusion" - 2nd Ed. Ellis Horwood (1974).
29. Patterson, G. K.
"Modelling complex chemical reactions in flows with turbulent diffusive mixing" - AICHE 70th Annual Meeting, New-York (Nov. 1977).
30. Patterson, G. K.
"Closure approximations for complex multiple reactions" - Second Symp. on Turbulent Shear Flow, London (July 1979).
31. Peters, L. K. and Richards, L. W.
"Extension of atmospheric dispersion models to incorporate fast reversible reactions" - Atm. Env. 11, 101-108 (1977).
32. Shu, W. R. e.a.
"A model of second-order chemical reactions in turbulent fluid" Part II. "Application to atmospheric plumes" - Atm. Env. 12, 1695-1704 (1978).
33. Snijder, W. H.
"Similarity criteria for the application of fluid models to the study or air pollution meteorology" - Bound. Lay. Met. 3, 113 (1972).
34. Spalding, D. B.
Combust. Sci. Techn. 13,3 (1976).
35. Tesche, T. W. e.a.
"Theoretical, numerical and physical techniques for characterizing power plant plumes" - EC-144 EPRI Project 572-2 (February 1976).
36. Tesche, T. W. e.a.
"Determination of the feasibility of ozone formation in power plant plumes " - EPRI EA-307 (November 1976).
37. Varey, R. H. e.a.
"The oxidation of nitric oxide in power station plumes, a numerical model" - CERL, Laboratory Note RD/L/N 184/78 (1978).

38. White, W. H.
"NO_x-O_3 Photochemistry in power plant plumes : Comparison of theory with observation" - Env. Sc. and Techn. 11, 10, 995-1000 (1977) - A, C2.
39. White, W. H. and Lamb, R. G.
Discussion in Atm. Env. 13, 1471 (1979).

DISCUSSION

W. B. JOHNSON

It is well known that ozone reacts rapidly when it comes in contact with surfaces, such as the bottom and sides of your windtunnel. Did you take this potentially large ozone deposition into account in your study and if not, is it significant enough to affect your results?

P. J. H. BUILTJES

Ozone deposition is detectable in the windtunnel, however, due to the relatively large air-resistance at these low velocities, this phenomenon is restricted to upto about 3-4 cm distance from the surface only, and will consequently not influence the results at stackheight.

M. K. LIU

Could you please comment briefly on the feasibility of using windtunnel to simulate slow reactions involving reactive hydrocarbons and NO_x leading to ozone formation?

P. J. H. BUILTJES

In principle you can take into account in your windtunnel experiment also other compounds, such as hydrocarbons The scaling of these reactants will be determined by the timescale of the reactions.

P. M. ROTH

What is the most significant limitation associated with the various scaling assumptions made in constructing the physical model?

P. J. H. BUILTJES

The most severe restriction is the assumption made that length scales smaller than about 0.25 m in the field don't have an influence on the reaction. Although this assumption seems physical logical, I cannot give a definite prove of this assumption.

M. SCHATZMANN

Wouldn't it be helpful to measure in the windtunnel not only the intensities and macroscales but the spectra of turbulence and to compare it with field measurements (in a non-dimensionalised form)?

P. J. H. BUILTJES

Turbulent spectra have been measured in the windtunnel, as shown in ref. (1).

A. BERGER

What are the values of K_z you have chosen. Are they constants? Have you tried to make them as functions of Z? Do you think that such a Z-dependence would change your conclusions?

P. J. H. BUILTJES

In the study no direct values of K_z have been used, only the dispersion coefficients σ_y and σ_z of the Gaussian plume model for a neutral stabilitiy have been used, according to Pasquill.

G. WHITTEN

It is not clear to me how the windtunnel be used to both suppress and enhance the "retardation" effect compared to the atmosphere.

P. J. H. BUILTJES

The retardation effect could be decreased or increased in the windtunnel by changing the ratio of $\tau chem/I_L$. The easiest way to do this would be the changing of $\tau chem$ by changing the O_3-concentration.

H. VAN DOP

You use in your experiments the chemical components NO and O_3. In order to study the effect of chemical reaction time scales in relation with turbulence time scales, would it not be useful to use species with (significantly) different reaction times?

P. J. H. BUILTJES

I agree with this remark. In principle all easily detectable species, with a not too slow and not too fast reaction, could be used.

A. VENKATRAM

Can you explain what you mean by "inhomogeneous" mixing? And why doesn't the Lamb/Shu model account for it?

P. J. H. BUILTJES

By "inhomogeneous" mixing I mean the process described by the term $k\langle O_3\rangle \langle NO\rangle$ in which $\langle O_3\rangle$ will be smaller than $\langle O_3\rangle$ environment, due to "inhomogeneo diffusion controlled, mixing. The Lamb/Shu model only calculates average concentrations across the plume, and consequently this inhomogeneous mixing is not taken into account.

M. K. LIU

(Comment:) I like to offer an explanatio to the apparent good comparison of both the Liu/Lusis model and the Lamb/Shu model Turbulent fluctuation is probably only important in the early stage of the plume evolution and the Liu/Lusis model might use a lower apparent reaction rate to accomodate the retardation.

A. VENKATRAM

You said that the NO-O_3 syste can be studied in the windtunnel. Anythin can be studied in the windtunnel. Explain your conclusions.

P. J. H. BUILTJES

I meant to say that the NO-plume in an O_3-environment can be studied in the windtunnel in such a way that the results are comparable to field situations and can be used as a data base to check reactive plume models.

J. L. WOODWARD

You showed that to satisfy scaling laws you need to increase reactio rates in a windtunnel by a factor of 35 over rates in the field. Could you not accomplish this with a uniformly heated windtunnel. Would it not be a better approach than increasing ozone concentration (because diffusion would take place proportionally)?

It is in principle not necessary to increase the reaction rates to satisfy the scaling laws; the scaling is correct by using the scaling laws shown leading to the fact that the time scales in the windtunnel are a factor 35 smaller than in the field. An increase of the time scale ratio or of τ_{chem} could be accomplished in the windtunnel by heating; however at this moment the windtunnel is not equipped with heating facilities.

Because $\tau_{chem} = \frac{1}{k < O_3 >}$ a decrease of τ_{chem} can also be accomplished by increasing $< O_3 >$, which is more easy at this moment.

AIR POLLUTANT REACTIONS IN MAJOR PLUMES

TRANSPORTED OVER THE NORTH SEA

Ir.T. Schneider

Director of the Environmental Hygiene Division of the
National Institute of Public Health
P.O.Box 1, Bilthoven, The Netherlands

INTRODUCTION

A study is being made of pollutant transport and transformation within a large number of major industrial and urban plumes. The North Sea to the west of The Netherlands was selected as the sampling area for the field experiments conducted during 1979 - 1981.

This paper deals with 3 types of transport conditions:

1. The general features of pollutant distribution with height over the North Sea during west-east transport circulation.
2. Air pollutant transformations within plumes over the North Sea as indicated by variations in pollutant concentrations with distance from the sources on the mainland.
3. Formation of air masses with high pollutant concentrations over the sea near the coast during semi-stagnant conditions prior to dispersion by transport.

EXPERIMENTAL PROGRAMME

A twin engine Piper Navajo Chieftain has been instrumented and is being used to determine the concentration variability and distribution of air pollutants (NO, NO_x, SO_2, O_3), both vertically and horizontally along flight tracks over the North Sea and the adjacent coastal area of The Netherlands, Belgium, France and the United Kingdom.

The flight tracks consisted of traverses perpendicular or at an angle to the mean wind direction at several altitudes. The number and specific flight levels were dependent on the weather conditions during the selected periods of steady wind direction.

The following instruments were used:

- for the measurement of the NO/NO_x concentrations a Bendix 8101 special and a Teco 14D analyser
- the ozone was measured with a Bendix 8002 analyser and
- the SO_2 concentrations by a Teco 43 pulsed fluorescent analyser
- air temperature was measured with a Rosemount total temperature sensor model 102BE
- the data system in the airplane consisted of a Monitor Labs 9400 datalogger, a King Radio Cooperation KDC 380 computer and a Perex 6300 cartridge recorder.

Continuous quality control during the experimental period in the consecutive seasons was maintained through regular laboratory calibration of the monitors and automatic checks of instrument performance before, during and after the measurement flights.

FLIGHT JULY 26, 1979

To the north of a high pressure area located between the south of England and south-eastern Europe (Balkan area), measurement flights were carried out on the track from Knokke (Belgium) due north for 180 km. Backtrajectories (850 mb) for 24 hours for the wes coast of The Netherlands are shown in Fig. 1. The air masses were moved over the flight track with a westerly wind of 4 - 5 m/s. The air had passed over the major source areas in the United Kingdom. Flight levels on the 26th of July were 150, 300, 600 and 1200 m.

At the higher levels the NO/NO_2 concentrations are as expected, very low (max. 5 µg/m^3). The SO_2 concentrations varied between 25 and 50 µg/m^3 (see Fig. 2 and 3 for SO_2 and NO_2). At the intermediate height of 300 m the concentration of SO_2 is 50 - 100 µg/m^3 and at the lowest level 50 - 75 µg/m^3 with a large plume producing maximum values of more than 200 µg/m^3. This plume is also evident in the NO_2 concentrations (up to 40 µg/m^3). The plur is shown in a negative sense in the O_3 concentrations at 150 m (see Fig. 4). Near the Belgium coast the ozone concentration range is 100 - 200 µg/m^3. Over the middle of the North Sea the concentrati is approximately constant at 100 µg/m^3 at all levels except for the 150 m level where values up to 150 µg/m^3 are found.

Using the results of the four flight levels we can construct a total "plume" profile by interpolation. For interpolation a "spline" function is used. The resulting plume profile is a two-dimensional matrix with isolines. The total gasburden, the total amount of pollution in a vertical column over a surface area, is defined here as the partial gasburden between the lowest and highest flight level. This gasburden in mg/m^2 follows from the plume profile matrix by integration over every vertical column of measurement values.

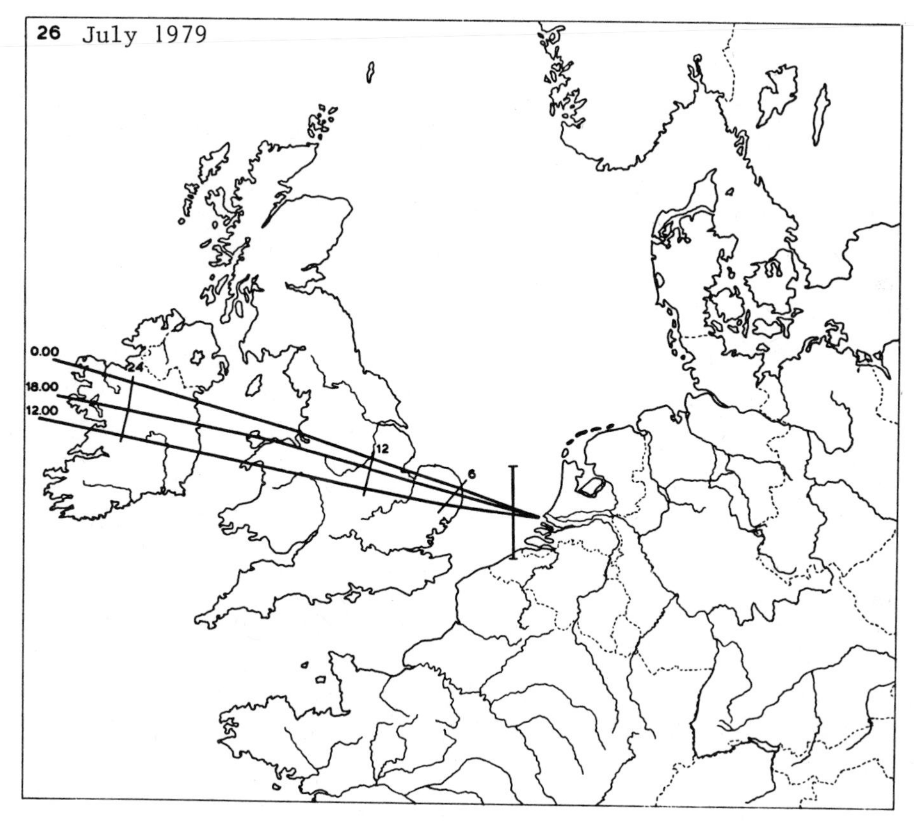

Fig. 1 Flight track and backtrajectories for the southern part of the North Sea

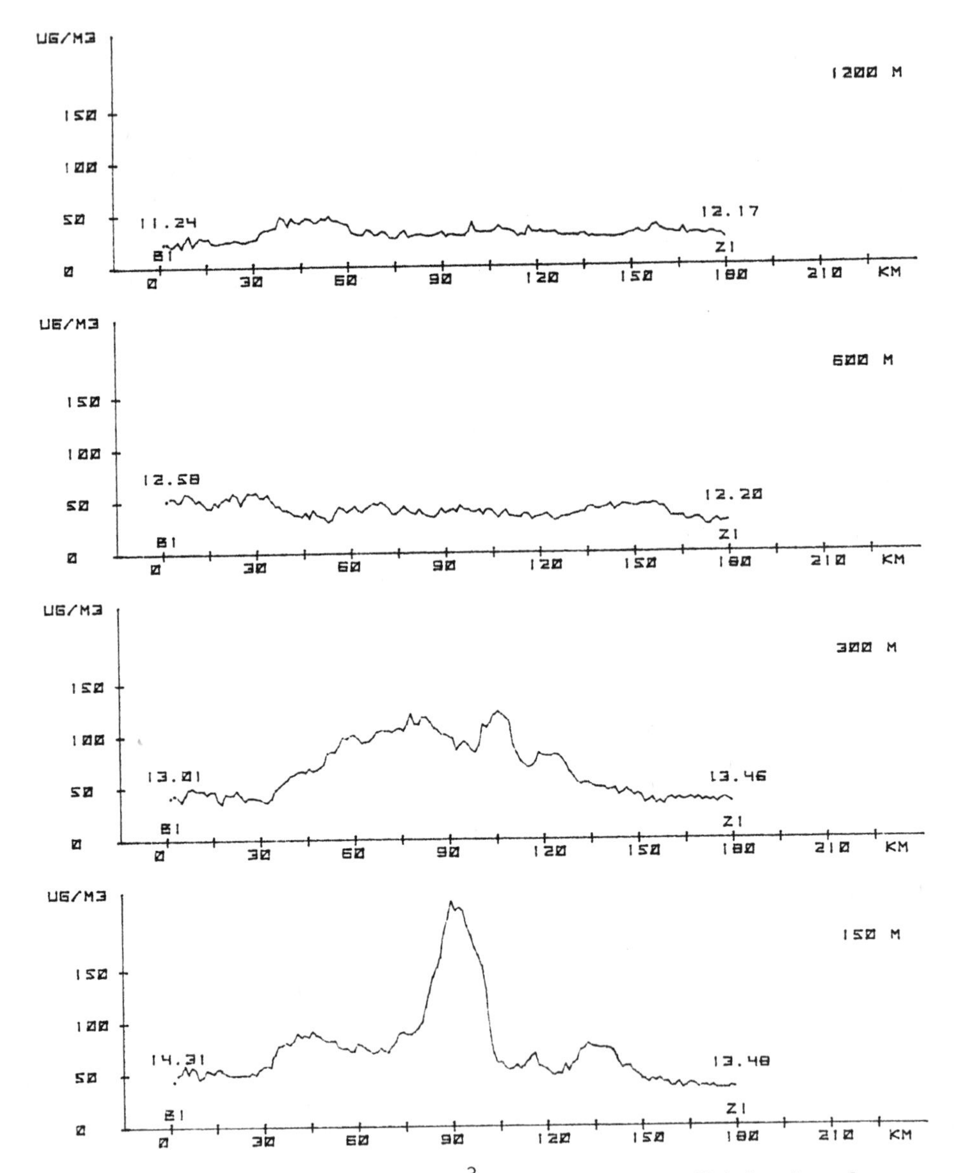

Fig. 2 SO_2 concentration ($\mu g/m^3$) at different flight levels over the North Sea, 26 July 1979

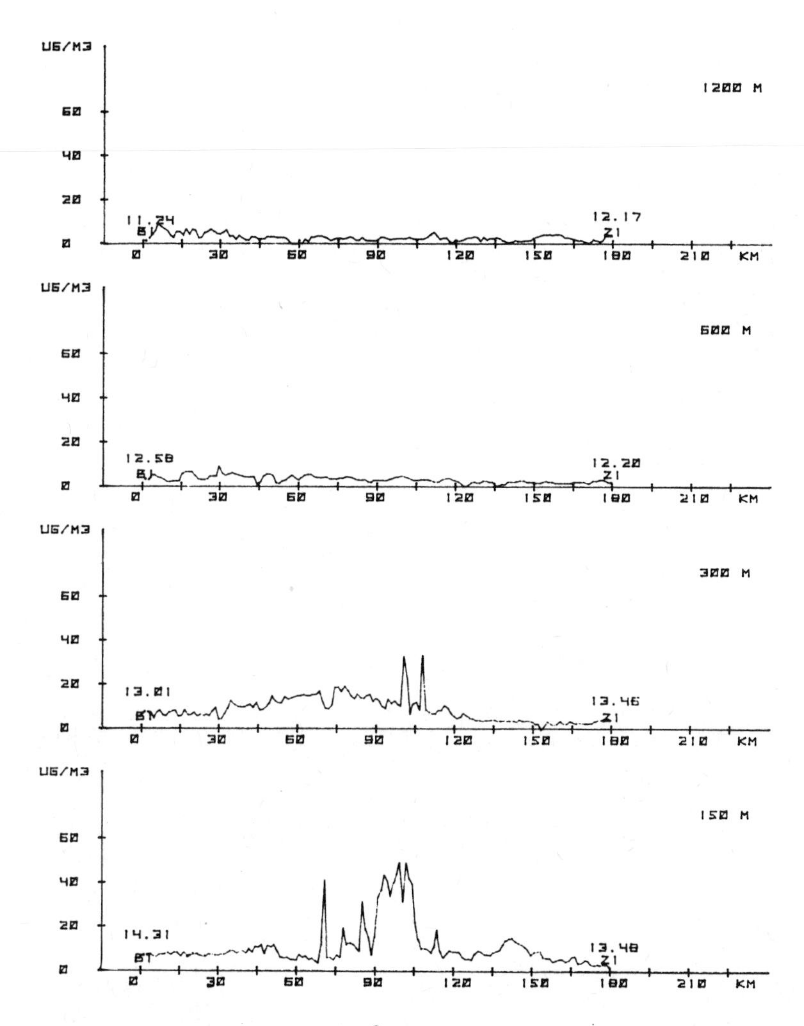

Fig. 3 NO_2 concentration (µg/m^3) at different flight levels over the North Sea, 26 July 1979

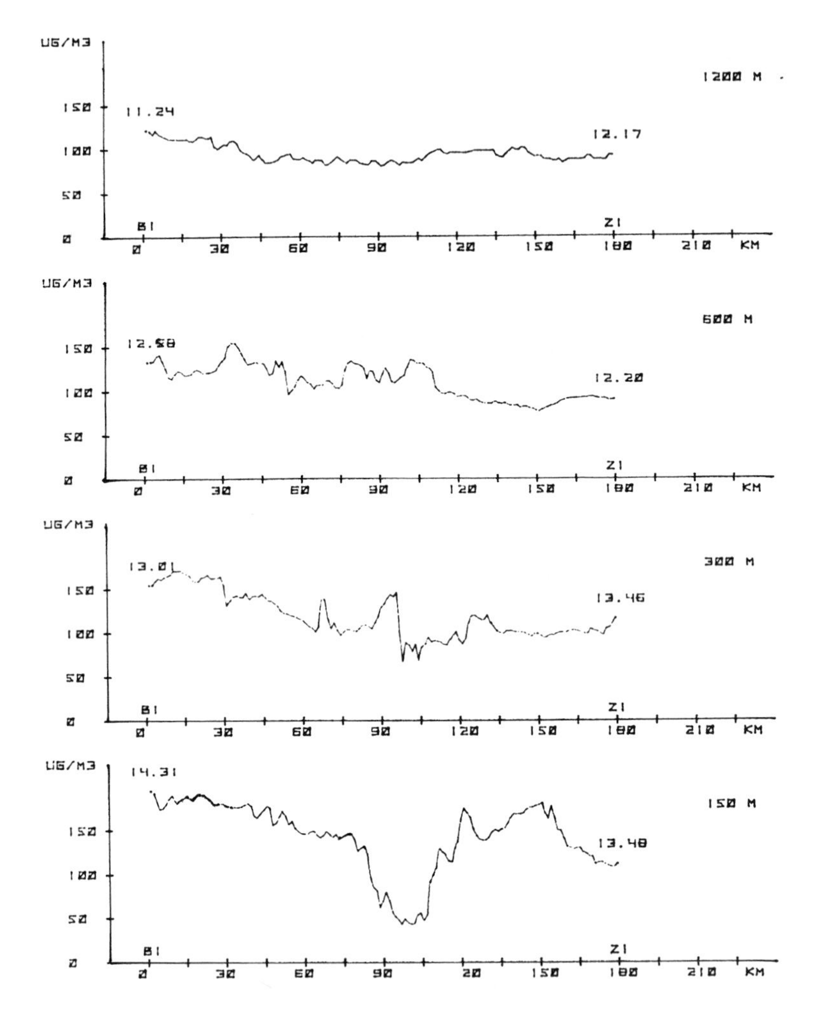

Fig. 4 O_3 concentration ($\mu g/m^3$) at different flight levels over the North Sea, 26 July 1979

The total flux of a pollutant over the flight track is given by multiplication of the gasburden with the windvector normal to the cross section (for the calculation the "ground wind speed" at the coast is used).

Calculated average fluxes for the period 11.30 a.m. to 2.30 p.m. on the 26th of July are:

SO_2 - 500 ton/h
NO_2 - 18 ton/h

Large variations in the calculations occur, primarily due to variations in the mean wind speed used. Also variations from day to day are found. On the 25th of July the mean wind direction was south-south west with mean fluxes of 250 ton/h for SO_2 and approximately 10 ton/h for NO_2.

Although valuable insights are obtained from the vertical distribution of the pollutant concentrations in the area over the sea, as was shown with the results of the transport calculations for the 26th of July, it is the change in pollutant concentrations with distance from the largest sources or source areas that is most useful in the analysis of the transformations of the individual pollutants. Therefore, an attempt was made to use flight patterns that enabled us to monitor the same plumes at different locations over the sea.

FLIGHT FEBRUARY 27, 1981

During a period with a high pressure system located over eastern Europe and a low pressure system above the Mediterranean, a steady easterly wind moved air masses from the continent towards the United Kingdom. On the 27th of February the transport took place with an average windspeed over the North Sea of approximately 8 m/s (winddirection was 110 - 115° at the Dutch coast).

The flight tracks are shown in Fig. 5. With this configuration of tracks individual plumes of large sources could be followed over the North Sea. At the flight tracks 1 and 2 respectively RC and CB, between 9 and 10 a.m. local time an inversion layer was indicated at a height of 260 m. Over land, flight track 3 and 4 (BRWDT), the inversion layer was found at a height of 300 - 350 m (at 1 - 2 p.m.). Above this inversion layer O_3 concentrations were approximately 90 $\mu g/m^3$, air temperature 4 - 5°C with a few $\mu g's/m^3$ of NO_x and SO_2

Over the middle of the North Sea inversion layers were later in the day (3.30 p.m.) found at 425 m and 900 m (see Fig. 6). The maximum O_3 concentrations were 100 - 110 $\mu g/m^3$ at the top of the lowest inversion layer. Below the inversion layers at a height of 200 m the pollution concentrations found were: SO_2: maximum 110 $\mu g/m^3$; NO_2: 110 - 115 $\mu g/m^3$; NO: maximum 25 $\mu g/m^3$ and O_3: 50 $\mu g/m^3$.

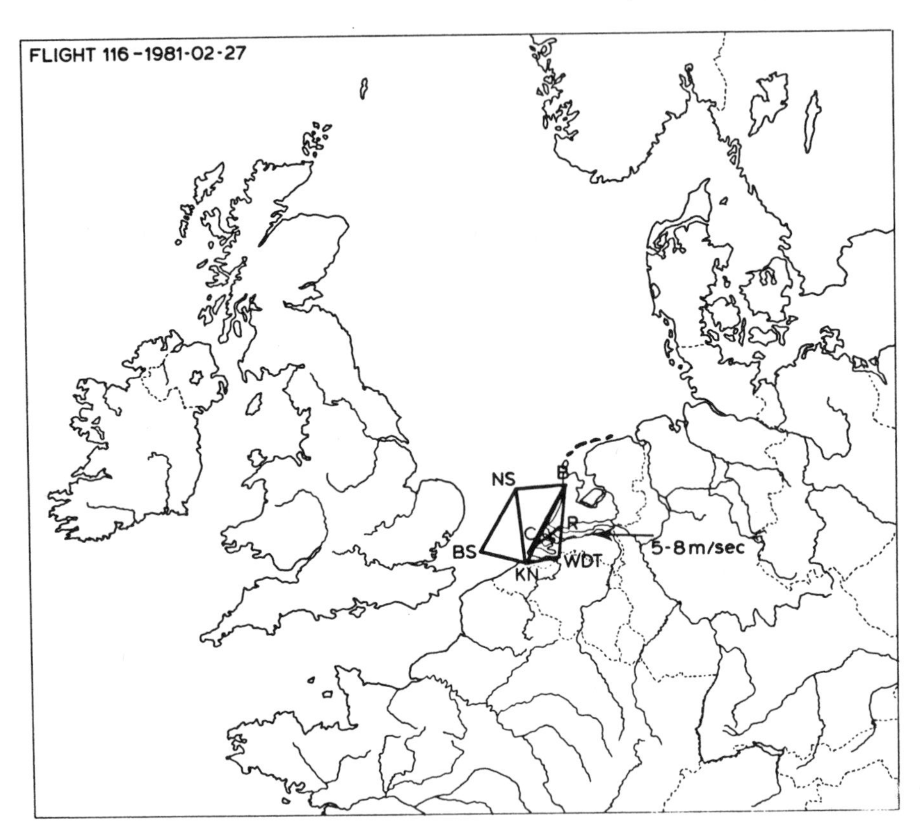

Fig. 5 Flight tracks over the west coast of The Netherlands and Belgium and over the southern part of the North Sea

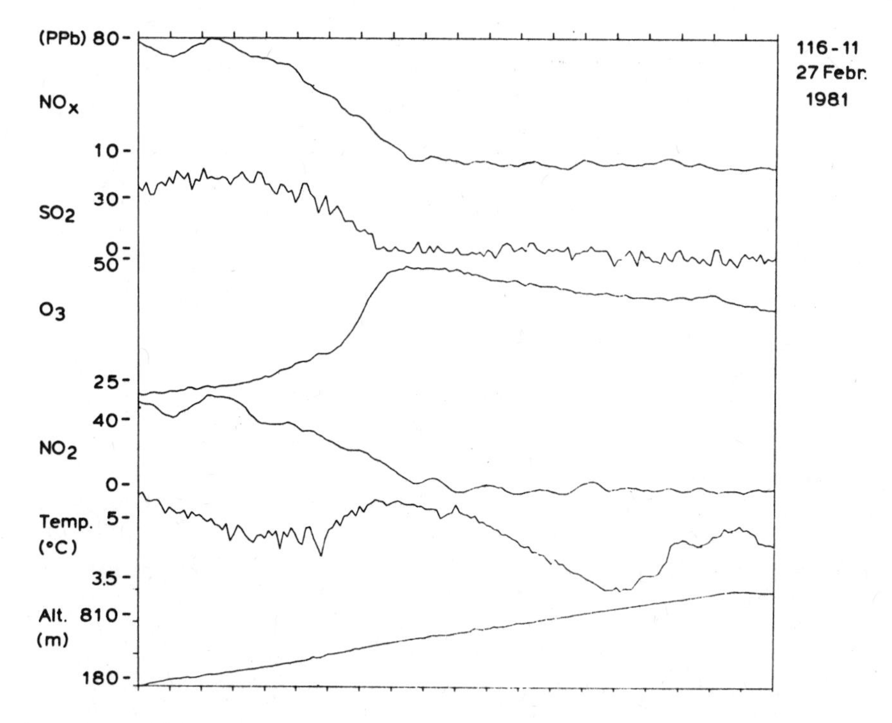

Fig. 6 Pollutant concentrations measured during an ascend above the North Sea at 3:30 p.m.

Table 1. Air pollutant concentrations (in $\mu g/m^3$) within the plume of a power plant at different distances from the source

flight track	13	09	10	12	07
distance from the source (km)	± 1	7	60	60	100
height (m)	335	330	390	350	410
NO_2	132	113	115	114	100
NO	200	35	33	22	32
SO_2	1010	430	310	306	210
O_3	25	36	44	45	40

Due to the steady mean windspeed and winddirection it was possible to study the variation in pollutant concentration with distance from the source, both from a single source (powerplant stack) and a source region (Rijnmond/Rotterdam area). Results are compared for the flight tracks 09 (D-KN), 10 (KN-NS), 12 (NS - KN) and 07 (BS-NS) located at respectively 7, 60, 60 and 100 km from the sources.

In Table 1 for comparison purposes also the results of a measurement near to the source (approximately 1 km) on flight track 13 is shown.

Several features can be noted from these results:

- There is a marked decrease of SO_2 concentrations with distance from the source. Assuming a steady mean windspeed of 8 m/s, the removal rate of SO_2 is on the order of 20%/hour within the plume
- For the nitrogen oxides on the other hand the concentration changes relatively slow, for the NO_2 concentrations within the plume the variation lies within the uncertainty in the measurements. Also the O_3 concentrations remain approximately constant, except for the flight track nearest to the source (09 and 13) were access NO is available.
- The rapid conversion of SO_2 within the plume of the single source is not comparable with the variation of SO_2 concentrations within the plume of a large industrial/urban source area (Rijnmond/ Rotterdam area, see Table 2).
- Both in the "Rijnmond/Rotterdam" plume as well as in the region

Table 2. Air pollutant concentrations (in $\mu g/m^3$) within the plume of the Rijnmond/Rotterdam area at different distances from the source area

flight track	09	10	12	07
distance from the source (km)	7	60	60	100
height (m)	335	330	390	350
NO_2	85	86	80	73
NO	24	27	19	24
SO_2	150	130	125	140
O_3	53	55	56	50

Table 3. Average air pollutant concentrations (in $\mu g/m^3$) outside the main plumes over the North Sea

flight track	09	10	12	07
distance from the source (km)	7	60	60	100
height (m)	335	330	390	350
NO_2	56	56	68	56
NO	10	12	6	12
SO_2	110	92	105	100
O_3	65	67	70	60

outside the main plumes (see Table 3), there is only a small reduction in the SO_2 and NO_x concentrations with growing distance from the source area. Also the O_3 concentrations show minor variations between the different tracks. (There is of course a clear reduction of the O_3 in the lower layer as a whole due to dry deposition as well as a sink where NO is sufficiently available - O_3 reduction and NO_2 production is evident in the plume near the source (see Table 1, track 13). The maximum "background" concentration over the sea of O_3 is 110 $\mu g/m^3$ at a height of approximately 1000 m.

Although in major plumes the overall decrease in O_3 concentration is the general rule, we have encountered a number of cases during the measuring flights where O_3 production within a plume was evident. One of these occurred during the flight on the 3rd of September 1980 in the track HSD (Schouwen-Duiveland) - Dover. The O_3 concentrations over the sea below the inversion height a approximately 800 m, reached a maximum of 130 - 140 $\mu g/m^3$. At lower layers the concentration varies, but outside plumes it is of the order 110 - 120 $\mu g/m^3$. At higher flight tracks up to 2 km, O_3 values of 100 $\mu g/m^3$ $\pm$ 20 $\mu g/m^3$ are found. At these high flight levels the SO_2 concentration is virtually nihil and NO/NO_2 concentrations are measurable up to 10 $\mu g/m^3$ NO_2 over the Dover area. Fig. 7 shows a plume measured at the height of 130 m that contains SO_2 (230 $\mu g/m^3$) and excess O_3 with a maximum of 175 $\mu g/m^3$. Remarkable is the fact that in this plume NO_2 and NO concentrations are relative low. The winddirection was south to south-east, windspeed 4 - 6 m/s, during a period with extensive sunny conditions. A further occasion of observed ozone production was found during the flight on the 16th of April 1981.

FLIGHT APRIL 16, 1981

The general weather circulation was determined by a high pressure system located over the northern part of the North Sea. Dry air was transported from Sweden over Germany into The Netherlands and over the sea. Winddirection was east to north-east with windspeed over the North Sea ranging from 5 - 8 m/s.

From 10.30 a.m. to 7.00 p.m. an extensive pattern of flight tracks was covered (see Fig. 8), including a number of ascends and descends near the coast of The Netherlands and the United Kingdom.

The profile over the North Sea (GY) indicated an inversion layer (approx. 6°C) between 525 and 575 m. Low pollutant levels were found for SO_2 and NO_2: 0 - 10 $\mu g/m^3$ above and below the inversion layer. O_3 concentration was approx. 80 $\mu g/m^3$ below the inversion layer and 95 $\mu g/m^3$ above, with a maximum of 110 $\mu g/m^3$. The features of the concentrations of pollutants below and above the inversion layer in the morning and late afternoon hours are demonstrated with the results of ascends and descends near the south-western coast of The Netherlands (see Fig. 9). Clear profiles in all pollutant concentrations are shown in the morning, in the afternoon the distribution below the inversion layer is approx. constant with height with a sharp decrease above the inversion layer for SO_2 and NO_2.

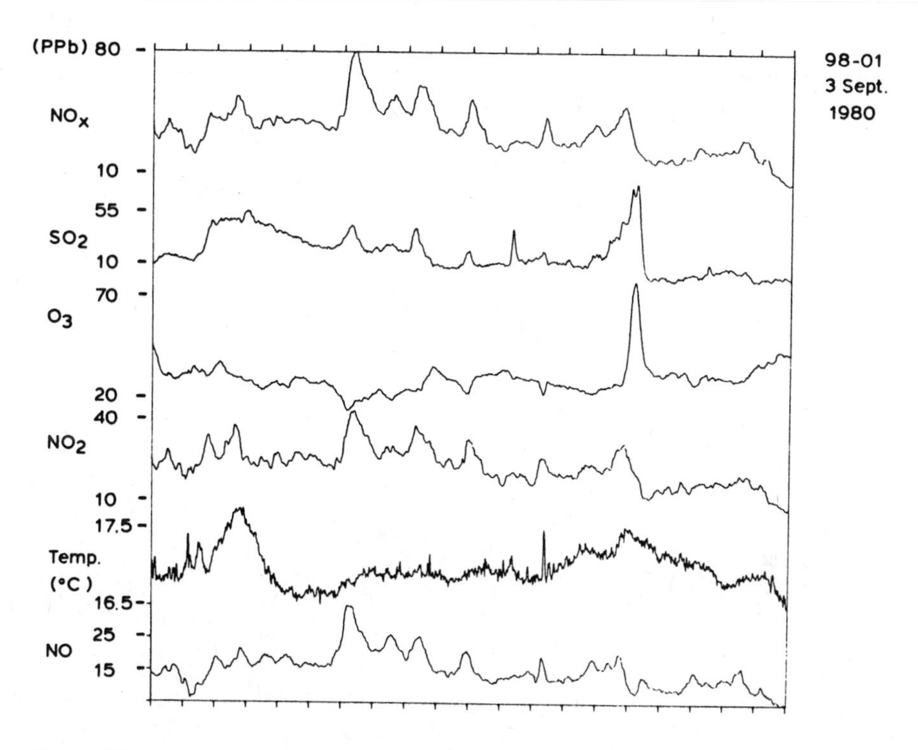

Fig. 7 Pollutant concentration above the flight track Haamstede (Neth.) - Dover (UK)

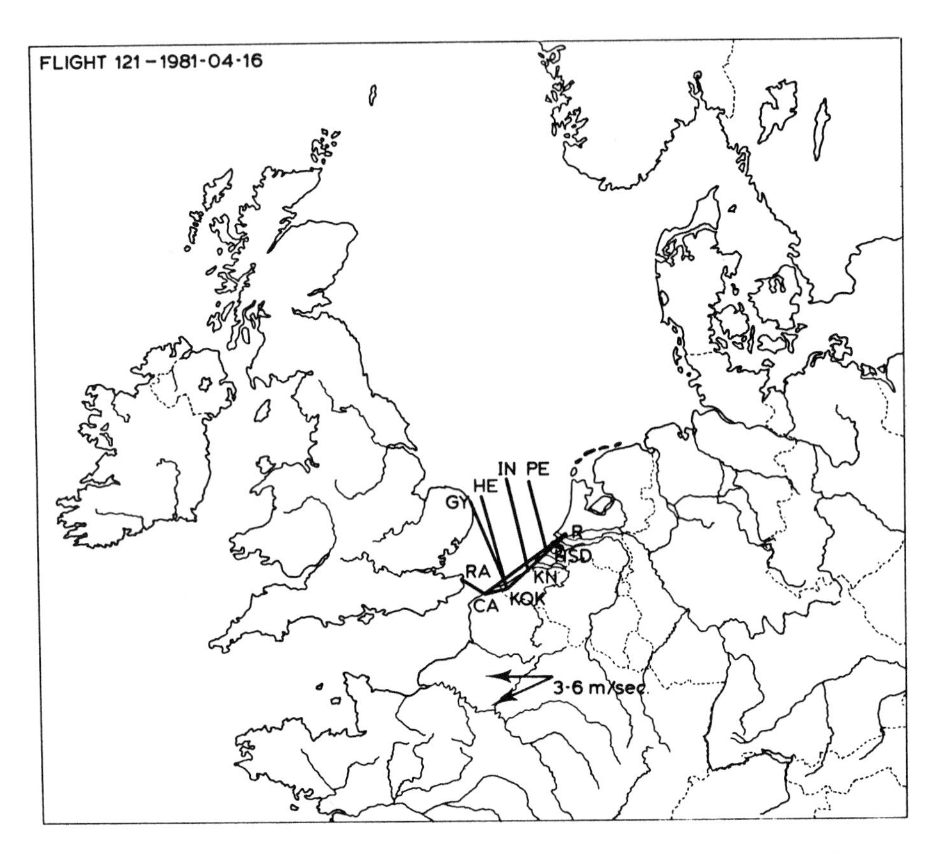

Fig. 8 Flight tracks over the southern part of the North Sea

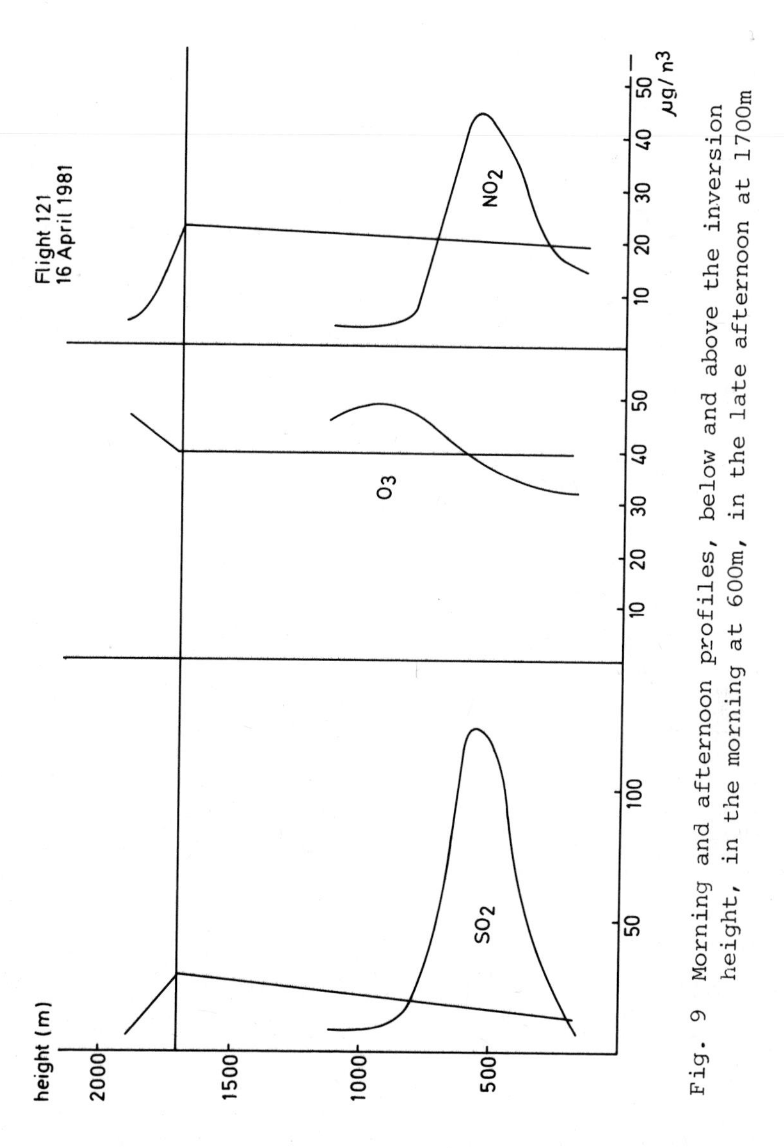

Fig. 9 Morning and afternoon profiles, below and above the inversion height, in the morning at 600m, in the late afternoon at 1700m

The O_3 values above the inversion layers are nearly constant over the whole area, for the coast of France (CA) the O_3 concentration was of the same order of magnitude (minimum 75 $\mu g/m^3$, maximum 115 $\mu g/m^3$ with a mean value of 100 $\mu g/m^3$).

Below the secondary inversion layer however, O_3 values were slightly higher due to O_3 formation. Clear indication of such O_3 formation was also found on the flight track between Rotterdam and Calais near the coast of Dunkerque and over the Westerschelde and the south-west area from the Rijnmonddistrict. Maximum values in the urban/industrial plume near Dunkerque were 120 $\mu g/m^3$ O_3 measured at a height of 370 m. In the same plume the SO_2 concentration was 240 $\mu g/m^3$ and the NO_2 concentration 80 $\mu g/m^3$.

Transport of pollutants originating from source areas on the mainland could be seen in the measurement results of the flight tracks over the sea (Table 4).

Again the nearly constant values of NO_2 at the individual heights are clearly marked. The SO_2 removal rate is 25-30%/ hour. This value is on the same order of magnitude as ratios found during earlier flights.

Table 4. Pollutant concentrations ($\mu g/m^3$) in a single industrial plume of a power station at several distances from the source and at several heights above the flight tracks

flight track	03	04	06	07	08	13
height (m)	260	500	165	300	555	185
distance from the source (km)	15	15	60	60	60	120
SO_2	100	250	35	40	110	-
NO_2	38	-	23	34	36	24
O_3 ("background")	54(80)	54(90)	68(82)	66(82)	88(104)	76(82)

FLIGHT JANUARY 30, 1981

During a number of days towards the end of January the general pressure distribution showed a high pressure system over eastern and middle Europe that was responsible for quiet weather with morning fog, relative few sunshine hours over land a very light (1 - 3m/s) southerly wind. The situation over the North Sea area was favourable for the build-up of high air pollutant concentrations in semi-stagnant air masses over the southern part of the North Sea.

A flight was carried out at two levels from Rotterdam via Clacton on Sea and Dover (UK) to northern France and Belgium (Antwerp) (see Fig. 10).

Measurements were made partly above and below the inversion layer located at approximately 500 m. Conditions above this layer over the sea show virtually no SO_2 or NO_x, 90 - 100 $\mu g/m^3$ O_3 with temperatures of 10 - 11°C. Also over land comparable results are found above the inversion layer as is shown in Fig. 11 in the second part of the flight track Clacton - Dover at a height of 660 m.

In the lower more polluted layers over the sea under the inversion layer, high concentrations are found in the major plumes at a height of 150 - 200 m. SO_2: 700 - 1200 $\mu g/m^3$; NO_2: 200 - 430 $\mu g/m^3$.

As a result of the scavenging of excess NO (approximately 150 $\mu g/m^3$), the O_3 concentrations within the plumes decreased to values ranging from 25 to 35 $\mu g/m^3$ in the lower layers.

That the region over the North Sea with relative high pollution values is covering a large area is shown with the results of the flight track Dover - KOK. At a height of 165 m over the sea the values are: SO_2: 100 - 150 $\mu g/m^3$; NO_2: 60 - 100 $\mu g/m^3$; NO: 50 - 90 $\mu g/m^3$; O_3: 30 - 40 $\mu g/m^3$ (see Fig. 12).

These pollution levels do not show plumes of individual sources in contrast to the values measured near the coast of France (Dunkerque) were plumes with very high concentrations are found at a height of 325 m with maxima for SO_2: 1120 - 1660 $\mu g/m^3$; NO_2: 120 - 180 $\mu g/m^3$ and a O_3 minimum of 10 - 15 $\mu g/m^3$.

Comparing the NO_2 versus SO_2 values found over land and over the sea during the different flight tracks above and below the inversion layer, it can be concluded that over land the NO_2 concentrations in plumes are significantly higher (in northern Belgium around the Antwerp region the maximum values at a height of 325 m were 200 - 250 $\mu g/m^3$), probably due to the fact that enough NO is available for reaction with O_3 to form the NO_2. O_3 reduction over land, from 80 - 90 $\mu g/m^3$ to a minimum of 10 $\mu g/m^3$,

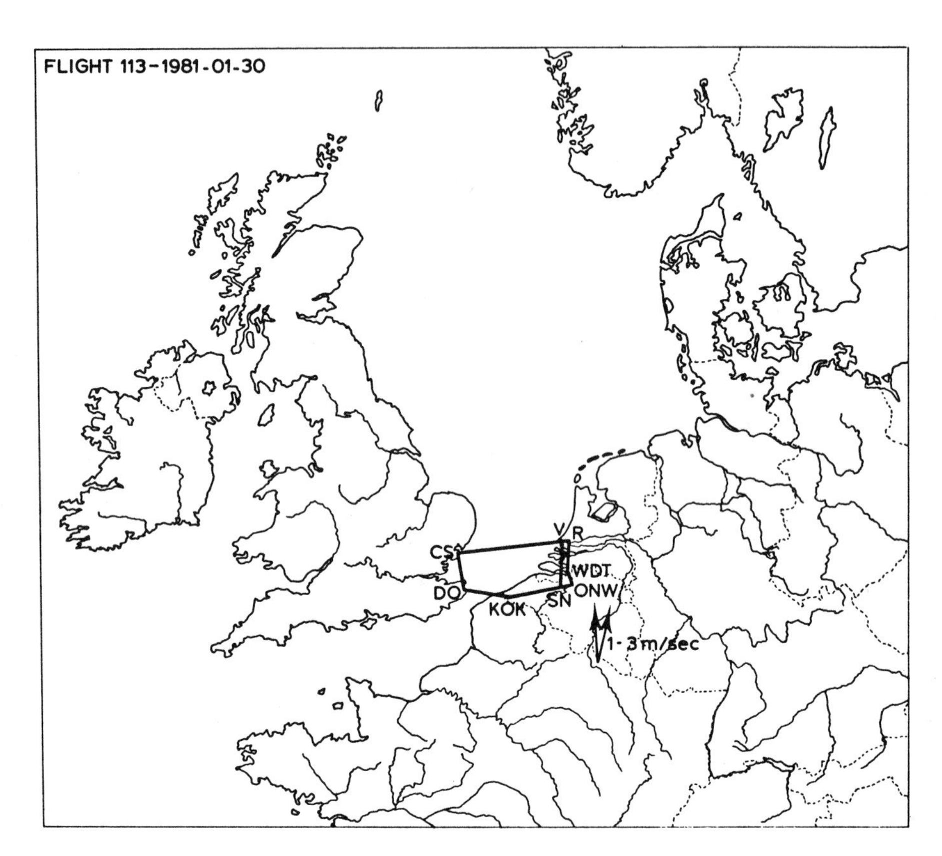

Fig. 10 Flight tracks over the southern part of the North Sea and the adjacent coastal area

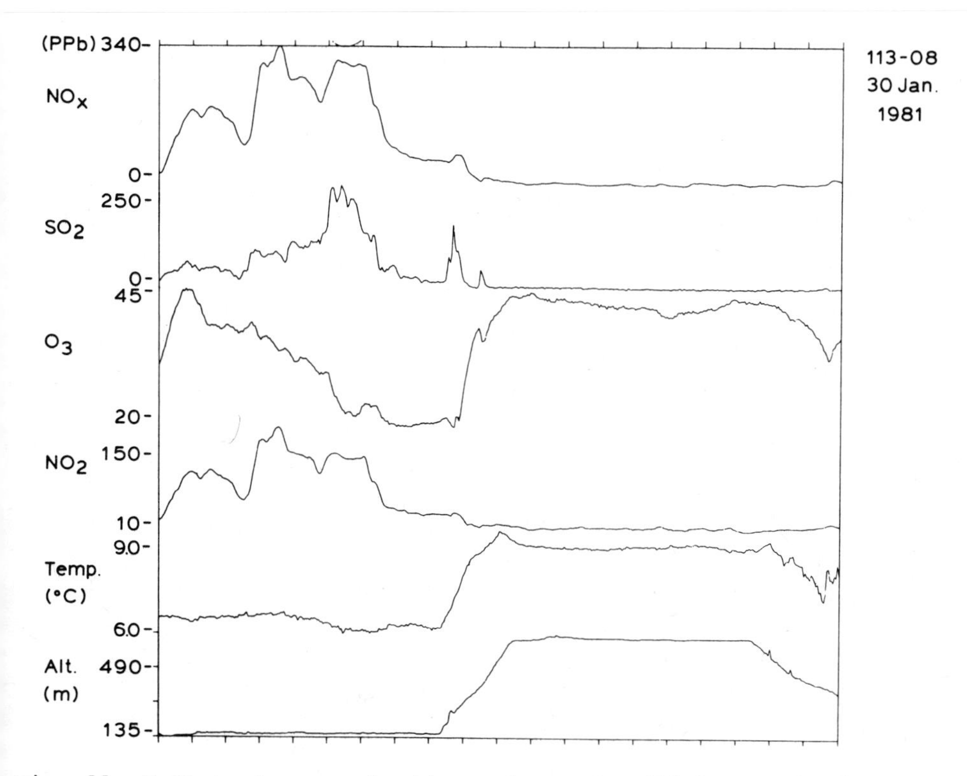

Fig. 11 Pollutant concentrations above the flight track Clacton on the North Sea - Dover (UK)

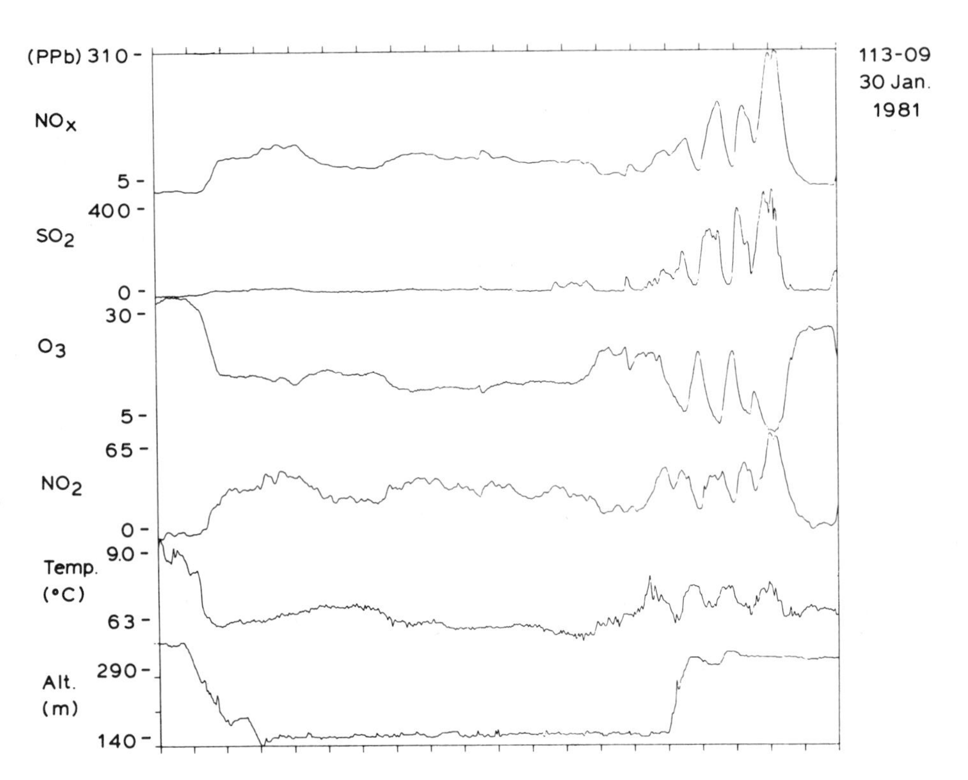

Fig. 12 Pollutant concentrations over the flight track Dover (UK)-KOK (Belgian coast)

is therefore larger than over the sea (reduction from 90 - 100 $\mu g/m^3$ to a minimum of 35 $\mu g/m^3$).

CONCLUDING REMARKS

A description has been given of experimental investigations designed primarily to obtain detailed measurements of pollutant concentrations during transport conditions over the North Sea.

The results discussed in the case histories described above document the variability that exists in the pollutant concentrations within major plumes originating from sources over land and transported over the sea. The data show also that significant amounts of pollution can exist and are transported in layers well above the earth surface. It is felt that this variability with height of specific pollutants strongly influences the mesoscale and long range transport and should therefore be considered in the transport models used.

On the basis of the results one can also conclude that there can be no unique conclusion based on single point or under most conditions also not on multi-stationary point measurements of single pollutant components or parameters. For a complete understanding and description of the phenomena that are occurring a three dimensional observation network must be used. It is obvious that such a three-dimensional observation network can only be used in special circumstances. Such case studies should therefore indicate the parameters that are important to be considered in transport modeling.

There is an urgent need to bring the two fields of investigation, model development and chemical and physical atmospheric measurement studies together in order to reach a more continuous interaction between both activities. Only then one can decide on a sound basis where transport models can be kept simple and straightforward and where details have to be inserted into the overall modeling to reach an acceptable level of accuracy in the model calculations.

ACKNOWLEDGEMENT

The cooperation of the Environmental Research Department of the KEMA, Arnhem, is gratefully acknowledged. The efforts of the field team, J.W.Viljeer and L.van de Beld, that performed the aerial measurements in a careful as well as in an efficient way are very much appreciated.

DISCUSSION

W.B. JOHNSON

Do you think it is possible that some of the pollutant peaks that appear to be plumes, as measured by the aircraft horizontal traverses, might actually be caused by small aircraft height variations in the vicinity of an inversion layer (where pollutant vertical gradients might be large), or alternatively, by waves or other "bumps" on the top of inversion layers ?

T. SCHNEIDER

This irregularity caused by height variations of 10 - 50 meters in the horizontal flight tracks did occur in our earlier flights. Since 1979, however, we recorded the flying height continuously and could discard the measurements made under such conditions. Flying through the inversion layer the variability in pollution concentrations is copied in the temperature fluctuations. In this case one can also take these variations into account.

W. KLUG

Were any of the measurements you showed connected with clouds or precipitation ?

T. SCHNEIDER

During the flights reported there were no periods with precipitation. Clouds were present on several flight tracks but results gathered while crossing a cloud are not used in the evaluation.

SOLUTIONS TO THE EQUATION FOR SURFACE DEPLETION OF A GAUSSIAN PLUME★

Robert J. Yamartino✱

GEOMET GmbH
Bundesallee 129
1000 Berlin 41

INTRODUCTION

The problem of surface depletion of a Gaussian plume has been formulated by Horst (1977). As in most models of deposition, the deposition flux, $F(x,y) = v_d\ C(x,y,z_d)$, at a given point in computed as the deposition velocity v_d times the pollutant concentration, $C(x,y,z_d)$, at a particular reference height z_d ; however, in Horst's model $C(x,y,z)$ is the concentration due to the initial, unabsorbed plume minus the sum of concentration deficits due to all upwind deposition. Thus, absorption is treated through the simultaneous ground reflection and dispersion of real particles plus the ground-level creation and dispersion of "anti-particles." The governing equation for this process is given by Horst as

$$C(x,y,z) = Q\ \frac{D(x,z,h)}{u}\ \frac{e^{(-y^2/2\sigma_y^2)}}{\sqrt{2\pi}\,\sigma_y} - \int_{-\infty}^{+\infty} dy' \int_0^x dx'\ v_d\ C(x',y',z_d)$$

$$\times \frac{D(x-x',z,0)}{u}\ \frac{1}{\sqrt{2\pi}\,\sigma_y(x-x')}\ \exp\left[-\frac{(y-y')^2}{2\sigma_y^2(x-x')}\right] \qquad (1)$$

★Sponsored in part by the Umweltbundesamt, Berlin.

✱Present address : ERT, Inc., 3 Militia Drive, Lexington, MA 02173.

and in crosswind-integrated form as

$$\bar{C}(x,z) = Q \frac{D(x,z,h)}{u} - r \int_0^x dx' \, \bar{C}(x', z_d) \, D(x-x',z,0) \tag{2}$$

where Q is the source strength, u is the wind speed, $r = v_d/u$, and D(x,z,h) expresses the vertical portion of the coupling between a point source at height h and receptor at height z and distance x downwind. For example, for a point source with ground reflection, D(x,z,h) would take the form :

$$D(x,z,h) = \frac{1}{\sqrt{2\pi}\,\sigma_z(x)} \left\{ \exp\left[\frac{-(z-h)^2}{2\sigma_z^2(x)}\right] + \exp\left[\frac{-(z+h)^2}{2\sigma_z^2(x)}\right] \right\}$$

Though the integral equation (2) may be solved by interative numerical techniques, this paper presents solutions that can be evaluated via a single integration. This simplification reduces the complexity and computation time associated with the surface depletion model to the same level as encountered in the source depletion model (Van der Hoven, 1968).

FORMAL SOLUTION OF THE PROBLEM

Equation 2 can be further reduced to the form

$$f(x,z) = r \int_0^x dx' \, [D(x',z_d,h) - f(x',z_d)] \, D(x-x',z,0) \tag{3}$$

by assuming a solution of the form

$$\bar{C}(x,z) = C(x,y,z) \cdot \sqrt{2\pi}\,\sigma_y(x) \exp\left[\frac{+y^2}{2\sigma_y^2(x)}\right]$$

$$= \frac{Q}{u} [D(x,z,h) - f(x,z)] \tag{4}$$

Where (Q/u) f(x,z) is the unknown, crosswind-integrated, "anti-particle" concentration to be solved for. As it is first necessary to determine $f(x,z_d)$ before finding the full solution f(x,z), Equation (3) is rewritten as

$$f(x,z_d) = r \int_0^x dx' \, [D(x',z_d,h) - f(x',z_d)] \, D(x-x',z_d,0) \tag{5}$$

Equation 5 is now in the form of a special type of the Volterra equation which is well-suited to integral transform methods+. Taking the Laplace transform of Equation 5 yields an algebraic equation

$$f(x,z_d) = r\left[\mathcal{L}D(x,z_d,h) - \mathcal{L}f(x,z_d)\right] \mathcal{L}D(x,z_d,0)$$

which has the following simple formal solution :

$$\mathcal{L}f(x,z_d) = \left[\frac{r\,\mathcal{L}D(x,z_d,0)}{1 + r\,\mathcal{L}D(x,z_d,0)}\right] \mathcal{L}D(x,z_d,h)$$

Thus,

$$f(x,z_d) = \int_0^x dx'\ D^{*}(x-x',z_d,0)\ D\,(x',z_d,h) \tag{6}$$

where D^* is given as $D^*(x,z_d,0) = \mathcal{L}^{-1}\left[\frac{r\,\mathcal{L}D(x,z_d,0)}{1 + r\,\mathcal{L}D(x,z_d,0)}\right]$ (7)

Though formal representation of D^* in terms of D is simple, the actual computation of D^* for subsequent use in equation 6 can be difficult and is deferred until the next Section.

The representation of D^* in terms of the Laplace transform is useful when the vertical coupling, D, is available as an analytic function of downwind distance x. This implies, of course, that $\sigma_z(x)$ is given as a simple function. If $\sigma_z(x)$ is available only in graphical form or if the Laplace operations cannot be evaluated,

---

+ It is only necessary to consider the Laplace transform

$$\mathcal{L}f(x) = F(s) = \int_0^\infty e^{-sx} f(x)dx, \text{ its inverse } \mathcal{L}^{-1},$$

and two important properties of this transform :

$$\mathcal{L}\int_0^x f(x')dx' = \frac{F(s)}{s} \text{ and } \mathcal{L}\int_0^x f(x-x')g(x')dx' = F(s)G(s)$$

$$= \mathcal{L}f(x)\,.\,\mathcal{L}g(x)$$

equation (7) may be recast as the integral equation

$$D^{*}(x,z_d,0) = r\left[D(x,z_d,0) - \int_0^x dx'\, D^{*}(x',z_d,0)D(x-x',z_d,0)\right] \quad (7a)$$

Although replacing one integral equation (5) with another (7a) seems somewhat academic , Equation (7a) is independent of source height h, so that for a fixed z_d (often taken as some multiple of surface roughness z_o), $D^{*}(x,z_d,0)$ must be evaluated as a function of x by iterative methods only once for each vertical stability class.

For many applications, the computation of $f(x,z_d)$ with Equation (6) is all that is needed, as it enables the near-ground-level concentration, $C(x,y,z_d)$, to be computed with Equation 4 and the surface deposition rate to be computed with the original relation, $F(x,y) = v_d\, C(x,y,z_d)$.

In some applications it will be of interest to know the amount of material Q(x) remaining in the plume. This is obtained by integrating Equation (4) over z, i.e.,

$$\frac{Q(x)}{u} = \int_0^\infty dz\, \overline{C}(x,z) = \frac{Q}{u}\int_0^\infty dz\left[D(x,z,h) - f(x,z)\right] \quad (8)$$

noting that $\int_0^\infty dz\, D(x,z,h) = 1$, substituting Equation (3) for f(x,z), and applying the Laplace transform operator. The resulting formal solution has the simple form

$$\frac{Q(x)}{Q} = 1 - r\int_0^x dx'\, D^{**}(x-x',\, z_d,0)D(x',z_d,h) \quad (9)$$

where

$$D^{**}(x,z_d,0) = L^{-1}\left[\frac{1/s}{1 + r\, L\, D(x,z_d,0)}\right], \quad (10)$$

and s is the transform variable. As with equation (7), Equation (10) can be expressed in the following h-independent, integral equation form

$$D^{**}(x,z_d,0) = 1 - r \int_0^x dx' \, D^{**}(x',z_d,0)D(x-x',z_d,0) \tag{10a}$$

Only in a few applications will the entire vertical concentration profile $\overline{C}(x,z)$ be required ; in many such cases, the zero deposition shape combined with the material $Q(x)$ in the plume, i.e.,

$$\overline{C}(x,z) \approx \frac{Q(x)}{u} D(x,z,h), \tag{11}$$

may provide the needed accuracy. This is similar to the approach taken in source depletion, except that the $Q(x)$ is based on the more realistic surface deposition approach.

The rigorously correct determination of $\overline{C}(x,z)$ can be obtained by plugging the solution for $f(x,z_d)$ (Equation (6)) into Equations (3) and (4) to obtain

$$\overline{C}(x,z) = \frac{Q}{u} \left\{ D(x,z,h) - r \int_0^x dx' D(x-x',z,0) \left[D(x',z_d,h) - \int_0^{x'} dx'' D(x'-x'',z_d,0) D(x'',z_d,h) \right] \right\} \tag{12}$$

While the presence of a double integral makes this expression appear correspondingly more difficult to compute, its nature is such that it is only necessary to accumulate the quantities $F(x',z_d)D(x-x',z,0)$ and $D(x-x',z,0)D(x',z_d,h)$ as $f(x,z_d)$ is being evaluated. Thus, it should not be too tedious to compute $\overline{C}(x,z)$ for several z values on a routine basis.

Finally, we note that applying the Laplace transform operator to equation (3) leads to the formal solution

$$\overline{C}(x,z) = \frac{Q}{u} \left\{ D(x,z,h) - \int_0^x dx' \, D^{***}(x-x',z:z_d,0) \, D(x',z_d,h) \right\} \tag{13}$$

where

$$D^{***}(x,z:z_d,0) = L^{-1} \left[\frac{r \;\; D(x,z,0)}{1+r \; LD(x,z_d,0)} \right] \tag{14}$$

This solution will not be further considered, however, because of the difficulty of evaluating $D(x,z,0)$ for $z \neq 0$ while using realistic forms of $\sigma_z(x)$; however, the integral equation formulation

$$D^{***}(x,z:z_d,0) = r\left[D(x,z,0) - \int_0^x dx' \, D^{***}(x',z:z_d,0) D(x-x',z_d,0)\right] \tag{14a}$$

appears more promising.

PRACTICAL SOLUTION OF THE PROBLEM

In the previous section, it was shown that the near-ground-level concentration and deposition rate is obtained through evaluation of Equation (6) and that the material, $Q(x)$, remaining in the plume is obtained through evaluation of Equation (9). Crucial to these evaluations is the ability to provide reasonable expressions for $D^*(x,z_d,0)$ (Equation (7)) and $D^{**}(x,z_d,0)$ (Equation 10)), respectively.

Fortunately, D^* and D^{**} need only be evaluated once for a given form of $\sigma_z(x)$; this will be done in this section.

The determination of either D^* or D^{**} requires first the assumed form of $D(x,z_d,0)$. Strictly speaking, D should depend on z_d, e.g.,

$$D(x,z_d,0) = \frac{\sqrt{2/\pi}}{\sigma_z(x)} \exp\left[\frac{-z_d^2}{2\sigma_z^2(x)}\right]$$

However, as the absorption occurs very near the ground, a null exponent can be assumed, provided $\sigma_z(x)$ does not go to zero for $x = 0$. Thus we assume

$$D(x,z_d,0) = D(x,0,0) \approx \frac{\sqrt{2/\pi}}{\sigma_z(x+x_o)} + 1/H \tag{15}$$

where $\sigma_z(x_o) = t$, the thickness of the layer from which the surface absorption occurs, and H is the mixing depth. Simple addition of the term $1/H$ is not rigorously correct but simplifies the problem and causes a maximum error of order t/H, an error that should not exeed a few percent for any x.

The ultimate focus of this section will be on σ_z of the form $\sigma_z(x) = a(x+x_o)^b$, but we will first consider $\sigma_z(x) = t\exp(a'x/t)$

because of mathematical ease and insights to be gained.★

Taking the Laplace transform of $D(x,z_d,0)$ when using the exponential form of $\sigma_z(x)$ yields the simple result

$$r\,\mathcal{L}\,D(x,z_d,0) = k/(s+a'/t) + r/(sH) \tag{16}$$

where s is the transformation variable, $r = (v_d/u)$ as before, and $k = \sqrt{2/\pi}\,r/t$. D^* and D^{**} then follow from equations (7) and (10) respectively and can be written as

$$\mathcal{L}\,D^*(x,z_d,0) = [s(k+r/H) + a'r/(tH)]/[(s+K_p)(s+K_m)] \tag{17}$$

and

$$\mathcal{L}\,D^{**}(x,z_d,0) = (s+a'/t)/[(s+K_p)(s+K_m)] \tag{18}$$

where the roots of the quadratic in s are

$$K_p = 1/2\,(a'/t+k+r/H) + 1/2\,[(a'/t+k+r/H)^2 - 4a'r/(tH)]^{1/2} \approx k + a't$$
$$K_m = \quad '' \quad - \quad '' \quad \approx r/H \tag{19}$$

Using a table of Laplace transforms or the Heaviside Expansion Theorem enables one to obtain the inverse Laplace transforms of Equations (17) and (18) as

$$D^*(x,z_d,0) = [(K_p(k+r/H) - a'r/(tH)\exp(-K_p x)$$
$$+ (a'r/(tH) - K_m(k+r/H))\exp(-K_m x)]/(K_p-K_m) \tag{20}$$

★Matching the value and derivative of the exponential and power law forms of $\sigma_z(x)$ at $x = 0$ leads to the choice $a' = bt/x_o$. This choice also follows from the comparison of the Laplace transformation forms as discussed in the text.

and

$$D^{**}(x,z_d,0) = [(K_p-a'/t)\exp(-K_p x) + (a'/t-K_m)\exp(-K_m x)]/(K_p-K_m) \quad (21)$$

Equations (20) and (21), when used in Equations (6) and (9) respectively, represent a complete solution to the surface depletion problem for the assumed exponential form of $\sigma_z(x)$.

Several interesting aspects of these last two equations are worth mentioning :

- Their exponential form does not follow from using an exponential form of $\sigma_z(x)$ but is rather a general property of such integral equation resolvants as recently proven by Jordan and Wheeler (1980).

- The fact that the only x dependence appears inside the exponential is advantageous, as the x dependence of $D(x-x',z_d,0)$ in (6) and $D^{**}(x-x',z_d,,0)$ in (9) can be taken outside of the x' integration with a considerable saving of effort. This advantageous situation can be expected for most physically reasonable forms of $\sigma_z(x)$.

- Assuming $D(x,z_d,h) \approx 1/H$, which might be a appropriate in long-range transport problems, gives a dominant dependence proportional to $\exp(-K_m x) \approx \exp(-v_d x/(uH))$, as may be expected intuitively or derived from source depletion.

Returning to Equation (15), we now consider the case of $\sigma_z(x) = a(x+x_o)^b$. Taking the Laplace transform yields

$$r\,\mathbf{L}D(x,z_d,0) = kx_o(sx_o)^{b-1}\exp(sx_o)\,\Gamma(1-b,sx_o) + r/(sH) \quad (22)$$

where $k = \sqrt{2/\pi}\,r/t$ and Γ denotes the incomplete gamma function. Although Equation (22) represents the exact Laplace transform of (15), it is difficult to utilize in this form. The following results assume the continued fractions expansion

$$\Gamma(1-b,y) = e^{-y}y^{1-b}\left(\frac{1}{y+}\;\frac{b}{1+}\;\frac{1}{y+}\;\frac{1+b}{1+}\;\frac{2}{y+}\cdots\right) \quad (23)$$

from Abramowitz and Stegun (1964).

Taking only the first two terms of the partial fractions expansion (23) yields

$$r\,\mathcal{L}D(x,z_d,0) \approx k/(s+b/x_o) + r/(sH) \qquad (24)$$

and comparison with Equation (16) shows that the two equations are identical if $a'/t = b/x_o$ is assumed. Thus, Equations (19) through (21) provide the required solution when b/x_o is substituted for a'/t; however, it is clear that this approximate form (24) cannot provide more realistic results than would the assumption of the exponential form of $\sigma_z(x)$.

Taking instead the first four terms of Equation (23) yields

$$r\,\mathcal{L}D(x,z_d,0) \approx k\,[s+(b+2)/x_o]/[s^2+2s(b+1)/x_o+b(b+1)/x_o^2] + r/(sH) \qquad (25)$$

If H is assumed infinite, then $\mathcal{L}D^*$ and $\mathcal{L}D^{**}$ can easily be computed, as were Equations (17) and (18), and inverted to give

$$D^*(x,z_d,0) = k\,[\,(K_p-(b+2)/x_o)\exp(-K_p x) + ((b+2)/x_o-K_m)\exp(-K_m x)\,]/(K_p-K_m) \qquad (26)$$

and

$$D^{**}(x,z_d,0) = [\,K_p^2-2K_p(b+1)/x_o+b(b+1)/x_o^2\,]\exp(-K_p x)/[K_p(K_p-K_m)] - [\,K_m^2-2K_m(b+1)/x_o+b(b+1)/x_o^2\,]\exp(-K_m x)/[K_m(K_p-K_m)] + b(b+1)/(x_o^2K_pK_m) \qquad (27)$$

where the quadratic roots are now

$$K_p = ((b+1)/x_o+k/2)+[(b+1)/x_o^2-k/x_o+k^2/4]^{1/2}$$

$$K_m = \quad '' \quad - \quad '' \qquad (28)$$

If H is assumed finite, LD^{*} and LD^{**} can be written as

$$LD^{*}(x,z_d,0) = k\left[s^2(1+\delta)+s((b+2)+2\delta(b+1))/x_o+\delta b(b+1)/x_o^2\right]/R(s)$$

$$= g_1(s)/R(s) \tag{29}$$

and

$$LD^{**}(x,z_d,0) = \left[s^2+2s(b+1)/x_o+b(b+1)/x_o^2\right]/R(s) = g_2(s)/R(s) \tag{30}$$

where

$\delta = \sqrt{\pi/2}\ t/H$ and R is cubic is s, that is,

$$R(s) = s^3+s^2\left[2(b+1)/x_o+k(1+\delta)\right]$$

$$+s\left[b(b+1)/x_o^2+k(b+2+2\delta(b+1))/x_o\right]+ k\delta b(b+1)/x_o^2 \tag{31}$$

In order to obtain $D^{*}(x,z_d,0)$ and $D^{**}(x,z_d,0)$ in the desired exponential form, R must be re-expressed in the form

$$R(s) = (s+K_1)(s+K_2)(s+K_3),$$

but this is extremely tedious and has not yet been accomplished exactly, though, $K_1 \approx K_p - k^2 x_o \delta/2$, $K_2 \approx K_m - k^2 x_o \delta/2$, and $K_3 \approx k\delta/(1+kx_o)$ might be used as preliminary estimates. The final solutions for D^{*} and D^{**} are then

$$D^{*}(x,z_d,0) = \sum_{i=1}^{3} g_1(-K_i)\exp(-K_i x)/R'(-K_i) \tag{32}$$

and

$$D^{**}(x,z_d,0) = \sum_{i=1}^{3} g_2(-K_i)\exp(-K_i x)/R'(-K_i) \tag{33}$$

where $R'(-K_i)$ denotes the derivative $dR(s)/ds$ evaluated at $s = -K_i$. Noting that $k\delta = r/H$ and $kx_o = (r/a^{1/b})\sqrt{2/\pi}\ t^{(1-b)/b}$, it is seen that, as in the lower-order approximations, the slowest decaying D^{**} term, $\exp(-K_3 x)$, resembles the decay, $\exp(-rx/H)$, from source depletion when uniform vertical mixing is assumed, except that $k_3 \approx (r/H)/(1+r/a^{b})$ for t on the order of a few meters.

Since $r/a^{1/b}$ increases with increasing stability and deposition velocity, K_3 will nearly equal the source depletion exponent, r/H, under unstable, low v_d/u conditions but can be several times less than r/H for stable, higher v_d/u conditions. The need for such a correction factor to source depletion has been indicated by several authors including Horst (1977) and Prahm and Berkowicz (1978) and is directly obtainable using the approach presented here. The need of determining the roots of higher-order polynomials in s does represent, however, a significant barrier to utilizing even more accurate approximations for Equation (22).

SUMMARY AND CONCLUSIONS

In this paper, exact formal solutions to the surface depletion formulation of deposition (Horst, 1977) are presented in Equations (6), (9), and (13). Practical representations of D^* and D^{**} for use in Equations (6) and (9) respectively are given for assumed exponential (see Equations (20) and (21)) and power law (see Equation pair (26) and (27) or pair (32) and (33)) forms of $\sigma_z(x)$, though these representations are only approximate in the power law case. Higher-order approximations in the case of the power law form would probably require readily available routines to obtain polynomial roots or numerical solutions of the alternate integral equation forms. The advantage is that, if required, such numerical techniques would need to be utilized far less than is the case in solving the entire surface depletion problem, as a set of source height independent equations are available.

Comparisons between the present method, exact numerical solutions of surface depletion, and the source depletion method are planned. This method, when fully evaluated on actual problems, will enable the user to compute deposition effects with the more accurate surface depletion techniques yet with the same level of effort currently expended on the source depletion method.

REFERENCES

ABRAMOWITZ, M. and I.A. STEGUN, 1964, "Handbook of Mathematical Functions," Applied Mathematics Series, Volume 55, National Bureau of Standards, Washington, D.C.

ERDELYI, A. and Staff of the Bateman Manuscript Project, 1954. "Tables of Integral Transforms - Volume I," McGraw-Hill Book Company, Inc., New York.

HILDEBRAND, F. B., 1952, "Methods of Applied Mathematics," (2nd Ed.), Prentice-Hall, Inc., Englewood Cliffs, NJ.

HORST, T. W., 1977, A Surface Depletion Model for Deposition from a Gaussian Plume, Atmos. Environ., 11:41-46.

JORDAN, G. S. and R. L. WHEELER, 1980, Structure of Resolvents of Volterra Integral and Integrodifferential Systems, SIAM J. Math. Anal., 11: 119-132.

PRAHM, L.P. and R. BERKOWICZ, 1978, Predicting Concentrations in Plumes Subject to Dry Deposition, Nature, 271, 232-234.

VAN DER HOVEN, I., 1968, Deposition of Particles and Gases, in : "Meteorology and Atomic Energy," D. SLADE, ed., USAEC, TID-24190, 202-208.

DISCUSSION

C. BLONDIN

How do you incorporate canopy-resistance and leave uptake in your model ?

R. YAMARTINO

The deposition velocity is usually modeled as the reciprocal of a sum of resistances associated with the atmospheric mixing, the surface layer, and the vegetative canopy itself. As the surface depletion model already incorporates atmospheric resistance, via diffusion of the deposition deficit, the deposition velocity would not need to include this resistance explicitly. This slightly modified, resistance - modeled deposition velocity could then be used in this surface depletion approach.

C. BLONDIN

As far as concentrations are concerned, the observation averaging time must be clearely defined because it influences drastically the σ values. The concept of surface deposition seems to involve short time scale eddies to explain the physics of the phenomenon. So, how can the use of such a concept be justified to compute one-hour averaging concentrations

R. YAMARTINO

While it is true that the deposition occurs on very short time scales, the deposition velocity would be based on longer time scales (i.e., the ratio of the hour average depositon flux to the hour average near-surface concentration). This is analogous to the situation with σ_y where the hour average σ_y used is not the same as the instantaneous plume width sigma. Therefore, I do not think the deposition modeling introduces fundamental approximations or assumptions not already implicit in the Gaussian plume approach.

B. VANDERBORGHT

Might not the maximum discrepancy between source and surface depletion occur at elevated points rather than at the surface ?

R. YAMARTINO

The surface and source depletion methods yield different vertical profiles as well as different downwind distance dependences of the amount of material remaining in the plume. I have not examined the issue you have raised but would guess that the answer would depend on downwind distance as well as the deposition velocity and rate of vertical diffusion.

MATHEMATICAL MODELING OF CHEMICALLY REACTIVE PLUMES IN AN URBAN ENVIRONMENT

Christian Seigneur
Philip M. Roth
Systems Applications, Inc.
San Rafael, California

Ronald E. Wyzga
Electric Power Research Institute
Palo Alto, California

ABSTRACT

Mathematical models used to simulate photochemical urban air pollution are generally based upon the solution of the atmospheric diffusion equation, which describes the rate of change of chemical species in each cell of a three-dimensional grid-mesh. In such models, point source emissions are assumed to be immediately dispersed into a grid cell. This approximation generally leads to an overestimation of plume dispersion and thus considerably affects the accurate treatment of photochemical kinetic processes.

A model has been developed that offers a realistic treatment of large point source emissions in urban areas. This Plume-Airshed Reactive-Interacting System (PARIS) is based upon the embedment of one or more reactive plume models in a conventional airshed model. These plume models describe the chemistry and dynamics of large point source plumes, including their interaction with the ambient environment. Plume trajectories are calculated from the gridded wind field used for the airshed model, and plume dispersion is computed according to the local temperature lapse rate. As the plume size becomes comparable with the grid cell size, the plume material is mixed into a grid of the airshed model with the background airshed material.

This paper presents some results of PARIS model simulations performed for the St. Louis urban area; in addition, the performance of the model is analyzed and implications regarding its use in the simulation of air quality are discussed.

INTRODUCTION

Mathematical modeling of urban air pollution has been the subject of considerable research in the past decade. The most comprehensive and scientifically sound approach to urban air quality modeling is based on the numerical solution of a set of partial differential equations--so-called atmospheric diffusion equations--that describe the dynamic behavior of polluants in a three-dimensional gridded domain.

The assumptions inherent to the atmospheric diffusion equation place some limitations on the spatial and temporal resolution of air quality models that can be summarized in the following manner : The horizontal and vertical averaging distances of the model should be greater than approximately 2000 and 20 m, respectively, and the averaging time should be greater than approximately 20 min.[1] Therefore, air quality models such as the Systems Applications, Inc. (SAI) Urban Airshed Model[2,3] that are based on the solution of the atmospheric diffusion equation involve variations in the velocity, concentration, and emission fields that are on scales greater than those of the spatial and temporal resolutions just given.

Clearly, the constraints that are thus placed on the concentrations and emission fields in urban air quality models are violated in the vicinity of strong emission sources such as roadways and power plants. These emission sources release large amounts of nitric oxide that react rapidly with the ambient ozone. In the basic formulation of urban air quality models, these emissions must be averaged over a grid cell, whereas the plumes of these emissions occupy a much smaller volume near the source, and do not expand to the size of a grid cell until they are several kilometers downwind of the source. Consequently, the mixing rate of these fresh emissions with the ambient air is overestimated; this can significantly affect the chemical rate processes involved in photochemical smog formation.

There has been little attempt to develop a systematic treatment of the modeling of large point source plumes, such as those emitted by power plants in urban airsheds. This aspect of urban air pollution modeling seems important in light of an increasing concern about the impact of large point sources on ambient nitrogen dioxide, ozone, and sulfate levels. The purpose of this paper is to present the development and application of an improved airshed model. This model considers the near-source microscale features of these large point source plumes by means of a submodel, the SAI Reactive Plume Model, which is incorporated into the SAI Urban Airshed Model.

FORMULATION OF THE MODEL

The Plume-Airshed Reactive-Interacting System (PARIS) is derived from the embedment of the SAI Reactive Plume Model in the SAI Urban Airshed Model. In the standard Urban Airshed Model, point source emissions are instantaneously mixed into the grid cell that corresponds to the plume height after calculation of the plume rise. The Plume-Airshed Reactive-Interacting System provides a more realistic treatment of large point sources because point source plumes are treated by means of the Reactive Plume Model until they reach the size of the Airshed Model grid cells. The plume constituents are then mixed into the gridded structure of the airshed. Point sources located outside of the airshed region are also taken into account in this system because their emissions can be carried into the airshed. Figure 1 presents a schematic representation of the Plume-Airshed Reactive-Interacting System. An overview of the model formulation is presented next. A more detailed description of the model can be found elsewhere[4].

The SAI Urban Airshed Model

The SAI Urban Airshed Model was developed to describe the three-dimensional dynamic behavior of reactive pollutants in a gridded airshed domain[2,3]. The mathematical basis of the model is the so-called atmospheric diffusion equation, which describes the advection, turbulent diffusion, chemical reactions, emissions, and surface removal of up to 20 reactive pollutants. These pollutants include nitrogen oxides (NO and NO_2), ozone, paraffins, ethylene, olefins, aromatics, aldehydes, peroxyacetylnitrate, carbon monoxide, sulfur dioxide, nitric acid and aerosol. Treatment of atmospheric processes by the model can be summarized in the following manner :

- Pollutant advection is treated by a three-dimensional wind field.

- Transport resulting from atmospheric turbulence is treated by means of vertical and horizontal eddy diffusion terms.

- Chemical reactions in an urban area are described by a 68-step kinetic mechanism--the so-called Carbon-Bond Mechanism (CBM)[5].--which has been extensively validated with smog chamber data.

- Source emissions are spatially and temporally distributed in the gridded region. In the case of large point sources, the total effective plume rise is calculated to enable the appropriate spatial placement of the emissions aloft.

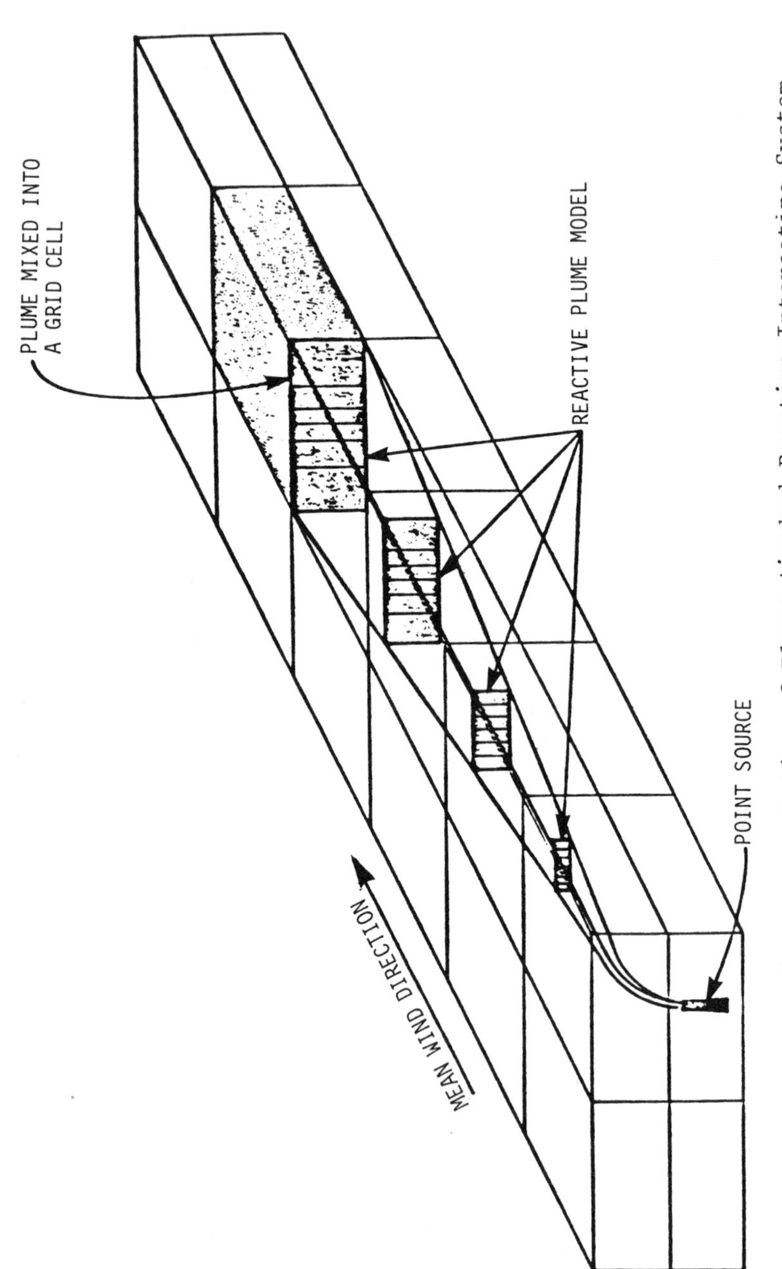

Figure 1. Schematic representation of Plume-Airshed Reactive-Interacting System.

- Removal of pollutants by surface uptake processes is treated in the ground-level boundary conditions.
- Initial and boundary conditions are defined from air quality data.

The governing model equations are solved numerically through the use of finite difference techniques on the three-dimensional grid that covers the urban airshed of interest[2].

The Reactive Plume Model

The Reactive Plume Model was developed to simulate pollutant concentration distributions across a plume emanating from a large point source[6]. This Lagrangian plume model is founded on mass balance and is composed of a fixed number of contiguous cells that expand in a prescribed fashion as the plume travels downwind. There is no vertical resolution of the chemical species concentrations in the plume, which is assumed to be vertically well-mixed. For an inert pollutant species, there is no net mass flux across the cell boundaries. An effective diffusion coefficient, founded on the relative concentration gradient, is used for reactive or background species to characterize the inter-cell mass fluxes.

The same chemical kinetic mechanism--the Carbon-Bond Mechanism--is used in both the plume and airshed models. In the Plume-Airshed Reactive-Interacting System, the plume released from a large point source is approximated by a plume with constant properties over a given time interval (i.e., steady-state is assumed for the plume trajectory, dispersion, and chemistry over a time interval of, say, 15 min). This steady-state approximation allows for the modeling of a continuous plume by means of a discrete series of puffs. Thus, puffs are released at constant time intervals from the point source. A puff is therefore defined as a Lagranian plume cross-section. At a given time, a plume is represented in the model by its puff concentrations at several downwind distances; these distances are separated by travel time increments equal to the puff release time intervals. The sensitivity of the model to the frequency of puff releas is analyzed in the section on model application. We consider next some specific aspects of the Plume-Airshed Reactive-Interacting System.

Calculation of the Puff/Plume Trajectories

The plume--or puff--trajectories are calculated according to the three-dimensional airshed wind field and are determined with respect to the Eulerian reference system of the Airshed Model.

Plume trajectories are calculated for point sources located inside as well as outside of the airshed region. The wind field of the Airshed Model is defined on an hourly basis. Thus, there may be some discontinuity in the calculations of the plume trajectories, since wind direction might change incrementally every hour. Should a wind field with better temporal resolution be used, the algorithm for plume trajectory calculations would tend towards the continuous solution of the plume transport equations.

Calculation of Plume Dispersion

The Urban Airshed Model used here consists of two grid layers in the mixing layer and two grid layers in the inversion layer. In the early morning, the mixing layer is usually at its minimum height; its depth increases during the day. Most emissions from tall stacks, therefore, are trapped in the inversion layer in the early morning hours and are entrained into the mixing layer later in the day as the mixing height increases. At night, the height of the bottom of the inversion layer decreases and the upper part of the plume becomes trapped in the inversion layer.

The dispersion rates, which are calculated according to Pasquill stability classes, have been derived from the McElroy-Pooler dispersion coefficients for an urban area and were adjusted for relative diffusion only. Plume meandering is taken into account when calculating time-averaged concentration. Stability classes are determined according to the temperature lapse rates.

The rate of dispersion of a plume depends on its size because, when the plume expands, more eddies are available to disperse the plume material. If the atmospheric stability class does not change, the dispersion rate can be expressed as a function of the downwind distance, since the plume size is uniquely related to this distance. However, when the stability class changes during plume transport, or when the plume extends over regions of different stabilities, this relationship is no longer possible and the dispersion rate must be determined from the plume size.

When the plume is included in a region characterized by a simple stability class, plume dispersion is calculated according to the dispersion curves from the stability class and the plume size. If W and D are the plume widths and plume depths, we have the following relationships :

$$W = 4\ \sigma_y(x) \tag{1}$$

$$D = 4\ \sigma_z(x) \tag{2}$$

Then, an effective downwind distance can be calculated that corresponds to the values of the horizontal and vertical dispersion coefficients, $\sigma_y(x)$ and $\sigma_z(x)$, given by equations (1) and (2) where x is the downwind distance. Clearly, if there has been no change in atmospheric stability since the release of the plume from the source, the effective downwind distance is identical to the actual downwind distance. If the atmosphere becomes more unstable, the effective downwind distance will be smaller than the actual downwind distance, whereas an increasing stability will lead to a larger effective downwind distance.

When the plume extends over regions of different stability (i.e., mixing and inversion layers), the plume is considered to be composed of two fractions, a plume fraction in the inversion layer and a plume fraction in the mixing layer.

The plume width is defined at the plume centerline and a single horizontal dispersion rate is calculated from the stability of the layer where the plume centerline is located. Different horizontal dispersion rates could be calculated for the two plume fractions if some vertical resolution were introduced into the plume model. Vertical plume dispersion, however, is calculated independently for these two plume fractions, according to the vertical size of each plume fraction.

Chemical Interactions between the Plumes and the Airshed

In the formulation of the Reactive Plume Model, the background air is entrained into the plume as it moves downwind and is dispersed. The background chemistry is computed by the Airshed Model. As the various puffs travel and are dispersed into the airshed, the corresponding airshed concentrations are entrained into the puffs. The puff locations in the airshed grid system, as well as the puff sizes, are known as a function of time from plume trajectory and dispersion calculations. Thus, it is possible to determine which grid cell concentrations should be used in the chemical interactions of the plumes and the airshed. Background air is entrained into the plume from the left and right edges as well as from the top and bottom of the plume. If the plume touches the ground, no entrainment occurs at the bottom of the plume.

In summary, the chemical species concentrations are given by the Airshed Model for the background, and by the puffs along the plume trajectories. Thus, a complete description of the chemistry of the modeled area is given, both at the gridscale and subgridscale levels.

Mixing of the Plume Material into the Airshed Grid System

Use of Reactive Plume Models is an appropriate means of overcoming the inability of the Urban Airshed Model to properly describe chemical processes that occur at the subgridscale level; however, its use is unnecessary once the spatial resolution of the plume becomes comparable with the Airshed Model gridscale. At this point, the behavior of the chemical constituents of the plume can be described appropriately by the atmospheric diffusion equation that governs the dynamics of chemical species in the Airshed Model. Therefore, the criterion for mixing of the plume material into the Airshed Model must involve a comparison of the plume size with the airshed grid cell size. In the Plume-Airshed Reactive-Interacting System, this criterion is based on the horizontal dispersion of the plume. When the plume width* reaches the grid cell horizontal size, the plume concentrations are mixed with the airshed grid cell concentrations. Usually, the plume depth will be larger than the vertical grid size when the plume width reaches the horizontal grid size.

The distance downwind from the source at which the plume material is mixed into the airshed depends on the rate of plume dispersion (i.e., on atmospheric stability). As shown in table 1, those downwind distances vary considerably according to the stability class. In the case of a stable atmosphere, plumes could well be carried out of the airshed domain without being mixed into the airshed grid cells.

Table 1

Downwind distances for plume material mixing into the airshed

(grid cell width = 4 km in this simulation)

Stability Category	Downwind Distance (km)
A	7.5
B	12
C	17
D	28
E	46
F	68

The procedure for connecting the Plume Model to the Airshed Model at the moment of the mixing of plume material is as follows : the position of the plume center is given by the plume trajectory model as just described. The plume material is then mixed in one single cell (for a given grid cell layer), that is, in the cell which corresponds to the location of the plume centerline when the instantaneous plume width equals the horizontal grid size. This corresponds to artificial movement of the plume so that it is located in only one grid cell. If the plume material were mixed into the two or three grid cells over which a plume can extend, dispersion of the plume material would be overestimated. Thus, the single cell approach for plume mixing appears to be more realistic .

* The plume width is defined as 4 σ_y, where σ_y is the horizontal dispersion coefficient. Similarly, the plume depth is defined as 4 σ_z, where σ_z is the vertical dispersion coefficient.

In the vertical direction, the plume material is mixed into the grid layers over which the plume extends, according to the location of the plume top and bottom. It is apparent that there can be some artificial diffusion, particularly in the grid cell corresponding to the top of the plume. However, in most cases, the plumes extend over the two grid cells of the mixing layer so that most of the material will be mixed into these cells and only a small amount of the plume material is affected by artificial diffusion. The plume concentrations are taken to be vertically uniform, since it is assumed in the Reactive Plume Model that the plume is vertically well mixed.

As the plume is mixed into a grid cell, it is treated as a volume source. The change in the airshed concentration as a result of the mixing of a plume in cell j is given by the following equation :

$$\frac{d\,{}^{c}c_i(t)}{dt} = \left({}^{P}c_i(t) - {}^{c}c_i(t) \right) \frac{4\,\sigma_y\,{}^{P}h_j\,u}{H^2\,{}^{c}h_j} \qquad (3)$$

where ${}^{c}c_i$ and ${}^{P}c_i$ are the airshed and plume concentrations of species i, respectively, t is the time, σ_y is the plume horizontal dispersion coefficient, ${}^{P}h_j$ is the plume depth contained in cell j, H and ${}^{c}h_j$ are the horizontal and vertical sizes of cell j, and u is the wind speed. This equation relates the Plume Model to the Airshed Model as the plume material is mixed into the general gridded structure of the airshed.

The mixing of the plume material from a given puff occurs as a volume source continuously for the time interval corresponding to the frequency of puff release so that the mass of the pollutant released from the point source is conserved.

Calculation of the Time-Averaged Concentrations

The predicted concentrations are calculated at the monitoring station sites for one-hour time averages in the following manner : if a plume impinges over the station during the hour, the plume concentration is taken into account for the time period corresponding to the plume impact. If several plumes impinge over the same station, an average plume concentration is calculated for NO, NO_2, and O_3 according to the photostationary state assumption[7]. Since concentrations represent instantaneous values, it is necessary to take plume meandering into account in the calculation of the time-average concentrations[8]. This is done in the following manner :

$$c_i = \left({}^{P}c_i + 0.5\,{}^{c}c_i \right) / 1.5 \qquad (4)$$

where c_i is the time-averaged concentration, ${}^{P}c_i$ is the plume concentration and ${}^{C}c_i$ is the corresponding airshed concentration. If no plume impinges over a station, the concentration at the station is simply the airshed concentration.

APPLICATION OF THE MODEL

The Plume-Airshed Reactive-Interacting System was applied to a simulation of air quality on 13 July 1976 in the St. Louis urban area. A map of the area showing the monitoring stations and the two major power plants--the Sioux and the Meramec-- is shown in figure 2. The Sioux and Meramec power plants represent 35 and 16 percent of the total NO_x emissions for the area, respectively. To evaluate the treatment of these two point sources by the plume models, simulations were carried out with the conventional SAI Urban Airshed Model and with PARIS. Puffs from the power plants were emitted at 15-minute time intervals for the base case. Model predictions were compared with data obtained during the Regional Air Pollution Study (RAPS) field program. Model performance for ozone levels above 1 pphm is shown in table 2. It appears that the overall performance of the PARIS and Urban Airshed Model simulations are comparable, since the absolute deviation varies from 42 to 37 percent when the Reactive Plume Model is used to simulate emissions of the two major power plants.

Table 2

Measures of Model Performance for Ozone Levels above 1 pphm

Performance Measure	Model	
	Plume-Airshed Reactive-Interacting System	Urban Airshed Model
Signed deviation	0.212	0.305
Absolute deviation	0.373	0.424
Ratio of predicted to measured station peaks	1.29	1.31
Peak station predictions	28.7 pphm at station 107	29.1 pphm at station 107
Peak station measurement	22.2 pphm at station 114	22.2 pphm at station 114

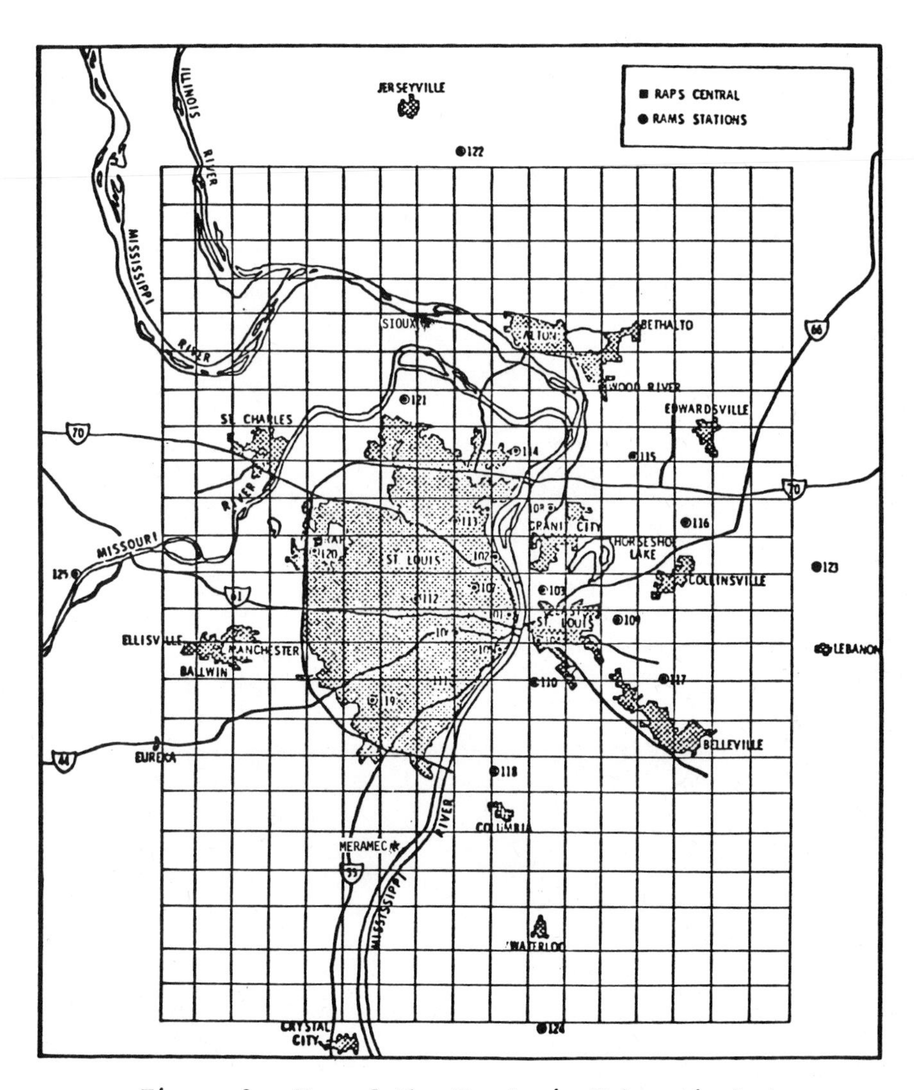

Figure 2. Map of the St. Louis Urban Airshed.

The signed deviation is defined as

$$\sum_{i=1}^{n} (C_i^p - C_i^m) / C_i^m$$

where

$$C_i^p \text{ and } C_i^m$$

are the predicted and measured concentrations, respectively at a given station and hour, and N is the total number of measurements of ozone concentration above 1 pphm. The absolute deviation is defined as

$$\sum_{i=1}^{n} \left| C_i^p - C_i^m \right| / C_i^m$$

On the average, both models tend to overpredict ozone levels. However, model performance is satisfactory and of a magnitude comparable with measurement uncertainties, which have been estimated to be on the order of 20 percent for the chemiluminescent technique[9] Predicted and measured ozone concentrations are shown in figure 3 for 4 monitoring stations--108, 109, 113, and 116. It appears that the plume of the Sioux power plant impinges on station 108 since, for PARIS, lower plume ozone concentrations are observed from 0900 to 1500 PST. The measurements tend to show better agremeent with the predictions when the plume is simulated using PARIS.

Station 109 shows lower ozone concentrations for PARIS than for the Urban Airshed Model. Model predictions compared well with the measurements at that station. It appears from the measurements that a plume impinged on station 113 around 0700 PST; this impingement appears slightly in the PARIS simulation. Both models overpredict the maximum ozone concentration at this station by about 50 percent.

There appears to be no direct effect from the power plant plume at station 116. PARIS predicts ozone concentrations that are slight ly lower than those predicted by the Urban Airshed Model because of the mixing of the power plant plumes within the grid cells adjacent to station 116.

In sensitivity studies carried out for PARIS, a change in the time interval for puff release from 15 minutes to 60 minutes led to an absolute variation in ozone predictions of 6 percent. This variation in model predictions occurs mainly because the frequency of puff release affects the time of plume impact at a given station.

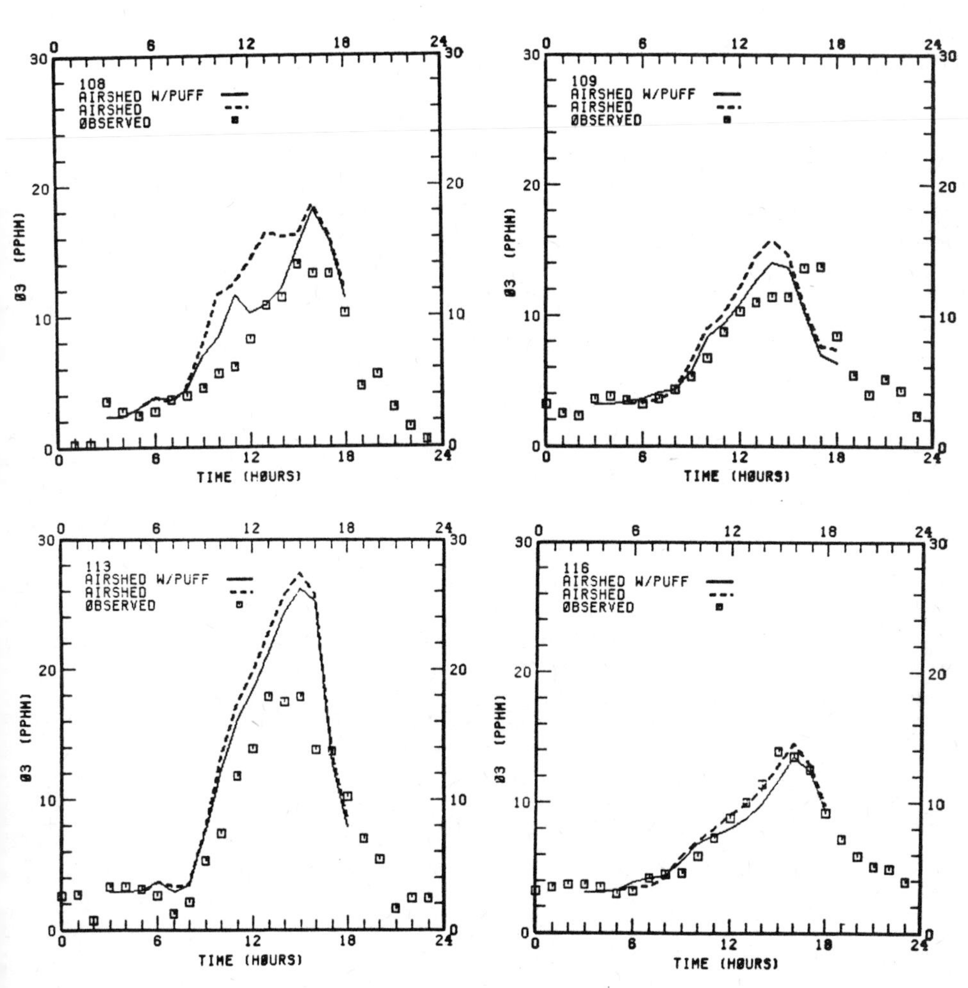

Figure 3. Comparison of measured and predicted ozone concentrations in the St. Louis area on 13 July 1976.

Simulating 16 additional point sources with the Reactive Plume Model led to a variation in ozone predictions of only 8 percent, therefore the increase in computational cost does not seem to justify the simulation of nonmajor point sources with the Reactive Plume Model. Finally, a 10 percent reduction in NO_x emissions from the Sioux and Meramec power plants led to an average increase in ozone levels of 2 percent.

CONCLUSION AND RECOMMENDATIONS

The Plume-Airshed Reactive-Interacting System (PARIS), which consists of one or several Reactive Plume Models imbedded into the SAI Urban Airshed Model, has been developed to provide a more realistic treatment of chemically reactive plumes emitted from large point sources in an urban area. Preliminary evaluation of the model show that model performance is modified by a few percent when the reactive plume treatment is used. This is within the uncertainty of the model predictions, which are on the order of 25 to 50 percent[10]. A more accurate assessment of the effect of this plume treatment on airshed model performance and of the usefulness of this more complex model for simulating urban air quality must await further model evaluation. The model is presently being applied to the Los Angeles basin for the simulation of a two-day air pollution episode[4].

The Plume Airshed Reactive-Interacting System appears to provide a valuable means for simulating the impact of large point-sources such as power plants or refineries located in urban areas. This model provides a unique approach to the description of the plume chemistry and its interactions with the urban background atmosphere and it is a promising tool for the prediction of maximum NO_2 concentrations downwind of major point sources in urban areas.

REFERENCES

1. R. G. Lamb and J. H. Seinfeld, "Mathematical Modeling of Urban Air Pollution--General Theory", Environ. Sci. Technol., 7, 253-261 (1973).
2. S. D. Reynolds, P. M. Roth, and J. H. Seinfeld, "Mathematicl Modeling of Photochemical Air Pollution I-Formulation of the Model", Atmos. Environ., 7, 1033-1061 (1973).
3. S. D. Reynolds, T. W. Tesche, and L. E. Reid, "An Introduction to the SAI Airshed Model and its Usage", EF79-31, Systems Applications, Inc., San Rafael, California (1979).
4. Sytems Application, Inc., "Impacts of NO_x Control upon Ozone Concentrations in Three Environments", Report to the Electric Power Research Institute, Systems Applications, Inc., San Rafael, California (1981).

5. G. Z. Whitten, J. P. Killus, and H. Hogo, "Modeling of Simulated Photochemical Smog with Kinetic Mechanisms", EPA-600/3-80-028a, U.S. Environmental Protection Agency, Research Triangle Park, North Carolina (1980)
6. D. A. Stewart and M. K. Liu, "Development and Application of a Reactive Plume Model", Atmos. Environ., 15, in press (1981).
7. W. H. White, "NO_x-O_3 Photochemistry in Power Plant Plumes : Comparison of Theory with Observation", Environ. Sci. Technol., 11, 995-1000 (1977).
8. F. Gifford, "Peak-to-Average Concentration Ratios According to a Fluctuating Plume Dispersion Model", Int. J. Air Poll., 3, 253-160 (1960).
9. C. S. Burton et al., "Oxidant-Ozone Ambient Measurement Methods : An Assessment and Evaluation", EF76-111R, Systems Applications, Inc., San Rafael, California (1976).
10. S. D. Reynolds and P. M. Roth, "The Systems Applications, Incorporated Airshed Model. A Review and Assessment of Recent Evaluation and Application Experience", submitted to J. Air Pollution Control Assoc. (1981).

ACKNOWLEDGMENTS

This work was supported by the Electric Power Research Institute under contract to Systems Applications, Inc., San Rafael, California. The authors wish to thank T. C. Meyers, R. A. Morris, A. B. Hudischewskyj, and D. A. Stewart for their assistance with the computer programs, and Dr. M. K. Liu for helpful discussion throughout the course of this study. We are also indebted to K. L. Schere of the Environmental Protection Agency for providing useful information and comments regarding the St. Louis simulation.

DISCUSSION

I. N. LEE — Can the model take into account terrain effects?

P. M. ROTH — The model takes into account terrain effects by means of the windfield calculation.

J. GOODIN — Did you analyze the sensitivity of your subgrid puff calculation to grid size?

P. M. ROTH — Up to this moment, no sensitivity analysis has been carried out with respect to the grid size.

CHEMICAL MODELLING STUDIES OF THE LONG-RANGE DISPERSION OF POWER-PLANT PLUMES

Alan T. Cocks, Ian S. Fletcher, and Anthony S. Kallend

Chemistry Division
Central Electricity Research Laboratories
Kelvin Avenue
Leatherhead, Surrey KT22 7SE, England

ABSTRACT

A comprehensive chemical model of gas-phase plume/ambient interactions using simple dispersion terms has been developed. Simulations have been carried out for an idealised source emitting into ambient air masses with different histories with particular reference to the formation of acidic species. Diurnal and seasonal effects together with the influence of ambient air composition and plume expansion rate have been investigated for dispersion times up to 19 hours.

The model predictions indicate that in general, greater chemical conversion of plume components occurs in material emitted in the late afternoon and overnight than in material emitted in the morning, and that chemical reactions are faster in Summer than in Winter. However, the nett quantitative environmental effect of the plume is a complex function of the physical and chemical parameters which cannot be determined a priori and further modelling studies are required before universally applicable simplifications to the chemical scheme can be made.

In addition to the general studies, the model has been modified to simulate, in an idealised manner, a specific case study in which a labelled power plant plume was tracked and analysed by instrumented aircraft on a trajectory from the U.K. over the North Sea. For these experiments conducted on a clear sunny day, satisfactory agreement between measured values for the important gaseous plume constituents and model predictions are obtained.

INTRODUCTION

The environmental impact of power plant emissions, particularly with respect to precipitation acidity, has been a subject of intense interest in recent years, in northern Europe[1] and North America[2].

Many relevant modelling studies have been undertaken to date, but the majority have concentrated on the physical aspects of dispersion and long range transport. For example, calculations of long term sulphur deposition in Europe have been performed using statistical distributions of meteorological parameters and emission inventories together with simple expressions for chemical conversion and removal by wet and dry deposition[3]. Detailed modelling of plume dispersion has been undertaken (e.g. ref 4) and coupling of detailed dispersion functions with simple kinetic expressions for single species of interest has also been carried out, particularly for the interpretation of near-field measurements (e.g. ref 5).

Environmental effects of gaseous power plant effluents depend on both the concentration and chemical nature of the species in the region of the receptor site which have been produced by the emitted material. Chemical reactions involving plume constituents are, therefore, important factors , particularly in the case of long-range transport where reaction times are long. Deposition processes during transit which likewise assume greater importance with longer times, are also influenced by the chemical nature of the plume-derived material.

Whilst the use of simple expressions to account for chemistry allows plume dispersion to be treated in a realistic manner without producing computational difficulties, such functions tend to be specific for a particular field situation. In addition, emphasis on a single species does not allow an assessment of the total effect of emissions on air quality. The complexity of the physico-chemical interactions produced by the dispersion of a multicomponent plume into multicomponent ambient air precludes the a priori extraction of simple universal rate expressions for important species.

The present work describes a simple idealised model for ambient emissions, atmospheric mixing, and plume dispersion, together with a more comprehensive gas phase chemical kinetic scheme which has been used to examine the major chemical and physical effects applicable to a dispersing power-plant plume over long distances and to predict important chemical features.

In addition,the model has been modified to simulate a field study of plume dispersion form a U.K. power plant over the North Sea using chemical tracers and followed by instrumented aircraft.

MODEL

General

A simple physical model comprising expansion of polluted "urban" air into a relatively clean "rural" atmosphere and the subsequent expansion of power-plant effluent into the dispersing urban plume was adopted. For the situations modelled in the present work the extent of the rural land mass is ca. 260 km, the width of a typical westerly wind trajectory passing over an area in the U.K. in which major power plants are situated. Beyond this distance, emissions and depositions appropriate to a marine environment are adopted. In the present study, the centre of the urban emissions is at ca. 160 km and the power plant ca. 190 km along the trajectory. These dispositions are characteristic of several major U.K. power plants.

Rural and Marine Atmosphere

The composition of the rural atmosphere is determined by emissions, depositions and reactions in air initially of oceanic composition (table 1) advected across the land mass and then over the sea.

Rural night-time emissions are given in table 2. They were derived from total annual U.K. emissions[13-15] by apportioning 50 % of the vehicle and non-industrial values to the total area of the U.K. Emission rates for reduced sulphur species were derived from global estimates[16,17].

Diurnal variation of SO_2, NO, CO, RH, and aldehydes was accounted for by assuming day and night emissions are in the ratio 2:1 and allowing a brief excursion to three times the night level at the morning and evening peak traffic times. Instantaneous uniform mixing of these emissions within the boundary layer, which for the present simulations was fixed at 1 km in depth (h), was assumed.

Table 1 : Concentrations (ppb) of Reactive Species in Oceanic Air

Species	NO[6]	NO_2[6]	O_3[7]	SO_2[7]	CO[8]	H_2[8]
Concentration	1.0	1.0	25	0.2	200	400
Species	CH_4[8]	HCL[9]	H_2S[7]	OCS[10,11]	CS_2[11]	N_2O[12]
Concentration	1000	3.0	0.2	0.4	0.2	300

Table 2 : Emissions

Species	Rural Emissions ($kg\ m^{-2}\ s^{-1}$)	Urban Emissions ($kg\ m^{-2}\ s^{-1}$)	Plume Emissions (ppm)
NO	1.5×10^{-11}	1.5×10^{-9}	500
SO_2	4.0×10^{-11}	6.0×10^{-9}	1500
CO	3.0×10^{-10}	1.0×10^{-8}	500
H_2	4.5×10^{-11}	1.5×10^{-9}	100
CH_4	2.0×10^{-10}	4.0×10^{-10}	1.0
RH	4.0×10^{-12}	3.0×10^{-10}	1.0
$RCH{=}CH_2$	7.5×10^{-12}	6.0×10^{-10}	0.1
HCHO	4.0×10^{-12}	3.0×10^{-10}	0.05
RCHO	3.0×10^{-12}	2.0×10^{-10}	0.05
CH_3OH	-	3.5×10^{-13}	-
ROH	-	4.5×10^{-10}	-
H_2S	6.0×10^{-13}	6.0×10^{-13}	-
OCS	1.0×10^{-13}	7.0×10^{-12}	-
CS_2	2.0×10^{-13}	1.5×10^{-11}	-

A constant water vapour partial pressure of 1×10^4 ppm, corresponding to a ca. 50 % relative humidity in Summer and close to 100 % in Winter, was adopted.

Only methane and H_2S were assumed to be emitted by the sea and emission rates identical to those for the "rural" environment were adopted.

Table 3 gives measured deposition velocities over land for important stable chemical species[18,19]. Water soluble and reactive gaseous species such as nitric acid and radicals, were assumed to have the same deposition velocity as SO_2 (1 cm s^{-1}) whereas other gases were assumed to have values similar to CO_2 (0.1 cm s^{-1}).

Only ozone deposition has been measured over the sea[18] and the velocity is about 7 % that over land. SO_2 deposition velocities of 40 % the land value have been measured over fresh water[20] but preliminary data from recent flights over the North Sea indicate a substantially lower value, possibly due to stratification and consequent decoupling of the sea surface. It has been assumed, therefore, that all deposition velocities over the sea are 10 % of the land values.

Table 3 : Deposition Velocities

Species	NO	NO_2	O_3	PAN	H_2S	SO_2	Aerosol Sulphate
Deposition Velocity ($m\ s^{-1}$)	6×10^{-4}	7×10^{-3}	6×10^{-3}	3×10^{-3}	2×10^{-4}	1×10^{-2}	1×10^{-3}

Variations of concentrations of species i, c_i, due to emission and deposition in rural or marine atmospheres are given by equation (1) where $(E_i)_{rural}$ is the emission rate and D_i is the deposition rate constant.

$$\left(\frac{dc_i}{dt}\right)_{rural} = (E_i)_{rural} - D_i(c_i)_{rural} \tag{1}$$

Urban Atmosphere

Emissions in the urban atmosphere were taken to be 50 % of vehicle and domestic emissions together with industrial non-power plant emissions. These are grouped into 11 urban conurbations 30 km x 20 km, which approximate to U.K. emission distributions allowing that some urban areas comprise 2 such conurbations. The urban plume is assumed to expand at a constant horizontal half-angle θ and to be fully mixed in the vertical and horizontal directions. Changes in concentrations due to emissions, deposition and rural entrainment in the urban plume are given by equation 2.

$$\left(\frac{dc_i}{dt}\right)_{urban} = (E_i)_{urban} - D_i(c_i)_{urban} + \frac{(c_i)_{rural} - (c_i)_{urban}}{t + t_o} \tag{2}$$

t_o is given by equation (3) where L is the town width and W is the wind speed.

$$t_o = L/(2W \tan\theta) \tag{3}$$

Plume Dispersion

Plume compositions, typical of U.K. coal-fired plant were used in the simulations and are shown in table 2. A flux typical of 2000 MW power plant was used and emission into the centre of the mixing layer (500 m a typical total plume rise) in an initially square cross-section plume of side B was assumed. The plume area was equated to the volume flux, at atmospheric temperature and pressure, divided by wind speed, W. The expansion of this plume was taken to be uniform with half-angle θ in the horizontal direction and ϕ in the vertical. Concentration changes in this plume due to expansion and entrainment up to the time, t_g, when the plume simultaneously makes contact with the ground and the top of the mixing layer are given by equation (4),

$$\left(\frac{dc_i}{dt}\right)_{plume} = \frac{(c_i)_{urban} - (c_i)_{plume}}{t + t_A} + \frac{(c_i)_{urban} - (c_i)_{plume}}{t + t_B} + \frac{(c_i)_{rural} - (c_i)_{urban}}{t + t_o + t_c} \tag{4}$$

where t_c is the distance between town and plume divided by W, and t_A and t_B are the back projected times to the hypothetical line sources for horizontal and vertical plume (equations (5) and (6)).

$$t_A = B/(2W \tan\theta) \tag{5}$$

$$t_B = B.t_g/(h-B) \tag{6}$$

For expansion beyond t_g, equation (7) is applicable.

$$\left(\frac{dc_i}{dt}\right)_{plume} = (E_i)_{rural} - D_i(c_i)_{rural} + \frac{(c_i)_{urban} - (c_i)_{plume}}{t + t_A} + \frac{(c_i)_{rural} - (c_i)_{urban}}{t + t_o + t_c} \tag{7}$$

Chemical Kinetic Scheme

The effects of chemical changes on equations (1), (2), (4) and

(7) are accounted for by the addition of a specific term, $\left(\frac{dc_i}{dt}\right)_{chem}$, for each species considered. These were derived from a detailed chemical scheme described in detail elsewhere[21].

Table 4 : Major Reactions

Reaction	$k/(ppm^{-1}s^{-1})$
1. $NO + O_3 \rightarrow NO_2 + O_2$	4.0×10^{-1}
2. $NO_2 + h\nu \rightarrow NO + O$	7.8×10^{-3} (a)
3. $NO_2 + O_3 \rightarrow NO_3 + O_2$	7.9×10^{-4}
4. $O + O_2 + M \rightarrow O_3 + M$	3.6×10^{-1} (b)
5. $O_3 + h\nu \rightarrow O_2 + O$	3.2×10^{-5} (a)
6. $O^* + M \rightarrow O + M$	7.0×10^{8} (c)
7. $O^* + H_2O \rightarrow 2OH$	3.0×10^{3}
8. $OH + RH \xrightarrow{O_2} RO_2 + H_2O$	5.0×10^{1} (d) (e)
9. $OH + CH_4 \xrightarrow{O_2} CH_3O_2 + H_2O$	1.6×10^{-1} (d)
10. $RO_2 + NO \xrightarrow{O_2} R'CHO + HO_2 + NO_2$	8.8×10^{1} (d)
11. $CH_3O_2 + NO \xrightarrow{O_2} HCHO + HO_2 + NO_2$	1.7×10^{2} (d)
12. $RO_2 + SO_2 \xrightarrow{O_2} SO_3 + R'CHO + HO_2$	6.0×10^{-2} (d)
13. $CH_3O_2 + SO_2 \xrightarrow{O_2} SO_3 + HCHO + HO_2$	1.2×10^{-1} (d)
14. $HO_2 + SO_2 \rightarrow SO_3 + OH$	2.2×10^{-2}
15. $NO_3 + SO_2 \rightarrow SO_3 + NO_2$	2.5×10^{-1}
16. $OH + SO_2 + M \rightarrow HSO_3 + M$	3.0×10^{1} (b)
17. $OH + NO_2 + M \rightarrow HNO_3 + M$	2.7×10^{2} (b)
18. $HO_2 + NO \rightarrow NO_2 + OH$	2.0×10^{2}

(a) First order rate constant (s^{-1}) for an overhead sun.
(b) Second order rate constant assuming M is air at atmospheric pressure.
(c) First order rate constant (s^{-1}) assuming M is air at atmospheric pressure.
(d) Second order rate constant for primary step. Subsequent reactions with O_2 are rapid.
(e) Rate constant for propane.

This chemical model has been modified to include reduced sulphur species and allow for different hydrocarbon reactivities. The main reactions are summarised in table 4 together with the appropriate critically evaluated rate constants.

Recent experimental work[22-25] has indicated that rate constants for peroxy radical attack on SO_2 (reactions 12-14) are uncertain and a consistent set of high values has been adopted for the present work. These may, however, be in error by more than an order of magnitude and further work is required to obtain unambiguous information on the rates and mechanisms of these processes. The rate of reaction of NO_3 with SO_2 (reaction 15) is also subject to uncertainty. SO_3 is presumed to react with water vapour to form H_2SO_4, and HSO_3 is presumed to form an acidic aerosol although the mechanism of this process is not firmly established[26].

Rate constants for photolytic reactions are those for an overhead sun and were modified for diurnal and seasonal variations by correcting for the solar zenith angle.

Modelling

The combined chemical and physical differential equations for the concentrations, c_i, form a stiff set which were solved numerically using a divided-difference Gear algorithm (NAGLIB DO2QBF) on an IBM 370 computer.

Simulations were carried out for a latitude 53.5°N (Northern England), a wind speed of 8 m s^{-1}, plume release times of 06:00, 12:00, 18:00 and 24:00 GMT in Summer (June 21st), and a value for Θ of 5° and ϕ of 2°. The angles are slightly less than average but well within the range of measured values. Simulations were also performed for a 12:00 plume release time in Winter (December 21s with the above plume parameters and also for a plume release at 12:00 in Summer with a value for θ of 10°.

In addition, the model was adapted to simulate a field study using an SF_6 labelled plume from a U.K. power plant monitored by an instrumented C130 aircraft of the Meteorological Research Flight on a trajectory over the North Sea. For this simulation, "rural" emissions of SO_2 were estimated from grid inventories along the back trajectory through the power plant and NO emissions were assumed to be 1/3 of these in volume. Urban SO_2 levels are similarly taken from a grid corresponding to a conurbation upwind of the power plant and non-vehicle NO was again estimated at 1/3 SO_2. All other emissions (vehicular NO, hydrocarbons etc.) were assumed to be identical to those used in the idealised "rural" and "urban" models.

The labelled power plant plume was one of three emitted from the 2000 MW plant during the measurement exercise and is situated close to two other 2000 MW plants. It was assumed that emission from all plants occurred from a single source and total fluxes were calculated from load factors. Concentrations of major effluents were measured by in-stack analysis at the plant of interest and similar compositions were assumed for the other plants.

The aircraft measurements indicated a value for θ of 6° and a mixing height of 1100 m. Ground measurements suggested a value for ϕ of ca 2° and the wind speed was approximately 5 m s^{-1} throughout the measurement period. All these measured values were used as input data for the model.

The in-flight measurements took place on June 18th 1980 under dry sunny conditions.

RESULTS

Predicted concentrations of OH radicals and nitric and sulphur acids in "rural", "urban", and plume atmospheres along a plume trajectory for an emission start time of 12:00 in Summer are shown in figs. 1-3.

OH radical concentrations show the expected diurnal variation (fig. 1) and peak concentrations are higher in the rural atmosphere and occur earlier than in the urban atmosphere. OH concentrations are substantially lower in the plume during the first day but exceed those in the ambient atmosphere at the beginning of the next day.

Peroxy radical and NO_3 concentrations show much later maxima (ca. 20:00) than does OH. Peak OH concentrations predicted for the range of conditions simulated are summarised in table 5.

Nitric acid concentrations in the urban atmosphere are higher than those in the rural atmosphere (fig. 2). However, for ca. 5 hours in the early stages of plume expansion, nitric acid concentrations in the plume are less than those in the ambient.

Sulphur acid concentrations are always greater in the plume than in the ambient atmospheres (fig. 3) and predicted concentrations are less than those for nitric acid. The relative importance of the OH and peroxy and NO_3 radical routes changes during the day. "H_2SO_4" produced via NO_3 and peroxy radical attack on SO_2 accounts for less than 5 % of the sulphur acid production in the early afternoon. The dominance of the "HSO_3" product from OH attack becomes less marked during the afternoon and falls to 85 % of the total at sunset. This decline continues during the hours of darkness and reaches ca. 65 % at dawn.

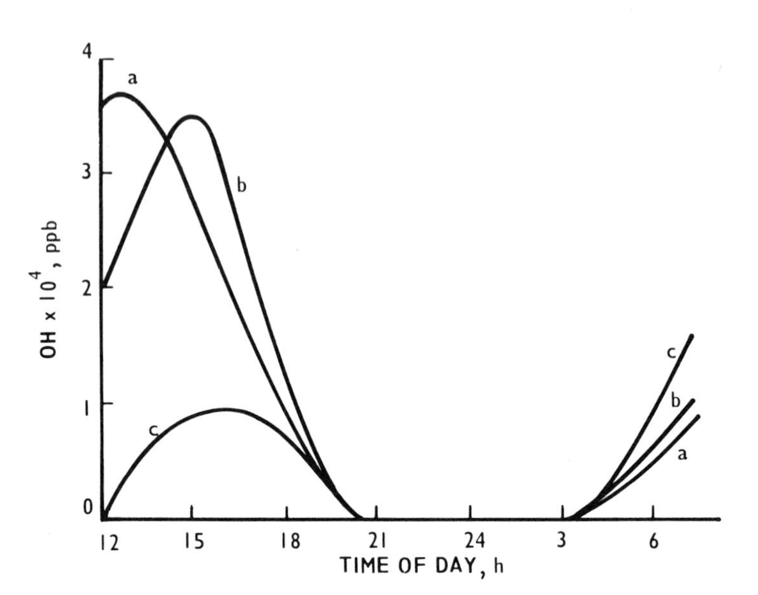

Fig. 1. Diurnal variation of OH concentrations in a "rural atmosphere" (a), an "urban atmosphere" (b) and a plume (c).

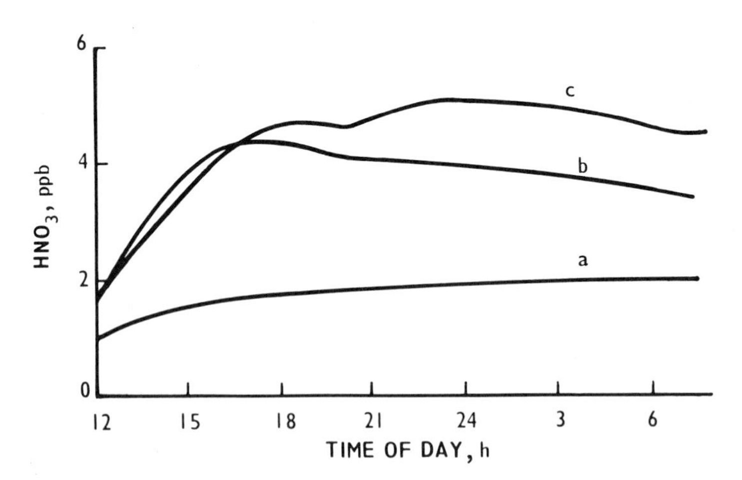

Fig. 2. Diurnal variation of HNO_3 concentrations in a "rural atmosphere" (a), an "urban atmosphere" (b) and a plume (c).

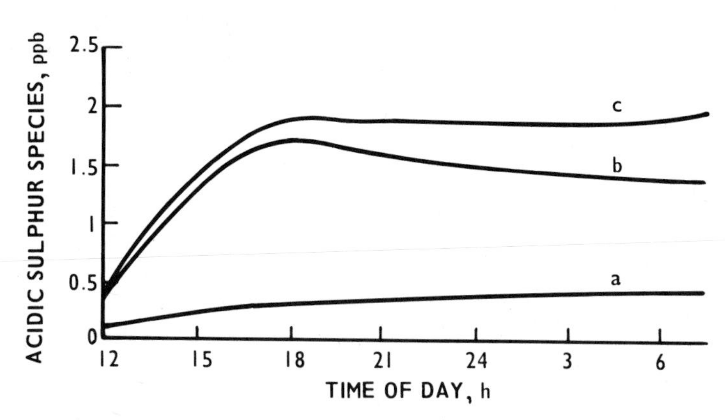

Fig. 3. Diurnal variation of the concentrations of sulphur acids in a "rural atmosphere" (a), an "urban atmosphere" (b) and a plume (c).

Table 5 : Peak Hydroxyl Radical Concentrations

Season	Plume Start	$\theta/^{o}$	[OH] max. (ppb)		
			Rural	Urban	Plume
Summer	06:00	5	3.4×10^{-4}	4.3×10^{-4}	1.7×10^{-4}
"	12:00	5	3.7×10^{-4}	3.5×10^{-4}	9.4×10^{-5}
"	12:00	10	3.7×10^{-4}	3.6×10^{-4}	1.4×10^{-4}
"	12:00	5★	3.7×10^{-4}	-	4.0×10^{-5}
"	18:00	5	2.4×10^{-4}	2.8×10^{-4}	3.5×10^{-4}
"	24:00	5	2.8×10^{-4}	3.6×10^{-4}	2.9×10^{-4}
Winter	12:00	5	4.6×10^{-5}	3.1×10^{-5}	7.2×10^{-6}

★ Plume Expansion into Rural Air

Table 6 summarises the predicted concentrations of SO_2 and acidic species in "rural", "urban" and plume atmospheres after 69000 s for the range of conditions simulated. Concentrations of all species increase with plume start time in the order 24:00 < 6:00 12:00 < 18:00. Nitric acid always exceeds sulphur acids and the greatest difference between plume and ambient acid concentrations occurs for the 18:00 start expanding into urban air. This is refle ted mainly in the greater sulphur acid difference. Expansion into a rural atmosphere results in lower total plume acid concentrations but the differences between plume and ambient are similar to those predicted for dispersion into "urban" air. Dispersion at 10° produces lower concentrations and smaller plume/ambient differences than 5° dispersion. Lower values are also predicted for Winter compared to Summer and a much higher proportion of "H_2SO_4" product is predicted for Winter conditions.

Under the conditions modelled, concentrations of photochemical smog products such as ozone and PAN are predicted to be always lower in the plume than in the ambient atmosphere for plume start times of 06:00 and 12:00. For start times of 18:00 and 24:00 small nett increases in plume concentrations of ozone and PAN are predicted towards the end of the simulation period.

Figures 4, 5 and 6 show predicted and measured profiles for SO_2, NO_x, and ozone at a downwind distance of 100 km in the labelled plume. Predicted NO_x results are 48 % NO, 43 % NO_2 and 9 % HNO_3 The agreement for SO_2 is reasonable but measured NO_x and ozone fluxes are higher than predicted. Measurements indicate that NO comprised ca. 50 % of the NO_x in good agreement with the simulation.

Predicted oxidant levels for the "rural" atmosphere in the flight simulation are generally lower than those calculated for the model cases. Maximum hydroxyl radical concentrations, for example, are about 1/4. "Urban" predicted values are similar to those for the model calculations, but plume concentrations are agai lower, with maximum OH concentrations of ca. 5×10^{-5} ppb.

Sulphur acid concentrations are predicted to be ca. 1.2 ppb and comprise 85 % "HSO_3" product. Filter samples collected during a cross-sectional transit through the plume but including a substantial period in the background air yielded a particulate sulphur concentration of ca. 1.6 ppb.

Table 6 : Concentrations (ppb) After Plume Dispersion for 69000 s

Season	Plume start	θ/°	Species	Rural	Urban	Plume
Summer	06:00	5	SO_2	0.6	1.5	5.1
"	"	"	Sulphur Acid	0.3	0.9	1.6
"	"	"	HNO_3	1.8	2.5	3.5
"	12:00	"	SO_2	0.8	2.7	6.4
"	"	"	Sulphur Acid	0.4	1.5	2.0
"	"	"	HNO_3	2.0	3.5	4.6
"	"	10	SO_2	0.8	2.0	3.8
"	"	"	Sulphur Acid	0.4	1.0	1.4
"	"	"	HNO_3	2.0	2.9	3.5
"	"	5★	SO_2	0.8		4.5
"	"	"	Sulphur Acid	0.4		0.9
"	"	"	HNO_3	2.0		3.0
"	18:00	"	SO_2	0.8	3.2	6.5
"	"	"	Sulphur Acid	0.3	1.2	2.2
"	"	"	HNO_3	1.9	3.6	4.9
"	24:00	"	SO_2	0.6	1.5	4.9
"	"	"	Sulphur Acid	0.2	0.7	1.5
"	"	"	HNO_3	1.7	2.4	3.5
Winter	12:00	"	SO_2	0.8	3.4	7.4
"	"	"	Sulphur Acid	0.1	0.3	0.5
"	"	"	HNO_3	1.7	2.7	3.3

★ Plume Expansion into Rural Air.

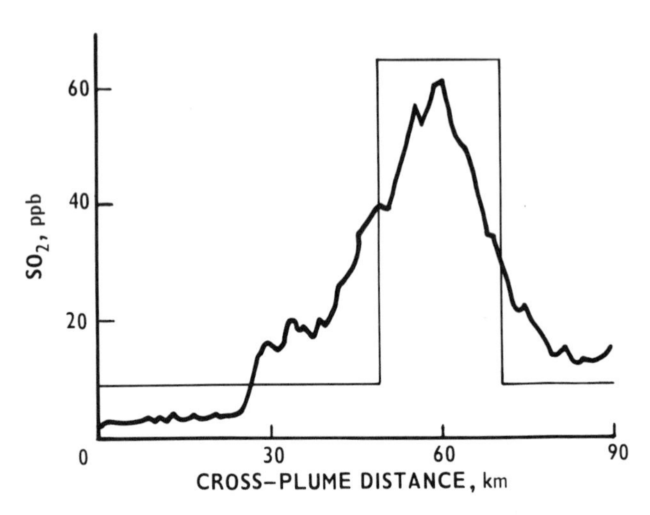

Fig. 4. Predicted and measured SO_2 plume profiles.

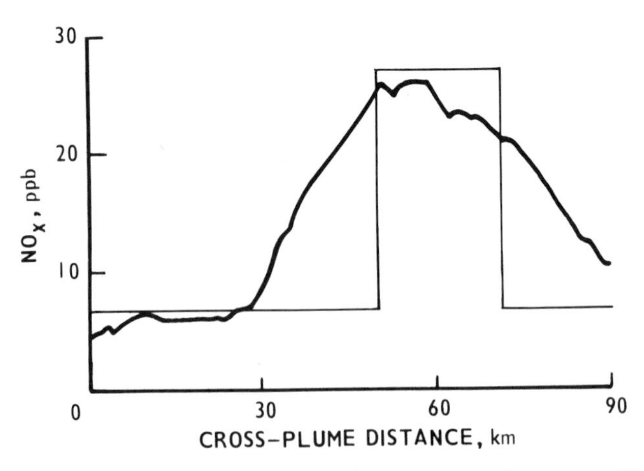

Fig. 5. Predicted and measured NO_x plume profiles.

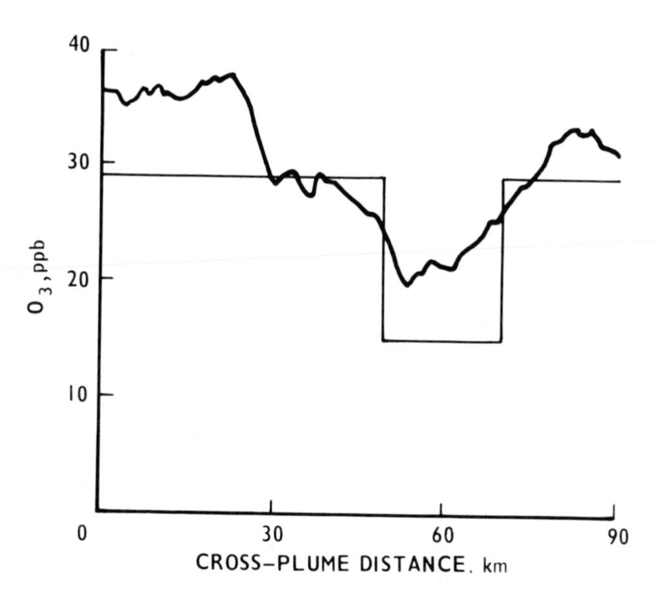

Fig. 6 Predicted and measured O_3 plume profiles

DISCUSSION

The major production routes for nitric acid and sulphur acid species involve attack by the hydroxyl radical on NO_2 and SO_2 (reactions 17 and 16). This radical is formed in an expanding power plant plume from entrained ambient material and the reaction pathways involved are affected by the much higher plume NO levels resulting in lower radical concentrations. Thus, greater expansion rates produce higher OH radical concentrations. However, there appears to be no simple relationship between the potential for oxidant production, measured by OH concentrations in unperturbed ambient air, and the effectivenss of OH formation from entrained ambient exposed to expanding plume concentrations. For example, for the model simulation with plume start time of 12:00, maximum OH concentrations in the "rural" and "urban" atmospheres are similar but the concentrations in a plume expanding into a "rural" ambient are one half those for one expanding into an "urban" ambient. The complexity of the physical/chemical interaction leading to OH production is further illustrated by the flight simulation predictions which indicate that for an "urban" ambient atmosphere more polluted than the model "urban", OH levels in the plume are similar to those for expansion into a model "rural" atmosphere.

Diurnal variations of OH for a 12:00 plume release may be rationalised in terms of insolation and NO concentration. NO inhibits the formation of "new" OH radicals by depressing the ozone concentration (reaction 1) whereas higher insolation increases OH production via greater ozone photolysis (reactions 5 and 7) and greater ozone production from NO_2 photolysis (reactions 2 and 4). However, NO can release OH radicals (reactions 10, 11 and 18) "stored" (reactions 8 and 9) as peroxy species. The higher NO levels in the "urban" and plume atmospheres thus account for the lower formation in the morning and greater formation in the afternoon and hence the shift in the maximum with respect to "rural" conditions. The effect of the greater concentration of reactive hydrocarbons in the "urban" ambient appears to be largely offset by the higher NO concentrations. The apparent increase in plume OH with respect to the ambient atmospheres after the period of darkness is a consequence of the more favourable NO_x/hydrocarbon ratio due to reaction of most of the NO. This is more clearly reflected in the higher OH concentrations produced on the following day by material released at 18:00 and 24:00. For these cases substantial dilution and NO reaction occur before photolysis commences.

Peroxy radicals persist after OH levels fall because of the lower NO levels. Ozone concentrations also peak in the late afternoon for the same reason and as high concentrations of ozone and NO_2 persist after dark, their interaction to form NO_3 is favoured.

Nitric acid is accumulated during the day along the plume trajectory. Initially, the decrease in OH within the plume volume off sets the higher concentration of NO_2 and lower conversions in the plume than in the "urban" atmosphere are predicted. The small increase in HNO_3 after dusk is due to the formation of N_2O_5, via the reaction of NO_3 and NO_2, which reacts with water to form nitric acid.

The amount of nitric acid produced over the 69000 s period is a function of precursor and OH concentration. Material emitted from the plume at 18:00 produces the highest daytime values. Hence the greatest HNO_3 concentration is predicted for the 18:00 plume start. Although the OH levels in the plume for the 24:00 release time are higher than for that at 12:00, NO_x levels in the former are determined by lower overnight emission rates and hence smaller concentrations of HNO_3 are predicted. Calculated concentrations of HNO_3 are lower than those measured in a polluted atmosphere[27].

The true environmental impact of the plume is measured by the difference between plume concentrations and ambient concentrations. For 5° dispersion in Summer, the nett effect of the plume on nitric acid concentrations is similar for all plume release times after 69000 s (ca. 1 ppb). If the extra acid is ascribed to initial plu NO, apparent first order rate constants for nitric acid formation,

k_{app}, may be calculated and these are summarised in table 7. Although the reaction is certainly not first order over the time considered, first order approximations are often made in simulations and the rate coefficients are useful comparative quantities. Values of k_{app} depend on differences in reaction time, oxidant concentrations and precursor concentrations between plume and ambient atmospheres and similar values can be produced by different combinations of circumstances (e.g. 06:00 and 24:00 plume release times give similar k_{app} values although the concentrations of the ambients are different and the time of greatest photochemical reactivity occurs at different stages of plume/ambient interaction). k_{app} does not reflect the true oxidation rate in the plume, however, as it depends on the rate of reaction of the entrained NO_x (up to 85 % of the plume total) with plume oxidants which will not normally produce an identical conversion to that in the ambient.

In the flight simulations, depression of OH is greater than increases in NO_2 and hence, predicted nitric acid concentrations are less (-0.6 ppb) in the plume than in the ambient.

Table 7 : Overall First Order Rate Constants for Acid Formation over 69000 s

Season	Plume Start	θ/°	HNO_3 k_{app}/% hr^{-1}	Sulphur Acid	
				k_{app}/% hr^{-1}	k_{eff}/% hr^{-1}
Summer	06:00	5	7.0	0.8	1.0
"	12:00	5	7.8	0.6	0.9
"	12:00	10	9.8	0.7	1.0
"	12:00	5*	6.6	0.6	0.6
"	18:00	5	10.7	1.4	1.4
"	24:00	5	7.2	1.1	1.3
Winter	12:00	5	2.8	0.3	0.3

* Plume Expansion into Rural Air

The diurnal varition in production of sulphur acids for a 12:00 plume release follows the OH concentration, modified by SO_2 accumulation, during the day and reaction 16 predominates. Towards sunset, however, the reactions of peroxy radicals, which persist after OH decreases, and particularly NO_3 become relatively more important. Although ozone levels are depressed in the plume, enhanced NO_2 concentrations result in greater amounts of NO_3 in the plume. Thus, an increase in the importance of the "H_2SO_4" product overnight is predicted reaching a value of ca 30 % of the total at dawn. However, the reaction routes to "H_2SO_4" have rate constants which are not well established and hence predicted sulphur acid concentrations at 69000 s may be in error by up to 30 %. As plume concentrations of SO_2 are in much greater excess over ambient compared to NO_x concentrations, acid sulphur products are always greater in the plume.

Concentrations of sulphur acids after 69000 s are dependent on oxidant and precursor concentrations in an analagous manner to nitri acid and hence follow the same pattern. The low levels of photochemically produced oxidants in Winter results in most of the sulphu acids being produced via NO_3 reactions. Differences between plume and ambient concentrations of sulphur acids whilst showing a simila qualitative pattern to those for nitric acid, reflect quantitative differences between accumulation of the acids in the ambient atmosphere. For example, whilst "urban" HNO_3 is similar for plume releas times of 12:00 and 18:00, sulphur acids are significantly less in th latter case. This results in a much greater effect on the nett increase in sulphur acid concentrations by the material emitted at 18:

Apparent first order rate constants for acid production, k_{app}, are listed in table 7. In the simulations, however, plume SO_2 was treated as a separate species and hence true effective first order rate constants k_{eff} for plume material can be calculated and are also shown in table 7. These are greater than k_{app} for the reasons outlined in the case of HNO_3, but for assessment of environmental impact, k_{app}, is more appropriate.

For the flight simulations, only a small amount of acid production is predicted as OH levels are low (k_{eff} ca. 0.2 % hr^{-1}). Additionally, as ambient SO_2 concentrations are greater than those in the model calculations, the plume/ambient differences are even lower (0.14 ppb) resulting in $k_{app} < 0.05$ % hr^{-1}. Thus, for the flight simulations a total decrease in acidity (0.14 ppb sulphur acid - 0.6 ppb HNO_3) in the plume is predicted at 100 km.

For plume release at 06:00 and 12:00, the depression of oxidant formation during most of the simulation time results in a nett decrease in plume ozone and PAN. However, the model calculations suggest that at longer times plume oxidants may exceed those in the ambient and hence, for very long range transport, enhanced levels of photochemical oxidants (and possibly higher acid concentrations) could be predicted for these plumes. Increased oxidant levels compared to ambient concentrations are predicted during the simulation period for plumes released at 18:00 and 24:00 but predicted increases in photochemical pollutants are modest and exist for only a short time.

The present model was developed primarily to examine the general effects of dispersion into reasonably realistic ambient atmospheres on the gas phase chemistry of power plant effluents allowing for the influence of emissions and depositions. Although gross simplifications of uniform mixing and constant mixing height have been adopted, reasonable agreement between the predictions for the modified model and flight measurements have been obtained particularly for SO_2 concentrations and NO/NO_2 ratios. The additional measured NO_x compared to that predicted is probably due in part to the channelling of an urban plume along an estuary into the vicinity of the measurement area. Also, the predictions were based on an estimate of NO_x emissions. The lower ozone levels predicted could be made to match the measurements by altering the estimated hydrocarbon emission rates. It would thus appear that uncertainties in the input data may introduce greater errors than approximations in the model. Our aircraft measurements show that the structure of the mixing layer is extremely complex and a more ambitious dispersion model based on theoretical functions, which would involve much greater computation times may not yield a commensurately more realistic description of pollutant concentrations.

An aim of all modelling work should be to produce as simple a description of the considered system within a universally applicable fundamental framework. The present study indicates that simplifications to the chemical scheme pertinent to plume/ambient interaction are not obvious. Unperturbed ambient behaviour does not indicate the resultant plume + ambient oxidant levels and hence all the important features of oxidant production must be maintained. Indeed, the scheme adopted in the present model which generalises the reactivity of non-methane hydrocarbons, alkenes and aromatics, and aldehydes other than formaldehyde, may be too gross a simplification for the accurate determination of oxidant levels. Steady state approximations do not appear to hold well in a plume/ambient interacting system and further modelling work, encompassing a wider range of conditions is required before any valid simplifications can be made.

The present work deals solely with gas phase processes but oxidation in aqueous solution within cloud droplets may constitute a significant route for acid production. A model incorporating cloud chemistry is currently under development and such considerations produce further restrictions on simplifications in gas phase chemistry as gaseous composition determines the concentration in cloud droplets. Hence reactions which do not affect gas phase acid production, but result in the formation of potential heterogeneous oxidants such as peroxides, cannot be neglected in a general predictive reactive plume model.

ACKNOWLEDGEMENTS

We thank our colleagues at C.E.R.L., the Meteorological Research Flight, and the Meteorological Office who provided the data from the C.E.R.L. flying chemistry programme. This work was carried out at the Central Electricity Research Laboratories, with partial funding from EPRI under contract RP 1311-1, and is published by permission of the Central Electricity Generating Board.

REFERENCES

1. OECD Report, Long range transport of air pollutants, OECD, Paris (1977).
2. G.E. Likens, R. F. Wright, J. N. Galloway and T. J. Butler, Acid Rain, Scientific American, 241:39 (1979).
3. B. E. A. Fisher, The Calculation of long term sulphur deposition in Europe, Atmospheric Environment, 12:489 (1978).
4. K. R. Szoc, Mathematical models for predicting the dispersion of atmospheric pollutants, Thesis, University of Florida (1976).
5. O. T. Melo, M. A. Lusis, and R. D. S. Stevens, Mathematical modelling of dispersion and chemical reactions in a plume-oxidation of NO to NO_2 in the plume of a power plant, Atmospheric Environment, 12:1231 (1978).
6. E. Robinson and R. C. Robbins, Sources, abundances and fate of gaseous atmospheric pollutants, NTIS report N71-25147 U.S. Department of Commerce, (1968).
7. J. Heicklen, "Atmospheric Chemistry", Academic Press, New York (1976).
8. I. M. Campbell, "Energy and the Atmosphere a Physical-chemical Approach", John Wiley, London (1977).
9. A. C. Stern, "Air Pollution Vol. 1 - air pollutants, their transformations and transport", Academic Press, New York (1976).
10. P. L. Hanst, L. L. Spiller, D. M. Watts, J. W. Spence, and M. F. Miller, Infrared measurement of fluorocarbons, carbon tetrachloride, carbonly sulphide, and other atmospheric trace gases, J. Air Pollution Control Assoc., 25 : 1220 (1975).
11. F. J. Sandalls and S. A. Penkett, Measurements of carbonyl sulphide and carbon disulphide in the atmosphere, Atmospheric Environment, 11:197 (1977).

12. K. A. Brice and S. A. Penkett, Measurements of nitrous oxide in the atmosphere via E.C.D. gas chromatography at Harwell, AERE report R8569, HMSO (1977).
13. A. J. Apling, C. J. Potter, and M. L. Williams, Air pollution from oxides of nitrogen, carbon monoxide, and hydrocarbons, Warren Spring Laboratory Report LR 306 (AP) (1979).
14. M-L Weatherley, Estimates of smoke and sulphur dioxide pollution from fuel combustion in the United Kingdom for the years 1976 and 1977, Clean Air, 9:8 (1979).
15. R. G. Derwent and Ö. Hov, The contribution from natural hydrocarbons to photochemical air pollution formation in the United Kingdom, in "Proceedings of the First European Symposium on the Physico-chemical behaviour of Atmospheric Pollutants", B. Versino and H. Ott eds., CEC (1980).
16. J. N. Galloway and D. M. Whelpdale, An atmospheric sulphur budget for eastern North America, Atmospheric Environment, 14:409 (1980).
17. R. P. Turco, R. C. Whitten, O. B. Toon, J. B. Pollack, and P. Hamill, OCS, stratospheric aerosols and climate, Nature, 283:283 (1980).
18. G. A. Sehmel, Particle and gas dry deposition : a review, Atmospheric Environment, 14:983 (1980).
19. J. A. Garland, Dry and wet removal of sulphur from the atmosphere, Atmospheric Environment, 12:349 (1978).
20. J. A. Garland, The dry deposition of sulphur dioxide to land and water surfaces, Proc. Roy. Soc. A, 354:245 (1971).
21. A. T. Cocks and I. S. Fletcher, Possible effects of dispersion on the chemistry of gaseous power plant effluents, Atmospheric Environment, in press.
22. W. A. Payne, L. J. Stief, and D. D. Davies, A Kinetics study of the reaction of HO_2 with SO_2 and NO, J. Amer Chem. Soc., 95:7614 (1973).
23. J. P. Burrows, D. I. Cliff, G. W. Harris, B. A. Thrush, and J. P. T. Wilkinson, Atmospheric reactions of the HO_2 radical studied by laser magnetic resonance spectroscopy, Proc. Roy. Soc. A, 369:463 (1979).
24. C. S. Kan, R. D. McQuigg, M. R. Whitbeck and J. G. Calvert, Kinetic flash spectroscopic study of the CH_3O_2-CH_3O_2 and CH_3O_2-SO_2 reactions, Internat. J. Chem. Kinetics, 11:921 (1979).
25. S. P. Sander and R. T. Watson, A kinetics study of the reaction of SO_2 with CH_3O_2, Chem. Phys. Letters, 77:473 (1981).
26. D. D. Davis, A. R. Ravishankara and S. Fischer, SO_2 oxidation via the hydroxyl radical : atmospheric fate of HSO_x radicals, Geophys. Res. Letters, 6:113 (1979).
27. G. J. Doyle, E. C. Tuazon, R. A. Graham, T. M. Mischke, A. M. Winer, and J. N. Pitts, Simultaneous concentrations of ammonia and nitric acid in a polluted atmosphere and their equilibrium relationship to particulate ammonium nitrate, Environ. Sci. and Technology, 13:1416 (1979).

A COMPARISON OF REGIONAL SCALE EFFECTS OF IN-CLOUD CONVERSION OF SO_2 TO $SO_4^=$ IN AN EIGHT LAYER DIABATIC MODEL WITH A SINGLE LAYER MODEL

W. E. Davis

Battelle, Pacific Northwest Laboratory

Richland, Wa. U.S.A.

A multilayer and single layer regional scale models have been used in a short term assessment during two frontal storms in October, 1977. A comparison with observed values has been made of results of the model assessments for $SO_4^=$ air concentrations using an in-cloud $SO_4^=$ wet deposition conversion, an in-rain conversion and no in-cloud or in-rain conversion in the multilayer model. Also an assessment using in-rain conversion using a single layer model was compared with observations. The results were mixed in that the multilayer model with no in-cloud or in-rain conversion yielded the best $SO_4^=$ air concentration patterns while the multilayer model using in-cloud conversion yielded the best fit to observed wet deposition of $SO_4^=$. Further work is necessary since only 35 % of the SO_2 emissions in the northeast United States were used in this study.

INTRODUCTION

In recent years, concern with the regional to long-range transport of pollutants has caused a significant effort in model development for assessment purposes. This modeling involves the simulation of transport, dispersion, transformation, and both wet and dry removal. Consideration of the complexity of these processes and their interactions have prevented a direct attack on the problem with time dependent, numerical solution techniques. One impressive approach to a portion of the problem, namely the transport and moisture budget, has been undertaken.[17] This model provides predicted three-dimensional velocity fields in frontal storms, as well as wet removal by precipitation scavenging.
However, the computer time required precludes use of this model as an assessment tool. The value of this type of model is to

provide insight and guidance in the development of the more practical diagnostic models.

The predominant type of diagnostic model used for studying atmospheric transport and dispersion involves the calculation of atmospheric trajectories from synoptic meteorological data. Trajectories are generated in the various air quality models in different ways. Some models[1] use the winds determined at a specific pressure level, e.g., 850 mb ; this is convenient when a large area of ocean is involved and standard level National Meteorological Center analyses are available. Over land areas where radiosonde date is available, some models use winds averaged through a layer, e.g., 100-1000 m above the surface.[2,11,15,24,28,29] Neither of these approaches considers the vertical component of the air motion Isentropic trajectories which do include the vertical component of the air motion have been used in determining pollutant transport on a case study basis.[6,7,9,22,23]

Isentropic analysis has not been used in long-term assessment modeling for two reasons. First the statistical nature of the results seemed to reduce the need for more accurate individual trajectories, especially in view of the more complex and expensive analysis required. Also, the diabatic effects near the earth's surface threaten the validity of this analysis. Condensation of water vapor also adds complexity and expense to the analysis technique. It is this condensation and subsequent removal of pollutants by precipitation scavenging which has generated new interest in more realistic diabatic air trajectories in long-term assessments.

In the middle latitudes of the northern hemisphere, frontal storms provide effective cleansing of most power plant pollutants.[5,13,18] The precipitation patterns associated with these storms are quite variable in space and time. To more realistically evaluate the effect of precipitation scavenging, the placement of pollutant-laden aire takes on a new importance. This is primarily true for models which are using real-time precipitation as opposed to models using average precipitation or none at all.

Recent work done by McNaughton[21] has looked at verification of the Pacific Northwest Laboratory (PNL) long-term single layer regional air pollutant transport model (RAPT) for the northeast United States. In this study tests were made to look at $SO_4^=$ air concentrations in the vicinity of a frontal storm during a period of October 10, 1977 through October 12, 1977. However, in the vicinity of fronts three dimensional motions become more important in determining transport and wet deposition.[10]

In an effort to compare the effects of 3-D motions with results using a single layer model, the eight layer model[10] was modified to handle SO_2 to $SO_4^=$ conversion in the same manner as another PNL regional scale assessment model.[19]

The purpose of this investigation was to compare the computed $SO_4^=$ fields with the observed fields as well as compare wet deposition results with field measurements of $SO_4^=$ in rain made during the same study period using two models. These results should indicate whether the use of more sophisticated models are necessary to describe the $SO_4^=$ air concentrations and wet deposition.

The Models

The two models being compared are actually a single model, the eight layer model. The single layer model fits within the framework of the eight layer model and thereby all features except those involving vertical motion calculation are identical. The only difference is that vertical motions are not allowed in the single layer mode. The eight layer lagrangian model was developed under the U.S. Environmental Protection Agency Multistate Atmospheric Power Production Pollution Study (MAP3S).

The models as used in this study, assumes hourly puffs. The puffs are advected in hourly steps using gridded layer averaged wind. Eight layers of average winds, mixing ratios, and potential temperature at 100-300 m, 300-500 m, 500-700 m, 700-1000 m, 1000-1500 m, 1500-2000 m, 2000-2500 m and 2500-3000 m above surface.

The following are assumptions made on atmospheric diffusion.

1. Horizontal diffusion is based on a formulation by Eimutis and Konicek[12] for distances <100 km. For distances greater than 100 km $\sigma_y=0.5$ t where σ_y is the horizontal diffusion in meters and t is the time in seconds since release.[14]

2. The vertical-turbulent diffusion parameterization is a function of stability in a specified diurnal cycle for all plume elements within the daytime mixed layer or the nocturnal stable layer.

3. The depth of the daytime mixed layer is represented by a diurnal cycle in which a daytime layer increased from a minimum depth at sunrise to a maximum depth in the afternoon. A nocturnal layer builds to a lower depth.

4. Depth of the boundary layer (mixed layer in daytime or nocturnal layer) for each hour determines which of two vertical dispersion regimes - mixed layer or above mixed layer - will be applied to a given plume element. Those elements released within the depth of the layer expand according to the stability

of the hour. Those released above the current depth of the layer expand as if the atmosphere were extremely stable. An exception to this method occurs when releases have expanded to the maxium depth of the mixed layer during the previous day ; in this case, their expansion continues according to a fixed function of travel time independent of stability or layer depth.

5. Depth of the mixed layer provides a vertical constraint for releases at heights within this layer. The depth of the nocturnal boundary does not serve as an analogous constraint. Carson[3] supports this assumption by depicting the mixed-layer depth as a function that increases during the daytime, then loses its definition. This picture is in relative agreement with modeling results by Venkatram and Viskanta[26] and with experiment results by Kaimal et al.[16] who find that about one hour before sunset, the convective layer disintegrates rather abruptly, although remnants of the capping inversion persist through the development of the nocturnal boundary layer.

6. When vertical motions of the puff occurs calculations of vertical diffusion are based on a Pasquill-Gifford class 6 stability.

7. At the present stage of model development, terrain affects are neglected.

For a more indepth description of the diffusion, wet and dry removal as well as advection the reader is referred to Eadie and Davis[11] and McNaughton and Powell.[19]

Hourly precipitation was used in the study. This was arrived at by analysis of hourly reporting stations which were then interpolated to a 1/2° latitude by 1/2° longitude grid. No Candadian data was used in this study.

A diurnal time change of SO_2 to $SO_4^=$, conversion rate, was used with a maximum of 2%/hour in the afternoon and 0% at night (see Table 1).

Three different modes of the model were used using different SO_2 to $SO_4^=$ conversions. For an in-cloud conversion test, a 10%/hour conversion of SO_2 to $SO_4^=$ was used whenever the relative humidity of the puff was calculated to be greater than 85%. A second conversion was used in rain.[20] In the in-rain conversion test a 10%/hour conversion of SO_2 to $SO_4^=$ was used whenever the puff encountered rain during the hourly time step. The last conversion used only the diurnal conversion shown in Table 1.

Table 1. Model Input Data

Time Period For Meteorological Data	Oct. 8/12Z-12/12, 1977
Advection Grid Spacing	2° longitude x 2° latitude
Grid Spacing For Hourly Precipitation Data And Sampling Of Results	1/2° latitude 1/2° longitude
Advection Grid Boundaries	19° latitude - 52° latitude 100° w longitude - 65° w longitude
Effective Stack Height	350 m
Mixing Height	Variable
Stability	Variable
Dry Deposition Velocities	
SO_2	1 cm/sec
Sulfates	0.1 cm/sec
Wet removal Coefficients	
SO_2	0.005 P*/hr
Sulfates	0.38 $P^{0.73}$*/hr
Transformation Rate of SO_2 to Sulfates	2% x sin ($\frac{T^{**}}{12}$ x π),✶ day 0%, night

*P is the precipitation measured in mm
**Is the time in hours after 7 a.m/ EDT
✶2% maximum during day, 0% at night (8-12 p.m. EDT, 1-7 a.m. EDT).

Case study

A period of four days were selected for a case study October 8, 12Z, 1977 to October 12, 127, 1977 for NE United States. The reason for selecting this period was based on the movement of two fronts through the region. The first an occluded front moved through the NE on the 8th and 9th of October. The second, a cold front moved through the region with associating rain on the 11 th and 12th of October. Measurements of sulfate in the rain were made near four locations, State College, Penn, Ithaca, New York, Virginia and Whiteface Mountain, NY.[4] Daily SO_2 and $SO_4^=$ measurements were obtained from daily data from an intensive study period by the Electric Power Research Institute in the Sulfate Regional Experiment (SURE).

The models were turned on October 8, 12Z, 1977 and allowed to run for 48 hours to build up a sufficient background before comparing the computed fields with the observed fields.

The eight layer model was run in three modes, 1) with in-cloud conversion for relative humidities > 85%; 2) with in-rain conversion for rain > 0.0 and 3) with no in-cloud or in-rain conversion. The single layer model was run in only one mode with in-rain conversion Table 1 shows the parameterization used in the model.

The number of sources used were the top 20 utilities based on a 1974 inventory. These represented approximately 35% of the SO_2 sources in the northeast (see Fig. 1).

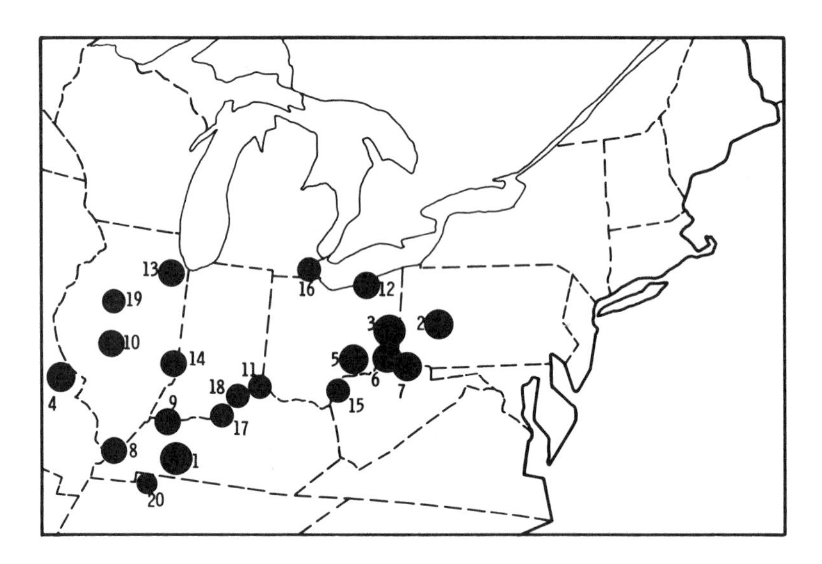

Fig. 1. Twenty Sources Used for Case Studies Size of Dot Indicates Relative Strength

Results

I. $SO_4^=$ Air Concentrations

The model results for $SO_4^=$ air concentrations along with the observed patterns are shown in Fig. 2-6 only for October 10, 1977, since the overall results for the other two days were generally the same. Discussion of results will be touching mostly on the broad scale features. The reader should keep in mind that only 35% of the SO_2 emissions were used for this study.

October 10

1. The model run with in-cloud conversion (Fig. 2) shows the computed maximum in the south shifted south of the observed maximum in the Alabama area (Fig. 6). There is agreement between observed and computed in the North and South Carolina region. Again a maximum is observed on the Pennsylvania, Deleware border which is not computed. The observed minimum in southwestern Ohio is computed. A minimum computed over Kentucky is not observed. The $SO_4^=$ computed values are too high.

Fig. 2. $SO_4^=$ Air Concentrations ($\mu g/m^3$) for 10/10/77 Using In-Cloud Conversion

2. The model run with in-rain conversion (Fig. 3) shows a better fit of the observed maximum in the south from Tennessee and Alabama to North and South Carolina (Fig. 6). A maximum is computed south of the observed maximum on the Pennsylvania, Deleware border. The observed minimum in southeastern Ohio is computed to be there. Also a minimum is computed over

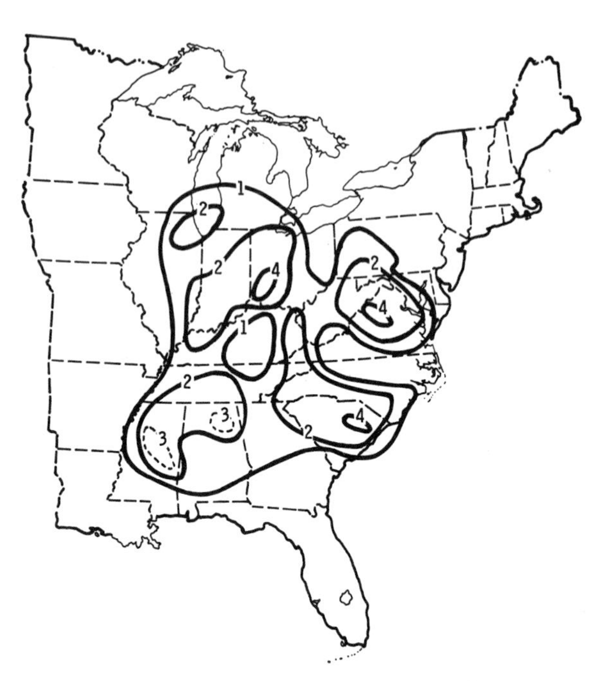

Fig. 3. $SO_4^=$ Air Concentrations ($\mu g/m^3$ for 10/10/77 Using In-Rain Conversion

Kentucky which is not observed. The $SO_4^=$ values computed are proportionately what they should be.

3. The model run with only the diurnal conversion (Fig. 4) compares well with the observed maximum in the south. Also, the values computed are proportionately what they should be for the observed. The maximum observed over Pennsylvania is computed to be south of the observed with it centered over the Pennsylvania, West Virginia, Virginia border area. The maximum observed near the Pennsylvania Deleware border is not calculated.

4. The model runs with in-rain conversion using a constant layer wind (Fig. 5) compares poorly with the maximum in the south being too far south (Fig. 6). Also, the maximum computed to be in Virginia does not compare favorably as the other modes with the observed maximum in Pennsylvania. This mode performed the poorest of all other modes run. The calculated values in the south are proportinately too low.

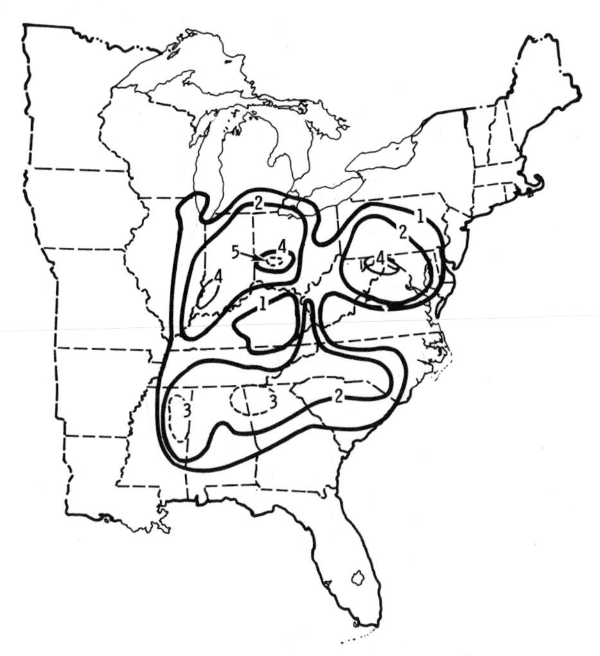

Fig. 4. $SO_4^=$ Air Concentrations ($\mu g/m^3$) for 10/10/77 Using Only Diurnal Conversions

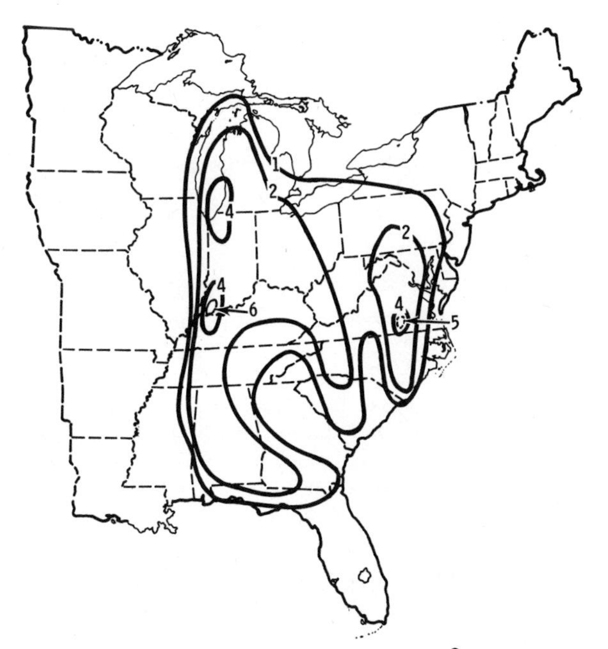

Fig. 5. $SO_4^=$ Air Concentrations ($\mu g/m^3$) for 10/10/77 for Constant Layer Flow and In-Rain Conversion

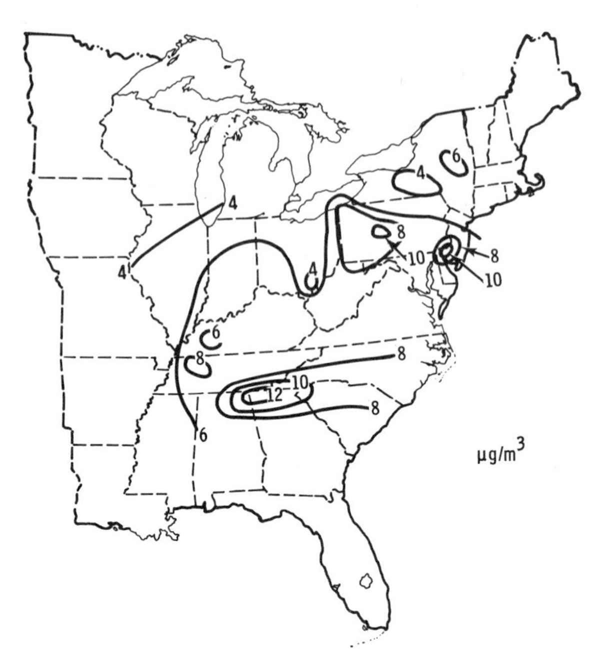

Fig. 6. Observed $SO_4^=$ Air Concentrations ($\mu g/m^3$) for 10/10/77

The overall general assessment for $SO_4^=$ air concentrations for the three days is that mode 1 (in-cloud) converts too much SO_2 to $SO_4^=$. The same is true for mode 2 (in-rain). The best results were for no in-cloud or in-rain conversion but using a diurnal variation in SO_2 to $SO_4^=$. The constant layer approach generally yielded the poorest comparison of maximums than the other three modes. It is possible if a different layer were used that better results may be obtained for the constant layer mode.

II. $SO_4^=$ Deposition In Rain

The results of the mass balance at October 11, 12Z, 1977 are shown in Table 2. This shows the dramatic difference in deposition in the no conversion compared with in-cloud or in-rain conversion.

The amount deposited at the three sites is also shown compared with measured quantities. As one can note that ∿35% of the sources were used, thus the amounts deposited should be considerably less by about a factor of 3. However, as noted in the table the values run anywhere from a factor of 20 lower for the in-cloud at Whiteface Mountain to about equal for Ithaca.

Tabel 2

. Mass Balance at 10/12/12Z/77

	SO_2 Air	SO_2 Deposit (Wet Deposit)	$SO_4^=$ Air	$SO_4^=$ Deposit (Wet Deposit)	SO_2 OFF	$SO_4^=$ OFF
Layer						
. In Cloud	21.4	24.9 (1.1)	17.9	13.1 (12.1)	12.2	10.6
. In Rain	26	27.4 (1.2)	12	12 (11.2)	17.4	5.7
. None	31.0	22.7 (1.1)	9.2	5.6 (5.1)	19.5	6.1
ingle Layer						
n Rain	24.9	44.4 (1.0)	11.1	11.5 (10.4)	3.9	4.1

. $SO_4^=$ Wet Deposit ($\mu g/m^2$)

	State College, Pennsylvania (1)	Ithaca, New York (2)	White Face Mountain, New York (3)	Virginia (4)
Layer				
. In Cloud	1×10^4	2×10^4	3×10^3	3×10^3
. In Rain	1×10^4	1×10^4	3×10^3	2×10^3
. None	4×10^3	7×10^3	6×10^2	1×10^3
ingle Layer				
n Rain	1×10^4	1×10^4	1×10^3	1×10^3
bserved	8×10^4	3×10^4	5.5×10^4	2.16×10^4

1) Cumulative for 8 Oct. 12Z - 12 Oct. 12Z
2) Cumulative for 8 Oct. 12Z - 10 Oct. 12Z
3) Cumulative for 8 Oct. 12Z - 11 Oct. 12Z
4) Cumulative for 8 Oct. 12Z - 10 Oct. 12Z

For the case of in-rain the deposition runs anywhere from a factor of 20 lower to about a factor 3 for Ithaca. For the case of no-incloud or in-rain, all estimates were low with a factor of 100 low for Whiteface Mountain to the best estimate with a factor of 4 low for Ithaca. The single layer in-rain produced a minimum of a factor of 60 at Whiteface Mountain to a factor of 3 low at Ithaca, New York.

The cumulative wet deposition patterns were computed for $SO_4^=$ for each of the techniques. The only comment is that additional sampling would have been necessary to tell if an observed pattern would match any one of the four. The only way to tell which of the techniques would be able to describe the observations is to do more case studies.

In general none of the techniques described in the paper adequately described the wet deposition at all sites. The one coming nearest was the in-cloud. The worst technique was where no in-clou or in-rain conversion was used. This leaves an obvious dilemna for this case study because the technique which produced the best $SO_4^=$ air concentration produced the worst estimate of wet deposition of $SO_4^=$. Also the technique which grossly overestimated the $SO_4^=$ air concentrations (in-cloud conversion) produced the closest wet deposition values. The reason for this was the conversion in air of the SO_2 to $SO_4^=$ in precipitation. This would not increase the air concentrations of $SO_4^=$ at the same time it would keep the depositio of $SO_4^=$ near those calculated.

CONCLUSIONS

A short term case study was carried out for October 8, 1977 to October 11, 1977 using an eight layer regional assessment model. The model was run in four different modes for 20 of the top utility sources (∿35 % emissions in northeast). The first mode was with in-cloud conversion or puffs with a relative humidity greater than 85% were assumed to convert SO_2 to $SO_4^=$ at 10% hour. The second mode was with in-rain conversion or puffs encountering rain were assumed to convert SO_2 to $SO_4^=$ at 10% hour. The third mode was with only standard conversion with a maximum of 2%/hour conversion of SO_2 to $SO_4^=$ in the daytime. The fourth and last mode was using a constant layer wind, averaged over the 300 m to 500 m above surface, and an in-rain conversion (see mode 2). Running the model in third mode yielded the best results for $SO_4^=$ air concentrations. The best results for wet deposition was from running the models in the first mode.

This apparent inconsistency in the result occurs because of the conversion of SO_2 to $SO_4^=$ in the air. A new technique using conversion directly to the precipitation is under way. Additional work is planned in an attempt to resolve these differences.

REFERENCES

BOLIN, B. and C. D. PERSSON. 1975. Regional Dispersion and Deposition of Atmospheric Pollutants with Particular Application to Sulfur Pollution Over Western Europe. Tellus XVII, 3, 281-310. (1)

BROOKHAVEN NATIONAL LABORATORIES. 1975. Regional Energy Studies Program. Annual Report, FY 1975, Phillip F. Palmedo, Ed. 46-61, 91-106. (2)

CARSON, D. J. 1973. "The Development of a Dry Inversion- Capped Convectively Unstable Boundary Layer," Quart. J. R. Met. Soc., 99 : 450-467. (3)

DANA, M. T. 1979. The MAP3S Precipitation Chemistry Network : Second Periodic Summary Report (July 1977-June 1978). PNL-2829, Pacific Northwest Laboratory, Richland, WA. (4)

DANA, M. T., J. M. HALES, W. G. N. SLINN, and M. A. WOLF. 1973. Natural Precipitation Washout of Sulfur Compounds from Plumes. Office of Research and Development, U.S. Environmental Protection Agency, EPA-R3-73-047, June 1973. (5)

DANA, M. T. and W. E. DAVIS. 1979. Redistribution of a layer of HTO by Rainfall. Preprint from the Institute of Environmental Sciences 25 th Annual Meeting, Seattle, WA. (6)

DANIELSON, E. F. 1974. Review of Trajectory Methods. Adv. in Geophysics, 18B, Academic Press, 73-94. (7)

DAVIS, W. E., W. J. EADIE and D. C. POWELL. Technical Progress in the Alternate Fuel Cycle Program. Pacific Northwest Laboratory 1977 Annual Report, PNL-2500 PT 3, UC-11, Part 3 Atmospheric Sciences, February 1978. (8)

DAVIS, W. E. and L. L. WENDELL. 1976. Some Effects of Isentropic Verical Motion Simulation in a Regional Scale Quasi Lagrangian Air Quality Model. Preprint from the Third Symposium on Atmospheric Turbulence, Diffusion and Air Quality. (9)

DAVIS, W. E. 1979. "Comparison of the Results of an Eight-Layer Regional Model Versus a Single Layer Regional Model for a Short-Term Assessment," World Meteorological Organization Symposium on Long-Range Transport of Pollutants, Sophia, Bulgaria. (10)

EADIE, W. J. and W. E. DAVIS. 1979. "The Development of a National Interregional Transport Matrix for Respirable Particulates," PNL-RAP-37, Pacific Northwest Laboratory, Richland, WA. (11)

EIMUTIS, E. C. and M. G. KONICEK. 1972. Derivations of Continuous Functions for the Lateral and Vertical Atmospheric Dispersion Coefficients. Atmos. Envir., 6, 863-869. (12)

ENGELMANN, R. J. 1970. Precipitation Scavenging. Proceedings of AEC Symposium Series 22, available through U.S. Atomic Energy Commission, Division of Technical Information. (13)

HEFFTER, J. L. 1965. The Variation of Horizontal Diffusion Parameters With Time for Travel Periods of One Hour or Longer, J. Appl. Meteor., Vol. 4, February 1965. (14)

HEFFTER, J. L. and G. A. FERBER. 1975. A Regional Continental Scale Transport Diffusion and Deposition Model, Part II : Diffusion-Deposition Models. NOAA Tech. Memo, ERL ARL-50. (15)

KRAIMAL, J. C., J. C. WYNGAARD, D. A. HAUGEN, O. R. COTE, Y. IZUMI, S. J. CAUGHEY and C. J. READINGS. 1976. "Turbulence Strucsture in Convective Boundary Layer. J of the Atm. Sci., 33:2152-2169. (16)

KREITZBERG, C. W., M. LUTZ and D. J. PERKEY. 1976. Precipitation Cleansing Computation in a Mesoscale Weather Prediction Model. Fate of Pollutants, John Wiley & Sons, Inc. (17)

Precipitation Scavenging. 1970. Editors, R. J. ENGELMAN and W. G. N. SLINN, U.S. AEC Division of Technical Information, CONF-700601. (18)

McNAUGHTON, D. J. and D. C. POWELL. 1980. RAPT - The Pacific Northwest Laboratory Regional Air Pollutant Transport Model : A Guide, PNL-3390, UC-11, Pacific Northwest Laboratory, Richland, WA, March, 1981. (19)

McNAUGHTON, D. J. and B. C. SCOTT. 1980. "Modeling Evidence of In-Cloud Transformation of Sulfur Dioxide to Sulfate". JAPCA, Vol. 30, No. 3. March 1980. (20)

McNAUGHTON, D. J. 1980. "Time Series Comparisons of Regional Model Predictions With Sulfur Oxide Observations From the Sure Program". Presented at 73rd Annual Meeting of the Air Pollution Control Association, Montreal Quebec, June 22-27, 1980. (2

REITER, E. R. and J. D. MAHLMAN. 1964. Heavy Radioactive Fallout Over the Southern United States, November 1962. Atmos. Sci. Tech., Paper No. 58, 21-49, Colorado State University. (22)

REITER, E. R. and J. D. MAHLMAN. 1965. Heavy Iodine-131 Fallout Over the Midwestern United States, May 1962. Atmos. Sci. Tech Paper No. 70, USAEC Report COO-1340-2, 1-53, Colorado State University. (23)

RODHE, H. 1974. Some Aspects of the Use of Air Trajectories for the Computation of Large-Scale Dispersion and Fallout Patterns. Adv. in Geophysics, 18E, Academic Press. (24)

SHEIH, J. 1976. Application of a Lagrangian Statistical Trajectory Model to the Simulation of Sulphur Pollution Over North Eastern United States. Proceedings of the Third Symposium on Atmospheric Turbulence, Diffusion, and Air Quality, October 1976. (2

VENKATRAM, A. and R. VISKANTA. 1976. Radiative Effects of Pollutants on the Planetary Boundary Layer, Research Triangle Park, NC, Office of Research and Development, U.S. Environmental Protection Agency Report EPA-600/4-76-039. (26)

WENDELL, L. L. 1972. Mesoscale Wind Fields and Transport Estimates Determined from a Network of Wind Towers. Mon. Wea. Rev., 100 565-578, July 1972. (27)

WENDELL, L. L. and T. D. FOX. 1976. Examination of the Wind Shear in Regional Scale Flow Layers Derived from Radiosonde Data. Pacific Northwest Laboratories Annual Report for 1975 to the USERDA Division of Biomedical and Environmental Research, Part Atmospheric Sciences, BNWL-2000 PT3, Richland, WA. (28)

WENDELL, L. L., D. C. POWELL and R. L. DRAKE. 1976. A Regional Scale Model for Computing Deposition and Ground Level Air Concentration of SO_2 and $SO_4^=$ from an Elevated Source. PNL Annual Report for 1975 to ERDA Division of Biomedical and Environmental Research, Part 3, Atmospheric Sciences, BNWL-2000 PT-3, UC-11, 218-223, March 1976.(29)

DISCUSSION

J.W.S. YOUNG

Scaling up 1/3 of an old inventory to compare to observed data is very dangerous

B. DAVIS

Of course any time you do not use the full data set you will make inherent errors. However the conclusions in this case are still valid that the constant layer results were considerable different from using the eight layers. The area it does effect is the comparison to the observed data.
Hopefully in the future we can use a larger source base and be able to answer this question.
I might add the lack of sources along the eastern coast may have been the reason for missing the peak observed in SE Pennsylvania.

J. SHANNON

Did you compare more than one single layer-model with the eight layer model?

B. DAVIS

We used a layer similar to the 2nd standard layer reported above surface. This layer in the past was used to track radioactive releases. In this case it was the 2nd layer the 300-500 m layer averaged winds. The real question is whether any constant layer can represent the shear that appears in a frontal storm. The answer to this is no..

L. NIEMANN

We have found the model evaluation of wet sulphur deposition is very important in the way the high resolution precipitation data are interpolated and extrapolated especially for grid squares contining rain gauges measuring data or no gauges at all. How do you of the three nearest station, the model will reach

B. DAVIS The method of griding the hourly data is described in a report by Fox et al at PNL. Basically each grid is filled by the use of the three nearest stations. If there is no data at the nearest station, the model will reach out to the next nearest. In the east the problem is not as bad as in the west. This is because there are more reporting stations in the east. Thus when one station is missing than there is usually another nearby to use. In the west larger erros will occur because the number of reporting stations are fewer and further apart.

J.K. WESTBROOK Can you comment on the time step you use?

B. DAVIS The model is set up to run on hourly steps.

THE WASHOUT OF HYDROGEN FLUORIDE

F.M. Bosch, P.M. De Keyzer

Laboratory for Inorganic Technical Chemistry
State University Ghent
Grote Steenweg Noord 12
B - 9710 Ghent (Belgium)

ABSTRACT

The washout of Hydrogen fluoride, emitted by a continuous point source has been considered as being an absorption process.

According to Hales (1972), the fundamental equations describing this process are developed and are discussed more in detail for 2 special cases; namely the irreversible and reversible washout. In both cases the influence on the fluoride concentration in a raindrop of stack height, distance from the source and raindrop radius has been discussed. Moreover, the mean concentration during a real rainshower has been calculated.

From the previous theoretical study a computer program has been developed in order to simulate the fluoride washout around a brickworks in Belgium. This program is able to predict very well the experimental washout values obtained from a precipitation sampling network set up in a radious of 2 km. around these brick works and using a sampling period of 2 weeks.

INTRODUCTION

Stack gases, resulting from the baking of bricks, may contain considerable amounts of hydrogen fluoride. So, emission concentrations have been noted in the Federal Republic of Germany ranging from 4 up to 120 mg F^-/m^3 (STP) with 30 mg F^-/m^3 (STP) as a mean (Mueller, 1976).

On account of the relative importance of the brick industry in Belgium (production in 1976 : 2,586,000 m^3 (Rebuffat, 1979))and because these brick-works are always near some residential sites, this being due to the high population density (400 inhabitants/km^2) an investigation program was set up to study the effect of these fluoride emissions on the environment.

As part of the program, this paper will discuss the washout of hydrogen fluoride emitted by a continuous point source. A mathematical model will be developed and applied to the simulation of the fluoride washout in the vicinity of a brick works.

FORMULATION OF THE MODEL

Assuming that

(i) the raindrops are perfectly spherical and falling vertically with their constant terminal velocity v (cm/s)
(ii) the raindrops do not coalescent nor fall apart in smaller ones
(iii) the gaseous pollutant's concentration remains constant with time
(iv) no reactions occur in the liquid phase involving the considered pollutant.

The pollutant concentration C_ℓ (mol/cm^3) in a raindrop with radius R (cm) falling with constant velocity v (cm/sec) along the vertical z-axis, can be expressed by

$$\frac{dC_\ell}{dz} = - \frac{3\ K}{R\ v} (C \quad - H\ C_\ell\ ) \tag{1}$$

$$C_\ell \quad = 0 \qquad z = \infty \tag{1'}$$

with C_g (mol/cm^3) the atmospheric pollutant concentration, H Henry' law constant (°) and K (cm/s) the overall gasphase masstransfer-coefficient (Hales, 1972).

(°) For species not obeying Henry's law, H is the slope of the equilibrium vaporpressure curve.

The overall gasphase masstransfercoefficient K contains a contribution from the masstransfer in the gas- and liquid phase :

$$\frac{1}{K} = \frac{1}{k_g} + \frac{H}{k_\ell} \tag{2}$$

were k_g (cm/s) and k_ℓ (cm/s) are the masstransfercoefficients referring respectively to the gas- and liquid phase.

The transportcoefficient for the gas phase k_g can be estimated from the semi-empirical Froessling equation (Froessling, 1938) :

$$Sh = 2. + 0.6\ Re^{1/2}\ Sc^{1/3} \tag{3}$$

$$Sh = \frac{2\ k_g\ R}{D_g} \quad \text{(Sherwood number)}$$

$$Re = \frac{2\ v\ R}{\nu} \quad \text{(Reynolds number)}$$

$$Sc = \frac{\nu}{D_g} \quad \text{(Schmidt number)}$$

where D_g (cm^2/s) is the diffusivity in air and ν (cm^2/s) the kinematic viscosity of air. Considering the resistance for transport in the liquid phase as being negligible (well mixed raindrop) in comparison to the resistance in the gasphase, the overall masstransportcoefficient K is reduced to

$$K = k_g \tag{2'}$$

The other possibility, the resistance for transport in the liquid phase being no more negligible, has not been considered, because referring to Hales (1972), the well mixed drop model satisfies with precipitation rates up to 2 $mm.h^{-1}$.

Assuming the gas washout an irreversible process as a special case, equation (1) is reduced to

$$\frac{dC_\ell}{dz} = - \frac{3\ k_g\ C_g}{R\ \ v} \tag{4}$$

$$C_\ell = 0, \qquad z = \infty \tag{4'}$$

v, the final velocity of a raindrop with radius R can be expressed by (Markowicz, 1976) :

$$v = 958\ (1 - \exp\ (-\ (R/0.0885)^{1.147})) \tag{5}$$

The concentration of gaseous pollutant C_g can be estimated using the bi-gaussian dispersion model with total reflection :

$$C_g = \frac{Q \exp(-0.5(y/\sigma_y)^2)}{2 \sigma_y \sigma_z \bar{u}} \times$$

$$(\exp(-0.5(\frac{z - h}{\sigma_z})^2) + \exp(-0.5(\frac{z + h}{\sigma_z})^2)) \quad (6)$$

with $\bar{u}$ (cm/s) the mean windvelocity, h (m) the effective stack height and σ_y and σ_z (m) the dispersion parameters both function of the atmospheric stability and the distance from the stack. The distribution of the raindrop dimensions can be approximated by (Best, 1950) :

$$f(R) = \frac{n}{a} (\frac{2 R}{a})^{n - 1} \exp(-(\frac{2 R}{a})^n) \quad (7)$$

with $a = 0.13 \ P^{0,232}$
$n = 2.25$
$f(R)\,dR$ = fraction of rain water comprised in drops with radius between R and R + dR
P = precipitation intensity (mm/h)

Derived from that distribution, the modus raindrop size is given by

$$R_p = (\frac{n - 1}{n})^{\frac{1}{n}} \ \frac{a}{2} \quad (8)$$

Furthermore, Laws & Parson's (1943) expression, giving the median radius, will also be used

$$R_{50} = 0.0404 \ P^{0,182} \quad (9)$$

NUMERICAL RESULTS

In order to obtain useful information about the variables affecting the washout of hydrogen fluoride, equations (1) and (4) have been solved.

The data and values of the constants used in the following calculations are summarised in table 1.

Irreversible washout

The solution of equation (4) leads to the following expression for the concentration C_ℓ (x,y,z) in a falling raindrop of radius R :

$$C_\ell\ (x,y,z,) = \frac{3\ kg\ Q\ \exp\ (-0.5(y/\sigma_y)^2)}{2\ \sqrt{2\pi}\ \ R\ v\ \bar{u}\ \sigma_y}\ \ x$$

$$(2 - \mathrm{erf}\ (\frac{z-h}{\sqrt{2}\ \sigma_z}) - \mathrm{erf}\ (\frac{z+h}{\sqrt{2}\ \sigma_z})) \qquad (10)$$

$$\mathrm{erf}(x) = \frac{2}{\sqrt{\pi}} \int_o^x e^{-t^2} dt$$

For the concentration at ground-level (z=0) expression (10) can be written as

$$C_\ell\ (x,y,0) = \frac{3\ k_g\ Q\ \exp(-0.5(y/\sigma_y)^2)}{\sqrt{2\pi}\ \ R\ v\ \sigma_y\ \bar{u}} \qquad (11)$$

In figure 1 the concentration (eq. 10) divided by

$$\frac{3\ k_g Q}{2\ \sqrt{2\pi}\ R\ \ v\ \ \bar{u}}$$

is plotted versus the vertical height, considering 2 different effective stack heights and different distances x. The concentration drops with increasing distances from the source. At the same height, the greater the effective stackheight is, the greater the concentration gets but the effect diminishes with increasing distance.

At ground-level, the effective stackheight has no more any influence but this is due to the assumption of irreversible washout and the dispersion model with total reflection.

Equation (11) was also used to evaluate the influence of the atmospheric stability. The expression $\frac{2}{\sigma_y}\ \exp\ (-0.5\ (y/\sigma_y)^2$ has been plotted versus the distances x for the 6 Pasquill stability classes A to F (fig. 2). The dispersion σ_y is function of the distance x :

$$\sigma_y = C_y x^{m_y}$$

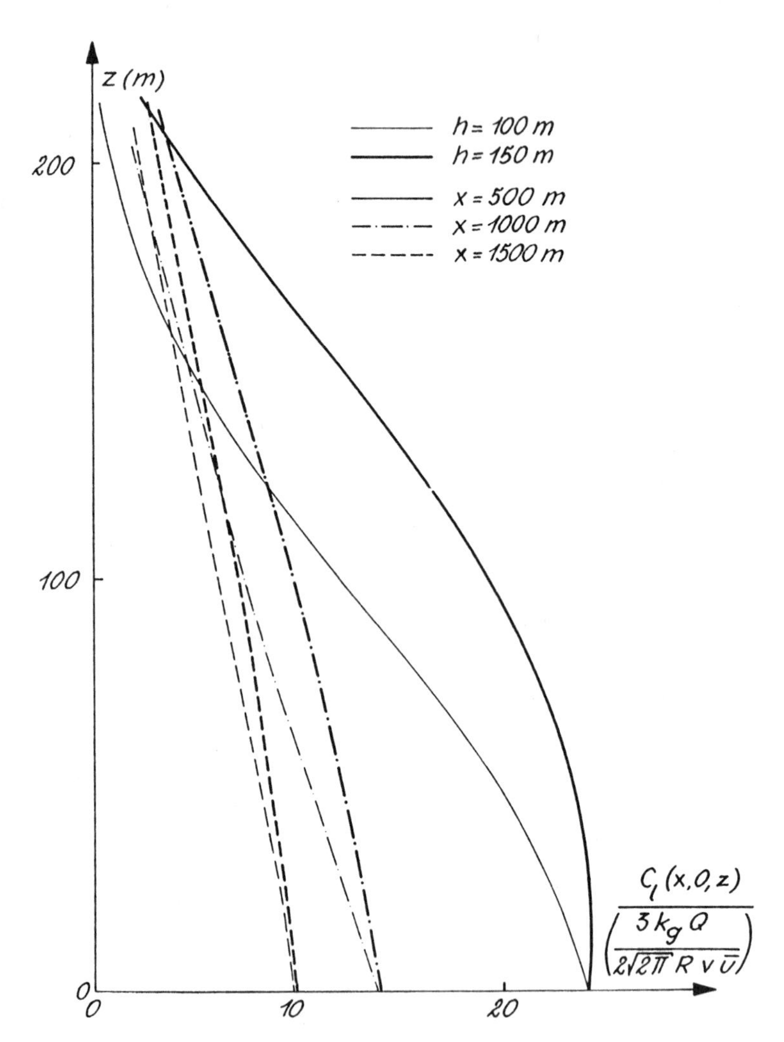

Fig. 1 : irreversible washout : concentration in function of height

Table 1. Data used in numerical calculations

Emission rate Q	: 1 mol F^-/s	(68.4 Kg/h)
Windvelocity $\bar{u}$	: 100 cm/s	
Dispersion σ_y	$= 0.586\ x^{0.796}$	(m)
σ_z	$= 0.700\ x^{0.711}$	(m)
cinematic viscosity of air (10 °C)	$= 0.133\ cm^2/s$	
diffusivity of HF in air (10 °C)	$= 0.13\ cm^2/s$	

The constants C_y and m_y are functions of the atmospheric stability. The values of these constants were determined at Mol (Belgium) by the Institute for Nuclear Research by means of a meterological mast of 120 m and were thought to be also representative for the considered region (Bultinck & Malet, 1972).

This function has a maximum that approaches the emission source with decreasing atmospheric stability. After the maximum, the concentration drops sharper as the atmospheric stability decreases. From this it can be concluded that the greater the atmospheric stability, the further away from the stack the influence of the emission on the rainwater composition will be remarked.

The value of the maximum is not directly function of the atmospheric stability. Nevertheless, according to the windvelocities connected with the different Pasquill stability classes, the maximum concentration is decreasing as the atmospheric stability increases (concentration is inversely proportional with wind velocity, cfr. eq. (11)).

Finally, the influence of the drop radius (and thus of the rain-intensity) is included in the term $\frac{3\ k_g}{v\ R}$. Increasing drop diameters will involve lower concentrations and the smaller the drop diameter is, the greater the influence on the concentration will be.

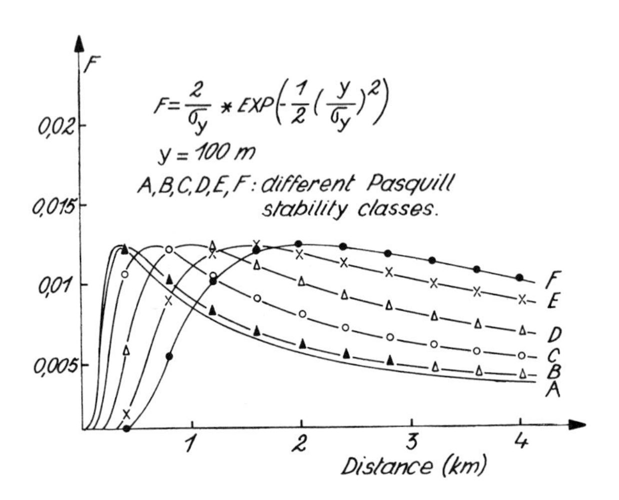

Fig. 2. : irreversible washout : influence of atmospheric stability

Reversible washout

The solution of equation (1) results in the following semi-analytical expression for the concentration at groundlevel in a raindrop with radius R (Hales, 1973) :

$$C_\ell\ (x,y,0) = \frac{\zeta Q\ e^{-\frac{1}{2}\left(\frac{y^2}{\sigma_y^2} - \frac{\xi^2\ \sigma_z^2}{2}\right)}}{2\ \sqrt{2\pi}\ \sigma_y\ \bar{u}} \quad x$$

$$\left[e^{-h\xi} \left(1-\mathrm{erf}\left(\frac{\xi\sigma_z^2 + h}{\sqrt{2}\ \sigma_z}\right)\right) + e^{h\xi} \left(1-\mathrm{erf}\left(\frac{\xi\sigma_z^2 + h}{\sqrt{2}\ \sigma_z}\right)\right) \right] \qquad (12)$$

$$\xi = \frac{3\ K\ H}{R\ V}$$

$$\zeta = \frac{3\ K}{R\ v}$$

In practice however, equation (12) is not very useful because the first exponential term may become great while the term between edged brackets becomes very small. Therefore, equation (1) has been solved numerically using the Hamming predictor - corrector method (Lapidus & Seinfield, 1971). The initial boundary condition (1') was taken as

$$C_\ell = 0 \text{ with } z = h + 5\ \sigma_z$$

The atmospheric stability was set as slightly unstable (Pasquill class C). In figure 3 the concentration in a 0.04 cm raindrop is given in function of the height for different effective stack heights, and distances from the emission source. In the case of smaller distance and greater effective stackheights, the reversibility of the washout process is very well illustrated : at a certain height the desorption becomes very clear (cfr. fig. 3.a). The influence of the effective stack height on the concentration at ground level is presented in figure 4. The results for an effective stack height of 130 m were deleted being very close to those obtained for a height of 100 m.

From this figure it is obvious that the effect of a higher stack will be practically negligible (at least with very soluble gases like hydrogen fluoride) omitting the smallest raindrops.

In figure 5, the concentration is plotted versus the distance from the stack with the effective stackheight and raindrop radius as parameters. The concentration drops sharply with increasing distance and the influence is greater in the case of the smaller raindrops. It is also clear that the effect of the effective stack height diminishes with increasing distance.

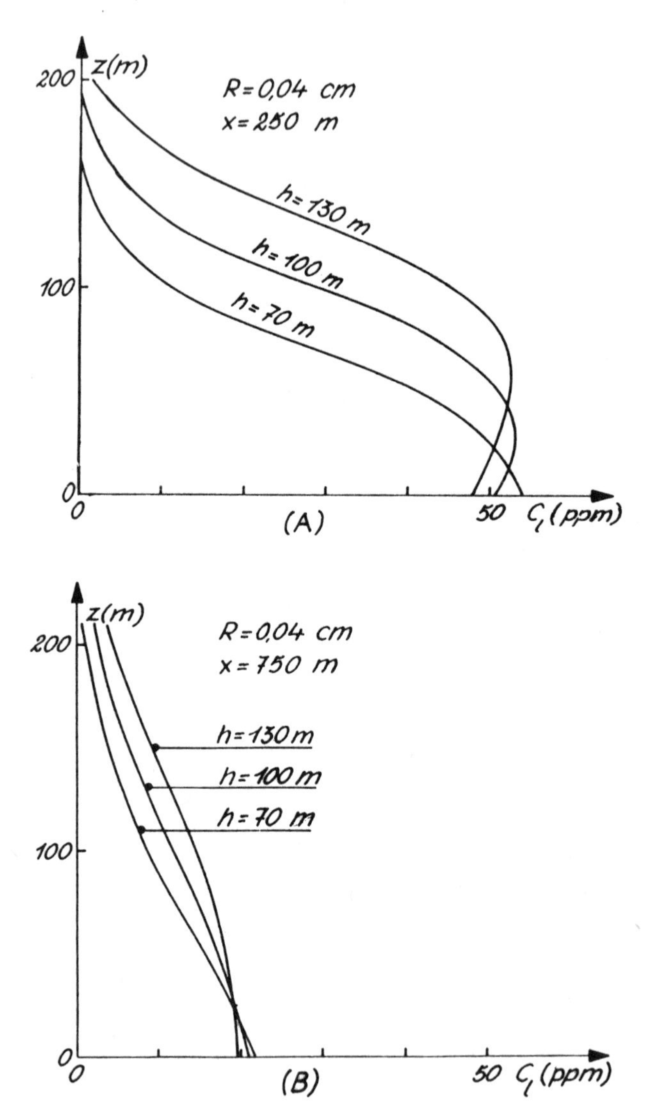

Fig. 3. : reversible washout : concentration in function of height

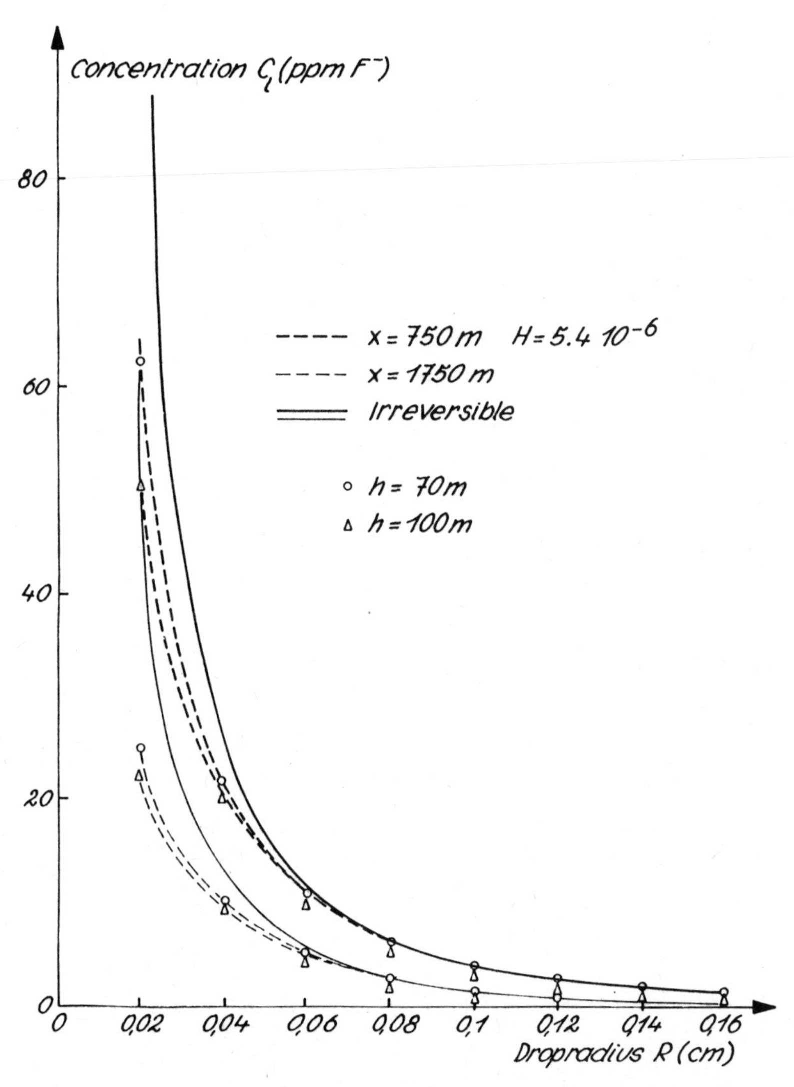

Fig. 4. : reversible washout : ground-level concentrations in function of effective stackheight and dropradius.

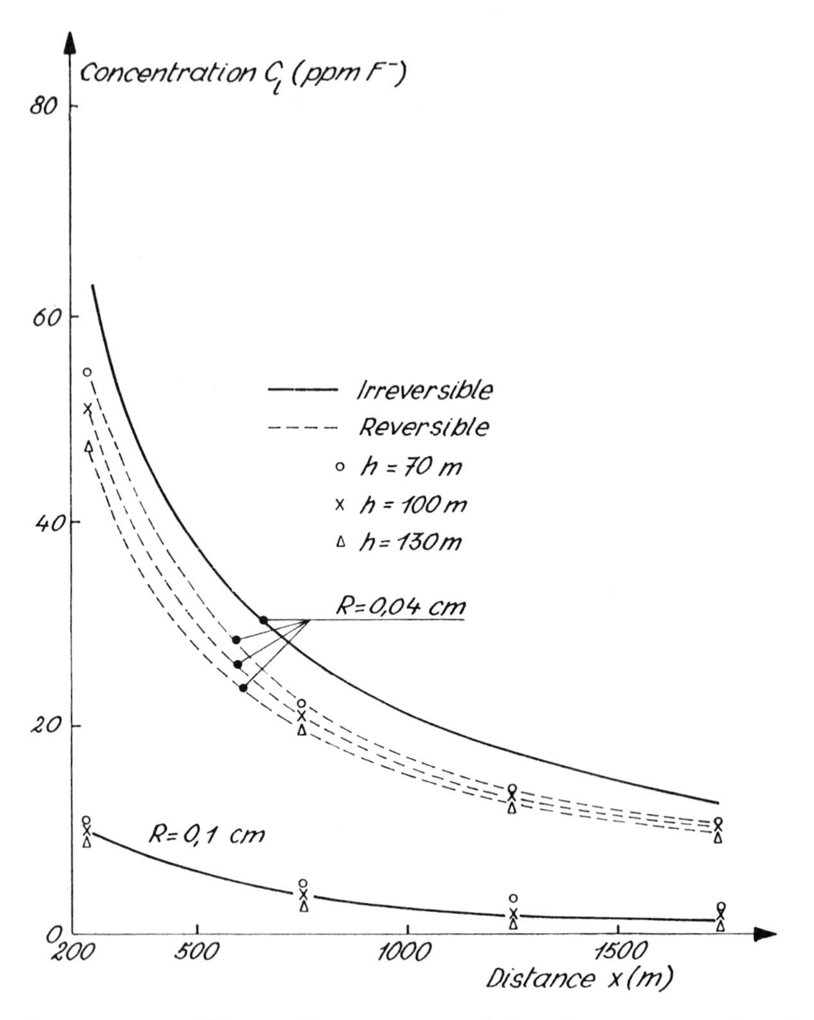

Fig. 5. : reversible washout : ground level concentration in function of distance from the source.

Further, similar to the case of the irreversible washout, a maximum will appear in function of the distance from the source when points are considered that are not exactly in line with the wind-direction axis.

Comparison : reversible - irreversible washout

Because of the greater simplicity in the computation of the irreversible washout concentrations, the difference between the computed reversible and irreversible washout concentrations and the factors affecting it, have been investigated.

In figure 6 the ratio $C_\ell(rev)/C_\ell(irr)$ is plotted versus the distance from the stack with the raindrop radius and effective stackheight as parameters. It is obvious that the raindropradius is the most important parameter : The difference between reversible and irreversible washout concentration drops sharply with increasing drop diameters. The difference becomes greater with increasing stack height and distance from the stack.

Nevertheless, for raindrop radii greater then 0.04 cm, the fault will never be greater than 25% by considering irreversible washout instead of reversible.

Mean washout concentration

In the preceding sections, the concentration was calculated for a singular raindrop. In a real shower of rain a whole range of raindrop dimensions will appear. The mean concentration $\overline{C}_\ell$ during a shower is given by

$$\overline{C}_\ell = \frac{\int_0^\infty C_\ell(R)\ v(R)\ f(R)\ dR}{\int_0^\infty v(R)\ f(R)\ dR} \tag{13}$$

Another way to calculate a "mean" concentration is by using a single dropradius, representative for the distribution of the raindrop dimensions, such as the modus or the median given by equations (8) and (9) respectively.

The difference between the "real" mean value $\overline{C}_\ell$ and the concentration obtained by using only the modus or median raindrop radius has been investigated

Therefore, the ratio $\overline{C}_\ell/C_\ell(\overline{R})$, given by

$$\frac{\overline{C}_\ell}{C_\ell(\overline{R})} = \frac{\int_0^\infty k_g(R)\ f(R)\ dR}{\int_0^\infty v(R)\ f(R)\ dR} \Big/ \frac{k_g(\overline{R})}{\overline{R}\ v(\overline{R})} \tag{14}$$

$\overline{R}$ = median or modus raindrop radius

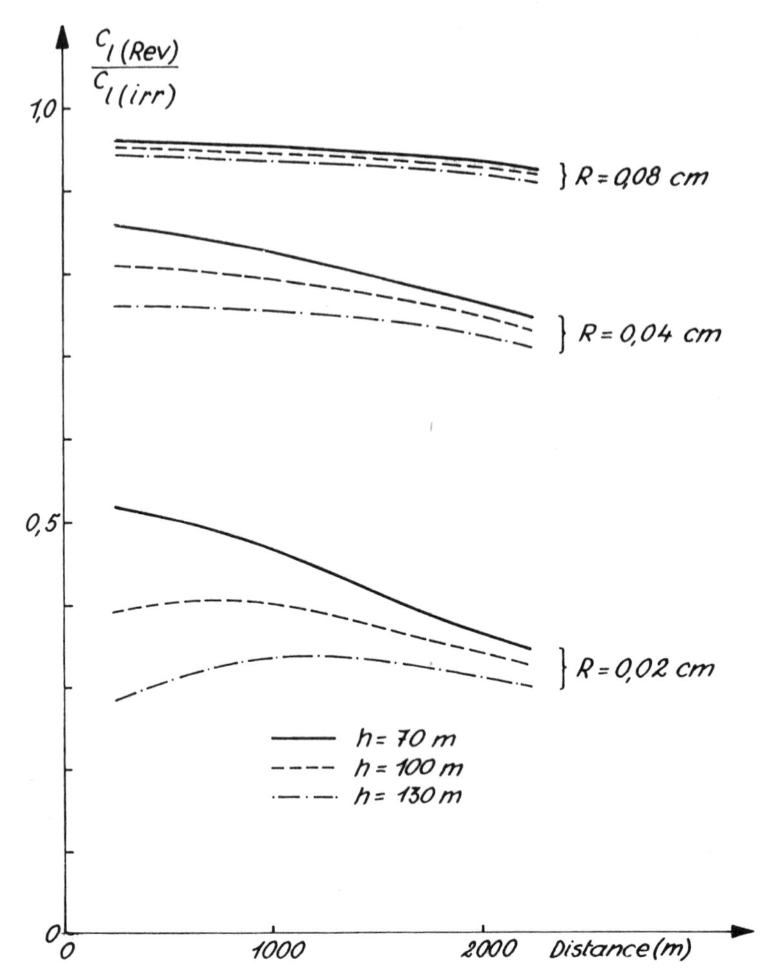

Fig. 6. : Comparison : reversible - irreversible washout.

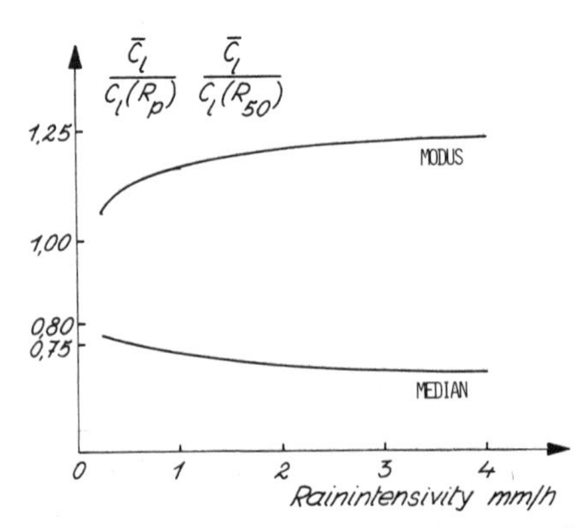

Fig. 7. : Mean concentration during real shower.

has been plotted versus the rainintensity in figure 7.

The use of the median, instead of the whole distribution leads to values too high. On the other hand, using the modus gives a too low value. In both cases, the difference increases with the rain intensity but the error will never exceed 20 - 30%, while using only one representative radius.

SIMULATION OF THE FLUORIDE WASHOUT

Since 1977, 24 precipitation samplers have been placed around a brickworks in Rumbeke (Belgium) at distances ranging from 0.3 to 2 km (figure 8). Most of the samplers are concentrated in the eastern zone of the brick works since western winds are predominant in this region.

The precipitation samplers consist of a polyethylene flask (5 or 10 l) equipped with a funnel (diameter 23,5 cm) and are collected every two weeks (15th and last day of the month). The fluoride concentration is determined using a fluoride specific electrode (Radiometer F 1052F) and the analytical procedure is given elsewhere (De Keyzer, 1980).

Taking into account the considerations of the previous sections a computer program was developed in order to stimulate the fluoride washout around the considered brick works. A simplified flow sheet of this program is given in Figure 9. The input to the program consists of meteorological data and the experimental washout values.

The meteorological data have been obtained from the weather station (Royal Meteorological Institute of Belgium) at Beitem, about 4 km away from the brick-works. We used the precipitation volumes recorded every 10 min. and the mean wind velocity and -direction over a two hour period.

The subroutine TZSUB calculates the fluoride washout in each of the 24 points using the irreversible washout model of the previous sections and the median drop radius. The mean fluoride concentration and the washout for a single shower are given by

$$\hat{C}_{\ell}(x,y,0) = \frac{3\ k_g Q \exp\left(-0,5\left(\frac{y}{\sigma_y}\right)^2\right)}{\sqrt{2\pi}\ \ R_{50}\ v\ \bar{u}\ \sigma_y} \qquad (15)$$

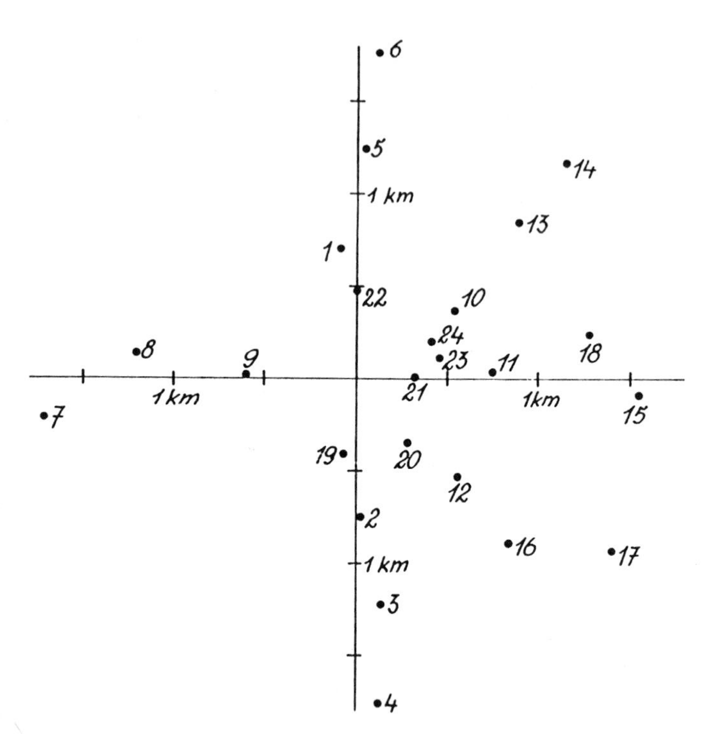

Fig. 8. : Precipitation sampling network around the brickworks at Rumbeke (Belgium).

$$W = 19\ 10^6\ PA\ \hat{C}_\ell\ (x,y,0) \tag{16}$$

with

$\hat{C}_\ell(x,y,0)$ = fluoride concentration calculated using the median dropradius (mol F^-/cm^3)

PA = precipitation amount (mm or l/m^2)

W = washout (mg F^-/m^2)

σ_y = $0.586\ x^{0.796}$ (Pasquill class C, slightly unstable)

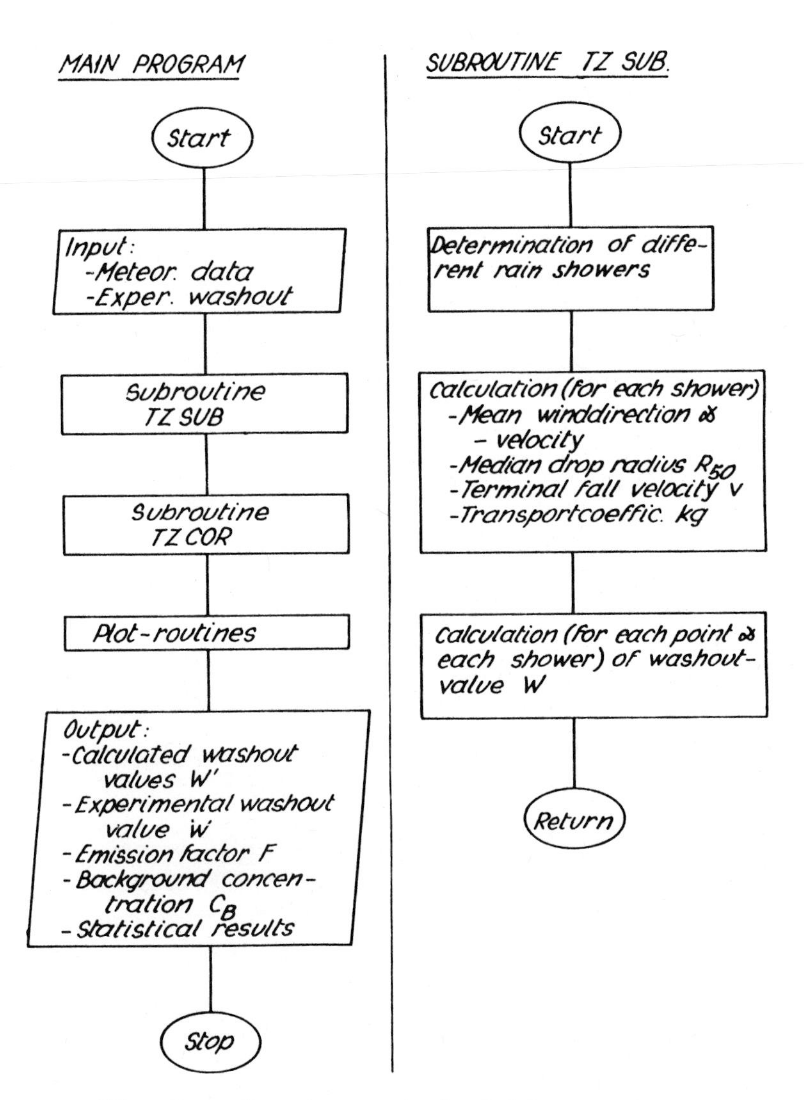

Fig. 9. : Structure of simulation program.

Because of the stack height of the brickworks is 60 m, the mean wind velocity $\bar{u}$ has been taken between 0 and 100 m. Assuming a power law

$$\frac{u(z_1)}{u(z_2)} = \left(\frac{z_1}{z_2}\right)^n \tag{17}$$

with n = 0,23 (slightly unstable, Pasquill class C) (Bultinck & Malet, 1972) and an anemometer height of 10, this leads to

$$\bar{u} = 1.38\ u_A \tag{18}$$

with u(z) the wind velocity at height z and u_A the anemometer wind velocity.
As the emission rate Q of fluoride is unknown, it is set arbitrarily at 1 mol F^-/s (68.4 Kg F^-/h).

At the 24 points the contributions to the half-monthly washout of the different showers are added, which leads to a first estimate W. These estimates do not agree yet with the experimental values for several reasons :

(i) the emission rate has been set arbitrarily at 1 mol F^-/s. From a few emission measurements and from literature (Müller, 1976), it is known that this value is certainly too high.
(ii) during rainfall, the gaseous fluoride concentration has been held constant with time. The decrease of the atmospheric concentration due to the gas washout has not been taken into account.
(iii) the initial fluoride concentration C_ℓ (z=∞) was set equal to zero.

With this in mind, a new washout value W' is calculated in the subroutine TZCOR, by a linear regression between the experimental washout values W_e and the first estimates W :

$$W' = PA \cdot C_B + F \cdot W \tag{19}$$

The physical meaning of C_B is a background-concentration and the factor F corrects the failures (ii) and (iii).

Besides the calculation of C_B and F, the program provides also a statistical analysis giving

- correlationcoefficient with significance level
- The 95 % confidence intervals of C_B and the calculated washout values W'.

The whole of the calculations is presented in a plot giving the experimental and calculated washout values (with confidence intervals). Moreover the contribution of the background washout is represented. As an example, in figure 10 the resulting plot is given for the first half of february 1978.

The simulation program contains also the possibility of performing a three-dimensional picture of the fluoride washout around the emission source. Therefore a grid of 4 x 4 Km is considered, containing 40 X 40 gridpoints (every 100 m one point). The gridpoints are on straight lines oriented north-south and east-west (figure 11).

The washout values at these 1600 points are calculated with the aid of the previous computer program and are stored on a sequential diskfile. These values are the input data for the plot routine, giving a three-dimensional view. The observer is in the south-east and looks to the north-west.

Figure 12 represents a three-dimensional picture of the washout in the first half of february 1980 in the surroundings of the brickworks. Although such plots provide a good picture of the directions in which the greatest washout has happened, the strong peaks in the centre (very near the stack) are to be relativated. Indeed, in reality the wind doesn't blow rigourously in the same direction and little fluctuations do occur. These fluctuations will have a great influence on the concentration close to the stack but further away, these fluctuations will have much less importance.

Finally, the results for the 24 periods in 1978 are given in table 2. For most of the periods the correlations coefficients of the linear regression, used to obtain the final washout values W' (cfr. eq. 19), are highly significant. Also the factor C_B, determined by regression does not fluctuate very much sustaining its physical meaning of a background concentration.

The reason for the less satisfactory results in some periods (1-15 June, 16-31 July...)is probably due to the fact that the meteorological data were not recorded in loco but a certain distance away from the considered region. Also the model would probably better fit the experimental data if a better (and unluckely a more complicated) dispersion model was to be used, in which direction we are working.

ACKNOWLEDGEMENT

We wish to thank the Ministry of Economics, the Ministry of Public Health, the State Secretary of Regional Economics of Flandres and the Belgian association of Clay Works who have made this investigation program possible.

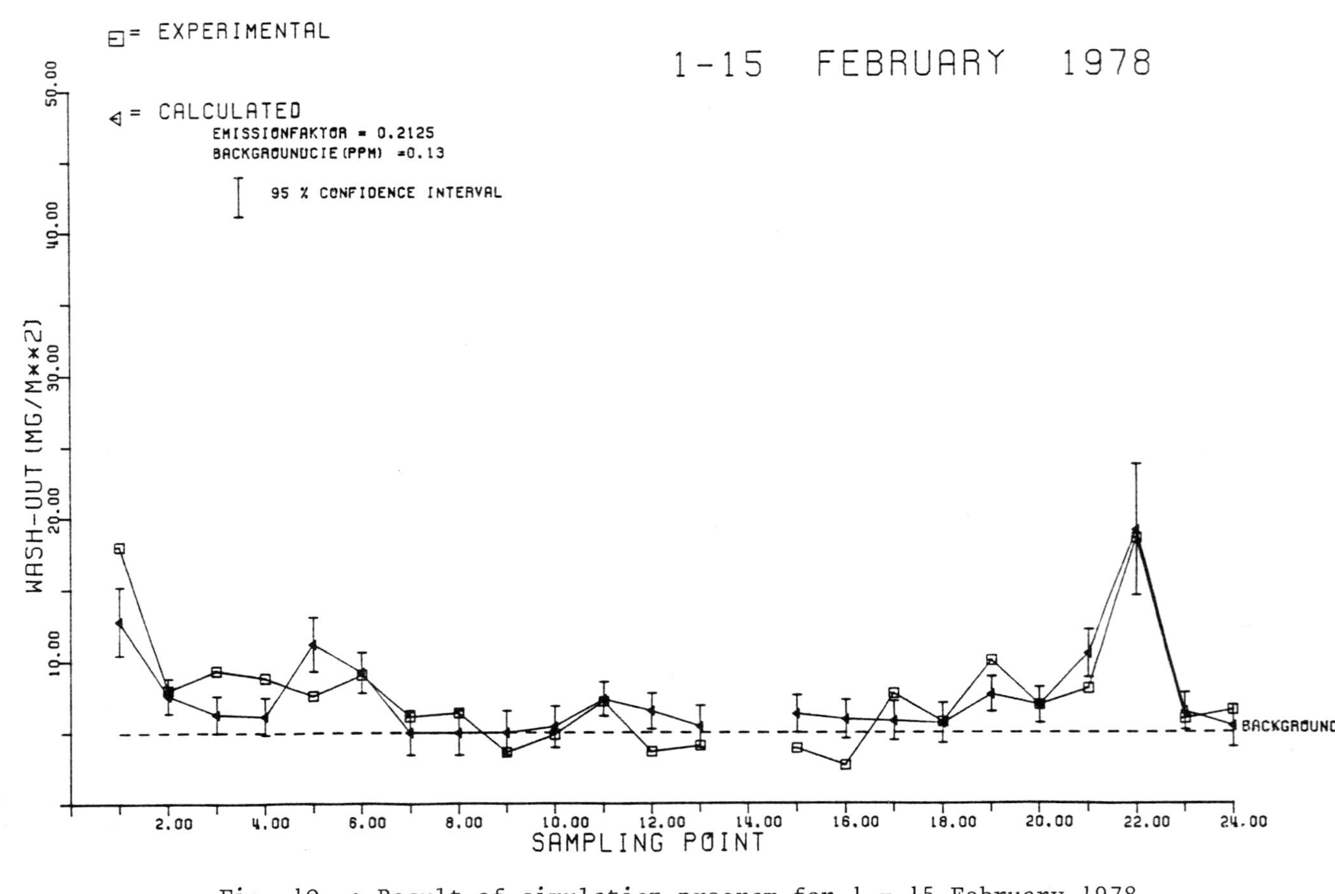

Fig. 10. : Result of simulation program for 1 - 15 February 1978.

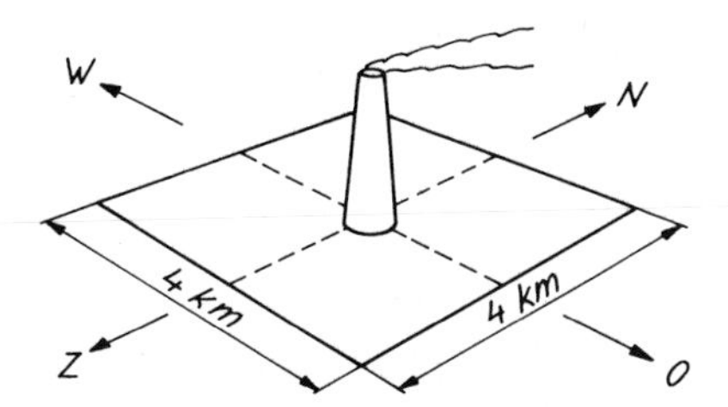

Fig . 11 " Orientation three-dimensional pictures.

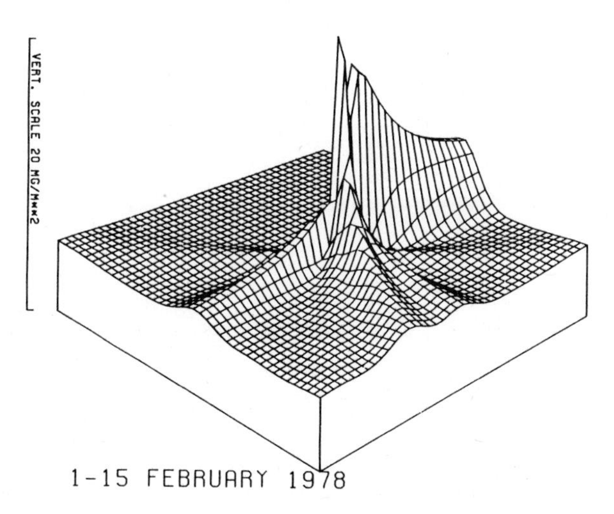

Fig. 12. : Three-dimensional representation of simulation program for 1-15 February 1978.

Table 2. Results of simulation program for 1978 period.

Period 1978	correlation coefficient	Emission factor	background concentration (ppm)
1-15 jan.	0.1431	0.1638	0.271
16-31 jan.	0.8775^{+++}	1.5845	0.309
1-15 febr.	0.8305^{+++}	0.2864	0.132
16-28 febr.	0.6637^{+++}	0.7138	1.306
1-15 march	0.4479^{+++}	0.4585	0.359
16-31 march	0.3648^{+}	0.3150	0.308
1-15 apr.	0.6099^{+++}	0.7950	0.331
16-30 apr.	0.3643^{+}	0.6675	0.235
1-15 may	0.8594^{+++}	0.2228	0.074
16-31 may	0.5212^{+++}	0.8726	2.037
1-15 june	0.0637	0.0163	0.307
16-30 june	0.2296	0.0478	0.130
1-15 july	0.8527^{+++}	0.3243	0.076
16-31 july	0.3270	0.2314	0.614
1-15 aug.	0.6499^{+++}	0.1993	0.172
16-31 aug.	0.5317^{+++}	0.3322	0.306
1-15 sept.	0.4070^{++}	0.0528	0.207
16-30 sept.	0.0074	-0.0040	0.322
1-15 oct.	0.7931^{+++}	0.6301	0.901
16-31 oct.	0.6667^{+++}	0.7424	0.322
1-15 nov.	0.7274^{+++}	1.0148	0.592
16-30 nov.	0.0073	0.0045	0.296
1-15 dec.	0.8039^{+++}	1.1718	0.220
16-31 dec.	0.0180	0.0051	0.145

+ 90 % significance level
++ 95 % significance level
+++ 99 % significance level

REFERENCES

Best, A. C., 1950 : The size distribution of raindrops, Q. Jl. R. Met. Soc., 76:16.
Bultynck H., Malet L., 1972 : Evaluation of atmospheric dilution factors for effluents diffused from an elevated continuous point source, Tellus, 24:455.
De Keyzer, P. M., 1980 : MS Thesis, State University Ghent
Froessling N., 1938 : The evaporation of falling raindrops, Beitr. Geophys., 52:170.
Hales J. M., Wolf M. A., Dana M. T., 1973 : A linear model for the prediction of the washout of pollutant gases from industrial plumes, AICHE, 19:292.
Hales J. M., 1972 : Fundamentals of the theory of gas scavenging by rain, Atm. Env., 6:635
Lapidus L., Seinfeld J. H., 1971 : Numerical solution of ordinary differential equations in : "Mathematics in science and engineering", 74:174, Academic Press, NY.
Laws J. O., Parson D. A., 1943 : The relation of raindrop-size to intensity, Trans. Amer. Geophys. Un., 24.452.
Mueller H. A., 1976 : Ziegelindustrie International nr. 12:552
Rebuffat P., 1979 : Le secteur briquetier, in : "Leefmilieu-Lucht, Eindrapport 1975-1978, nr. 7 - Economische studies", Prime Minister's Services, Brussels.

PERFORMANCE EVALUATION OF THE EMPIRICAL KINETIC MODELING APPROACH (EKMA)

J. Raul Martinez, Christopher Maxwell, Harold S. Javitz, and Richard Bawol

SRI International
Menlo Park, California 94025

ABSTRACT

The EKMA is a Lagrangian photochemical air quality simulation model that calculates ozone from its precursors: nonmethane hydrocarbons (NMHC) and nitrogen oxides (NO_x). This study evaluated the performance of the EKMA when it is used to estimate the maximum ozone concentration that can occur in an urban area and its environs. The evaluation was conducted using data for five U.S. cities. This paper reports the results for St. Louis, Missouri.

A novel statistical evaluation procedure was developed to measure the accuracy of the EKMA ozone estimates. The accuracy parameter is defined as the ratio of observed to estimated ozone. Associated with this ratio is an accuracy probability, which is defined as the probability that the ratio lies within a predefined percent (e.g., ±20 percent) of unity, a unit value of the ratio denoting perfect agreement between observation and prediction. Equations were derived that express the ratio as a function of NMHC and NO_x. The evaluation procedure thus uses NMHC and NO_x as inputs to calculate the accuracy probability of the EKMA ozone estimate. The full range of accuracy probabilities associated with the EKMA ozone estimates is displayed in graphical form on the NMHC-NO_x plane.

INTRODUCTION

The Empirical Kinetic Modeling Approach (EKMA) is a Lagrangian photochemical model that calculates ozone (O_3) as a function of nonmethane hydrocarbons (NMHC) and oxides of nitrogen (NO_x). The EKMA has been extensively documented and we will forego discussing the technical details of the model (EPA, 1977, 1978; Dodge, 1977; Trijonis and Hunsaker, 1978; EPA, 1980; Whitten and Hogo, 1978).

This paper describes the results of a study that evaluated the performance of the EKMA when it is used to estimate the maximum O_3 concentration that could occur in an urban area and its environs. The study measures the EKMA's ability to predict maximum O_3 and defines conditions under which O_3 estimates can achieve specific accuracy levels.

The study encompassed five cities in the United States. Becau of space limitations this paper will focus on the results obtained for St. Louis, Missouri. The complete results for all the cities ar described in a report by Martinez et al. (1981).

METHODOLOGY

The evaluation was conducted using air-quality, emissions, and meteorological data for St. Louis, Missouri. The evaluation procedure consisted of three steps:

- Define evaluation data set
- Obtain O_3 estimates using EKMA
- Conduct a statistical evaluation of EKMA performance as a predictor of maximum O_3.

Evaluation Data Set

Briefly, the evaluation data set consisted of observed 0600-09 local time (LT) NMHC and NO_x, and the measured O_3 maximum for each day. The 0600-0900 NMHC and NO_x were averaged over a group of site located in the subregion of the urban area where their respective sources are concentrated. The EKMA considers the 0600-0900 (LT) NMHC and NO_x to be the precursors of the maximum O_3 that occurs down wind of the emission subregion later in the day. The observed daily O_3 maximum is defined as the highest hourly average O_3 measured from 1200 to 1700 (LT) at any monitoring station in the area. The timing of the O_3 maximum is dictated by the EKMA's assumptions. Thus, each evaluation data set contains one entry per day. The data set for St. Louis contains 100 days covering the period May-October 1976. The data set is listed in Martinez et al. (1981).

Reduced to essentials, the criteria for selecting days for the evaluation data set consisted of: data availability, the time of occurrence of the daily O_3 maximum, and the prevalence of meteorological conditions that are necessary, but may not be sufficient, for the O_3 maximum to be at least 100 parts per billion (ppb). The 100 ppb cutoff ensures that all days when the O_3 national ambient air quality standard (NAAQS) of 120 ppb is exceeded are included in the evaluation data set. There are two reasons for choosing such days for the evaluation. First, EKMA is based on worst-case assumptions and therefore cannot be expected to perform well for days that do not meet these conditions. Second, because we are interested in maximum O_3 potential with respect to the NAAQS, days with conditions associated with low O_3 are not important to our evaluation.

EKMA O_3 Estimates

Basically, the EKMA takes two forms: standard and city-specific. The standard EKMA is based on conditions that prevail in the Los Angeles area, whereas the city-specific version uses model inputs for a particular city. The performance evaluation was conducted using both forms of EKMA for each city.

For both standard and city specific EKMA, the inputs used to estimate O_3 were the 0600-0900 (LT) NMHC and NO_x, spatially averaged over a specific set of monitors located in the emissions region of the city.

Statistical Evaluation of EKMA Performance

The assessment uses the ratio R = OBS/EST, where OBS denotes the observed daily O_3 maximum and EST the EKMA estimate, as the performance measure. Using this ratio, three accuracy regions were defined as follows:

- Region 1: $1.2 < \text{OBS/EST}$
- Region 2: $0.8 \leq \text{OBS/EST} \leq 1.2$
- Region 3: $\text{OBS/EST} < 0.8$

Region 1 contains ratios representing cases of substantial underprediction. We consider ratios in Region 2 to represent the most accurate predictions that can be made because the error is at most 20 percent, which is consistent with measurement precision. Region 3 contains ratios representing cases of substantial overprediction.

To evaluate EKMA performance we calculated the probability that a given EKMA estimate falls in one of the three regions defined above. This was done by finding a multiple linear regression equation that describes the ratio R as a function of NMHC, NO_x, and possibly other variables such as maximum daily temperature. The equation for R was then used to estimate the probabilities $P(R > 1.2)$, $P(0.8 \leq R \leq 1.2)$, and $P(R > 0.8)$, based on the assumption that the residuals of the regression fit are Gaussian. The accuracy probabilities were displayed on graphs showing the probability as a function of NMHC and NO_x. In addition, probability isopleths were plotted on the NMHC-NO_x plane. The plots for the various cases are shown in the next section.

RESULTS FOR ST. LOUIS

For the standard-EKMA estimates, the regression equation for the ratio R = PBS/EST had $1/NO_x$, 1/NMHC, background O_3 and maximum daily temperature as the independent variables. The multiple regression coefficient of the fit was 0.86. Setting the background O_3 and the temperature equal to their respective mean values of 78 ppb and 30.4°C yields

$$R = 0.3196 + 8.68/NO_x + 99.23/NMHC \ , \tag{1}$$

where NO_x is in ppb and NMHC is in ppbC.

Figure 1 shows a plot of the three probability regions as a function of a new variable $Z = 8.68/NO_x + 99.23/NMHC$, which is the right hand side of Equation (1) without the constant term. Using Figure 1, one can estimate the probability associated with each accuracy region for a given pair of (NMHC, NO_x) coordinates. In general, Figure 1 shows that there is a high probability of overprediction [i.e., $P(R < 0.8)$] when NMHC and NO_x are large. Conversely, the figure indicates that there is a high probability of underprediction [i.e., $P(R > 1.2)$] for low NMHC and NO_x. Accurate predictions are most probable in the shaded region, where $0.56 \leq Z \leq 0.80$ and $P(0.8 \leq R \leq 1.2)$ varies between 0.70 and 0.83, the maximum probability of 0.83 occurring in the neighborhood of $Z \doteq 0.68$.

In Figure 1, a value of Z is uniquely associated with a particular accuracy probability. Hence, a constant-Z curve defines an accuracy probability isopleth on the NMHC-NO_x plane. A plot of constant-Z curves in the NMHC-NO_x plane is shown in Figure 2. The figure also displays the evaluation region for NMHC and NO_x, which defines the area of the plane containing all the NMHC and NO_x concentrations present in the evaluation data set. The shaded area of Figure 2 corresponds to that of Figure 1, and defines the (NMHC, NO_x) combinations that are most likely to yield accurate predictions. A high probability of overprediction occurs in the region where $Z \leq 0.3$, and this region is seen in Figure 2 to be linked to high NMHC and NO_x. A high probability of underprediction occurs for $Z \geq 1.0$, and Figure 2 shows that this is a very small area within the evaluation region.

For the city-specific EKMA, a multiple regression equation was derived for R as a function of 1/NMHC, daily maximum temperature, and background O_3. For mean temperature and background O_3, the equation is

$$R = 0.572 + 118.39\ (1/NMHC) \tag{2}$$

where NMHC is in ppbC.

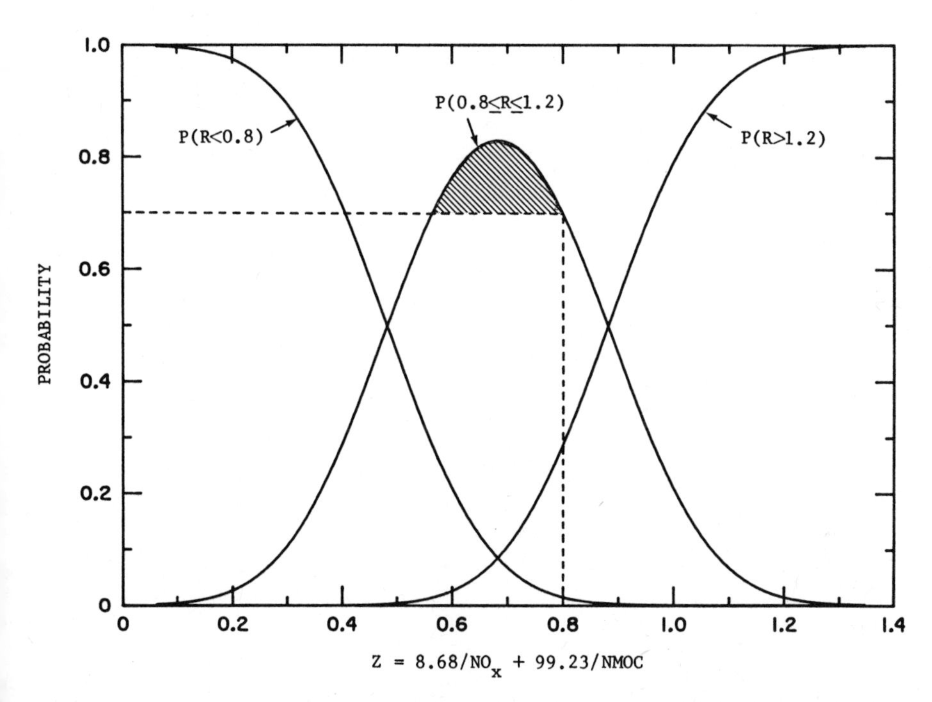

'ig. 1. Accuracy Probability Plot for Standard-EKMA Ozone Estimates for St. Louis. Mean values are assumed for background ozone and maximum daily temperature.

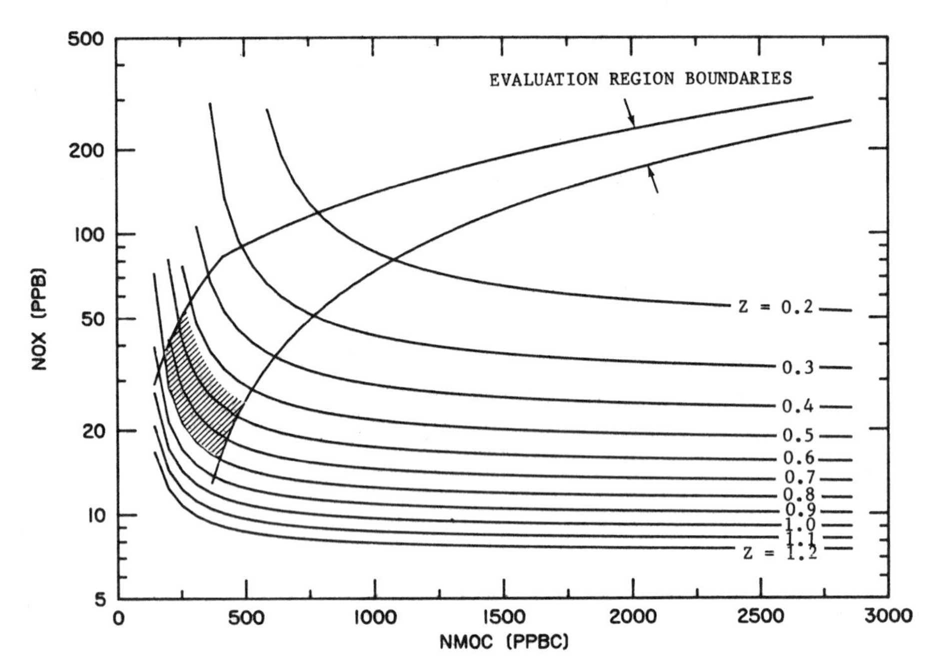

Fig. 2. Plot of Constant-Z Curves on NMOC-NO_x Plane for Standard-EKMA Estimates for St. Louis. Shading denotes area where ozone estimates are most accurate, assuming average conditions for temperature and background ozone.

The multiple regression coefficient is 0.74, and the standard error of the regression is 0.182. In contrast to Equation 1, this time the variable $1/NO_x$ does not appear in the regression. The stepwise regression procedure attempted to include $1/NO_x$, but its coefficient was not significant at the 0.05 level ; NO_x was also tried with the same result.

Using Equation 2, we plotted the accuracy probabilities as a function of Z = 118.39/NMHC ; the plot is shown in Figure 3. Complementing Figure 3, a family of lines for several values of Z is displayed in Figure 4, which also shows the evaluation region in the NMHC-NO_x plane.

The shaded area of Figure 3 is the region where the probability $P(0.8 \leq R \leq 1.2)$ attains its highest values, hence it is associated with ozone estimates that are most likely to be accurate. Within the shaded area, the variable Z varies from 0.35 to 0.50, and $P(0.8 \leq R \leq 1.2)$ varies from 0.70 to 0.73, reaching its maximum value of 0.73 in the neighborhood of Z = 0.43. The shaded area of Figure 4 corresponds to that of Figure 3. Inside the shaded area of Figure 4, NMHC is defined by $237 \leq NMHC \leq 338$ ppbC, and NOx is limited by the boundaries of the evaluation region.

Comparing Figures 1 and 3 shows that the maximum value of $P(0.8 \leq R \leq 1.2)$ in Figure 1 is higher than that in Figure 3,the respective maxima being 0.83 and 0.73. This discrepancy is largely due to the fact that the standard deviation for Equation 1 is smaller than that for Equation 2 (s = 0.146 and s = 0.182, respectively). The smaller standard deviation causes the probability curve for Equation 1 (cf. Figure 1) to be less spread out and thus taller, than that for Equation 2 (see Figure 3).

Comparison of Figures 2 and 4 shows that the location of the shaded area has been shifted to the left in the evaluation region of Figure 4. As a result, the values of NMHC and NO associated with the ozone estimates most likely to be accurate can be higher for the standard EKMA (Figure 2) than for the city-specific case (Figure 4). The relative placement of the shaded area in Figures 2 and 4 also is indicative of the difference in the predictive performance of standard and city-specific EKMA. Thus, overprediction is more common for the standard than for the city-specific EKMA, and this is reflected in the fact that the region above the shaded area in Figure 2 is larger than that in Figure 4. This situation is reversed in the case of underprediction, since the region below the shaded area in Figure 2 is smaller than in Figure 4.

CONCLUSIONS

The analysis showed that, within certain bounds, the standard EKMA can be used to provide realistic estimates of maximum ozone

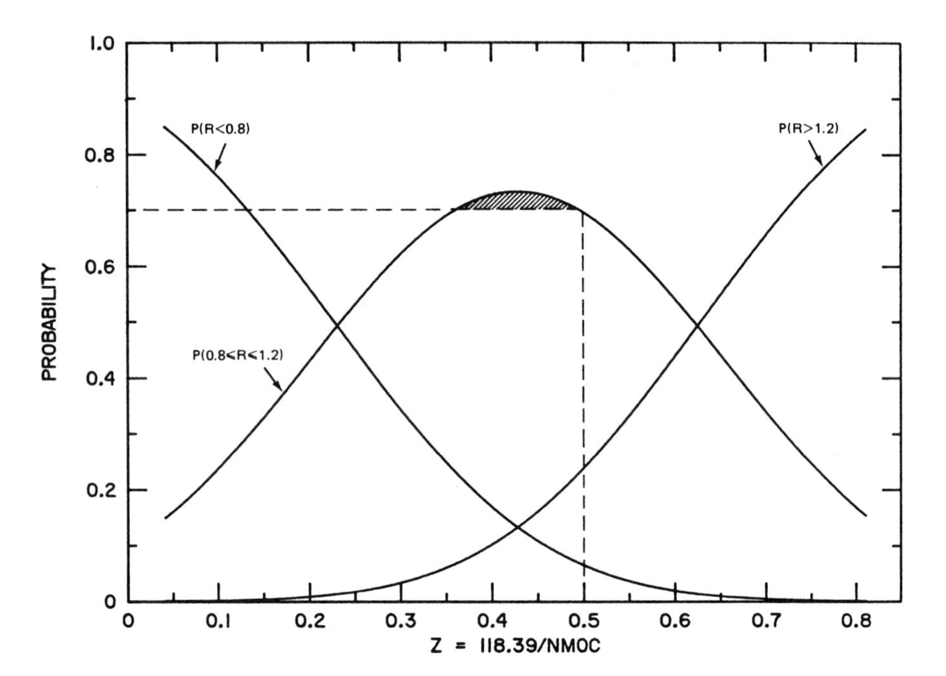

Fig. 3. Accuracy Probability Plot for City-Specific EKMA Ozone Estimates for St. Louis. Mean values are assumed for background ozone and maximum daily temperatures.

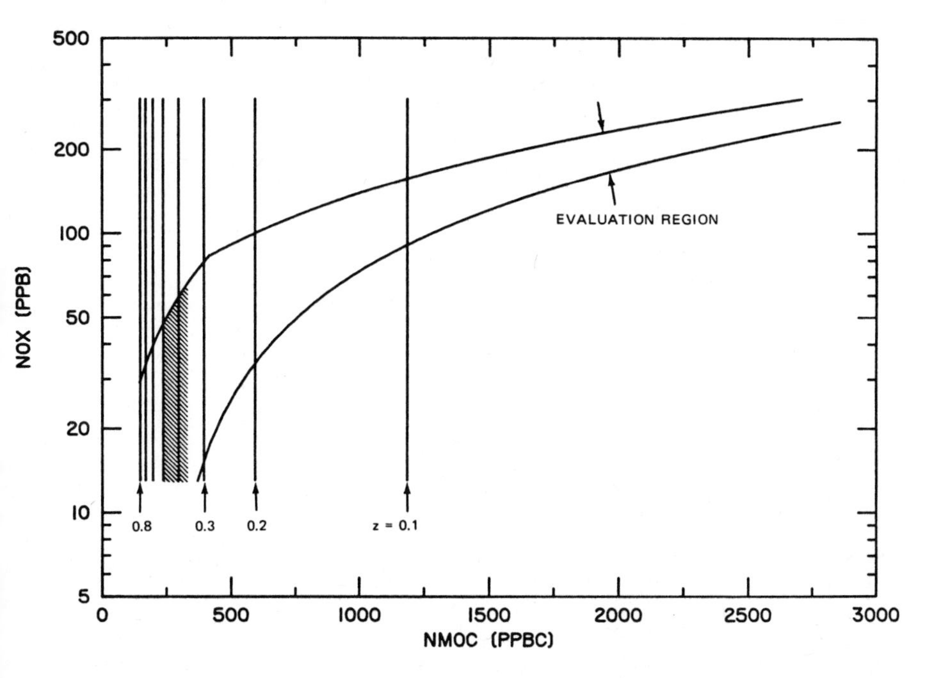

Fig. 4. Plot of Constant-Z Lines on NMOC-NO_x Plane for City-Specific EKMA Estimates for St. Louis. Shading denotes area where ozone estimates tend to be most accurate, assuming average conditions for maximum temperature and background ozone.

levels for the St. Louis area. Three regions of predictive accurac were defined wherein the ratio of observed to estimated ozone is greater than 1.2 between 0.8 and 1.2, and less than 0.8. Of the 100 estimates, 78 fell in the third region, which is indicative of substantial overprediction and was expected. Eighteen estimates fe in the second region, which is where EKMA is accurate to within ±2(percent. This shows that reasonably credible estimates can be obtained with the standard EKMA. Four cases fell in the first regi where an underestimate occurs. The latter suggests that the standa EKMA could be used as a screening tool with the expectation that there is a low probability that its predictions would be exceeded i actuality.

The city-specific EKMA ozone estimates were generally more accurate than the standard EKMA's. Thus, although the city-specifi EKMA also produced upper bounds for the observations, 36 percent of the city-specific estimates fell in the region where 0.8 $\leq$ observed estimated $\leq$ 1.2, compared to 18 percent for the standard-EKMA estimates. However, the city-specific EKMA displayed a greater ten dency to underestimate the observations than did the standard EKMA (8 percent compared to 4 percent, respectively).

The accuracy regions for the standard-EKMA estimates were show to be functions of NMHC, NO_x, background ozone and maximum daily temperature. Thus, to define the accuracy regions on the NMHC-NO_x plane it is necessary to assign a value to background ozone and temperature. The example presented used mean values for these two quantities (cf. Figures 1 and 2). When applied to a particular problem, it is desirable to tailor the accuracy regions to the specific prevailing background ozone and temperature. Having defin the values of these two variables, it is a simple matter to generat a plot similar to Figure 1. Similar considerations apply to the accuracy regions associated with the city-specific estimates.

ACKNOWLEDGMENT

The work reported here was sponsored by the U.S. Environmenta Protection Agency. The views expressed in this report are the authors' and do not necessarily represent those of the sponsor.

REFERENCES

Dodge, M.C., 1977, "Effect of Selected Parameters on Predictions of a Photochemical Model," EPA-600/3-77-048 (June).

Martinez, J.R., C. Maxwell, H.S. Javitz, and R. Bawol, 1981, "Evalu tion of the Empirical Kinetic Modeling Approach (EKMA)," Draft Final Report, SRI Project 7938, SRI International, Menlo Park, California (July).

J.S. Environmental Protection Agency, 1977, "Uses, Limitations, and Technical Basis of Procedures for Quantifying Relationships Between Photochemical Oxidants and Precursors," EPA-450/2-77-021a (November).

J.S. Environmental Protection Agency, 1978, "Procedures for Quantifying Relationships Between Photochemical Oxidants and Precursors : Supporting Documentation," EPA-450/2-77-021b (February).

J.S. Environmental Protection Agency, 1980, "Guideline for Use of City-Specific EKMA in Preparing Ozone SIPs," EPA-450/4-80-027 (October).

Trijonis, J. and D. Hunsaker, 1978, "Verification of the Isopleth Method for Relating Photochemical Oxidant to Precursors," EPA-600/3-78-019 (February).

Whitten, G.Z. and H. Hogo, 1978, "User's Manual for Kinetics Model and Ozone Isopleth Plotting Package," EPA-600/8-78-014a (July).

DISCUSSION

G. WHITTEN

Have you used the new algoritm for the mixing height? The way you use EKMA is not the way in which the EPA advices to use EKMA, can you comment on that?

J. R. MARTINEZ

The work described was completed about one year ago, before the new mixing height algorithm was published. Consequently, the model did not use the new mixing height algorithm.
The method in which EKMA was used in this study differs from that recommended by EPA. Our intent was to investigate new methods of using EKMA, and our results indicate that the EKMA can be used to estimate maximum ozone levels.

R. M. STERN

What are you doing if you are ending at "refine the analysis" in the flowchart 2?

J. R. MARTINEZ

The question refers to a flowchart presented at the meeting, but not included in the paper; the flowchart is reproduced below. The flowchart describes how to use the results of the study so that the EKMA can be applied to asse the effectiveness of ozone control stra gies. In essence, the approach is to use the EKMA prediction of maximum ozon in conjunction with the accuracy probab lities calculated in our study. Becaus the determination of the strategy's effectiveness is based on probabilities there are some cases in which the answe obtained is inconclusive. The question refers to the actions taken in such circumstances, and refining the analysi is one course of action that could be profitably taken. By refining the analysis, we mean redefining the accuracy regions by increasing their number and reducing their width. This allows us to obtain a more accurate estimate of t difference between the ozone estimate a the ambient standard, in the hope that the more accurate estimate can resolve the ambiguity associated with the coars analysis.

Ph. ROTH

Did you take the shape of the city into account?

J. R. MARTINEZ

The shape of the city was take into account in obtaining city-specifi ozone estimates. This was done in the step in which trajectories were calculated to obtain estimates of the post -0800 emissions load.

W. GOODIN

The rate of the increase of th depth of the mixed layer differs in the five cities that you have studied, yet EKMA uses one procedure for this calculation. Could this have affected your results?

J. R. MARTINEZ

Two modes of EKMA operation we evaluated. One was the standard EKMA, i which the rate of increase of the depth of the mixed layer is assumed to be the same for all areas. The other is the city-specific EKMA, in which individual rates for each city are used. The results for both EKMA modes were genera ly different, indicating that the mixin height assumption affected the results. However, we did not measure the effect of mixing height alone, and the effects of other inputs such as post-0800 emissions and initial NO_2/NO_x ratio may have been more important than the effect of mixing height.

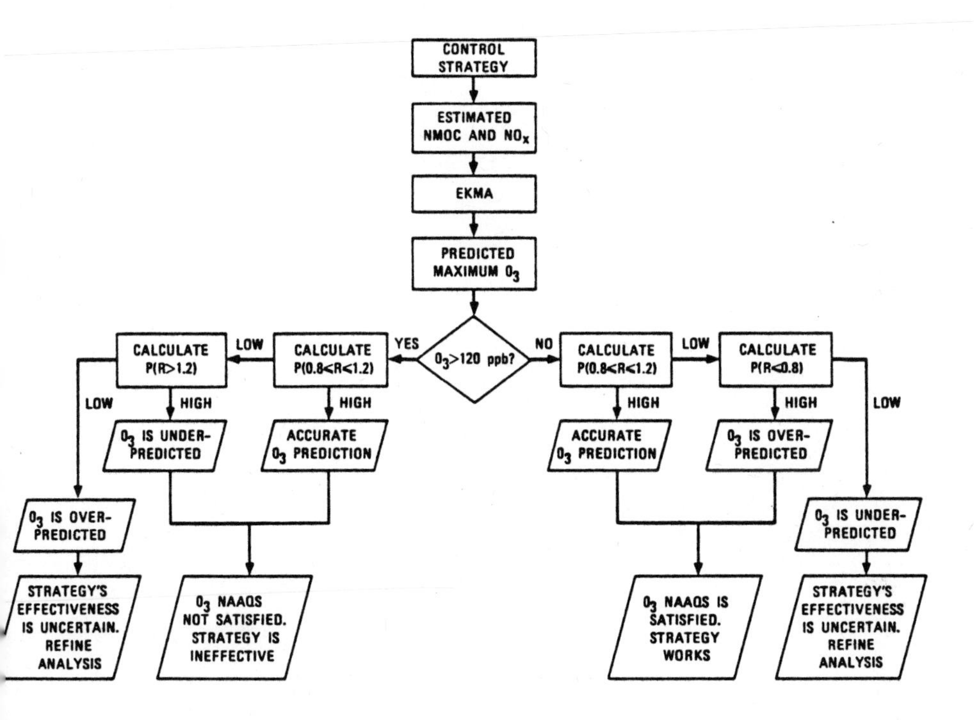

URBAN VISUAL AIR QUALITY : MODELLED AND PERCEIVED

Paulette Middleton
Robin L. Dennis
Thomas R. Stewart

National Center for Atmospheric Research *
P.O. Box 3000
Boulder, Colorado, 80307 USA

ABSTRACT

Comparisons of modelled and perceived urban visual air quality are presented for an average visual air quality day in the winter of 1981 for Denver, Colorado. The field study design for capturing human judgments of visual air quality and the three dimensional visual air quality simulation model are outlined. The comparisons illustrate the feasibility as well as the difficulties associated with predicting human judgments of visual air quality with an air quality simulation model.

The research for this project was facilitated by the Center for Research on Judgment and Policy, Institute of Behavioral Science, University of Colorado. The project was supported in part by BRSG Grant RR07013-14 awarded by the Biomedical Research Support Program, Division of Research Resources, NIH.

* The National Center for Atmospheric Research is sponsored by the National Science Foundation.

INTRODUCTION

Although visual air quality is an important issue in regional air quality management, the ability to forecast effects of potenti air quality management plans on visual air quality is currently quite poor, particularly for urban areas. No widely accepted metri for measuring visual air quality has been developed, nor appropriat techniques for linking potential policy actions with such an index. The ultimate goal of the present research effort is to develop a visual air quality index that (1) is based upon human judgments of visual air quality, (2) can be related to easily monitored properties of the physical environment, and (3) can be used in forecastin levels of visual air quality expected from alternative air quality management plans.

Our method for obtaining a visual air quality index involves (1) calculating perceptual cues using a three dimensional visibility model which relates emissions to perceptual cues for a set of charactertistic days in the urban area being studied; (2) validating the model by comparing the calculated cues with observer judgments for a variety of different atmospheric conditions; (3) deriving a simple index from parameterization of the visibility model runs for the characteristic days; and (4) producing a weight index for the year/season based on estimates of the average freque of occurence of each characteristic day over a year or a season.

Central to our method is the belief that development of an appropriate air quality index will require improved understanding of the relationships among emissions changes and the four types of variables important to urban visual air quality: (1) individual's judgments of overall visual air quality; (2) perceptual cues used in making judgments of visual air quality; (3) measureab physical characteristics of the visual environment; and (4) concentrations of visibility-reducing pollutants and their precursors (see Figure 1). The relations among emissions, physical characteristics, pollutant concentrations and perceptual cues can be descri with a three dimensional dispersion-radiative transfer model and/o a parameterization of such a model. The relationship between the human judgments and the physical environment variables whether calculated or measured has yet to be described in any theoretical model. Our approach to achieving a complete model link between emissions and human judgments for a particular day -- steps 1 and 2 of our 4-step method -- requires two types of information. Firs the visual air quality simulation model is needed to calculate perceptual cues from emissions. Validation of the model's predictive ability requires an extensive data base of human observations. Prior to our study, neither the visual air quality simulation model nor the observer data base were available. Thus, major research efforts have been focused on obtaining an appropriate model and data base.

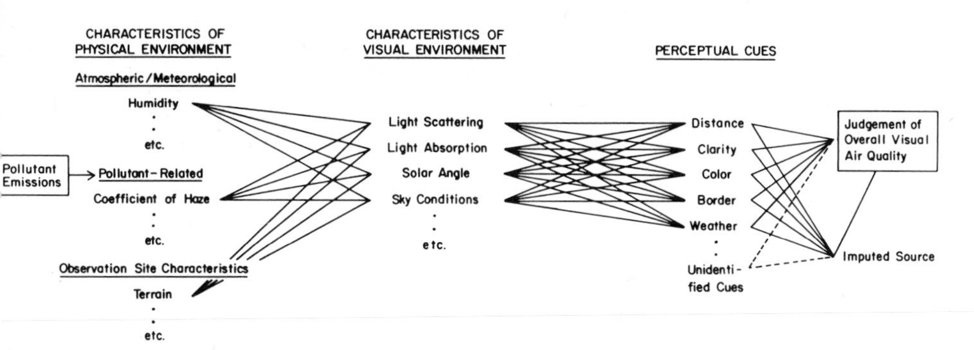

Figure 1. Conceptual Framework for the Denver Regional Visual Air Quality Study.

The purpose of this paper is three fold: (1) to describe our approach to analyzing the human component of visual air quality assessment; (2) to describe our visual air quality simulation model; and (3) to present a comparison of modelled and perceived cues for Denver, Colorado. The comparison illustrates the first two steps in our method for obtaining a visual air quality index and the difficulties in using a simulation model to predict human judgments.

ANALYZING THE HUMAN COMPONENT

The judgment of visual air quality contains elements of both perception (seeing) and cognition (thinking). Such judgments are best studied under natural conditions -- conditions that are as close as possible in all important respects to those we wish to generalize to. In research on urban visual air quality, the focus of research is the impact of air quality on the visual environment as experienced by people who live and/or work in urban areas. As the conditions under which people experience visual air quality vary, so must the conditions represented in the research. A single vista, time of day, or season cannot adequately represent the range of "urban visual air quality".

Since its inception in 1979, NCAR's urban visual air quality study (see Mumpower et al, 1981, for complete description of field program) has been carried out with attention to the principle of "representative design" (Brunswik, 1952; 1956). That is, the research has been designed to be representative of the environment of interest. Operationally, this has meant field observations rather than ratings of photographs, many observation sites rather than one, several views at each site, numerous untrained observers, and observations taken at several times of day, both summer and winter. Representative design ensures that the results are not produced by a specific set of conditions that have been created for purposes of research.

Visual air quality has been measured using a standard 7-point rating scale (see Figure 2). Observations are not instructed to make their judgments in a particular way, but rather are encouraged to use the rating scale to express their subjective impression of the visual air quality at a given time. Thus, individual differences in judgments of visual air quality are allowed to emerge.

Figure 2

Judgment Scales Used to Rate Visual Air Quality

1. What is your overall impression of the visual air quality in the metropolitan area at this hour?

1	2	3	4	5	6	7
Extremely Poor						Extremely Good

In a specific direction:

D1. Rate the visual air quality in the present direction:

1	2	3	4	5	6	7
Extremely Poor						Extremely Good

A person's judgment of visual air quality may be based on a number of different kinds of information, including

a. the effect of the air mass on the scene (e.g. clarity of objects, air color, presence of a border).
b. characteristics of a specific view (e.g. elevation of site, presence of distant targets, features in the foreground, mountains),
c. context of the observation (e.g. cloud cover, time of day, sun angle) and
d. background information (e.g. other scenes viewed earlier in the day, news reports, chance conversations).

The observer integrates these different types of information into a single judgment of visual air quality. When some of the information conflicts, the judgment made depends on the weight or relative importance that the observer places on each piece of information. The judgment also depends on the form of the functional relation between the information and the judgment, e.g. linear or nonlinear. Finally, the judgment depends on the organizing principle by which many items of information are combined into a single judgment, (e.g. additive or multiplicative). The concepts of weight, unction form, and organizing principle have been found useful for escribing the judgment process and for characterizing differences mong people in judgment research. For a full discussion, see ammond, Stewart, Brehmer and Steinmann, 1975.

There is also a component of error, or inconsistency, in udgment. Under identical conditions, the same person will often ot make identical judgments. Numerous studies of judgment have ound that this error component generally constitutes 20 to 40 per-ent of the variability in judgments (Slovic and Lichtenstein, 1973). he amount of error depends on the person and the characteristics f the judgment problem. In spite of this inconsistency, judgment nalysts have been able to use statistical techniques successfully o uncover the consistent basis for inconsistent judgments.

To summarize, the variation in judgments of visual air quality an be due to a number of influences including

a. differences among views,
b. differences in the background information,
c. differences in the context of the observation,
d. differences in the air mass, and its effect on the scene,
e. differences among individuals with regard to the process for integrating information into a judgment and,
f. inconsistency.

The study is primarily concerned with the effect of d) above, s mediated by c), on judgments of visual air quality. Considerable ttention is also given to individual differences among observers e). Inconsistency (f) cannot be directly studied because it is mpossible to make two field observations under identical conditions. irect consideration of (a) and (b) would require introduction of a arge number of variables, quickly exhausting the information in he data.

bservational Procedure

Paid observers were sent to specified sites at prearranged imes during the study period. At the site, they were instructed o observe views in several different directions and to give their mpression of the visual air quality in each direction. At the ame time, they rated the clarity with which designated targets n each direction could be seen. Generally, two target objects ere associated with each view. One was designated the "distant arget" and the other was the "intermediate target". These targets, hich were usually buildings or mountains, were between .8 and 117 m from the observation sites.

Other judgments made by the observers for each of the direc-ional views included the color of the air and sky (as compared to

Munsel color charts) the sharpness of a border between discolored air and clean air and the imputed source of visibility degredation-natural or human. The clarity of target object, color, and border judgments relate to aspects of the views which have been found to be important cues to visual air quality (Mumpower, et al, 1981).

MODELLING THE PHYSICAL ENVIRONMENT

The three-dimensional visual air quality simulation model includes dispersion of fine particles in the Denver airshed and interaction of the particles with solar radiation. Three dimension concentration fields are produced using a dispersion model. Perceptual cues -- visual range, color. contrast and border -- are calculated for observation points in the field using a radiative transfer model. The impact of NO_2 is not included in the model since it has been shown that the influence of NO_2 on the color of Denver's air is small compared to the influence of particles (Waggoner and Weiss, 1980).

The atmospheric dispersion model used to advect and diffuse the fine particles is a multibox, numerical model. Finite differen techniques are used to solve the turbulent diffusion equation for a Eulerian (fixed) grid system. The eddy diffusivity in the vertical direction is a function of height. For stable and neutral conditio the algorithms suggested by Shin and Shieh (1974) are used. For unstable conditions, the polynomial developed by Lamb (1975; Reynolds et al, 1979) is used. The wind is assumed to be homogeneous across the urban area and the wind shear is explicitly accounted for. The model incorporates a surface layer for superadiabatic convection and follows the base of the inversion height through the day. The top of the model is always above the base of the inversion. The model of Tennekes (1973) is used to determine the time of surface inversion fill-in base on the a.m. radiosonde readings and the rate of rise of the inversion in the morning. The inversion height in the p.m. determines the afternoon inversion height.

Fine particle emissions are based on the Colorado State emissi inventory for Denver. In the model results presented here aerosol formation and growth is not included. The ability of this type of air quality model to predict observed aerosol concentrations is expected to be significantly reduced due to the neglect of aerosol growth.

The interaction of aerosols with solar radiation is described with a model based on inhomogeneous radiative transfer theory (Chandrasekhar, 1960). To calculate perceptual cues which are to be compared to human observer ratings, first the observation time and observer location and direction of view in the three dimensional field are prescribed. Cues are then calculated for that particular observation.

The definitions of the calculated cues are chosen to match as closely as possible the descriptions of cues used by the human observers in their ratings. For the winter study used in the comparisons described here, observers rated clarity of targets at different distances and color and border perceived at the horizon. The calculations for clarity are based on contrast which is defined as the difference between light intensity of the sky and the assigned target in the concentration field divided by the intensity of the sky. Calculated color is defined as the ratio of horizon air intensity weighted by the eye's sensitivity to red color to the intensity weighted by the eye's sensitivity to blue color. Calculated border is defined as the ratio of contrast of air and sky background at the horizon along the horizonal--0° viewing angle--to the contrast at a 45° viewing angle.

The aerosol concentrations used are taken directly from the dispersion model calculations every hour. The calculations of phase functions and light extension parameters used in the intensity calculations are carried out using Mie theory (van de Hulst, 1957). Values for direct sunlight are taken from reported measurements (Bolle, 1977). Eye sensitivity functions are taken from Land and McCann (1971). The angle between the sun and the observer's direction of sight are calculated given the time of year, month and day and observation direction.

COMPARISON OF PERCEIVED AND MODELLED CUES

Judgments of clarity of targets, color and border are compared with modelled values for a day judged to have intermediate visual air quality for the winter 1981 study in Denver, Colorado (January 9). The comparison illustrates the ability of a physical model to predict variations in observer perceptions and judgments. It should be stressed that these are preliminary calculations. Analysis for only one day is presented. The comparison between calculated and perceived cues will vary from day to day. Thus, the results presented should be viewed as an illustration of the analysis, not the final conclusion regarding human judgment prediction by physical models.

Table 1 shows the correlations obtained between each of the modelled cues and the corresponding judgments by observers. Judgments that were replicated by more than one observer have been averaged over observers. The resulting 31 distinct observations taken during the day were correlated with the model calculations for the corresponding site, time, and direction. The table shows the values of r (Pearson product-moment correlation), n (sample size) and the significance level for each correlation. The sample size for the distant target is only 29 because two of the views did not contain distant targets. The significance levels are not strictly applicable in this situation because the observers do not represent a random sample from a population (although, in a limited sense, they might be thought of as a random sample of all the possible observa- tions on January 9). Significance levels are reported here only as an aid in interpretting the correlation coefficients. They indicat that, for the sample sizes given, the correlations for distant targ and intermediate target exceed the level that would be required to reject the possibility that they are due to chance, and the correla tions for color and border do not exceed that level.

Table 1

Correlations between judgments averaged over observers and modelled cues.

	Correlation	N	Sig. level
Distant target	.42	29	.01
Intermediate Target	.37	31	.02
Color	.11	31	NS
Border	.23	31	NS

In a classic paper in psychological measurement theory, Campb and Fiske (1959) have shown that a good measure should not only co relate with what it is supposed to measure, but also should not co relate with what it is not supposed to measure. Table 2 shows that for the case of the calculations for the targets, this criterion is met. The appropriate target calculation correlates with the appro- priate target judgment, even though the two target judgments them- selves are correlated .64. In the case of color and border, the criterion is not met. In fact, the highest correlation is between the calculation for color and judgment of border. This indicates a need for further work on the definitions of border and color and the methods of calculating these measures.

Table 2

Intercorrelations among modelled cues and judgments.

	Judgments	
Modelled Cues	Distant Target	Intermediate Target
Distant Target	.42	.14
Intermediate Target	.29	.37

	Judgments	
Modelles Cues	Color	Border
Color	.11	.37
Border	- .11	.23

Table 3 provides a comparison of the relations of three different sets of variables and visual air quality. The first row shows the multiple correlation between visual air quality judgments and the four calculated measures. The squared multiple correlation (R^2) of .19 indicates that, in this sample of observations, 19 percent of the variation in judgments of visual air quality is associated with variation in the calculated measures. The second row provides a similar calculation for a set of measured characteristics of the atmosphere the time the observations were made. A number of environmental measures were taken and then reduced to 8 factors by factor analysis.

Table 3

Multiple correlations between visual air quality judgments and modelled cues, factor scores and judged cues.

	R	R^2	N
4 modelled cues *	.44	.19	29
8 factor scores **	.47	.23	31
4 judged cues *	.90	.80	29

* Cues are listed in Table 1

** See Table 4

Table 4 lists the physical measures and shows how they were grouped into factors. The multiple correlation between this extensive set of physical measurements taken simultaneously with the observations is only slightly higher than the correlation with the model observations.

Table 4

Physical measurement factors*

Factor 1	Factor 2	Factor 3	Factor 4
CO(A)	Temperature (E)	CO(C)	Windspeed (B)
CO(D)	Temperature (A)	NO(C)	Windspeed (D)
NO(E)	Relative Humidity (A)	THC(C)	Windspeed (E)
		CH_4(C)	Windspeed (F)
SO_2(C)	Visibility Range (A)	COH(C)	Windspeed (A)

(continued)

Table 4 (continued)

Factor 5	Factor 6	Factor 7	Factor 8
0 (B)	Skytype(A)	NO (C)	CO(F)
0 (C)	Skycover (A)	NO (ED)	
0 (D)	Opacity(A)	TSPB(C)	
0 (E)		SO (E)	
0 (F)			

* Factor groupings based on patterns of correlations among measures over daylight hours between 12/29/80 and 1/25/81.

Measuring Stations

A - Stapleton Airport
B - Arvada
C - Camp
D - Carih
E - Welby
F - Highland

The final row in table 3 shows the relation between judgments of distant and intermediate target, color, and border, and visual air quality. This correlation is somewhat inflated because the judgments were not made independently of one another, but the high R^2 does indicate that adequate measures of the four dimensions should be capable of accounting for a substantial proportion of the variation in visual air quality.

DISCUSSION

Two important questions are brought out by these comparisons: (1) What is the greatest proportion of the variations in judgments of VAQ that can be accounted for by measured or calculated differences in the visual environment or, stated another way, are the predictions presented here as good as can be expected? (2) How can the physical model and/or design of the observational study be improved to increase the physical model's ability to predict human judgments.

The proportion of variations in judgment that can be accounted for by calculated differences in the visual environment must be less than 1.0 because some non-trivial sources of variation in the judgments are left uncontrolled either intentionally (differences among views and background information) or unavoidably (error). The exact upper limit on the proportion of variability in visual air quality that we could expect to be accounted for by environmental measurements or theoretical calculations is indeterminant. However, since the upper bound is of some interest in interpretting our results, we will make a rough estimate.

Inconsistency

The consistency of VAQ judgments will be moderate to low because a) the judgment is complex b) observers were not trained to make judgments according to a prescribed system and c) judgments are distributed over time rather than being concentrated in one session. For a moderately consistent judgment, 25% of the variance is due to inconsistency. This corresponds to a test-retest reliability coefficient of .87. This leaves a maximum of 75% of the variance to be potentially accounted for by a model based on characteristics of the environment.

Views and Background

Since the views included in the study were diverse and the circumstances preceding each observation could also vary substantially, these factors could easily account for 10% or more of the variation in visual air quality. Assuming that this variation is independent of the variation due to inconsistency, we are left with .75 x .90 = .675 of the variance in visual air quality that could be accounted for by a model.

Individual Differences

Correlations between different individuals rating the same views have averaged .60 to .70. If we adopt the higher figure and correct for inconsistency, we can estimate that 65% of the consistent variation in judgements of visual air quality is shared among individuals (35% is due to individual differences). Since a model based index is an index of visual air quality averaged over individuals, it is not expected to account for variation due to individual differences. Assuming independence, we could therefore expect a visual air quality index to account for at most .675 x .65 = .44 of the variation in visual air quality.

Of course, this is a very crude estimate meant only to illustrate that, due to multiple influences on visual air quality judgments, we should not expect a visual air quality index to achieve perfect predictability. In fact, a model that accounted for 60-70% of the variation in individual judgments or group average judgments or a model that accounted for 40-50% of the variation in judgments aggregated across observers would be doing very well.

The results presented in Tables 1, 2, and 3 therefore illustrate the feasibility of predicting judgments from model calculations, but it may well be possible to obtain higher levels of predictability through further work. As our work progresses, we hope to make a better determination of the limits of predictability for judgments of visual air quality.

To achieve increased predictability several improvements in th visual air quality model and the observer field design are recommended.

Model Modification

The model modifications are based on the fact that the major uncertainties in the model calculations are associated with the definition and calculation of the cues. The horizon and distant target may not always be considered comparable by the observers whereas for simplicity the model assumes they are. The assumption that calculated border is adequately defined with respect to the same two model layers in every observation may also be an over simplification. The use of a red to blue light intensity ratio focuses more on the amount of differences between two colors rather than the absolute color. This fact is supported by the higher correlation of calculated color with perceived border than perceive color. For both color and border calculations it is assumed, again for simplicity, that observers are basing their judgments only on the air directly in front of them. It is quite possible that observers are subtending a much larger angle when they look at the horizon as instructed and make judgments about color and border.

In future calculations several changes are suggested which would possibly make the definitions and calculations more comparable to the observations. (1) The horizon will be defined as the point at which a black object can just barely be distinguished against the background air. In directions where there are terrain obstacles before the horizon -- for Denver terrain barriers are mountains or ridges that occur to the west, southwest and northwest -- horizon will be defined as the distance to the terrain barrier. In the calculations presented here the distant targets satisfy the definition of horizon 50% of the time. (2) the definition of border will be changed to the largest gradient in color perceived in the 0° to 45° viewing angle range. (3) The definition of color will be changed to the amount of redness in the air rather than the red to blue color ratio. (4) Calculations of color and border will be done assuming that observers subtend a 30°-45° rather than a 0°-10° angle when making their judgments.

Field Study Modifications

In order to adequately understand a judgment such as visual ai quality, it is important to have a) multiple observers, selected to be representative of the general population, judging the same views under the same conditions and b) each individual observer judging identical views on at least two different occasions. The first condition is desirable in order to assess the extent of individual differences in judgments of visual air quality, and to develop an index that represents a meaningful "average" of many people's impression of visual air quality. The second condition is necessary in order to assess the reliability of visual air quality judgments.

Unfortunately, in field studies such as ours, both conditions are difficult to achieve. The problem with having multiple observers at the same site is primarily logistic. Each observation requires about one hour, so it is unlikely that volunteers could be found to make more than a few observations. Paying a group of 20 to 30 observers to make a large number of observations is beyond the means of most research projects. Even if it were possible, significant difficulties would be caused by having a large group of observers at a site trying to make observations simultaneously. Having observers repeat observations under identical conditions is impossible because atmospheric conditions vary so much that two sets of identical conditions rarely, if ever, occur.

Our response to the first problem has been to have observations replicated by two and sometimes three observers and to have the observers interview passersby at the scene. It would clearly be desirable to have more observers at the scene and more passersby interviewed. It would be desirable to conduct a future study focussed specifically on individual differences in which a group of observers would make a series of observations together. If possible the group of observers should be representative of the general population.

The impossibility of directly measuring the reliability of judgments is inherent in field studies. A partial solution is to gather information about reliability from other sources. This is possible because, in the long run, the reliability of a judgment is an upper bound on its correlations with other variables that are measured independently. Therefore, the correlation between two observers' judgments cannot exceed the product of their reliabilities (of course, sample values of correlations can exceed reliabilities by chance if the number of observations is few). Thus correlations between judments of different people and between judgments and physical measures or calculations, can be used to establish a lower bound on reliability. Another source of information about reliability is the reliability of judgments of visual air quality made from photographs. The reliability of slide judgments should be at least as great as for field judgments because judgments of photographs are made under more controlled conditions. The reliability of judgments of photographs might, therefore, indicate an upper bound for the reliability of field observations. We are currently studying the relation between photographic and field observations and we would recommend that judgments of photographs be obtained along with field observations in any study of judgments of visual air quality.

SUMMARY

The comparisons presented strongly indicate that prediction of human judgments of visual air quality with an air quality simulation model is possible. Although the correlations for the case study day are not as high as could be expected, they are still considered good for predictions of human judgments given the research design constraints imposed by the principle of representative desig Recommended modifications in the physical model are expected to improve the model's predictability. Improved analysis of the human component using photographs is also expected to enhance our understanding of visual air quality judgments. Research is presently underway in our program on both improved modelling and analysis of visual air quality judgments.

REFERENCES

Bolle, H.J. (1977) : Proceedings of the Symposium on Radiation in the Atmosphere, Science Press, xiv.

Brunswik, E. (1952) : Conceptual Framework of Psychology. University of Chicago Press, Chicago, IL.

Brunswik, E. (1956) : Perception and Representative Design of Experiments. University of California Press, Berkeley, CA.

Campbell, D.T. and D.W. Fiske (1959) : "Convergent and discriminant validation by the multitrait-multimethod matrix". Psychological Bulletin, 56, pp. 81 - 105.

Chandrasekhar, S. (1960) : Radiative Transfer, Dover, New York.

Hammond, K.R., T.R. Stewart, B. Brehmer, and D.O. Steinmann (1975). "Social judgment theory". In Human Judgment and decision processes
M.F. Kaplan and S. Schwartz (Eds.). New York : Academic Press Inc.

Lamb, R.G., W.H. Chen,and J.H. Seinfeld (1975) : "Numerico-Empirical Analysis of Atmospheric Diffusion Theories". J. of the Atmos. Sci., 32, 1794 - 1807.

Land, E.H., and J.J. McCann (1971) : "Lightness and Retinex Theory" Optical Society of America, 61, 1 - 10.

Mumpower, J., P. Middleton, R.L. Dennis, T.R. Stewart and V. Viers (1981) : "Visual air quality assessment : Denver Case Study". Atmos. Sci., in press.

Reynolds, J.D. Reynolds (1979) : Photochemical Modelling of transportation Control Strategies - Vol. 1. Model Development, Performance Evaluation and Strategy Assessment. Report EF79-Federal Highway Administration, prepared by SAI.

Shin, C.C. and L.J. Shieh (1974) : "A Generalized Urban Air Pollution Model and its Application to the Study of SO_2 Distributions in the St. Louis Metropolitan Area". J. Appl. Meteor., 13, 185 - 204.

Slovic, P. and S. Lichtenstein (1973) : "Comparison of Bayesian and regression approaches to the study of information processing in judgment". In Human Judgment and Social Interaction, L. Rappoport and D. Summers (Eds.). New York : Holt, Rinehart and Winston.

Tennekes, H. (1973) : "A model for the Dynamics of the Inversion Above a Convective Boundary Layer." J. of Atmos. Sci., 30, 558 - 567.

van de Hulst, H.C. (1957) : Light Scattering by Small Particles, Wiley, New York.

Waggoner, A.P. and R.E. Weiss (1980) "The Color of Denver Haze", in 73rd Annual Meeting of the Air Pollution Control Association, Paper 80-58.5

DISCUSSION

D. SKIBIN

In your definitions of clarity and border :

1) If the intensity of the object is zero the clarity is 1. Isn't this odd?
2) In the definition of border there is the intensity of object at an elevation angle of 45°. Do you have objects for observation at such high elevations?

P. MIDDLETON

1) Our model definition of clarity is identical to the standard definition of contrast ground in discussions of radiative transfer. On the limit of very small object intensity relative to sky intensity, clarity (contrast) does approach one. This is not particularly odd. It is, however, not a "typical"limit in the type of study reported here : a detection of targets in a polluted atmosphere.

2) In our field studies, there are no targets at a viewing angle of 45°. In the model calculations hypothetical black objects, are assumed to be located at the model defined horizon at the 0° and 45° viewing angles. A more realistic model definition of border is being developed. Details are discussed in the paper.

2: AIR TRAJECTORY MODELS FOR AIR POLLUTION TRANSPORT

Chairmen: W. B. Johnson
L. Niemeyer
H. E. Turner

Rapporteurs: N. D. Van Egmond
C. De Wispelaere
A. Venkatram

A GAUSSIAN TRAJECTORY MODEL FOR HAZARD EVALUATION FOR PROLONGED RELEASES FROM NUCLEAR REACTORS

E. Doron and E. Asculai

Nuclear Research Centre - Negev
P.O.Box 9001, Beer-Sheva, Israel

ABSTRACT

A model for mesoscale hazard evaluation, along a trajectory, with time/distance varying meteorological conditions, has been developed. For release periods longer than one hour, multiple superimposed trajectories and meteorological conditions are utilized.

The centreline concentration for each hour's release is assumed to move along a trajectory, calculated from actual wind measurements or climatological average data. The model utilizes the Gaussian distribution formulae for the calculation of the concentration at preselected grid points near the ground. The wind speed and stability category may vary continuously along the trajectory. At these grid points the instantaneous (one hour average) and the total integrated concentrations are calculated. Ground deposition is calculated assuming a constant settling velocity.

The final results are given as radiation fields and total integrated doses, computed by the methods suggested by WASH-1400 and the ICRP-26 report.

Three experiments are presented in the work and the main advantages of the present method for the chosen cases are discussed.

INTRODUCTION

The environmental hazard assessment of either routine or accidental releases of radioactive material from nuclear reactors has usually proceeded according to strict and preset procedures. With the advance of nuclear technology the use of these rules has become more and more unrealistic. While the use of most rules tend to make the results of the hazard evaluation more conservative, it can, in some cases, achieve the opposite.

As a case in point we may take the common assumption of a constant wind speed.

Since the source term has become so large with the advent of large reactors, the distance at which a release is still potentiall hazardous has become larger. Instead of a small exclusion radius outside of which relatively harmless doses occure, one must calculate hazards for distances of tens of miles or more.

Thus, the assumption of a constant wind speed may, for the greater distances, distort the whole picture, predicting a rapid passage of radioactive cloud over close-by areas but contaminating a far away area much sooner then actually possible - when utilizing an above average wind speed in the calculations.

The distortions which may result from keeping the wind direc- constant or the stability conditions unvarying are, of course much greater.

The motivation for the present work is thus the need to know, in advance, and from meteorological data, what the potential hazar from a power reactor really are, and to provide an important tool for real time hazard evaluation in case of an actual release.

In the present work we utilize wind speed and direction vary- ing in time and space (i.e. a different trajectory for different release times), varying stability conditions and, of course, vary- ing release rates.

THE GAUSSIAN TRAJECTORY MODEL

The total dose calculations are based on the summation of th contributions from a series of instantaneous puffs.

The trajectory of the centre of mass of each individual puff is calculated using a method described in Doron (1979). Following Drexler (1977) it is assumed that the concentration distribution around the centre of mass is Gaussian in the horizontal plane to- gether with a vertical mixing formulation.

For the present experiments it was assumed that the variation of the horizontal standard deviation σ_y with distance from the source is given by the well known Pasquill-Gifford curves (see Turner, 1970) for the case in which stability does not change in time. For long travel times this assumption is usually not valid. In order to allow for a stability category change the following method is used:

From the Pasquill-Gifford curves we may deduce an approximation $\sigma_y = ax^\alpha$ where α is the same for all stability categories and a is stability dependent. Since it is obvious that for the dispersion process the important parameter is time of travel and not the distance travelled, and utilizing the fact that in the Pasquill-Gifford scheme each stability category may be assigned a definite wind speed we may write $\sigma_y = bt^\alpha$, where b is stability dependent.

On the other hand it is also obvious that the growth rate $\frac{d\sigma}{dt}$ is dependent both on the stability and σ at a certain time t_0. Since $\sigma(t_0)$ is a result of the whole past history of the puff it cannot be represented in the above form by simply adjusting the value of b. In order to overcome this difficulty we utilize a virtual time of origin at each time step in the following way:

First, assign a value for b according to the stability category. Secondly calculate a virtual time from start t_* from the relation $\sigma_y{}^t = bt_*^\alpha$ using the current value of σ and the assigned value of b. Finally calculate a new valve of σ_y for the next time step from:

$$\sigma_y{}^{t+\Delta t} = b(t_*+\Delta t)^\alpha$$

For the vertical diffusion representation a vertical grid was chosen and a K theory formulation applied. For the experiments described in the present work twenty levels, from the surface to 2000 metres, were employed, with a resolution varying from 25 m at the surface to 200 m at the uppermost level.

The values of K_z according to stability and their dependence on z are the same as in Drexler (1977). Since, in the present version, the stability depends only on time and not on location, there is a complete decoupling between the advective part and the diffusive part of the calculations.

Since the advective part is Lagrangian, the maximum time step will be determined by the diffusive terms. With a non-constant K_z the time step is thus determined by the maximum value of K_z during the integration at the lowest levels, where we have the highest resolution.

The dependent parameters that are calculated by the model are, thus, the position coordinates of the centre of the puff, the horizontal standard deviation σ_y and the normalized concentration (the concentration for a unit puff released) at the centre of the puff - for all levels.

The updating of the concentrations is performed in the following way: As a first step, new concentrations are calculated at all levels due to vertical diffusion only, from:

$$C_{z*}^{t+\Delta t} = C_z^t + \frac{\partial}{\partial z}\left(K_z \frac{\partial c}{\partial_z}\right) \cdot \Delta t,$$

Next, a new horizontal standard deviation $\sigma_y^{t+\Delta t}$ is calculated using the procedures outlined above, and, finally, all C values are adjusted using the equation

$$C_z^{t+\Delta t} = \left(\frac{\sigma_y^t}{\sigma_y^{t+\Delta t}}\right)^2 \cdot C_{z*}^{t+\Delta t} .$$

The upper and lower boundary conditions for the solution of t vertical diffusion equation include a no-flux condition at the uppe boundary: $K_z = \partial c/\partial z = 0$ at $z = H$, and a deposition velocity formul tion (Vander Hoven, 1968) at the lower boundary.

DOSE CALCULATIONS

The Source Term

The source term is either internally calculated for one of the WASH-1400 (1975) release categories, or externally supplied by a computer code for an arbitrary choice of:

1. Irradiation time.
2. Cooling time.
3. Release rates and fractionation from reactor core to containment building.
4. Release rates and fractionation from the containment to the atmosphere.

The nuclear data is externally supplied and the program is virtuall unlimited in the number of isotope chains and isotopes it can handl

Cloud Depletion and Concentrations Calculations

Cloud depletion by radioactive decay and fallout is calculate in the following way:

1. A time step Δt is set.
2. After each puff is released, the radioactive isotopes' inventory in it is decayed for the time step Δt.
3. The fallout term is calculated by multiplying the activit of each isotope j (A_j) by a constant deposition velocity

which is specific for the isotope (e.g. $Vd_j = 0$ for noble gases).

4. The fallout term is subtracted from the cloud puff inventory.

   Air and ground concentrations are calculated in the following way:

5. From the trajectory calculations, the puff's centre of gravity is moved from (x_i,y_i) to (x_{i+1},y_{i+1}) in the time step Δt.
6. A dilution factor $F_{i+1} = C_{i+1}/C_i$ is calculated for the puff's centre of gravity.
7. The air and ground inventories are multiplied by F_{i+1}.
8. At the completion of the calculation of the updated inventory at the centre of the puff, the integrated activity at the points of a regular grid is updated in the following way: The activity at the grid points $A_{i,k}$ is calculated using the Gaussian relationship, with the distances from (x_{i+1},y_{i+1}) to grid points i,k being utilized.

   In order to save computer time the updating is performed for the grid points within a distance of $n.\sigma_y$ from the centre of the puff $(2 \leq n \leq 4,\ n=n(\sigma))$.

   For the concentration calculation from prolonged releases the following scheme is utilized:

9. A time interval $\Delta\tau$ between puff releases is set.
10. Steps I through 8 are repeated for each individual puff.
11. At each grid point where the ground inventory is not zero, the inventory is decayed for the time interval $\Delta\tau$.
12. The integrated air and ground concentrations are calculated by

$$I^{t+\Delta t}_{i,k} = I^{t}_{i,k} + \Delta t \cdot A_{i,k}$$

Dose Calculations

Four exposure types were calculated.

- The dose to the thyroid from inhalation of contaminated air during the cloud passage.
- The whole-body equivalent dose from inhaling contaminated air.
- Whole-body external irradiation from immersion during the cloud passage.
- Whole-body external irradiation from contaminated ground.

In all four types of exposure a person was assumed to be at the grid point during the whole incident, and for the relevant exposure following, unsheltered.

The specific doses (rem/μC_i inhaled) were taken from appendix VI of WASH-1400, and equivalent whole-body doses were calculated according to the ICRP-26 scheme.

For the external gamma doses, all energies with an abundance of more than 1% were considered and new (Rad/hr)/C_i/m^3) and (Rad/hr)/(C_i/m^2) were calculated. For the dose calculations the time integrals of those dose rates were calculated and stored for a series of preset times (1,2,3,6 etc. hours, following the first release). An option provides the dose rates at preset times.

Another option, incorporated in the program, is the introduction of protective measures such as sheltering, filtration etc. and the computation of their effectiveness in the reeducation of the exposure to the population, as an aid to decision making.

EXPERIMENTS WITH THE MODEL

In order to illustrate the capabilities of the model three experiments are presented. In the first, the model was run with stationary and constant conditions over the whole domain. This was done in order to demonstrate the capability to reproduce the conventional curves and also to serve as a basis for comparison with experiments in which these restrictions were relaxed. For the remaining two experiments, the conditions prevailing at the Mediterranean coastal plane in Israel during the summer were utilized. These meteorological conditions are dominated by the land - and see-breeze circulation, which is a daily phenomenon, in the absence of synoptic-scale disturbences. One of the main features of the sea and land breeze circulation is the rapid change of the wind direction with the time of the day. It is thus a particularly useful situation in which to demonstrate the superiority of a trajectory model over the usual constant wind speed and direction assumptions.

For the purpose of the present experiments the average wind vectors for the month of July for 35 meteorological stations in central Israel were analyzed, and trajectories calculated according to the scheme proposed by Doron (1979). 20 such trajectories for the hours from $0^{\underline{00}}$ to $19^{\underline{00}}$ L.S.T. are presented in Fig. 1.2. The point of origin is the proposed site for a nuclear power station at Zikim, on the Mediterranean coast. As can be seen, the constant wind speed and direction assumption is only valid between 1000 hrs. and 1700 hrs. L.S.T., but may lead to very erroneous results at any other time of the day.

In order to illustrate the effect of the curved trajectory, the model has next been run with the 0000 hrs. trajectory, assuming that these conditions prevail during the whole pollution incident.

As a final example for the present work the model was utilize for a prolonged release, between 1430 hrs. and 1730 hrs. Here the effects of both changing trajectories and stagnation areas in individual trajectories should be apparent.

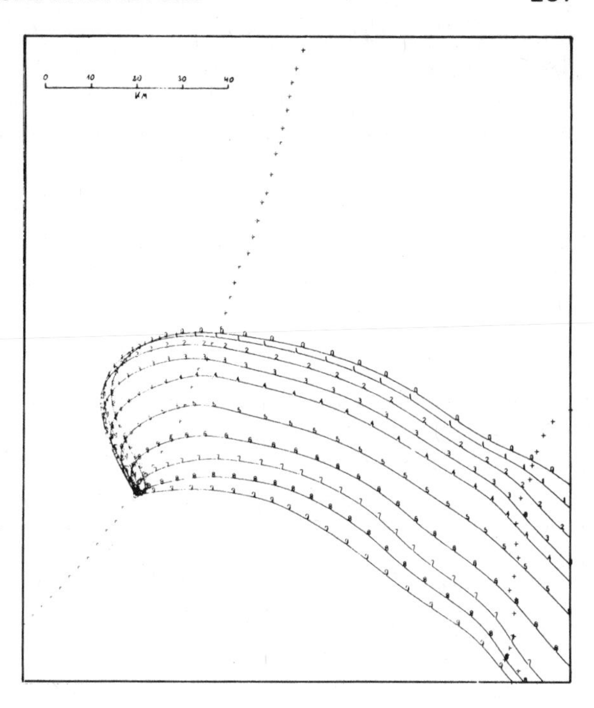

'ig. 1. : Trajectoriesfor uly from 00^{00} (0) to 9^{00} (9) (Interval be- ween numbers - 30 inutes)

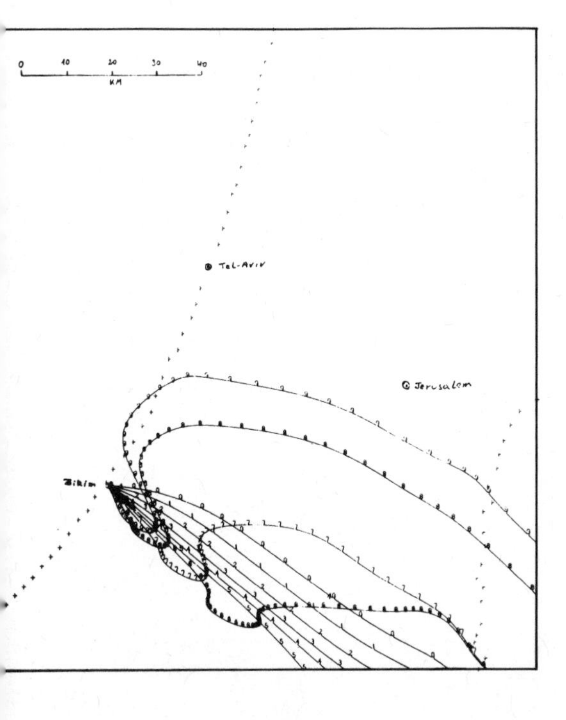

Fig. 2. : Trajectories for july from 10^{00} (0) to 19^{00} (9) (Interval between numbers - 30 minutes)

For the source term in these examples, a variation on the PWR-4 Accident Category (WASH-1400) has been utilized.

The basic assumptions for our source term are:

- Reactor type and power: PWR, operating at 3200 MW(t)
  Inventory: as in WASH-1400, App. VI, table VI, 3-1 (fission products only)
  Start of release: 2 hours after shutdown.
  Duration of release: 3 hours.
  Rate of release: exponential, involving 95% of the core inventory

Fractionation: all of the noble gases, 10% of the volatile fission products are assumed to be available for release to the atmosphere.

Height of release: as this accident category postulates a low rate of release of sensible energy, ground level emission was assumed.

RESULTS AND DISCUSSION

Experiment I: As mentioned in the previous section this experiment was run with stationary and homogeneous meteorological conditions, i.e. stability category D and a constant wind speed of 5 m/sec. The vertical diffusion coefficient K_z was kept constant at a maximum of 20 m^2/sec with the vertical variation as mentioned above. Sample integrated dose maps (in Rad) at various stages of the exposure are presented in Figs. 3,4. Each line represents a change by an order of magnitude. In order to avoid the cluttering of maps with lines of very low exposure a maximum was artificially set at $9x10^{-6}$ Rad so that the outermost line represents a value of 10^{-5} Rad.

A comparison of the resulting doses at the centerline of the cloud with the values obtained from the Pasquill-Gifford curves (Turner 1970) indicates a quite good agreement - the values agree to within a factor of two in all cases. Nevertheless, in the present case, the decrease of the doses with distance is somewhat more rapid. This is attributed to two main differences between the two models: First, in the present case, depletion of the cloud through deposition on the ground and radioactive decay is included and secondly,the vertical distribution in the present model is not Gaussian due to the change of K_z with elevation and the finite depth of the boundary layer. Taking all those differences into account it is felt that the results for this simple case are quite good and justify our confidence in the results for the more complicated cases for which the Pasquill-Gifford curves are not relevent at all

Experiment II: In the second experiment a 3 hour release beginning at midnight during the nocturnal land breeze was simulated.

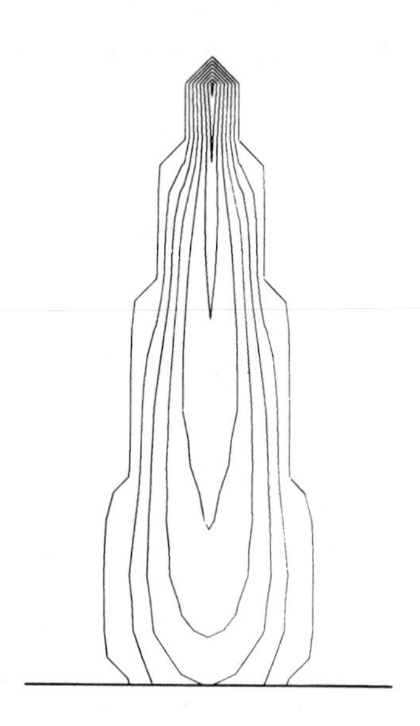

Fig. 3. : Log of ground deposition dose (rad) 6 hours from start to release

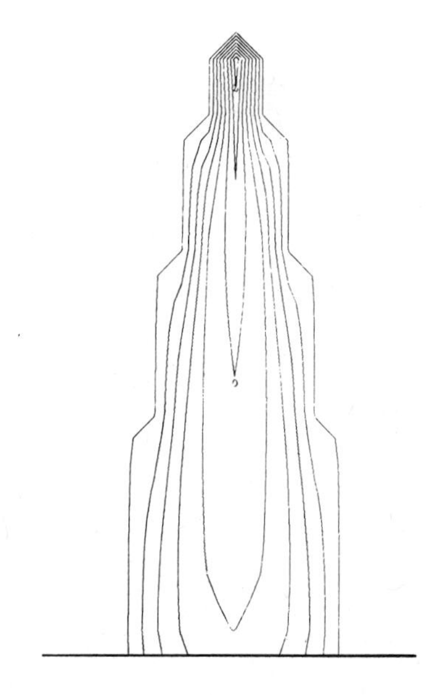

Fig. 4. : Log of ground deposition dose (rad) 18 hours from start of release

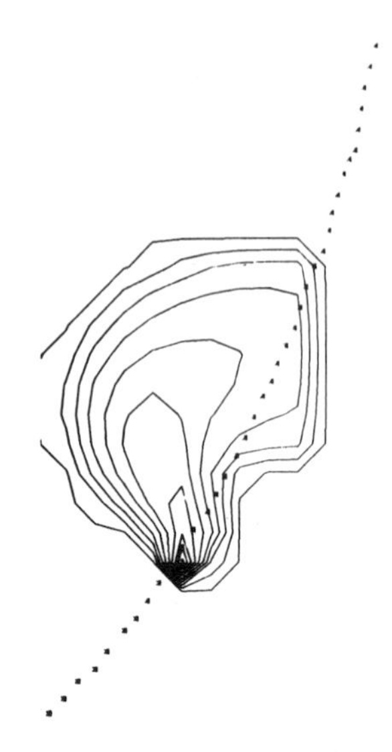

Fig. 5. : Log of ground deposition dose (rad) 12 hours from start to release

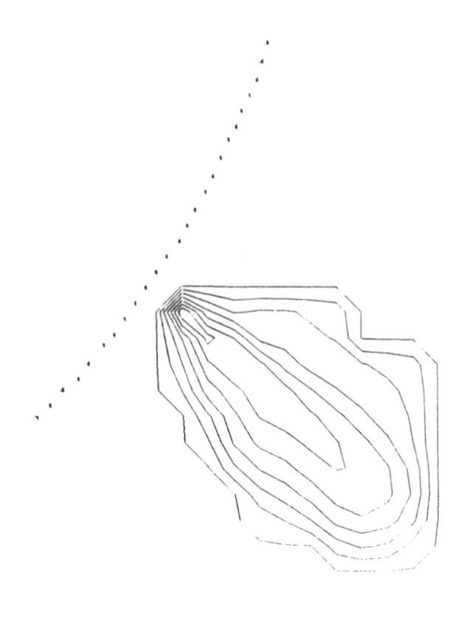

Fig. 7. : Log of cloud dose in rad 12 hours from start of release

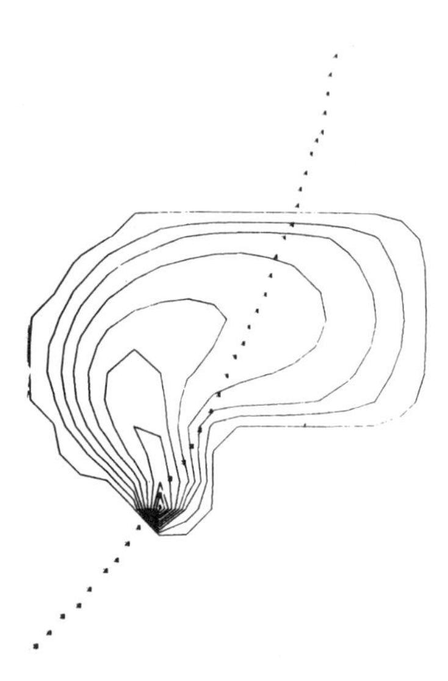

Fig. 6. : Log of ground deposition dose (rad) 18 hours from start of release

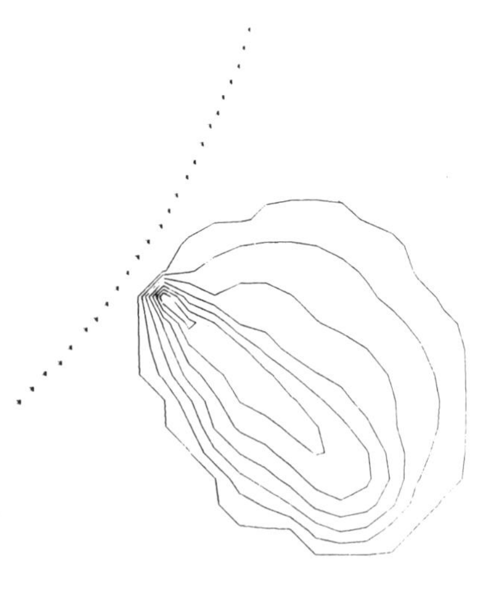

Fig. 8. : Log of cloud dose in rad 24 hours from start of release

The meteorological conditions were assumed to be the same for all releases. The exposure maps at various stages and to various organs are presented in Figs. 5,6. The main feature of the exposure maps is the fact that even though at the time of release the wind is towards the sea, and it looks as if inland areas will not be affected, in fact, the turning of the wind later on returns the cloud toward the land. In our specific case release from the aforementioned power station might endanger the cities of Ashdod and even Tel-Aviv. Another interesting feature is the relative narrowness of the cloud in the beginning, during the stable conditions, and the rapid widening during the daytime hours, again emphasizing the importance of changing stability conditions.

Experiment III: This experiment was performed for a prolonged release during a period of rapid changes in both wind conditions and stability category. Sample dose maps for the case are presented in Figs. 7,8. Examining the maps, two main features became apparent: The much larger width of the exposed area as compared to the previous experiment and the very marked aasymmetry around the centerline. The larger width is of course a result of the meandering trajectories - this is a well known effect and is already taken into account in the Turner workbook for example, where multiplication factors are assigned to the calculation of standard deviations according to the length of the sampling time. The advantage in the present case is of course the fact that no artifical average factors are assigned to all cases and each case is treated individually. The apparent asymmetry, i.e. the bulging of the cloud towards the north is a result of the fact that, with time, the trajectories meander northwards and develop stagnation areas resulting in much higher exposures in the northern part.

CONCLUSION

A model utilizing the Gaussian-Trajectory formulation is presented. Several sample calculations were performed. The results of these runs clearly demonstrate the advisability and advantages in the use of such models in hazard evaluations as compared with the commonly used, straight line, steady conditions Pasquill-Gifford method.

ACKNOWLEDGEMENT

Most of the computations were carried out at the Ben-Gurion University Computer Centre.

REFERENCES

Doron, E. 1979 : "Objective analysis of mesoscale flow fields in Israel and trajectory calculations", Isr. J. Earth Sci. 28 pp 33-41.

Draxler, R.R. 1977 : A Mesoscale Transport and Diffusion Model, NOAA Tech. Memo, ERL ARL-64.

ICRP, 1977 : Recommendations of the International Commission on Radiological Protection, ICRP publication 26, Pergamon Pres

Turner, D. B. 1970 : Workbook of Atmospheric Dispersion Estimates, U.S. Dept. of Health Education and Welfare, Environmental Health Service, PB 191482.

U.S.N.R.C. 1975 : Reactor Safety Study : An Assessment of Accident risks in U.S. Commercial Nuclear Power Plants, Appendix VI. Wash-1400, U.S. N.R.C. (NUREG-75/014).

Van der Hoven, I. 1968 : "Deposition of particles and gases", in Meteorology and Atomic Energy, D.H. slade ed., pp 202-208. U.S.A.E.C. Division of Technical Information, TID-24190.

DISCUSSION

P. HECQ

We fully agree with your approach of σ_y and K_z but how do you compute this K_z and why do you say that you have no problem with it ?

E. ASCULAI

The $K_z(H)$ are those suggested by DREXLER (1977), for the various stability categories, and are only time dependent(and not x,y coordinates dependent).
Since we are puff-integrating for continuous releases we can deal with each puff separately and utilize the appropriate K_z's.

A.J.H. GODDARD

You have used a sea-breeze situation as your main example, but this is strictly a 3-dimensioned problem. What limits does this impose upon your simulation ?

E. ASCULAI

This model is a two dimensional one, not taking into account wind-shear and other 3-d phenomena. However, we feel that this is not too restricting - especially in the more severe neutral/stable and stable cases.

A. VENKATRAM

For a hazardous release air quality model the validity of the concentration predictions is obviously very important. How does your model address this problem ?

C. ASCULAI The most inaccurate factor, we feel, in our model is the source-term, which is very controvertial. There are widely opposed views, e.g. the WASH-1400 and the EPRI approaches. Otherwise, I would be quite happy with the model which could be applied to and checked for conventional power plant emissions, even including some chemical reactions.

PPLICATION OF TRAJECTORY MODEL TO

EGIONAL CHARACTERIZATION

Boris Weisman

MEP Company
Downsview, Ontario
Canada

NTRODUCTION

The heightened awareness among scientists in Europe and North merica of the opotential dangers to the ecosystem posed by the nput of acidic substances to lakes, soil, and vegetation through he atmospheric medium has spurred the development and application f models to describe the long range transport of pollutants. ost long range transport models are intended for use in charact-rizing regional patterns over long averaging period, thus allowing ignificant simplifications in the level of detail required for he treatment of transport an dispersion.

Such implementations of regional modeling are not adequate to dress questions of relative impact of source regions with ufficient confidence to provide answers which will be needed by egulatory authorities for developing abatement strategies.

The objective in the development of the TRANS model is to rovide sufficiet flexibility to address more adequately the new pplications of long range transport models. To permit his flex-bility, the model is based on the Langrangian trajectory approach n which long-term averages are obtained by sequential simulation f short time periods.

The model can be used to evaluate single source contributions n the meso-scale and regional scale, as well as individual and otal contributions from multiple sources on a regional basis. It an utilize detailed time-varying emissions data for SO_2 and NO_x, as well as primary sulphate and nitrate and generates fields of concentration, dry deposition and wet deposition of the primary and secondary species over short and long averaging periods.

This paper outlines the structure of the model and examines the model sensitivity to parameter variation. The results of the application of TRANS to simulation of total eastern North American sources of SO_2 and NO_x for the year 1978 are then presented and discussed.

STRUCTURE OF TRANS MODEL

TRANS is a Lagrangian trajectory plume segment model intended for use in sequential simulation of individual source plumes. The structure of the model incorporates two levels of detail for the dispersion component in order to provide the modeling capability for single source meso-scale detail as well as practicability for regional-scale multi-source simulations. The sequential simulation approach allows the use of the model for short-term predictions as well as seasonal and annual averages.

Transport and Dispersion

The plume location at a given time is determined from the trajectories of air parcels released from the source point at regular intervals. The transport wind is determined from observed or analyzed pressure surfaces using the geostrophic approximation. These two-dimensional winds are optionally adjusted to respresent the transport layer winds using an empirically derived regression equation. To facilitate accurate integration of the trajectories, space and time orthogonal polynomials are used to provide the transport wind specific to the given point at the required time. (See Weisman, 1980 ans Sykes and Hatton, 1976).

The effect of small scale eddies in dispersing the pollutant with time is incorporated as a plume spread according to the normal distribution. For the longer travel times, the spread is assumed to be linear in time (Heffter and Ferber, 1975).

Vertical dispersion is assumed to proceed instantaneously within the mixed layer. In order to accomodate a diurnally varying mixing height, the maximum extent of the seasonal mixing height is made up of four layers with heights of 0.1, 0.25, 0.5 and 1.0 times the maximul height. As the mixing height cycles diurnally through the levels in sequence, the total material in the layers below the given level is assumed to be uniformly mixed in the vertical and contributing to the ground level concentration. Material in the layers above the mixing level undergoes transport with the same advecting winds, but is not mixed into the surface layer.

Dry Deposition

The flux of material to the ground surface is modeled by means of the deposition velocity concept. The removal of material from the plume segment is then treated as a source depletion process, only the mixed layer material undergoing depletion at the surface.

The deposition velocity is not modeled in the present version of TRANS, but is allowed to vary diurnally along the pattern of the mixing height to account for turbulence variation. Seasonally varying values of deposition velocity can be specified for each pollutant species.

Wet Deposition

The washout and rainout of sulphate particulate is empirically parameterized through a bulk washout scavenging coefficient which is proportional to precipitation rate. The precipitation rate for a given segment is determined from observations of intensity at the given time as a mean of several neighboring reporting stations. All layers are assumed to participate in the wet removal process.

The wet removal of SO_2 is modeled in terms of a washout ratio, using the approach suggested by Barrie (1981). The washout ratio is expressed in terms of the rate constants for the solution of SO_2 in the water droplets and the rainwater pH.

The washout of both gaseous and aerosol nitrogen oxides is modeled by a scavenging coefficient formulation.

Chemical Transformation

The chemical conversion of SO_2 to SO_4 is assumed to follow a first order process. The rate of conversion is seasonally variable, and is assumed to follow a diurnal variation similar to the mixing height to reflect the importance of photochemical reaction components in the oxidation process.

The nitrogen chemistry is similarly treated as a first order conversion of NO_2 to NO_3, the conversion of NO to NO_2 being assumed to be instantaneous for the purpose of long range transport modeling. It is generally recognized that the nitrogen chemistry is more critically dependent on the concentration of reactive radicals; however, the approach adopted seems to produce a useful first approximation to the complex processes involved.

MODEL SENSITIVITY

The model described in the previous section contains free parameters to account for conversions rate, dry deposition flux, and wet removal processes. These are variable in both space and time on the regional scale, necessitating the use of concentration and loadings observations to select a set of parameters which gives the optimum correspondence of predictions with observations. In order to carry out such an adjustment procedure, the behaviour of the model needs to be documented over the likely range of variation of the parameters.

Plume Budget

The plume sulphur budget and nitrogen budget are shown in Figure 1 for a point source plume emitted in the morning hours with no precipitation and the following choice of parameters :

i) initial sulphur al SO_2, nitrogen as NO_2
ii) transformation rates : sulphur .01 hr^{-1}, nitrogen .05 hr^{-1} (average)
iii) deposition velocities : SO_2 .75 cm/s, SO_4 . 25 cm/s, NO_2.50 cm/s, NO_3.50 cm/s (average)
iv) mixing height 700 m (average)

The decay time for SO_2 is 24 hrs, while that for sulphur is 48 hrs, with a maximum of 17 % of sulphur converted to sulphate within 60 hours. For NO_2, the decay time is 16 hrs, total nitrogen decay time being 72 hrs. A maximum of 46 % of nitrogen is converted to nitrate within 36 hours.

Table 1 shows the sensitivity of the plume budget to parameter variation. The sulphur decay time ranges from a low of 10 hrs for a 2 cm/s SO_2 deposition velocity to 150 hrs for a 0.1 cm/s SO_2 deposition velocity. Sulphur converted to sulphate ranges from 2 % for the 0.1% hr^{-1} conversion rate to 36 % of the sulphur for the 0.1 cm/s SO_2 deposition velocity. Pollutants emitted at the time of maximum mixing height show a 50 % increase in decay times and maximum amount of sulphur converted.

The nitrogen decay time ranges from 15 hours for a 2.0 cm/s NO_2 deposition velocity to 160 hrs for a 0.1 cm/s NO_3 deposition velocity. The converted fraction ranges from 10 % for the 0.5 % hr conversion rate to 60 % for the 0.1 cm/s NO_3 deposition velocity.

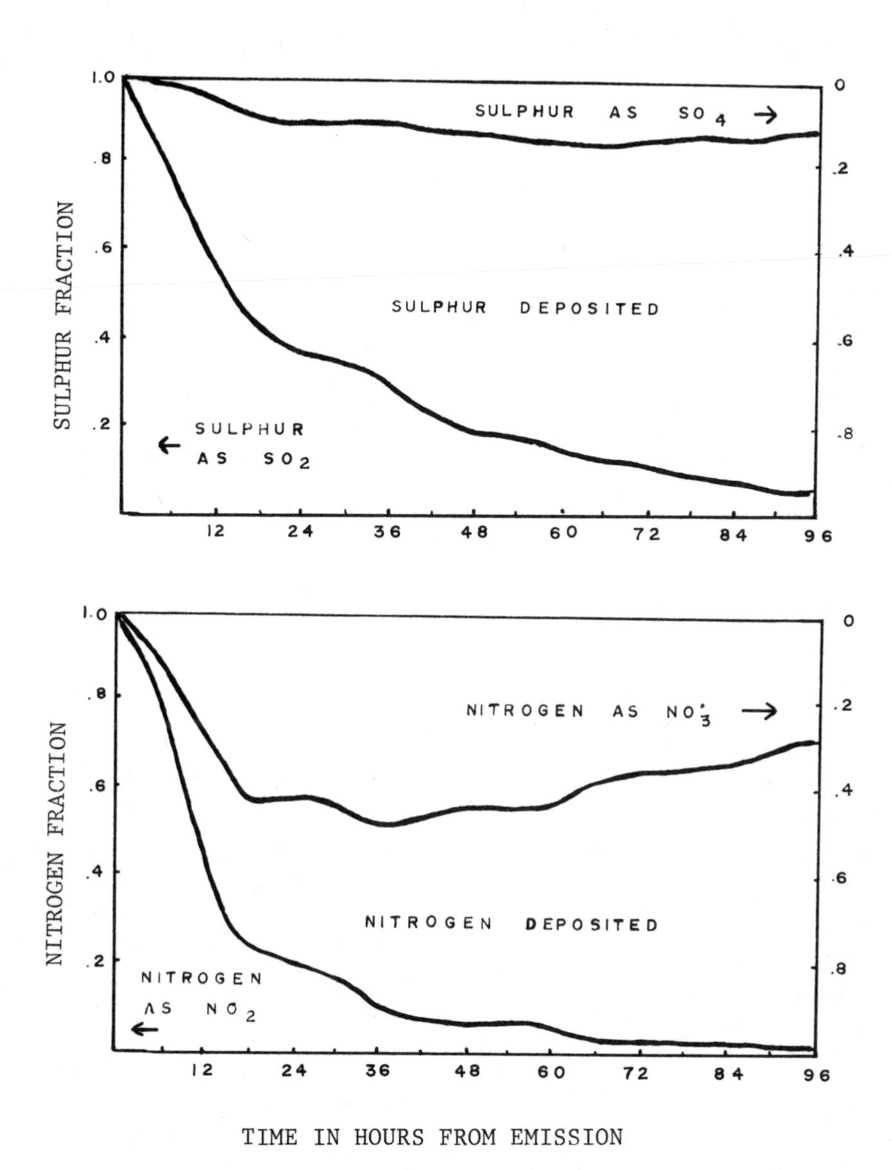

Figure 1 Plume Sulphur (top and Nitrogen (bottom) Budget for Standard Values of Parameters.

TABLE 1. Variation of Plume Budget for Range of Parameter Variation

	SULPHUR			
	Decay Time (hrs)		Max SO_4	
	SO_2	Sulphur	Fraction	Time
Standard	24	48	.17	60 hrs
Noon Time Emission	36	72	.22	75
SO_2 Dep. 2.0 cm/s	9	10	.06	48
0.1 cm/s	86	150	.36	90
SO_2 Transf. Rate .025 hr^{-1}	16	66	.33	48
.001 hr^{-1}	40	42	.02	36
SO_4 Dep. 2.0 cm/s	24	40	.07	36
0.1 cm/s	24	86	.20	90
Mixing Height 1200	40	120	.30	90
500	15	30	.12	40

	NITROGEN			
	Decay Time (hrs)		Max NO_3	
	NO_2	Nitrogen	Fraction	Time
Standard	16	72	.46	36 hrs
Noon Time Emission	21	96	.56	30
NO_2 Dep. 2.0 cm/s	10	15	.20	18
0.1 cm/s	18	120	.62	40
NO_2 Transf. Rate .10 hr^{-1}	10	70	.58	18
.005hr^{-1}	48	80	.10	66
NO_3 Dep. 2.0 cm/s	16	38	.28	16
0.1 cm/s	16	160	.60	40

Concentrations and Depositions

The sensitivity of concentration and deposition predictions to parameter variation has been determined by performing calculations for a full year (1978) for a single source at 75 receptor locations, and varying the parameters individually within the ranges indicated in Table 2. The results of the analysis are presented in Tables 3 and 4 for sulphur and nitrogen respectively. The first entry in the table lists the calculated concentrations and depositions for short range and long range from the source. The results for the short range are the average over all receptors within 100 km of the source, while the long range values are the averages for the range of 1000 to 1500 km.

The other entries in the table give the percentage change from the standard values for each choice of parameters as indicated in the table. All parameters other than the one being varied are kept at the standard values listed in Table 2.

It can be seen from the table that the annual concentrations and loadings in the short range generally fall within 100% of the standard values, although the range of parameter variation is generally over a factor of 20. The variability in the values at long range is considerably greater than for short range.

The complex dependence of concentrations and depositions on the several parameters is shown by reversals in sign of the change with distance, as in the case of sulphate wet deposition. Lowering the wet removal rate for SO_4 reduces the wet deposition in the short range, but increases it considerably in the long range as more sulphate is left in the plume for long-range transport.

MODEL RESULTS FOR EASTERN NORTH AMERICA

The TRANS model has been used to generate predictions of annual average concentrations and depositions resulting from emmissions of sulphur and nitrogen oxides in eastern North America for the year 1978.

Emissions Data

Emissions data for the model run consisted of gridded SO_2 and NO_x emissions on a 127 km grid as prepared by Environment Canada (Voldner and Shah, 1979). For purposes of this run, each 127 km square was treated as a virtual point source, all squares with emissions in excess of 200 tonnes/yr being modeled; a total of 366 sources comprising 31.8 Million Tonnes of SO_2 (99.7 % of the inventory) and 18.1 Million Tonnes of NO_2 (98.9 % of the inventory).

TABLE 2. Range of Parameter Variation for Sensitivity Analysis

	Sulphur						Nitrogen					
	Summer			Winter			Summer			Winter		
Parameter	Low	Std.	High	Low	Std.	High	Low	Std.	high	Low	Std.	High
Prim Deposition Vel. (cm/s)	0.1	0.8	2.0	0.1	0.3	2.0	0.1	0.5	2.0	0.1	0.3	1.0
Sec. Deposition Vel. (cm/s)	0.1	0.4	2.0	0.1	0.3	1.0	0.1	0.5	2.0	0.1	0.3	1.0
Transformation Rate (% hr^{-1})	1.0	2.0	4.0	0.5	1.0	2.0	3.0	5.0	10.0	1.5	2.0	4.0
Primary Washout Rate (hr^{-1})	Not varied						0.01	0.04	0.1	0.01	0.04	0.1
Secondary Washout	0.04	0.5	0.8	0.04	0.5	0.8	0.04	0.5	0.8	0.04	0.5	0.8

TABLE 3. Sensitivy of Model Predictions of Sulphur Concentrations and Loadings to Parameter Variation.
Each entry indicates the percentage change from the standard values indicated in the first line.

		Concentrations				Dry Depositions				Wet Depositions			
		SO_2		SO_4		SO_2		SO_4		SO_2		SO_4	
		SR	LR	SR	LR	SR	LR	SR	LR	SR	LR	SR	LR
Predicted Values for Standard Parameters		8.3★	.01	.76	.02	438.6★	7.0	30.0	.6	180.4	.4	42.3	.2
Transf.Rate	Low	4	69	-47	-30	7	78	-47	-29	4	59	-47	-23
	High	-6	-59	80	16	-11	-63	79	15	-7	-54	80	-2
Dep Vel of SO_2	Low	16	191	19	100	-77	-48	20	102	14	148	18	140
	High	-27	-85	-24	-61	98	-54	-25	-61	-29	-83	-26	-72
Dep Vel of SO_4	Low	NC	NC	10	43	NC	NC	-71	-62	NC	NC	5	22
	High	NC	NC	-30	-68	NC	NC	240	70	NC	NC	-16	-40
Wet Removal Rate of SO_4	Low	NC	NC	18	185	NC	NC	15	232	NC	NC	-52	433
	High	NC	NC	-3	-6	NC	NC	-3	-7	NC	NC	3	-21

★concentrations in $\mu g/m^3$ (annual average), depositions in $mg\text{-}S/m^2$/year.

TABLE 4

Sensitivity of Model Predictions of Nitrogen Concentrations and Loading to Parameter Variation. Each entry indicates the percentage change from the standard values indicated in the first line.

		Concentrations				Dry Depositions				Wet Depositions			
		NO_2		NO_3		NO_2		NO_3		NO_2		NO_3	
		SR	LR	SR	LR	SR	LR	SR	LR	SR	LR	SR	LR
Predicted Values for Standard Parameters		3.3★	0	0.6	.01	74.0★	.06	19.0	.22	45.0	.06	21.0	.04
Transf. Rate	Low	5	83	-30	2	10	98	-30	2	7	69	-28	14
	High	-11	-72	64	-10	-18	-75	59	-12	-15	-68	61	-26
Dep Vel of NO_2	Low	11	89	10	40	-71	-45	10	41	12	79	11	57
	High	-28	-85	-26	-59	156	-44	-27	-59	-31	-81	-28	-70
Dep Vel of NO_3	Low	NC	NC	12	65	NC	NC	-76	-61	NC	NC	6	37
	High	NC	NC	-28	-67	NC	NC	190	36	NC	NC	-16	-43
Wet Removal	Low	8	381	3	166	7	304	3	137	-58	116	14	182
	High	-8	-60	- 3	-30	-7	-57	-3	-30	32	-66	-15	-51
Wet Removal of NO_3	Low	NC	NC	17	586	NC	NC	14	597	NC	NC	-52	646
	High	NC	NC	-3	-8	NC	NC	-3	-9	NC	NC	3	-25

★concentrations in μg/m^3 (annual average), deposition in mg N/m^2/

The emission density is illustrated in Figure 2 and 3.

Trajectory Calculations

In order to reduce computational time, the number of trajectory calculations was limited to a total of 70 points corresponding to grid squares of high emission rate. The same trajectories, suitably translated, were then used to represent neighbouring grid square.

For each trajectory origin point, 5 day trajectories at 3 hourly steps were derived for the entire year. Each trajectory is used to determine the individual plume location and precipitation intensity for each segment of plume, all of this data being stored for subsequent processing.

Concentration and Deposition Calculations.

The trajectory information for each source square is used as input to the full model which calculates average concentrations and dry/wet depositions for the four species over the entire period of simulation for a network of 75 receptor points distributed over eastern North America. The contribution of each source to each of the 75 receptor points is stored separately for further analysis.

The parameter choice for the 1978 simulation is shown in Table 5, and was based on the sensitivity analysis and values cited in the literature. No tuning of these parameters has been attempted to-date.

TABLE 5 Parameter Choice for 1978 Model Simulation

Parameter	Sulphur		Nitrogen	
	Summer	Winter	Summer	Winter
Primary Deposition Velocity (cm/s)	0.8	0.3	0.5	0.3
Secondary Deposition Velocity (cm/s)	0.1	0.1	0.1	0.1
Transformation Rate (% hr^{-1})	2.0	1.0	5.0	2.0
Primary Washout Rate (hr^{-1})	-	-	0.04	0.04
Secondary Washout Rate (hr^{-1})	0.3	0.3	0.3	0.3
Mixing Height (m)	750	500	750	500

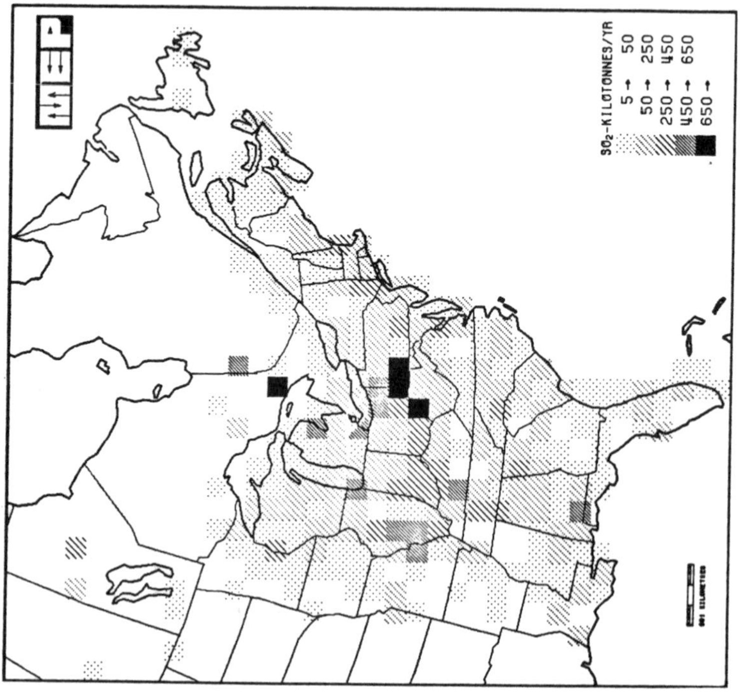

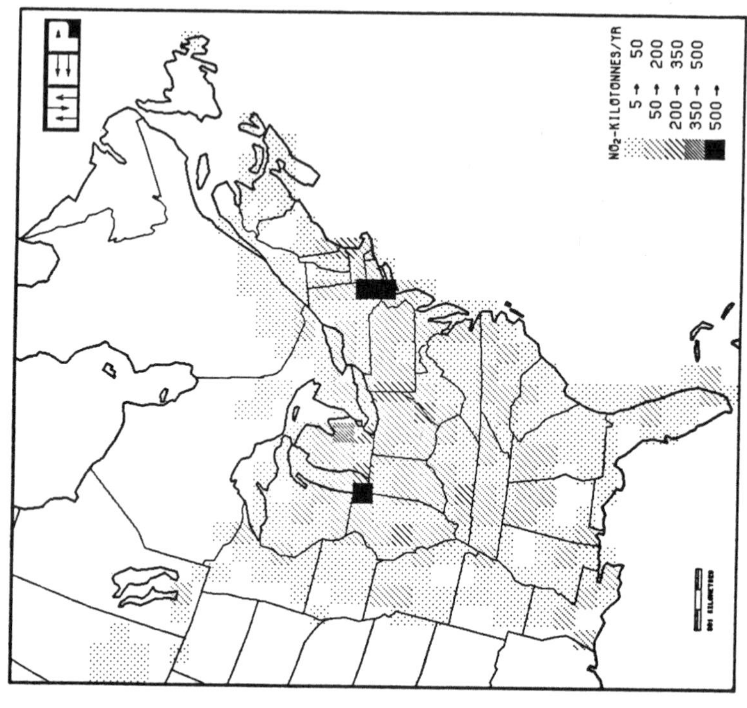

Figure 3 Distribution of NO_2 Emissions in 127 km grid prepared by AES.

Figure 2 Distribution of SO_2 Emissions in 127 km grid prepared by AES.

Model Predictions

Predicted annual average concentrations are shown in Figure 4. Due to the low density of receptor locations, the contours are considerably smoothed by comparison to observed concentration fields.

The SO_2 concentrations range from less than 1 $\mu g/m^3$ in Northern Canada to a high of 50 $\mu g/m^3$ coinciding with the densest source regions in Ohio and Pennsylvania. Sulphate concentrations range from less than 1 $\mu g/m^3$ to a peak of 25 $\mu g/m^3$, the contour gradient being much smoother than for SO_2.

The NO_2 and NO_3 contours show several areas of high concentration corresponding to the different distribution of high density NO_2 sources as seen in Figure 3.

Figure 5 shows the wet deposition of secondary species as well as total wet depositions. Peak wet sulphur deposition of 2.5 g-S/m^2/year occurs over Ohio and Southwestern Ontario, with a large area of eastern North America receiving more than 1 g-S/m^2/year.

DISCUSSION

Comparison data for model verification is very limited for 1978, so that only qualitative comparisons have been possible to date.

Figure 6 reproduces a three year average (1975-1977) map of SO_4 (US-Canada MOI, 1981) derived from AQCR measurements. The peak SO_4 of 20 $\mu g/m^3$ occurs in Ohio, with values of 18 $\mu g/m^3$ in Pennsylvania and 12 to 14 $\mu g/m^3$ in Southwestern Ontario. The corresponding model predictions are 25 $\mu g/m^3$, 20 $\mu g/m^3$ and 18 $\mu g/m^3$ respectively.

These results indicate a tendency to overpredict the SO_4 concentrations, likely attributable to the choice of conversion rate for sulphur.

The wet sulphur despositions have been compared to a one year compilation of CANSAP and NADP data for the year April 1979 to March 1980 as reported by NADP (1981). This shows a peak wet sulphur loading of 1.5 g-S/m^2/year with a significant area of the eastern states receiving more than 1 g-S/m^2/year.

The model would seem to overpredict the wet loading in the high source density region and underpredict the loading in the low density regions of eastern Canada and southeastern US.

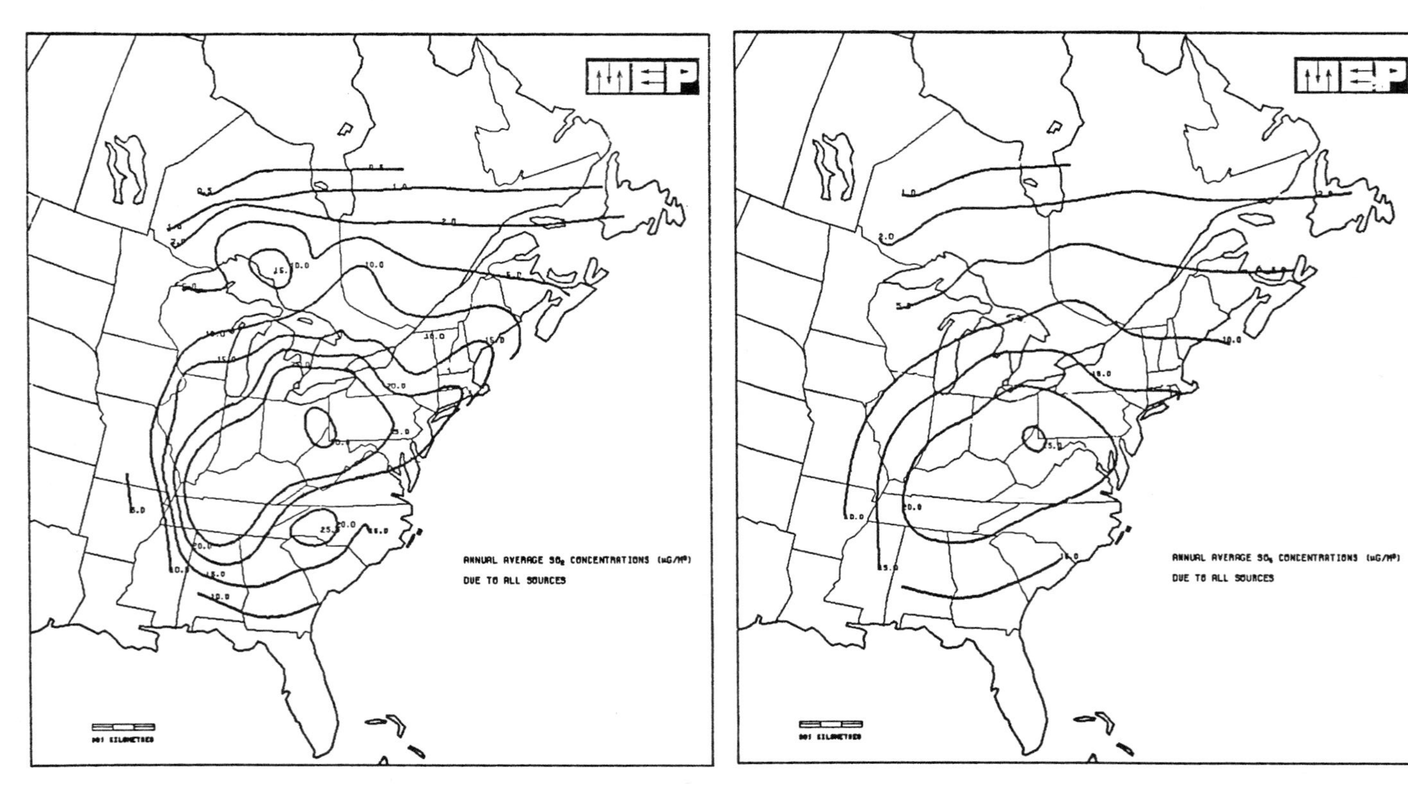

Figure 4a Annual average predicted concentrations of SO_2 and SO_4 for 1978.

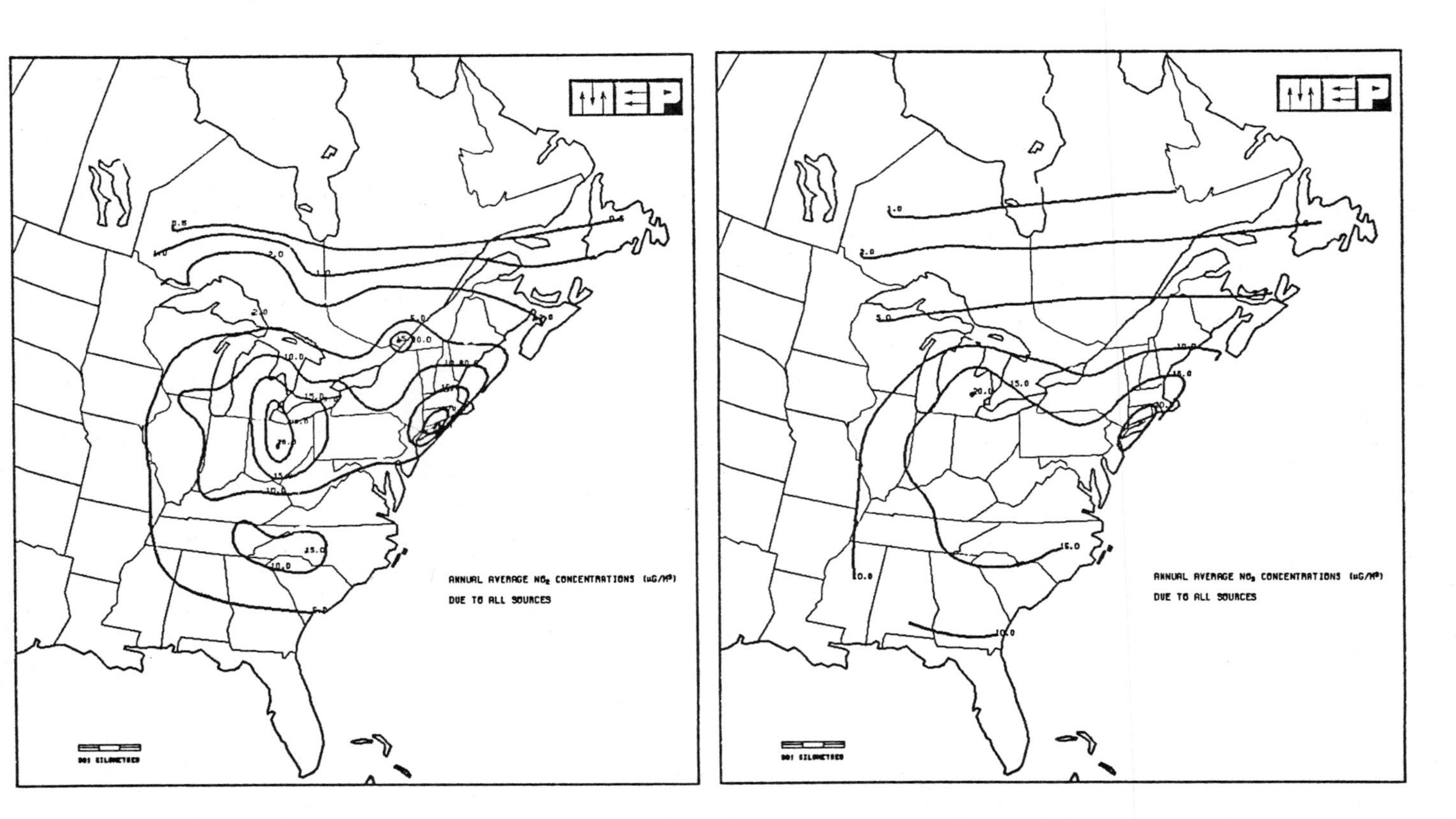

Figure 4b Annual average predicted concentrations of NO_2 and NO_3 for 1978.

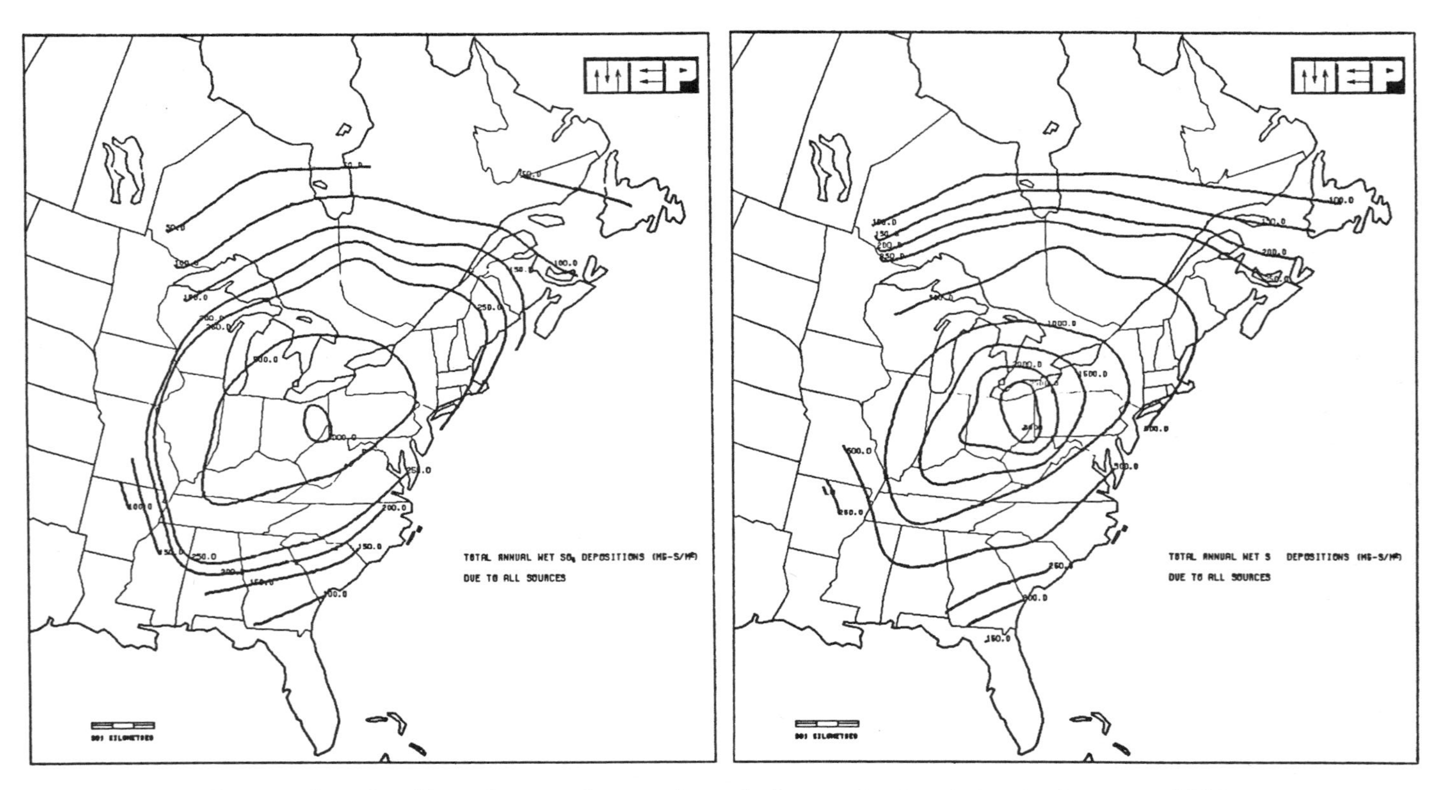

Figure 5a Predicted annual wet depositions of SO_4 and total S for 1978.

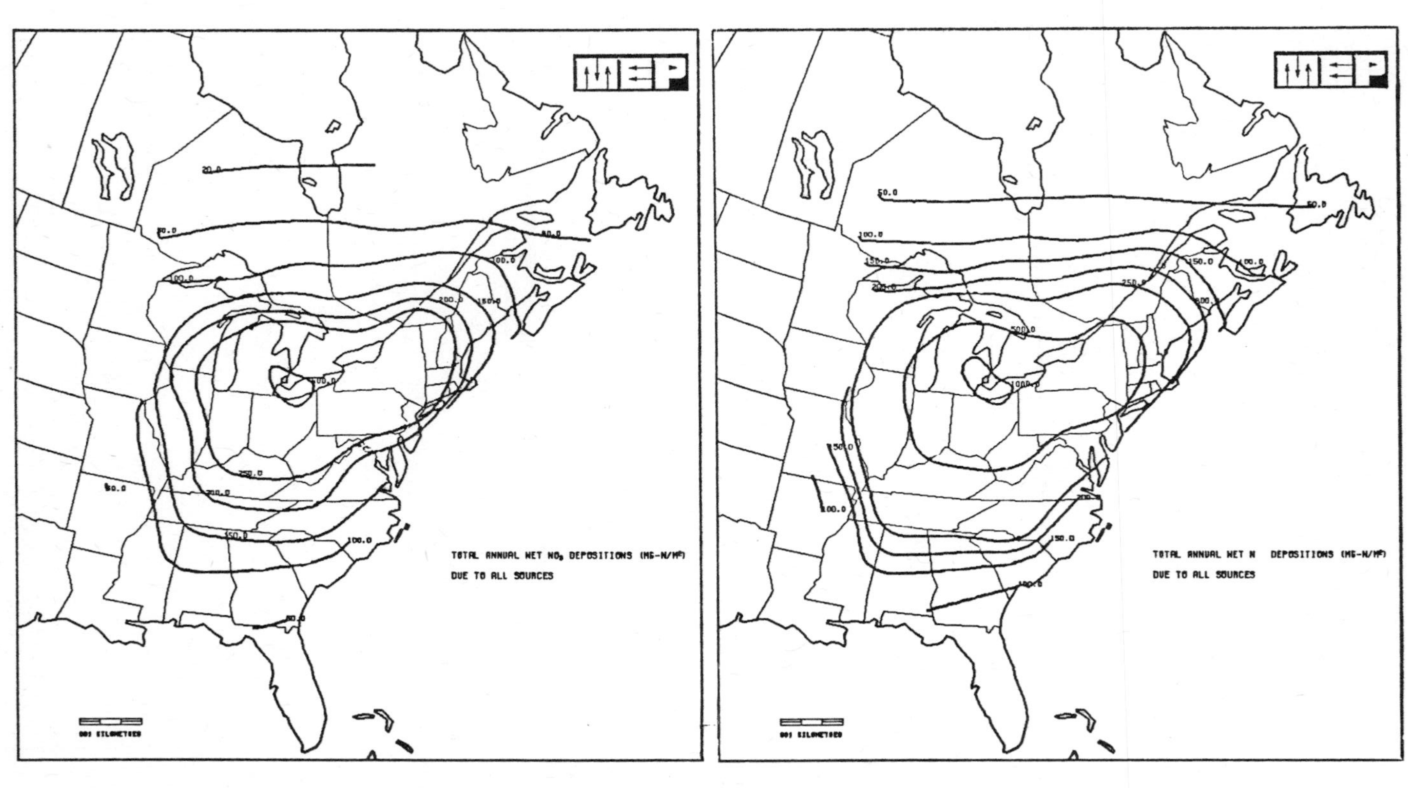

Figure 5b Predicted annual wet depositions of NO_3 and total N for 1978.

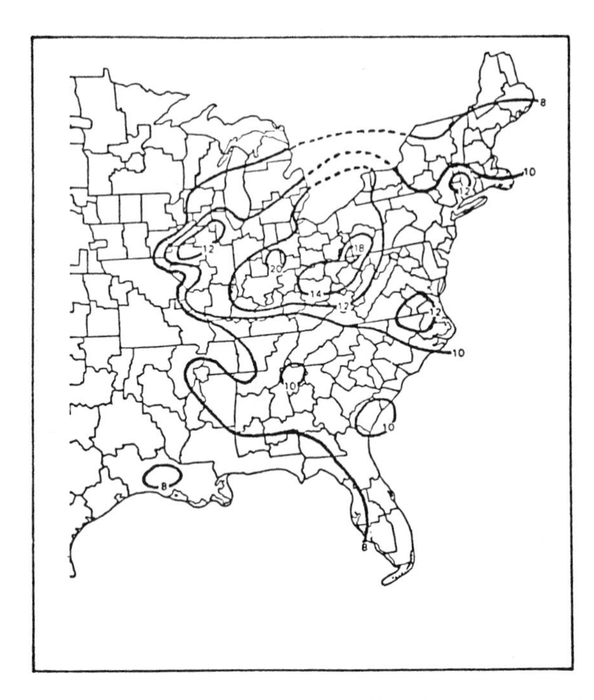

Figure 6 Three-year average (1975-1977) of AQCR average sulphate concentrations ($\mu g/m^3$). Reproduced from US-Canada MOI (1981).

TABLE 6 Comparison of Predicted and Observed Sulphur Wet Deposition in 11 Ecologically Sensitive Areas.

		Model Predicted Wet S Deposition for 1978 (kg-S/ha/yr)	Observed Wet S Deposition NADP (1981)
1	Boundary Waters	1.3	2.5
2	Algoma	6.0	3.5
3	Muskoka	9.5	6.0
4	Que.-Montmorency	5.8	5.0
5	Nova Scotia	2.7	3.0
6	New Hampshire	8.0	7.0
7	Whiteface	8.7	7.5
8	PennState	16.1	12.0
9	Appalachians	7.3	6.5
10	Florida	1.5	5.0
11	Arkansas	1.8	7.0
	Mean	6.2	5.9
	Correlation Coefficient r = 0.80		

Table 6 shows a comparison of model prediction to monitored wet loading in 11 ecologically sensitive areas in the US and Canada. The monitored data are extracted from a map presented in the NADP report cited above. The means over these stations differ by 5%, with a spatial correlation coefficient of 0.80.

Data on nitrogen oxides are even more fragmentary than for sulpur, and no evaluation of model performance for NO_x can be made at this stage.

ACKOWLEDGEMENTS

Part of this study was sponsored by the Atmospheric Environment Service, who also supplied the emissions inventory. I would like to acknowledge the dedicated computational support of M. Minuk, T. Speakman and W. McCormick.

REFERENCES

Barrie, L.A., 1981 : The Prediction of Rain Acidity and SO_2 Scavenging in Eastern North America, Atmos. Env., 15, 31.

Heffter, J.L. and G.J. Ferber, 1975 : A Regional-Continental Scale Transport, Diffusion, and Deposition Model, Part II : Diffusion-Deposition Models, NOAA Technical Memorandum ERL ARL-50, Air Resources Laboratories, Silver Spring, Maryland.

NADP, 1981 : Annual Report of Co-operative Regional Project, NC-141-NADP.

Sykes, R.I. and L. Hatton, 1976 : Computation of Horizontal Trajectories Based on the Surface Geostrophic Wind, Atmos. Environ. 10, 925.

United States-Canada Memorandum of Intent on Transboundary Air Pollution, 1981 : Interim Report.

Voldner, E. C. and Y. Shah, 1979 : Preliminary North American Emissions Inventories for Sulfur and Nitrogen Oxides and Total Suspended Particulate - A Summary, AES Report, LRTAP 79-4.

Weisman, B., 1980 : Long Range Transport Model for Sulphur, 73rd Annual Meeting of the Air Pollution Control Association, Montreal.

DISCUSSION

J.L. WOODWARD

What effect does your use of diurnally-varying parameters (particularly reaction rate fore SO_2 conversion) have as opposed to using constant average values.

B. WEISMAN

The magnitude of the effect of using diurnally varying parameters depends on the choice of mean parameter values and the time of release ; however, for the standard choice of parameters, up to a factor of two in the sulphur lifetime is evident, the lifetime being higher for diurnally varying parameters and noon-time releases.

W. GOODIN

You indicated that your model is applicable to short-term (episode) and long-term (annual periods). Aren't the assumptions different for these two approaches ? Have you validated your model for episode periods ?

B. WEISMAN

The sequential structure of the model allows the use of the model in both modes ; parameters can be chosen for the specific conditions. The model is currently being exercised on a 24 hour regional basis, but no results are available at this time. The model is also being used to predict wet depostion episodes at an observational station in Eastern USA.

REFINED AIR POLLUTION MODEL FOR CALCULATING DAILY REGIONAL PATTERNS AND TRANSFRONTIER EXCHANGES OF AIRBONE SULFUR IN CENTRAL AND WESTERN EUROPE

C. M. Bhumralkar, R. M. Endlich, K. Nitz, R. Brodzinsky,
J. R. Martinez, and W. B. Johnson

SRI International
Menlo Park, California 94025

INTRODUCTION

Previous studies at SRI International (Johnson et al., 1978 ; Bhumralkar et al., 1979 and 1981) developed a puff trajectory model for regional-scale air pollution modeling.★ The SRI model exists in several forms : The simpler form (EURMAP-1) can be used to obtain long-term averages ; the more complicated form (EURMAP-2) is used for short-term pollution episodes.

FEATURES OF THE EURMAP-2 MODEL

EURMAP-2A (an early version) and EURMAP-2B (the most recent version) have the following features in common : both use a 50-km grid, use emissions from 19 countries, introduce pollution puffs at 6-hour intervals, and use tracking increments of 1 hour. Standard daily weather observations at the surface and 850 mb are used. The transport wind, which is based on both surface and 850-mb winds, can be computed in EURMAP-2B using the pollution mass within sublayers as a weighting function. Horizontal diffusion is based on deformation of the horizontal wind field.

The two versions of the EURMAP-2 model differ in the following respects : In EURMAP-2B the mixing layer is divided into three sublayers as shown in Figure 1. The total emissions are divided

---

★ There are numerous similar studies for Europe (for example Fisher, 1978 and Ottar, 1978), for the United States (Shieh, 1977 and Perhac, 1978), and for other countries.

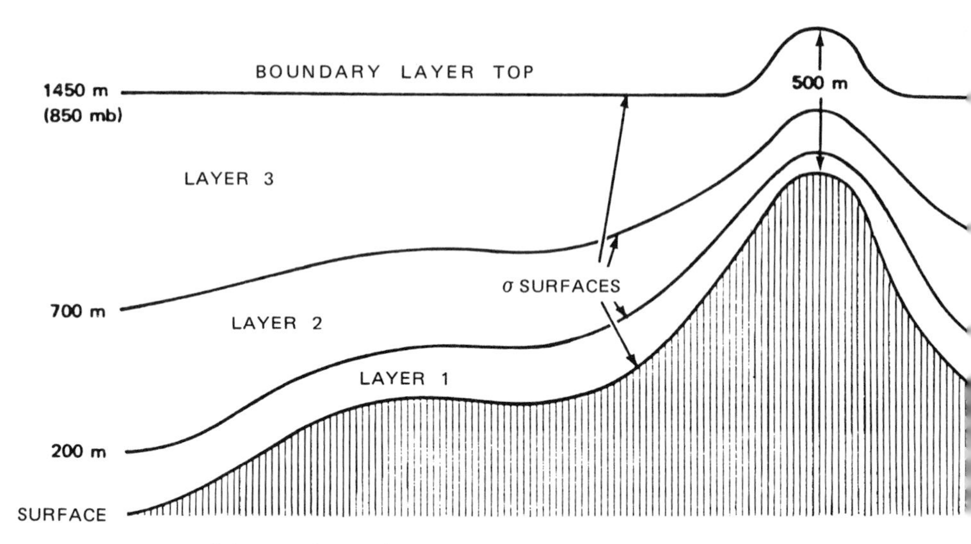

Figure 1. Division of the Boundary Layer into Three Parts Using σ Surfaces

between sublayers 1 and 2 to separate near-ground and tall-stack emissions. Vertical diffusion between the sublayers has been introduced in EURMAP-2B to replace uniform mixing within puffs. The vertical diffusion is controlled by diffusion coefficients that are computed from wind shear and stability. The transformation rate for SO_2 to $SO_4^=$ was 0.01 (one percent) per hour in EURMAP-2A: In EURMAP-2B the rate has been made dependent on solar radiation using parameters that vary with season and latitude. The new rate (based on Carmichael and Peters, 1979) is considerably larger than the old and varies from approximately 0.03 in winter to 0.08 summer. The wet deposition rates for SO_2 and $SO_4^=$ in EURMAP-2A were constant values (0.28 and 0.07 respectively) multiplied by the precipitation rate. In EURMAP-2B the wet deposition rates depend on precipitation type, i.e., cold-season (Bergeron) processes versus summertime convection. The wet deposition rates for SO_2 are substantially reduced in EURMAP-2B for both winter and summer. For $SO_4^=$ the new rate is reduced in winter and increased in summer.

To separate the sublayers in EURMAP-2B we define a coordinate σ as

$$\sigma \equiv \frac{z - h(x,y)}{H(x,y) - h(x,y)} \quad ,$$

where h(x,y) is the height of the terrain above a certain reference level (e.g., sea level) and H(x,y) is the corresponding height of the boundary layer top, as shown in Figure 1. The top of the boundary layer is selected as the 850-mb level (where radiosonde data are available). However, over high terrain the boundary layer top is taken to be 500 m above the terrain, so that the minimum depth of the boundary layer is 500 m. The boundary layer has been subdivided into three parts to permit the use of vertical diffusion computations (as mentioned above) instead of the previous assumption of uniform mixing within puffs. The layers are defined in σ coordinates so that they are proportional parts of the total thickness. The top of the lowest layer is currently set at approximately 200 m above-ground. Winds on the σ surfaces are interpolated from the surface and from 850-mb values as a function of the logarithm of height to give smooth profiles, with the largest shear of speed and direction near the ground.

As mentioned earlier, pollution emissions from near-ground sources go into the bottom layer. Pollution from higher stacks, and allowing for the rise of heated plumes, goes into the second layer. The model has a parameter to control the percentage of the total emissions introduced into each layer. (Unfortunately, there is currently little information to guide the selection of these percentages.)

The initial pollution placed into sublayers 1 and 2 (see Figure 1) is mixed vertically among the three layers. A small amount (controlled by an appropriate coefficient) is mixed upward from sublayer 3 into the overlying air. The diffusion computation is made using the standard expression

$$\frac{\partial C(i)}{\partial t} = \frac{\partial}{\partial Z}\left(K_Z(i,t)\ \frac{\partial C(i)}{\partial Z}\right) ,$$

where

$C(i)$ = pollutant concentration in the ith layer,
Z = vertical coordinate,
t = time,
$K_Z(i,t)$ = diffusion coefficient at time t.

$K_Z(i,t)$ varies with height as well as with horizontal position and time. It is defined in different areas of the modeled region, and the puff uses the value corresponding to the region it is traversing at time t.

The corresponding finite difference equation is formulated for layers of variable depth. All three values of concentration in a column after mixing for a given time increment are obtained simultaneously by using the Crank-Nicolson method (see Crank, 1967). The system of equations yields a tridiagonal matrix, which is solved using recursion formulas given by Young (Ch. 11 in Todd, 1962).

RESULTS

The effects of terrain features on the flow were computed by making the winds on σ surfaces nondivergent using the method of Endlich (1967). This calculation is based on the reasonable assumption that the flow is parallel to the σ surfaces. Then pollution transport was modeled using the nondivergent winds and compared to computer runs with uncorrected winds. For the cases studied, which had predominantly west-east flow over the Federal Republic of Germany (FRG), the differences were very minor. We concluded that terrain influences would be important only for cases where the wind traveled from main pollution sources over the Alps (for example, southward from the FRG). Because such cases are infrequent, we have omitted the terrain correction from further computations.

For the pollution episode of 25 and 26 August 1974, we computed the pollution transport using the EURMAP-2B model. (The computations shown below include only the emissions from the FRG.) As mentioned above, the transformation rate from SO_2 to $SO_4^=$ is approximately 4 times larger than that used previously, thus increasing the proportion of $SO_4^=$ in concentrations and depositions. Lacking knowledge of the proportion of near-ground versus tall-stack emissions, we divided the emissions equally (50/50 percent) between sublayers 1 and 2, as seems reasonable. If the proportion is changed (to, say, 75/25 percent) the computed patterns change markedly, with higher values close to the sources. Thus, the proportion is an important parameter in the computations. The vertical diffusion process between the three sublayers appears to mix the pollution in a realistic manner.

The results are illustrated for emissions from the FRG on 25 and 26 August 1974. Figure 2 shows the airborne SO_2 and $SO_4^=$ concentrations when the emissions have been divided initially 50/50 between sublayers 1 and 2. The highest concentrations are close to the sources for SO_2 and displaced slightly downwind for $SO_4^=$. Figure 3 shows that the concentrations in the surface layer are much higher if the emissions are divided 75/25 between sublayers 1 and 2 (corresponding to a predominance of near-ground emissions). In contrast, Figure 4 shows lowered concentrations for emissions divided 25/75 between sublayers 1 and 2 (implying a predominance of tall-stack emissions). It is clear that the initial division of the emissions has major effects on the pollution patterns; information on the amounts of near-ground and tall-stack emissions is thus needed.

The effect of increasing the transformation rate of SO_2 to $SO_4^=$ is shown in Figure 5. Compared to Figure 2, the SO_2 concentrations are lower and the $SO_4^=$ concentrations increase. Similar computation are currently being made for 19 emitter countries.

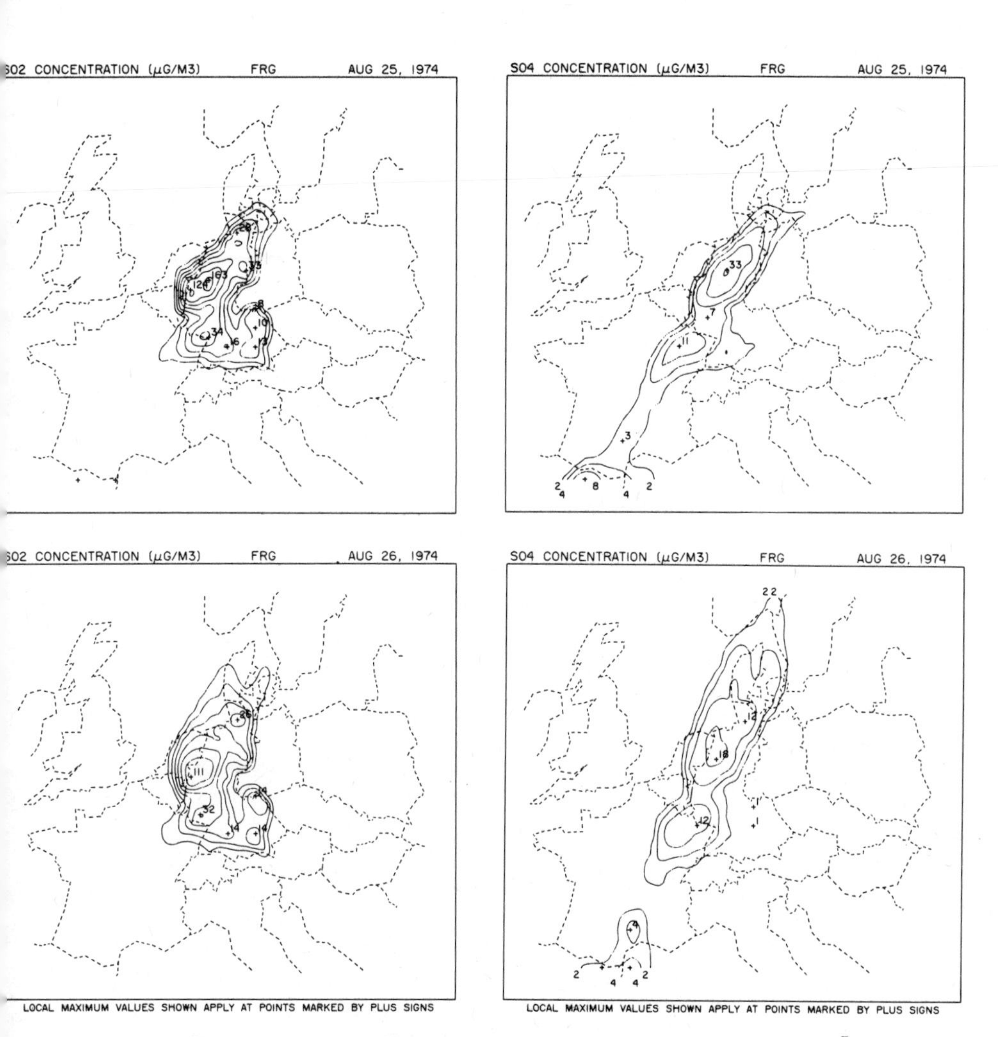

Figure 2. SO_2 and $SO_4^{=}$ Pollution Concentrations ($\mu g/m^3$) Calculated Using Three Layers, with Emission 50 Percent into Layer 1 and 50 Percent into Layer 2

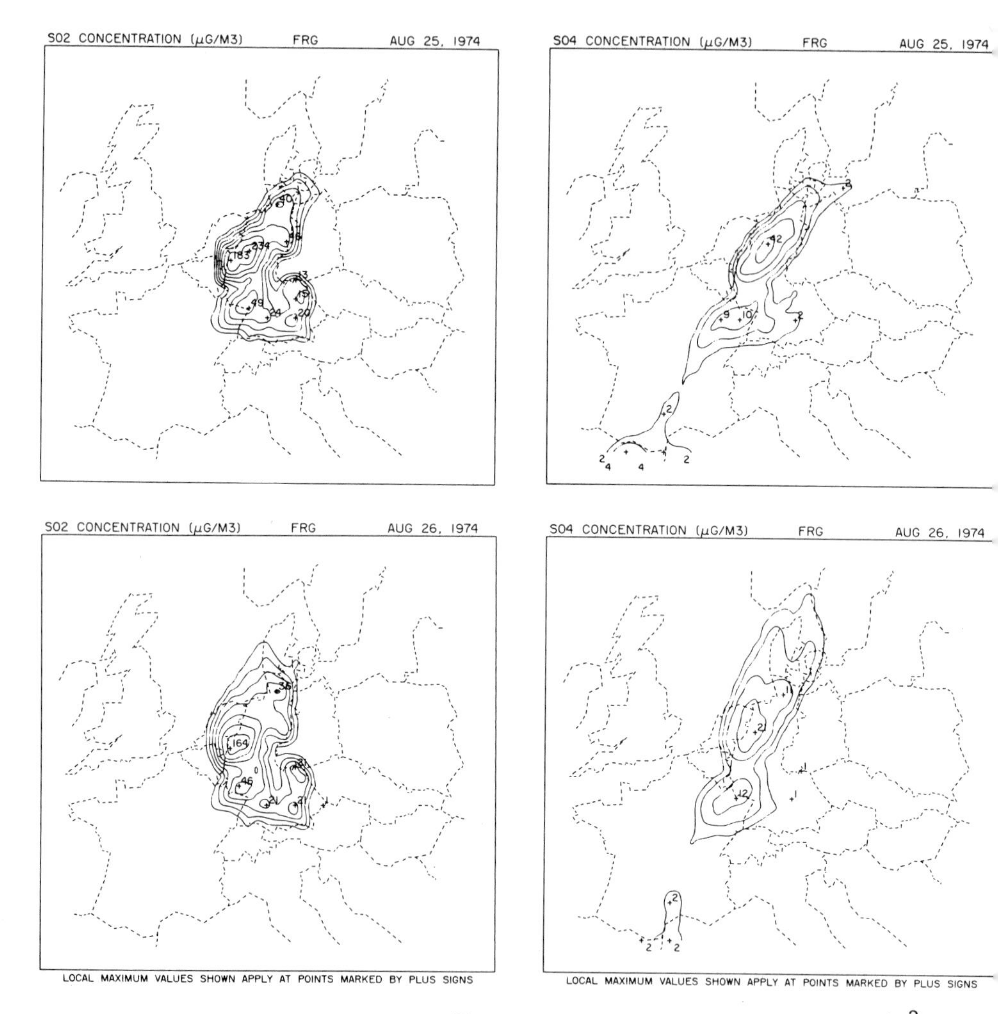

Figure 3. SO_2 and $SO_4^=$ Pollution Concentrations ($\mu g/m^3$) Calculated Using Three Layers, with Emission 75 Percent into Layer 1 and 25 Percent into Layer 2

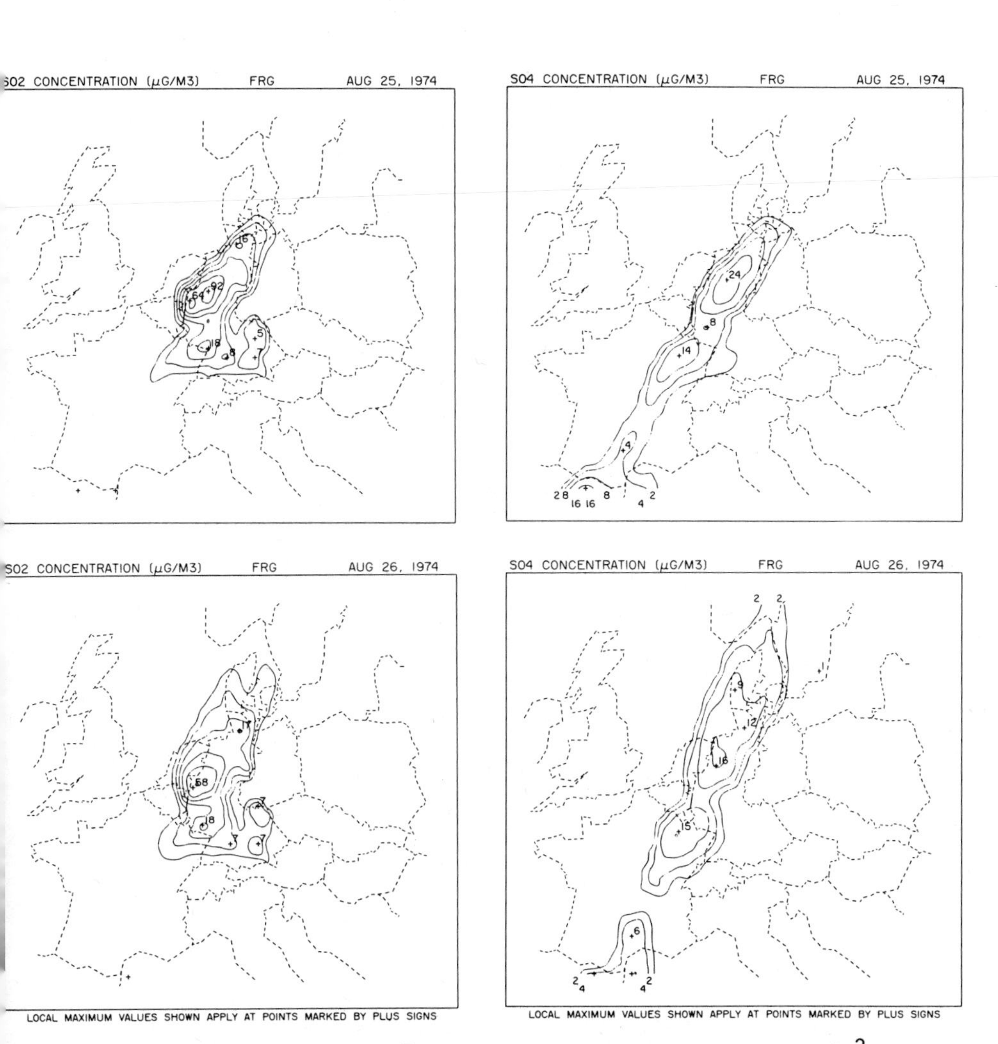

Figure 4. SO_2 and $SO_4^=$ Pollution Concentrations ($\mu g/m^3$) Calculated Using Three Layers, with Emission 25 Percent into Layer 1 and 75 Percent into Layer 2

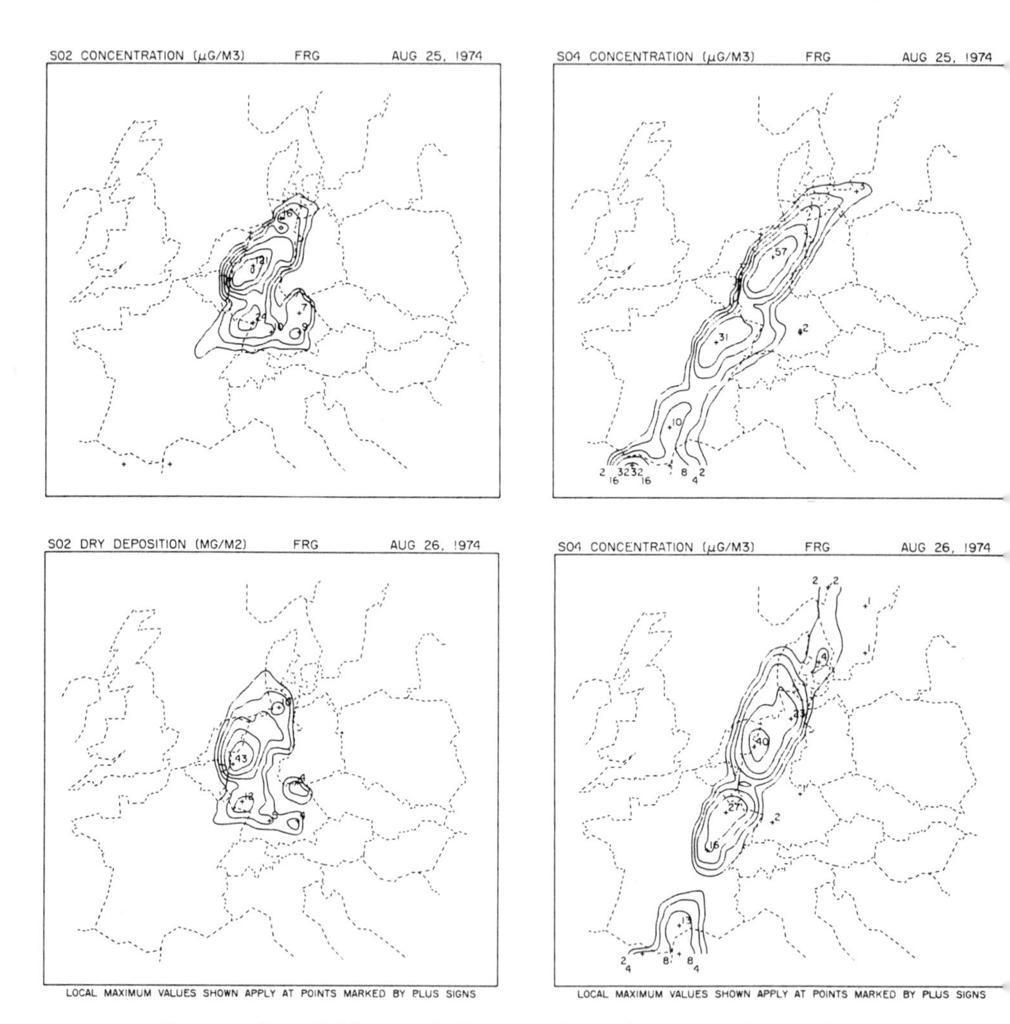

Figure 5. Effect of Increasing the Transformation Rate of SO_2 to $SO_4^=$

SUMMARY

The long-range pollution transport model EURMAP-2B has several new features. A subroutine permits correction of winds for complex terrain. Vertical mixing is carried out using three sublayers within the mixing layer. The transformation rate for SO_2 to $SO_4^=$ and the wet deposition rates have also been changed to conform to recent published values. In practical application the correction of winds for terrain does not appear to be important, because in Europe the predominant wind flows do not pass from major emission sources directly over the Alps or Pyrenees. On the other hand, the use of sublayers is advantageous in permitting pollution emission from near-ground sources and tall stacks to be placed into sublayers 1 and 2, respectively. The computations are sensitive to the proportions of total emissions assigned to each sublayer. Information is needed on the actual proportions for major pollution source regions. The computations are also sensitive to the amount of total emissions. For episode studies the emissions should be known more accurately than one can compute using annual emission totals prorated for each day. The new transformation rates markedly affect the deposition patterns and affect the concentration patterns to a lesser extent.

It is difficult to determine whether the EURMAP-2B model is more accurate (on an absolute scale) than EURMAP-2A because of the lack of actual measurements of depositions (which account for most of the sulfur emission). The new model provides greater flexibility than does the old, however, and we believe it will be a valuable tool for further studies of pollution transport over large areas.

ACKNOWLEDGMENTS

This work was conducted under contract to the Federal Environmental Agency (Umweltbundesamt) of the Federal Republic of Germany (FRG), under the supervision of Dr. D. Jost and Dr. J. Pankrath.

REFERENCES

Bhumralkar, C. M., Johnson, W. B., Mancuso, R. L., and Wolf, D. E., 1979, Regional patterns and transfrontier exchanges of airborne sulfur pollution in Europe, Final Report to the Umweltbundesamt (FRG), Project 4797, SRI International, Menlo Park, California, 110 pp.

Bhumralkar, C. M., Mancuso, R. L., Wolf, D. E., and Johnson, W. B., 1981, Regional air pollution model for calculating short-term (daily) patterns and transfrontier exchanges of airborne sulfur in Europe, Tellus, 33:142.

Carmichael, G. R., and Peters, L. K., Numerical simulation of the regional transport of SO_2 and sulfate in the eastern United States, Preprint Volume, Fourth Symposium on Turbulence, Diffusion, and Air Pollution, pp. 337-344 (American Meteorological Society, Boston, Massachusetts).
Crank, J., 1967: "The Mathematics of Diffusion," Oxford University Press, London, 347 pp.
Endlich, R. M., 1967, An iterative technique for altering the kinematic properties of wind fields, J. Appl. Meteor., 6:837.
Fisher, B.E.A., 1978, The calculation of long term sulphur depositi in Europe, Atmos. Environ., 12:489.
Johnson, W. B., Wolf, D. E., and Mancuso, R. L., 1978, Long term regional patterns and transfrontier exchanges of airborne sulf pollution in Europe, Atmos. Environ., 12:511.
Ottar, B., 1978, The OECD study on long range transport of air pollutants (LRTAP), Atmos. Environ., 12:445.
Perhac, R. M., 1978, Sulfate regional experiment in northeastern United States: The SURE program, Atmos. Environ., 12:641.
Shieh, C. M., 1977, Application of a statistical trajectory model to the simulation of sulfur pollution over the northeastern United States, Atmos. Environ., 11:173.
Todd, J., 1962, "Survey of Numerical Analysis," McGraw-Hill Book Co New York, 589 pp.

DISCUSSION

B. WEISMAN

Are you recommending the higher value of transformation rate of SO_2 to SO_4 and on what is this consideration based ?

R.M. ENDLICH

The higher transformation rates are based on recent literature. We have not yet verified them by comparing the results to observations.

C. HAKKARINEN

What are the initial SO_2 and SO_4 concentrations that you assume for your study region ?

R.M. ENDLICH

We initialize the model by beginning an episode study with data for the five previous days.

M. SCHATZMANN

One of your conclusions was t terrain effects seem to be of minor influence on your model predictions. How reliable is this statement and doesn't it only reflect your model assumption concerning streamline deterioration in the vicinity of hills ?

..M. ENDLICH We were referring to the fact that the usual wind patterns carry the pollution eastward from the major sources. In occasional cases when the winds travel from an industrial region directly over the Alps the terrain effects may be significant. Overall we judge that the terrain effects are not large.

. SKIBIN When treating such large scale there are cases (like your example) of high pressure systems with divergence and downward motion of order 1 cm/s (200 m in 6 hours). Do you think that this effect is important and do you take it into account in the computation of the air concentration of the pollution.

..M. ENDLICH In such cases the horizontal winds are light and do not give much transport of the pollution. We do not specifically account for the effects of the sinking motion.

..N. LEE The wind information used in the model is essentially based onthe information at the surface and 850 mb. Since themodel is divided in 3 layers in Z-coordinate, the mid-layer wind information should be incorporated into the model to take into account the shear effects.

..M. ENDLICH We assume that the internal winds can be interpolated (versus the logarithm of height) from the surface and 850 mb values.

..J. VOGT On which plume rise model is your separation of the sources with respect to the lower and upper release layer based ?

..M. ENDLICH It is empirical and not based on a plume rise model. Further information on the proper separation is needed.

ON A METHOD OF EVALUATION OF PERFORMANCE OF A TRAJECTORY MODEL FOR LONG-RANGE TRANSPORT OF ATMOSPHERIC POLLUTANTS

John L. Walmsley and Jocelyn Mailhot

Atmospheric Environment Service
4905 Dufferin Street
Downsview, Ontario, M3H 5T4 Canada

INTRODUCTION

During the 1960's and early 1970's observations of increasing acidity of lakes in southern Norway and in the Adirondacks of New York gave an indication that pollutants were being transported great distances from their point of origin. The long transport times were sufficient to allow chemical transformations to occur, adding further complexity. Thus in the early 1970's the phenomenon of long-range transport of air pollution began to be examined with increasing interest. Studies of transport, transformation and deposition mechanisms have been undertaken and attempts have been made to model these processes. Eliassen (1980) has presented a review of these modelling efforts.

At Atmospheric Environment Service (AES) a project on the Long-Range Transport of Air Pollutants (LRTAP) was initiated during the mid-1970's encompassing all of the above mentioned research activities. In particular, a computerized trajectory model (Olson et al., 1978) was developed for application to the LRTAP project.

The present study describes a method of direct evaluation of the AES trajectory model by using specified wind fields rather than those obtained by objective analysis of observational data. The wind fields we use are determined by specifying the stream function as a simple, yet fairly realistic, function of latitude and longitude. This method enables us to write differential equations for trajectories which we solve numerically, for a given starting position with high accuracy. Trajectories produced in this manner may be used for verifying the AES trajectory model in a number of sensitivity tests. This method is applicable to the evaluation of other LRTAP models.

SPECIFICATION OF WIND FIELDS

First we define a stream function, ψ, which is a function of longitude, λ, and latitude, ϕ:

$$\psi(\lambda,\phi) = \cos^2\phi \left[A+B \cos n \lambda + C \cos k(\lambda+\alpha)\right] - D \sin\phi. \tag{1}$$

Here A, B, C, D, n, k and α are specified parameters which are described below. If we then define the eastward- and northward-wind velocity components (u',v') as:

$$u' = -u_o \frac{\partial\psi}{\partial\phi} , \tag{2}$$

$$v' \cos\phi = +u_o \frac{\partial\psi}{\partial\lambda} , \tag{3}$$

where u_o is a specified parameter, we obtain a non-divergent wind field, as shown in Walmsley et al., (1981).

The (u',v') winds in (λ,ϕ) coordinates are transformed to (u,v) components in a (x,y) grid system, as shown in Walmsley et al (1981) and the resulting trajectory equations to be solved by the trajectory model are:

$$\frac{dx}{dt} = \frac{mu}{\Delta X} , \tag{4}$$

$$\frac{dy}{dt} = \frac{mv}{\Delta X} , \tag{5}$$

where m is a map scale factor, ΔX is the grid spacing and t is time Methods of solution of these equations are described below.

SELECTION OF PARAMETERS

In Equation (1), parameters A and D establish a zonal circulation and the position of a jet stream and/or sub-tropical high pressure regions. Values for two basic flows are given in Table 1a By an appropriate choice of parameters B, C, n, and k one can simulate short waves (synoptic scale) and long waves (Rossby waves) superimposed on a zonal circulation as are often observed on constant pressure charts.

The principal function of the different parameters is indicated in Table 1a. By choosing $k = 4$ and $\alpha = 55^o$ a circulation marked by a major trough over the centre of the North American continent (100^oW) is obtained. The choice of $C = 0.2$ is intended

Table 1a. Parameters used for the Sensitivity Tests

PARAMETER	PRINCIPAL FUNCTION	FLOW 1	FLOW 2
u_o	Mean Speed	10 m s^{-1}	10 m s^{-1}
A and D	Zonal circulation and mean latitude of jet stream and/or sub-tropical high pressure zone	0.25 1.33	3.5 -3.5
n	Wavenumber of short waves	10, 20, 40	10, 20, 40
B	Maximum amplitude of short-wave zonal wind component	0.1,0.05,0.03	0.1,0.05,0.03
k	Wavenumber of long waves	4	4
C	Maximum amplitude of long-wave zonal wind component	0.2	0.2
α	Phase of long waves	55°	55°

to ensure that the long-wave amplitude of wind velocity perturbations remains within reasonable limits. Sensitivity of the model to synoptic-scale systems may be evaluated by using n = 10, 20 and 40. Corresponding values of B are set to 0.1, 0.05 and 0.03, respectively. Wavelengths of the short-waves are 36, 18 and 9 degrees longitude for n = 10, 20 and 40, respectively. At 60^oN this would mean wavelengths of 2000, 1000 and 500 km and at 45^oN, 2830, 1415 and 708 km, respectively. It should be recalled that the grid spacing of the wind data is 381 km so that the n = 40 case, in particular will be poorly resolved by the grid.

Some properties of the resulting flows are given in Table 1b, the three values of short-wave amplitudes corresponding to n = 10, 20 and 30, respectively.

EXACT SOLUTIONS

For comparison with model solutions we may obtain accurate solutions by another means. Since

$$u' = R \cos\phi \frac{d\lambda}{dt} \tag{6}$$

and

$$v' = R \frac{d\phi}{dt} \quad , \tag{7}$$

where R is the radius of the Earth, we may use Equations (1) - (3) to derive:

$$\frac{d\lambda}{dt} = \frac{u_o}{R} \quad [2 \sin \phi (A+B \cos n\lambda + C \cos k (\lambda+\alpha)) + D] , \tag{8}$$

$$\frac{d\phi}{dt} = \frac{-u_o \cos \phi}{R} \quad [n B \sin n\lambda + k C \sin k (\lambda+\alpha)] \quad . \tag{9}$$

Given a starting position (λ_o, ϕ_o), this coupled system of equation is solved numerically for $\lambda(t)$ and $\phi(t)$ using the Hamming Predictor Corrector model (Ralston and Wilf, 1960). The method is a stable fourth-order procedure. The size of the time increment is automatically adjusted to keep local truncation error within specified limits.

Solutions obtained by the above method will be referred to as analytic or "exact" solutions. Strictly speaking they are numerica or approximate, but they can be determined to any reasonable degree of accuracy required. Hence they form a standard against which trajectory model results may be compared.

Table 1b. Properties of the Flow

PROPERTY	FLOW 1	FLOW 2
Mean latitude of Jet Stream	17.8°N	57.5°N
Mean latitude of Sub-Tropical High	none	30°N
Short-wave amplitudes at 45°N:		
u'/u_o	0.10,0.05,0.03	0.10,0.05,0.03
v'/u_o	0.71,0.71,0.85	0.71,0.71,0.85
Long-wave amplitudes at 45°N:		
u'/u_o	0.20	0.20
v'/u_o	0.57	0.57
Mean zonal wind speed at 45°N	1.19 u_o	1.03 u_o
Mean zonal wind speed at mean latitude of jet stream	1.41 u_o	1.29 u_o

MODEL SOLUTIONS

The simplest finite-difference approximation to Equation (4) is:

$$x(t_1) = x(t_o) + \frac{m(t_o)\ u(t_o)}{\Delta X} (\Delta t) \tag{10}$$

where $(\Delta t) = t_1 - t_o$ is the timestep. The approximation to Equation (5) is similar. For a given trajectory of several timesteps, the total error may be shown to be proportional to (Δt) (see Walmsley et al., 1981).

A somewhat better finite-difference approximation to Equation (4) is given by:

$$x(t_1) = x(t_o) + \frac{(\overline{mu})}{\Delta X} (\Delta t) \tag{11}$$

where $(\overline{mu}) = \frac{1}{2} [m_o u_o + m_1 u_1]$ is a mean value of (mu) during the timestep. It is evaluated by an iterative process. In Walmsley et al. (1981) we show that the total error for several timesteps with this "constant acceleration" model is proportional to $(\Delta t)^2$.

Other errors arise from the need to interpolate wind velocity components from grid-points to positions needed by the finite-difference approximation to the trajectory equation. These are discussed in Walmsley et al., (1981) where theoretical estimates and results obtained from model calculations are given.

MODEL RESULTS

An example of the model results is given in Figure 1 where root-mean-square position errors (in gridlengths of size 381 km) are given as a function of timestep (logarithmic scale) for two different flow situations. Errors were computed by comparing model solutions with exact solutions each timestep for five trajectories of 42-hour duration. A constant acceleration model (Equation 11) was used. In cases without interpolation error (referred to as "exact" in Figure 1), the curves have a slope of +2, indicating that the error in Equation (11) is proportional to $(\Delta t)^2$.

When wind data are assumed to be known only at grid points, an interpolation scheme is needed to obtain winds at positions require by the trajectory model. Errors due to linear and cubic interpolation schemes, in both cases added to the error arising from the use of Equation (11), are also shown in Figure 1. Interpolation errors clearly dominate those due to the finite-difference approximation. The cubic interpolation scheme affords some improvement over the linear scheme, especially in the case with wavenumber n= 1

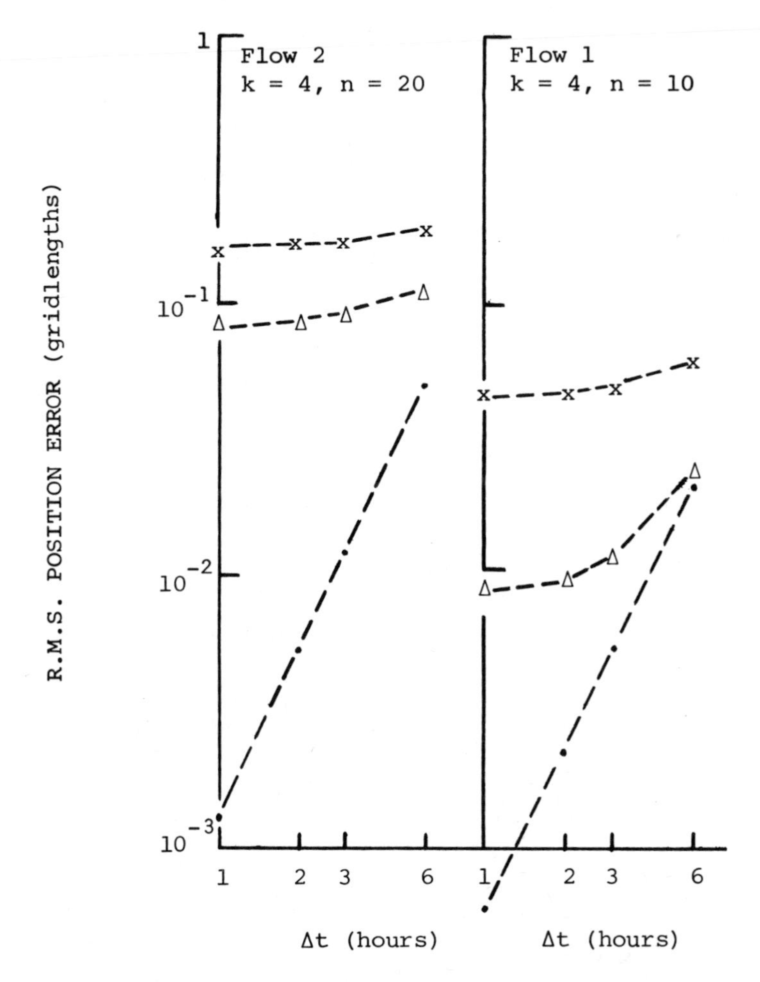

Fig. 1. Root-mean-sqaure position error as a function of timestep for the constant acceleration model with specified wind fields. Five trajectories of 42-hour duration were computed.

CONCLUSIONS

A method of direct evaluation for the accuracy of trajectory models has been described in which non-divergent, quasi-realistic wind fields are specified and model calculations may be compared with highly accurate solutions for trajectories in those analytic fields.

An example of results obtained from this method has been illustrated. More details on the method and further results appear in Walmsley et al., (1981).

A possible extension of this project would be the specification of a stream function which is also time- and/or height-dependent, in order to evaluate the accuracy of temporal and vertical interpolation schemes.

REFERENCES

Eliassen, A., 1980, A Review of Long-Range Transport Modelling. J. Appl. Meteor., 19, 231-240.

Olson, M.P., Oikawa, K.K., and MacAfee, A.W., 1978, A Trajectory Model Applied to the Long-Range Transport of Air Pollutants. Rep. LRTAP 78-4. Atmos. Environ. Service, Downsview, Ontario, 24 pp.

Ralson, A., and Wilf, H.W., 1960, "Mathematical Methods for Digital Computers," Wiley, New York, 293 pp.

Walmsley, J.L., Mailhot, J., and Hopkinson, R.F., 1981, Sensitivity Tests with a Trajectory MOdel. Rep. AQRB-81-027-L, Atmos. Environ. Service, Downsview, Ontario, 49 pp.

DISCUSSION

H.N. LEE — Can the model take into account the variation of eddy diffusivity ?

A.D. CHRISTIE — The study does not address the diffusion problem, only the location of the centre of gravity of a parcel following a trajectory consistent with the stream function in λ, ψ coordinates.

THE PSEUDO-SPECTRAL TECHNIQUE IN THE COMPUTATION OF SEASONAL AVERAGE CONCENTRATIONS AND ITS COMPARISON WITH THE TRAJECTORY TECHNIQUE

B. D. Murphy

Computer Sciences
at the
Oak Ridge National Laboratory*

LONG-RANGE ATMOSPHERIC TRANSPORT

Since there is now a greater awareness of the chronic effects of exposure to low levels of atmospheric pollutants, the long-range transport of atmospheric material is receiving increasing attention in recent years. Furthermore, there is the concern about the effects of successor species which come about by, for instance, chemical transformation or radioactive decay of the primary species during transport in the atmosphere. Concentrations of successor species may build up slowly and consequently transport processes may need to be studied over long distances.

Quite a few codes have been designed to study the long-range movement of material in the atmosphere. The ORNL code RETADD (Begovich, Murphy and Nappo, 1978) is an example of such a code. The National Coal Utilization Study (R. M. Davis et al., 1978) and the work of Murphy, Parzyck and Raridon (1978) are examples of the application of the code. The RETADD code is based on one by Heffter, Taylor and Ferber (1975). Other well-known long-range transport models are described in the work of Egan and Mahoney (1972), Start and Wendell (1974) and Rao et al. (1976). Further examples of the use of long-range transport calculations can be found in Husar, Lodge, and Moore (1978). Eliassen (1980) is a good review of various approaches to long-range transport modeling. Bass (1980) discusses various U.S. models with particular emphasis on regulatory applications. Both the Eliassen and Bass papers contain many references giving detailed information on different modeling efforts.

---

*Operated by Union Carbide Corporation, Nuclear Division under contract W-7405 eng 26 with the U.S. Department of Energy.

Codes such as RETADD operate by following wind trajectories which advect material from a source location and which superimpose a diffusion rate upon the advection process. By storing the results of many trajectory passes, average patterns of ground level concentration are deduced; and by repeating the process, the average pattern resulting from a combination of a number of sources can be deduced. Of course, this procedure becomes time-consuming when a large number of sources is involved. Many current problems (e.g. the NCUA study referenced above) require the analysis of the fate of atmospheric effluents associated with a given technology. Such effluents may arise from many sources and a code which can handle a large number of sources efficiently would therefore be useful.

An approach which uses an Eulerian grid and follows the time evolution of concentration patterns is better suited to handling a large number of sources since it avoids the time-consuming problem of serially following trajectories from each of the sources. A number of such approaches exist, e.g. Egan and Mahoney (1972), Rao et. al. (1976), Shannon (1979). One approach which has recently gained popularity is the pseudo-spectral method (Orszag, 1971; Christensen and Prahm, 1976; Prahm and Christensen, 1977; Mills and Hirata, 1978). This technique has been shown to have good accuracy when compared to other advective techniques (Mahlman and Sinclair, 1977; Chock and Dunker, 1981). In this paper we describe a pseudo-spectral code which has been designed to deal with multiple source assessment problems involving transport over continental scale distances and which uses readily available upper-air wind data. This code is compared with a trajectory code which also uses the same upper-air data to interpolate trajectory segments. It is true, of course, that no long-range transport code exists which has been fully validated. Although we compare the pseudo-spectral results with those of a trajectory code we are not implying that the latter has been validated. However, such codes have come to be accepted and are probably better understood than most.

The pseudo-spectral code described here will be referred to as PHENIX (Murphy, 1981) and the trajectory code with which it is compared is known as RETADD (Begovich et al., 1978).

The Pseudo-Spectral Technique

The advection and diffusion of a pollutant species in the atmosphere are governed by the equation of continuity for a binary mixture (Bird, Stewart and Lightfoot, 1960)

$$\frac{\partial(\rho x)}{\partial t} = -\nabla.(\rho\bar{V}x) + \nabla.(\rho D\nabla x) + Q - S \quad , \tag{1}$$

where ρ is the density of the mixture and x is a fraction such that $\rho x = C$, the pollutant concentration,
$\bar{V}$ is the wind vector,
D is the diffusion coefficient,
Q is a source strength, and
S is a sink strength.

The first term on the right-hand side represents advection and the second term represents diffusion. If one is concerned only with the two-dimensional problem, one can assume a constant value for D. Furthermore, the assumption of incompressibility implies a constant value of ρ and a non-divergent $\bar{V}$ field. Thus, the equation simplifies to

$$\frac{\partial C}{\partial t} = -\bar{V}.\nabla C + D\nabla^2 C + Q - S \quad . \qquad (2)$$

The present approach, which is two-dimensional here, computes the spatial derivatives using spectral techniques and then uses these derivatives together with source and sink strengths to solve the ensuing differential equation in time. Specifically, the concentration is expressed as a Fourier expansion and the Fourier expansion coefficients (i.e. in spectral space) are used to derive the spatial derivatives, whereas the differential equation in time is solved in physical space. Hence the description of the process as pseudo-spectral. Further details on the pseudo-spectral technique may be obtained in the references cited above.

OUTLINE OF THE PHENIX CODE

The code described here and used in these tests is two-dimensional and employs a spatial grid which has an upper limit of 32 x 32 cells. The number of cells in the x direction (N_x) need not equal the number in the y direction(N_y). Furthermore N_x and N_y need not be limited to being powers of 2 because this model employs a mixed radix fast Fourier transform (FFT) algorithm (Singleton, 1969).

Output from the model consists of two concentration fields on the $(N_x)x(N_y)$ grid of cells. In the current configuration of the code, these two concentration fields are intended to represent a primary and a derived species, e.g., SO_2 and sulfate or a parent and daughter radionuclide pair. The code requires one to specify emission source strengths and locations for all primary and secondary (derived) species. The code also requires one to specify transfer rates between species and the rates at which other loss mechanisms operate. Important loss mechanisms are deposition to ground either in the form of dry deposition or wet deposition following precipitation scavenging.

Frequency Cutoff

The maximum frequency which can be resolved by the grid (the Nyquist frequency) is that corresponding to the spacing of two grid intervals. In the present work the higher frequencies (i.e. from about half the Nyquist frequency upwards) were removed. This was done partly to produce a smoothing effect and partly because of a concern about higher frequency components introduced by the products on the right hand side of Eq. (2). The details of how this frequency cutoff was performed are discussed below.

Periodic Boundary Conditions

Because of the implied periodicity of the discrete Fourier expansion, the boundary conditions can give rise to an inconvenient phenomenon. Namely, material which is advected or which diffuses across a boundary will reappear at the opposite boundary. The present code treats this problem in the same manner as do Christensen and Prahm (1976). A decay term is employed for cells adjacent to the boundaries and all sources occurring in such cells are suppressed. Other techniques exist for treating this problem, e.g., Roache (1978). The Christensen and Prahm treatment of the problem has been found to be efficient and is currently being used although it may be modified at a later date.

Wind Field Generation

In order to proceed with the solution to Eq. (2), a wind field must be constructed for the $(N_x)x(N_y)$ grid. The present formulation of the model uses upper-air wind data available from rawinsonde and pibal observation in the U.S. and Canada. These wind data have been used in models such as RETADD (Begovich, Murphy and Nappo, 1978) and are available on a six-hourly basis.

The wind field is currently interpolated using the same technique as in RETADD for the interpolation of trajectory segments. The wind observations at any station are at first averaged throughout the transport layer (depth specified by the user). An interpolated wind vector is deduced for each grid cell from a weighted average of the surrounding (observed) layer-averaged wind vectors. The weighting has a distance component which is of the form $1/r^2$ and an alignment factor of the form $1 - 0.5\ (\mathrm{Sin}|\Theta|)$ where Θ is the angle between the station-averaged wind vector and the line joining the point of interpolation to the point of observation.

Time Integration

Having calculated the spatial derivatives and having the source strengths, sinks and interspecies transfer rates, one sees from Eq. (2) that the formulation now becomes an initial value problem in time. Mostly this is solved using a third-order Adams-Bashforth technique; however, in getting the process started, first-and second-order techniques are used.

Time steps must be chosen for the time integration such that the process remains stable. The stability of these procedures is discussed in detail by Shampine and Gordon (1975), and it is discussed in the context of the current problem by Prahm and Christensen (1977). Generally speaking, instabilities are more likely to arise via the advection terms than via the diffusion terms. The criterion for stability which pertains to the advection process is

$$\Delta t \leq 0.22 \frac{\Delta L}{v} \qquad (3)$$

where ΔL is the grid spacing and v the maximum wind velocity. There is also a criterion to ensure stability in the diffusion process; however, this is usually satisfied if (3) is satisfied. The current code takes note of both criteria, however, and chooses Δt by satisfying the more demanding of the two.

Long-Range Atmospheric Diffusion

The PHENIX code is here concerned only with lateral diffusion since uniform vertical mixing is assumed for the case being studied. Gifford (1977) discusses various proposed forms of lateral plume spread as a function of time and reports on the existing observational data. The approach in the PHENIX code is to recognize that, when simulating long time periods (e.g. months), the large-scale eddies of the wind field will contribute the large-scale component of the turbulence. The smaller scale local turbulence for individual plumes is supplied by the diffusion coefficient D (see eqs.1 and 2).

In summary, we employ a diffusion coefficient which takes care of diffusion over distances on the order of a typical grid spacing. We point out that this parameter is known only very roughly. However, when the model is used as intended, i.e. to produce long-term average concentrations, the turbulence scales of interest will be contained in the wind data used to drive the model. This idea is contained, for instance, in the approach advocated by Fay and Rosenzweig (1980) which, however, requires one to know the mean wind vectors for pollutants originating at various sources of interest together with dispersion rates about the mean wind. The present approach allows one to calculate such things for any given averaging period; and, furthermore, the calculations can accommodate a large number of sources without added effort.

Source and Sink Terms

The concentration function with which this model operates is complex. Thus one has the freedom to follow two components of the concentration. As structured, the model considers the real part of the concentration as being a primary emitted pollutant (e.g. SO_2) and the imaginary part is used to follow a derived species (e.g. sulfate).

When modeling the evolution of a primary and a derived species, one must specify a term which gives the rate of decay of the first species and the consequent rate of generation of the second species. Then, at each time step those rate terms are used to account for decay and growth of the respective species for the duration of the time step.

Sink terms for both species are also required in order to describe deposition to ground. These are entered as deposition velocities (dry + wet), for each species.

Non-Negativity Constraints

As explained, some higher frequencies are removed from the Fourier expansion. This, however, has a penalty associated with it which is apparent in the form of oscillations (Gibbs' phenomenon) in the data when the inverse transformation is performed. The disadvantage of such oscillations is that in those places where the concentration is near zero, they tend to produce alternating positive and negative values. Such a situation is not always a serious error considering the overall accuracy of long-range transport models. But, at the very least, the problem is a nuisance from the point of view of data display.

The PHENIX code sets all negative values to zero at each time step. Since the pseudo-spectral technique conserves material it is necessary to reduce the positive total by the same amount. This is done at each positive cell in an amount proportional to the local value of the concentration. This procedure is likely to introduce some diffusive effects and it is invoked here primarily out of convenience.

THE RETADD CODE

The trajectory code RETADD computes trajectories originating from each source four times per day. These trajectories are composed of three hour segments interpolated from the upper-air wind data base. Horizontal diffusion of the form σ_y=0.5t (with σ_y in meters and t in seconds) is employed to calculate the lateral contribution to the diffusion process. For the distances involved in the present calculations the assumption of a uniformly mixed boundary layer applied over most of the travel distances. However, in the early stages of travel a vertical diffusion process applies. The model uses a vertical diffusion coefficient specified by the user and one can specify an arbitrary height for the emission source. This is important from the point of view of deposition loss during the early stages of plume travel. Although not important in the present context the model can account for the decay, growth and deposition loss of a parent and up to seven daughter products (Murphy, Nelson and Ohr, 1980). This model will be reported on in more detail at a later date.

COMAPRISON OF PSEUDO-SPECTRAL AND TRAJECTORY MODELS

For comparison purposes a grid containing 21 cells in the X direction (East-West) and 15 cells in the Y direction (North-South) was used. This grid extends from 85° to 106° W and from 30°N to 40°N. The study area is shown in Fig. 1.

In the case of RETADD, the model was run for this actual area described above. In the case of PHENIX, the model was run for a larger area. This larger area consisted of a grid of 25 by 21 cells. The central area of 21 x 15 cells was therefore obtained by discarding two cells at each end in the X direction and three cells at each end in the Y direction. This larger area was used in the calculations so as to avoid possible edge effects due to the boundary conditions.

A group of 10 sources was located within the study area. These were considered as emitting a primary pollutant with source strengths between 4440 and 7376 units per second. Figure 2 shows the location of the 10 sources superimposed on the 21 x 15 grid.

Comparisons were made using a test run of both models for the month of January 1975. The ten sources emitted a primary species which decayed to a secondary species with a decay rate of 1.0 x 10^{-5}/s (corresponding to a half-life of about 19 hours). No other loss mechanisms were considered (e.g. no loss by ground deposition). A mixing height of 1000m was employed. In the case of the pseudo-spectral model the lateral diffusion coefficient was chosen as 40,000 m^2/s. Keep in mind that in the case of both models the lateral diffusion is determined mainly by the structure of the wind field. The rather large value of 40,000 m^2/s chosen for D in the pseudo-spectral model may have the effect of alleviating some of the Gibbs phenomena oscillations since it more quickly removes the high frequencies as the plumes migrate away from the source cells.

Discussion of Comparisons

Figures 3 and 4 show respectively the trajectory model results for species 1 and 2. These should be compared with figures 5 and 6 which show respectively the pseudo-spectral model results for species 1 and 2. Figures 5 and 6 each have two parts a) and b). These show the results of running the pseudo-spectral model under varying circumstances; a) were obtained with all frequencies greater than half the Nyquist frequency being set to zero, and b) were obtained when those frequencies above half the Nyquist frequency were set to zero gradually by employing a Gaussian filter as illustrated in Figure 7. We refer to these two frequency cutoff processes as "abrupt filtering" and "smooth filtering" respectively.

An inspection of figures 3, 4, 5 and 6 shows the following general results:

(i) The pseudo-spectral results for species 2 are more alike to those of the trajectory model than are the results for species 1.
(ii) The use of a smooth filter when performing the frequency cutoff tends to produce better agreement with the trajectory model.

It is likely that these results come about for the following reasons:
(i) The concentration function for species 2 is much smoother than for species 1 (which approximates a delta function in places). Thus the Fourier expansion for species 2 converges more rapidly.
(ii) The smoothing of the cutoff in frequency space reduces the magnitude of Gibbs phenomena in physical space.

None of these results is very surprising. They do, however, point out the problems and limitations of the pseudo-spectral technique and the extent to which these problems are tractable.

The results of these simulations were also compared on the basis of the total amount of material in the study area. Table 1 shows the results of these comparisons. From the average monthly concentrations the average amount of material within the mixing

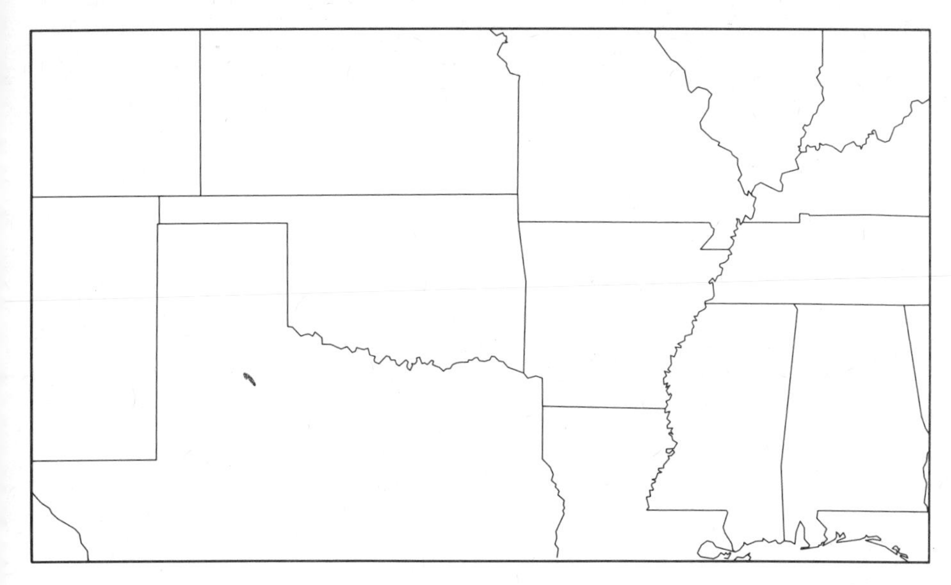

Figure 1. Study Area

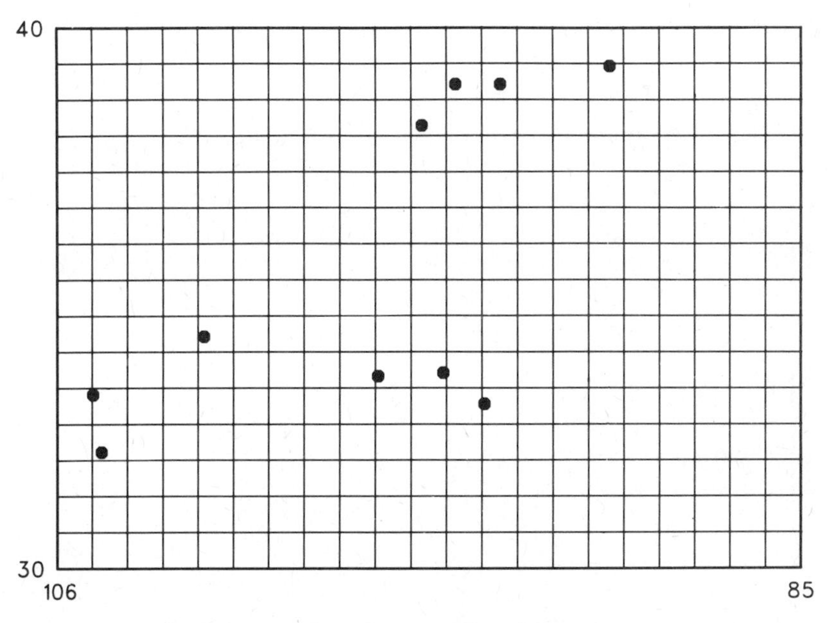

Figure 2. Source Locations

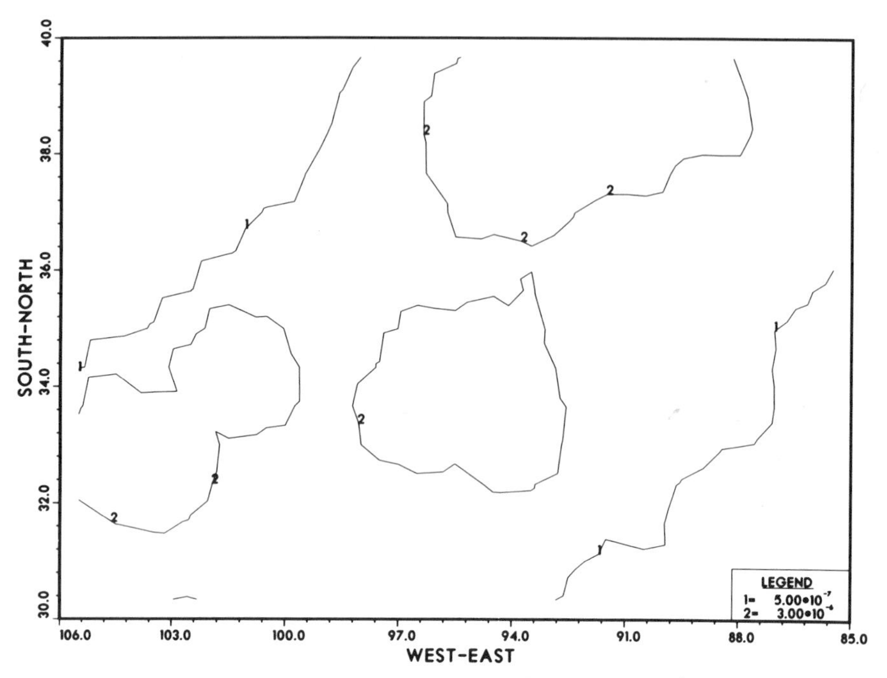

Figure 3. First Species (Trajectory)

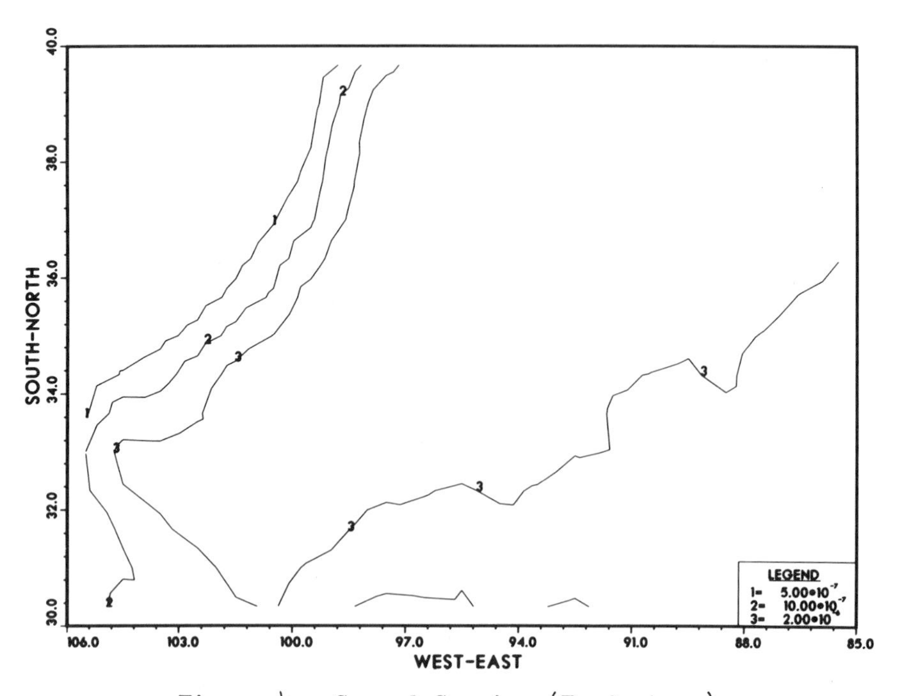

Figure 4. Second Species (Trajectory)

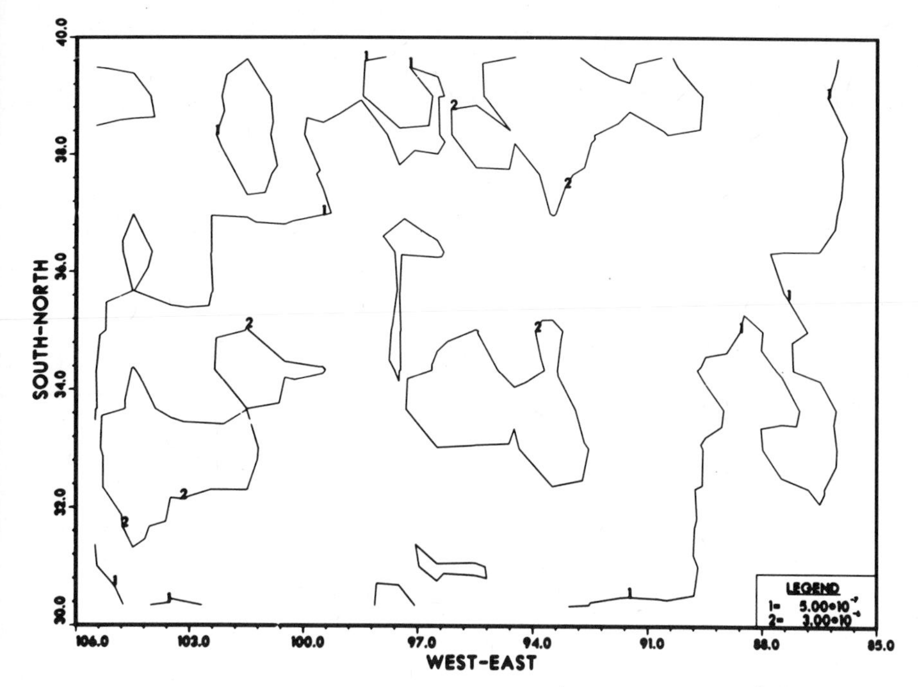

Figure 5a. First Species (Abrupt Filter)

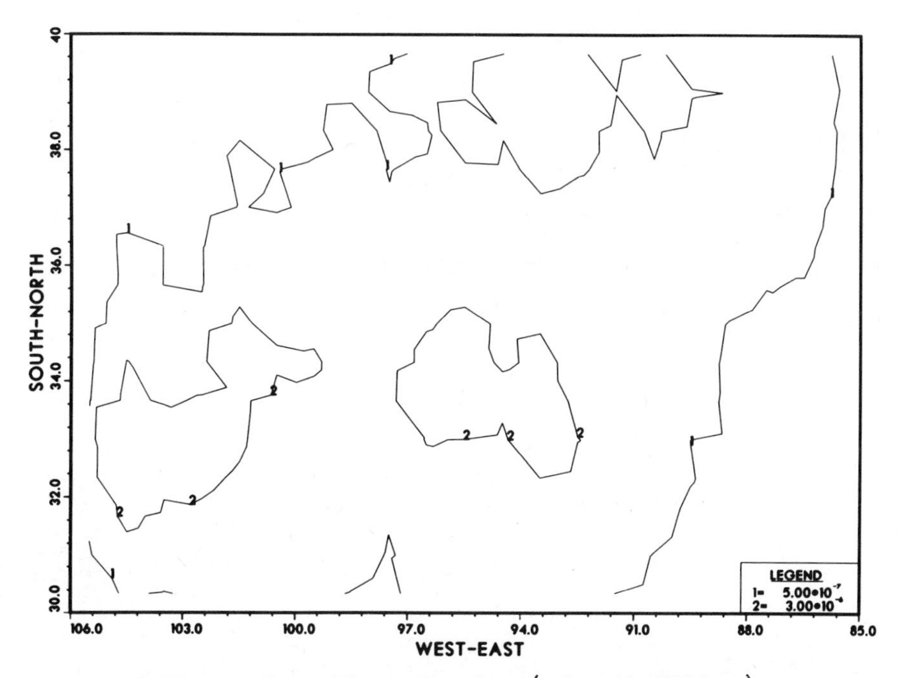

Figure 5b. First Species (Smooth Filter)

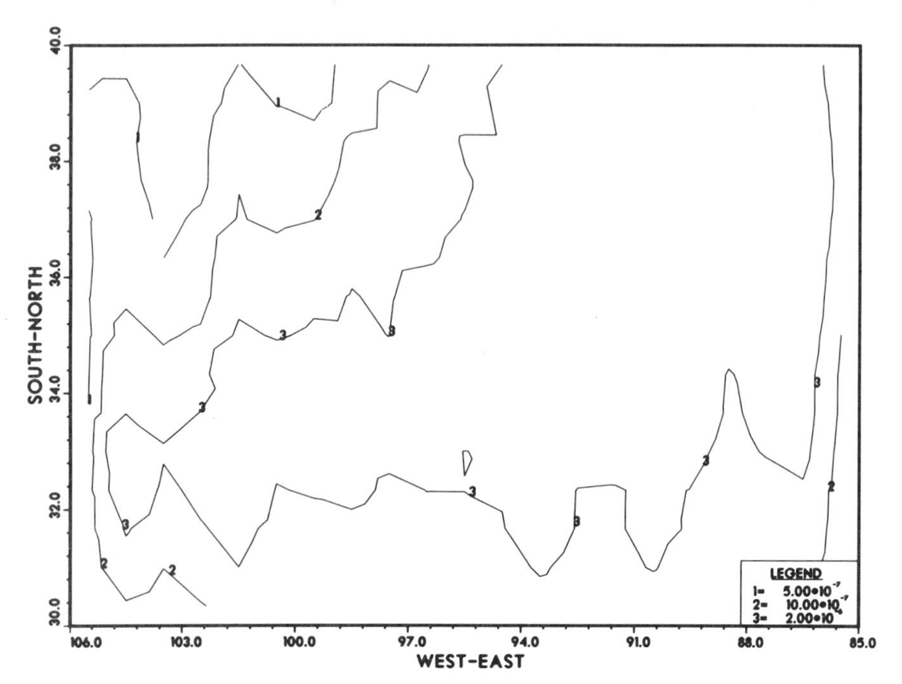

Figure 6a. Second Species (Abrupt Filter)

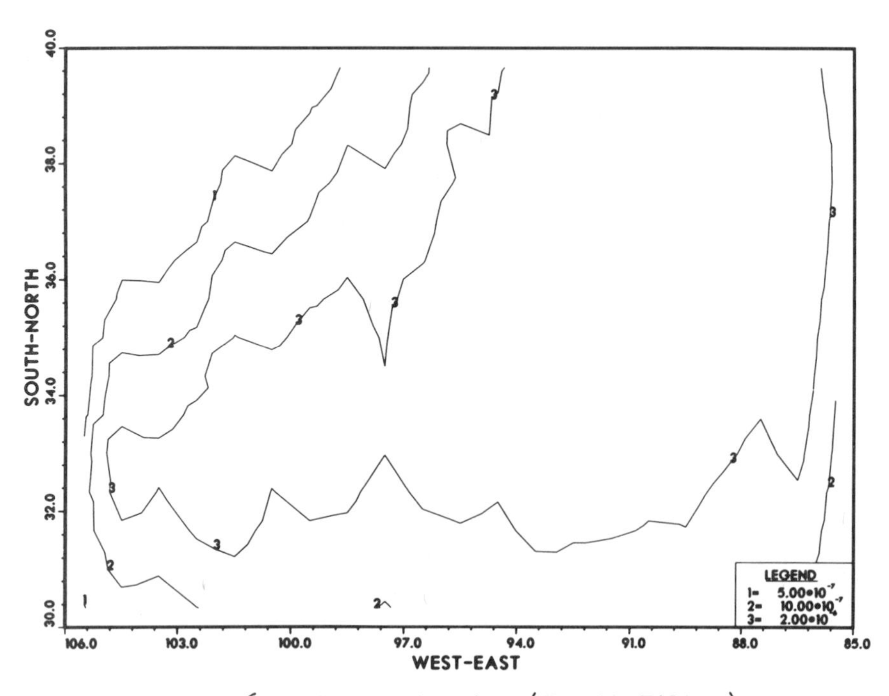

Figure 6b. Second Species (Smooth Filter)

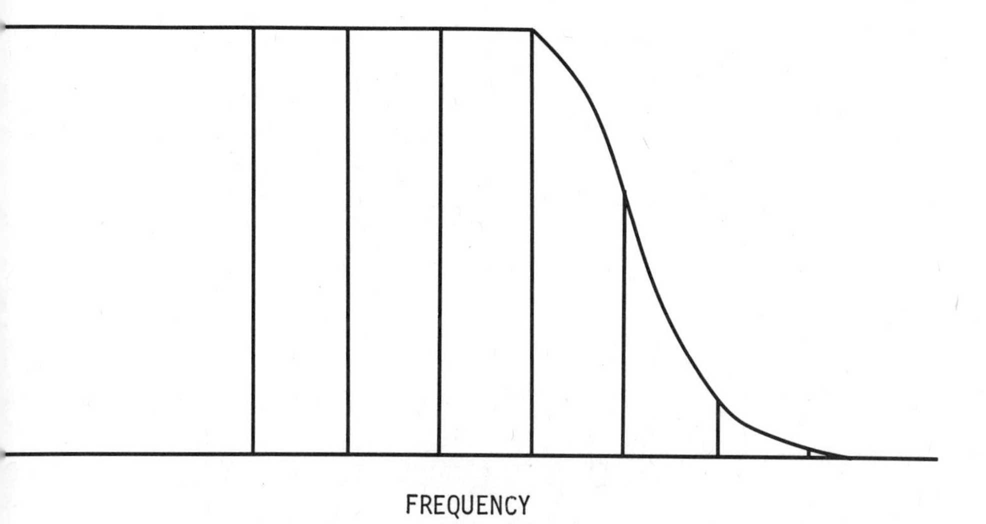

Figure 7. Frequency Filter

layer and within the 21 x 15 grid has been calculated. Table 1 gives the total deduced from the PHENIX calculations as a percentage of the corresponding values calculated with RETADD.

TABLE 1
Total Material in Study Area
(Percentage of RETADD Total predicted by PHENIX)

	SPECIES NO. 1	SPECIES NO. 2
ABRUPT FILTER	68	98
SMOOTH FILTER	72	99

It can be seen from the table that there is good consistency between the models in the case of the second species, however, the models are not as consistent as regards the first species.

In general the models are in better agreement in the case of the second species but they are reasonably in agreement for both species.

The version of PHENIX which has been described above is a preliminary one and some modifications are currently being considered. In particular, the handling of the periodic boundary conditions and techniques for the elimination of negative quantities are areas which can probably be improved upon to a significant extent.

EFERENCES

ass, A., Modeling Long Range Transport and Diffusion, AMS/APCA Second Joint Conference on Applications of Air Pollution Meteorology, New Orleans, Louisiana (March 24-27, 1980).

egovich, C. L.; Murphy, B. D.; and Nappo, C. L., RETADD: A Regional Trajectory and Diffusion-Deposition Model, ORNL/TM-5859 (June 1978).

ird, R. B.; Stewart, W. E.; and Lightfoot, E.N., Transport Phenomena, John Wiley and Sons (1960).

nock, D. P. and Dunker, A. M., A Comparison of Numerical Methods for Solving the Advection Equation, AMS Fifth Symposium on Turbulence Diffusion and Air Pollution, Atlanta, Georgia (March 9-13, 1981).

hristensen, O. and Prahm, L. D., "A Pseudo-Spectral Model for Dispersion of Atmospheric Pollutants", J. Appl. Meteor. 15: 1284-1294 (1976).

avis, R. M., et al. National Coal Utilization Assessment. A Preliminary Assessment of Coal Utilization in the South, ORNL/TM-6122 (Oct. 1978).

gan, B. A. and Mahoney, J. R., "Numerical Modeling of Advection and Diffusion of Urban Area Source Pollutants", J. Appl. Meteor. 11: 312-322 (1972).

liassen, A., "A Review of Long-Range Transport Modeling", J. Appl. Meteor. 19: 231-240 (1980).

ay, J. A. and Rosenzweig, J. J., "An Analytical Diffusion Model for Long Distance Transport of Air Pollutants", Atmospheric Environment 14: 355-365 (1980).

ifford, F. A., "Tropospheric Relative Diffusion Observations", J. Appl. Meteor. 16: 311-313 (1977).

effter, J. L.; Taylor, A. D.; and Ferber, G. J., A Regional-Continental Scale Transport Diffusion and Deposition Model, NOAA Tech. Memo. ERL ARL-50 (1975).

usar, R. B.; Lodge, J. P. and Moore, D. J., editors, "Sulfur in the Atmosphere", Proceedings of the International Symposium held in Dubrovnik (September, 1977), Pergamon Press Ltd. (1978) (also published as volume 12, number 1/3 of Atmospheric Environment).

ahlman, J. D. and Sinclair, R. W., Tests of Various Numerical Algorithms Applied to a Simple Trace Constituent Air Transport Problem, Fate of Pollutants in the Air and Water Environments, Part 1, Vol. 8, I. H. Suffet, Ed., John Wiley and Sons (1977).

ills, M. T. and Hirata, A. A., A Multi-Scale Transport and Dispersion Model for Local and Regional Scale Sulfur Dioxide/Sulfate Concentrations: Formulation and Initial Evaluation, NATO/CCMS Air Pollution Pilot Study, Toronto (August 1978).

urphy, B. D.; Nelson, C. B. and Ohr, S. Y., A Consistent Treatment of Ground Deposition Together with Species Growth and Decay During Atmospheric Transport, Symposium on Intermediate Range Atmospheric Transport Processes and Technology Assessment, Gatlinburg, TN, October 1-3, 1980.

urphy, B. D.; Parzyck, D. C. and Raridon, R. J., Estimated Pollutant Loading Patterns and Population Exposures Associated with Power Plant Siting in the South, Air Pollution Control Association, Houston, Texas (June 1978).

Murphy, B. D., PHENIX - A Pseudo-Spectral Model of Long-Range Atmospheric Transport, ORNL-5761 (July 1981).

Orszag, S. A., "Numerical Simulation of Incompressible Flows within Simple Boundaries: Accuracy", J. Fluid Mech. 49 Part 75-112 (1971).

Prahm, L. P. and Christensen, O., "Long-Range Transmission of Pollutants Simulated by a Two-Dimensional Pseudo-Spectral Dispersion Model", J. Appl. Meteor. 16: 896-910 (1977).

Rao, K. S.; Thomsen, I. and Egan, B. A., Regional Transport Model of Atmospheric Sulfates, NOAA, ATDL Contribution No. 76/13 (1976).

Roache, P. J., A Pseudo-Spectral FFT Technique for Non-Periodic Problems, J. Computational Phys. 27, 204-220 (1978).

Shampine, L. F. and Gordon, M. K., Computer Solution of Ordinary Differential Equations, W. H. Freeman and Company, San Francisco (1975).

Shannon, J. D., "A Gaussian Moment - Conservation Diffusion Model" J. Appl. Meteor. 18: 1406-1414 (1979).

Singleton, R. C., "An Algorithm for Computing the Mixed Radix Fast Fourier Transform", IEEE Trans. Audio and Electroacoustics AU-17: 93-103 (1969).

Start, G. E. and Wendell, L. L., Regional Effluent Dispersion Calculations Considering Spatial and Temporal Meteorological Variations, NOAA Technical Memorandum, ERL ARL-44, Idaho Fall Idaho, 1974.

DISCUSSION

J. K. WESTBROOK

How can you use the term "improvement" to describe the result of the pseudo-spectral method in comparison to the trajectory model when the sources are hypothetical?

B. D. MURPHY

The use of the term "improvement" was unfortunate. I sought only to indicate the general behavior of the two models and illustrate weaknesses and strengths of the pseudo-spectral model.

As used "improvement" means that when the smooth filter was employed the pseudo-spectral results approached more closely those of the trajectory model. Since no long-range transport model exists which has been validated, I have used a trajectory model as a standard against which to judge the pseudo-spectral model. However, I am well aware that the trajectory model may also have shortcomings.

A MESOSCALE AIR POLLUTION TRANSPORT MODEL : OUTLINE AND PRELIMINARY RESULTS

H. van Dop and B.J. de Haan

Royal Netherlands Meteorological Institute
P.O. Box 201
3701 AE De Bilt, The Netherlands

INTRODUCTION

An air pollution transport model is developed. It will be applied in an area which covers the Netherlands and (parts of) the surrounding countries (500 x 500 km^2). Transport is described by

$$\partial c/\partial t + u\partial c/\partial x + v\partial c/\partial y = \partial/\partial z\ (K\partial c/\partial z) + S. \qquad (1)$$

Here, u and v are the horizontal components of the average wind velocity, and $K = (K(x,y,z,t)$, is the eddy diffusivity. The term S contains the sources and sinks. Horizontal diffusion and average vertical velocity are neglected.

The magnitude of the diffusive term in Eq. (1) varies strongly both in time and in space. As a consequence the nature of the transport process can be either strongly dilutive in the vertical direction, or mainly (horizontally) advective. Therefore, we will only apply the full equation (1) only when advection and diffusion are equally important, and use adequate approximations otherwise.

Much attention is given to the construction of the meteorological input of the model. The wind field and eddy diffusivity data are obtained by interpolation procedures which are extensively discussed in a previous paper (Van Dop et al., 1980). The input data are deduced from mainly synoptic data so that the model can be applied in any region with a standard meteorological network.

The required synoptic data are (i) the 10 m wind velocity, (ii) the surface pressure, (iii) the pressure at a specified elevated level (850 mbar), and (iv) the surface temperature. In addition it is desirable (but not required) to have some knowledge of the topography in order to determine some relevant surface characteristics (roughness length, heat flux, deposition velocity).

A second field of interest is the numerical treatment of the transport equation (1). In an earlier paper (De Haan, 1980) compared some surrent finite different schemes for the solution of Equation (1). He points out that the so called "pseudo-spectral" method is one of the most efficient and accurate procedures. However, using this method some care should be taken with the numerical treatment of the (horizontal) boundaries. (Roache 1975). This method, combined with the Crank-Nicolson procedure for the diffusive part of the equation, is the numerical basis for the solution of the transport equation.

In the next section the model concept is treated, and in the third some results of the interpolation procedures of the meteorological input are given.

MODEL CONCEPT

An important meteorological parameter in air pollution transport modelling is the diffusivity $K(x,y,z,t)$. The parameter varies over orders of magnitude both in time and in height. This determines the magnitude of the diffusive term in Eq. (1). As a consequence the nature of this equation changes alternatively from an equation where advection is dominated by diffusion and vice versa.

The basic approach here is, that only where advection and diffusion are equally important the full transport equation Eq. (1) is numerically solved. When advection dominates diffusion only advection is considered, and we solve the equation neglecting the diffusive term. When on the other hand diffusion is very effective, we assume that the (vertical) concentration distribution is homogeneou

These assumptions save a considerable amount of computation time, but divide the space-time domain in which we are interested i several sub domains, according to the magnitude of the diffusion coefficient K. This coefficient is determined by the turbulence which is created by wind shear and buoancy. Therefore, large values are attained in the boundary layer only. Above this layer K values are generally small. We distinguish the following domains.

1. the boundary layer, $0 < z < z_i(t)$•
2. the region above the boundary layer $z > z_i(t)$.

• We assume that the variation of the boundary layer height as a function of spatial coordinates may be neglected.

The boundary layer region, $z < z_i$

Within this layer the turbulence may vary from moderate to strong, depending on heat flux and wind velocity. Also, the height of the boundary layer itself may vary, proportional to the generated turbulence. Based on these considerations we make another distinction : (i) moderate turbulence, and (ii) strong turbulence.

In the former case, which is usually encountered during the evening, night and early morning, we assume that the boundary layer height does not exceed a certain fixed value, conveniently set at 600 m •). In this case the full transport equation (1) is applied.

In the latter case we assume that vertical mixing is so vigorous that the concentration distribution is homogeneous above say 25 m. Then only advection is considered. The dilution which occurs due to the increase of the boundary-layer height can be simply incorporated in the advection equation :

$$\partial c/\partial t + u\partial c/\partial x + v\partial c/\partial y = -(c/z_i)\, dz_i/dt + S. \quad (2)$$

In Eq. (2) $c = c(x,y)$ denotes the average boundary layer concentration.

The region above the boundary layer, $z > z_i$

In this region diffusion will be neglected, so that Eq. (1) reduces to the simple advection equation,

$$\partial c/\partial t + u\partial c/\partial x + v\partial c/\partial y = S, \quad (3)$$

for which the exact solution are streak lines. Physically this solution can be interpreted as a plume axis. The streak lines can be simply evaluated from the (known) wind field.

The model concept is summarized in Fig. 1

Numerical procedures

For the numerical evaluation all data are presented on a grid. The horizontal gridpoint distance is 20 km and the vertical grid-

---

•) This value is equal to the vertical extension of the model. This particular value was chosen because it was assumed that no important sources or influx were present above that height. Further, vertical exchange at that height can be neglected most the time. Only in strongly convective conditions this process should (and will) be considered.

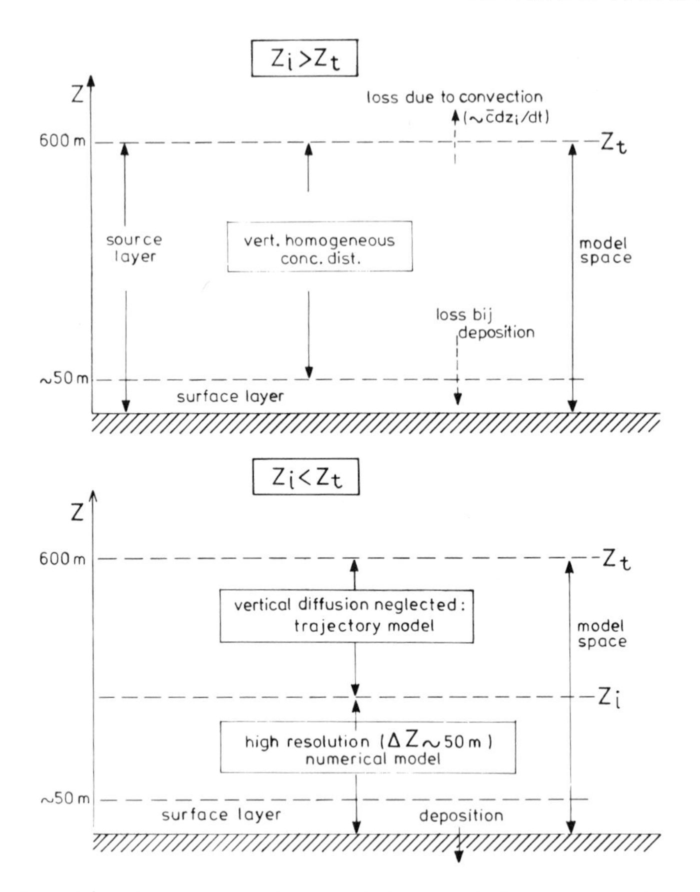

Fig. 1 Graphical presentation of the model concept (vertical cross section). Two cases are distinguished : (a) the inverstion depth (or mixed layer height) z_i, does not exceed the (fixed) upper boundary of the model, z_t (at present 600 m), and (b) the inversion height exceeds the upper boundary, z_t. The latter situation usually occurs during daytime conditions.
All sources are assumed to be contained within the model space.

point distance is fixed on 50 m. The model contains now 17 x 17 x 12 gridpoints, with a maximum extension to 25 x 25 x 12 gridpoints. The lowest gridlevel represents the surface layer.

It should be noted here that the full solution (Eq. (1)) is only required for a fraction of the time and in a limited space, because Eq. (1) is solved in the region $z < z_i(t)$ only,with the additional constraint on the time domain given by $z_i(t) < 600$.

At the boundary-layer height, z_i, we neglect vertical tran-
port :

$(K\ \partial c/\partial z)_{z=z_i} = 0$, and at the lowest level we assume that

$K\ \partial c/\partial z = v_g\ C)_{z=25\ m}$. The deposition velocity v_g at that level
s determined by the of the inverse sum of the turbulent resistan-
es of the layer 1-25 m and the surface resistance 0-1 m. The
ormer value is determined from surface layer similarity expres-
ions, while the latter are obtained from literature (Van Dop, 1981)

At the lateral inflow boundary external boundary values are
upplied. At the outflow side no boundary conditions are forced
De Haan, 1981).

For the solution of the full equation we use a time splitting
ethod (Marchuk, 1969),

$$\partial c/\partial t + u\partial c/\partial x + v\partial c/\partial y = 0 \qquad (4)$$

$$\partial c/\partial t - \partial/\partial z\ (K\ \partial c/\partial z) = S. \qquad (5)$$

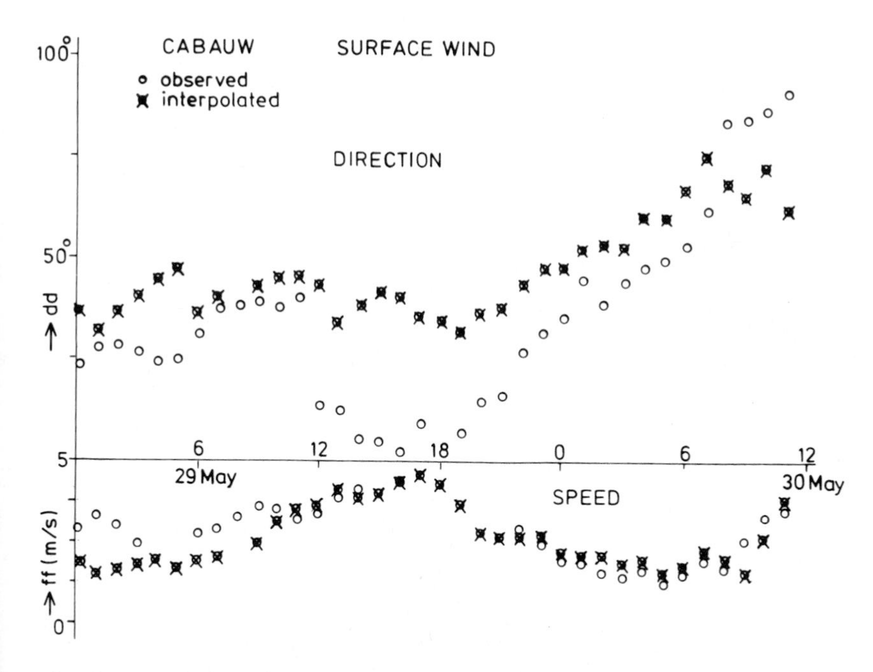

ig. 2 Observed and interpolated wind at Cabauw at 10 m height,
29, 30 May, 1978.

Eq. (4) and (5) are stepwise subsequently solved. Eq. (4) is solved by a pseudo spectral technique (Orszag, 1971), in combination with the 4th order Runge Kutta method (Gear, 1971) for the time integration. Special attention has been paid to the boundary conditions (Roache, 1978, De Haan, 1981).
Eq. (3) is solved by a Crank Nicolson scheme (Richtmyer and Morton 1967). The time step used is approximately 15 minutes.

METEOROLOGICAL INPUT

The wind, diffusivity filed and boundary-layer height are the basic meteorological input of the model. It appears from the previous section that wind and eddy diffusivity data are required up to a height of 600 m. In this section we will give a summary of the results of some interpolation procedures.

Synoptic data

Parameters which are required for the construction of the win

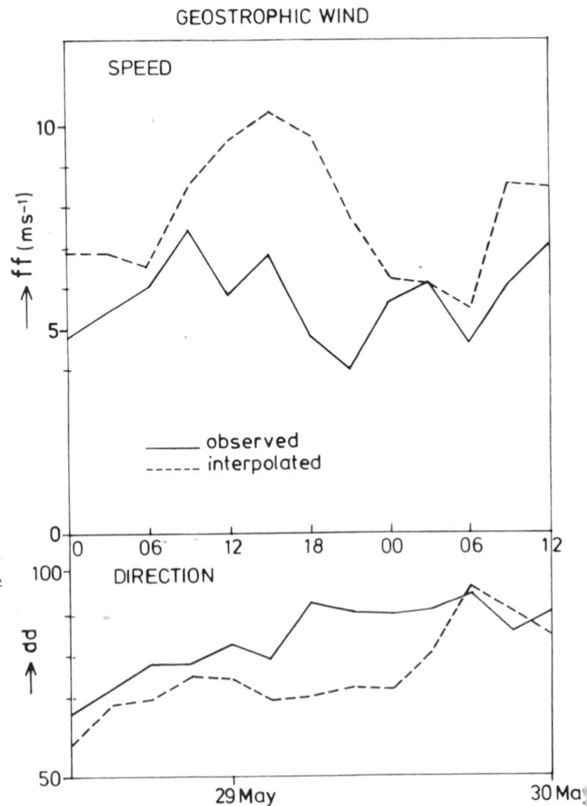

Figure 3
Interpolated ground level geostropic wind speed and direction at Cabauw. The observed 850 mbar radiosonde analysis is given by the solid line.

field are the observed 10 m and 850 mbar wind, the geostrophic wind at ground level and, when available, wind measurements at towers. The determination of the vertical wind profile requires also the knowledge of atmospheric stability (expressed as Obukhov length, L), the friction velocity u , the surface roughness, z_o, and the boundary-layer height, z_i. A discussion of the horizontal and vertical interpolation procedure will be given.

Horizontal interpolation

a. 10 m wind.
In order to test the interpolation schemes a data base was assembled from 36 subsequent hours of meteorological observations (00 GMT, 29 May, 1978 to 12 GMT, 30 May 1978). For each hour approximately 50 synoptic observations were available in the region. They were interpolated by means of the optimum interpolation method (Cats, 1980). Corrections of varying roughness of the terrain were included (Van Dop et al., 1980).
At Cabauw, where the experimental platform of the Institute is situated, we were able to compare observed data with interpolated ones (Fig. 2). The maximum difference in direction amounts to nearly 40°. In 70 % of all cases, however, it is less than 15°. The interpolated wind speed fits the observed data fairly well. The differences are in general well beyond 0.5 ms^{-1}.

b. Ground level geostrophic wind.
Surface pressure was interpolated and differentiated with respect to the horizontal coordinates x,y. Though no measurements were available with which interpolated values could be compared, we have plotted the geostrophic wind together with the 850 mbar radiosonde data, in order to have some reference (Fig. 3).

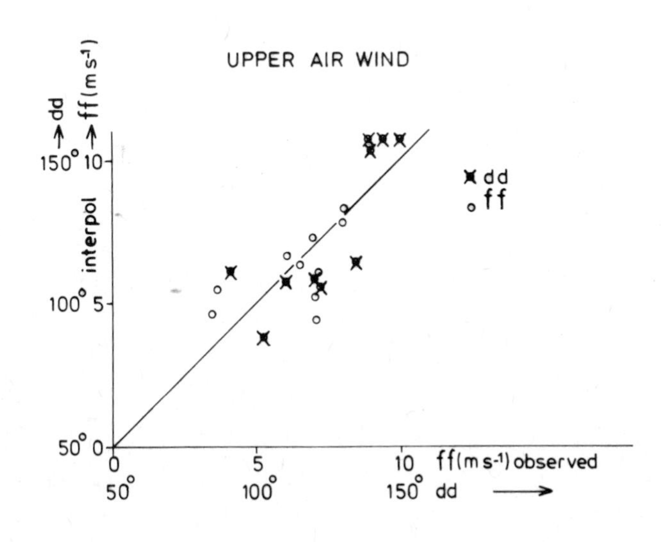

Figure 4.
Observed and interpolated wind speed and direction at Cabauw, at a height of 1500 m.

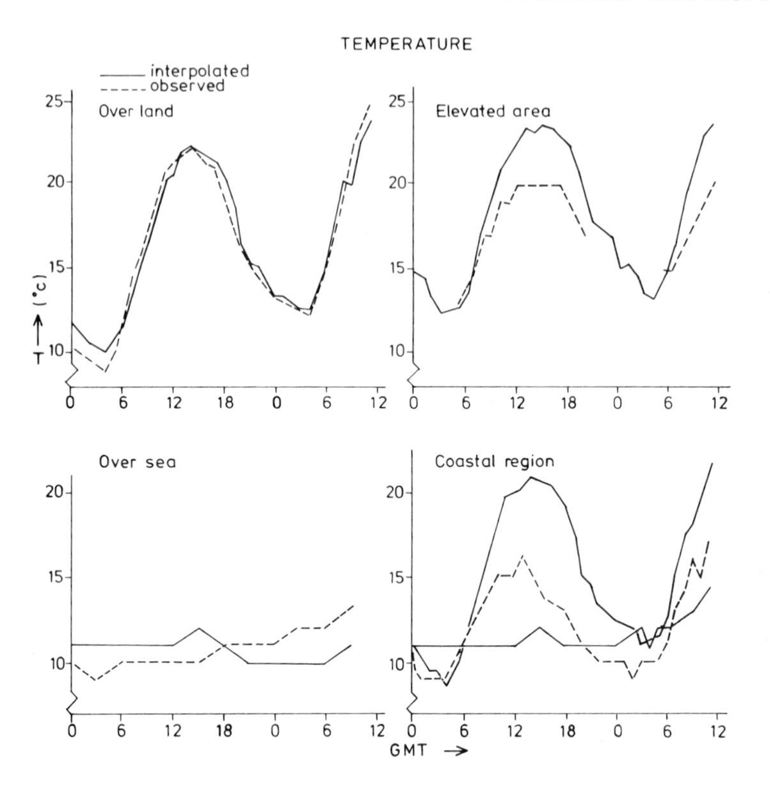

Fig. 5 Observed and interpolated temperature at 4 sites.

c. 850 mbar wind.
The analysis program of the Institutes numerical weather forecast model (Heijboer, 1977) interpolates radiosonde data on grindpoints covering Western-Europe and a large part of the Atlantic Ocean (gri distance ∿360 km at 50° latitude). Nine gridpoints were selected, which were situated in or near our model region. Wind data at 1500 m height •) were interpolated to our fine mesh grid (grid distance 20 km), by a standard r^{-2} method. At Cabauw the data were compared with in situ balloon measurements. Fig. 4 shows that -the inaccuracy of the data itself considered - the agreement both in speed and direction is excellent.
d. Temperature.
Temperature (observation height ∿1.50 m) was also interpolated by means of the optimum interpolation method. A distinction was made between land and sea surface : the temperature at gridpoints over land was obtained from synoptic observations over land only. The air temperature over sea was taken from the temperature, which

•) In fact at the 750 mbar level, which for simplicity was set equa to 1500 m.

were measured at light vessels situated in the North Sea. The results are given in Fig. 5. At four different sites the interpolated temperatures were compared with observations. The figure indicates that the interpolation is inaccurate in coastal regions and at elevated sites. The temperature interpolation is required to calculate the surface energy balance and atmospheric stability.

e. Obukhov length and friction velocity.
The parameters L and u_* are calculated according to procedures given by Holtslag et al. (1980), Nieuwstadt (1977) and Venkatram (1980). Besides the above mentioned data also data on cloud cover were required in order to determine L and u_*. Cloud cover at gridpoints was estimated by r^{-2} interpolation. The accuracy, of the procedures followed was tested at Cabaux (cf. also Holtslag, 1980). In the considered period special turbulence measurements were carried out from which L and u_* could be determined directly. In Fig. 6 the diurnal cycle is depicted. We observe a reasonable agreement between observed and interpolated data. The calculated friction velocity is slightly overpredicting, presumable due to the relatively high z_o-values used. We should note here that each gridpoint represents an area of 20 x 20 km^2 and that the roughness length is assigned to that particular area rather than to the close and relatively smooth vicinity of the meteorological tower at Cabauw.

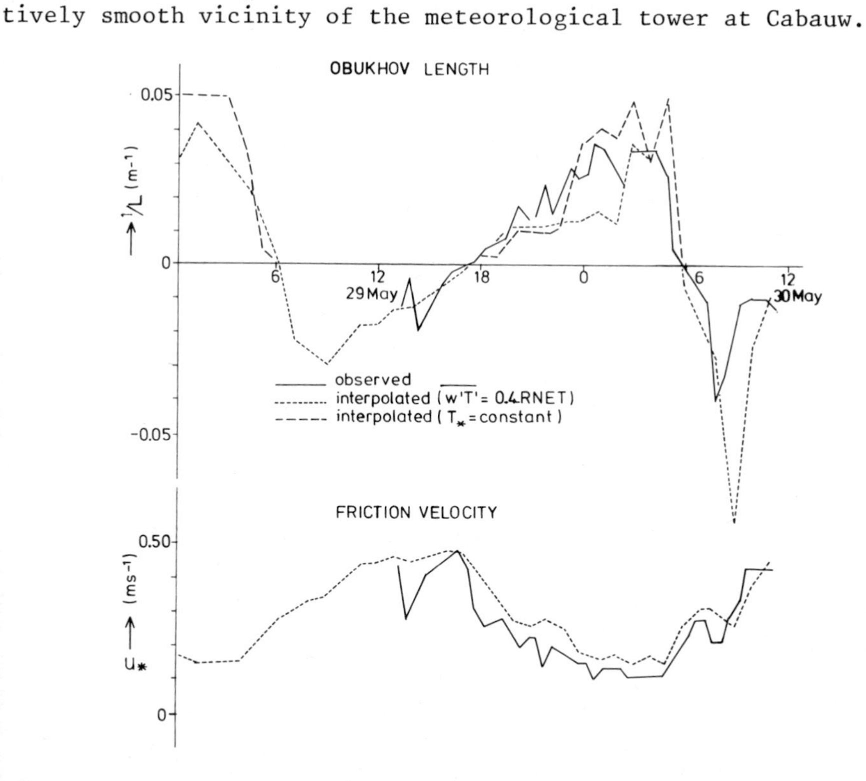

Fig. 6 Interpolated turbulence parameters (L and u_*) at Cabauw.

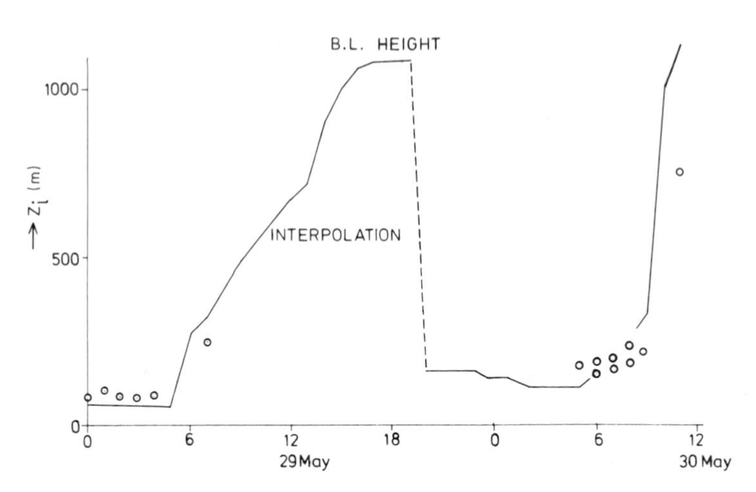

Fig. 7 Calculated inversion height at Cabauw. Sodar and radiosonde observations are indicated by open circles.

The Obukhov length in stable conditions was calculated in two different ways. In the first case proportionality with the net radiation was assumed (Holtslag, et al., 1980). In the second case, T_* was held constant during nighttime conditions (Venkatram, 1980). The data in the considered period, however, do not enable us to draw decisive conclusions on the quality of both methods.

f. Roughness length.

The surface properties play an important role in atmospheric transport models. The surface roughness, characterized by z_0, influence the wind field in the lowest 50 m, the diffusion and deposition. For each grid square the roughness was estimated by inspection of topographical land maps (Van Dop, 1981). The procedures followed are similar to those use by Smith and Carson (1977).

g. Boundary-layer height.

During nighttime the boundary-layer height was determined by a steady state expression adopted from Nieuwstadt (1981). During daytime (in fact when $L < 0$), an inversion rise expression was used (Tenneke 1973). Again at Cabauw calculated values are compared with balloon and Sodar observations (Fig. 7).

The agreement with the sparse observations is reasonable, though in view of the importance of this parameter a more thorough validation including more varying meteorological conditions, is desired. The determination of the boundary-layer height over sea is still more uncertain. In this particular application, however, the sea will be mainly at the outflow region of the model region, and meteorological circumstances there will play a minor role in the determination of ambient concentrations over land. Therefore, the boundary-layer height is assumed to be uniform over the whole area and time dependent only.

h. Eddy diffusivity.
For the vertical eddy diffusivity an expression of Brost and Wyngaard (1978) is used, which depends on L, u_*, z_i and z. In the surface layer it approaches the surface layer expressions for the eddy diffusivity of heat (Businger, 1973). At half the boundary-layer height its value attains a maximum. It vanishes at the boundary-layer height.

Vertical interpolation

The above calculations yield a horizontal field of hourly values of the :
- 10 m wind velocity,
- 1500 m wind velocity,
- the ground level geostrophic wind velocity,
- the turbulence parameters, L and u_*,
- the boundary-layer height, z_i,
- the surface roughness, z_o.

From these data vertical wind profiles can be constructed at any desired horizontal coordinate. The method which was already outlined in an earlier paper (Van Dop et al., 1980) will be briefly summarized here.

a. In the surface layer a wind profile was used based on the 10 m wind levocity, L, u_* and z_o. Similarity expressions were used to obtain the profile $u_b(z)$.•).
b. Above the boundary-layer height a geostrophic wind profile, $u_g(z)$, was constructed. It was obtained by linear interpolation of the 850 mbar wind velocity and the ground level geostrophic wind velocity.
c. At each tower a wind profile was constructed by fitting a polynominal through the measured data. These profiles were horizontally interpolated (r^{-2} interpolation), and resulted in the tower wind profile, $u_m(z)$.

The final wind profile was obtained by weighted interpolation of the three profiles,

$$u(z) = f_b(z).u_b(z) + f_g(z)\, u_g(z) + f_m(z)\, u_m(z), \qquad (6)$$

with $f_b + f_g + f_m = 1$. The weighting functions were chosen such that each profile got its maximum weight in the layer where it is thought valid.

•) For simplicity we denote only one component.

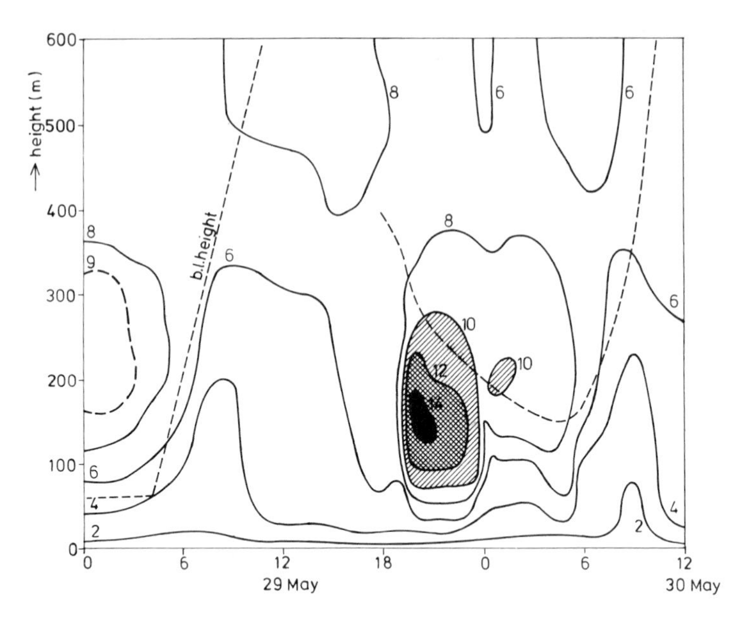

Fig. 8 Time height cross section of the interpolated wind speed at Cabauw. Isoplethes are drawn every 2 ms^{-1}. On 29 May also the 9 ms^{-1} isopleth is depicted in order to depict the low level jet.

An example of vertical interpolation is given in Fig. 8. Here a time-height cross section of iostaches is shown (Cabauw). During day time the wind speed gradient is weak. During both nights low level jets are observed. The wind profile measurements at Cabauw up to 200 m were, as expected, well reproduced.

REFERENCES

Brost, R.A., Wyngaard, J.C. (1978), A model study of the stably stratified planetary boundary. J. Atmos. Sci. 35, 1427-1440.

Businger, J.A. (1973), In : Working on micrometeorology. Haugen, D.A. (Ed.) American Meteorological Society.

Cats, G.J. (1980), Analysis of surface wind and its gradient in a mesoscale wind observation network, Mon. Wea. Rev., 108, 1100-1107.

Van Dop, H., DeHaan, B.J. and Cats, G.J. (1980), Meteorological input for a three dimensional medium range air quality model. Preprint of the procs. of the 11th ITM, Nato-CCMS, Amsterdam 24-27 November 1980.

Van Dop, H. (1981), Terrain classification for interregional transport models. To be published (preprint available).
Gear, C.W. (1971), Numerical initial value problems of ordinary differential equations,Prentice Hall Englewood Cliffs N.J.
De Haan, B.J. (1980), A comparison of finite difference schemes, describing the two-dimensional advection equation. Preprint of the procs. of the 11th ITM, Nato-CCMS, Amsterdam, 24-27 November 1980.
De Haan, B.J. (1981), Two numerical methods for the solution of the advection equation including cross-boundary flow. To be published (preprint available).
Heijboer, L.C. (1977), Design of a baroclinic three-level quasi frontal waves. Thesis, Royal Netherlands Meteorological Institute.
Holtslag, A.A.M., Van Ulden, A.P., de Bruin, H.A.R. (1980), Estimation of the sensible heat flux from standard meteorological data for stability calculations during daytime. Preprint of the procs. of the 11th ITM, Nato-CCMS, Amsterdam, 24-27 November.
Marchuk, G.I. (1969), Lectures on numerical short range weather prediction. Hydrometeoizdat Leningrad, 92-162.
Nieuwstadt, F.T.M. (1977), The dispersion of pollutants over a water surface. Procs. of the 8th ITM, Nato-CCMS, Louvain la Neuve 20-23 september.
Nieuwstadt, F.T.M. (1981), The steady-state and resistance laws of the nocturnal boundary layer : theory compared with observations. Bound. Layer Meteor. 20, 003-017.
Orszag, S.A. (1971), Numerical simulation of incompressible flows within simple boundaries : accuracy. J. Fluid. Mech., vol. 49, part 1, 75-112.
Richtmyer, R.D., Morton, K.W. (1967), Difference methods for initial value problems. Interscience New York
Roache, P.J. (1975), Recent developments and problem areas in computational fluid dynamics. Lecture notes in mathematics, No. 461, Springer Verlag Berlin.
Roache, P.J. (1978), A pseudo spectral FFT technique for non-periodic problems. Journal of comp. phys. 27, 204-220.
Smith, F.B. and Carson, D.J. (1977), Some thoughts on the specification of the boundary layer relevant to numerical modelling. Bound. Layer Meteor. 12, 307-330.
Tennekes, H. (1973), A model for the dynamics of the inversion above a convective boundary layer. J. Atmos. Sci. 30, 558-567.
Venkatram, A. (1980), Estimation of turbulence velocity scales in the stable and the unstable boundary layer for dispersion applications. Preprint of the procs. of the 11th ITM, Nato-CCMS, Amsterdam, 24-27 November.

DISCUSSION

K. VOGT

How do you take account of the difference of the deposition processes and of the deposition velocity in the source layer and in the surface layer ?

H. van DOP

The lower boundary condition of the model is

$$(K\ \partial c/\partial z = Vd.C)_{Z = Zi}$$

The deposition velocity $Vd(Z_i)$ is derived from the "normal" deposition velocity at a reference height of 1 m by adding to it the turbulent resistance of the layer 1 - Z.

W. GOODIN

How do your results compare with those from previous studies performed to construct accurate wind fields and advection-diffusion schemes ?

H. van DOP

There are two main topics, where our model differs considerably from previous ones :

1) In order to describe the turbulent diffusion, the Pasquill Stability scheme is abandoned. Instead turbulence is described in terms of Obukhov length and friction velocity, using surface and mixed layer similarity
2) Special care has been taken at the numerical methods to describe advection, so that numerical errors do not exceed 10 %.

J. TAFT

1) What was the grid resolution used in the advection diffusion test ?

2) Does your model formulation include terrains ?

H. van DOP

1) The horizontal gridspacing Δx amounts to 20 km, the vertical to 50 m.

2) Yes. Inspection of landmaps resulted in a terrain classification scheme, from which roughnesslength, deposition velocity and energy balance properties could be derived.

INITIAL EVALUATION OF CANADIAN AND UNITED STATES REGIONAL MODELS FOR TRANSBOUNDARY AIR POLLUTION

J. W. S. Young
B. L. Niemann*

Atmospheric Environment Service
Downsview, Ontario
*Environmental Protection Agency
Washington, D.C.

SUMMARY

In accordance with the U.S.-Canada Memorandum of Intent on Transboundary Air Pollution, Work Group 2 and its Modeling Subgroup are charged with describing the transport and transformation of air pollutants from their source regions to final deposition, especially wet sulfur deposition in sensitive ecological regions. The Modeling Subgroup has organized a comprehensive inter-comparison and evaluation of eight regional air quality simulation models for eastern North America. The initial inter-comparison and evaluation was based on transfer matrix elements for 11 source regions and 9 targeted sensitive areas. So far, no model has emerged as clearly superior or inferior to the others from the first of three rounds of model evaluation. The first round of evaluation has primarily served to reveal (1) the deficiences in the monitoring data bases, (2) the need for some changes in input parameters for some of the models, and (3) the need to use at least one more year of independent data for model evaluation. The future plans for completing the comprehensive inter-comparison and evaluation are described.

INTRODUCTION

Transboundary air pollution problems include (1) local situations where emissions from an identified source in one country adversely effect human health or welfare in another country within a few tens of kilometers from the source, (2) mesoscale situations where one or several sources or an urban area in one country produce adverse effects in another country up to many tens of kilometers distant, and (3) regional and long range transport situations where many sources in one or both countries can in combination produce a regional air pollution problem that crosses an international border resulting in acid deposition or visibility impairment.

Transboundary air pollution problems can be dealt with using the air quality management approach which involves the use of air quality simulation models and other environmental assessment tools. However, this requires that the countries involved in transboundary air pollution problems reach agreement through a negotiation process on the nature and extent of the problems and the most effective means to mitigate them. One of the most critical elements in the scientific part of the process is to establish the level of confidence which can be placed in model results and to recommend how best those results can be used in the negotiation process.

The purpose of this paper is to briefly describe the general structure which Canada and the United States have established to reach agreement on the scientific aspects of the problem, specifically in the area of simulation modeling, and to describe the initial results from a bilateral evaluation of Canadian and United States regional models.

OVERVIEW OF THE MOI AND WORK GROUP 2

A Transboundary Air Pollution Agreement, a Memorandum of Intent (MOI), was signed by United States and Canada on August 5, 1980, in Washington, D.C. The MOI includes the following major elements :

- Establishment of a formal coordinating committee to begin negotiations by June 1, 1981[1] ;
- Vigorous enforcement of existing air pollution controls ;
- Advance notification on industrial or regulatory developments relating to the acid rain problem ; and
- Exchange of scientific information from research and development on the problem.

An annex to the MOI set forth the structure of Work Groups to assist in the preparation for and the conduct of the treaty negotiations. The five Work Groups are Impact Assessment (# 1),

[1] The negotiations opened on June 23, 1981, in Washington, D.C.

Atmospheric Modeling (# 2), Strategies Development and Implementation (# 3A), Emissions, Costs, and Engineering Assessment (# 3B), and Legal, Institutional Arrangements and Drafting (# 4). Each government has appointed members to the 5 Work Groups and the individual bilateral Work Groups have worked cooperatively to produce interim reports (February 1981, except for Work Group 4) in their assigned areas. All the Work Groups are expected to interact informally with one another and formally through the Coordinating Committee and Work Group 3A.

Work Group 2, which was called Atmospheric Modeling in Phase I (September 1980 to January 1981) and is now called Atmospheric Sciences and Analysis in Phases II (February to June 1981) and III (July 1981 to January 1982), has the following specific task as set forth in the Annex to the MOI :

"Provide information based on cooperative atmospheric modeling activities leading to an understanding of the transport of air pollutants between source regions and sensitive areas, and to prepare proposals for the Research, Monitoring, and Modeling element of an agreement."

The principal objective of the modeling activity is to calculate emission reductions required from source regions to achieve proposed reductions in pollutant concentrations and deposition rates which would be necessary to protect sensitive areas.

During Phase II the Work Group 2 activities have been structured into three activity areas as follows :

(1) Atmospheric Sciences Review - assess (a) the appropriateness of the parameterizations used in models to quantify source-receptor relationships, (b) the issue of trends in acid deposition, and (c) global and western North American precipitation pH measurements ;

(2) Regional Modeling - document, evaluate, intercompare, and apply available practical models ; and

(3) Data Analysis Review - use data to independently establish the usefulness of models and the existance of source-receptor relationships or the lack thereof.

In Phase III, the third activity area will be called Monitoring and Interpretation and a fourth area called Local Source Analysis will be started. The Regional Modeling Subgroup consists of Work Group 2 members, the participating modelers who are directly involved in the operation of selected models and individuals interested in advancing the science of model evaluation and intercomparisons.

PARTICIPATING REGIONAL MODELERS (PHASES I AND II)

A considerable effort was made to inventory and document regional air quality simulation models by the Atmospheric Transport and Deposition Modeling Subgroup of the United States-Canada Research Consulation Group (RCG) on LRTAP during the previous year (Smith, et al., 1981). The RCG inventory was reviewed and adopted by Work Group 2 for inclusion in its Phase I report.

Work Group 2 agreed upon criteria for selection of regional models to be used in Phase I and applied them to the RCG subgroup inventory. Due to the large amount of preparatory work required to provide adequate short-term (episode) modeling results, Work Group 2 decided that the annual time period should be the primary focus for modeling source receptor relationships in Phases I and II. The selection criteria in Phase I were that the models :

- Be fully operational
- Be numerically practical
- Be capable of expansion as the knowledge base increases
- Be capable of use over the geographical and temporal time scales of interest (all of Eastern North America)
- Been at least partially evaluated through comparison with measurements
- Be currently funded and accessible to the governments

The results of the selection process were to use five regional models in Phase I - two models developed in Canada and three models developed in the U.S. It was also agreed that additional Canadian and/or U.S. developed models could be added or replace the initial group of five models in Phase II. In fact, three additional models (two developed in the U.S. and one developed in Canada) joined the inter-comparison and evaluation effort in Phase II.

A very brief description of each of the eight participating modelers is given (see next page).

Work Group 2 built upon the initial documentation in the RCG Subgroup report for the five Phase I models and provided more complete documentation along with the available evaluation results in Appendix 5 of their Phase I report. In Phase II, Work Group 2 decided to provide even more comprehensive documentation for the eight models in the form of individual Model Profile reports which contain model descriptions, sensitivity analyses and evaluation results. These Model Profile reports are summarized in the Phase II report of the Modeling Subgroup.

Country	Phase	Name of Organization	Model Acronym	Type of Model	Variation
Canada	I	Atmospheric Envir. Service	AES-LRT	Lagrangian	Day-by-Day
U.S.	I	Argonne Nat'l Laboratory	ASTRAP	Langrangian	Statistical
U.S.	I	SRI International/EPA	ENAMAP	Langrangian	Day-by-Day
Canada	I	Ontario Ministry of the Environ.	OME-LRT	Langrangian	Statistical
U.S.	I	University Of Illinois	RCDM-2	Eulerian	Analytical
U.S.	II	Washington Univ.-CAPITA	MCARLO	Lagrangian/ Eulerian	Statistical
Canada	II	MEP, Ltd.	MEP-TRANS	Lagrangian	Day-by-Day
U.S.	II	University of Michigan	UMACID	Lagrangian	Statistical

INITIAL MODEL EVALUATION

The techniques for regional model evaluation and intercomparison have not been fully developed, but some progress has been made as a result of recent workshops on Regional Models (Pt. Deposit, Maryland, November 1979) and Short Range Dispersion Model Performance (Woods Hole, Massachusetts, September 1980).

During Phase I, the Modeling Subgroup of the Research Consulation Group (RCG) on Long Range Transport of Air Pollution and Work Group 2 prepared a plan (Smith and Whelpdale, 1981) for (1) the further development of model evaluation and intercomparisons using the mechanism of regular workshops and for (2) the actual conduct of the activity to meet the specific needs of Work Group 2 in Phases II and III. Since the RCG, the MAP3S/RAINE Program, and Work Group 2 all had directives and specific needs for intercomparison and evaluation of regional models, the work plan developed by Work Group 2 provided for participation of all these groups so that there would be an integration and coordination of their efforts during 1980-1981. An example of this was the adoption of the same periods (January and July 1978) used in the MAP3S/RAINE model evaluation for the first round of Work Group 2 model evaluation.

The Modeling Subgroup of Work Group 2 developed formal definitions and terms of reference for Phase II during a series of monthly workshops (see Compilation of Attendance, Agenda, and Minutes in Work Group 2 Document No. 2-16). Some of the most essential definitions are as follows :

- Regional Air Quality Simulation Model - a relationship between a group of emission sources and the concentration or deposition over a receptor area some distance away (more than 100 km).

  Evaluation - a procedure of assessing the validity and sensitivity of a model. Usually the validity is ascertained by comparison with measurements, and the sensitivity through a series of model runs in which input parameters are altered in sequence.

- Model Sensitivity - the ability of a model to distinguish between small changes in an input parameter.

- Model Intercomparison - a procedure of comparing the results of several models which have been run on specified data bases and with (usually) pre-specified parameter values.

So far, most of the evaluations of regional models against monitoring data, including those in the Phase I report of Work Group 2 (Work Group Document 2-1) consist of very simple comparisons between simulated and observed values. These comparisons have taken a variety of forms as can be seen in Appendix 5 of the Phase I report or in the general literature. Since the main objective of Work Group 2 is to apply regional models to simulate current and future sulfur depositions in targeted sensitive areas, the Phase I model evaluation consisted of comparing simulations with available observation in the 11 Phase I targeted sensitive areas. The variations in results among the five Phase I models at the 11 targeted sensitive areas were due principally to differences in emission inventories, periods of meteorological data, and parameterizations. In general, the simulations were within a factor of two of the observations and the Work Group felt that this uncertainty could be narrowed during Phase II with the use of unified data bases and the discussion of differences between the models in the workshops.

In order to standardize and further quantify the model evaluation in Phase II, Work Group 2 adopted the confidence interval statement as recommended by the AMS/EPA Woods Hole Model Performance Workshop and agreed to compute the residuals between model simulations and observations. The four specific model evaluation criteria agreed upon were :

- Bias - mean of the residual between pairs of simulated and observed values (using the 95% confidence interval) ;
- Deviation - standard deviation of the residuals between pairs of simulated and observed values ;
- High end of the distribution - 5 highest observed values versus the simulated values at the locations of the highest observed ; and
- Reproduction of the spatial distribution - square of the spatial correlation coefficient.

Furthermore, an agreed upon list of evaluation sites, parameters (SO_2, $SO_4^=$, and wet sulfur deposition) and averaging periods (January and July 1978 and all of 1978) was used (Niemann, 1981).

It was also agreed that the models would be intercompared using the normalized values (per Teragram of sulfur) to minimize some remaining differences in the emission inventories used and would include seven parameters as follows :

(1) SO_2 concentration in $\mu g/m^3$/(TgS/year) ;
(2) $SO_4^=$ concentration in $\mu g/m^3$/(TgS/year) ;
(3) Wet SO_2 deposition (kgS/ha/year)/(TgS/year) ;
(4) Wet $SO_4^=$ deposition (kgS/ha/year)/(TgS/year) ;
(5) Dry SO_2 deposition (kgS/ha/year)/(TgS/year) ;
(6) Dry $SO_4^=$ deposition (kgS/ha/year)/(TgS/year) ;
(7) Total S deposition (kgS/ha/year)/(TgS/year) ;

for the 11 source regions and 9 receptor areas used by the Canadian modelers in Phase I (see Figure 4.1 in the Phase I report).

The Phase I and II model simulations were compared to calculated and interpolated wet sulfur deposition values at the nine targeted sensitive areas (see Table 1). The estimates by the OME-LRT and AES-LRT models did not change between Phase I and Phase II. In contrast, the ASTRAP, ENAMAP, and RCDM model estimates changed significantly due to the use of 1978 meteorology and the Phase II SO_2 emissions inventory and some changes to the input parameters. In general, the Phase II estimates of wet sulfur deposition by the OME-LRT, MEP-TRANS, and MCARLO models are within $\pm$ 50 % of the observed values, while the estimates for the AES-LRT and RCDM models are within about $\pm$ 75 % of the observed values. The ASTRAP and ENAMAP models estimates are generally much higher than the observed values at all the targeted sensitive areas. All the models show a general tendency to overpredict the observed wet sulfur depositions in the targeted sensitive areas except for the MCARLO model. These preliminary evaluation results will be used by the modelers to refine their input parameter and check their model results before starting the second round of model evaluation (see Future Plans).

Another interesting inter-comparison that was made is the relative contribution of source regions at various distances from the targeted sensitive areas to the total concentration or depositions. Table 2 shows some preliminary (since parameterizations are not final) numbers for the relative contributions and shows that long range transport makes a significant contribution in most of the sensitive areas.

PRELIMINARY ANALYSES OF TRANSFER MATRICES

In Phase I, wet sulfur deposition transfer matrices were generated by the AES-LRT, OME-LRT, ASTRAP, ENAMAP and RCDM models using different years of meteorological data and somewhat different emission inventories. In order to intercompare the individual transfer matrix elements, the values were normalized using a unit emissions rate of 1 Teragram of sulfur per year. The normalized matrix elements for 11 emission regions (8 in the eastern U.S. and 3 in south-eastern Canada) and 9 targeted sensitive areas (5 in the eastern U.S. and 4 in eastern Canada) were presented in an addendum to the Phase I report. The normalized matrix elements showed generally the largest variations for the impact of Ontario emissions on the Muskoka sensitive area and the smallest variations for the receptors farthest from the major source regions (i.e. Boundary Waters and the Canadian Atlantic Provinces). The standard deviations over the 5 model values of wet sulfur deposition in the transfer matrix, a convenient measure of the variability among the models, ranged from 0.01 to 4.32 $kg.S.ha.^{-1}yr.^{-1}$. The ENAMAP model results were primarily responsible for the largest standard deviations. The standard deviations over the four model values for wet sulfur deposition in the transfer matrix, excluding ENAMAP, ranged from 0.01 to 1.75 $kg.S.ha.^{-1}yr.^{-1}$.

The present contributions of each source region on each targeted sensitive area in terms of wet sulfur depositions were also computed and displayed in a transfer matrix for the five Phase I models (see Addendum to Appendix 8 of the Phase I report). There was generally much better agreement for the percent contributions among the five models than for normalized values and the standard deviations over the five model values in the transfer matrix ranged from 0.01 % to 22.7 % with an average value of about 5 %. Again the ENAMAP model results were primarily responsible for the largest standard deviations. These variations are attributable to the differences in the modelers parameterized the physical processes and treated the available input data - emissions, winds, precipitation, etc.

Table 1. Preliminary Model Estimates and Observations of Annual Wet Sulfur Deposition ($kg.S.ha.^{-1}yr.^{-1}$) at the Nine Targeted Sensitive Areas

Sensitive Areas	Obs.[1] Values	Phase I Canadian OME*	Phase I Canadian AES	Phase I United States RCDM	Phase I United States ASTRAP	Phase I ENAMAP	Phase II Canadian OME*	Phase II Canadian AES	Phase II United States MEP	Phase II United States ASTRAP	Phase II United States ENAMAP	Phase II United States RCDM	Phase II United States MCARLO
1 B. Waters	3	3	2	16	7	10	3	2	1	2	0.7	4	0.8
2 Algoma	3	5	10	20	16	6	5	10	6	17	5	8	3
3 Muskoka	6	7	18	23	21	19	7	18	10	25	7	15	5
4 Quebec	5	6	9	17	17	1	6	9	6	24	13	10	5
5 S.N. Scotia	3	7	6	10	12	11	7	6	3	5	0	7	1
6 Vt., N.H.	7	8	13	15	15	4	8	13	8	21	19	9	5
7 Adirondacks	7	8	16	19	19	3	8	16	9	31	27	13	6
8 Pennsylvania	12	17	34	27	25	21	17	34	16	52	71	21	7
9 Smokies	7	7	17	18	17	8	7	17	7	36	35	12	6

*Background of 2 $Kg.S.ha.^{-1}yr.^{-1}$ added to MOE values

(1) Source : Interpolated from a map prepared by the National Atmospheric Deposition Program (1981) based on data for March 1979 to March 1980.

Table 2. Percentage Contribution to Wet Sulfur Deposition and Sulfate Concentrations (in parentheses) in Sensitive Areas as a Function of Separation Distance Between Source Areas and the Sensitive Areas (Phase II)

Sensitive Areas	Model	Separation Distances (km) 0-300*	301-1000	> 1000
New Hampshire	ASTRAP	14(16)	45(54)	41(30)
	ENAMAP	19(6)	45(45)	35(49)
	RCDM	7(4)	58(47)	35(49)
Adirondacks	ASTRAP	6(4)	72(78)	22(18)
	ENAMAP	13(4)	73(74)	14(22)
	RCDM	9(3)	74(69)	17(28)
Pennsylvania	ASTRAP	57(50)	41(45)	2(5)
	ENAMAP	42(28)	57(67)	1(5)
	RCDM	37(20)	58(68)	5(12)
So. Appalachia	ASTRAP	61(61)	38(38)	1(1)
	ENAMAP	50(26)	49(69)	1(5)
	RCDM	27(15)	72(78)	1(6)
Florida	ASTRAP	56(68)	38(33)	6(9)
	ENAMAP	75(26)	15(44)	10(30)
	RCDM	32(18)	51(49)	17(33)
Arkansas	ASTRAP	23(19)	72(74)	5(7)
	ENAMAP	13(5)	81(70)	6(25)
	RCDM	5(1)	83(78)	12(21)
Boundary Waters	ASTRAP	16(39)	67(49)	16(12)
	ENAMAP	6(2)	91(94)	3(4)
	RCDM	1(1)	63(46)	36(53)
Ontario	ASTRAP	35(42)	56(50)	9(8)
	ENAMAP	25(20)	72(61)	3(19)
	RCDM	18(9)	62(73)	10(18)
Quebec	ASTRAP	8(10)	74(80)	18(10)
	ENAMAP	5(3)	75(66)	20(31)
	RCDM	0(0)	77(66)	23(34)
So. Nova Scotia	ASTRAP	4(13)	35(41)	61(46)
	ENAMAP	--	--	--
	RCDM	3(2)	41(22)	56(76)

* Subgrid scale resolution

Transfer matrices using the Phase II data bases and agreed-pon conditions have been generated by seven of the eight partici-ating modelers (see Table 3). The standard deviations over the model values of wet sulfur deposition values in the transfer ma-rix ranged from 0.01 to 4.75 $kg.S.ha^{-1} yr.^{-1}$. The largest standard eviations were generally in the western Pennsylvania sensitive area. nterestingly the use of standardized inputs in Phase II did not educe the range of variation among models in some of the transfer atrix elements. However, since additional refinements will be ade to most of the models and a new set of source-receptor regions ill be used during Phase III, it is premature to draw any general onclusions at this time. Although the desirability of a single, nified transfer matrix was recognized, the Modeling Subgroup has eservations about the generation and application of a unified ransfer matrix at this time because no matter how it is generated ts interpretation is subject to some question.

Transfer matrices generated by regional models can be used or analyzing emission limits for selected source regions given he desired wet sulfur deposition objectives at targeted sensitive eceptors. Transfer matrices are based on the assumption that the verage concentration or deposition of a pollutant in any receptor reas is a linear combination of emissions or its precursors in very source region. Using matrix notation, the relationship be-ween sources and deposition may be expressed as

$$D_i = \sum_{j=1}^{N} E_j T_{ij} \quad , i = 1, \ldots . M \tag{1}$$

here N is the total number of source regions and M the total num-er of receptors. T_{ij}, the normalized transfer matrix may be com-ined with selected emission vectors E representing various emis-ion scenarios to assess the impact on the deposition array D.

There are uncertainties in both the emissions and the transfer atrix coefficients. Work Group 2 is expending considerable effort o quantify these uncertainties.

When using the transfer matrix method, the aggregation of ources selected turns out to be very important. In addition, the elation between sources and receptors may be non-linear and the xtend of any non-linearary is being examined.

Table 3. Phase II Transfer Matrix of Annual Wet Depositon of Sulfur ($kg.ha^{-1}.yr^{-1}$) per unit emission ($Tg.S.yr^{-i}$)

			Receptor Areas								
Source Regions	Models	Emiss. (Tg.S)	B. Waters (1)	Alg. (2)	Musk. (3)	Que. (4)	S. N.Sc. (5)	Vt. NH. (6)	Adir. (7)	Penn. (8)	Smokies (9)
1 Mich	MOE	0.784	0.07	0.43	0.92	0.33	0.37	0.54	0.83	1.66	0.12
	AES	0.973	0.21	2.36	3.19	1.03	0.31	0.72	1.13	1.75	0.10
	ASTRAP	1.194	0.06	1.47	3.44	1.10	0.20	1.03	2.33	1.08	0.07
	ENAMAP	1.194	0.04	0.71	1.40	1.32	0.00	0.73	1.35	1.56	0.79
	RCDM	1.194	0.30	0.94	1.37	0.73	0.28	0.35	1.57	0.69	0.18
	MEP		0.14	1.24	1.27	0.18	0.08	0.25	0.42	0.66	0.11
	MCARLO		0.06	0.50	0.61	0.33	0.06	0.31	0.40	0.27	0.20
2 Ill Ind.	MOE	2.538	0.06	0.23	0.31	0.15	0.18	0.22	0.30	0.74	0.47
	AES	1.937	0.05	1.24	1.14	0.31	0.10	0.26	0.36	1.14	0.77
	ASTRAP	2.077	0.07	0.61	1.02	0.71	0.12	0.70	1.16	1.04	0.40
	ENAMAP	2.077	0.00	0.11	0.58	0.59	0.00	0.46	0.35	1.12	0.56
	RCDM	2.077	0.25	0.37	0.50	0.28	0.12	0.17	0.28	0.66	0.54
	MEP		0.05	0.32	0.31	0.08	0.02	0.12	0.13	0.17	0.38
	MCARLO		0.01	0.23	0.28	0.20	0.04	0.17	0.28	0.32	0.43
3 Ohio	MOE	1.983	0.04	0.16	0.23	0.18	0.28	0.31	0.46	1.99	0.24
	AES	2.381	0.00	0.25	1.85	0.46	0.21	1.01	1.34	4.75	0.25
	ASTRAP	2.163	0.02	0.31	0.75	0.92	0.25	0.83	1.79	3.91	0.24
	ENAMAP	2.163	0.00	0.00	0.02	0.75	0.00	0.99	0.90	7.24	0.23
	RCDM	2.163	0.11	0.31	0.91	0.47	0.23	0.38	0.65	1.88	0.42
	MEP		0.01	0.14	0.95	0.23	0.08	0.33	0.47	1.97	0.29
	MCARLO		0.00	0.08	0.33	0.33	0.05	0.27	0.47	0.61	0.23
4 Penn.	MOE	1.021	0.03	0.13	0.29	0.21	0.39	0.38	0.57	5.42	0.12
	AES	1.028	0.00	0.29	1.26	0.68	0.29	1.75	2.24	7.88	0.00
	ASTRAP	0.990	0.00	0.14	0.48	0.74	0.52	1.94	2.15	12.16	0.04
	ENAMAP	0.990	0.00	0.00	0.00	0.73	0.00	1.55	3.91	12.82	0.24
	RCDM	0.990	0.04	0.16	0.84	0.54	0.37	0.79	1.37	1.37	0.13
	MEP		0.01	0.08	0.60	0.45	0.20	0.83	1.28	6.46	0.16
	MCARLO		0.00	0.03	0.20	0.34	0.05	0.36	0.50	0.68	0.05
5 N. York to Maine	MOE	1.143	0.02	0.08	0.22	0.26	0.98	0.59	0.83	0.40	0.05
	AES	1.204	0.00	0.17	0.50	1.33	1.99	2.24	2.41	0.42	0.00
	ASTRAP	1.208	0.00	0.02	0.12	1.38	0.59	3.02	2.94	0.31	0.01
	ENAMAP	1.208	0.00	0.01	0.04	0.060	0.00	1.63	3.46	3.37	0.06
	RCDM	1.208	0.02	0.10	0.53	0.47	0.53	1.30	1.71	0.46	0.06
	MEP		0.00	0.05	0.66	1.24	1.06	2.29	1.89	0.63	0.01
	MCARLO		0.00	0.01	0.17	0.35	0.08	0.52	0.29	0.22	0.01

(1) Annual deposition matrix elements for ASTRAP and ENAMAP models are based on the mean of the equivalent January 1978 and July 1978 elements.

6 Kent. Tenn.	MOE	1.202	0.03	0.10	0.14	0.09	0.13	0.13	0.17	0.45	1.64
	AES	1.418	0.00	0.14	0.71	0.07	0.07	0.21	0.42	1.48	3.10
	ASTRAP	1.473	0.03	0.22	0.40	0.14	0.04	0.52	0.96	1.30	8.17
	ENAMAP	1.473	0.00	0.00	0.03	0.05	0.00	0.12	0.17	1.16	3.91
	RCDM	1.473	0.14	0.18	0.28	0.16	0.07	0.11	0.19	0.61	1.29
	MEP		0.01	0.05	0.09	0.03	0.01	0.04	0.06	0.18	2.01
	MCARLO		0.00	0.04	0.10	0.11	0.01	0.07	0.16	0.26	0.75
7 W.Virg. to N.C.	MOE	1.703	0.03	0.08	0.16	0.13	0.27	0.22	0.29	0.91	0.18
	AES	1.223	0.00	0.00	0.33	0.33	0.25	0.90	1.14	3.52	0.49
	ASTRAP	1.610	0.00	0.05	0.25	0.37	0.26	0.84	1.03	5.98	0.13
	ENAMAP	1.610	0.00	0.00	0.00	0.38	0.00	0.83	1.77	7.73	2.73
	RCDM	1.610	0.06	0.15	0.51	0.30	0.18	0.37	0.60	2.79	0.48
	MEP		0.01	0.02	0.16	0.09	0.05	0.22	0.28	1.25	0.75
	MCARLO		0.05	0.01	0.09	0.14	0.02	0.15	0.20	0.42	0.17
8 Rest of (USA) Fld to Mo. to Minn.	MOE	1.196	0.09	0.38	0.33	0.15	0.15	0.20	0.25	0.39	0.99
	AES	3.743	0.24	0.61	0.27	0.05	0.03	0.08	0.13	0.53	2.46
	ASTRAP	4.012	0.24	0.38	0.37	0.12	0.03	0.15	0.26	0.32	2.74
	ENAMAP	4.012	0.12	0.15	0.02	0.08	0.00	0.12	0.15	0.19	3.05
	RCDM	4.012	0.37	0.25	0.18	0.12	0.05	0.07	0.11	0.25	0.81
	MEP		0.23	0.34	0.12	0.02	0.01	0.03	0.05	0.05	0.29
	MCARLO		0.06	0.11	0.09	0.05	0.01	0.05	0.09	0.08	0.20
9 Ontario	MOE	0.906	0.08	0.62	1.63	1.00	0.55	1.04	1.14	0.54	0.05
	AES	0.985	0.10	1.83	3.35	1.73	0.61	1.62	2.03	1.22	0.00
	ASTRAP	0.949	0.04	5.43	6.89	2.15	0.16	1.18	2.18	0.25	0.02
	ENAMAP	0.949	0.02	1.45	1.41	1.61	0.00	1.20	1.75	0.53	0.25
	RCDM	0.949	0.17	1.30	1.91	1.52	0.56	0.64	0.95	0.39	0.06
	MEP		0.07	2.17	2.81	0.72	0.14	0.89	1.56	0.99	0.05
	MCARLO		0.20	0.47	0.61	0.44	0.08	0.37	0.32	0.17	0.09
10 Quebec	MOE	0.595	0.06	0.20	0.38	1.58	0.71	2.99	0.65	0.13	0.03
	AES	0.519	0.00	0.19	0.38	2.89	0.96	3.28	1.54	0.19	0.00
	ASTRAP	0.464	0.00	0.19	0.65	7.71	0.25	1.54	2.11	0.12	0.00
	ENAMAP	0.464	0.00	0.00	0.00	0.90	0.00	3.06	1.35	0.03	0.03
	RCDM	0.464	0.03	0.12	0.30	0.59	2.33	0.81	0.53	0.13	0.03
	MEP		0.03	0.34	0.49	3.53	0.57	3.15	1.87	0.16	0.00
	MCARLO		0.01	0.09	0.28	0.56	0.21	0.20	0.17	0.10	0.01
11 -Atlantic Provinces	MOE	0.187	0.01	0.03	0.06	0.18	0.75	0.19	0.12	0.05	0.02
	AES	0.235	0.00	0.00	0.00	0.43	2.55	0.00	0.00	0.00	0.00
	ASTRAP	0.453	0.00	0.00	0.00	0.02	0.87	0.03	0.00	0.00	0.00
	ENAMAP	0.453	0.00	0.00	0.00	0.07	0.00	5.78	5.39	0.35	0.09
	RCDM	0.453	0.00	0.01	0.04	0.10	0.93	0.33	0.12	0.02	0.01
	MEP		0.00	0.01	0.06	0.30	1.42	0.19	0.10	0.07	0.00
	MCARLO		0.00	0.00	0.01	0.05	0.17	0.04	0.01	0.00	0.00

INTERIM CONCLUSIONS AND FUTURE PLANS

Conclusions

While there is still no general agreement within the modeling community as to (1) the proper method and (2) the statistics to be used for inter-comparison and evaluation of models, the Modeling Subgroup of Work Group 2 has selected a common basis for performing these tasks for the eight participating modelers. The uncertainti in model predictions can be quantified from the differences betwee model predictions and observations (the residuals). In Phase II, a complete set of evaluation statistics was computed by only one mo del, while the monthly and annual residuals were computed at 9 to 20 sites by four of eight models. So far, no single model has e- merged as clearly superior or inferior to the others from this first of three rounds of model evaluation. The evaluation has primarily served to reveal (1) the deficiences in the monitoring data bases, (2) the need for some changes in input parameters for some of the models, and (3) the need to use at least one more year of independent data for model evaluation.

Future Plans

The Work Group 2 objectives for Phase III are as follows :

- Check and consolidate the Phases I and II work
- Integrate the work by the Atmospheric Science Review, Moni- toring and Interpretation, and Regional Modeling Subgroups
- Expand the Atmospheric Science Review to prepare state-of- the-art statements for other pollutants
- Complete the regional sulfur model evaluation work by Decem- ber 1981
- Complete the technical peer review of the Phase II working reports by December 1981
- Produce a high quality final report by January 15, 1982.

With regard to completing the regional sulfur model evaluatio work, the Regional Modeling Subgroup has agreed upon the following basis steps :

(1) develop terms of reference for model evaluation on easter North America ;

(2) run the models using the agreed-upon data bases for the year 1978 ;

(3) evaluate the models against the 1978 monitoring data using the agreed-upon statistics (first round) ;

(4) analyze and intercompare the 1978 model evaluation statistics and transfer matrices ;

(5) refine and improve the models based on the first round of model evaluation ;

(6) review and possibly revise the terms of reference for the second round of model evaluation ;

(7) evaluate the models against the 1979 monitoring data using the final parameterizations and other specifications and using the agreed-upon statistics (second round) ;

(8) analyze and intercompare the 1978 model evaluation statistics and transfer matrices ;

(9) develop terms of reference for model evaluation on Western Europe ;

(10) run the participating models using the agreed-upon data bases from Western Europe ;

(11) evaluate the models against the Western European monitoring data using the agreed-upon statistics (third round) ;

(12) analyze and intercompare the Western European evaluation statistics and transfer matrices ;

(13) prepare a statement on the state-of-the-art of LRT models based on the three rounds of model evaluation for inclusion in the Phase III final report to the Coordinating Committee.

It should be noted that steps 9-12 are optional depending upon he time and resources available to the participating modelers.

.CKNOWLEDGEMENTS

The authors are deeply indebted to many colleagues and research orkers who have participated in the Research Consulation Group on RTAP and the MOI Work Group 2 for the work described in the paper. he authors especially acknowledge the contributions of Barbara Ley nd Ed Pechan to the analyses presented and to Carolyn Acklin for ord processing.

REFERENCES

1. Atmospheric Modeling, Interim Report by Work Group 2, Prepared Under the Memorandum of Intent on Transboundary Air Pollution Signed by United States and Canada on August 5, 1980, January 14, 1981.

2. Atmospheric Sciences and Analysis, Phase II Working Report by Work Group 2, Prepared Under the Memorandum of Intent on Transboundary Air Pollution Signed by United States and Canada on August 5, 1980, July 10, 1981.

3. Altshuller, A.P., and McBean, G.A., 1980 : Second Report of the United States-Canada Research Consulation Group on the Long-Rang Transport of Air Pollutants, U.S. State Department and Canada Department of External Affairs, November 1980, 40 pp.

4. Compilation of Attendance, Agenda, and Minutes form the Phase II Workshops and Meetings of the Modeling Subgroup of Work Group 2 (Atmospheric Sciences and Analysis), Document No. 2-16, July 10,

5. NIEMANN, B.L., 1981 : Initial Data Bases for the Intercomparison of Regional Air Quality/Acid Deposition Simulation Models - Description and Applications, Paper 81-46.4 Presented at the 74th Annual Meeting of the Air Pollution Control Association, June 21-26, Philadelphia, PA.

6. SMITH, L.F., Whelpdale, D.M., 1980 : Proposed Plan for the Intercomparison and Evaluation of Regional Air Quality Simulatio Models in Support of Work Group 2 in Phase II) December 15 (revised January 14, 1981), 9 pp.

7. SMITH, L.F., WHELPDALE, D.M. SHAW R.W., and NIEMANN, B.L., 1980 Atmospheric Transport and Deposition Modeling : Inventory , Analysis, and Recommendations. Report to the United States-Canada Research Consulation Group on LRTAP. December (revised June 1981).

8. YOUNG, J.W.S., and NIEMANN, B.L., 1981 : Status Report on LRTAP Modeling Activity by the Atmospheric Sciences and Analysis Work Group # 2, Prepared for the EMEP-Cooperative Programs for the Monitoring and Evaluation of Long Range Transmission of Air Pollutants in Europe (ECE), March 30 - April 1, Reading, UK, 30 pp.

DISCUSSION

J.S. TOUMA — Canadian emission into the U.S. are high in the West? What are your plans to study the impact on the western U.S.?

J.W.S. YOUNG — During Phase 3 (May 15, 1981 to January 15, 1982) our models and analyses are being expanded to deal with the whole of North America since the treaty will cover the entire border between the two countries.

W. KLUG — Which were the criteria you used to select the eight models in the first place?

J.W.S. YOUNG —

1) Fully operational
2) Numerically practical
3) Capable of expansion as the knowledge base increases
4) Capable of use over the geographical and temporal scales of interest
5) At least partially evaluated through comparison with measurements
6) Currently funded and accessible to the governments.

J.L. WOODWARD — We saw this morning that modellers are continually changing the parameter values they use. Will you make any attempt to freeze the model parameters at a certain time? Will you attempt to standardize model parameter values where possible.

J.W.S. YOUNG — Modellers will be allowed to make their own choices of parameters values. After the first round of model evaluation (1978 data), the modellers will be encouraged to finalize their parameter values. The performance of the models will then be evaluated against the 1979 north American data set.

I don't think that any standardization will be attempted but we may examine that possibility.

COMPARISON OF AN EULERIAN WITH A LAGRANGIAN TYPE NUMERICAL AIR POLLUTION DISPERSION MODEL

N.D. van Egmond and H. Kesseboom

National Institute of Public Health
P.O.box 1 Bilthoven
the Netherlands

INTRODUCTION

For the quantitative interpretation of the measurement results of the Dutch national air pollution monitoring network two retrospective numerical models were developed. Based on the same meteorology both models only differ in numerical treatment of air pollution transport. In the Eulerian GRID-model advection is simulated by the pseudo-spectral advection scheme as given by Christensen and Prahm (1976), over a 32x32 grid with gridpoint-distances of 15 km, covering the 450x450 km^2 surroudings of the Netherlands. In the Lagrangian PUFF-trajectory model the 15x15 km^2 source areas are aggregated to puffs which are advected according to the same wind fields as used in the GRID-model. This dual approach was chosen as both models have specific advantages which are relevant at intensive applications within the framework of an operational monitoring network:

GRID-model:
- chemical processes can be modelled by applying chemical submodels to the pollutant concentrations within every grid-cell. In trajectory models it is impossible of 'mixing' two chemical components which are not emitted at the same time and place, for example NO and O_3.
- at larger numbers of sources the model is more efficient than the trajectory model as the numerical procedure only depends on the grid-size.

PUFF-model:
- spatial resolution can be increased strongly in specific sub-areas of interest by decreasing the grid-distances of the Eulerian grid to which the puffs are finally projected and by extending the individual puffs to plume segments. In the fully Eulerian GRID-

model the highest spatial resolution is given by the grid-distance (15 km).
- at smaller numbers of sources and/or source areas the required computation time is less than for the Eulerian model as the number of computations is proportional to the number of puffs, and consequently the number of sources.
Moreover the numerical procedures for the PUFF-model are more simple than for the GRID-model.

In this paper a short description of the common meteorological treatment and of the two numerical procedures will be given. The two models will be compared with respect to the mentioned advantages and disadvantages and with respect to computational efficiency. As the meteorology is the same and both models are run on the same HP-1000F, model 45 minicomputer system, this efficiency can be considered in dependence of the number of sources to be treated. The current computations are restricted to SO_2; extension to SO_4 is effectuated and extension to NO, NO_3 and O_3 is foreseen. The computations only apply to dry episodes; wet deposition is not (yet) taken into account.

MODEL DESCRIPTIONS

Meteorology

Vertical stratification

As the models are to be used within the framework of the operational network the treatment of meteorology is directed to the parameters which are acquired on a routine basis i.e.:
- wind speed, - direction and - deviation at 40 stations at 10 m level and at 5 tv-towers at levels between 150 and 300 m (hourly values)
- radiation at 3 stations (hourly values)
- diurnal variation of mixing height as obtained by an acoustic sounder.

Vertical stratification is given by three layers, as shown in figure 1:
- mixing layer, assumed to be constant during the night and increasing during daytime with a speed given by the acoustic sounder measurement,
- reservoir layer in which the pollution of higher sources is emitted, especially during the night,
- surface layer of 50 m in which the vertical turbulent diffusion is restricted by interactions with the earth surface, resulting in a limited mass transfer from the mixing layer to the ground.

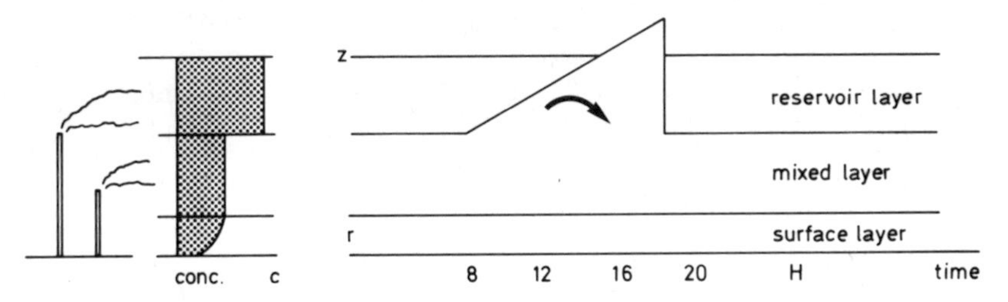

Figure 1. Vertical stratification in surface-, mixing- and reservoir layer and diurnal variation of mixing height.

During the morning hours the mixing height increases and the pollution in the reservoir-layer will be mixed downward to the mixing-layer, resulting in higher concentrations (fumigation). After passing the top of the reservoir layer (14.00 h) the increasing mixing height will result in an overall decrease of concentrations during the afternoon. As such the models describe the major diurnal variation.

Surface layer: dry deposition

The interaction of the earth surface limits the vertical mixing in the lowest layer, resulting in a concentration gradient over this surface layer and implicitely a reduction of dry deposition. As the category of low sources in the emission inventarisation has effective heights of 50 m, the surface layer is considered as the lowest 50 m of the atmosphere. The dry deposition is obtained as (Weseley and Hicks, 1977)

$v_g = (r_a + r_s + r_c)^{-1}$ with

aerodynamical resistance $r_a = (ku_*)^{-1}(\ln {}^z/_{z_0} - \phi_c)$

surface resistance $r_s = 2.6\ (ku_*)^{-1}$

canopy resistance $r_c = 70$ s/m and

ϕ_c is the stability correction given by

$\phi_c = \exp\ [0.598 + 0.39\ \ln(-z/L) - 0.09\ (\ln(-\ z/L))^2]$ at $L < 0$

and $\phi_c = -5\ z/L$ at $L > 0$ (stable conditions).

Obukhov length L and friction velocity u_* are computed from the 10 m-wind speed and, during daytime, the measured radiation, which is assumed to be proportional to the sensible heat flux and a linear function of albedo. During the night this iterative solution of L and u_* is impossible as $R_g = 0$ and Venkatram's (1980) empirical relation $L = 1100\ u_*2$ is used.[8]

Wind fields: vertical and horizontal interpolations

For vertical in - and extrapolation of wind speeds, measured at 10 m level and at the tv-towers (150-300 m) the exponent of the power law is computed:

$$u(z) = u(10)\ (z/10)^p$$

The winds at the representative heights of mixing and reservoir layer are computed from the spatial average < p >.

Wind direction is vertically interpolated by fitting the Ekman spiral wind profile

$$\alpha(z) = \sin(zq)/(e^{zq} - \cos(zq))$$

with $q = \sqrt{\lambda/2K_{zm}}$ and $\lambda = 2\ \omega \sin 52^o$

Wind directions (deviations $\alpha(z)$ from the geostrophic wind) are calculated at representative heights z with spatial mean bulk diffusivities $< K_{zm} >$.

Horizontal interpolations are made by means of a negative exponential weighting scheme. Horizontal divergence is reduced by application of the iterative scheme of Endlich (1967), resulting in an average divergence lower than $1 \times 10^{-6}\ s^{-1}$.

II.2 Emission and chemical transformations

The emissions are given as time-independent 15 x 15 km^2 averages; temperature dependency is not (yet) taken into account. The emissions for both high and low sources are given in figure 2.

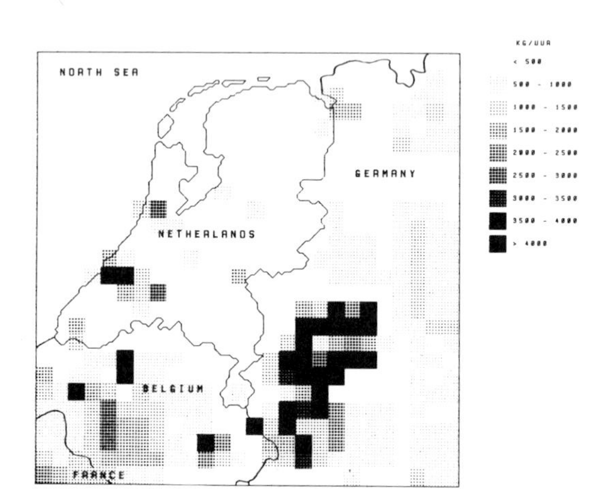

Figure 2. SO_2-emissions in the 450 x 450 km^2 model-area; total emission is 521 tons/h excluding point sources in the Netherlands.

Apart from area-average emissions a number of 22 larger point sources within the Netherlands are taken into account. According to their effective height the emissions are allocated to the mixing or to the reservoir layer.

Chemical transformation of SO_2 to SO_4 is modelled by

$$S_c = -(c_1 + c_2 R_g) C$$

with constants c_1 and c_2, resulting in transformations between 1% h^{-1} at night and 2.5% h^{-1} during daytime.

Emission and chemical transformation is treated the same way for the GRID- as for the PUFF-model.

GRID-model

The Eulerian GRID-model is based on the numerical integration of the continuity equation

$$\frac{\partial C}{\partial t} = -\left(u \frac{\partial C}{\partial x} + v \frac{\partial C}{\partial y}\right) + K_H \left(\frac{\partial^2 C}{\partial x^2} + \frac{\partial^2 C}{\partial y^2}\right) + Q_c - S_d - S_c$$

The two-dimensional equation applies to both the mixing and the reservoir layer. The wind components u and v are given by the respective wind fields as described above. Emission Q_c and dry deposition S_d applies to the mixing layer and chemical decay S_c to both layers. The spatial gradients $\partial C/\partial x$ and $\partial^2 C/\partial x^2$ are obtained by the pseudo-spectral method given by Cristensen and Prahm (1976). The (one dimensional) concentration field is written as

$$C(x) = \sum_k A(k) \exp(j\ 2\ \Pi \frac{k}{N\Delta x}\ x)$$

with N = 32 gridcells, Δx = 15 km and A (k) the spectral representation of the spatial profile C(x). The gradients then simply are obtained as

$$\frac{\partial C(x)}{\partial x} = \sum_h j\ 2\ \Pi \frac{k}{N\Delta x}\ A(k) \exp(j\ 2\ \Pi \frac{k}{N\Delta x}\ x),$$

thus by multiplication of the Fourier components A(k) by the arguments $k/N\Delta x$ times $2\Pi j$.

To avoid numerical bias, generated by high frequencies (wave numbers) in the concentration field, a Lanczos-Sigma filter is applied to the fields every 3 model-hours. Time integeretation is performed by a leap-frog scheme.

For every time step (about 6 min) and every layer, 2 x 32 = 64 Fast Fourier Transformations of 32 elements have to be performed to compute the spectral representation A(k) in both x- and y-direction. Neglecting turbulent diffusion (K_H = o) the derivatives $\partial C/\partial x$ are obtained by inversed transformation of the multiplied spectrum, resulting in a total of 128 FFT's per layer and per time step. So the

two layer model requires for one model-hour about 2 x 10 x 128 = 25
Fast Fourier Transformations.

II.4 PUFF-model

The Lagrangian PUFF-model is based on the transport of a large number of Gaussian puffs according to the above described wind fiel

$$C\ (x,y) = f \frac{M}{2\Pi\ \sigma_y^2\ h} \exp\left[\frac{-(x^2+y^2)}{2\sigma_y^2}\right]$$

where the symbols have their usual meaning and f is a factor which describes the concentration distribution in z-direction.

At smaller distances the value of f is given by the convention Gaussian plume model, accounting for effective plume height and reflections at ground- and inversion level. The required values for σ_z are obtained from the conventional empirical Pasquill functions where the Pasquill stability class is derived from Obukhov length L from the scheme given by Golder (1972).

At larger distances f attains the above mentioned value of C_4/C_{50}. The puffs are advected with the wind field of the (mixing-reservoir) layer in which the puff mass M is the largest. During fumigation pollution is transported from the reservoir part of the puff to the mixing layer part. The vertical concentration distribution within the puffs is given by figure 3.

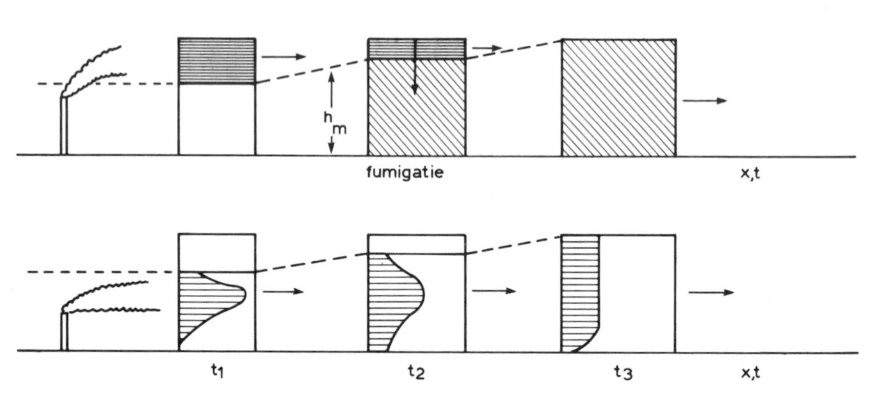

Figure 3 Vertical concentration distribution in the PUFF-model for reservoir (above) and mixing layer (below)

The horizontal distribution is given by σ_y and the increase of σ_y is found as

$$\sigma_{y\ t+\Delta t}^{2} = \sigma_{y\ t}^{2} + 2\ K_H\ \Delta t$$

K_H is the time dependent apparent diffusivity and given by

$$K_H\ (t) = \overline{v_m^2} \int_o^t R_L\ (\zeta)\ d\zeta$$

From the Lagrangian correlation function $R_L(\zeta) = e^{-\zeta/t_L}$ and the measured cross wind turbulence $\overline{v_m^2}$, $K_H(t)$ is computed for every puff-travel time t. The puffs are generated with initial $\sigma_{y\ o}$ such that the source area diameter is 4 x $\sigma_{y\ o}$. The puffs are released with inter-puff distances of 10 km. At the final projection of the individual puffs to the Eulerian mapping grid, additional puffs are generated in the direction of the wind vector, which is assigned to every puff at the time and location of emission. As such the puffs actually represent plume segments over 10 km intervals and together form a continuous plume.

The plume-segment approach enlarges the number of parameters which has to be handled for every puff, but on the other side reduces the number of required puffs and thus reduces computation time. Finally the following 10 parameters are assigned to every puff:

- x - position
- y - position
- σ_y - standard deviation of puff
- t total time of travel
- f fraction of concentration affecting ground level.
- Mass in reservoir layer
- Mass in mixing layer
- x - component plume-segment vector (wind at time of emission)
- y - component plume-segment vector
- effective height H of initial emission

PERFOMANCE AND INTERCOMPARISON OF MODELS

As might be expected from the use of common meteorological-, emission- and decay-routines, both models compute about the same overall concentration levels. Especially at south easterly circulations when the receptor area (the Netherlands) is not affected by nearby sources, the results of both models are much the same. However at winds from south-westerly directions the concentrations in the receptor area are dominated by the Dutch Rijnmond-Rotterdam area and the nearby Antwerpen-area in Belgium (figure 2). In that case a pronounced discrepancy is observed. As an example the results for 23 november 1979 are summarized; the concentration fields for 14.00 hours as modelled by PUFF and GRID are presented in figure 4. The diurnal profiles of modelled and measured spatial average concentrations (within the Netherlands) and the correlations between the modelled and measured hourly concentration fields

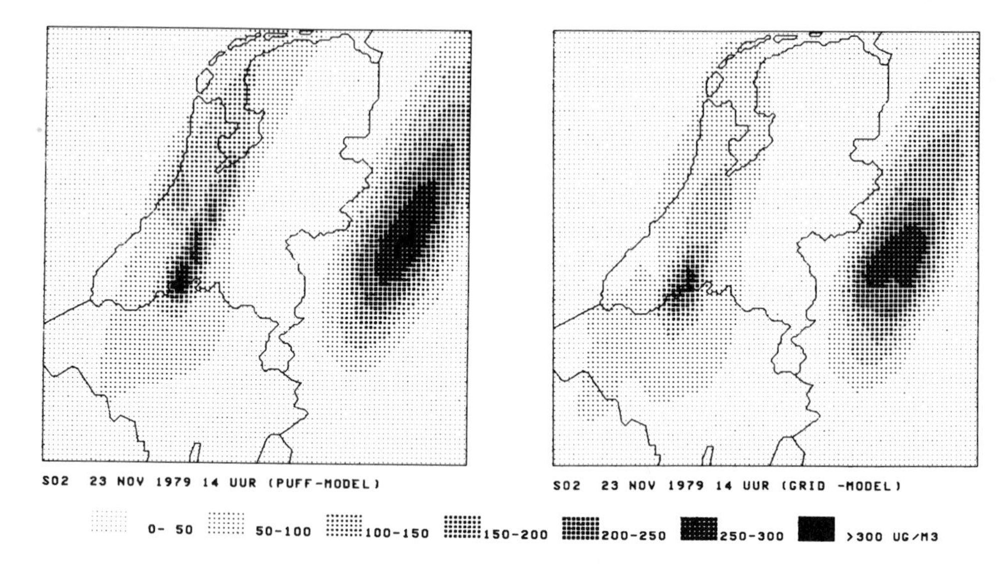

Figure 4 SO_2-concentration fields for 23 November 1979, 14.00 hour as modelled by PUFF (left and GRID (right).

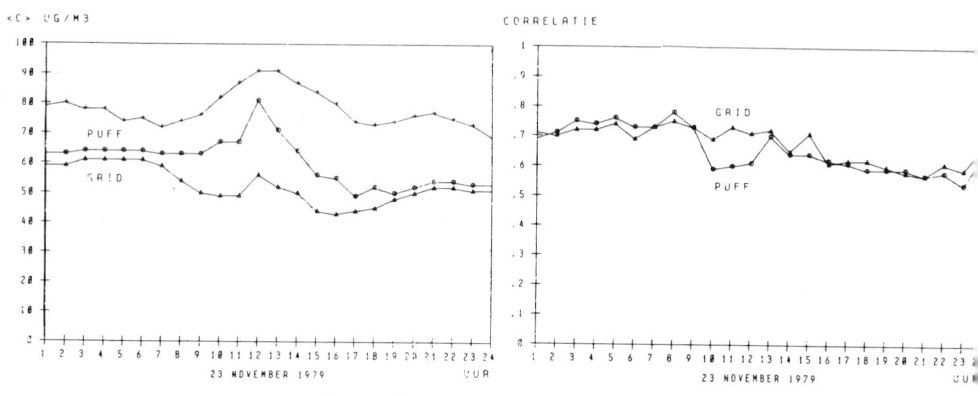

Figure 5 Diurnal profiles of measured and modelled spatial average concentrations (left) and correlations of modelled with measured hourly concentration fields (right).

are given in figure 5. The mixing height at 8.00 hours initially was 100 m and increased till 16.00 hours with 15 m/h, according to the scheme of figure 1.

The discrepancy between the modelled and measured overall levels is partly explained by the unaccounted background concentration of about 20 $\mu g/m^3$. Further the measured fields indicate emission zones which were strongly underestimated in the emission inventarisation data (figure 2). Improvement of these data seems the most effective factor to obtain higher correlations. In the third place the lower level of concentrations computed by the GRID-model are explained from the unrealistic instantaneous dilution of pollution within the 15 x 15 km^2 grid cells. The PUFF model gives a better description of horizontal and vertical dispersion and consequently computes higher local concentrations.

A second very typical discrepancy results from the simplification of describing pollutants in the reservoir and mixing layer after the fumigation period still by the same puff. A puff which represents emission from a high source in the reservoir layer is transported according to the reservoir layer wind. As soon as the mixing layer mass of the puff becomes larger than the reservoir layer part, due to fumigation (figure 3), the puff suddenly is advected by the mixing layer wind. However in the GRID-model pollution which is mixed downward, immediately is advected with the mixing layer wind which implies a correct description of vertical wind shear. An example is given in figure 6 for a high source in the Ruhr area, emitting a puff of one hour SO_2-emission between 7.00 and 8.00 hours. After going through the fumigation process at 14.00 hours the puff in the GRID-model is dispersed by wind shear, while in the PUFF-model the original (reservoir layer) dimensions are maintained. This effect of improper modelling shear dispersion lowers the correlations for the PUFF-model during the fumigation period (figure 5).

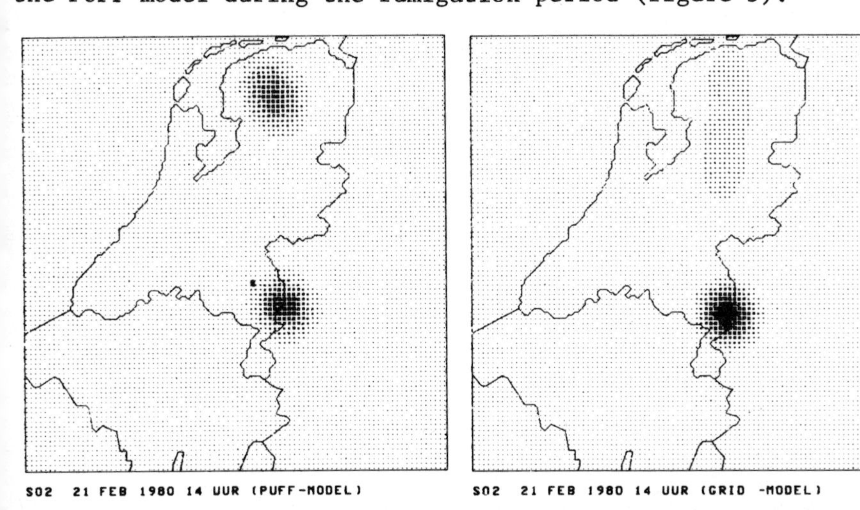

Figure 6 Concentrations at 14.00 hours resulting from SO_2-emission between 7.00 and 8.00 hours for the PUFF-(left) and GRID-model (right).

A striking advantage of the PUFF-model is its ability to focu on a specific area of interest. While computing the transport of po lutant over the total 450 x 450 km^2 model area, the detailed concer tration field in for example the Dutch Rijnmond-area can be compute by projecting the puffs, especially those of point sources, on a hi resolution Eulerian grid. From this 3 km-grid the concentration fie is mapped. The plume-segment procedure generates as much as secondary puffs along the plume-segment (axis)-vectors as necessary. An example is given by figure 7 where the Rijnmond concentration field for 23 november 1979, 14.00 h are presented; the map has to be considered as an enlargement of the indicated area of figure 4. Individual (Gaussian) plumes can be delineated, enabling an interpretation of local concentrations along with computed large scale back ground contributions.

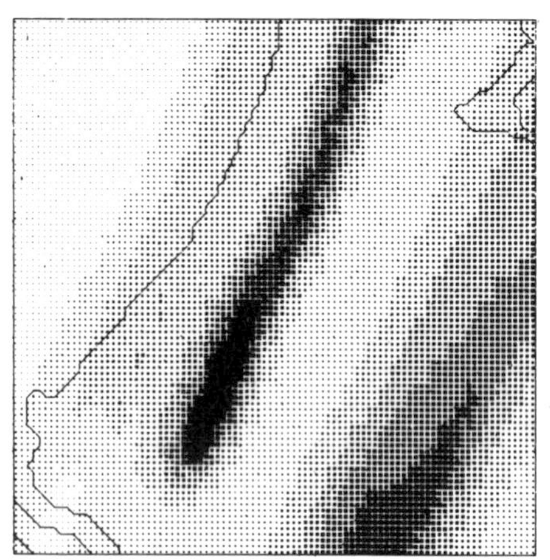

Figure 7 SO_2-concentration fields for 23 november 1979, 14.00 hour in the urban industrial area of the Netherlands as modell by PUFF (left) and as measured (right). (detail of figure 4).

IV COMPUTATIONAL INTERCOMPARISON

Both models were run on the same mini-computer configuration (HP-1000F). The GRID-model consists of 3 segments of totally 62 K bytes, requiring a maximum memory size of 52 K and additionally 80 K array-area. The PUFF-model is not segmented and requires 52 K bytes. The additional array area is mainly used for the 10 puff parameters (integers); at 5000 puffs this amounts 100 K. With 20 K for other arrays the total size is 172 K. So the required memory capicities for the GRID (142 K) and PUFF (172 K) model are about the same.

Computation time for the GRID-model is dependent on the number of required time steps and consequently on wind speed as for maintaining numerical stability the advection per time step is limited to 1/5 grid cell = 3 km. The representative time step is 6 min (at a reservoir layer wind speed of about 9 m/s). The computation time in the PUFF-model is, apart from the time consumed by the meteorological subroutines, dependent on the number of sources and consequently the number of puffs. Computation times for one model-hour, as a function of the number of sources are given in figure 8 for both GRID and PUFF-model.

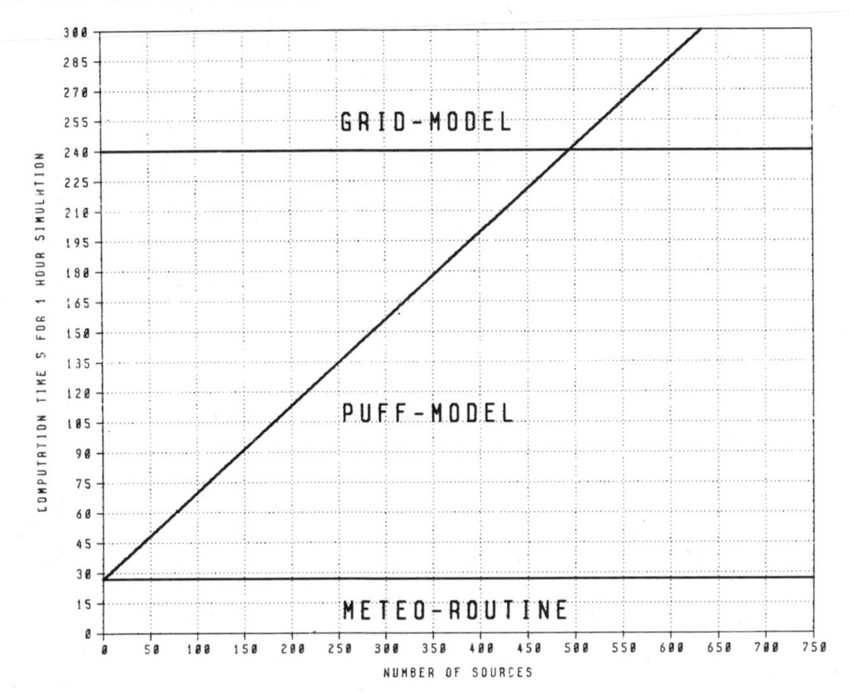

Figure 8 Computation times (CPU) for one hour simulation as a function of the number of sources for the PUFF- and the GRID-model (HP-1000 F).

At 500 sources the PUFF-model consumes as much as time as the GRID-model for one hour pollutant transport simulation. In the present application the number of sources in the PUFF-model, as aggregated from the 15 x 15 km^2 emissions presented in figure 2, amounts 160 and it can be concluded that the PUFF-model is far more efficient than the GRID-model under fully comparable performance conditions.

Of course the detailed computations for a specific sub-area, an example of which was given in figure 7, require about 100% more time as a second "mapping-array" of the same size has to be com-

At 500 sources the PUFF-model consumes as much as time as the GRID-model for one hour pollutant transport simulation. In the present application the number of sources in the PUFF-model, as aggregated from the 15 x 15 km^2 emissions presented in figure 2, amounts to 160 and it can be concluded that the PUFF-model is far more efficient than the GRID-model under fully comparable performance conditions.

Of course the detailed computations for a specific sub-area, an example of which was given in figure 7, requires about 100 % more time as a second "mapping-array" of the same size has to be computed. Even then the PUFF-model is not slower than the GRID-model, although its performance is substantially larger.

CONCLUSIONS

Under fully comparable performance conditions the trajectory PUFF-model is more efficient than the Eulerian GRID-model. The model performance only differs in the unadequate simulation of wind shear diffusion by the PUFF-model. Both models give good descriptions of the SO_2-concentration fields in het Netherlands resulting from mesoscale transport of SO_2 from inland and foreign source areas in the 450 x 450 km^2 surroundings. The PUFF-model allows the computation of spatially detailed concentration fields in specific sub-areas. However, the PUFF-model cannot be extended to other chemical species like NO_2 NO and O_3.

ACKNOWLEDGEMENT

The authors are grateful to dr.L.Prahm for his support with respect to the application of the pseudo-spectral advection scheme.

REFERENCES

Christensen O. and Prahm L.P., 1976, A pseudo-spectral model for dispersion of atmospheric pollutants, J. Appl. Meteor. 15, pp. 1284-1294

Endlich R.M., 1967, An iterative method for altering the kinematic properties of wind fields, J. Appl. Meteor. 6, pp. 837-844

Golder D., 1972, Relation among stability parameters in the surface layer, Boundary Layer Meteor. 3, pp.47-58.

Venkatram A., 1980, Estimating the Monin-Obukhov length in the stable boundary layer for dispersion calculations, Boundary Layer Met. 19, pp. 481-485

Wesely M.L. and Hicks B.B., 1977, Some factors that effect the deposition rates of sulfur dioxide and similar gases on vegetation, J. Air Poll. Contr. Ass. 27 no. 11, pp. 1110-1116.

DISCUSSION

B. WEISMAN

In view of the small extent of the modelling region, how do you account for SO_2 flux into the modelling region ?

N.D. VAN EGMOND

The model area is not that small : sources over an area of 450 x 450 km^2 are taken into account.
Background levels for SO_2 were not taken into account. For the present NO_x-O_3 modelling by means of the GRID-model, backgrounds are simulated by introducing additional emission sources on the 3 rows of grid-cells at the 2 upwind sides of the grid.

B. VANDERBORGHT

What is the information source for the emissions of Belgium, Germany and France ?

N.D. VAN EGMOND

Emission data were obtained from the TNO-emission inventorisation (TNO, Rekensysteem, Luchtverontreiniging, 1979). These data are partly based on the OECD-LRTAP program data, and the German emission inventarisation as published by the Bundesministerium des Innern.
The emission data for Germany and Belgium were updated on the basis of the results of numerous remote sensing - SO_2 gasburden scans over the major in-flux zones from the relevant source areas in these countries.

P.J.H. BUILTJES

Do you think it is possible to combine the pseudospectral solution method with non-linear chemical transformations? You might run into severe convergention problems.

N.D. VAN EGMOND

I don't know; you are probably right. So far we did not introduce non-linear chemistry in the model. Present NO_x-O_3 modelling of NO_x and O_x and analytical derivation of the individual components NO, NO_2 and O_3.
Due to aliasing errors, negative concentrations will occur in intermediate steps of the advection scheme and this already will give rise to substantial problems in the treatment of non-lineair chemistry. Indeed it is further expected that additional conversion of the (biased) concentration profiles by chemical reactions will result in numerical instabilities (non-convergence).

ANALYSIS OF PARTICLES TRAJECTORIES DURING A LAND-SEA BREEZE CYCLE USING TWO-DIMENSIONAL NUMERICAL MESO-SCALE MODELS

Christian Blondin and Gérard Therry

Etablissement d'Etudes et de Recherches Météorologiques
73-77 Rue de Sèvres
92100 Boulogne-Billancourt (France)

INTRODUCTION

It is now a current trend for scientists engaged in studies connected with atmospheric pollution to try to link closer and closer the respective knowledges in the conventional pollution and in the meteorology research domains. For a great number of problems, this is widely due to the very strong relationship between thermodynamic properties of the atmosphere and the physico-chemical behaviour of atmospheric pollutants.

Referring to the medium range transport and dispersion of pollutants, taking place mostly in the atmospheric planetary boundary layer, the greatest need concerns the evolution of the turbulence in this region of the atmosphere.

The operational numerical models developped with a view to treat this problem often choose very simple ways to simulate the turbulents processes, loosing more of less reliability or realism in their results.

It is the advantage and the purpose of research modelling to take the best profit of the large increase in computing facilities and to try to improve the various methods in the field. More specifically, in the case of atmospheric numerical meso-scale models in which the activity of the turbulent transfers are parameterized, one can examine the influence of the introduction of a higher order closure in the turbulence equations.

The aim of this paper is to show the discrepancies observed on computed trajectories in a two-dimensional numerical meso-scale model using two different approaches to calculate the vertical diffusion coeficients appearing in the classical thermodynamic equations applied to the atmosphere (Blondin, 1980).

In the first version of the model, we parameterized the vertical turbulent fluxes using the exchange coefficients proposed by Louis (1979). In this first order closure model, the vertical upwards propagation of the sensible heat flux is possible only if the vertical potential temperature gradient is negative. This implies that the warming of the planetary boundary layer by the convection occurs in the regions of absolute instability.

We shall now pay more attention to the second version of the model in which we added a new variable : the turbulent kinetic energy.

A CLOSURE WITH THE USE OF THE TURBULENT KINETIC ENERGY

The $\bar{e}$ equation

Kolmogorov proposed in 1942 the following relation between the turbulent kinetic energy $\bar{e}$:

$$\bar{e} = \frac{1}{2} (\overline{u'^2} + \overline{v'^2} + \overline{w'^2})$$

and the diffusion coefficient K :

$$K = C_K \, L \, \bar{e}^{\frac{1}{2}}$$

C_K is a constant and L is an appropriate mixing length which must be parameterized to close the turbulence equations.

In its one-dimensional version, the equation of the $\bar{e}$ evolution is :

$$\frac{\partial \bar{e}}{\partial t} = \frac{\partial}{\partial z} \left(K \frac{\partial \bar{e}}{\partial z}\right) + K \left(\left(\frac{\partial \bar{u}}{\partial z}\right)^2 + \left(\frac{\partial \bar{v}}{\partial z}\right)^2 - \alpha_T \, \beta \left(\frac{\partial \theta}{\partial z} - \gamma\right) \right) - C_\varepsilon \frac{\bar{e}^{\frac{3}{2}}}{L}$$

and we have :

$$\overline{u'w'} = - K \frac{\partial \bar{u}}{\partial z} \qquad\qquad K = C_K \, L \, \bar{e}^{\frac{1}{2}}$$

$$\overline{v'w'} = - K \frac{\partial \bar{v}}{\partial z} \qquad\qquad K_\theta = \alpha_T \, K$$

with

$$\overline{w'\theta'} = - K_\theta \left(\frac{\partial \bar{\theta}}{\partial z} - \gamma \right) \qquad\qquad K_e = C_e \, K$$

$$\overline{w'e} = - K_e \frac{\partial \bar{e}}{\partial z} \qquad\qquad \beta = \frac{g}{T_o}$$

It is generally admitted that the C_K and the C_ε constants are linked in such a way that the $\bar{e}$ equation gives the correct solution in the surface flux layer for the neutral and stationary case :

$$u_*^2 = C_K \, L \, \bar{e}^{\frac{1}{2}} \, \frac{u_*}{kz}$$

In the neutral surface flux layer, L is proportional to kz and so :

$$\bar{e} = \left(\frac{1}{C_K} \right)^2 u_*^2$$

If $\frac{\partial \bar{e}}{\partial t} = 0$ and if we neglect the transfer term, we have :

$$0 = - \overline{u'w'} \, \frac{\partial \bar{u}}{\partial z} - C_\varepsilon \, \frac{\bar{e}^{\frac{3}{2}}}{L} \quad \text{and} \quad \bar{e} = \left(\frac{1}{C_\varepsilon} \right)^{\frac{2}{3}}$$

Experimental values give :

$$\bar{e} = 5 \, u_*^2$$

and so :

$$C_K = (0.2)^{\frac{1}{2}} \qquad C_\varepsilon = (0.2)^{\frac{3}{2}}$$

The other constants are :

$$\alpha_T = 1.35 \qquad C_e = 1.2$$

The mixing length problem

The Kolmogorov formulation of the exchange coefficient K is based on the assumption that K is a function only of one characteristic speed and one length, characteristic of the most energetic eddies and which is the same as the one appearing in the dissipation term.

The simulations carried out with a mixing length taking account of the experimental values of the dissipative length L_ε show systematic defects in the convective situations. This can be explained by the fact that, in this case, the vertical and horizontal velocity variances have neither the same magnitude nor the same production processes. Then, two characteristic speeds seem to be relevant :

$$\left(\overline{w'^2}\right)^{\frac{1}{2}} \quad \text{and} \quad \left(\overline{u'^2} + \overline{v'^2}\right)^{\frac{1}{2}}$$

The degree of complexity we want to face does not allow us to solve two separate evolution equations for these quantities.

We propose an other way to increase the degrees of freedom of our model by differentiating the lenghts appearing in the dissipation term and in the exchange coefficient

This approach is justified by the following expression of the turbulent heat flux $\overline{w'\theta'}$, which is valid in a convective regime :

$$\overline{w'\theta'} = - \frac{1}{c_\varepsilon \; c_6} \; L_\varepsilon \left(\frac{\overline{w'^2}}{\bar{e}} \right) \bar{e}^{\frac{1}{2}} \left\{ \left(\frac{\partial \theta}{\partial z}\right) - (1 - c_7) \; \beta \; \frac{\overline{\theta'^2}}{\overline{w'^2}} \right\}$$

This expression comes from the evolution equation of $\overline{w'\theta'}$ (Deardorff, 1972) written with the parameterization formulation used in André et al (1978).

If we adopt a classical formulation for the exchange coefficient, there is a multiplicative factor : $(\overline{w'^2} / \bar{e})$ between the dissipative mixing length L_ε and the length L_K present in the exchange coefficient.

With the same analytical formula, similar to the one proposed by Bodin (1979), but with other limit values, we can represent approximately the two lengths and so their ratio induced by the variability of this multiplicative factor with height and thermal regime :

$$\frac{1}{L} = \frac{1}{kz} + \frac{c_1}{z_i} - \left(\frac{1}{kz} + \frac{C_2}{z_i}\right) \cdot m_1 \cdot m_2$$

with :

$$m_1 = \frac{1}{1 + c_3 \dfrac{z_i}{kz}} \qquad \text{and} \qquad m_2 = \frac{1}{1 - c_4 \dfrac{L_{Mo}}{z_i}} \quad \text{if} \quad L_{Mo} < 0$$

$$m_2 = 0 \qquad \text{if} \quad L_{Mo} > 0$$

by giving different numerical values to the c_i coefficients :

for L_ε $\quad c_1 = 15 \quad ; \quad c_2 = 5 \quad ; \quad c_3 = 5\ 10^{-3} \quad ;$

for L_K $\quad c_1 = 15 \quad ; \quad c_2 = 11 \quad ; \quad c_3 = 2.5\ 10^{-3} ;$

Moreover, to find a closer fit with experimental values, we multiply in the upper part of the convective layer the two lenghts by a linear in z decreasing function.

The γ gradient

From a fundamental point of view, the advantage of the introduction of the γ gradient term is to allow a better representation of the vertical stratification of the planetary boundary layer.

Experimental evidences show that a fully convective layer is not always completely well mixed and neutral but can present the schematic following temperature profile : in the lower part, the layer is slightly unstable, and, in the upper part, the gradient is weakly positive. But, in this upper part, it is clear that the sensible heat flux coming from the ground is still propagating upwards.

So, and only if the kinematic sensible heat flux Q is positive, it is a convenient artifice to introduce this pseudô-gradient in the whole convective layer.

According to André *et al* (1978), and taking their notations and their expression of $\overline{w'\theta'}$, we can estimate the γ term by :

$$\gamma = (1 - c_7)\beta \frac{\overline{\theta'^2}}{\overline{w'^2}}$$

Lenschow et al (1980) have proposed the analytical formula for $z < 0.8\, z_i$:

$$\frac{\overline{\theta'^2}}{\overline{w'^2}} \alpha \left(\frac{z}{z_i}\right)^{-\frac{4}{3}} \left(1 - 0.8 \frac{z}{z_i}\right)^{-2} \frac{\vartheta_*^2}{w_*^2}$$

$$w_* = (k\ \beta\ Q_\circ)^{\frac{1}{3}} \qquad \vartheta_* = \frac{Q_\circ}{w_*}$$

Referring also to other experimental data collected during the French Voves Experiment (Weill, 1980), it appears that the approximation :

$$\frac{\overline{\theta'^2}}{\overline{w'^2}} = C_{\theta w} \frac{\theta_*^2}{w_*^2}$$

is correct in the range $0.3\, z_i - 0.8\, z_i$. So:

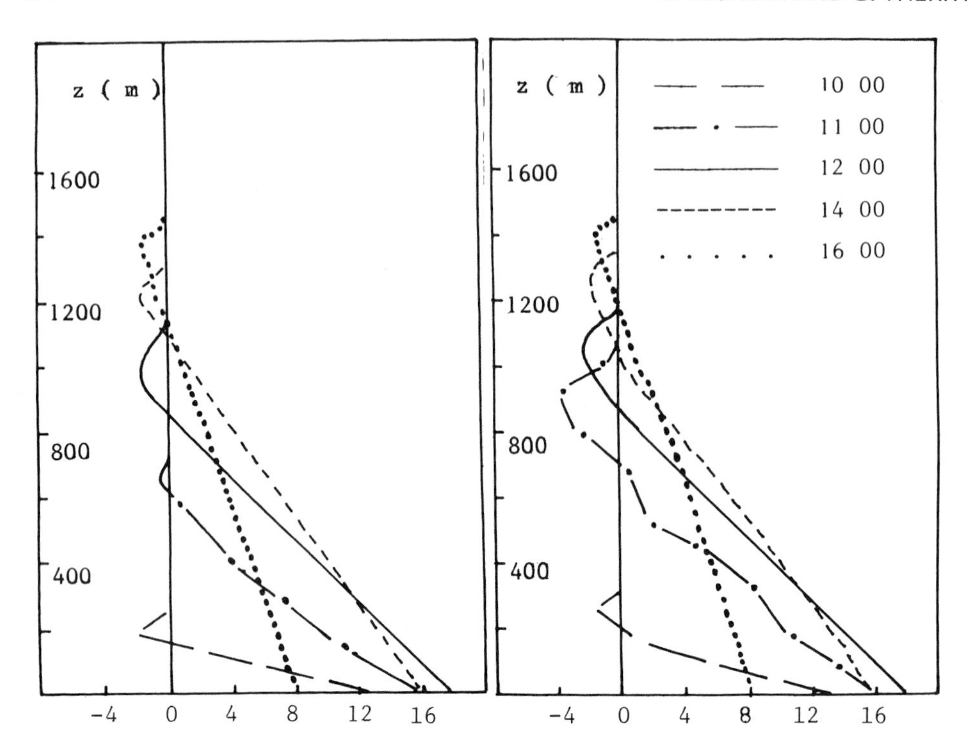

Fig. 1. Comparison between the $\overline{w'\theta'}$ (in cm K s^{-1}) evolution between our model (left) and André's model (right) for the simulation of the Day-33 of the Wangara Experiment

$$\gamma = (1 - c_7)\beta \frac{\overline{\theta'^2}}{\overline{w'^2}} = (1 - c_7)\, C_{\theta w}\, \beta \frac{\theta_*^2}{w_*^2}$$

$$(1 - c_7) \simeq 0.6 \qquad \text{so} \qquad \gamma \simeq 5\, \frac{Q_o}{w_* z_i}$$

The factor 5 is connected to the altitude where the potential temperature gradient comes to zero. With the assumed value, the change in the gradient occurs at approximately the half of the convective boundary layer.

We compared, using the data set of the Day-33 of the Wangara Experiment, the evolution of the diurnal boundary layer with this formulation of the turbulence closure and with the high-order closure model presented in André et al (1978).

The very good agreement on the global evolution of the temperature and on the budgets of the source and sink terms between the two simulations seemed to be a convincing result and we introduced this turbulence parameterization in the two-dimensional version of our meso-scale model. Of course, a more complete equation was then applied to take into account the advection and horizontal diffusion terms.

APPLICATION TO A LAND-SEA BREEZE CYCLE SIMULATION AND TRAJECTORIES STUDY

We carried out a comparative run with the two-dimensional model for a land-sea breeze cycle simulation, with a view to exhibit the discrepancies in the dynamic (wind and temperature) fields. As a more practical evidence of these differences, we computed, using a Langragian scheme, the trajectories of particles emitted just above the coastline at regularly spaced time steps during the day.

Conditions of the experiment

The model has a grid of fifty points on the horizontal direction, with a mesh of five kilometers, and twenty levels on the vertical direction, with ten layers of 100 m depth in the lower part of the domain. The coast-line is placed so that we have twenty points on the sea (left hand side of the grid) and thirty points on the land (right hand side of the grid).

The sea and the land present different physical characteristics, the sea temperature remains constant during the simulation, and, for the land surface temperature, we use the evolution equation proposed by Bhumralkar (1975). The land roughness length is five times greater than the sea one.

As we simulate here a dry atmosphere, with no latent heat flux at the ground, and to provide a reasonable sea-land temperature difference evolution during the daytime , the astronomic parameters for the sun course are those of the 27 October (day 300 in the year).

At the beginning of the simulation, which corresponds to the sunrise (06 31 a.m), the temperature field is horizontally homogeneous, the vertical profile showing a 2000 meters neutral layer with an inversion layer above, up to 2750 m, and finally a neutral layer, up to 3000 m which is the top altitude of the model domain. The sea and the land temperature are the same ($17^{\circ}C$) and equal to the first layer air temperature.

We do not impose a large scale pressure gradient, so that there is no wind in the initial state.

From now, we shall speak of the K-model for the version using the Louis coefficients and of the $\overline{e}$ model for the version running with the turbulent kinetic energy equation.

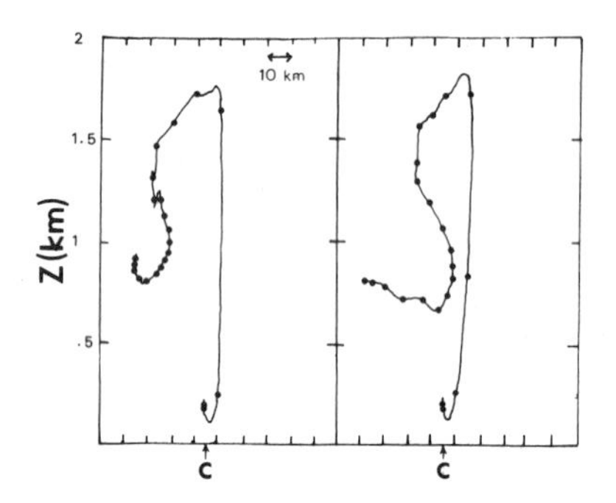

Fig. 2. Comparison between the computed trajectory n° 1 of a particle emitted at 07 31 a.m , 200 m high above the coast-line (C) in the K-model (left) and the $\overline{e}$ model (right)

Analysis of particles trajectories

We integrate the model equations on a 23-hour simulation run starting from the sunrise (the sunset occurs at 04 38 p.m). The particles are emitted each half-hour from 07 31 a.m to 05 31 p.m, from a point source located 200 m high above the coast line.

The trajectories are numbered from 1 to 21, according to the emission order of the particles, the number 1 corresponding to the trajectory of the first particle emitted.

We shall comment on three particular trajectories. Each following picture (2, 3, 4) is divided in two parts :
the left hand side represents the trajectory in the K-model ;
the right hand side refers to the trajectory of the particle emitted at the same time in the $\overline{e}$ model. The interval between two points on a trajectory corresponds to a one-hour travel.

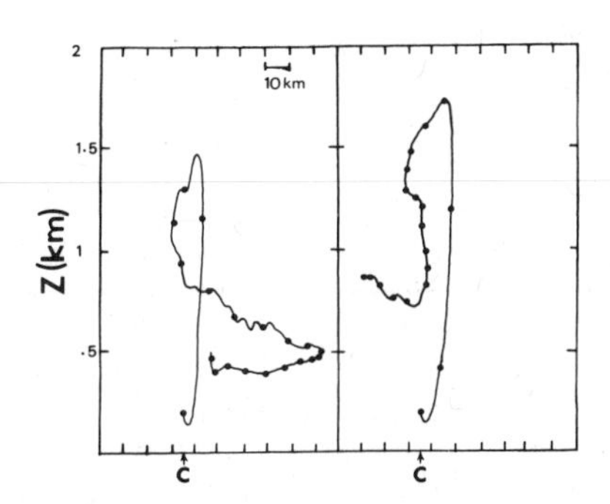

Figure 3. As in Fig. 2 but for the Trajectory n° 7 (emission 10:31 a.m.)

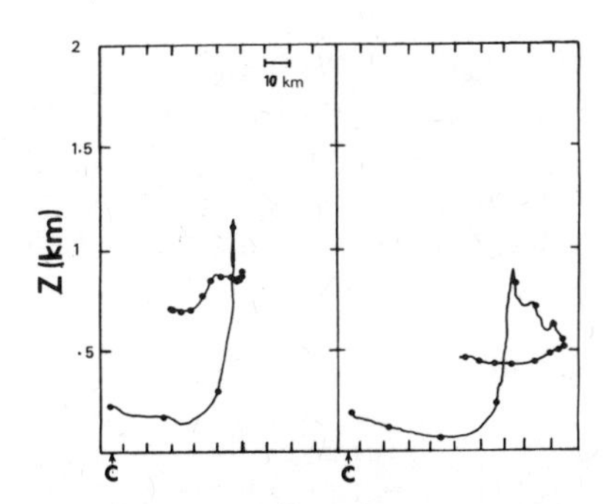

Figure 4. As in Fig. 2 but for the trajectory n° 14 (emission 0.2:01 p.m.)

In each simulation run, we reproduce a land-sea breeze cycle circulation, with the apparition, a few hours after the sunrise, of a convergence zone, indicated by important upward vertical wind velocities. This breeze front advances in-land till a bit later than the sunset. So, during the daytime, we have a wind blowing from the sea towards the land, behind the front and in the first 1500 meters, with significant wind speeds (up to 8 m/s), and a weaker return flow above.

When the convection regime stops, the wind speed comes close to zero in the whole domain, and then appears a light land breeze in the stable radiative inversion layer and above up to 1200-1500 m but without convergence areas inducing important vertical velocities.

Inside these common features that can be obviously noticed on the trajectories pattern, the models run with different evolutions due to the upward propagation of the sensible heat flux : we checked that the kinematic surface heat flux Q_o show a very similar behaviour in the two models.

In the $\bar{e}$-model, Q_o is propagating in a thicker layer than in the K-model because of the γ term and the temperature of the first layers in the K-model increases more rapidly. The breeze activity begins with a stronger intensity.

It can be seen on the trajectory n° 1 that it takes approximately one hour to the particle to reach the 2000 meter level in the K-model instead of more than two hours in the $\bar{e}$ model.

The trajectory n° 7 of the particle emitted at 10 31 a.m prooves that the front breeze advances more quickly in the K-model at the beginning of the convective period. This tendance is reversed at the end of the day because of the warming of the air above the land which becomes more intense in the $\bar{e}$ model. The reason is that the vertical temperature gradient is always very weak in the two models, but, in the $\bar{e}$ model, the supplementary contribution of the γ term becomes significant.

We can notice on the trajectory n° 14 of the particle emitted at 02 01 p.m that, finally, the front breeze penetrates deeper inland in the $\bar{e}$ model. Comparing the trajectories n° 7 and n° 14, we can estimate that the end of the sea breeze penetration occurs two hours later in the $\bar{e}$ model than in the K-model.

CONCLUSION

Of course, the aim of this experiment is not to show a spectacular improvement in the breeze understanding or simulation, primary because of the fundamental three-dimensional nature of this phenomenon. But we have shown that the theoretical advantages of a more sophisticated treatment of the turbulence in the planetary boundary layer can be transfered in the numerical modelling of the atmosphere.

This is one step towards a more reliable simulation of the atmospheric properties, with all the possible consequences for the realism of pollution studies.

REFERENCES

ANDRE, J.C., DE MOOR, G., LACARRERE, P., THERRY, G., and du VACHAT, R., 1978, Modeling the 24-hour evolution of the mean and turbulent structures of the planetary boundary layer, J. Atmos. Sci., 35:1861-1883.

BHUMRALKAR, C.M., 1975, Numerical experiments on the computation of ground surface temperature in a atmospheric general circulation model, J. Appl.Meteor., 14:1246-1258.

BLONDIN, C., 1980, Interest of an atmospheric meso-scale model for air pollution transport studies over medium distances, in : Proc. 11th I.T.M on Air Pollution Modeling and its Application , Amsterdam, The Netherlands.

BODIN, S., 1979, A predictive numerical model of the atmospheric boundary layer based on the turbulent energy equation, in : SMHI report n° 13, Norrköping, Sweden.

DEARDORFF, J.W., 1972, Theorical expression for the countergradient vertical heat flux, J.Geophys.Res., 77:5900-5904.

LENSCHOW, D.H., PENNELL, W.T., and WYNGAARD, J.C., 1980, Meanfield and second-moment budgets in a baroclinic, convective boundary layer, J.Atmos.Sci., 37:1313-1326.

LOUIS, J.F., 1979, a parametric model of vertical eddy fluxes in the atmosphere, Bound.Layer.Meteor., 17:187-202.

WEILL, A., 1980, measuring heat flux and structure functions of temperature fluctuations with an acoustic doppler sodar, J.Appl.Meteor., 19:199-205.

DISCUSSION

J. TAFT

Does your model contain enough generalism to be directly applicable to problems in the first 100 m of the boundary layer. If no, what formulas or parameters, if any, need to be modified.

C. BLONDIN

The formula proposed in the paper for l_K and l_ε have been improved by the introduction of a new length l_s which satisfies the following criteria

$$\frac{1}{l_s} = \begin{cases} 0 & \text{if } \overline{w'\theta'}(z) \geq 0 \text{ or } \frac{\partial\bar\theta}{\partial z} \leq 0 \text{ (unstable local stratification)} \\ \left[\frac{\beta\frac{\partial\theta}{\partial z}}{\bar e}\right]^{1/2} & \text{if } \overline{w'\theta'}(z) < 0 \text{ or } \frac{\partial\bar\theta}{\partial z} > 0 \text{ (stable local stratification)} \end{cases}$$

This allows to say that the model can be applied to treat the problem of the first 100 meters in neutral or fairly stable or unstable situations. The problem of strong stability conditions, (typically the nocturnal inversion layer) is still open and the basic difficulty to model mathematically and numerically this problem is due to the intermittent character of the turbulence in this case. For this case, I do not know any appropriate solution.

Concerning unstable conditions or strong convective conditions, the $\bar e$ behaviour is not satisfying for two reasons:

- the first one, which is less important, is connected to the ground boundary condition for the $\bar e$ value. In the present study, we use the following condition: $\frac{\partial\bar e}{\partial z} = 0$. In fact, it seems more reasonable to stand that $\bar e\ (z=z_0) = \bar e\ (u_\star, w_\star)$ and we actually try this approach.
- the second one is more important and fundamental: experimental measurements show the following vertical profiles for $\overline{w'e}$ and $\bar e$.

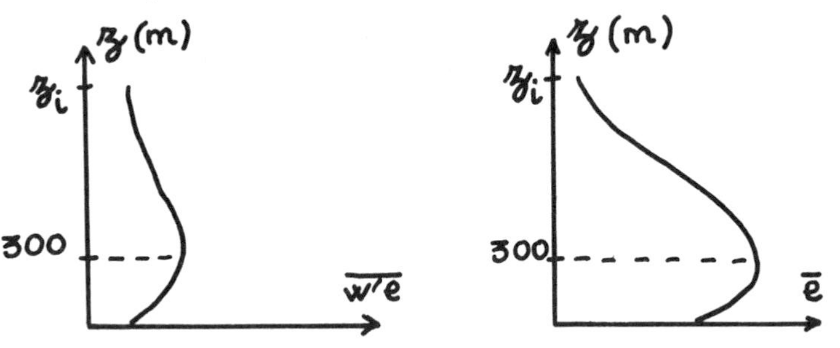

This implies that, in the first hundred meters, where $\overline{w'e}$ is positive, you cannot express this quantity by $- K\frac{\partial\bar e}{\partial z}$ which is obviously negative in the layer under the $\bar e$ maximum height.

So the current simulation gives rise to too large, but reasonable value of $\bar e$, in the first hundred meters. The introduction of a counter gradient term is less easy to justify for $\overline{w'e}$ than for $\overline{w'\theta'}$ but we search for a convenient solution.

VERIFICATION OF A THREE-DIMENSIONAL TRANSPORT MODEL USING TETROON DATA FROM PROJECTS STATE AND NEROS

T. T. Warner
R. R. Fizz

Department of Meteorology
The Pennsylvania State University
University Park, PA 16802

INTRODUCTION

No general procedure is available for calculation of parcel trajectories that demonstrates consistent accuracy over a broad range of conditions. Many techniques exist (Pack et al., 1978; Hoecker, 1977; Peterson, 1966), but most have limitations such as being site-specific, being applicable under only certain meteorological conditions or times of day, relying on the synoptic-scale rawinsonde network for wind data or being applicable to only specific levels in the atmosphere. Peterson (1966) provides particularly interesting comparisons of trajectories calculated using a number of standard diagnostic techniques. In general, procedures that are relatively successful in some situations, provide poor trajectory estimates in other situations. It is also difficult to determine a priori which procedure will be best in a given case. Thus, there is a need for a general and reliable technique for atmospheric transport computations.

Warner (1981) discusses and demonstrates the use of a 3-D dynamic model for providing winds for parcel trajectory calculations Even though the model's predictive skill has been demonstrated for many meteorological applications (Warner et al., 1978; Anthes and Keyser, 1979), its ability to accurately predict transport in situations of interest to air pollution meteorologists is still being evaluated. As a preliminary test of a basic version of the model, Warner (1981) performed transport simulations for a period during August 1978 when the Environmental Protection Agency (EPA) conducted a major field study at the Cumberland Steam Plant of the Tennessee Valley Authorty. This study, known as the Tennessee Plume Study, was conducted as part of EPA's Sulfur Transport and

Transformation in the Environment (STATE) Project. The field experiments included the release and tracking of tetroons from Cumberland during numerous intervals within the period of the study On 15 August, 10 tetroons were released, traveling distances ranging from less than 25 km to in excess of 200 km. The tetroons' position data were compared with 3-D kinematic trajectory predictions from the 3-D regional-scale dynamic model. The 3-D model produced forecasts of horizontal and vertical velocities for period where tetroon trajectory data were available for model verification The predicted winds, available at each model time step, were then used to produce 3-D kinematic trajectories. Given a starting point in space and time for the model trajectory, corresponding to the observed tetroon position after it had reached its balasted level, a prediction of the tetroon path was produced. These 3-D kinematic trajectories were then compared to the observed tetroon displacements. The average angular error was found to be 7° and the averag displacement error was 9% of the observed path of the tetroon. These results were encouraging, however the winds over the domain were fairly strong and the flow was relatively straight--conditions which favor accurate trajectory predictions by any method.

As a followup to this study, we have used the same 3-D dynamic model to predict trajectories during a period of the Northeast Regional Oxidant Study (NEROS). One tetroon flight occurred in a particularly interesting meteorological situation where a strongly diffluent wind field in the vicinity of the tetroon created a situation where potentially large trajectory forecast errors were possible. Trajectory predictions were produced by using different techniques and meteorological data sets. Specifically, trajectorie computed from winds produced by the 3-D model are compared with trajectories that were based on routinely observed winds. The sensitivitity of the trajectories produced by both of these procedures to the type of meteorological wind data used is also illustrated.

THE THREE-DIMENSIONAL PROGNOSTIC DYNAMIC MODEL

The 3-D dynamic model is described in detail by Anthes and Warner (1978) and will only be summarized here. It is based on the primitive equations of atmospheric motion and is applicable over a wide range of scales. Calculation of convective and non-convective precipitation is optional. Either a bulk planetary boundary layer (PBL) parameterization or a high vertical resolution treatment of the PBL can be used. In the former, the lowest model layer represents the PBL, where in the latter, the moisture, temperature and wind structure of the PBL can be resolved by many model layers (Anthes and Warner, 1978). Surface heating, which must be modeled correctly if thermally forced circulations such as coastal breezes or mountain valley circulations are to be properly simulated, is computed using a technique developed by Blackadar (1976). It involves calculation of a ground temperature from a surface energy equation. This requires the specification of the surface characteristics as well as the time of day and the date. The model's vertical coordinate is σ which represents the local pressure normalized by the surface pressure. It has the advantage of being well-suited for applications over complex terrain. The specific characteristics of the version of the Penn State model used for these transport calculations are provided in Table 1.

Table 1. Characteristics of the three-dimensional dynamic model.

Characteristics	Description
Number of vertical levels	6
Horizontal mesh size	31 x 31
Grid increment	25 km
Initialization	Observed data (no balancing)
Planetary boundary layer	Bulk formulation (Anthes and Warner, 1978)
Surface heating	None

CASE CHOSEN FOR STUDY

This particular tetroon flight was chosen for study because it coincided with a challenging meteorological situation for trajectory calculations. The tetroon flight extended from 2320 GMT 30 August to 2100 GMT 31 August 1979 and maintained an elevation between 1600 m (840 mb) and 2400 m (760 mb). Figure 1 shows an analysis of the 850 mb wind field for each of the three synoptic times that encompass the tetroon flight. The observed tetroon path is shown on the analyses as well as the position of the tetroon at the time the analyses apply. The wide solid line shows the extent of the model domain and the border of the figures corresponds to the boundary of the domain on which the meteorological analysis was performed. The diffluence line on the analysis for 1200 GMT 31 August is clearly identifiable as it separates the northerly flow over Virginia from the easterly flow over West Virginia. A small error in the predicted position of the trajectory early in the flight has the potential for producing a large separation in the endpoints of the predicted and observed trajectories.

A COMPARISON OF TRAJECTORIES BASED ON OBSERVED AND MODELED WINDS

Most trajectory calculation procedures use winds obtained from conventional synoptic observations. A wide variety of such techniques exists, where various assumptions are used to obtain the winds at the location of the trajectory endpoint. In order to give some

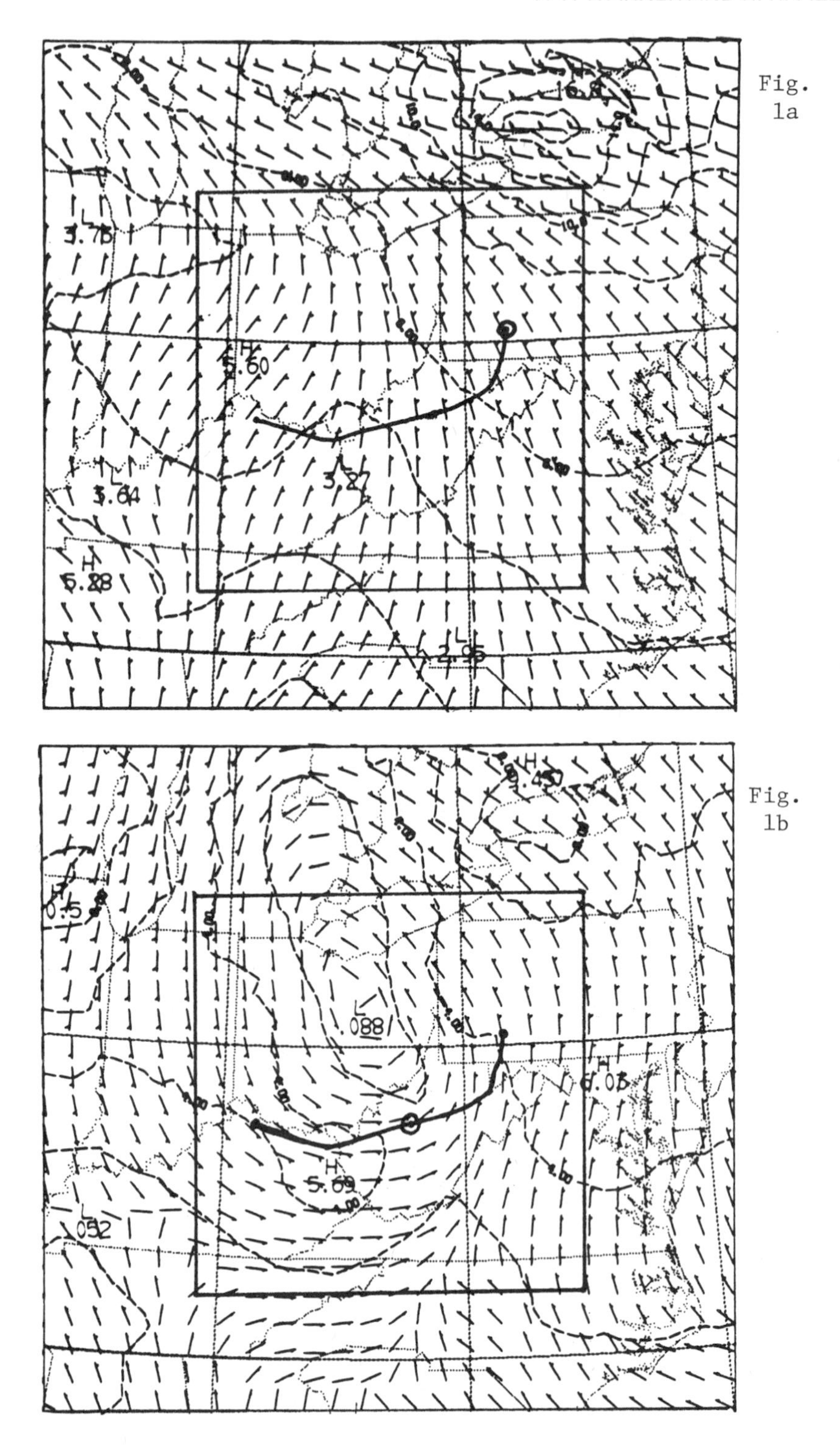
Fig. 1a
3.78
H
5.60
3.27
3.64
H
5.28
2.96
Fig. 1b
H
0.5
.088
H
6.03
H
5.69
.052

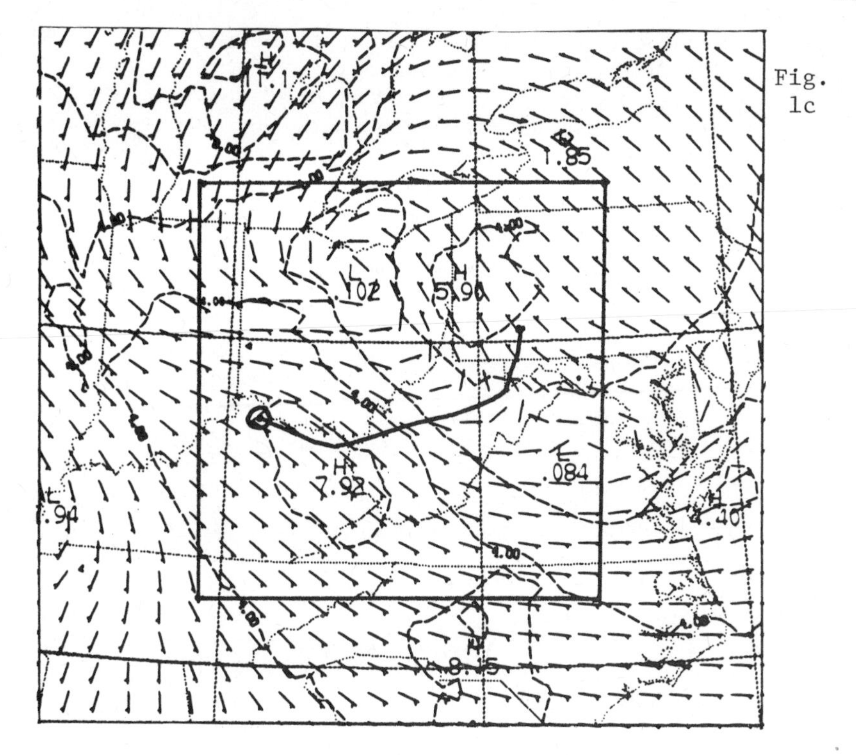

Fig. 1. Objective analysis of winds at 850 mb for 00 GMT 31 August 1979 (a), 12 GMT 31 August 1979 (b), and 00 GMT 1 September 1979 (c). Dashed lines are isotachs labeled in ms^{-1} and wind direction symbols have a barb on the upwind end that is proportional to the wind speed. The observed tetroon trajectory is shown as a solid line on which the position of the tetroon at the time of the analysis is indicated with a circle.

insight into the question of whether trajectories obtained from model-predicted winds are superior to those obtained from these conventional techniques, a simple procedure has been developed for calculating the trajectories based on 12-hourly National Weather Service (NWS) rawinsonde data.

Parcel trajectories calculated from observed wind data

In this procedure, horizontal wind data are obtained from conventional NWS rawinsonde observations made every 12 h. The horizontal winds are analyzed to a three-dimensional grid, where the vertical coordinate is σ, using a Cressman (1959)-type objective analysis procedure. For convenience and to permit a controlled comparison of results, this three-dimensional grid is the same as the one used for the dynamic model. Vertical velocities are then obtained from the observed winds through the use of a continuity equation. These three-dimensional arrays of wind vectors, availabl every 12 h, then provide the wind vector at the end point of a trajectory through an interpolation in space and time. This wind vector is then used to calculate a short displacement of the end point. The four-dimensional interpolation of the wind field to the endpoint and its displacement are repeated until the trajectory is the desired duration. If we wish to compare a calculated trajector with a tetroon trajectory (which does not reflect nonturbulent vertical velocities) we normally cause the computed trajectory to maintain the same altitude as the observed trajectory. That is, the vertical displacement of the computed trajectory is based on observations instead of the computed vertical velocity.

Parcel trajectories calculated from winds predicted by the dynamic model.

The model is initialized using the data obtained from the analysis of rawinsonde observations to the model grid. Now, howeve horizontal winds are available on the model grid every time step (i this case, every 5 minutes) after the initialization. Vertical velocities are calculated in the same way as noted above and the space and time interpolation of the winds to the trajectory endpoint is again performed.

Results of the comparison

Using the winds from rawinsonde observations as input to the trajectory calculation has potential disadvantages compared to using the windfield from the dynamic model. The rawinsonde observations have an average spacing of about 400 km. When the observations are analyzed to the gridpoints, only large-scale variations i the wind field appear on the 25 km grid which is capable of resolving much greater detail. Also, the interpolation of the grided win vectors between 12 hourly observation times is linear. The trajectories based on observed winds are therefore only responding to large-scale flow features and do not reflect any temporally nonline changes. The dynamic model, however, can generate dynamically consistent small-scale features in the winds even though the model is initialized with large-scale data. In addition, rapid temporal variations in the flow can be resolved. Figure 2 compares the trajectory predictions based on modeled and observed winds (curves 2 and 3 respectively). Even though the model-predicted trajectory do not compare expecially well with the observed trajectory, it still is clearly better than the one based on observed winds. This impro ment seems to be a result of the fact that the model winds undergo shift in direction from NNW to NNE over the Virginia-West Virginia border during the first 12 h, more rapidly than do the winds that a linearly interpolated between 00 GMT and 12 GMT 31 August. This allows the trajectory to penetrate more deeply into the easterly flow over West Virginia during the second 12 h of the simulation.

A COMPARISON OF TRAJECTORIES BASED ON RAWINSONDE DATA AND LARGE-SCALE ANALYSES

In the previous comparison of trajectories based on modeled and observed winds, the grided meteorological data were obtained by analyzing the NWS rawinsonde observations. In the procedure which used the model, the grided data served to define the initial state for the model as well as the lateral boundary conditions during the simulation. The procedure which was based on the linear time interpolation between grided data used the grided wind values at the 12 h synoptic times in the interpolation process.

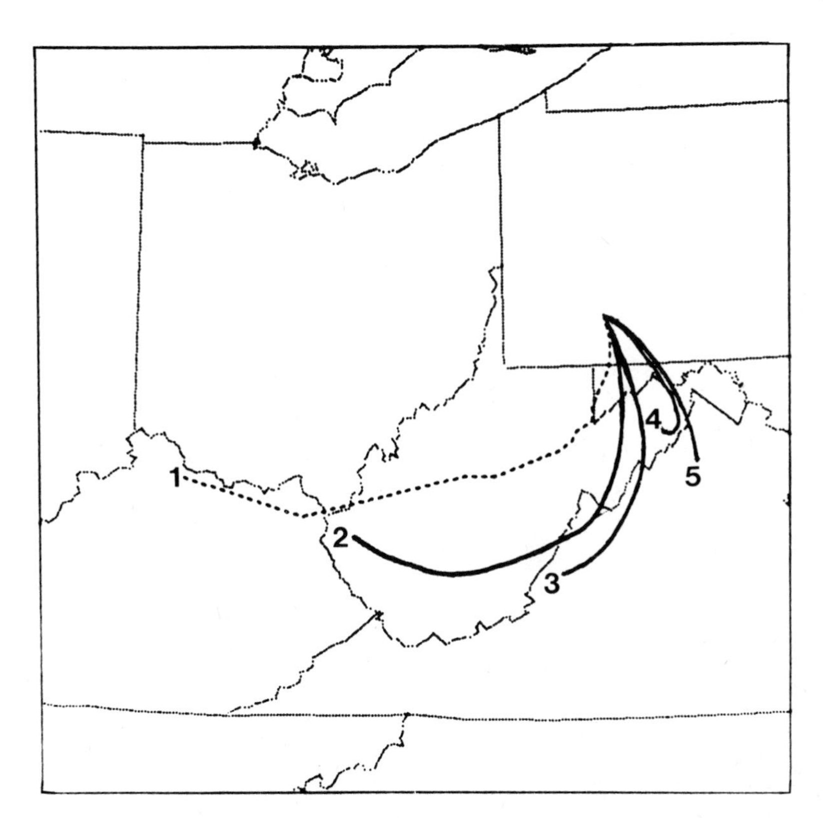

Fig. 2. Observed tetroon trajectory (1), trajectories predicted by the 3-D model based on objectively analyzed rawinsonde winds (2) and winds from the NMC analysis (5), and trajectories diagnosed by temporally interpolating winds obtained from an objective analysis of rawinsonde winds (3) and from the NMC analysis (4).

Because the analysis procedure which defines grid-point values from irregularly spaced observations requires computer resources an programming time, it may be tempting to some to use existing very large-scale analyses of the rawinsonde observations. An example is the readily available, global analysis of the National Meteorologic Center (NMC) available at 12-hourly synoptic times. This analysis (Bergman et al., 1974) has many desirable features and is convenien to use, however because of its smooth character, it does not always conform closely with observations. We will illustrate how trajectories produced using meteorological data from this existing large-scale analysis differ from those produced using data obtained from the aforementioned careful Cressman (1959)-type objective analysis of the rawinsonde data. Figure 2 shows the trajectories obtained from the NMC analyses (curve 4) and from the Cressman-type analyses of rawinsonde observations (curve 3). In addition, the trajectorie based on the model-forecast winds are compared for the two sources of initial wind data (curves 2 and 5). In both cases the use of th Cressman-type objective analysis leads to trajectories that are closer to the observed trajectory than the trajectories based on winds from the NMC analyses. In order to further illustrate differences in the wind analyses obtained from the two different data sources, 24 h trajectories from 00 GMT 31 August to 00 GMT 1 September 1979 have been calculated from various locations at 835 mb over the grid (Figure 3). The winds for these trajectory calculations are obtained from the temporal interpolation of the data and not fr the model so that differences in the trajectories based on the NMC and Cressman analyses more directly reflect differences in the actu analyses themselves. The NMC analyses were clearly more uniform ov the domain and had slower wind speeds than did the analyses from rawinsonde data. Because the NMC analyses and the rawinsonde observations that serve as the basis for the Cressman analyses are both available every 12 h, both procedures use data with the same temporal resolution.

SUMMARY

These results illustrate how certain sources of error can contribute to incorrect diagnosis or prediction of parcel trajectories. Use of rawinsonde-scale data that has been averaged, smoothed or filtered, such as the NMC analysis in this case, may be convenient but has clear limitations in terms of its ability to resolve even rawinsonde-scale features of the flow field. By comparison, a careful objective analysis of rawinsonde data can preserve meteorologic features that are potentially essential for good trajectory forecas The use of a 3-D dynamic model can also help provide more resolutio in time and space. Wind data are available at every model time ste and grid point; in this case that improves the temporal resolution from 12 h to 5 min and the spatial resolution from ~ 400 km to 25 k

Nonlinear variations in the windfield in space and time can therefore be better represented.

It should be recalled that the version of the dynamic model used here had relatively coarse vertical resolution and a bulk formulation for the planetary boundary layer. Work is underway to determine if the use of higher vertical resolution and a more realistic planetary boundary layer parameterization will provide further improvement in the trajectory prediction.

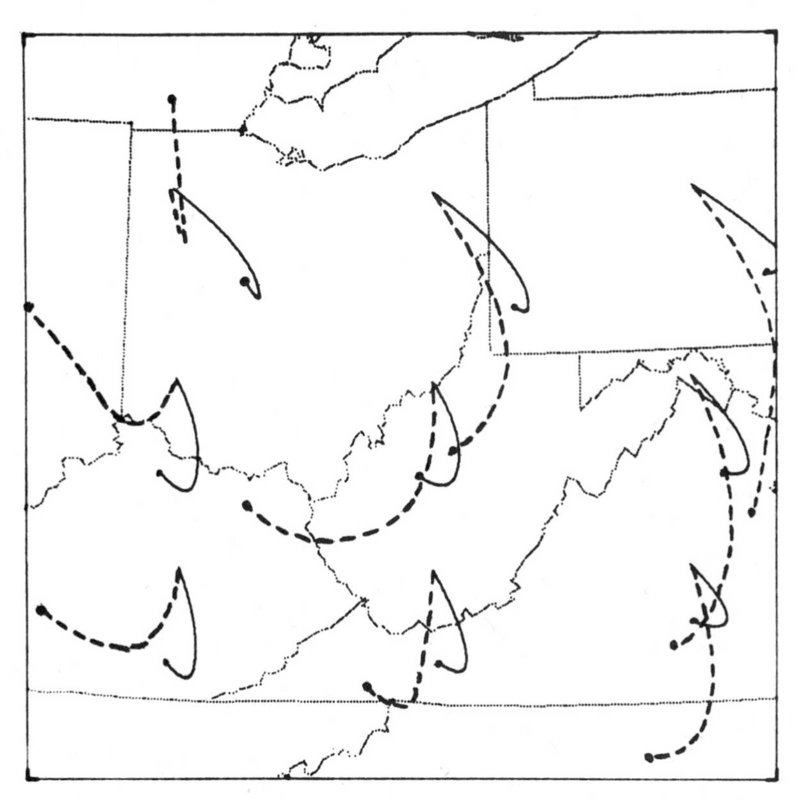

Fig. 3. Twenty-four hour trajectories with initial points at 835 mb at various locations over the domain. The trajectories were based on the temporal interpolation of the winds between the 12-hourly synoptic times where the NMC analyses (solid lines) and the objective analyses of the rawinsonde data (dashed lines) were used.

ACKNOWLEDGEMENTS

This research was performed under the sponsorship of EPA Grant R-805659. Computing support was provided by the National Center fo Atmospheric Research which is sponsored by the National Science Foundation. The tetroon trajectory data were obtained from the Meteorology Division, Environmental Sciences Research Laboratory, Environmental Protection Agency.

REFERENCES

Anthes, R. A. and Keyser, D., 1979, Tests of a fine-mesh model over Europe and the United States, Mon. Wea. Rev., 107:963-984.
Anthes, R. A. and Warner, T. T., 1978, The development of mesoscale models suitable for air pollution and other mesometeorological studies, Mon. Wea. Rev., 106:1045-1078.
Bergman, K., McPherson, R., and Newell, J., 1974, A description of Flattery global analysis method--No. 1, Technical Procedures Bull. No. 90, National Weather Service, 9 pp.
Blackadar, A. K., 1976, Modeling the nocturnal boundary layer, Third Symp. on Atmospheric Turbulence, Diffusion and Air Quality, Raleigh, NC, Oct. 9-12, Preprints, Amer. Meteor. Soc. Boston, pp. 46-49.
Cressman, G. P., 1959, An operational objective analysis system. Mon. Wea. Rev., 87:367-374.
Hoecker, W. H., 1977, Accuracy of various techniques for estimating boundary-layer trajectories, J. Appl. Meteor., 16:374-383.
Pack, D. H., Ferber, G. J., Heffter, J. L., Telegadas, K., Angell, J. K., Hoecker, W. H., and Machta, L., 1978, Meteorology of long-range transport, Atmos. Envir., 12:425-444.
Peterson, K. R., 1966, Estimating low-level tetroon trajectories, J. Appl. Meteor., 5:553-564.
Warner, T. T., 1981, Verification of a three-dimensional transport model using tetroon data from project STATE, Atmos. Envir., 15:in press.
Warner, T. T., Anthes, R. A., and McNab, A. L., 1978, Numerical simulations with a three-dimensional mesoscale model, Mon. Wea. Rev., 106:1079-1099.

DISCUSSION

A. VENKATRAM: Does an "observed" trajector correspond to the motion of a single tetroon ? If so, explain your assumption.

T.T. WARNER: Certainly a multiple tetroon release would have been more desirable. The sing tetroon data, provided by the U.S. EPA, did coin-cide with an interesting meteorological situation however, so we used it in spite of the limitations of using a single tetroon trajectory to represent the nonturbulent transport. When satisfactory 3-D data from chemical tracers become available for long to medium-range transport verification, we will employ them.

P. MICHAEL

Are you using a prognostic tool for diagnosis - I have a two part question - 1) How good is your forecast ? 2) Are there means of pushing the forecasts towards later observations ?

T.T. WARNER

1) The forecast meteorological variables compare very favourably with the observations after 24 h.

2) During the next two years we will be testing a "four-dimensional data assimilation" system whereby the forecast variables, that coincide approximately in space and time with observations applicable during the forecast, are nudged towards the observations.

L. NIEMEYER

While I agree that several tetroons would give a better approximation of the trajectory of an air parcel, experience in conducting tracer tests over ranges of up to 100 miles using fluorescent particles and SF_6 gas showed that a single tetroon used to tag the leading edge of an air parcel was an effective measure of the mean wind in the mixing layer. The tetroon arrived at a sampling area at the same time the tracer cloud arrived. The tetroon is displaced above and below its predetermined flight altitude by turbulent impulses which result in the tetroon being an effective integrating device which closely approximates the mean transport wind trajectory.

W. GOODIN

Over what range of grid sizes is your model appropriate ?

T.T. WARNER

Our applications of the model with real data have employed grid increments from 2 km to 140 km. The horizontal grid mesh generally ranges from a 30 x 30 point mesh to a 50 x 50 point mesh.

A.D. CHRISTIE

It appears inappropriate to

me to give so much weight to the tetroon observed trajectory. If tetroon trajectories are to be used as track criteria a large enough number (in both space and time dimensions) should be tracked to be consistent with the wind field data used to derive computed trajectories.

T.T. WARNER Same answer as for the first question.

J.D. TAFT How many rawinsondes did you use to generate the wind fields.

T.T. WARNER The model forecast domain contained about 5 observations while the larger domain on which initial data were analyzed contained about 10 rawinsonde obervations.

A NEW TRAJECTORY MODEL AND ITS PRACTICAL APPLICATION FOR ACCIDENT AND RISK ASSESSMENTS

K.J. Vogt, H. Geiß, J. Straka

Kernforschungsanlage Jülich GmbH

INTRODUCTION

Since the conventional straight-line Gaussian diffusion model is only applicable to environmental exposure calculations under constant meteorological diffusion conditions and as the so-called advanced diffusion models are, as a rule, not suited because

- they are not (or insufficiently) calibrated and validated
- the required input parameters such as vertical wind and temperature profiles are difficult to obtain
- a classification according to diffusion categories is problematic
- the diffusion and weather march statistics required for important ranges of application are lacking
- the evaluation requires a sophisticated numerical integration,

we have extended the Gaussian diffusion model to cover non-stationary diffusion conditions in a so-called volume source model, which permits calculation of the environmental exposure due to released pollutant puffs taking account of the change in wind velocity, wind direction, diffusion category and precipitation intensity during transport, using customary diffusion parameters and the diffusion as well as weather march statistics available for many sites.

The computation expenditure required in this connection is small, since the necessary integrations can be carried out analytically or may be attributed to error functions, so that the diffusion calculation for the majority of tasks can be performed on medium-sized computers or on process computers.

The spatial application range of the model, which in its present form uses meteorological data measured at the site as input parameters, is dependent on the topography. In the case of even terrain, sufficiently precise environmental exposure values can still be calculated up to a source distance of about 50 km. Rough estimates are even still possible at larger distances, in which case, however, inhomogeneities of the large-area weather situations and influences of the Coriolis forces will increasingly augment the error. For this reason, the use of meteorological data from a network of observation stations would be preferable for larger source distances, in which case the trajectory would result e.g. from wind field measurements. The computation model would have to be adapted to the adoption of input parameters dependent on the measurement network.

VOLUME SOURCE MODEL AND RELEASE CONSEQUENCES MODEL

The basic features of the volume source model were already presented by us at the meeting in Rome (1). The meteorological and mathematical foundations have meanwhile been published in (2).

First of all, it must be assumed that the meteorological diffusion conditions at successive time intervals are known and represented e.g. by measured hourly or 10-minute mean values. The diffusion factor of an instantaneously released puff, i.e. the time-integrated activity concentration in the n-th time interval normalized to the unit of the source strength, may then be represented after superposition of the contributions of the volume elements regarded as point sources at the end of the (n-1)th time interval by a modified Gaussian diffusion function:

$$\hat{\chi}_n(x,y,z) = \frac{\exp\left|-\frac{\eta_n^2}{2\sigma_{y\,eff}^2}\right|\left\{\exp\left|-\frac{(z-H)^2}{2\sigma_{z\,eff}^2}\right| + \exp\left|-\frac{(z+H)^2}{2\sigma_{z\,eff}^2}\right|\right\}}{2\pi\,\sigma_{y\,eff}\,\sigma_{z\,eff}\,u_n} \quad (1)$$

with H : effective release height

u_n : wind velocity at release height during the n-th time interval

x,y,z : downwind, crosswind and vertical coordinate (oriented according to the wind direction in the 1st time interval).

In this case,

$$\eta_n = \frac{1}{u_n}\left\{u_{xn}\left(y-\sum_{k=1}^{n-1} u_{yk}\,\Delta t\right) - u_{yn}\left(x-\sum_{k=1}^{n-1} u_{xk}\,\Delta t\right)\right\} \quad (2)$$

with u_{ik} : wind velocity components in the k-th time interval

$t=t_k-t_{k-1}$: travel time of the puff in the k-th time interval

is the crosswind distance from the trajectory in the n-th time interval and

$$\sigma^2_{i\ eff} = \sum_{k=1}^{n} \{\sigma^2_{ik}(t_k)-\sigma^2_{ik}(t_{k-1})\} \quad (3)$$

is the relation for determining the effective diffusion parameters $\sigma_{i\ eff}$ (i=y,z) from the diffusion parameters σ_{ik} of the successive intervals (k=1...n).

The waste air plume can be described accordingly by a succession of puffs (possibly with varying source strength), which will all follow their own trajectory in the case of a change in wind direction. On the simplifying assumption that all of the puffs constituting the waste air plume will follow the trajectory of the first puff(in single file), eq. (1) may also be used for the description of waste air plumes (for cases with constant release rate during each time interval). The error of this approximation is negligible, in particular if the changes in wind direction are insignificant.

The following effects have meanwhile been additionally incorporated into the basic equation of the volume source model eq. (1):

- limitation of the vertical distribution according to the mixing heigth dependent on the diffusion category
- consideration of pollutant conversion (e.g. due to radioactive decay)
- consideration of the depletion in the waste air plume on account of pollutant losses due to fallout and washout effects (cf.Appendix).

During programming operations, a further important modification of eq. (1) turned out to be necessary. Instead of carrying out the time integration over the waste air concentration from $-\infty$ to $+\infty$ passing the point of reference, as is common practice when using the straight-line stationary Gaussian function, integration in the case of the volume source model may always be effected only for the duration of the respective time interval for reasons of correctness, in which case the limited interval can be represented via the Gaussian function by a difference of error functions. This takes account of the fact that e.g. a puff whose center of gravity has just reached the end of its trajectory in the (n-1)th time interval, will be assessed correctly with respect to the environmental exposure caused by such puff in the (n-1)th and the n-th time interval. The first half of the puff has passed the point of reference at the time t_{n-1}

and will be treated using the diffusion conditions of the (n-1)th time interval with regard to time integration, whereas the rest of the dose will result when passing the rear part of the concentratio field under the meteorological conditions of the n-th time interval

The thus corrected diffusion equation in the n-th time interva then reads:

$$\chi_n(x,y,z) = \hat{\chi}(x,y,z)\ \frac{1}{2}\ \{\mathrm{erf}(\frac{u_n(t_n - t_{n-1}) - \xi_n}{\sqrt{2}\ \sigma_{y\ eff}}) - \mathrm{erf}(\frac{-\xi_n}{\sqrt{2}\ \sigma_{y\ e}}$$

with $\mathrm{erf}(x) = \frac{2}{\sqrt{\pi}} \int_0^x e^{-t^2}\, dt$: error function (5)

ξ_n, η_n: downwind and crosswind coordinates at the point of reference related to the trajectory in the (n-t)th time interval.

Based on the volume source model which, in essence, represents a trajectory model considering non-stationary diffusion conditions, an environmental exposure model was developed, permitting to calculate the contributions of the different exposure pathways, i.e.

- the inhalation of contaminated respiratory air
- the ingestion of contaminated food (on account of pollutant deposition on the ground due to fallout and washout)

and (for radioactive releases) additionally

- the β-and γ-dose from the waste air plume (β- and γ- submersion)
- the γ-dose due to depositions on the ground (γ-ground radiation

The environmental exposure, i.e. the dose commitment in the case of radioactive releases, will then result for each release pathway by multiplication of the source strength by the diffusion factors (and dose factors) depending on the exposure pathway and nuclide involved. Furthermore, the consumption rate is reflected additionally in the case of ingestion, and the breathing rate in the case of inhalation.

The diffusion equation eq. (4) represents the diffusion factor for the inhalation and the β-submersion as well as for the fallout contribution to the ingestion and γ-ground dose when multiplied by the deposition rate. The washout diffusion factors are obtained by vertical integration of the concentration (3) multiplied by the washout coefficients, whereas the γ-submersion diffusion factor results from a spatial integration over the contributions of the volume elements of the waste air plume, taking account of the distance, absorption and build-up factor (4)(cf. Appendix).

APPLICATION OF THE VOLUME SOURCE MODEL FOR ACUTE ACCIDENT AND RISK ANALYSIS

The special benefits of the volume source model become apparant in connection with the release consequences analysis which requires a non-stationary diffusion calculation due to the range of environmental exposure. Under the project on safety analyses for waste disposal (PSE - Projekt Sicherheitsanalysen Entsorgung), we have therefore developed a computer code MUSE with which the following environmental exposure values can be calculated:

1) for individual concrete accidents considering the actual weather march
   - the individual dose in the vicinity of the source
   - the collective dose depending on the population distribution

2) for risk analyses of postulated types of accidents considering weather march statistics
   - the frequency distribution and the mean value of the individual dose at the measuring point observed
   - the collective dose averaged over all weather marches
   - the mean dose density distribution of the population.

As an example of the application for concrete accidents, Figure 1a-c shows the development of the isopleths of the diffusion factor in terms of time as a measure of environmental exposure at different times after the release of pollutants for a predefined weather march. In the case of an acute accident, the meteorological data measured during successive time intervals (e.g. the 10-minute mean values) can be given into the computer as input parameters, possibly on-line. The curves in Figure 1d demonstrate the completely unrealistic course which would be obtained using the conventional Gaussian model for stationary diffusion conditions, i.e. maintaining the meteorological data available at the time of release.

A forecast on the probable environmental exposure situation for the coming time interval will be possible when applying the volume source model by keeping the diffusion conditions measured in the last time interval constant. The error in dose forecasts is generally governed more strongly by an inexact knowledge of the source term than by the inaccuracy of the diffusion model.

While in the case of a concrete accident, the environmental exposure has to be calculated according to the actual weather march during the diffusion process, all of the weather marches occurring or, at least, a representative portion, must be covered for a risk analysis. In the case of nuclear installations, the risk involved in early damage results from the dose density distribution of the population (to be assessed with the dose/effect relation), i.e. from

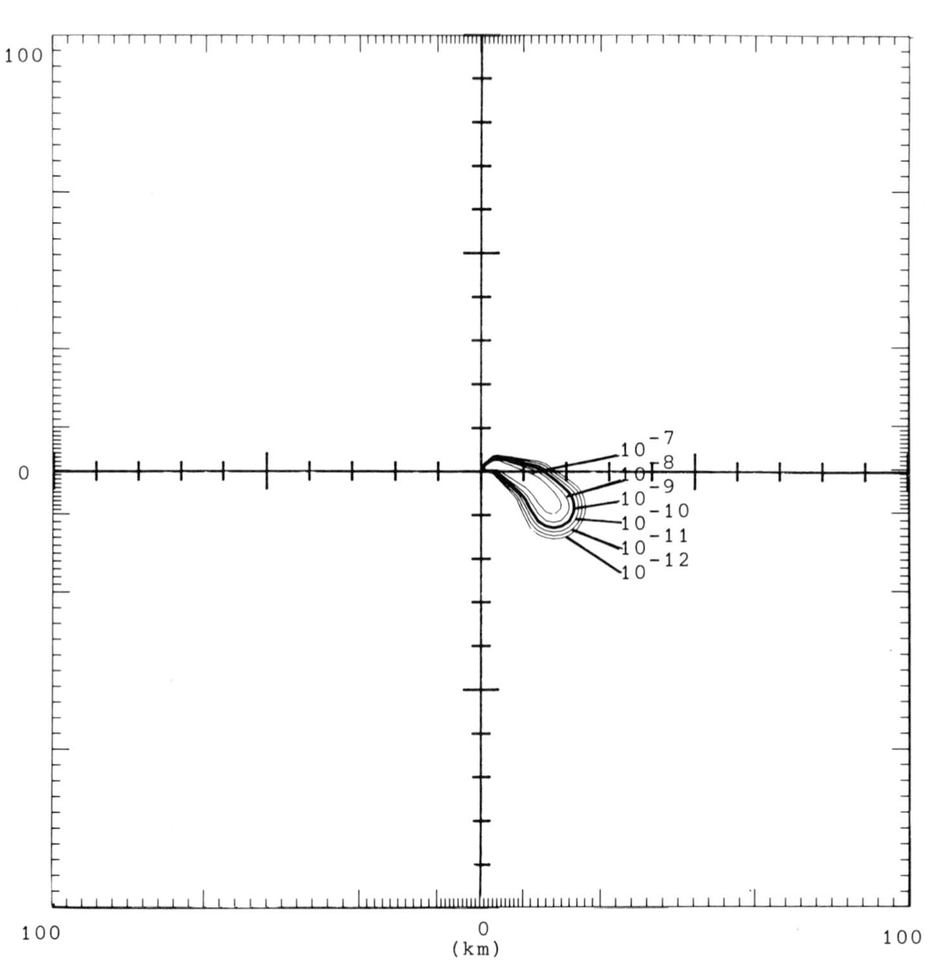

Fig. 1a Isopleths of diffusion factor χ for a given weather sequen
3 hours after release.

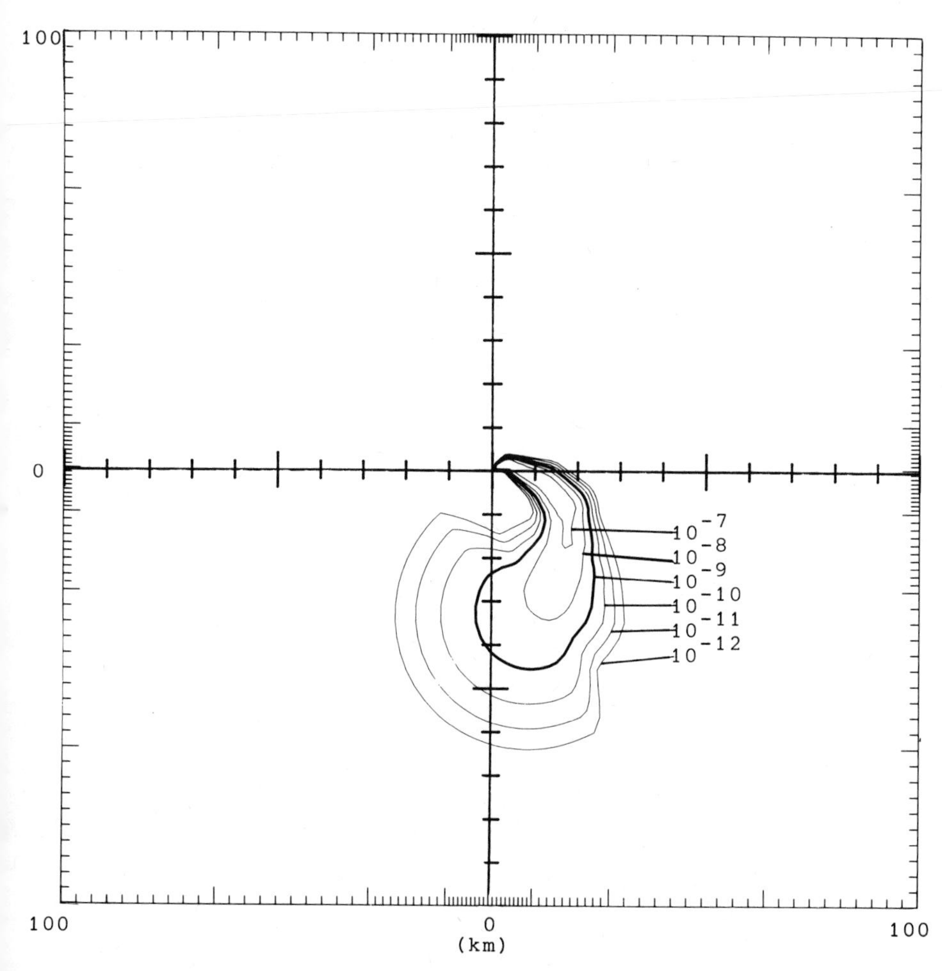

Fig. 1b Isopleths of dissusion factor χ for a given weather sequence 7 hours after release.

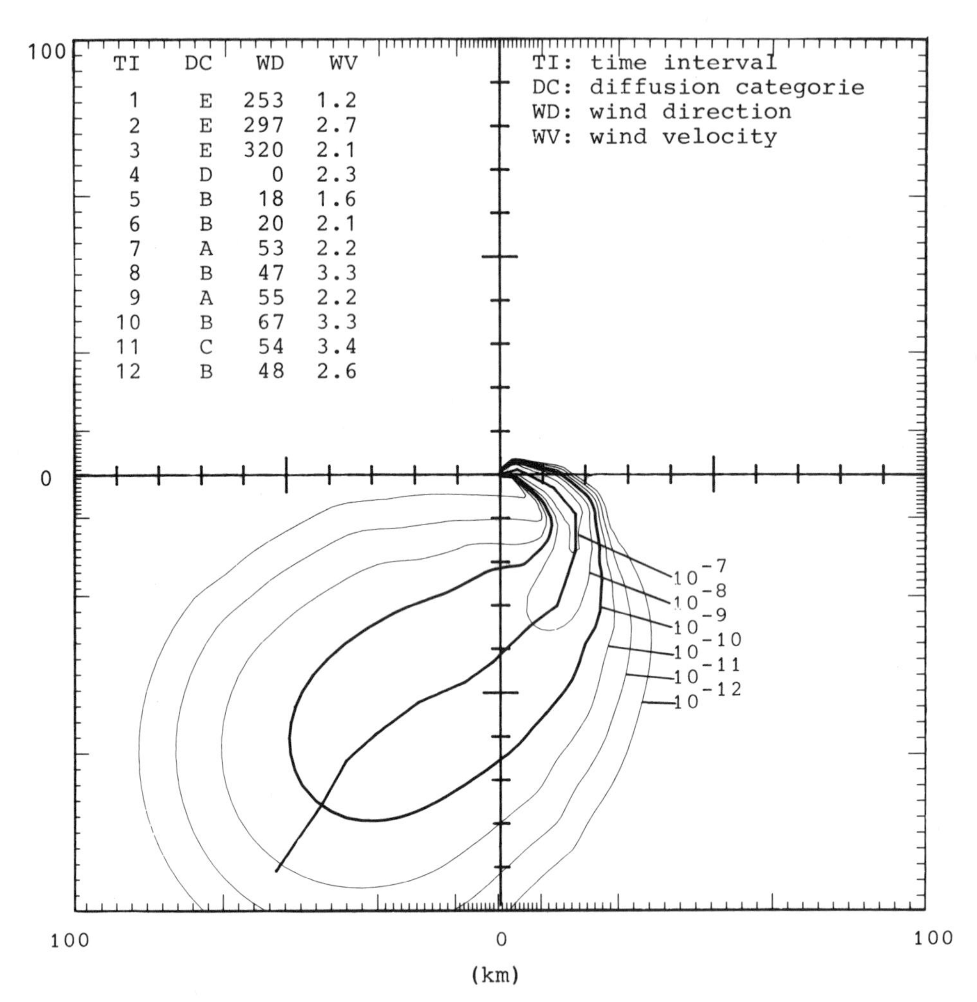

Fig. 1c Isopleths and trajectory of dissusion factor χ for a given weather sequence 12 hours after release

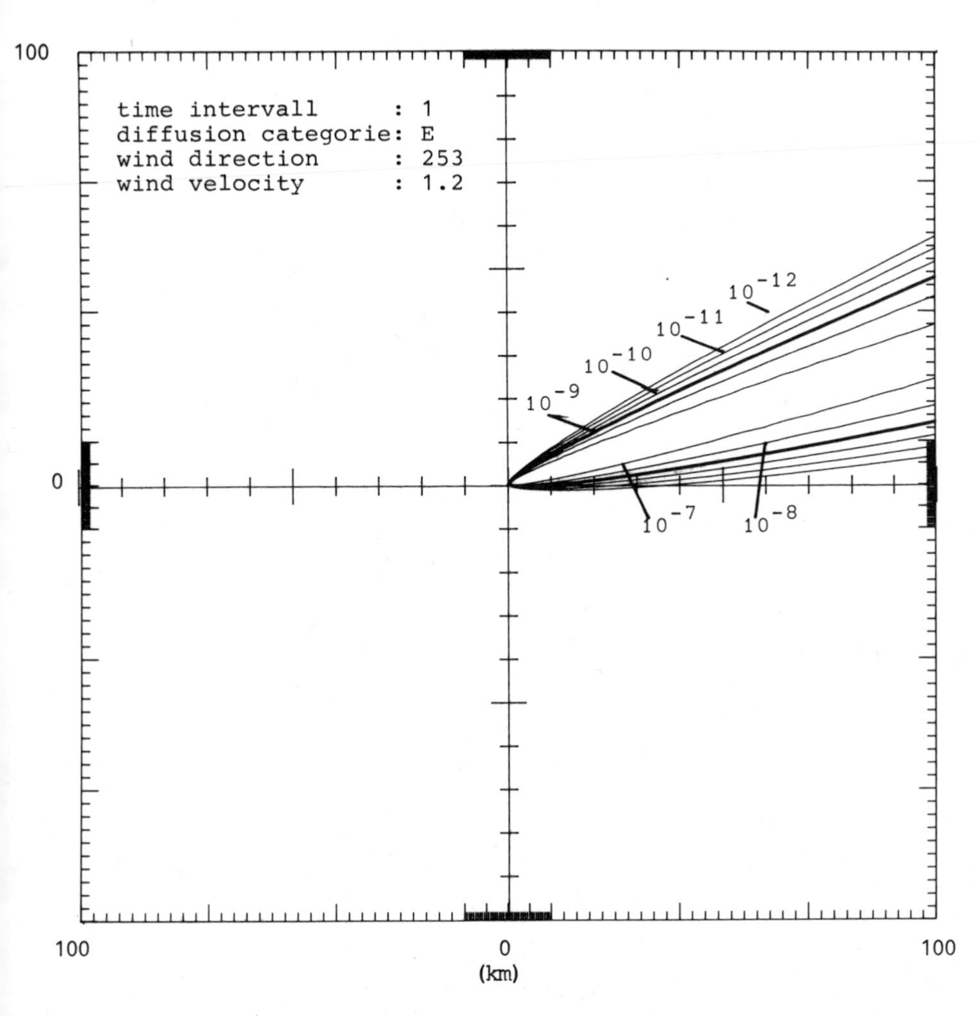

Fig. 1d Isopleths of dissusion factor χ obtained using the conventional Gaussian model for stationary diffusion conditions 1 hour after release.

statistics indicating how many persons are exposed to certain doses. Early damage is only possible, if the activity volumes released are so large that the resultant dose values exceed 200 rem. For most of the nuclear sources (probably including the reprocessing plants) no such large source strengths are to be expected, so that the overall risk is determined by late injuries, which are proportional to the collective dose, i.e. to the sum of the individual doses of the population. Their derivation must be based on the mean diffusion factors for the relevant exposure pathways χ, F, W and χ_{γ} whose isopleths are shown in Figure 2a - c. 2500 weather marches were incorporated into the mean value formation (cf. example in Figure 1), which were taken statistically out of the Jülich weather data. While the diffusion factors for the concentration near ground level and for the fallout F exhibit the same distribution pattern (Figure 2a) which, among other things,reflects the wind direction distribution, the diffusion factor for washout (Figure 2b) is distinctly dependent on the precipitation wind-rose (precipitations in Jülich occur preferentially with west or south-west winds). The isopleths of the diffusion factor for γ-submersion (Figure 2c) show a course depending least on the wind-rose at short source distances due to the range of γ-radiation.

For the most unfavourable point of impact in the main wind direction according to Figure 2a, Figure 3 shows the complementary cumulative frequency distribution of the diffusion factors calculated for 2500 weather marches for the concentration χ near ground level, the depositions on the ground due to fallout and washout (F + W) and the γ- submersion χ_{γ} . While in the case of γ-submersion, the values only vary by a factor of approx. 10 about the mean value, a far higher variation range results in the case of concentrations near ground level and, in particular, for the activity deposited (here especially on account of the precipitation statistics). While in the case of γ-submersion, the 99% percentile is only by a factor of six above the 70% percentile, this amounts to more than two orders of magnitude for concentration near ground level and for surface contamination.

A last example of the application for risk analyses is shown in Figure 4, where the normalized collective dose is plotted as a function of the radius of the observed region for all exposure pathways which are governed by the concentration χ near ground level. In this figure, source strength and dose factor were equated to one, and a homogeneous population distribution comprising 250 persons/km² was assumed. The curve then virtually indicates the diffusion factor weighted with the population, averaged over all wind directions and over 2500 weather marches. It may be seen that the collective dose still increases even after a source distance of 100 km, although less than linearly due to the depletion effects ($v_g = 10^{-3}$ m/s were assumed for fallout, and $\Lambda = 2.65 \times 10^{-5} \times I$ for washout). A complete risk analysis for a concrete accident type will be published in the near future (5).

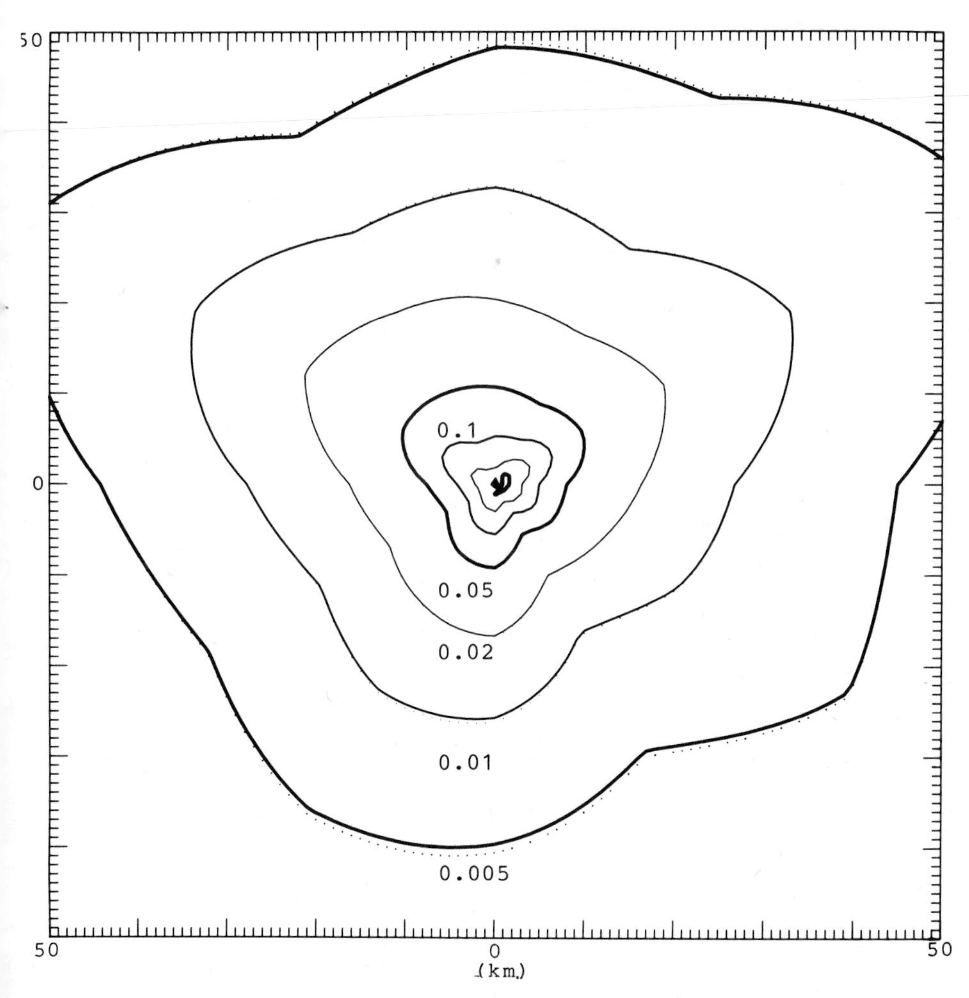

Fig. 2a Isopleths of the diffusion factor χ (averaged over 2500 weather sequences).

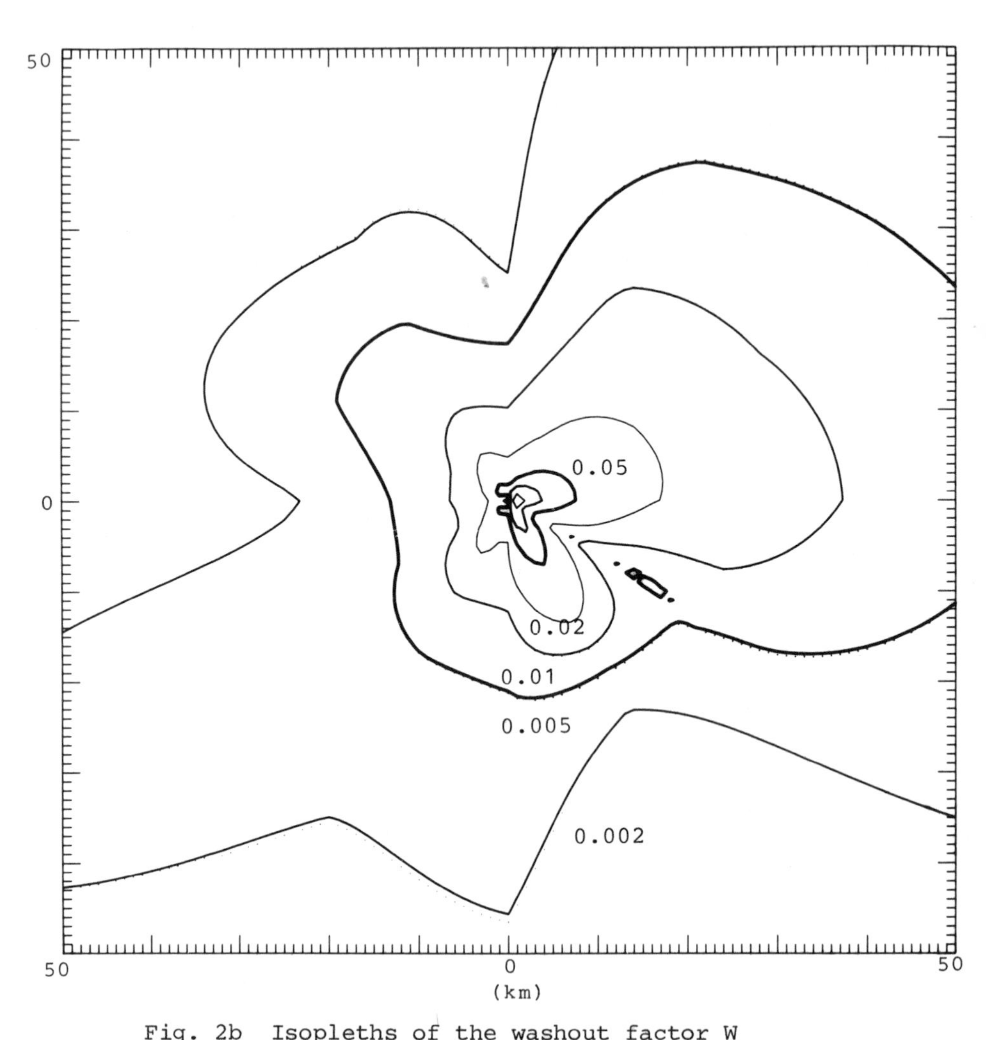

Fig. 2b Isopleths of the washout factor W (averaged over 2500 weather sequences).

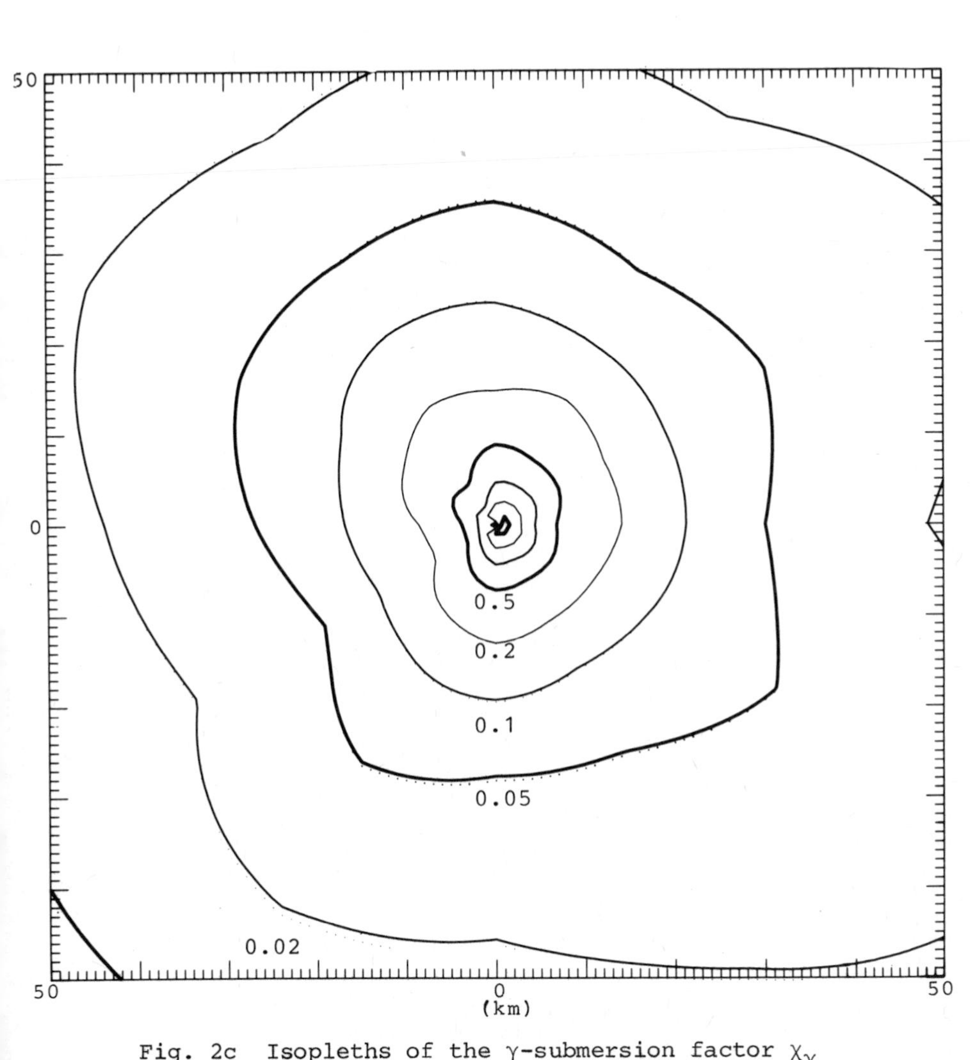

Fig. 2c Isopleths of the γ-submersion factor χ_γ (averaged over 2500 weather sequences).

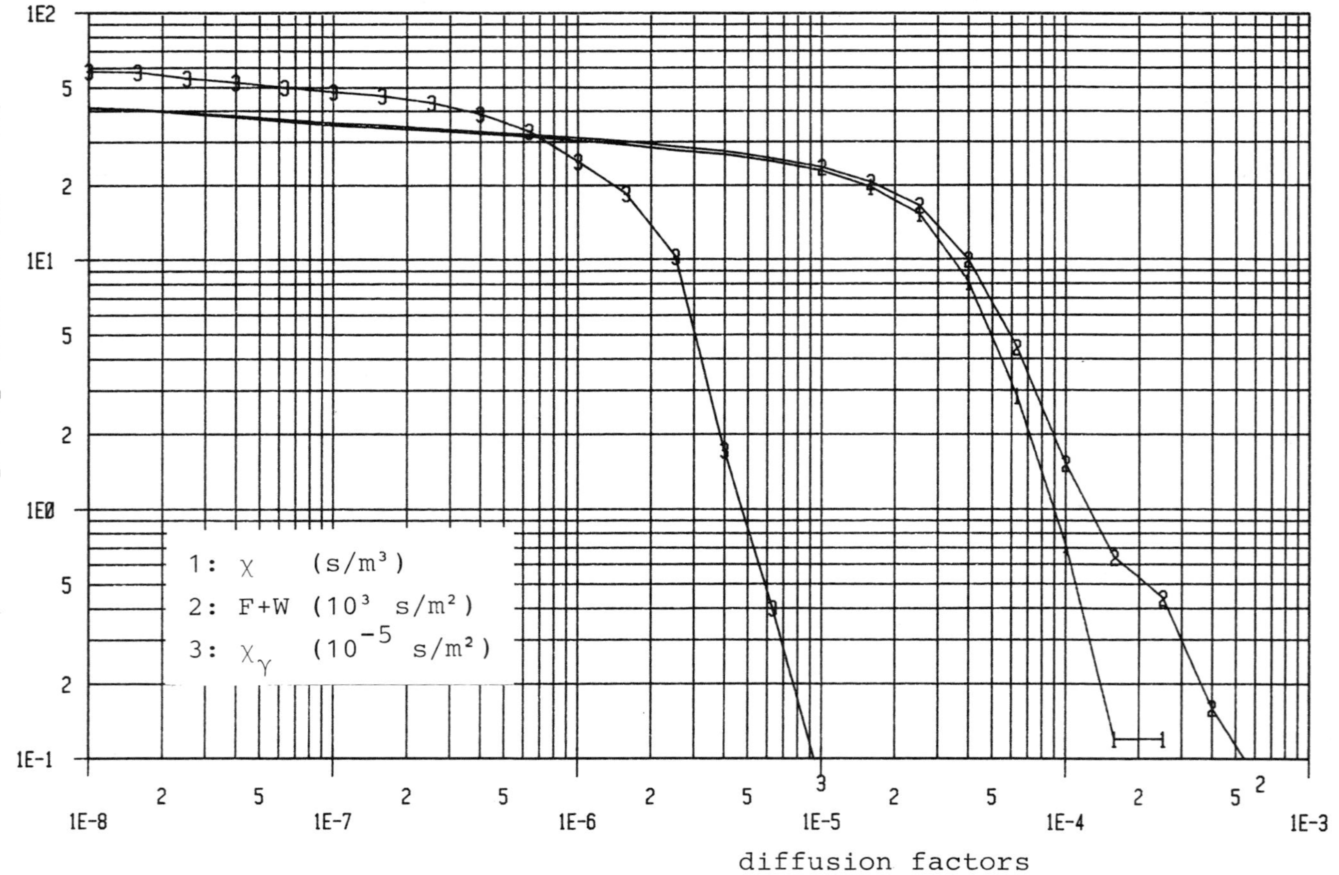

Fig. 3 Complementary cumulative frequency distribution of the diffusion factors χ, (F+W) and $\chi\gamma$ for the different exposure pathways at the most unfavourable point of impact (calculated for 2500 weather sequences)

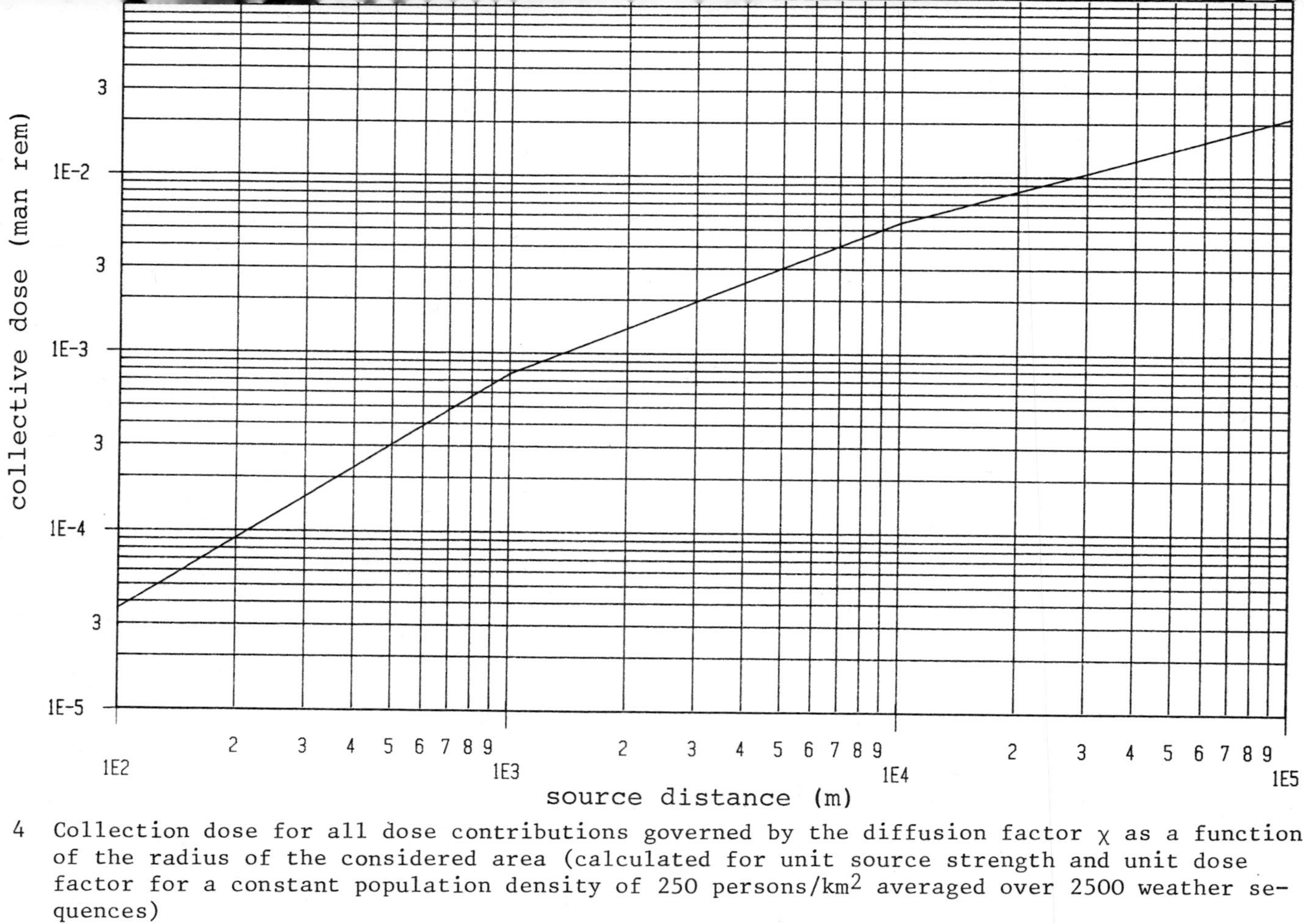

Fig. 4 Collection dose for all dose contributions governed by the diffusion factor χ as a function of the radius of the considered area (calculated for unit source strength and unit dose factor for a constant population density of 250 persons/km^2 averaged over 2500 weather sequences)

LIST OF SYMBOLS

D_β, D_μ, D_g, D_B, D_γ	dose contributions (cf. appendix)
g_β, g_μ, g_g, g_B, g_γ	dose factors (cf. appendix)
Q	source strength
λ	decay constant
Λ	washout coefficient
v_g	deposition velocity
χ, F, W, χ_γ	diffusion factors for different pathwa
B	build-up factor
μ	attenuation coefficient
r	distance between (X,Y,Z) and (x,y,0)
x,y,z	coordinates of reference point
X,Y,Z	coordinates of volume elements
ξ_n, η_n	coordinates relative to trajectory
H	effective release height
σ_{yk}, σ_{zk}	diffusion parameters in k-th time int
$\Delta t=t_n-t_{n-1}$	time interval
u_n	wind velocity in release height in n-th time interval
u_{xk}, u_{yk}	wind velocity components in k-th time interval
$\sigma_{y\ eff}$, $\sigma_{z\ eff}$	effektive diffusion parameter in n-th time interval
V	inhalation rate
K	consumption rate
R	radioecological transfer factor
I	precipitation rate

APPENDIX: Dose contributions from exposure pathways (equations)

ß-submersion : $D_{\beta} = g_{\beta} \sum_{\nu=1}^{n} Q_{\nu}\, \chi_{\nu}\, (x,y,0)$

inhalation : $D_h = g_h\, V \sum_{\nu=1}^{n} Q_{\nu}\, \chi_{\nu}\, (x,y,0)$

ingestion : $D_g = g_g\, R\, K \sum_{\nu=1}^{n} Q_{\nu}\, \{F_{\nu}(x,y) + 0{,}2\, W_{\nu}(x,y)\}$

γ-surface : $D_B = g_B \dfrac{1-e^{-\lambda t_n}}{\lambda} \sum_{\nu=1}^{n} Q_{\nu}\, \{F_{\nu}(x,y) + W_{\nu}(x,y)\}$

γ-submersion : $D_{\gamma} = g_{\gamma} \sum_{\nu=1}^{n} Q_{\nu}\, \chi_{\gamma\nu}\, (x,y)$

$Q_{\nu} = Q \prod_{k=1}^{\nu} f_k$: airborne activity

$f_k = f_{Rk}\, f_{Wk}\, f_{Fk}$: total depletion

$f_{Rk} = \exp\,(-\lambda\, \Delta t)$: decay depletion

$f_{Wk} = \exp\,(-\Lambda_k \Delta t)$: washout depletion

$f_{Fk} = \exp\,\left(-v_g\, \sqrt{2/\pi} \int_{t_{k-1}}^{t_k} \dfrac{e^{-H^2/2\sigma^2_{z_k\mathrm{eff}}}}{\sigma_{z_k\mathrm{eff}}}\, dt\right)$: fallout depletion

$F_{\nu}(x,y) = v_g\, \chi_{\nu}\, (x,y,0)$: fallout deposition

$W_{\nu}(x,y) = \Lambda_{\nu} \int_0^{\infty} \chi_{\nu}\, (x,y,z)\, dz$: washout deposition

$\chi_{\gamma 1}(x,y) = \iiint_{-\infty}^{\infty} \chi_1\, (X,Y,Z)\, \dfrac{B}{r^2}\, e^{-\mu r}\, dX\, dY\, dZ$ and

$\chi_{\gamma\nu}(x,y) = c\, \chi_{\nu}(x,y,z)\, ,\ \nu \neq 1$: γ-submersion factor

REFERENCES

K.J. Vogt, J. Straka, H. Geiss, An Extension of the Gaussian Plume Model for the Case of Changing Weather Conditions, 10th International Technical Meeting on Air Pollution Modeling and its Application, NATO/CCMS Air Pollution Pilot Study, 23./26.10.1979, Rome.

J. Straka, H. Geiss, K.J. Vogt, Diffusion of Waste Air Puffs and Plumes under Changing Weather Conditions, Beitr.z.Phys.d.Atm. 54:207 (1981).

H.D. Brenk, K.J. Vogt, The Calculation of Wet Deposition from Radioactive Plumes, Nuclear Safety 22:362 (1981).

F. Rohloff, E. Brunen, H.D. Brenk, H. Geiss, K.J. Vogt, LIGA, ein Programm zur Berechnung der lokalen individuellen Gammasubmersionsdosis durch Abluftfahnen aus kerntechnischen Anlagen, Jül-1577 (1979).

Abschlussbericht von Phase I des Projekts Sicherheitsstudien Entsorgung, PSE-Report (in preparation).

DISCUSSION

B. RUDOLF

The measurements of wind direction and velocity, which you apply for calculations of the trajectory, are measured at one single point near the source or do you respect more than one point ?
I think, it is necessary to respect more than one

K.J. VOGT

The input parameters requirements depend on the problem and on the range to be dealt with. Our model has originally been developed for local and medium range applications up to approximately 50 km. Under these conditions near source measurements of the meteorological conditions seem to be sufficient if the terrain has no significant orographic characteristics For larger distances the introduction of trajectories, calculated from measurements of a grid of meteorological stations, may be necessary, for which the model has to be adapted.

J. YOUNG

You compare your model with an ordinary Gaussian model using the first time step. This is not a fair comparison - you should use something like the vector average conditions over the 12 hours!

K.J. VOGT For the comparison the meteorological diffusion conditions during the release of the puff, i.e. during the first time interval, have been taken as input parameters, because this is the usual way of application of the conventional Gaussian diffusion model.

A FINITE ELEMENT FLOW MODEL OVER THE ALSACE PLAIN

P. Racher(1), F.X. Le Dimet(1), J.F. Roussel(1),
P. Rosset(1) and P. Mery(2)

(1) LAMP, Univ. Clermont II, BP 45, 63170 Aubière, France
(2) EDF, Etudes et Recherches, Division MAPA, 6 quai Watier, 78400 Chatou, France

INTRODUCTION

In a previous paper (Racher et al.,[1]), we have applied to the mid-Rhine valley a mass-consistent wind field model based on a variational technique for solving the incompressible continuity equation originally developed by Sherman[2]. Taking advantage of this variational approach which directly inserts the collected experimental information into the integration procedure itself, we depart from reference 2 mainly on three points. First, we use a terrain-following vertical coordinate (σ-coordinate), thus avoiding the problems of intersecting the topography. Secondly, as will be seen in next paragraph, we have generalized the boundary conditions so as to include the effects of inflow and outflow across the boundaries. And thirdly, we have designed an initialization procedure (Roussel[3], Mac Lain[4]) which aims at retaining most of the meteorological information collected on an irregularly-distributed network of stations. This is an important point which deserves further discussion, since the final relaxed solution is rather strongly dependent upon the initial first-guess wind field. Our initialization scheme avoids too much smoothing in the measured wind field through local expansion of the solution for each horizontal wind component in terms of a polynomial basis function within every triangle obtained by joining the stations. The values thus obtained are then redistributed over the grid points of a regular grid covering the domain under study. In doing so, we retain some of the fine-scale atmospheric structures reflected into the local wind measurements. These structures involve radiative and thermal as well as dynamic effects at many spatial and temporal scales : the adjustment implied through the continuity equation bears only upon the wind components, though accomodating in part the non-dynamic effects which contribute to them. One can assume that a larger number of constraint equations would lead to a reduction in the crucial role played by the initialization procedure.

A further improvement in our present finite difference version of the model lies in the local imbedding of a finer mesh grid within the larger one : it thus appears possible to treat with an increased resolution of the secondary circulations developing in some restricted areas of the domain. This is particularly true of the Alsace Valley between the Vosges and the Black Forest massifs, where strong topographic gradients determine singularities in the air flow such as strong confluence-diffluence near Belfort and Basel or possible flow detachment near Colmar according to the wind speed.Nevertheless, we were confronted at this stage with the problem of interfacing between the two grids of different resolutions. We then decided to recast our model into a finite element formulation more apt to retain internal coherence and consistency all over the domain in spite of a variable spatial and temporal grid resolution required in the presence of relief.

In the following, we give some general considerations about an alternative in posing the variational problem of solving the continuity equation. We then conclude with our own choice and with the design of an automatic mesh generation over a complex terrain.

ALTERNATIVE VARIATIONAL FORMALISMS

In variational objective analysis, one generally defines :

a) an observation ϕ_0 of the meteorological variables (a set of data collected at a number of observing stations) ;

b) a set of prognostic and/or diagnostic equations which govern the meteorological fields ϕ to be evaluated ;

c) a cost function for measuring the departure between the observed ϕ_0 and computed ϕ fields.

More specifically, for a mass-consistent model of the Sherman-type, the field of concern is the wind field in its three scalar components (i.e. $\phi_0 = \begin{Bmatrix} u_0 \\ v_0 \\ w_0 \end{Bmatrix}$ and $\phi = \begin{Bmatrix} u \\ v \\ w \end{Bmatrix}$).

After Sasaki[5] and Sherman[2] and after addition of a further term accounting for non-zero adjustement of the tangential velocity on the lateral boundaries (Γ_V), one defines the following cost function $J(\vec{V})$:

$$J(\vec{V}) = \frac{1}{2} \int_{(\Omega)} (\vec{V} - \vec{V}_0)^T \; W \; (\vec{V} - \vec{V}_0) \; d\Omega + \frac{\beta}{2} \int_{(\Gamma_V)} (\vec{n} \; (\vec{V} - \vec{V}_0))^2 \; d\Gamma_V \quad (1)$$

under the constraint of zero divergence ($\vec{\nabla}.\vec{V} = 0$) applying in the domain (Ω) bounded by the frontier (Γ) and where n is the oriented normal to (Γ). Apart form the part (Γ_V) of the frontier (Γ), the boundary condition $\vec{n}.\vec{V} = 0$ applies (no flow-through boundary). In equation (1), [W] is a diagonal weighting matrix and the symbol T stands for the transpose of the vector $(\vec{V} - \vec{V}_0)$.

We then state the variational analysis problem in terms of determination of $\vec{V}^* \varepsilon$ U such as :

$$J(\vec{V}^*) \leq J(\vec{V}), \; \forall \; \vec{V} \; \varepsilon \; U \quad (2)$$

where U is the set of all possible solutions satisfying the imposed constraints. This is a problem of optimization under constraint for which there are two classes of methods of solution(Le Dimet)[6,7]

1) Method of duality

It is a strong constraint method in which $\vec{V}$ is required to verify exactly the condition : $\vec{\nabla}.\vec{V} = 0$. In order to solve (2), the Lagrangian L $(\vec{V},\lambda)$ is defined such as :

$$L(\vec{V},\lambda) = J(\vec{V}) + (\lambda, \vec{\nabla}.\vec{V}), \quad (3)$$

where λ is a Lagrange multiplier and $\vec{\nabla}.\vec{V}$ the constraint term. The solution of (3) is obtained through determination of the saddle point (V^*, λ^*) of $L(\vec{V},\lambda)$. This leads to the two general following optimum conditions :

$$L(\vec{V}^*,\lambda^*) = \min_{\forall \vec{V} \varepsilon U} L(\vec{V},\lambda^*) \quad (4)$$

and

$$L(\vec{V}^*,\lambda^*) = \max_{\forall\lambda} L(\vec{V}^*,\lambda) \quad (5)$$

In our particular case, equations (4) and (5) lead to the following system to be solved for λ and then for $\vec{V}$:

$$\vec{\nabla}.(W)\ \vec{\nabla}\lambda = -\ \vec{\nabla}.\vec{V}_0 \qquad (6\text{-}1)$$

$$[W]\ (\vec{V} - \vec{V}_0) = \vec{\nabla}\lambda \qquad (6\text{-}2)$$

in (Ω), and :

$$\beta\ \vec{n}\ .\ [(W)\ \vec{\nabla}\lambda\ ] + \lambda = 0 \qquad (6\text{-}3) \text{ on } (\Gamma_V),$$

and $\vec{n}\ .\ [(W)\ \vec{\nabla}\lambda] = -\ \vec{n}.\vec{V}_0$ $\qquad (6\text{-}4)$ on $(\Gamma) - (\Gamma_V)$.

2) Method of penalization

It is a weak constraint method in which $\vec{V}$ is required to satisfy $\vec{\nabla}.\vec{V} = 0$ only approximately. One defines a penalized functional such as :

$$J_\varepsilon(\vec{V}) = J(\vec{V}) + \frac{1}{\varepsilon}\ \|\vec{\nabla}.\vec{V}\|^2 ,$$

where $\|\vec{\nabla}.\vec{V}\|^2$ is the squared norm of the wind divergence and where ε, a scalar comprised between 0 and 1, is the penalty function. The final solution is then obtained through satisfaction of the following (a) and (b) requirements :

(a) $\vec{V}^*_\varepsilon$ must verify $J_\varepsilon(\vec{V}^*_\varepsilon) = \min J_\varepsilon(\vec{V})$,

or alternatively,

$$\vec{\nabla}^*_{\vec{V}_\varepsilon}[J_\varepsilon(\vec{V}^*_\varepsilon)] = \vec{\nabla}^*_{\vec{V}_\varepsilon}[J(\vec{V}^*_\varepsilon)] + \frac{1}{\varepsilon}\ \vec{\nabla}_{\vec{V}^*_\varepsilon}\ [\|\vec{\nabla}.\vec{V}_\varepsilon^*\|]^2 = 0 \qquad (7)$$

and (b) $\vec{V}^*_\varepsilon \longrightarrow \vec{V}^*$, when $\varepsilon \longrightarrow 0$.

RESOLUTION STRATEGIES

In each of the two preceding variational formalisms, there are two possible algorithms for solution :

a) a direct solution method for systems (6) or (7) ;

b) use of an iterative method of the conjugate gradient type, directly applied to the functional itself.

The previous finite difference version of our model (Racher et al.[1]) was implemented with the above direct (a) algorithm as applied to a duality formalism. The same is true of Sherman[2] and of Tuerpe et al.[8], in their finite difference as well as finite element formulations. Unfortunately, especially the finite element form appears to be very heavy in its implementation and is much computer time-consuming. It is why, from the start of our new finite element recasting of the model, we have decided to apply an interative (b)-method to a penalty algorithm in order to maintain the computing cost at an acceptable level. Moreover, our choice has been dictated too by the capability of the penalty formalism to incorporate rather easily and simply new other constraints in addition to the zero divergence one (e.g. the momentum and thermodynamic equations). This is due to the fact that, in contrast to the duality method where new constraints are reflected in further unknowns and more equations to be solved, the penalty method only introduces new penalty parameters with no subsequent change in the number of unknowns.

DISCRETIZATION IN OUR FINITE ELEMENT FORMULATION

In a three-dimensional domain, we use a distorted eight-node parallelipied as the base element, in which the basis function N_i associated with every mode is trilinear. The spatial coordinates X and the wind variable $\vec{V}$ are given similar forms :

$$X = \sum_{i=1}^{N} N_i X_i, \text{ with } X = \begin{Bmatrix} x \\ y \\ z \end{Bmatrix}$$

and $$\vec{V} = \sum_{i=1}^{N} N_i \vec{V}_i, \text{ with } \vec{V} = \begin{Bmatrix} u \\ v \\ w \end{Bmatrix}$$

Related to these field descriptions, the topography Z(x,y) takes on the form :

$$Z = \sum_{i=1}^{N} N_i Z_i,$$

where Z_i denotes the ground height at the nodes of the lower boundary in the finite element mesh.

Aiming at developing a general operational objective analysis procedure, we have designed an automatic mesh generation scheme with quadrilateral elements at the ground surface. This scheme is first based upon the following requirements :

(a) the observation stations are nodes ;

(b) when desired, even flat regions can be given fine meshing in order to depict small scale circulations (e.g. the Comar region in the Alsace Plain).

Furthermore, the nodes are defined so as the topography is optimally meshed : the number of nodes is proportional to the ground slope with gradual increase in the mesh size away from the highest terrain elevations. This optimal meshing is iteratively selected in order to satisfy the following inequalities :

$$\max \left| \frac{Z_p - (\Sigma_i N_i Z_i)p}{Z_p} \right| < \varepsilon_1 \tag{8}$$

and

$$\frac{\Sigma_P (\Sigma_i N_i Z_i)_P}{\Sigma_P Z_P} < \varepsilon_2 \tag{9}$$

where Z_P is issued from direct topographic survey at point P(x,y) and where $(\Sigma_i N_i Z_i)_P$ is the ground height recomputed at the same point P from our relief description in finite elements; ε_1 and ε_2 are convergence parameters.

An academic example of automatic meshing is displayed in Figure 1, where the relief is constituted of three Gaussian mountains sited as indicated. One can observe the economy in the grid point number as opposed to what would be needed in case of finite difference meshing, together with their close spacing near the mountain tops.

CONCLUSIONS

We have developed some general basic considerations about variational objective analysis, in order to select at best our own procedures for recasting a previous finite difference mass-consistent wind field model into a finite element code. In contrast to other authors, we have chosen to use a penalty method in association with an iterative resolution algorithm. After some information about the elements and the basis function, we then put the stress on an automatic mesh generation scheme, a crucial part in the code implementation. The whole process is now being applied to the Alsace plain and the mid-Rhine valley.

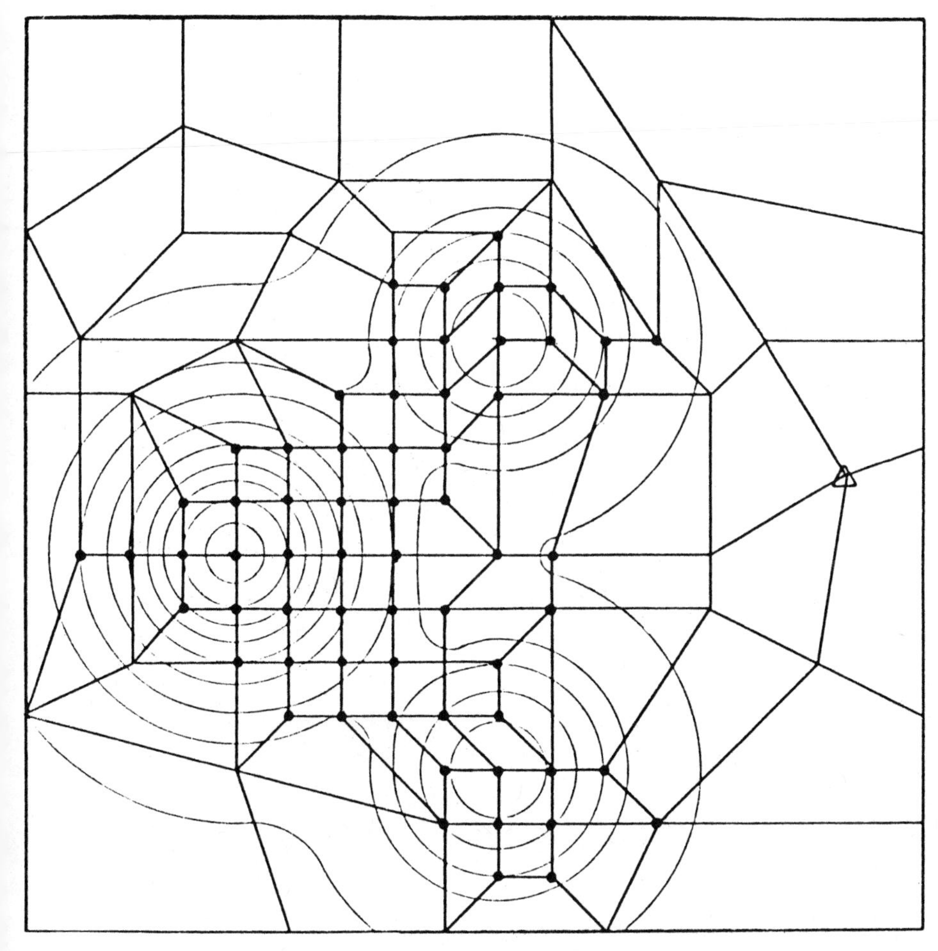

Figure 1 : An example of automatic meshing. The heavily dotted points are the chosen initial points near the tops of the three Gaussian mountains displayed too with thin isocontour lines. The triangle denote the observation station.

REFERENCES

1. P. Racher, R. Rosset and Y.Y. Caneill, A mass-consistent wind field model over the mid-Rhine valley, Proc. 11th NATO/CCMS International Technocal Meeting on Air Pollution Modeling and its Application, Amsterdam, The Netherlands, 24-27 nov. 1980.

2. C.E. Sherman, A mass-consistent model for wind fields over complex terrain, Journal Appl. Met., 17:312 (1978).

3. J.F. Roussel, Une méthode d'analyse objective des champs basée sur une interpolation locale par fonctions de base, Rapport Scientifique n° 31 du LAMP, Univ. de Clermont II, janvier 1981.

4. D.H. Mac Lain, Two-dimensional interpolation from random data, Comput. Journal, 19:178 (1976).

5. Y. Sasaki, Some basic formalisms in numerical variational analysis, Month. Weath. Rev., 98:875 (1970).

6. F.X. Le Dimet, Formalisme et applications de l'analyse objec-variationnelle, Rapport Scientifique n° 33 du LAMP, Univ. de Clermont II, avril 1981.

7. F.X. Le Dimet, A global approach of variational objective analysis, Internal Report of LAMP, Univ. of Clermont II,1981.

8. D.R. Tuerpe, P.M. Gresho and R.L. Sani, Variational wind field adjustement over complex terrain using finite element techniques, Proc. A.M.S. Conf. on Sierra Nevada Meteo., South Lake Tahoe, Ca., June 19-21 1978.

DISCUSSION

A. VENKATRAM — How do you determine dispersion conditions using a sodar ? Also, I don't see how dispersion conditions increase with height.

P. MERY — Perhaps I used a wrong term. In fact by dispersion conditions I meant stability conditions.
The index 1 I use corresponds to stable conditions (temperature gradient > - 0,5° C/100 m).
The index 2 corresponds to neutral and unstable conditions (temperature gradient < - 0,5 °C/100 m).

The definition of this index results from comparisons between sodar data and radiosounding data.

M. WILLIAMS You measure your inversion height at the position of the maximum echo - what do you do when multiple layers are present when the maximum echo may not be the limiting inversion ?

P. MERY I determine an inversion height from the position of the maximum echo and from the vertical shape of the vertical distribution of the echo.
If the first inversion layer is very strong, the sodar cannot determine the other inversion layer located above, but in many cases, we can have two maximum of echos and we can determine the position of two inversion layers. In these cases, inversions are not very strong or too thick.

MODIFICATIONS OF TRAJECTORY MODELS NEEDED FOR POLLUTANT SOURCE-RECEPTOR ANALYSIS

Paul Michael and Gilbert S. Raynor

Brookhaven National Laboratory

Upton, New York 11973, U.S.A.

INTRODUCTION

A most important application of air pollution modeling is the prediction of the changes in the airborne concentrations and surface deposition of pollutants as a function of changes in source emissions. Both the evaluation of the impact of new sources and of the effectiveness of control strategies are issues that are addressed by model calculations. Of particular interest, for some time in Europe and more recently in North America, is the incorporation of pollutants into precipitation, the so-called "acid rain" problem. One aspect of this problem which emphasizes the need for accurate source-receptor relationship is the concern over the long-range transport of pollutants from sources in one political jurisdiction to precipitation in another jurisdiction; see, for example, the following papers: Johnson, et al. (1978), Bhumralkar, et al. (1981), and Shannon (1981), which use two-dimensional trajectory methods.

The purpose of this note is to discuss some of the difficulties one should expect to encounter when using the usual trajectory models to calculate source-receptor relationships for the incorporation of pollutants into precipitation. Section II discusses sources of difficulty; Section III gives the results of trajectory calculations that were performed to test the sensitivity of source-receptor relationships to transport layer assumptions; Section IV presents some suggestions for future developments.

GENERAL DISCUSSION

The primary concern over the applicability of trajectory models to the problem of determining source-receptor relationships for the incorporation of pollutants into precipitation is in the area of regional or long-range transport. It is recognized that the scale of the problem and the limitation due to spacing between upper air soundings preclude very precise source-receptor definitions; the resolvable scale is probably of the order of a few tens of kilometers to about 100 kilometers.

The situation for which it is felt that special care should be exercised is when the precipitation bearing pollutants are associated with a warm front, and the particular concern is to be able to determine the origin of material that is scavenged in cloud, i.e., the material that is "rained out" as compared to the material that is "washed out" by precipitation below clouds. In general, clouds and precipitation are formed when air moves from lower levels behind the surface front, and then rises over the cold air. It is during this lifting that material is incorporated into cloud water. During the ascent, the direction of motion changes, and the air above the front can be moving in a different direction than the air below. In most applications of trajectory models the motion of near-surface air, or the mean motion in a layer between the surface and the mixing height is used. This is, of course, correct if one wishes to calculate airborne concentrations of pollutants that impact upon people, that are deposited by dry processes, or that are washed out below clouds. It would seem to be incorrect for in-cloud processes in the neighborhood of fronts.

For situations in which the pollutants are incorporated into clouds during convective, or air mass storms, the standard method of trajectory calculations would seem to be adequate; even though very complicated three-dimensional flows accompany convective storms the scales are smaller than those that one would hope to resolve in a regional-scale model.

The extent to which one should be concerned over incorporating more details on the air motion in the vicinity of fronts depends upon the importance of frontal storm precipitation to impacted areas and the relative importance of in-cloud vs below-cloud scavenging. The former is dependent upon locality; the latter may well depend upon the pollutant of concern. In particular, analysis of data by Raynor and Hayes (1980) has determined that rain associated with warm front delivers 35 to 40% of the major pollutants deposited by precipitation at Brookhaven National Laboratory which is on the East Coast of the United States. For sulfates in-cloud processes are generally deemed of great importance (e.g., Scott 1980).

When considering the motion of pollutant-bearing air parcels that travel for time intervals longer than one day, one should take into account day-to-day variations in the depth of the mixing layer. If on a later day of transport the mixing layer is shallower than on the day before then material can be left aloft and be transported in a direction quite different from the material in the mixed layer of the second day. If this effect is not included in the calculation erroneous results can occur. Most models used in applications either assume a fixed mixing height (Shannon, 1981, Bhumralkar, et al. 1981) or calculate strictly on the basis of current conditions.

EXAMPLES OF TRAJECTORY CALCULATION RESULTS

The sensitivity of source-receptor relationships to the assumptions made in the calculation of trajectories will be illustrated in this section by showing the results of pollutant transport calculations for two events in which precipitation associated with frontal systems was collected and chemical composition measured.

The model used is a version of the Air Resources Laboratories' Atmospheric Transport and Diffusion Model (ARL-ATAD) (Heffter, 1980), that has been modified for interactive use on the Brookhaven Computer (Benkovitz and Heffter, 1980). This widely used model calculates trajectories from computed wind fields that result from the interpolation between upper air soundings.

The precipitation chemistry results are from the MultiState Atmospheric Power Production Pollution Study (MAP3S) Regional Acidity of Industrial Emissions (RAINE) program, (Pacific Northwest Laboratory, 1980) and from additional measurements made at Brookhaven National Laboratory (Raynor and Hayes, 1979).

Fig. 1 shows the calculated trajectories of air parcels arriving at the MAP3S/RAINE precipitation chemistry network stations at Cornell University, Ithaca, New York (ITH) and Brookhaven National Laboratory, Upton, New York (BNL) on May 9, 1978 at 0600 GMT (0100 EST). This time was the approximate midpoint of an event that delivered 38 mm of rain at Brookhaven and 8.5 mm at Ithaca. This precipitation preceded a warm front that moved from the Kentucky-Tennessee region on 8 May through New York on 9 May. As indicated on the figure, three different averaging layers were used for the calculation of the transport winds. The trajectories marked with the solid dots used the ARL-ATAD "default" methodology which uses the layer from the surface to the first inversion during daylight hours. At nighttime the transport layer is determined by the vertical standard deviation of the plume. The calculated top of the transport layers ranged from 1000 to 1900 meters for the trajectory

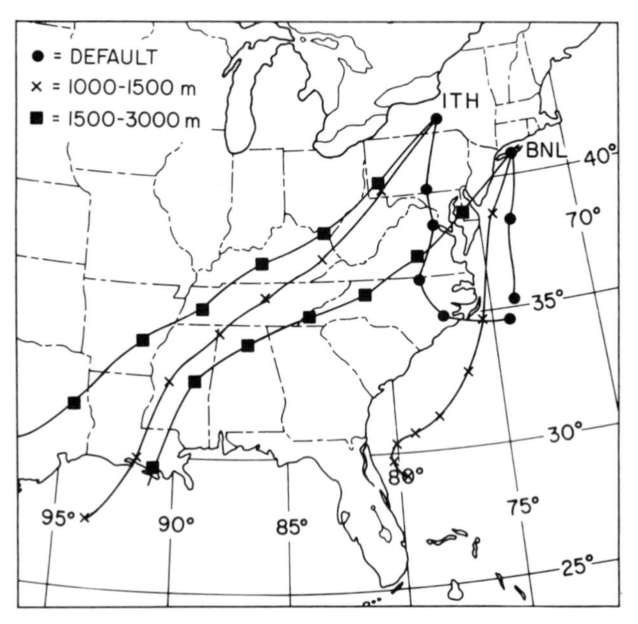

Fig. 1. Calculated air parcel trajectories reaching Cornell University, Ithaca, NY (ITH) and Brookhaven National Laboratory, Upton, NY (BNL), at 0600 GMT, 9 May 1978. Differing transport layer assumptions were used as indicated. Symbols are plotted at six-hour intervals.

to BNL. The top of the transport for the trajectories to ITH was somewhat lower, 1100 meters just before the parcel reached ITH, 500 to 1000 meters during the previous 24 hours and ranging up to 1500 meters prior to that. The trajectories marked with crosses used a transport layer of 1000 to 5000 meters--these are at the approximate level of the cloud base. The trajectories marked with solid squares used a 1500 - 3000 meter transport layer--these being an attempt to simulate air motions well within the cloud layer.

The concentration of major pollutants in the rain is listed in Table 1. They are close to the average springtime values in the Northeastern U.S. Thus, this is an event for which one would wish to be able to determine the location of source regions that are emitting the pollutants that become incorporated into the precipitation. We see that different algorithms for the selection of the transport layer can lead to different conclusions as to source location. The "default" or "mixing layer" algorithm would imply that the source of pollutants in the precipitation at both Ithaca and BNL came from a southerly direction; central Pennsylvania, Maryland, Virginia and North Carolina being the source region for Ithaca, and the Atlantic Ocean being the source region for BNL. The use of a 1000 - 5000 meter transport layer would imply that the trajectory to Ithaca came more from the southwest, western Pennsylvania, West Virginia, Kentucky, and central Tennessee; the trajectory to Upton would still be calculated to be over the ocean but sufficiently close enough to shore so that the coastal region could be considered to be the source region. The 1500 - 3000 transport layer gives a trajectory to Ithaca that travels through the Ohio River Valley from Kentucky and western Tennessee; the trajectory to Upton is along the populated East Coast--New Jersey, near Philadelphia, Pennsylvania; Maryland, near Washington, D.C., and through Virginia.

Table 1. The Concentration of Major Pollutants in Precipitation at Ithaca, NY (ITH) and Brookhaven National Laboratory, NY (BNL)

	Concentration (μ mole/liter) 9 May 1978		9 April 1979
Species	ITH	BNL	BNL
H^+	126 (pH = 3.90)	62 (pH = 4.21)	65 (pH = 4.19)
$SO_4^=$	44	17	69
NO_3^-	40	10	103
NH_4^+	18	4	50

The results of another group of trajectory calculations are shown in Fig. 2. Shown here are trajectories reaching BNL on 9 April 1979 at 0600 GMT. This arrival time is in the middle of an event that delivered 22.3 mm of rain to BNL. The precipitation preceded a warm front that accompanied a storm that moved nearly straight eastward starting at 0600 GMT on 8 April from southwestern Iowa and reaching the east coast at 0000 GMT on 10 April. Table 1 also gives the concentration of pollutants in the precipitation. The pH was close to the seasonal mean; the $SO_4^=$ and NO_3^- concentrations were more than a factor of two greater than the seasonal mean; the NH_4^+ was similarly high, likely keeping the pH up.

As in the previous example, different methodologies were used to determine the transport layer. We see that the results fall into two broad categories: (1) the "default" method for calculating the transport layer and those trajectories calculated with the top of the transport layer at 1000 meters give trajectories that reach BNL from the northwest--air parcels traveling through northeastern Pennsylvania, extreme western New York, Michigan and Canada, and (2) those trajectories which are calculated using transport layers above 1800 meters from southwest through southern Pennsylvania, West Virginia, and the Ohio River Valley. This example is from a more complete analysis of the rain event (Raynor and Hayes, 1981). In that study, particular attention was given to day-to-day variations in the mixing height, the concern being that material well mixed on one day will be transported above a lower mixing layer on the next day. For this example, it was decided that the 500 - 2000 meter transport layer best represented the long-distance transport. The high top to the transport layer also is more appropriate for input to in-cloud processes.

In this example, we again see that the choice of the transport layer strongly influences any conclusions one may make concerning source-receptor relationships.

In both cases, there is the implied suggestion that those trajectories that extend above the mixed layer more realistically represent the flow of pollutants into precipitation. This is because it is felt that in-cloud processes are likely the most important, that the over-ocean trajectories in the first example would seem unlikely to deliver very much pollution, and that in general the trajectories that are calculated using deeper transport layers travel over regions with more anthropogenic sources. (This latter point is further discussed by Raynor and Hayes (1981).) However, the methods used to calculate the trajectories in these examples are too approximate to be the basis for firm conclusions. The important point is that in most applications the lower layers are heavily weighted in the calculation of the transport wind field, and that the selection of alternate transport layers can drastically change the path of a trajectory.

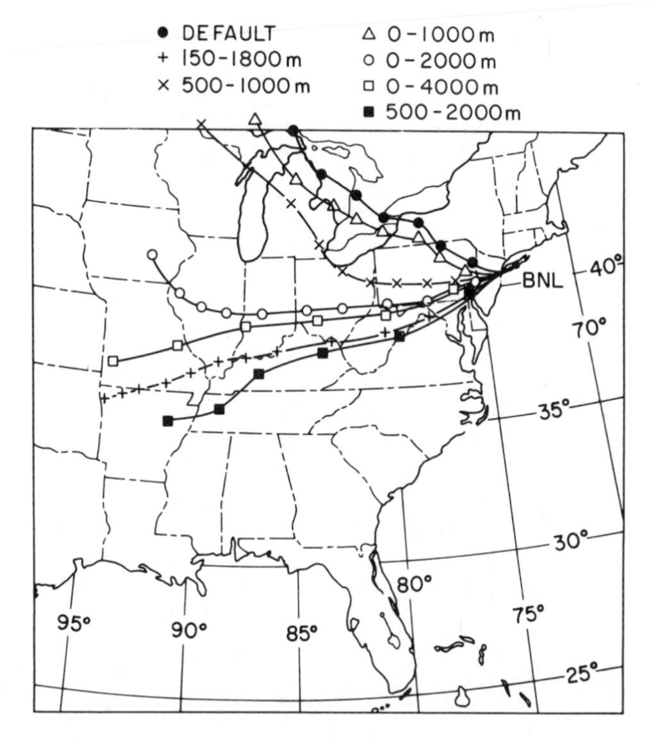

Fig. 2. Calculated air parcel trajectories reaching Brookhaven National Laboratory, Upton, NY (BNL) at 0600 GMT, 9 April 1979. Differing transport layer assumptions used as indicated. Symbols are plotted at six-hour intervals.

RECOMMENDATIONS

The examples in the preceding section illustrate the fact that extreme care should be taken when using trajectory calculations in an attempt to make source-receptor identifications. The examples do not use particularly sophisticated methodologies; they are useful in that the sensitivity to assumptions is displayed and that questions are raised, although not answered. In this section some suggestions are made as to what ought to be included in trajectory models used for source-receptor investigations.

Diurnal and day-to-day variations in transport layers and shear in general should be taken into account. This could be done by one of a number of methods: allowing a plume to fission into secondary plumes when new transport layers are created, calculating higher spatial moments that take into account plume tilting (Tyldesley and Wallington, 1965), or increasing the lateral plume standard deviation to account for shear (Samson, 1980). In any event, one would need to run source-oriented calculations rather than the back trajectory calculations of the type discussed above. (Most models used in applications are indeed operated in source-oriented mode.)

In order to compute the flow fields in the presence of frontal systems, three-dimensional trajectories should be used so that the input to clouds can be determined. Isentropic analysis, using the wet bulb potential temperature, which remains invariant whether or not evaporation or condensation is occurring could be the best method for determining three-dimensional trajectories (Mason, 1971; Palmen and Newton, 1969; Kreitzberg, 1964). It remains to be determined whether sounding data has enough resolution to allow for such analysis.

If input to clouds could be calculated by analysis such as indicated above, one could merge the results with the results from more conventional analysis which provides pollutant concentrations below cloud. The models would then require separate parameterization for in-cloud rainout and sub-cloud washout. Though this distinction is recognized in many discussions of the mechanisms of precipitation scavenging, commonly used models do not disaggregate these processes.

The above considerations, as has been mentioned above, apply to situations when the precipitation of interest results from frontal storms; if such precipitation is significant, as it is for much of the year in the northeastern U.S., then models that include the phenomena discussed here must be used if one is to have confidence in the resulting source-receptor relationships.

ACKNOWLEDGMENTS

The assistance of Carmen Benkovitz, who modified the ARL-ATAD model and has maintained the needed meteorological data base, is gratefully acknowledged. Janet Hayes performed a good deal of the analysis using the trajectory model; this assistance is also gratefully acknowledged.

This work is part of the MAP3S/RAINE research effort performed under the auspices of the U.S. E.P.A. under Contract No. 79-D-X-0533, and the U.S. Dept. of Energy under Contract No. DE-AC02-76CH00016. Accordingly, the U.S. Government retains a nonexclusive royalty-free license to publish or reproduce the published form of this contribution or allow others to do so, for U.S. Government purposes.

REFERENCES

Benkovitz, C. M. and J. L. Heffter, 1980, User's Guide to the Heffter Interactive-Terminal Transport Model (ARL-HITTM), Informal Report BNL 27801, Brookhaven National Laboratory, Upton, NY 11973.

Bhumralkar, C. M., R. L. Mancuso, D. E. Wolf, W. R. Johnson, and J. Pankrath, 1981, Regional air pollution model for calculating short term (daily) patterns of transfrontier exchanges of airborne sulfur in Europe, Tellus, 33:142-161.

Heffter, J. L., 1980, Air Resources Laboratories Atmospheric Transport and Dispersion Model (ARL-ATAD). NOAA Tech. Memo ERL ARL-81, Air Resources Laboratories, Silver Spring, MD.

Johnson, W. D., D. E. Wolf, and R. L. Mancuso, 1978, Long-term regional patterns of transfrontier exchange of airborne sulfur pollution in Europe, Atmos. Environ. 12:516-527.

Kreitzberg, C. W., 1964, The structure of occlusions as determined from serial ascents and vertically directed radar. AFCRL-64-26, Air Force Cambridge Research Labs., L. G. Hanscom Field, MA.

Mason, B. J., 1971, "The Physics of Clouds," 2nd Ed., Oxford University Press, London (see p. 302).

Pacific Northwest Laboratory, 1980, The MAP3S precipitation chemistry network: Third periodic summary report (July 1978-December 1979). Report PNL 3400, UC-11, Pacific Northwest Laboratory, Richland, WA.

Palmen, E. and C. W. Newton, 1969, Atmospheric Circulation Systems: Their Structure and Physical Interpretation. Academic Press, New York, NY (see p. 376).

Raynor, G. S. and J. V. Hayes, 1979, Analytical summary of experimental data from two years of hourly sequential precipitation samples at Brookhaven National Laboratory. Report BNL 51058, Brookhaven National Laboratory, Upton, NY.

Raynor, G. S. and J. V. Hayes, 1980, Variation in chemical wet deposition with meteorological conditions, BNL 28706R, (to be published).

Raynor, G. S. and J. V. Hayes, 1981a, Effects of varying air trajectories on spatial and temporal precipitation chemistry patterns. BNL 29623. (to be published).

Raynor, G. S., J. V. Hayes, and D. M. Lewis, 1981b, Testing of the ARL-ATAD trajectory model on cases of particle wet deposition after long distance transport from known source regions. BNL 29702. (to be published).

Samson, P. J., 1980, Trajectory analysis of summertime sulfate concentrations in the Northeastern United States. J. Appl. Meteor., 19:1382-1395.

Scott, B. C., 1980, Predictions of in-cloud conversion of SO_2 to SO_4 based upon a simple chemical and dynamical model. Proc. Second Joint Conf. on Applications of Air Pollution Meteorology, p. 389-416, American Meteorological Society, Boston, MA.

Shannon, J. D., 1981, A model of regional long-term average sulfur atmospheric pollution, surface removal and net horizontal flux. Atmos. Environ., 15:689-701.

Tyldesley, J. B. and C. W. Wallington, 1965, The effect of wind shear and vertical diffusion on horizontal dispersion. Qt. J. Roy. Met. Soc., 91:158-174.

DISCUSSION

A. CHRISTIF

The use of isentropic trajectories in preferance to horizontal trajectories in the lower troposphere may be desirable but is impracticable. In a joint study of pollen deposition in arctic regions carried out with Dr. PITCHIE, it was found that though the Montgomery stream function could be calculated from observed data the stability was generally consistent with almost zero vertical gradient of wet bulb potential temperature so little confidence could be placed in the vertical (θ_w) specification. In practise better source-receptor relations were found with horizontal trajectories

P. MICHAEL

In a warm front situation it should be possible to determine the motion of the warm air as it is overriding the cold air because of the fact of large scale stability that exists.

A. VENKATRAM

Could you explain what you mean by a "source-receptor" relationship ?

P. MICHAEL I mean the quantitative identification of the amount of interest delivered to a receptor region by each source region. (In this paper I have not calculated source-receptor relationships ; I have simply attempted to point out difficulties and to raise some questions).

W.E. DAVIS I am not sure any of those trajectories you have shown will describe the flow because of the tendency of the trajectory to turn with height as it moves over the front.

P. MICHAEL You are correct ; the constant layer trajectories are presented here in order to illustrate the sensitivity of the trajectories to the transport layer assumptions ; I am not recommending that these trajectories be used for actual source-receptor calculation.

SENSITIVITY ANALYSIS OF A

CLIMATOLOGICAL AIR TRAJECTORY MODEL

Perry J. Samson, Randy J. Fox and Richard A. Foltman

Department of Atmospheric and Oceanic Science
University of Michigan
Ann Arbor, Michigan 48109 U.S.A.

INTRODUCTION

The modeling of regional-scale transport and deposition of atmospheric pollutants is complicated by the considerable uncertainties involved in atmospheric process parameterizations. Beyond the uncertainties involved in parameterizing dispersion over relatively short distances (say less than 50 km) is the need to estimate dispersion over long time scales, usually including multi-diurnal cycles. Moreover, other atmospheric processes, usually ignored, must be considered. Dry deposition plays an important role as a removal process over long time scales and must be parameterized in a physically realistic manner. The removal of pollutants by wet deposition is an extremely complex process parameterized at best to be a simple non-linear function. Chemical conversion of primary pollutants to secondary is also poorly understood but has been parameterized for some pollutants, notably sulfur dioxide to sulfate, as a first-order chemical conversion. The purpose of this paper is to examine the sensitivity of individual trajectory model runs to variations in model parameterizations. This information is intended to help define the confidence in which we can hope to characterize the effects of anthropogenic emissions on receptor locations far downwind of the source.

This analysis has used the Atmospheric Contributions to Interregional Deposition (ACID) model developed to investigate probable source regions for measured sulfate concentrations in remote regions of the northeastern United States. The model incorporates horizontal dispersion due to vertical wind shear, first-order chemical transformation, spatially and temporally varying dry deposition, and precipitation scavenging at each time step. It is not intended for use in the near-field range (travel times less than six hours) since the trajectories are generated from upper-level winds measured at twelve hour increments using the technique of Heffter (1980).

The diagnoses of the downstream path of emitted material requires an accurate description of the motion of air in time. Data input errors or inaccurate interpolation schemes will cause deviations in the location of the calculated trajectory from the actual path of the air. Additionally, the calculated trajectory represents the mean motion of the total layer and should be thought of as a centerline of possible transport. The vertical wind velocity shear through the mixed layer, coupled with the vertical mixing will disperse the emitted material away from the center line.

The use of ensembles of trajectories is an attempt to minimize the effects of random errors in individual trajectory analysis. The main factors which could hinder an exact evaluation of the trajectory path using a constant layer model such as Heffter's are:

1. large-scale vertical motions,

2. the relative sparsity of observations in both time and space,

3. the inaccuracies of observations, and

4. inaccuracies in the time and space interpolation schemes employed.

Clearly, large scale vertical motions do pose a serious threat to the accuracy of a constant layer model. A sustained vertical motion of 0.5 cm sec^{-1} (possible in the vicinity of a warm front, for example) over the course of one day of trajectory motion would lead to a vertical displacement of over 400 m. In a layer model with an average mixed layer depth of about 1500 m this shift represents a sizable departure from the layer being averaged. If a particular wind direction was climatologically associated with a particular large-scale vertical motion, then this could lead to systematic errors in the trajectory results.

It is assumed in the use of ensembles of trajectories that this is not the case. The errors resulting from vertical motion, interpolation schemes, observational errors, and sparsity of data are assumed to be random. Thus the ensemble analysis of trajectories over reasonably long time periods will yield climatologically meaningful results about the transport of material from the source.

Horizontal Dispersion

The use of a mixed-layer trajectory model assumes that the material being traced is moving with the mean motion of the mixed-layer. This assumption will only be true if the material is, in fact, well-mixed through the layer and there is no shear in the layer to disperse the material away from the mean flow.

Over sufficiently long travel times (> 3 hours) the dispersion of atmospheric admixtures will be dominated by the shear of the wind velocity in the vertical. The climatology of this dispersion for long travel times has been calculated by Samson (1980) from the divergence of trajectories for sublayers of the mixed-layer. The distribution of the probability of contributing to receptors downstream was found to be normally distributed about the mixed-layer trajectory. The broadening of the probability field was estimated to be a linear function of travel time as

$$\sigma_x = \sigma_y = 5.4t \tag{1}$$

where σ_x and σ_y are the along and cross-trajectory standard deviations of sublayer displacements, respectively, at time t hours away from the origin.

The ACID model uses this parameterization to describe the potential contribution of a source to downstream receptors. Additionally, the dispersion over long sampling times due to the meander of plume centerlines is explicitly included in the model through the use of trajectory integrations at six hour time steps through the period of interest.

Dry Deposition

The ACID model uses a scheme for calculating dry deposition which follows the method of Shieh et. al. (1979). Dry deposition is allowed to vary as a function of time of day, season of the year, and location. To simulate the variations due to location, the estimates of Shieh et. al. for varying land use are stored for each grid point in the model. These values of deposition velocity are stored in BLOCK DATA subroutines with different values for each season.

To simulate the variation in dry deposition due to changes in atmospheric stability, the gridded deposition velocities are forced to vary diurnally according to the form

$$v_d(X,Y;t) = 2\bar{v}_d(X,Y) \sin\left\{\frac{\pi(t_{ss}-t_{sr})}{t_{ss}-t_{sr}}\right\} + v_{dn} \tag{2}$$

for $t_{sr} < t < t_{ss}$, and

$$v_d(X,Y,;t) = v_{dn} \tag{3}$$

for $t_{sr} > t$, $t > t_{ss}$

where v_d is the deposition velocity as a function on longitude, X; latitude, Y; and time, t. The deposition velocity varies as a sine wave from sunrise, t_{sr}, to sunset, t_{ss}, with an average value of $\overline{v}_d(X,Y)$ as obtained from Shieh et. al. The deposition velocity during the nocturnal hours is kept at a small value, v_{dn}, to simulate the reduction in removal expected because of the increased stability of the lowest layer of the atmosphere.

Removal by dry deposition is thought to be as important as wet removal over long averaging times in removing material from the atmosphere. Wet removal is, however, much more efficient that dry removal once precipitation has commenced.

Chemical Transformation

The reaction rates associated with the chemical transformation of pollutants will change the probability of affecting downwind receptors. A receptor very close to the source will have little chance of being impacted by the secondary pollutants of the emitted gas if the conversion rate is small. The rate of chemial conversion to secondary pollutant competes with the rate of dispersion of the primary pollutant. If dispersion is slow and chemical conversion is slow, one would expect the maximum probability of secondary impact to be considered downwind of the source. If, on the other hand, the dispersion is relatively large compared to the conversion rate, then the area of highest relative impact may well be close to the source.

The ACID model assumes a diurnally varying transformation rate, as suggested by Gillani et. al. (1978) with a form similar to dry deposition. The conversion rate, κ_t, from sulfar dioxide to sulfate species is assumed to behave as:

$$\kappa_t(t) = 2\overline{\kappa}_t \sin\left\{\frac{\pi(t-t_{sr})}{(t_{ss}-t_{sr})}\right\} + \kappa_{tn} \tag{4}$$

for $t_{sr} \leq t \leq t_{ss}$, and

$$\kappa_t(t) = \kappa_{tn} \tag{5}$$

for $t_{sr} > t, \quad t > t_{ss}$

where $\bar{\kappa}_t$ is an average afternoon value and κ_{tn} is a low night time value.

The system of equations to be solved along each trajectory result from mass continuity for both sulfur dioxide and sulfate. For sulfur dioxide, the equation expressing the rate of change of probability of impact is

$$\frac{dp}{dt} = - \kappa_t p - k_d p - k_w p \qquad (6)$$

where dp/dt expresses the rate of change along the trajectory of the probability of impact, p. The equation for the rate of change of probability of impact for sulfate is likewise written as

$$\frac{ds}{dt} = \kappa_t p - \kappa_d s - \kappa_w s \ . \qquad (7)$$

The terms k and κ refer to the removal rates for sulfur dioxide and sulfate, respectively, and the subscripts d and w refer to dry and wet removal.

Wet Deposition

A number of methods have been proposed for parameterizing the loss of gas and aerosol material from the atmosphere due to precipitation. Rates have been put forth by Dingle and Lee (1973), Engelmann (1970), and Scott (1978). The wet removal of sulfur dioxide in ACID is accomplished by assuming a first order loss rate with respect to pollutant concentrations. The loss rates for sulfur dioxide is given by $k_w = 0.005\ P(t)$ (Dana et. al., 1975) where P(t) is the precipitation rate in millimeters per hour. Wet removal of sulfate aerosol is parameterized to be $\kappa_w = 0.232\ P(t)^{0.625}$.

Precipitation data collected hourly at stations in the United States and eastern Canada have been obtained for the years from 1976 to 1978. The data, available in station form (three years of data for each station, station after station) was sorted and reassembled in a synoptic format (all stations for each hour, hour by hour) for use by the ACID model. This precipitation data was summed over three hour increments, checking for data validity, and gridded in the same form as dry deposition parameters.

The ACID model is a sequential puff trajectory model. An initial puff is generated at the first timestep as a two-dimensional normal distribution of probability of sulfur dioxide impact. The domain of this puff is interogated to determine the underlying location for dry deposition, the time of day for diurnal processes, and the existence of precipitation.

For analysis of individual episodes the ACID model allows partial losses from the puff without affecting the probability in adjacent areas of the puff. After the first time step, the remnant puff after deposition may well be non-symmetric about the segment endpoint. A moving grid is formed along the axis of the first trajectory segment as shown in Figure 1. The distances to the existing fixed grid points within a radius of 3 σ_y are computed in terms of the new axes and stored with the remnant probabilities at that time step. This set of points if moved to the next segment endpoint and placed in the same relative locations from this end-point as shown in Figure 2. These values are then interpolated and smoothed to the new underlying grid points in a manner which approximates the dispersion expected due to vertical wind shear.

The moving grid approach is designed for analysis of ensembles of trajectories for a month or less. For longer periods, the detail of irregular deposition is not justified. For that analysis ACID simply uses the dispersion of trajectories over time to simulate climatological dispersion.

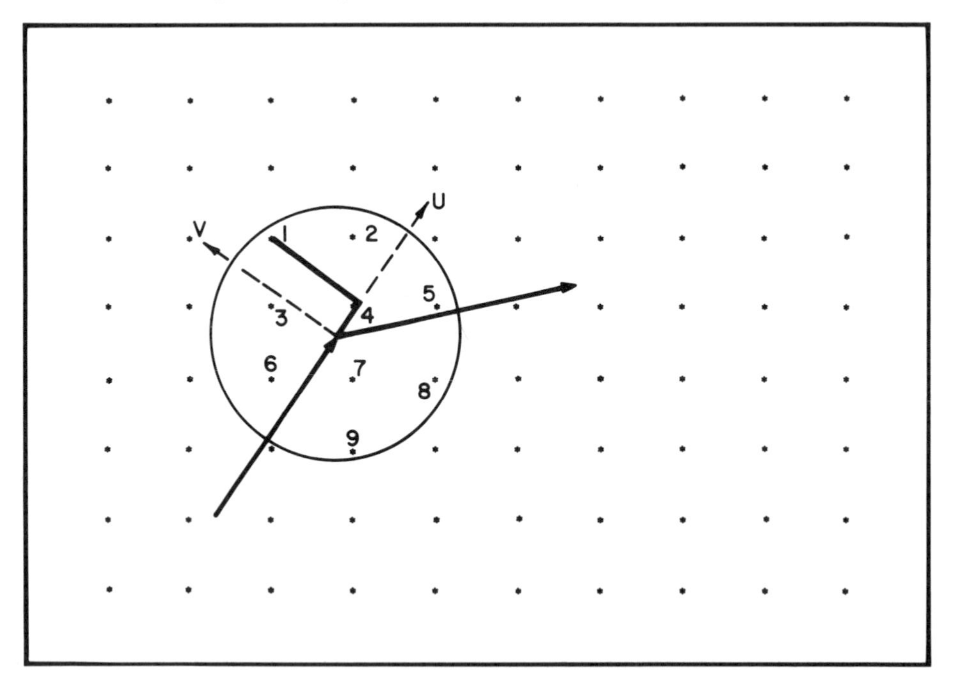

Fig. 1. Initial puff showing the trajectory axis orientation for computation of distances to the points in the domain.

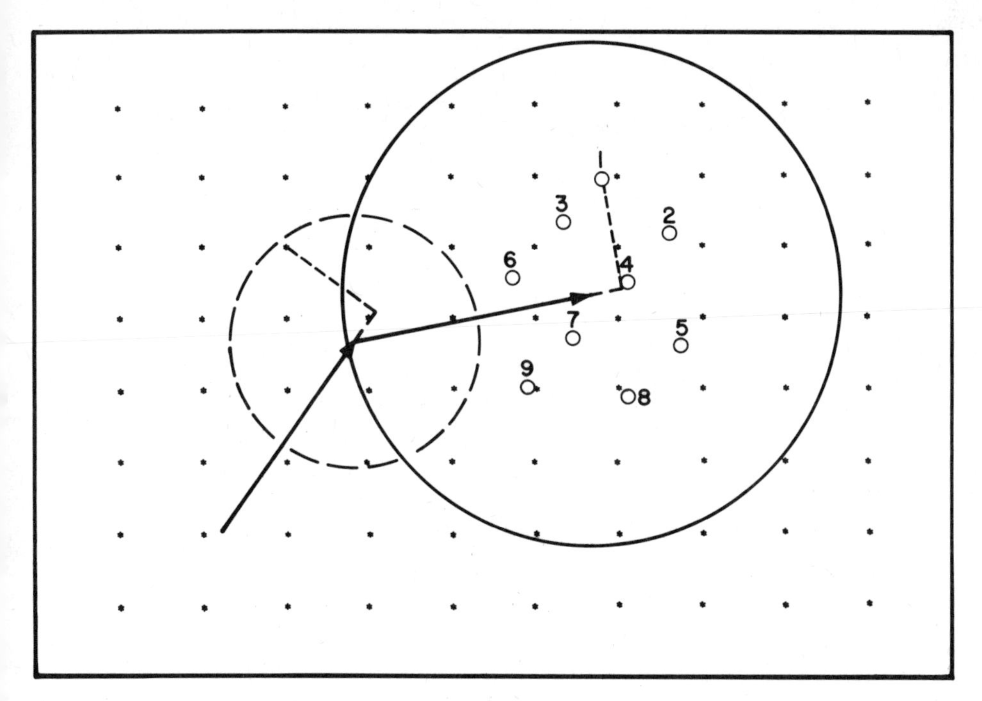

Fig. 2. Transposition of initial points to domain of the second puff. Initial data is smoothed and interpolated to new domain to form its initial guess field.

RESULTS

The model was run for one year, 1978, from a location near Ann Arbor, Michigan. The number of trajectories possible for the period were 1460 (365 days x 4 trajectories/day). Figure 3 shows that the number of trajectories available for analysis dropped rapidly from 1300 at three hours downwind to less than 200 at 72 hours downwind. Trajectories were prematurely terminated if (1) they ran outside of the area of data availability, (2) they ran into a grid cell in which precipitation data was not available, or (3) they ran outside the region arbitrarily chosen as the boundaries of analysis (50.5°N, 90.9°W, 35.6°N, 65.9°W). The dominant reason for loss of trajectories was the lack of observed winds at latitudes north of about 50° latitude and over the oceans.

By tabulating the SO_2 and SO_4 which has not been removed at each time step we can observe the percentage of total material emitted which is still available within the grid. Figure 3 shows the decay of SO_2 mass downwind of the source. As expected, the SO_2 drops rapidly due to losses to day and wet deposition and chemical conversion to SO_4=. The average SO_4= mass increases slowly and

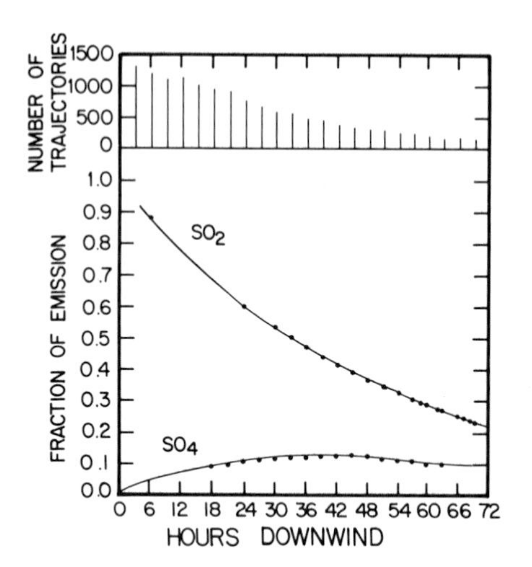

Fig. 3. The drop in the number of trajectories available as a function of time for the 1978 run. The drop in SO_2 an SO_4 derived from the available trajectories for the period.

appears to be highest 36 to 48 hours downstream. Because individual plume dispersion will mix the mass through a growing volume it is unlikely that any particular plume would exhibit a similar concentration increase downstream. The generation of $SO_4^=$ weighted agains the rate of dispersion will determine the time downstream of maximum concentration.

It is interesting to note that the percentage of emitted material still available after 72 hours is about 30 % with roughly 8 % as $SO_4^=$ and 22 % as SO_2. This material will go unexplained by the model, either leaving the domain of the model of remaining withi the grid, assumed to be randomly distributed about the grid.

In fact, the ability of the model to explain the deposition of emitted material is highly dependent upon the parameterizations used for deposition and conversion rates. Table 1 shows the percentage of unexplained material owing to the unavailability of data for the 1978 run. In the "baseline case" where parameterization was as described earlier, the unexplained deposition was over

Table 1. Percent of emissions not deposited by model due to premature termination of trajectories.

Condition	Remnant SO_2	Remnant SO_4	total Remnant
Baseline Case	53.6%	7.3%	60.9%
0.5 κ_t	59.0	3.4	62.4
2.0 κ_t	43.4	14.5	67.9
0.5 k_d (SO_2)	61.0	8.1	69.1
2.0 k_d (SO_2)	40.6	5.8	46.4
0.5 κ_d (SO_4)	53.6	8.3	61.9
2.0 κ_d (SO_4)	53.6	5.1	58.7
0.5 k_w (SO_2)	55.1	7.4	62.5
2.0 k_w (SO_2)	51.2	7.1	58.3
0.5 κ_w (SO_4)	53.6	7.9	61.5
2.0 κ_w (SO_4)	53.6	6.8	60.4
0.5 of all k and κ	70.3	4.8	75.1
2.0 of all k and κ	33.9	7.4	41.3

60%* indicating that, in addition to the 30% of material which was not explained by having a lifetime longer than 72 hours, another 30% is unaccounted for because many trajectories terminated prematurely.

The amount of unexplained deposition varies with varying rates of deposition and conversion. It is most sensitive to variations in the dry deposition rate of SO_2. If the general value of k_d were doubled the amount of explained deposition would increase to about 55% while halving k_d yields only about 30%.

Of course, not all trajectory models are going to suffer from these limitations. Trajectories based on analyzed winds from numerical model grids can be used to calculate trajectories for even longer durations without loss of data. However, the accuracy of the calculations for such long periods and the relevance of isobaric or constant level trajectories to the layer of pollutant

*This number should be lower for most other years. The observed winds data for 1978 had a extraordinary amount of missing values.

transport limits the confidence of such results.

The important question which must be studied next is where and how is the undeposited material eventually deposited. It is possible that overall spatial patterns of annual $SO_4^=$ and SO_2 dry deposition may not be significantly influenced by the remnant emissions. However, the spatial distribution of wet deposition is dependent on the juxtaposition of pollutant puffs and precipitation areas and could, conceivably, be altered substantially from our estimates based on only 40% of the emitted material.

REFERENCES

Dana, M.T., Hales, J.M., and Wolf, M.A., 1975, Rain scavenging of SO_2 and sulfate from power plant plumes, J. gepohys. Res., 80:4119.

Dingle, A.N. and Lee, Y., 1973, An analysis of in-cloud scavenging, J. Appl. Meteor., 12:1295.

Engelmann, R.J., Scavenging prediction using ratios of concentration in air and precipitation, AEC Symp. Ser. Precipitation Scavenging (1970), 475.

Gillani, N.V., Husar, R.B., Husar, J.D., Patterson, D.E., and Wilson W.E., 1978, Project MISTT: Kinetics of particulate sulfur formation in a power plant plume out to 300 km. Atmos. Environment, 12:589.

Heffter, J.L., 1980: "Air Resources Laboratories Atmospheric Transport and Dispersion Model (ARL-ATAD)," NOAA Tech. Memo. ERL-ARL-81, Silver Springs, MD.

Samson, P.J., 1980, Trajectory analysis of summertime sulfate concentrations in the northeastern United States, J. Appl. Meteor., 19:1382.

Scott, B.C., 1978, Parameterization of sulfate removal by precipitation, J. Appl. Meteor, 17:1375.

Shieh, C.M., Wesley, M.L., and Hicks, B.B., 1979, Estimated dry deposition velocities of sulfur over the eastern United States and surrounding regions, Atmos. Environment, 13:1361.

DISCUSSION

J.W.S. YOUNG

The percentage of material lost is an artifact of your model since the domain is restricted. By using a larger domain, this will not be a problem.

P.J. SAMSON

No, the largest percentage is due to (1) the finite time duration of trajectories (12 hours in our model) leaving undeposited materials within the grid and (2) premature termination of trajectory calculations because of data limitations within the grid

J. SHANNON One rarely has confidence in any single trajectory followed for 12 hours, but a trajectory ensemble out to 5 or more days has meaning, sufficient individual trajectories make up the ensemble.
Trajectories produced by numerical methods that reproduce closed circulations over data sparse areas should be used to investigate recurvature into the southern U.S.A.

P.J. SAMSON I agree.

W. GOODIN Can you clarify the concept of "loss of emissions" associated with your trajectory model?

P.J. SAMSON I am referring to the percentage of emissions which are not deposited by the model and remain unexplained.

3: ADVANCED MATHEMATICAL TECHNIQUES IN AIR POLLUTION MODELING

Chairmen: E. Runca
H. Van Dop
J. C. Oppenau

Rapporteurs: C. Blondin
D. Eppel
J. Van Ham

A MODEL FOR POLLUTANT CONCENTRATION PREDICTION IN COMPLEX TERRAIN

H. N. Lee, W. S. Kau and S. K. Kao

Department of Meteorology
University of Utah
Salt Lake City, Utah 84112

INTRODUCTION

In modeling the distributions of atmospheric pollutants, the study of the effects of the topography on the pollutants is of vital importance. As the terrain is flat, the Gaussian dispersion model is the basic method used to calculate pollutant concentrations. However, the Gaussian dispersion models are found to over-predict the concentration over mountainous terrain (Start, et. al., 1977). In order to better understand dispersion over complex terrain, the sophisticated dynamic model based on second-order closure schemes (Yamada, 1979) has been constructed. Due to the complexity of dynamic model, an alternative approach with the use of statistical method for analyzing the dynamic relationship between the geostrophic wind, vertical heat flux and surface wind over complex terrain was proposed (Kau et. al., 1981). The results were encouraging. It was found that the correlations between the calculated and observed surface wind speed were high for all time periods of the day and night. The surface wind speed was dependent primarily on the slope wind, cross isobaric angle, surface thermal stability and geostrophic wind. The wind direction was dependent primarily on the geostrophic wind direction, aspect angle of the topography, upcanyon direction and cross isobaric angle. In this paper, extending the works of Kau et. al., (1982) intend to predict the pollutant concentration over complex terrain.

Statistical - Dynamic Dispersion Model

In order to predict the surface pollutant concentration, it is necessary to obtain the wind speed and wind direction at the surface. The predicted surface wind speed V(t) is taking the following form (Kau et. al., 1981)

$$\frac{V(t)}{G(t)} = a_1 \left| V_s(t) / G(t) \right| + a_2\alpha_0(t) + a_3\mu(t) + a_4 \qquad (1)$$

where $V_s(t)$ is the slope wind speed which is calculated based on relationships developed by Petkovsek and Hocevar (1971) and Ryan (1974), α_0 (t) is the cross isobaric angle, G(t) is the geostrophic wind speed, μ (t) is a nondimensional thermal stability parameter and a_i are constant regression coefficients.

The predicted surface wind direction D(t) for the mountain and valley stations are as follows:

At Mountain Station

(a) $D_g > Asp$ $\begin{cases} D(t) = D_g - 0.32(D_g + \alpha_0 - Asp) & \text{for daytime periods} \\ D(t) = D_{uc} - 0.50(D_g - Asp) & \text{for evening and early morning periods} \end{cases}$

(b) $180° < D_g \leq Asp$, we have $D(t) = D_g + 0.32\ (D_g + \alpha_0 - Asp)$ for all periods

(c) $D_g \leq 180°$, we have $D(t) = D_g + \alpha_0$ for all periods

At Valley Station

(a) $D_g > Asp - \alpha_0$ $\begin{cases} D(t) = D_g - 0.50(D_g + \alpha_0 - Asp) & \text{for daytime periods} \\ D(t) = D_{uc} & \text{for evening and early morning periods} \end{cases}$

(b) $180° - \alpha_0 < D_g \leq Asp - \alpha_0$ $\begin{cases} D(t) = Dg + 0.50(D_g + \alpha_0 - Asp) & \text{for daytime periods} \\ D(t) = D_{uc} - 0.80(D_g + \alpha_0 - Asp) & \text{for evening and early morning periods} \end{cases}$

(c) $D_g \leq 180° - \alpha_0$, we have $D(t) = D_{uc}$ for all periods

where D_g is the geostrophic wind direction, Asp is aspect angle of the topography and D_{uc} up-canyon wind direction.

With the information of predicted surface wind speed, predicted surface wind direction and geostrophic wind speed, the surface pollutant concentration C(t) can be predicted by use of following form

$$C(t) = b_1 \mu(t) + b_2 W_D(t) + b_3 V(t)/G(t) + b_4 W_D(t)G(t)/V(t) + b_5 \quad (2)$$

where b_i are constant regression coefficients and $W_D(t)$ is the wind direction function which takes the following form

$$W_D(t) = \cos\left(\frac{\pi}{32} \times J(t)\right) \quad (3)$$

with the integer value J(t). By seeking the relations between the observed surface concentration and surface wind direction, J(t) is determined in the order of their importance of each sector in which the wind direction is divided into 16 different wind vectors.

Equation (2) should contain an emission source term which unfortunately, are not available. However, for simplicity we could assume constant emission rate which have been absorbed into the coefficient b_5.

Model Computation and Results

The meteorological data collected near Anderson Creek, California, from August 1 to August 20, 1977, were used to compute the coefficients a_i in Eq. (1) and b_i in Eq. (2). The coefficients should be determined from a larger good observation data set covering a wider range of weather conditions. Unfortunately, they are not available, particularly from a mountain-valley site.

The technique for determining coefficients is a method of stepwise multiple regression analysis which is a statistical technique for analyzing a relationship between a dependent variable and a set of independent variables and for selecting the independent variables in the order of their importane. The values of coefficients a_i have been computed and presented in the paper of Kau et al., (1982). In this study the values of coefficients b_i for the mountain station (SRI - 1) and valley station (SRI - 5) as shown in Fig. 1 for the eight time periods (three hours for each period) since midnight are listed in Tables 1 and 2.

Table 3 shows the correlation coefficients between the observed and calculated surface pollutant concentrations for the eight time periods in the computed range (from August 1 to August 20, 1977). It is seen that the correlations are generally high for most of time periodes of the day and night.

Fig. 2 and Fig. 3 show the observed surface H_2S concentrations (solid lines) based on SRI International measurement data in 1977 and the predicted surface H_2S concentrations (dashed curves) with the use of Eq. (2) at the mountain and valley stations for the periods in the predicted range (from August 24 to August 28, 1977). It is found that the predicted surface pollutant concentrations

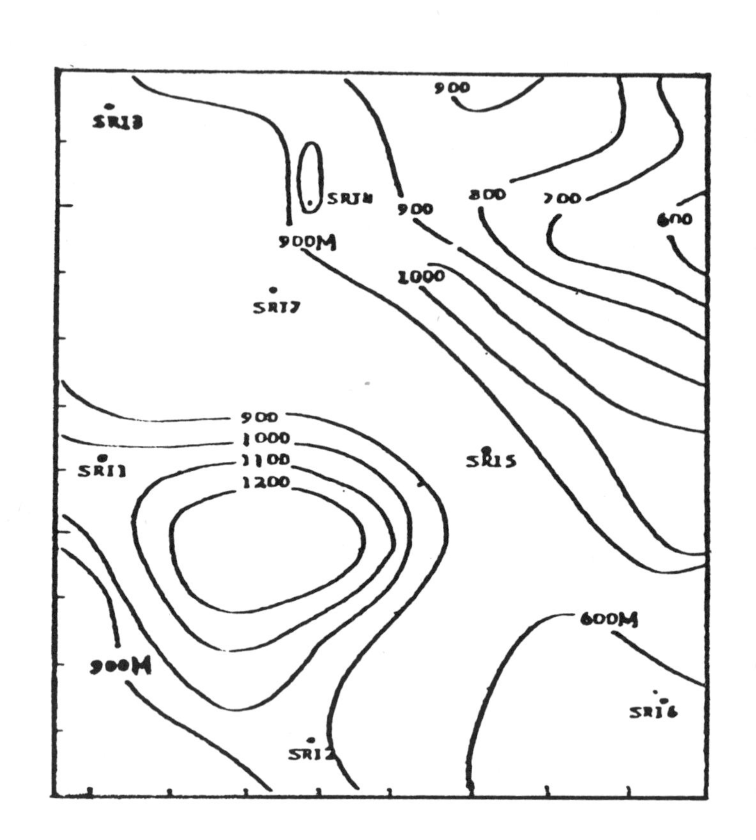

Figure 1 Topographic configuration near Anderson Creek, California

Table 1. Values of coefficients b_i for the mountain station at various time periods

Time Periods

	1	2	3	4	5	6	7	8
b_1	-0.004	0.105	0.008	0.046	0.108	-0.120	0.241	-0.123
b_2	1.899	3.590	7.189	51.328	25.249	62.312	-39.199	-6.658
b_3	9.804	0.528	3.528	-4.002	-6.750	34.884	12.350	23.561
b_4	-0.022	0.501	0.043	-0.500	4.834	3.482	2.461	1.158
b_5	-0.702	-7.416	-2.538	-32.209	-31.967	-58.085	-25.439	-2.498

Table 2. Values of coefficients b_i for the valley station at various time periods

Time Periods

	1	2	3	4	5	6	7	8
b_1	0.118	-0.008	0.024	-0.001	0.050	0.060	-0.009	-0.029
b_2	7.327	10.033	-3.884	0.947	-0.838	-17.687	-1.564	-11.880
b_3	37.312	60.291	3.605	1.004	-6.081	30.325	27.832	111.200
b_4	0.362	0.077	0.057	-0.081	0.169	4.654	-0.307	1.205
b_5	-13.175	-11.293	0.681	-0.459	-0.113	-12.522	6.718	-7.756

Table 3. Correlation coefficients between the observed and calculated surface pollutant concentration in the computed range

Time Periods	1	2	3	4	5	6	7	8
Mountain St.	0.703	0.920	0.904	0.666	0.840	0.811	0.903	0.628
Valley St.	0.769	0.963	0.985	0.780	0.848	0.820	0.650	0.899

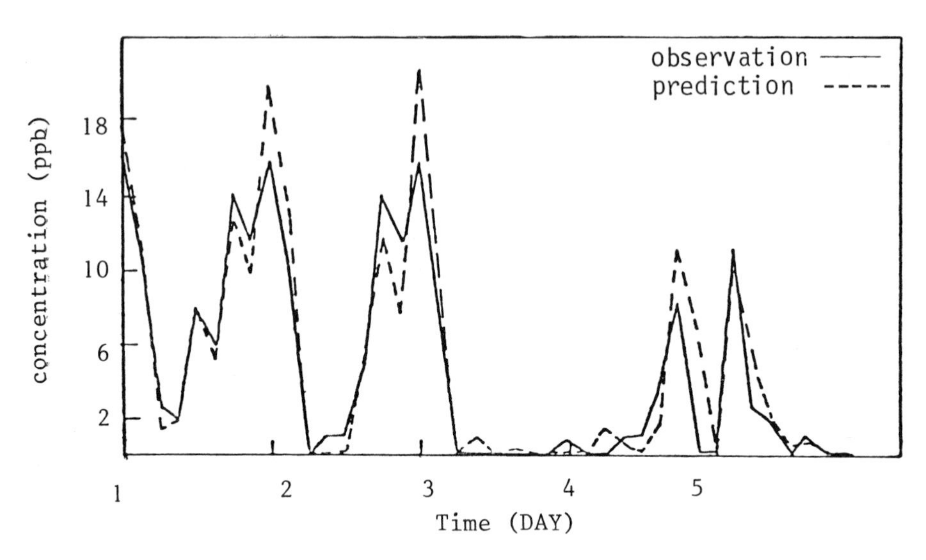

Figure 2 Observed and predicted H_2S concentration at the mountain station

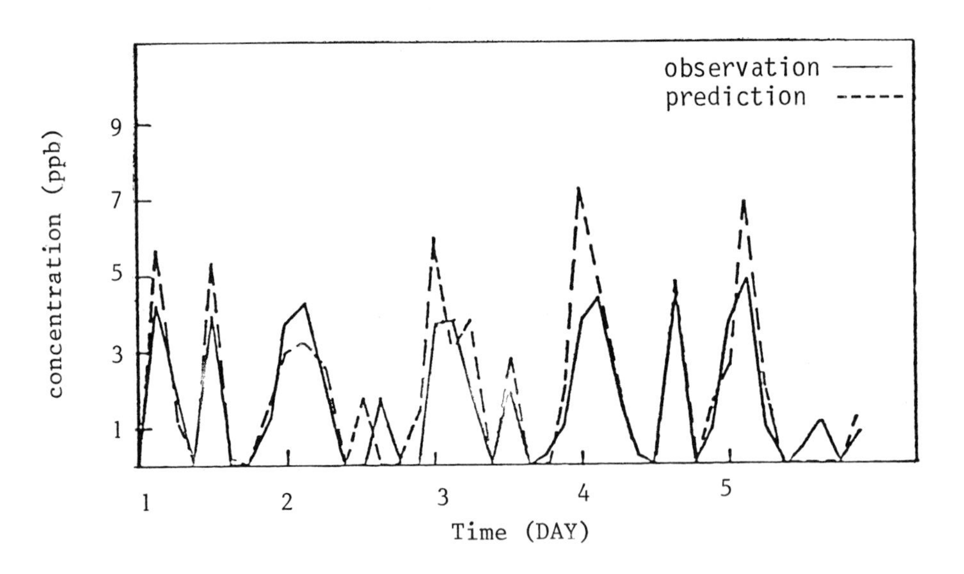

Figure 3 Observed and predicted H_2S concentration at the valley station

agree well with the observed; their correlations for 5 days periods at the mountain and valley stations are 0.927 and 0.838, respectively. It may be concluded that the surface pollutant concentration at the mountain and valley stations depends primarily on the surface wind speed, surface wind direction, geostrophic wind and surface thermal stability.

Conclusions

An atmospheric transport-dispersion model for surface pollutant concentration prediction in complex terrain has been presented. The model is statistical and treats the wind speed and wind direction in a more dynamic manner than the conventional dispersion approach. The predicted surface pollutant concentrations are high. It is found that the surface pollutant concentration at the mountain and valley stations depends primarily on the surface wind speed, surface wind direction, geostrophic wind and surface thermal stability.

Acknowledgement

This research was supported by the Office of Health and Environmental Research, U.S. Departement of Energy under Contract DE-AS02-76EV02455.

References

Kau, W.S., LEE, H.N. and Kao, S.K., 1982 : A Statistical Model for Wind Prediction at a Mountain and Valley Station Near Anderson Creek, California, accepted for publication in Journal of Applied Meteorology, Jan. issue, 1982.

Petkovsek, Z., and Hocevar, A., 1971 : Night Drainage Winds, Arch. Meteoro. Geophys. Bioklimatol., Ser. A, 20, 355-350.

Ryan, B.C., 1974 : A Mathematical Model for Diagnosis and Prediction of Surface Winds in Mountainous Terrain, in Proceeding of the sixth conference on weather forecasting and analysis, May 10-14, 1976, Albany, New York, 146-151.

Start, G.E., Dickson, Cr. R., and Ricks, N.R., 1977 : Effluent dilution over mountainous terrain and within mountain canyons. Symposium on Atmospheric Diffusion and Air Pollution, Amer. Meteor. Soc., 226-232.

Yamada, T., 1979 : A Numerical Study of the Effects of Complex Terrain on Dynamics of Airflow and Pollutant Dispersion, Fourth Symposium on Turbulence, Diffusion and Air Pollution, Reno, Nevada, Jan. 15-18, 1979, 213-216.

DISCUSSION

A. VENKATRAM

1. Is your model a fit to data?

2. Why doesn't your regressi equation for concentration contain an emission term?

H. N. LEE

1. No, it isn't. We compute regression coefficients based on history data bases.
Then we use them to predict future concentrations based on the model equations described in the paper. We compare the predicted results with the observations which are available.

2. In this case, we assume the emission rate is constant, i.e., the constant emission rate all the time at a station, which is a reasonable assumption. Under this assumption the emission source term could be combined with the coefficient b_5.

E. RUNCA

Was your model for the concentration prediction an autoregressive one?

H. N. LEE

Yes.

K. SCHAEDLICH

Do the time periods in table 1-3 refer to 3 hour averages?

H. N. LEE

Yes.

POLLUTANT DISPERSION IN THE EKMAN LAYER

USING MOMENT-REDUCED TRANSPORT EQUATIONS

D. Eppel, J. Häuser, and
H. Lohse
Institut für Physik
GKSS Forschungszentrum
2054 Geesthact, FRG

F. Tanzer

1. Institut für Physik
Universität Giessen
6300 Giessen, FRG

ABSTRACT

The program MODIS (Moment Distribution) is designed to extend the Gaussian plume model up to the range of about 10^2 km. Numerical experiments with stationary solutions are reported to show reliability and accuracy (essentially dependent on the grid Peclét number) of the results. The influence of the Coriolis force on the ground centre line of the plume is shown to be considerably smaller than estimates based on analytical models.

INTRODUCTION

Motivated by the necessity for the reduction and interpretation of air pollution data to be obtained in the near future by a remote sensing device called DIAL (differential absorption and scattering lidar)[1] we set up a code based on the transport equation and designed as complement and extension to the Gaussian plume model[2]. An extension, insofar as the range of this model should reach up to distances where the original plume is homogeneously distributed over the depth of the boundary layer. For a velocity $u \sim 5 - 10$ m/s and taking the mixing time to be of the order of 10^4 seconds[3] we get a range of 10^2 km. A complement, as the model should incorporate the possibility of using height dependent exchange coefficients and velocities and, because of its expected range, Ekman-type wind fields. Moreover we would like to have investigated removal processes, bottom deposition rates, and effects associated with a finite sink velocity. A more practical point which led us to the use of moment-reduced transport equation was the restriction to low computational expense. Certainly, all these requirements cannot be fulfilled equally well. However, we hope to show that for a wide range of applications our model may be used with confidence.

MODEL DESCRIPTION

To show the physical content of the model we start with the transport equation for a non-buoyant concentration $c = c(x,y,z,t)$:

$$\frac{\partial}{\partial t}c + \vec{\nabla}((\vec{U} - \vec{v}_s)c) = \vec{\nabla}K\vec{\nabla}c - Rc + q. \qquad (1)$$

$\vec{\nabla}$ denotes the 3-dimensional gradient operator, $\vec{U} = (u,v,0)$ is the horizontal velocity field, $\vec{v}_s = (0,0,v_s)$ a finite sink velocity, and K represents the turbulent exchange tensor. Removal processes are described in linear order by $-Rc$ where R is the removal parameter, and sources are summarized in q. The coordinate system is oriented with the positive x-axis parallel to the main wind direction. y and z are used as transverse and height coordinates. Eq. (1) is supplied with the boundary conditions:

$$c = 0 \text{ or } \frac{\partial}{\partial z}c = 0 \text{ or } c \text{ continuous} \quad \text{at } z = H$$

and (2)

$$[-v_s - (K_{xz} + K_{yz} + K_{zz})\frac{\partial}{\partial z}]c = -v_d c \quad \text{at } z = 0$$

where v_d is a prescribed deposition velocity. At the left (upwind) boundary we set $c = 0$, and at the right boundary we require c to be continuous.

To approximately solve eq. (1) we perform a moment expansion, i.e. we replace the transport equation by the first three members of the well-known hierarchy of moment equations[4]

$$\frac{\partial}{\partial t}c^{(n)} + \frac{\partial}{\partial x}(uc^{(n)}) - \frac{\partial}{\partial z}(v_s c^{(u)}) =$$

$$\frac{\partial}{\partial z}(K_z\frac{\partial}{\partial z}c^{(n)}) - Rc^{(n)} + n\, vc^{(n-1)} + n(n-1)K_y c^{(n-2)} + q^{(n)};$$

$$n = 0,\ 1,\ 2. \qquad (3)$$

The moments are defined with respect to the y-axis:

$$c^{(n)}(x,z,t) = \int_{-\infty}^{+\infty} dy\ y^n c. \qquad (4)$$

The restriction to the first three moments implies the plume to be Gaussian in y-direction. The Gauss Parameters, normalization Q, centre y_o, and width σ_y, are related to the moments by

$$c^{(o)} = Q; \; c^{(1)} = Qy_o; \; c^{(2)} = Q(y_o^2 + \sigma_y^2). \qquad (5)$$

Motivated by numerical tests with stationary solutions to eq. (3) we have retained from the eddy exchange tensor only the y- and z-diagonal parts, $K_y = K_{yy}$ and $K_z = K_{zz}$, as the left-out quantities are of minor importance compared to the effects resulting from the numerical scheme used. Similar to other works[5] we expect the x-diagonal and off-diagonal exchange coefficients as well as higher moments to have an appreciable influence on non-stationary situations such as the transport and diffusion of an instant puff. In such cases the methods employed here are not appropriate, and other techniques must be used[6].

The numerical solution of eq. (3) is done within the framework of elementary discrete element analysis. Technical details can be found in many textbooks[7]. The vertical solution area is covered with a rectangular grid which may be non-equidistant. Space-derivatives occuring in the diffusive part of the equation are approximated by central differences as well as the convective term associated with the sink velocity v_s. The convective part in x-direction is approximated by upstream differencing. The result is an equation system for the moment values at the grid points (x_i, z_j) first order in time. This system is integrated by a fourth-order Runge-Kutta scheme[8].

The different treatment of x- and z-convection is motivated by the vertical motion beeing usually diffusion-dominated, whereas horizontal motion is in the problems considered here advection-dominated.

Apart from ease of programming and favourable stability properties of equations containing upstream-differenced terms, the size of numerical diffusion associated with this method has led many people to look for more sophisticated discretization methods. However, as has been shown[9], upwind-differencing yields accurate results, namely for stationary situations with one dominant flow direction.

NUMERICAL EXAMPLES

To approximately determine the range of validity of the model we have performed some test calculations. However, we are aware that a systematic sensitivity analysis should be done[10].

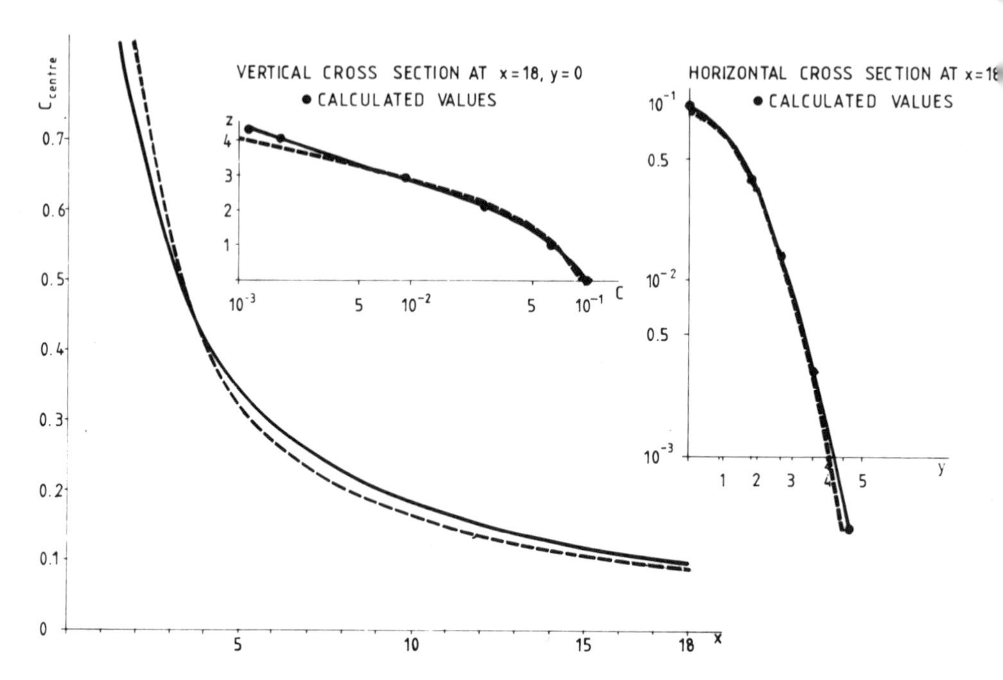

Fig. 1: Comparison between numerical and analytical solution for $K_y = K_z = 0.05$ ($P_\Delta = 20$). The dashed line is the exact result.

For the numerical investigations an equidistant grid of unit spacing in x- and z-direction is used with (10 x 20) grid points. We take $u = 1.0$, $v = 0$, and $K_y = K_z$, together with a unit source located at grid point (1,5). As a typical example, fig. 1 shows a comparison between calculated and exact concentrations for the plume centre line where the largest errors occur. Except within about five space units distance from the source the calculated concentrations are always higher than the exact values, an effect which is also present in other numerical schemes[11]. The error decreases with increasing distance from the source.

Dependence on Peclét Number

The most severe errors induced by the numerical scheme stem from the incorrect interplay between diffusion and advection. As can be seen from eq. (3) a sensitive parameter for investigating the behaviour of the result as a function of advection and diffusion is the grid-Peclét Number $P_\Delta = u\Delta x/K_z = 1/K_z$. Errors induced by K_y are negligible, as K_y only determines the source strength of the second moment equation. In fact, for $K_z = 0$ ($P_\Delta = \infty$) the numerical result is, apart from rounding errors, identical to the analytical solution. In fig. 2 the concentration error is given as a function of the Peclét Number P_Δ. The curve has been determined for the

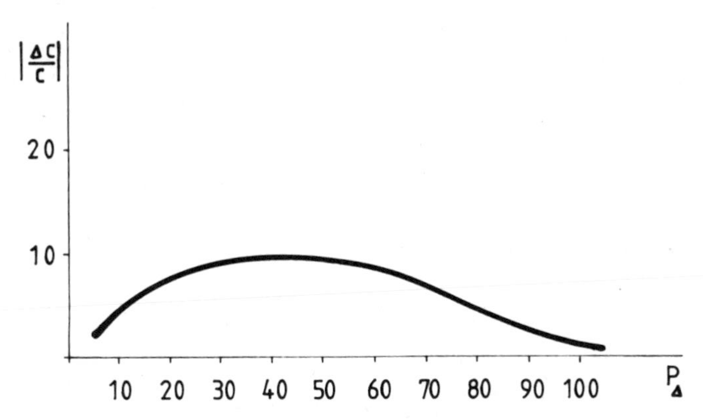

Fig. 2: Concentration error as a function of Peclét Number for the plume centre line.

plume centre line for points more than 15 space units away from the source. As the centre line deviates most from the exact result, fig. 2 actually shows the upper error bound. Outside the plume's centre line the agreement is considerably better. In our opinion, the errors induced by a finite Peclét number are the main limitation on the application of the model. In order to obtain a result with an error better than 5 % we must have $P_\Delta \sim 65$. For $u \sim 5$ m/s and $K_z \sim 10\ m^2/s$ we cannot obtain a grid resolution better than $\Delta x \geq P_\Delta K_y/u \sim 130$ m.

For the sake of completeness it should be noted that there is also a slight dependence on the z-grid spacing. This influence becomes prominent when Δx is made considerably smaller than Δz, a situation which is not encountered in problems considered here. We have therefore chosen $\Delta x = \Delta z$ which could perhaps occur as most unfavourable application case. Noting that a reduction of Δz by a factor of three improves the result by about 1 % we neglected this influence altogether.

Dependence on Boundary Conditions

Lower boundary

In eq. (2) we specified a flux condition with a prescribed deposition velocity. In fig. 3, the change in bottom concentration is shown when the no-flux condition is relaxed. In applications typical values range from $v_d \sim 0.01$ m/s (elementary iodine) to $v_d \sim 0.001\ \frac{m}{s}$ (aerosol) so that an incorrect treatment of the lower boundary may have an non-negligible effect.

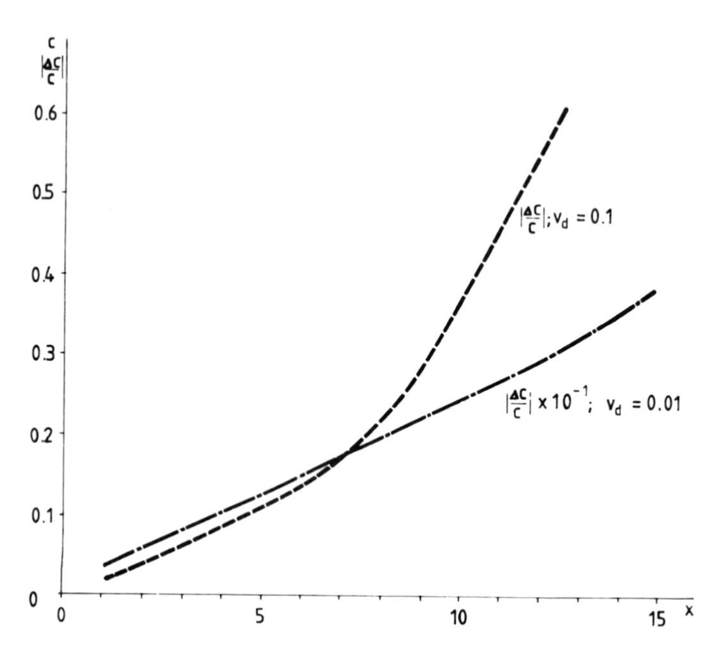

Fig. 3: Change of bottom concentration for $K_y = K_z = 0.5$ as a function of source distance for two deposition velocities. The reference concentration corresponds to $v_d = 0$ (Gauss).

Upper boundary

The conditions, complete absorption $c = 0$, and no-flux $\partial c/\partial z =$ have a physical justification only in that case in which the soluti area contains the full boundary layer. In situations where the exchange coefficient is still finite these boundary conditions are incorrect. Introducing a finite removal velocity[12] similar to the treatment of the lower boundary is no way out since this parameter can in practice not be measured. In our opinion, the only reasonable requirement one should apply at open boundaries is the continuity of the quantity to be calculated. With this prescription, no unphysical mass accumulation is possible. Fig. 4 shows the difference in concentration for the three cases. The reference concentration calculated with 'c continuous' is practically identical to the analytical solution. Depending on the size of the exchange coefficient concentrations calculated with improper boundary conditions may be unrealistic.

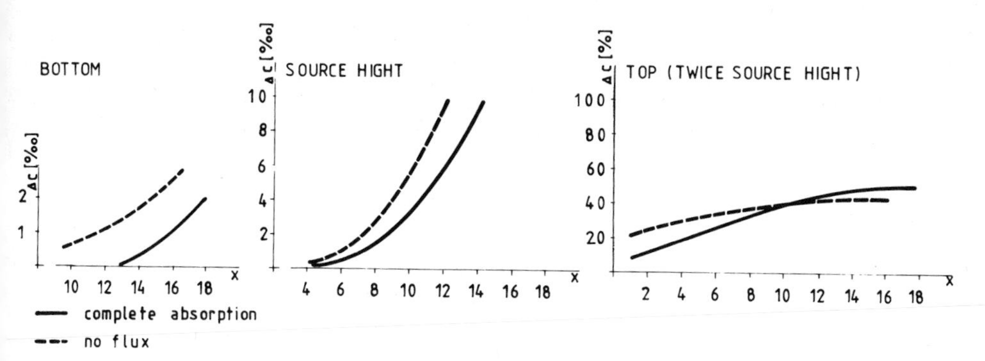

Fig. 4: Influence of the top boundary conditions on the concentration field at three heights for $K_y = K_z = 0.5$. The reference concentration is calculated with the boundary condition 'c continuous'.

Dependence on Removal Parameter and Sink Velocity

In fig. 5 we have shown the effect of the removal term on the accuracy of the solution for a value $r_\Delta = R\Delta x/u = R = 0.05$, which is the appropriate quantity. Except for the immediate vicinity of the source, where the finite approximation to the point source is felt, the error is negligible. For smaller values of r_Δ the error is further reduced. In applications, rain-out coefficients are of the order $10^{-4}(\frac{1}{s})$ (elementary iodine). Taking $\Delta x \sim 200$ m, $u \sim 5$ m/s we obtain $r_\Delta \sim 5 \cdot 10^{-3}$, which is small enough to neglect the errors introduced by the removal term.

Fig. 6 shows the bottom concentration for sink velocities v_s ranging from 0.005 to 0.1. The concentration line for $v_s = 0.005$ is nearly identical to the Gauss solution with $v_s = 0$. A slight shift away of the point of maximum ground concentration from the source

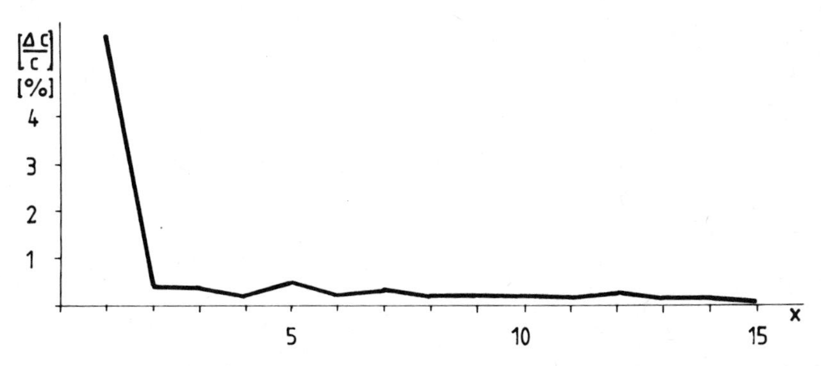

Fig. 5: Concentration error of the plume centre line for R = 0.05, $K_y = K_z = 0.1$. ($P_\Delta = 10$). The reference concentration is the analytic solution.

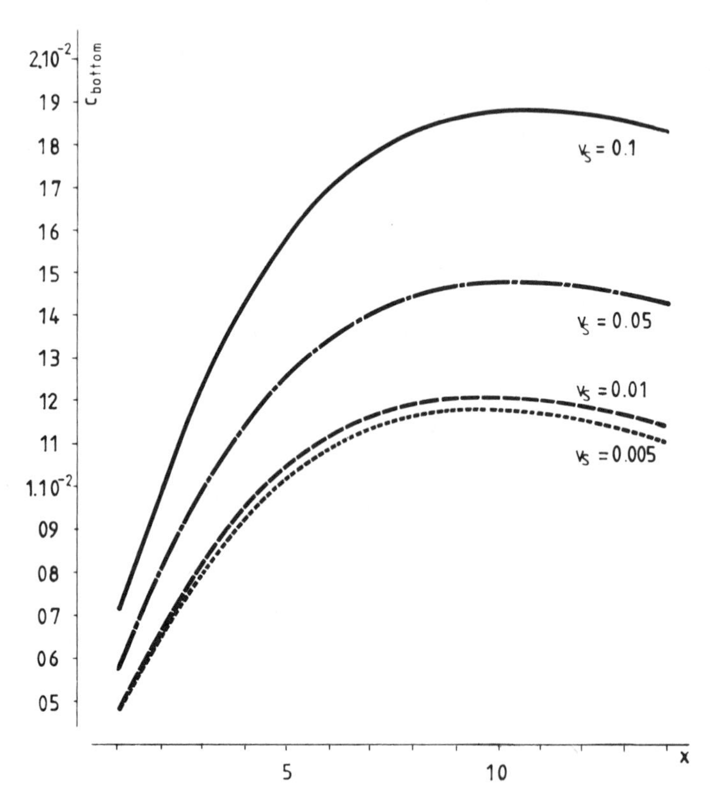

Fig. 6: Bottom concentration as a function of the source distance for different sink velocities. The calculations are done with $K_y = K_z = 0.5$ and $v_d = 0$.

with increasing sink velocity is observed. For particles with a size ~ 2 μm v_s ~ $5 \cdot 10^{-4}$ m/s. The differences to the distributions we have calculated with and without sink velocity, are for such cases well within the error limits given by the other approximations.

Comparison with Gaussian Plume Model

In order to simulate a Gaussian calculation we have created x-dependent exchange coefficients according to

$$\frac{1}{2}\frac{d}{dt}\sigma_y^2 = K_y \ , \quad \frac{1}{2}\frac{d}{dt}\sigma_z^2 = K_z$$

where the widths are given by

$$\sigma_y = Fx^f \quad ; \quad \sigma_z = Gx^g \ .$$

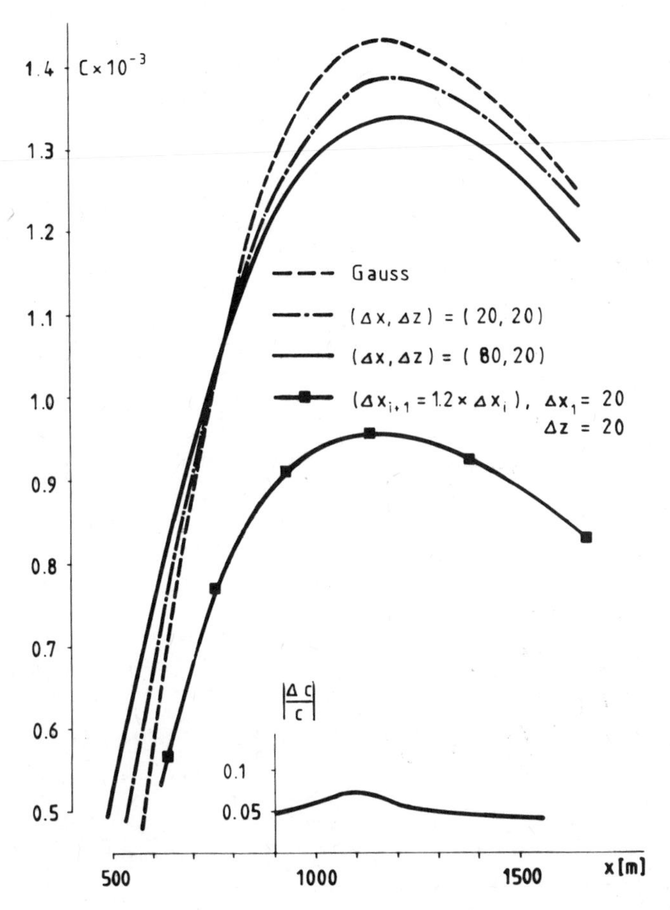

Fig. 7: Comparison of bottom concentrations as function of source distance obtained by the Gaussian model and by numerical calculation using identical input parameters given in the text. Concentration is measured in $\mu Ci/m^3$. The lower curve is the result of a non-equidistant grid calculation. The relative error shown at the bottom refers to the equidistant grid calculation with $\Delta x = 80$.

The result was compared to the corresponding Gaussian model solution for the following input parameters: source strength $\overline{Q} = 1.108 \cdot 10^3$ µC effective source height h = 100 m, average velocity $\bar{U}_{zh} = 6.606$ m/s (corresponding to an exponent m = 0.2666 for stability class D), F = 0.195, f = 0.968, G = 0.240, and g = 0.797. These parameters are determined from experiment 15,1 of the Karlsruhe field study[13]. For the calculations we used an equidistant grid with 20 m grid-spacing in the vertical and a horizontal spacing from 20 m to 80 m. The height of the solution area was 200 m (twice the source height). Fig. shows the calculated bottom concentrations together with the Gauss-solution. A linear and strongly stretched scale is used in order to show the differences. Whereas the location of maximum bottom concentration is insensitive to the grid spacing, the value of the maximum concentration is reduced by less than 5 % when increasing Δx by a factor of four. In both cases the agreement with the analytical solution is better than 10 %, the largest error occurring at the point of maximum ground concentration. Fig. 8 shows the ground concentration pattern for the Gaussian and grid model. The deviations near the source are due partly to the deficiencies of the interpolation routine and partly to the crude resolution of the grid near the point source.

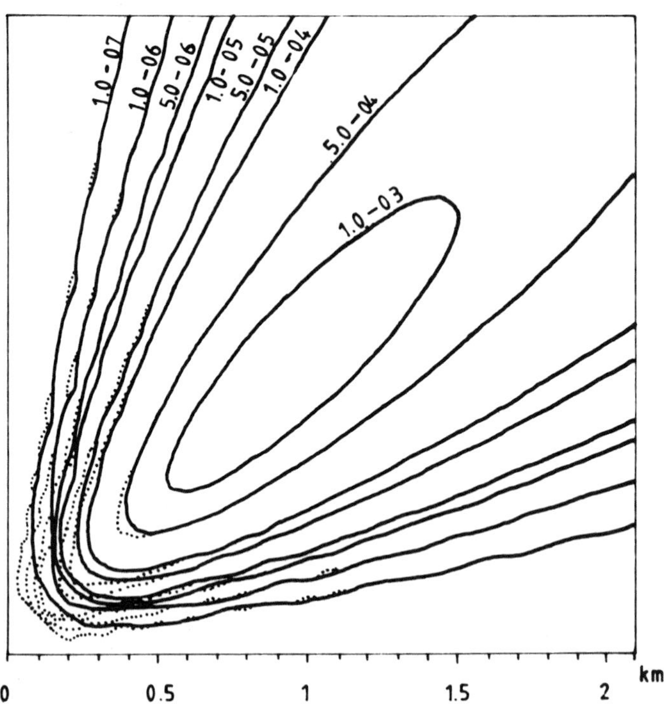

Fig. 8: Bottom concentration pattern of Gaussian solution. The dotted lines are the result of the calculation with Δx = 80. The source is located in the origin and the x-axis lies diagonal. Concentrations are given in µCi/m^3.

Influence of Coriolis Force

The numerical investigation showed a comparatively strong error dependence on the transverse velocity. To retain an accuracy better than 10 % the crosswind velocity must not exceed 0.5 · u.

In order to get an impression of the effect the Coriolis force we have used in the calculation a wind field together with a diffusion coefficient $K = K_y = K_z$ obtained by the Pielke model[14] (see rhs of fig. 9) with $u_* = 0.3$ m/s, $z_o = 0.01$ cm/s for neutral stability and a boundary layer height of 1400 m. The calculations have been checked by runs with different grid spacing ranging from $\Delta x = 500$ m up to $\Delta x = 2$ km. Fig. 1o shows the plume's ground deviation from the direction of the surface wind. The deviation calculated here is considerably smaller than that obtained by an investigation based on the analytic Ekman layer solution[15]. The lhs of fig.9 shows two vertical cuts through the plume at 5 and 10 km distance. The tendency towards homogeneous mixing is clearly seen.

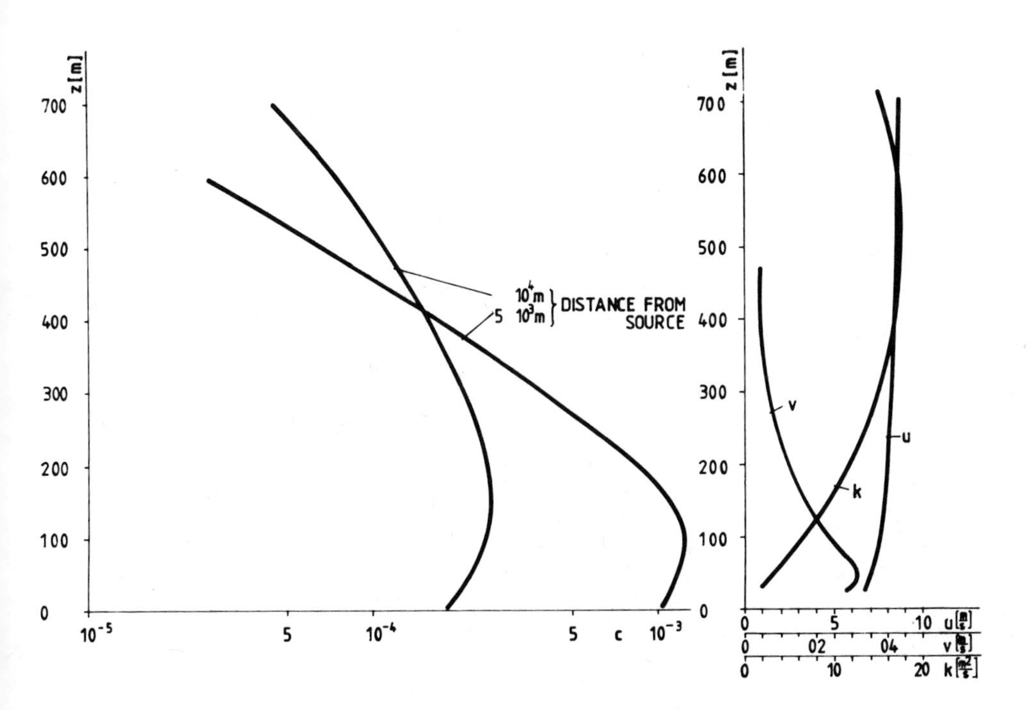

Fig. 9: Lhs: Vertical cut through the plume at 5 km and 10 km.
rhs: Velocities and exchange coefficient used in the calculation.

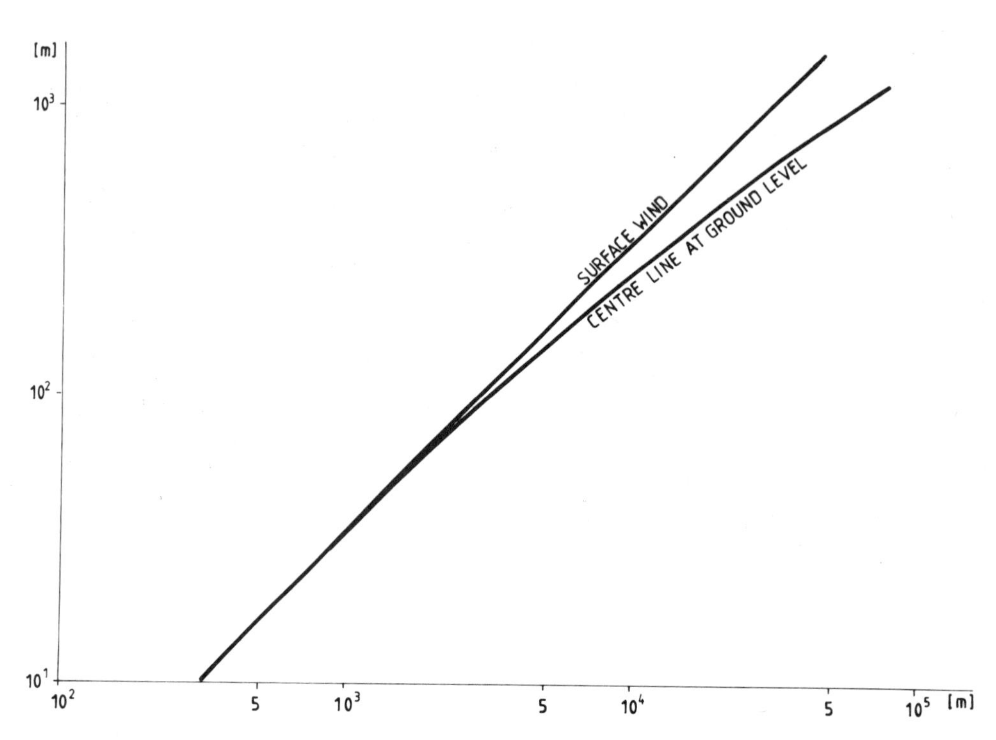

Fig. 10: Deviation of the ground centre line of the plume from the surface wind direction.

CONCLUSION

We have presented a sequence of numerical tests for the stationary version of the transport model MODIS. The tests have been performed for pollutant transport from point sources. It may be noted that, due to the treatment of the lower boundary and due to the moment reduced version of the transport equation, it is also possibl to consider extended area and volume sources. The main restriction of the model results from the dependence of the error on the Peclét Number. Within restrictions obtained in this work, which is the case for many applications, the model yields remarkably accurate results. Discrepancies with experiments, when using the model in connection with velocities and exchange coefficients created from meteorological measurements, can then, hopefully, be attributed to the data.

A note should be made concerning the use of non-equidistant grids. While a non-equidistant grid-spacing represents - within limits - no problem, a variable grid in flow direction may have disastrous effects on the result (c.f. non-equidistant grid result in Figure 7). When the grid-spacing is increased more and more lower fourier components of the solution get lost in an uncontrollable manner. We therefore think it is necessary for obtaining reliable results to use an equidistant grid along the main wind direction.

We are gratefully indebted to A. Müller for numerous discussions. A comment of E.J. Plate stressing the importance of reliability and sensitivity investigations is also acknowledged.

REFERENCES

1. C. Weitkamp, H.-J. Heinrich, W. Herrmann, W. Michaelis, U. Lenhard, R.-N. Schindler, Measurement of hydrogen chloride in the plume of incineration ships, 5th Int.Clean Air Congress, Oct.2o-26, 198o, Buenos Aires.
2. D. Eppel, J. Häuser, A. Müller, Three-dimensional simulation model MODIS for the propagation of air pollutants, 5th Int.Clean Air Congress, Oct.2o-26, 198o, Buenos Aires.
3. M. Benarie, How do tall stacks influence on long range transport, Proceedings of the 1o. CCMS Int.Meeting on air pollution modelling and its applications, Oct.22-26, 1979, Rome.
4. M.J. Manton, On the dispersion of particles in the atmosphere, Boundary Layer Met., 17 (1979) 145-165.
5. M. Mulholland, Simulation of tracer experiments using a numerical model for point sources in a sheared atmosphere, Atm. 14 (198o) 1347-136o.
   S.K. Kao, J.C. Doran, Turbulent diffusion in linear shear flow, Pageoph, 114 (1976) 357-363.
   P.G. Saffman, The effect of wind shear on horizontal spread from an instantaneous ground source, Q.J.R. met.Soc. 88 (1962) 382-393.
6. D. Eppel, J. Häuser, to be published
7. P.G. Roache, Computational fluid dynamics, Hermosa Publ., Albuquerque 1976.
8. L.F. Shampine, R.C. Allen, Numerical computing: an introduction, W.B. Saunders, Philadelphia 1973.
9. G.D. Raithby, A critical evaluation of upstream differencing applied to problems involving fluid flow, Comp.Meth.Appl. Mech.Eng. 9 (1976) 75-1o3.
1o. M. Koda, A.H. Dogru, J.H. Seinfeld, Sensitivity analysis of partial differential equations with application to reaction and diffusion processes, J.Comp.Phys. 3o (1979) 259-282.
11. K.W. Ragland, R.L. Dennis, Point source atmospheric diffusion model with variable wind and diffusivity profiles, Atmosph. Env. 9 (1975) 175-189.
12. G.R. Carmichael, D.K. Yang, C. Lin, A numerical technique for the investigation of the transport and dry deposition of chemically reactive pollutants, Atm.Env. 14 (198o) 1433-1438.

13. K. Nester, P. Thomas, Im Kernforschungszentrum Karlsruhe experimentell ermittelte Ausbreitungsparameter für Emissionshöhen bis 195 m, Staub-Reinhaltung Luft 39 (1978).
P. Thomas, W. Hübschmann L.A. König, H. Schüttelkopf, S. Vogt, M. Winter, Experimental determination of the atmospheric dispersion parameters over rough terrain, Part 1, Juli 1976, KfK 2285

14. R.A. Pielke, Y. Mahrer, Verification analysis of a three-dimensional mesoscale model predictions over South Florida for July 1, 1973, Report No. UVA-ENV SCI-MESO-1976-1, Dept. of Env. Sciences, University of Virginia, Charlottesville, Virginia 22 9o3.

15. G.T. Czanady, Turbulent diffusion in the environment, D. Reidel Publ. Comp., Dortrecht 1973.

DISCUSSION

P. HECQ

In your figure 6 where you analyse the influence of v_s you consider the case of $v_d = 0$. What does that mean ?

D. EPPEL

It means that the particles which sink down are not removed but accumulate near the ground to the extent shown in the figure. The size of the effect depends upon the relative magnitude between sink velocity transport and vertical diffusion transport. In reality there exists a finite deposition velocity, so that this effect is reduced.

M. SCHATZMANN

I wonder to what extend your error estimates reflect just the accuracy of your own model. Can it really be generalized that models which use non-equidistant grid distances in each direction provide concentration predictions up to 50% below the true value ?

D. EPPEL

If you require no better accuracy than 50% you will certainly find some crude rule. In my opinion, as long as one is not forced to change the grid-spacing to save storage in region where precise results are desired, one should use an equidistant grid when integrating the transport equation of a passive contaminant.

A SIMPLE MODEL FOR DISPERSION OF NON-BUOYANT PARTICLES INSIDE A CONVECTIVE BOUNDARY LAYER

P. K. Misra

Atm. Mod. Dev. Unit, Air Resources BR. OME, 880
Bay St., 4th Floor, Toronto, Ontario, Canada

INTRODUCTION

High ground level concentrations are very often observed for plumes dispersing in convective boundary layers. As has been pointed out by Lamb (1978, 1979), Willis and Deardorff (1975, 1978) and Venkatram (1980), conventional Gaussian plume methods are inadequate for treating dispersion inside a convective boundary layer. This is because, convective boundary layers consists of large scale eddies, the updrafts and downdrafts. Transport processes due to these eddies cannot be expressed by simple gradient transfer hypothesis for which the Gaussian plumes are valid.

In an earlier paper, Misra (1981) (hereafter referred to as M1), has shown that the plume inside a convective boundary layer can be divided into an updraft plume and a downdraft plume. It was shown also that for a stack height, $Z_s \sim 0.24Z_i$, where Z_i is the mixed layer height, the ground level concentration field was explained entirely by the downdraft plume. A reason for this plume behaviour is that the updraft plume probably hits the top of the mixed layer before descending to the ground. With a characteristic updraft, velocity, w_*, this implies that the updraft plume touches the ground at a downwind distance of approximately $Z_i u/w_*$ form the stack, were u is the mean wind speed. The maximum ground level concentration, on the other hand, occurs at a distance 0.39 $Z_i u/w_*$ (Willis and Deardorff, 1978). The concentration levels at a distance $Z_i u/w_*$ are low relative to the maximum concentration.

For taller stacks, however, the updraft plume hits the top of the mixed layer quicker, whereas the downdraft plume takes longer to reach the ground. Consequently the maximum ground level concentration is observed farther downwind from the stack and the contribution of the updraft plume to the ground level concentration field is no longer negligible.

Recently Willis and Deardorff (1981) have published water tank data for dispersion of non-buoyant plume from a stack height 0.49 Z_i. In this paper the model of M1 is applied to a stack height, 0.49 Z_i and the model predictions are compared with the water tank data.

The theory in M1 is based on the short diffusion time theory and should be applied with caution for very tall stacks. We never-the-less obtain good comparison between the model predictions and water tank data for $Z_s \sim 0.49\ Z_i$. Unlike the case of, $Z_s \sim 0.24\ Z_i$, the contribution of the updraft plume to the ground level concentration field is found to be significant.

DESCRIPTION OF THE MODEL

It is assumed that the downdraft velocities are distributed normally given by the following expression

$$P(W_d) = \frac{1}{\sqrt{2\pi}\ \sigma_{W_d}} \exp\left[-\frac{(W_d - \bar{W}_d)^2}{2\ \sigma_{W_d}^2}\right] \tag{1}$$

This is based on the observation of Lenschow (1970). Here, $\sigma_{W_d}^2$ is the variance of the downdraft velocities W_d, $\bar{W}_d$ their mean, and $P(W_d)$ the probability density function. It is assumed that $\sigma_{W_d}^2$ is approximately 1/3 σ_W^2 where σ_W^2 is the total variance of the vertical velocity (Lenschow and Stephens (1980)). 2/3 σ_W^2 is contributed by the updrafts to the total variance.

Using the short diffussion time theory (Hinze, 1976) we can relate the probability density function for particle position in the yz plane at a fixed downwind distance, x to P (W_d) as follows :

$$P\ (Z,Y)\Big|_x = P(W_d)\left|\frac{dW_d}{dZ}\right| \frac{1}{\sqrt{2\pi}\ \sigma_y} \exp\left[-\ Y^2/2\ \sigma_y^2\right] \tag{2}$$

We have assumed a normal density function in the y direction in confirmity with field and laboratory data.

After a few steps of mathematical manipulations the concentration field, due to the downdraft plume, $C_d(x,y,z)$, is obtained to be the following :

$$C_d(x,y,z) = \frac{F_i f_i(Z_s) Q}{2\pi\ f_i(Z)\ \sigma_{W_d}(ZS)\ \sigma_y x} \times \exp\left[-\frac{Y^2}{2\ \sigma_y^2} - \frac{1}{2\ \sigma_{W_d}^2(ZS)} \left\{ \frac{g(Z) f_i(Z_s) w}{x} - \bar{W}_d(Z_s) \right\}^2 \right] \tag{3}$$

Here F_i is the fraction of the plume in downdraft, Z_s is the stack height and

$$f_i(Z) = -(Z/Z_1)^{1/3}\ (1 - 1.1\ Z/Z_i) \tag{4}$$

and

$$g(Z) = \int_{Z_s}^{Z} \frac{1}{f_i(Z)}\ dZ \tag{5}$$

The reader is referred to the original paper for the mathematical details.

For the updraft plume we assume a well mixed condition in the vertical. Thus, the concentration field due to the updraft plume is given by the following :

$$C_u(x,y,o) = \frac{(1 - F_i)Q}{\sqrt{2\pi}\ Z_i\ \sigma_y\ u} \exp\left[- {Y^2}/{z\ \sigma_y^2} \right] \tag{6}$$

The applicability of (6) depends on the stack height. In this paper, (6) is applied in the range $W\ x/_{Z_i u} > 0.8$.

COMPARISON WITH OBSERVATIONS

The model predictions are compared with data generated in the water tank experiments (Willis and Deardoff, 1978, 1981). The following parameters are used for $Z_s/Z_i = 0.24$ and $Z_s/Z_i = 0.49$.

A. $Z_s/Z_i = 0.24$

$\sigma_{W_d} = 0.37\ W_*$
$\bar{W}_d = 0.11\ W_*$
$F_i = 0.6$
$\sigma_y = 0.4\ W_* \frac{x}{u}$

B. $Z_s/Z_i = 0.49$

$\sigma_{W_d} = 0.39\ W_*$
$\bar{W}_d = 0.1\ W_*$
$F_i = 0.6$
$\sigma_y = 0.4\ W_* \frac{x}{u}$

The above values are chosen from experiments in the atmospheric convective boundary layer (Lenschow and Stephens, 1980). We assume that the kinetic energy associated with turbulent fluctuations in the vertical velocity are divided in the ratio 2:1 between the updrafts and the downdrafts.

Tables 1 and 2 show the comparison of the non-dimensional concentration, $\bar{C} = CZ_i^2\ u/Q$ versus non-dimensional downwind distance, $W_*\ x/Z_i u$ between model predictions and water tank experiments. In Table 1, which is for the stack height 0.24 Z_i, it is seen that the downdraft plume alone explains the downwind variation of the concentration field. The magnitude and location of the peak ground level concentration are predicted well by the model. It is noted that Willis and Deardorff (1978) quote an experimental analysis error of about 7 % in the magnitudes of concentrations at the peak region. The contribution due to the updraft plume is negligible at virtually all downwind locations. Of course, the updraft plume would have to be treated in detail if one is interested in the vertical profiles of the concentration field.

Table 2 shows the same simulation for $Z_s = 0.49\ Z_i$. Once again, it is noted that the position of the peak ground level concentration is predicted well by the downdraft plume. The magnitudes of the concentrations near the peak region, as predicted by the downdraft plume are substantially lower than the observed values. Adding the contribution of the updraft plume brings the magnitude of the peak closer to the observed values. The updraft plume contributes about 25 % to the magnitude of the peak concentration for a stack height, $Z_s = 0.49\ Z_i$. The good comparison between model predictions and observations beyond $Z_i u/W_*$ from the stack may be fortuitous as the theory is not applicable in this range. Further, the simple approach of our study is unable to reproduce the complex plume behaviour, as observed by Willis and Deardorff (1981), beyond $Z_i u/W_*$.

CONCLUSIONS

We have demonstrated that the simple modelling approach for dispersion of plumes in a convective boundary layer compares favourably with laboratory observations for prediciton of ground level concentrations. This paper deals with non-buoyant particle release. When the particles have buoyancy, their trajectories assume a complex form in the x-z plane. This is not amenable to the simple approach undertaken here. The works of Lamb (1978) and Weil (1981) are similar to our approach.

Table 1

Comparison of Observed and Predicted Concentration (Ground Level) for a Stack Height, $Z_s/Z_i = 0.24$

Non-Dimensional Downwind Distance	Non-Dimensional Predicted Conc. Downdraft	Non-Dimensional Predicted Conc. Updraft	Non-Dimensional Observed Conc.
0.1	0	0	
0.2	0.7	0	0.9
0.3	4.1	0	4.9
0.38	5.37	0	5.9 ± 0.4
0.40	5.45	0	5.9 ± 0.4
0.50	4.9	0	5.1
0.60	4.1	0	4.0
0.70	3.4	0	3.2
0.80	2.85	0.5	2.2
1.00	1.95	0.4	1.6
1.50	0.91	0.27	0.9

Table 2

Comparison of Observed and Predicted Concentration (Ground Level) for a Stack Height, $Z_s/Z_i = 0.49$

Non-Dimensional Downwind Distance	Non-Dimensional Predicted Conc. Downdraft	Updraft	Non-Dimensional Observed Conc.
0.1	0	0	
0.2	0	0	
0.3	0.1	0	
0.4	0.75	0	0.7
0.5	1.47	0	1.5
0.6	1.88	0	2.95
0.7	1.99	0	3.00
0.8	1.94	0.5	3.00
0.9	1.81	0.44	2.90
1.0	1.67	0.4	2.4
1.5	1.03	0.27	1.5
2.0	0.67	0.2	0.7
2.5	0.48	0.16	
3.0	0.36	0.13	
3.5	0.28	0.11	
4.0	0.23	0.1	

REFERENCES

Hinze, J. O. (1976), Turbulence, McGraw Hill, Second Edition, pp. 7

Lamb, R. G. (1978), A Numerical Simulation of Dispersion from an Elevated Point Source in the Convective Boundary Layer, Atmospheric Environment, 12, 1297-1304.

Lamb, R. G. (1979), The Effects of Release Height on Material Dispersion in the Convective Boundary Layer, Pre-Print Volume, Fourth Symposium on Turbulence, Diffusion, and Air Pollution, Reno Nevada, 15-18 January 1979, 27-33.

Lenschow, D. H., and P. L. Stephens (1980), The Role of Thermals in the Convective Boundary Layer, Boundary Layer Meteorology, Volume 19, No. 4, 509-532.

Lenschow, D. H. (1970), Air Plane Measurements of Planetary Boundary Layer Structure, J. Appl. Meteor., 9, 874-884.

Misra, P. K. (1981), Dispersion of Non-Buoyant Particles Inside A Convective Boundary Layer, Atmospheric Environment, in Press.

Venkatram, A. (1980) Dispersion from an Elevated Source in Convective Boundary Layer, Atmospheric Environment, 14, 1-10.

Weil, J. C. and W. F. Furth, (1981), A Simplified Numerical Model of Dispersion from Elevated Sources in the Convective Boundary Layer, Pre-Print Volume, Fifth Symposium on Turbulence, Diffusion, and Air Pollution, March 9-14, 1981, AMS, Atlanta.

Willis, G. E., and j. W. Deardorff (1975), Computer and Laboratory Study of Modelling of the Vertical Diffusion of Non-Buoyant Particles in the Mixed Layer, Adv. Geophys., 18B, 187-200, Academic Press New York.

(1978), A Laboratory Study of Dispersion from an Elevated Source Within a Modelled Convective Planetary Boundary Layer, Atmospheric Environment, 12, 1305-1312.

(1981), A Laboratory Study of Dispersion from a Source in the Middle of the Convectively Mixed Layer, Atmospheric Environment, 15, 109-117.

DISCUSSION

D. SKIBIN

1. Observations of the plume shows sometimes upward motion, sometimes downward motion to the ground. Reverse cases also exist. How does this fit the authors divisions of plumes to descrete "upward" and "downward?"

2. Does the author feel that a neutral parcel or balloon can be continuously raised (or drifted to the ground) over travel distances of tens of km, in a convectively PBL?

P. K. MISRA

1. A byoyant plume would initially rise relative to a downdraft. You are probably referring to a buoyant plume. My paper deals with non-buoyant particles only.

2. Observations in convective boundary layers show "chinks" of plume to continue a downward motion to the ground. I would, therefore, think that a neutral parcel would follow the vertical motions in updrafts and downdrafts as it is advected downwind.

B. WEISMAN

How do you determine the fraction F_i of the downdrafting component of plume?

P. K. MISRA

Data (both numerical simulation (Lamb, 1978) and field experiments (Lenschow, 1980)), on the average areas occupied by updrafts and downdrafts at different levels within the convective boundary layer are available. I have chosen F_i from these data.

DISPERSION FROM FUGITIVE SOURCES

C. Demuth, P. Hecq, A. Berger

Institut d'Astronomie et de Géophysique G. Lemaître
Université Catholique de Louvain
B-1348 Louvain-la-Neuve, Belgium

ABSTRACT

Emissions from fugitive sources can be estimated from observed concentrations around an industry. The mathematical technique designed and calibrated for this application is a K-analytical diffusion model. The influence of buildings upon the nearby concentration field is also analysed, by a numerical two dimensional model for wind and concentration fields; this allows to locate monitoring stations to avoid these effects.

INTRODUCTION

The continuous recording of Sb atmospheric conentrations around an industry in Beerse (Belgium), has shown pollution levels much higher than expected from the emission rate of the main source (stack of 80m). This lead to a careful study of the influence of possible fugitive sources in the plant.

As no quantitative information exist about the strength of these fugitive sources, it has been necessary to design a mathematical technique in order to estimate them on the basis of diffusion models, provided their geographical location is known.

We shall present in this paper, a description of the model used, its calibration based on a tracer experiment and finally the computation of the fugitive emissions. The last part of the study will be related to the influence of buildings on the observed concentration field.

DESCRIPTION OF THE DISPERSION MODELS

The basic purpose of a diffusion model is to link a given emission Q to any immission χ around the source, by taking into account a maximum of physical relevant phenomena, as precisely as possible. The model used for this experiment is mainly based on the principle of mass conservation and on the theory of the atmospheric boundary layer (ABL).

On the basis of vertical and horizontal length scale ($\sim$1 and 10 km) and for a flat surface, all the mean parameters like the wind profile and the diffusivities, are assumed to be homogeneous in the horizontal plane. On that basis, it is possible to express the concentration field χ as the product of three terms : the emission Q, a lateral concentration field Y and a vertical field Z.

This Y function is generally simple because, as there are no lateral boundaries, χ can go to zero at large cross wind distances; in most cases this Y function is given by a gaussian law with a standard deviation σ_y. On the opposite, the Z function strongly depends upon the vertical scale in the ABL, which means that the turbulent diffusion, the sedimentation, the ground deposit and the physico-chemical interactions can not be represented any more by a simple law.

This Z function must in fact take into account :

- the transport by the wind, the vertical profile $\bar{u}(z)$ being given by a power law, $\bar{u}(z) = u_1 z^m$ where m is a function of the ABL stability.
- the vertical turbulent diffusivity K_z. The definition of its verticale profile is difficult mainly because it can not be deduced directly from experimental data. It may either be deduced from a parameterization or from a non stationnary model of the ABL. In the first case, meteorological data along a tower allows to define $K_z(z)$; in the second case, vertical profiles of wind, temperature and humidity are needed. In any case, a realistic approach of K_z profiles is given by the following power law

$$K_z(z) = K_1 z^n \tag{1}$$

- the sedimentation and wet deposition. In this particular case, the Sb particles have a diameter of 0,2 to 11 μm and the corresponding maximum sedimentation velocity v_s is 0,002 m s^{-1}. The dry deposit caracterised by a deposition velocity v_d will take into account the adsorption of antimonium by the ground and this velocity v_d will be considered as increasing with the wind speed.

- the interactions. This term include the removal of effluent by chemical reactions or by wash-out with a given coefficient Λ. In this latest case the effluent is found back on the ground.

The concentration χ at a given point (x,y,z) will thus be given by :

$$\chi(x,y,z) = Q.Y(x,y).Z(x,y) \qquad (2)$$

with :

$$Y(x,y) = \frac{1}{\sqrt{2\pi}\,\sigma_y(x)} \exp\left[-\frac{y^2}{2\sigma_y^2(x)}\right] \qquad (3)$$

Z(x,z) is the solution of the following stationnary problem :

$$\begin{cases} u\,(z)\,\dfrac{\partial Z}{\partial x} = \dfrac{\partial}{\partial z}\left[K_z(z)\,\dfrac{\partial Z}{\partial x} + v_s Z\right] - \Lambda\,Z \\ Z\,(x=0,z) = \dfrac{1}{u(h_e)}\,\delta(z-h_e) \\ K_z\,\dfrac{\partial Z}{\partial z} + v_s\,Z = v_d\,.\,Z \text{ for } z = 0 \\ K_z\,\dfrac{\partial Z}{\partial z} + v_s\,Z = 0 \quad \text{for } z = z_i \end{cases} \qquad (4)$$

where z_i is the thickness of the ABL.

For most of the practical cases analysed, the problem (4) has either an exact analytical solution or at least an approximate one sufficiently accurate for the present study if the influence of the buildings is not considered. Otherwise, it is necessary to solve the fundamental equations of fluid mechanics, as it will be shown later.

CALIBRATION OF THE K-MODEL IN BEERSE

The calibration and the validation of the model presented above may be done through a tracer experiment during which emission, immission and meteorological factors are carefully measured.

The calibration was based on the experiment carried out on December 4, 1979 during which the conditions were the following :

- two-hours release (12h30-14h30) of SF_6
- stack height : 75m
- stack diameter : 3m
- gaz flow rate : 6,5 g/s

- gaz velocity : 5,66 m/s
- gaz temperature : 20°C
- 13 receptors were under the influence of S-W winds

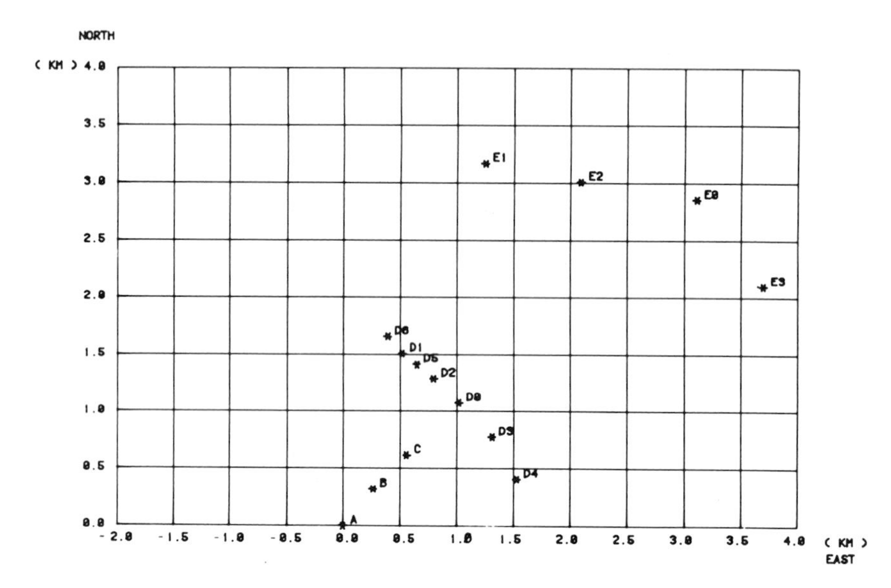

Fig. 1 Position of the recptors in Beerse. 'A' represents the source and 'DO' the meteorological tower.

Two solutions of the model have been investigated : one analytical[1] with a power law for K_z and the other semi-numerical with a simulated profile for K_z. These models need the knowledge of the wind field and of the turbulent diffusivities over the investigated region.

The wind profile is simulated by a power law of z, on the basis of half-hourly observations at a tower 30m high located close to the receptors ; the thickness of the dispersion layer was estimated to 200 to 300m.

The diffusion parameter σ_y was determined according to Bultynck-Malet[2] as a function of the stability. The diffusion coefficient K_z is deduced from in situ measurements and through a model of the ABL[3]. The vertical profile of K_z for each half hour is presented in fig. 2 which shows the existence of a maximum around 150m and fast decreasing towards very low values at the level of the inversion layer.

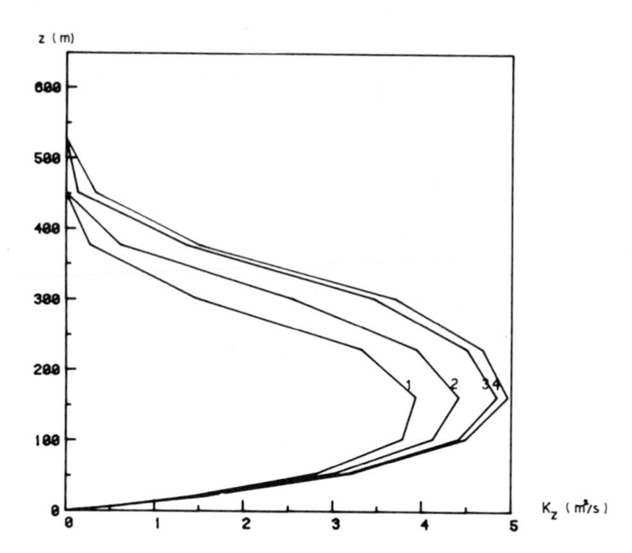

Fig. 2 Mean profile of the vertical diffusion coefficient K_z as a function of z for each half-hour of the experiment : 1(12h30-13h00), 2(13h00-13h30), 3(13h30-14h00), 4(14h00-14h30).

The values used for all the different parameters of the two models are the following

- semi-numerical models :
  . the exponent of the power law for the wind profile : 0, 33 (neutral stability)
  . the wind speed at 30m high is given in table 1 for every half-hour
  . the thickness of the diffusion layer is z_i=400m
  . the horizontal step along x is Δx=50m
  . the vertical step along z corresponds to 1/50 of the length of ABL

- analytical model : as the effective height h_e of the release is always lower than the level of the maximum of K_z, the power law for K_z has been adjusted with the rising part of the profile given by the ABL model (table 2); the power law for the wind profile is the same as for the semi-numerical model.

Table 1. Meteorological data measured at the Beerse and Mol (Belgium) towers: mean wind speed, direction, temperature

Averaging period	Beerse: z_a=30m ff (m/s)	dd (°)	Temperature at Mol (°C) z_a=24m	z_a=78m	z_a=114m	Stability
12h30-13h00	4,1	199	9,6	8,5	8,5	Neutral
13h00-13h30	3,6	190	9,9	8,9	8,8	Neutral
13h30-14h00	4,8	178	9,9	9,0	8,8	Neutral
14h00-14h30	4,9	179	10,0	9,4	9,3	Neutral

Table 2. Definition of the parameters caracterising the vertical diffusion K_z of the analytical model. Angle ϕ for the fluctuations in the wind direction.

Averaging period	K_1	n	ϕ
12h30 - 13h00	1,11	0,228	10
13h00 - 13h30	1,62	0,180	12,5
13h30 - 14h00	1,12	0,257	2,5
14h00 - 14h30	1,12	0,260	2,5

The registration of the wind direction in Beerse shows that during one half-hour, the direction was varying with independant long and short periods. As the high frequency fluctuations have a zero mean over half an hour and are already include through σ_y, only the slower variations must be taken into account : this effect is simulated by introducing the mean of the concentration calculated over an angular sector 2ϕ, where ϕ is given in table 2.

Table 3 shows the concentrations in SF_6 ($\mu g/m^3$) measured and computed at the principal receptors; there are only results for the 13h-14h30 period because the transient period 12h30-13h can not be treated by a stationnary model. The measured ground concentrations (M) have been given by the Nuclear Center of Mol[5] on the basis of measurements made by CEN/SCK(C) and Center of Ispra (I).

Table 3. SF_6 concentration measured (μg/m^3) and computed during the period 13h00-14h30, 4/12/79 in Beerse.

RECEP.	TIME											
	13h00-13h30				13h30-14h00				14h00-14h30			
	M		SNM	AM	M		SNM	AM	M		SNM	AM
	C	I			C	I			C	I		
B	0.08	0.12	0.001	0.006	0	0	0	0	-	0	0	0
C	0	0	0	0	0	0	0	0	-	0	0	0
D_0	0	0	0	0	0	0	0	0	-	0	0	0
D_1	6.68	5.70	5.71	8.21	0	0	0	0.01	-	0	0.01	0.01
D_2	0.20	0.15	0.22	0.34	0	0	0		-	0	0	0
D_5	-	-	2.49	3.68	0.3	-	0	0	-	-	0	0
D_6	-	-	7.44	10.25	0.16	-	0.14	0.25	0	-	0.23	0.41
E_0	0	0	0	0	0	-	0	0	-	0	0	0
E_1	-	5.58	3.44	3.76	-	0.21	0	0	-	0.08	0	0
E_2	-	0	0.002	0.008	-	0	0	0	-	0	0	0

M : concentration given by the CEN/SCK from the measurement by CEN (C) and Ispra (I)
SNM : results of the semi-numerical model
AM : results of the analytical model

For the period considered, the results of both the semi-numerical and analytical models are in agreement with the experimental results, except for point B near the source. The maximum concentration is located between points D_1 and E_1 alike the measurements. The differences between the analytical and semi-numerical models are very small with a tendancy of the analytical model to give values a little bit higher.

QUANTITATIVE ESTIMATION OF THE FUGITIVE EMISSIONS

The manufactory of Sb treatment in Beerse is caracterised, from the emission point a view, by the existence of a number of fugitive sources having a non-negligible effect on the atmospheric pollution level. Indeed, if only the emission of the main source (stack of 75m) is considered, the concentration computed by the models at different receptors are significantly lower than the measured concentrations. Therefore, it has been necessary to estimate the flow rate of the fugitive sources. Since it is not easy to measure accurately fugitive emissions, these flow rates have been computed by diffusion models using the concentrations measured at the receptors close to the manufactory. This implies an inverse application of the computation of an effluent concentration nearby one or several sources.

The research of a single unknown flow rate $\hat{Q}$ is rather easy if one notes that the relation between the emission and the immission is linear in Q. So, at a given receptor, the ratio

$$\chi_{ir}^{mes} / \chi_{ir}^{cal}(1) \tag{5}$$

of the measured concentration to the computed concentration for a unit emission, gives the flow rate $\hat{Q}$ when all the parameters are kept identical. This argument is still true for nr receptors and leads to nr estimations Q_{ir} of the flow rate Q of the fugitive sources. In order to deduce $\hat{Q}$ from all this estimations, it is necessary to introduce a constraint over this set of Q_{ir}. In our case, $\hat{Q}$ has been chosen as the value that minimise the square of the difference between measured and computed values ; in other words, $\hat{Q}$ is given by the minimum of :

$$\sigma^2 = \frac{1}{nr} \sum_{ir=1}^{nr} \left[Q\, \chi_{ir}^{cal}(1) - \chi_{ir}^{mes} \right]^2 \tag{6}$$

If n_p fugitive sources have been detected, the above method may easily be generalised : the fugitive flow rates Q_{ip} are defined as the values that minimize :

$$\sigma^2 = \frac{1}{nr} \sum_{ir=1}^{nr} \left[\sum_{ip=1}^{np} Q_{ip}\, \chi_{ir}^{cal}(1) - \chi_{ir}^{mes} \right]^2 \tag{7}$$

with the constraint $Q_{ip} > 0$. The minimization(7) has an explicit solution.

In the particular case of one fugitive source, the value $\hat{Q}$ that minimize σ^2 defined by (6) is simply given by :

$$\hat{Q} = \frac{\sum_{ir=1}^{nr} \chi_{ir}^{mes}\, \chi_{ir}^{cal}(1)}{\sum_{ir=1}^{nr} \chi_{ir}^{cal}\, \chi_{ir}^{cal}(1)} \tag{8}$$

This method of estimation has been applied for the experiment conducted on June 17, 1980 in Beerse. One source has been selected (H21) during a period starting at 12h50 and ending at 16h ; SF_6 and Sb have been measured at 8 receptors in the North-East sector during 4 periods of half-an hour. Table 4 gives the coordinates of the source and of the receptors and figure 3 shows their location and the mean wind direction for each period T_m. For each period T_m, Table 5 provides the wind speed and direction, the temperature, the humidity, the stability as deduced from measurements made in Beerse and Mol.

Table 4 Lambert coordinates of the receptors and of the source (H21) with their height above the ground level at the manufactory site.

RECEPTOR	LAMBERT COORDINATE X(m)	Y(m)	HEIGHT (m)
A2	180 431	224 232	3,0
R1	180 454	224 255	3,0
S6	180 469	224 248	3,0
S7	180 458	224 338	5,5
R3	180 475	224 333	5,5
S8	180 487	224 332	5,5
R9	180 499	224 330	5,5
R4	180 514	224 327	3,0
SOURCE	180 391	224 171	18,0

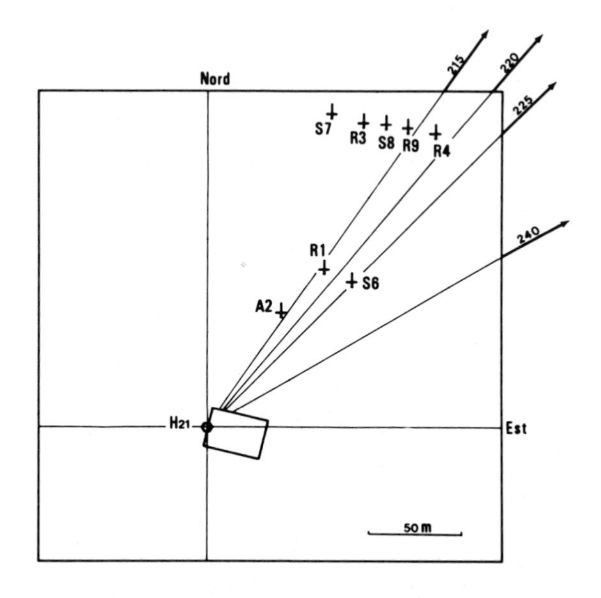

Fig. 3 Receptor localisation and the mean wind direction during the four half-hour periods June 17, 1980 in Beerse.

Table 5 Meteorological data measured at an altitude z_a on June 17, 1980 at the Beerse and Mol towers. The mean vertical turbulent diffusivity is deduced from a parameterization of the ABL.

Time		Beerse: z_a=30m		Temperature at Mol		Stability	$K_z=K_1 z$	
T_m		ff (m/s)	dd (°)	z_a=8m	z_a=24m	(*)	K_1	
I	13h15-13h45	4,5	215	18,3	17,5	4	1,09	0,
II	14h00-14h30	5,5	220	18,8	17,9	4	1,06	0,
III	14h45-15h15	7,5	225	19,0	18,2	4	1,05	0,
IV	15h30-16h00	3,5	240	17,7	17,4	3	0,07	1

(*) Bultynck-Malet

The model used for the estimation of the concentration near the source is based on an analytical solution of the diffusion equation when the following assumptions are made : stationnarity, flat terra power law for wind and K_z profiles.

This solution is given by :

$$\begin{cases} \chi(x,y,z) = Q\, Y(x,y)\, Z(x,z)\,. \\ Y(x,y) = \dfrac{1}{\sqrt{2\pi}\,\sigma_y(x)} \exp\left[-\dfrac{y^2}{2\sigma_y^2(x)}\right] \\ Z(x,z) = \dfrac{h_e}{2qK_z(h_e)x} \left(\dfrac{z}{h_e}\right)^p \exp\left[-\left[1+\left(\dfrac{z}{h_e}\right)^{2q}\right] \dfrac{\bar{u}(h_e)h_e^2}{4q^2K_z(h_e)x}\right] \\ \qquad I_{\mu-1}\left[2\left(\dfrac{z}{h_e}\right)^q \dfrac{\bar{u}(h_e)h_e^2}{4q^2K_z(h_e)}\right] \end{cases} \tag{9}$$

where :

x,y,z	:	coordinates of the measuring point	[m]
χ	:	mean concentration during T_m	[kg.m^{-3}]
Q	:	flow rate at the source	[kg.s^{-1}]
Y	:	part of the concentration due to crosswind diffusion	[m^{-1}]
Z	:	part of the concentration due to vertical diffusion	[s.m^{-2}]
h_e	:	effective height of the source	[m]
$\bar{u}(z)$	:	mean profile of the wind $\bar{u}(z) = \bar{u}_1 z^m$	[m.s^{-1}]

$K_z(z)$: mean profile of vertical diffusion $[m^2\ s^{-1}]$
$K_z(z) = K_1\ z^n$

$\sigma_y(x)$: standard deviation of crosswind distribution $[m]$

with the parameters

$2q = m-n+2$
$2p = 1-n$
$\mu = (1+m)/2q$
I_ν : modified Bessel function

The parameters m and $\sigma_y(x)$ are defined in function of the boundary layer stability[2]. K_z is computed on the basis of the data of table 1.

The results of the computation for SF_6 for one single source (H21) and with the hypothesis of no background are presented in table 6. Except for the last period, the estimated flow rates are in good agreement with the experimental conditions. During the last period, it seems that the buildings have a non negligible influence on the wind field (this fact will be carefully analyze in the next section). If the receptor A2 (the nearest to the source) is not taken in the minimization procedure to avoid the buildings effects the agreement is still better. For each period T_m, table 7 presents these new estimations with the corresponding values of the ratio $(\sigma/\sigma_t)^2_{min}$.

For the SF_6, the K-analytical model gives flow rate estimates of the order of 6,7 kg/h, 10,3 kg/h and 8,8 kg/h when the measured flow rates where 9,1 kg/h, 9,6 kg/h and 8,3 kg/h. On the basis of such results, it seems that the proposed methodology can be successfully applied to other periods and other pollutants.

Simultaneously to the controlled SF_6 emission, there were uncontrolled Sb emissions and Sb measurement at receptors A2, R1, R3, R9 and R4. The application of the above method leads to the following results : 2,7 kg/h, 5,6 kg/h and 2 kg/h ; the computation being done without the receptor A2.

The use of a stationnary bi-gaussian model, instead of K-analytical one, has also be tested 6,7 and definitely gives less accurate results for the SF_6 experiment.

Table 6 Estimated SF_6 flow rate (kg/h) and concentrations (µg/m^3) during 4 periods with the criteria of minimization applied to 7 or 8 receptors.

T_m	13h15-13h45		14h00-14h30		14h45-15h15		15h30-16h00	
	OBS.	EST.	OBS.	EST.	OBS.	EST.	OBS.	EST.
Flow rate	9,1	7,1	9,6	12	8,3	9,6	8,7	5,3
CONC.								
A2	203	160	311	183	105	71	55	5
R1	181	178	208	229	92	102	58	9
S6	-		-		-		65	78
S7	48	85	68	63	7	15	1	0
R3	107	130	113	121	29	38	10	0
S8	158	148	113	156	51	57	24	0
R9	168	155	144	185	80	77	38	1
R4	109	143	181	199	87	95	59	5
$(\sigma/\sigma_t)^2_{min}$		0,30		0,54		0,19		2,35

Table 7 Estimated SF_6 flow rates (kg/h) and concentrations (µg/m^3) during 4 periods with the criteria of minimization applied with 6 or 7 receptors and without receptor A2.

T_m	13h15-13h45		14h00-14h30		14h45-15h15		15h30-16h00	
	OBS.	EST.	OBS.	EST.	OBS.	EST.	OBS.	EST.
Flow rate	9,1	6,7	9,6	10,3	8,3	8,8	8,7	5,06
CONC.								
A2								
R1	181	168	208	198	92	94	58	10
S6	-		-		-		65	75
S7	48	81	68	54	7	14	1	0
R3	107	123	113	104	29	35	10	0
S8	158	139	113	135	51	52	24	0
R9	168	146	144	160	80	71	38	1
R4	109	135	181	171	87	88	59	5
$(\sigma/\sigma_t)^2_{min}$		0,24		0,09		0,03		1,88

INFLUENCE OF BUILDINGS ON THE DIFFUSION

In the previous sections the influence of the building has already been mentionned but was not taken into account, the hall where the source is situated modifying the wind field and generating a turbulence that the K-analytical or the bi-gaussian models can not take into account. This complex situation which is reflected both in the wind field and in the dispersion field will be simulated by a numerical model (two-dimensional in this study). A simulation is shown on figure 4 ; the selected configuration is the one of a wind blowing along a direction perpendicular to the obstacles ; these being the hall 21 and a little store situated 45m behind it.

For the wind field around the hall 21, the numerical model was based on the resolution of the stationnary equation of continuity and of the 2-D mean flow equation of motion with the Boussinesq approximation (this model is similar to the one used by Hecq et al., 1981). If the x-axis is oriented along the wind and the z-axis along the vertical, these equations are the following :

$$\begin{cases} \dfrac{\partial \bar{u}}{\partial x} + \dfrac{\partial \bar{w}}{\partial z} = 0 \\[2ex] \dfrac{\partial \bar{u}}{\partial t} + \dfrac{\partial}{\partial x}\,\bar{u}^2 + \dfrac{\partial}{\partial z}\,\overline{wu} = -\dfrac{\partial}{\partial r}\dfrac{\bar{p}}{\rho} + \nu\nabla^2\bar{u} + \dfrac{\partial}{\partial x}K_{xx}\dfrac{\partial \bar{u}}{\partial x} + \dfrac{\partial}{\partial z}K_{xz}\dfrac{\partial \bar{u}}{\partial z} \\[2ex] \dfrac{\partial \bar{w}}{\partial t} + \dfrac{\partial}{\partial x}\,\overline{uw} + \dfrac{\partial}{\partial z}\,\bar{w}^2 = -g - \dfrac{\partial}{\partial z}\dfrac{\bar{p}}{\rho} + \nu\nabla^2\bar{u} + \dfrac{\partial}{\partial x}K_{zx}\dfrac{\partial \bar{w}}{\partial x} + \dfrac{\partial}{\partial z}K_{zz}\dfrac{\partial \bar{w}}{\partial z} \end{cases} \qquad (10)$$

where $\bar{u}$, $\bar{w}$ are the mean components of the wind, $\bar{p}/\rho$ the mean pressure divided by the air density, g the gravity, K_{ij} the turbulent diffusion coefficients and ν the coefficient of molecular diffusion; in this simulation, K_{xz} and K_{zx} have been neglected and K_{xx} and K_{zz} are considered as constant with values respectively of 0,1 m^2/s and 2 m^2/s. The mean initial wind is defined by :

$$\begin{cases} \bar{u}(z) = 3{,}5\ (z/30)^{0,33} \\ \bar{w}(z) = 0 \quad ; \end{cases} \qquad (11)$$

The initial pressure $\bar{p}$ is given by the hydrostatic equation

$$\frac{\partial}{\partial z}\frac{\bar{p}}{\rho} = -\ g \qquad (12)$$

The boundary conditions are the classical one : zero wind speed along solid boundaries and zero wind speed gradients along open boundaries. The numerical solution of the system (13) by the method of the donnor cell gives the wind field represented in figure 4 .

The analysis of this figure shows :

- a large recirculation area between hall 21 and the store and little one behind the store ;
- a wind field influenced by obstacles over distances extending to 120m from the source ;
- only the concentration at the receptor A2 is influenced by the wind field perturbations due to obstacles.

The concentration field has also been computed by a numerical model. If Z(x,y,t) is the mean pollutant concentration (kg/m^2) in the plane (x,z), the conservation law may be expressed by :

$$\frac{\partial}{\partial t} Z + \bar{u} \frac{\partial Z}{\partial x} + \bar{w} \frac{\partial Z}{dz} = \frac{\partial}{\partial x} K_x \frac{\partial Z}{\partial x} + \frac{\partial}{\partial z} K_z \frac{\partial Z}{\partial z} \qquad (14)$$

where K_x and K_z are supposed to be identical to K_{xx} and K_{zz} in the previous model. The boundary conditions are :

- at the source : $Z(x_s, h_e, t) = \frac{Q}{u(x_s, h_e, t)}$ (15)

- on the ground, on the obstacles and at the top of the mixing layers (reflection) :

$$\begin{cases} K_z \frac{\partial Z}{\partial z} = 0 \\ K_x \frac{\partial Z}{\partial x} = 0 \end{cases} \qquad (16)$$

- at the left boundary (continuity of the flux) : $\frac{\partial Z}{\partial x} = 0$ (17)

The numerical method used to solve the problem is a two-dimension generalisation of the method by Crowley[5]. This leads to the results presented in fig.5 which shows the vertical profile of the concentration Z at four distances downwind of the source (x = 30, 45, 75 and 120 m) which emits 0,75 g/s of Sb at a height of 18,75 m.

For a useful comparison of the results with or without buildings, it is necessary to note two important properties of the concentration field without obstacles (fig. 5) :

- the maximum of concentration (if any) is always at the level of the source ;
- the downwind effect of the vertical mixing is rather slow : the bell shaped curve (+) near the source becomes flatter with the distance.

If we now look to the concentration field with obstacles it appears that :

- because of the recirculation area, the maximum concentration is above the source level ;
- turbulence enhances the homogenisation of the concentration field : the first three profiles are near uniform and induce high concentration at the ground.

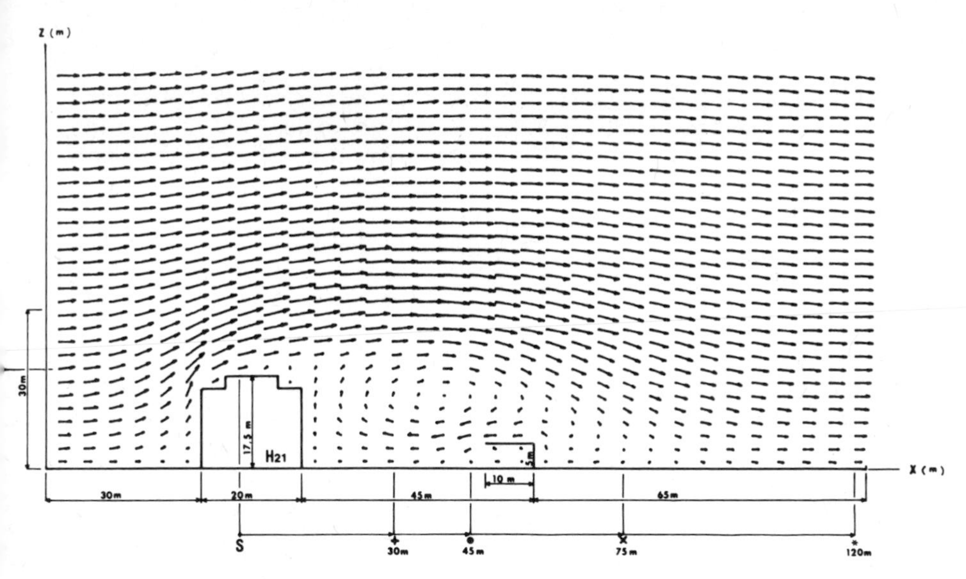

Fig. 4 Two-dimensional flow around the hall 21 and the store for a wind speed of 3,5 m/s at 30m ; the atmospheric layer being supposed to be slightly unstable with a $K_z = 2\ m^2/s$.

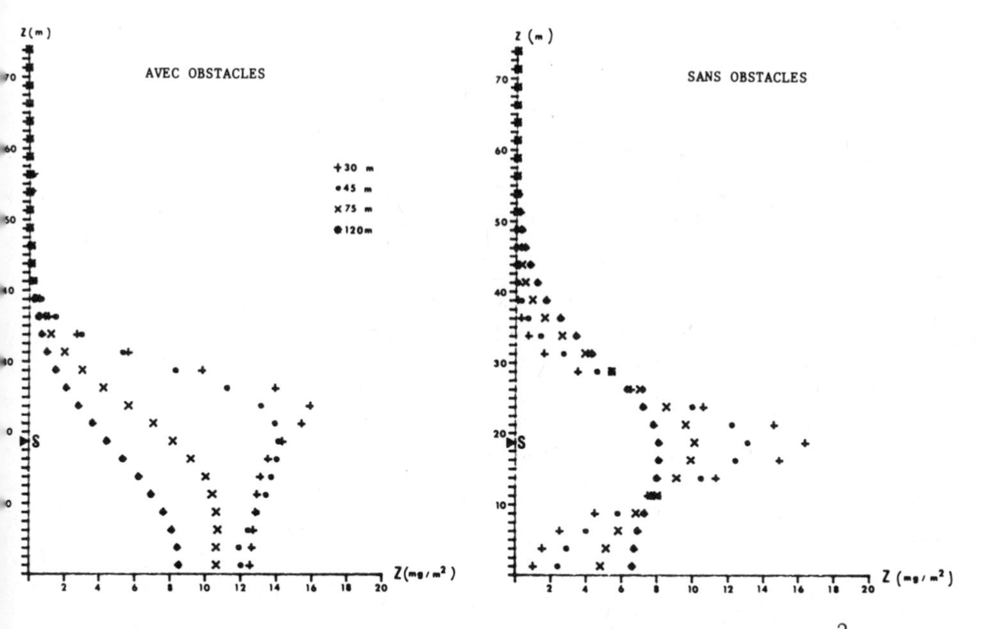

Fig. 5 Vertical distribution of the concentration Z (mg/m^2) at different distances downwind of the source; the Sb flow rate is 0,75 g/s and the source height 18,75m. The left figure shows the distribution in presence of buildings and the right one assumes no buildings at all.

CONCLUSION

For continuous release of pollutants during several hours, a stationary semi-numeric diffusion model associated to a simple atmospheric boundary layer model provide an adequate global model for pollution impact study at a local scale.

The application of the numerical model to estimate fugitive emissions is not absolutely necessary since the K-analytical model gives a good approximation for the unknown rate $\hat{Q}$.

The consideration of building effect has shown that any receptor within a distance of 120m downwind the source is under the influence of the turbulence generated by the building. This implies the need of a model based on numerical solution.

ACKNOWLEDGMENT

This research was financed by the Prime Minister Office for Science Policy within the framework of the National Program "R & D" Environment Air. We wish to acknowledge this organisation for its support during this project.

REFERENCES

1. Cl. Demuth, A contribution to the analytical steady solution of the diffusion equation for line sources, Atm. Env. 12 : 1255-1258 (1978).
2. H. Bultynck, L. Malet, Diffusion turbulente des effluents émis dans l'atmosphère par une source élevée à émission continue en relation avec la stabilité de l'air, S.C.K./C.E.N., BLG 434 (1969).
3. G. Schayes and M. Cravatte, Diffusivity profiles deduced from Synoptic data, in : 11th International Technical Meeting on Air Pollution Modelling and its Application, NATO/CCMS, Amsterdam (1980).
4. H. Bultynck, Meteorologsiche waarnemingen te Beerse en Mol op 4 december 1979 (SF_6 - proefneming), werkdokument, A.L./4.2 1980/2, National R-D Program on Environment Air, Services for Science Policy Programming, Brussels (1980).
5. J. Kretschmar, C. Ronneau, T. Rymen, W. Slegers and M. Termonia, Vergelijkende SF_6-metingen te Beerse (4/12/79) en te Dersel (7/12/79). Werkdokument, A.L./4.2-1980/1. National R-D Program on Environment Air, Services for Science Policy Programming, Brussels (1980).
6. Cl. Demuth and P. Hecq, Estimation du débit massique de certaines sources fugitives à Beerse le 4-5 décembre 1979. Progress Report 1980/5, Institute of Astronomy and Geophysics, Catholic University of LOuvain-la-Neuve, Research Unit AL/4.1 National R-D Program on Environment Air, Services

for Science Policy Programming, Brussels (1980).
7. Cl. Demuth and P. Hecq, Estimation du débit massique d'une source fugitive à Beerse le 17 juin 1980. Scientific Report 1980/8. Institute of Astronomy and Geophysics, Catholic University of Louvain-la-Neuve, Research Unit AL/4.1. National R-D Program on Environment Air, Services for Science Policy Programming, Brussels (1980).
8. Ph. Gaspar, Modèles mathématiques de dispersion des polluants dans la couche limite planétaire. Mémoire d'ingénieur civil en mathématiques appliquées (directeur A. Berger), Institut d'Astronomie et de Géophysique, Université Catholique de Louvain-la-Neuve (1980).
9. P. Hecq, Cl. Demuth and A. Berger, Physical and mathematical simulation of Pb dispersion in and around a Pb industry in a urbanized area. To be published in "12th NATO/CCMS ITM on Air Pollution Modeling and its Application", Palo Alto, August (1981).

DISCUSSION

J. W. S. YOUNG

Have you tried to estimate fugitive emissions directly by using the methods of Cowherd et al. and a simple area or volume source model?

P. HECQ

No.

H. N. LEE

Have you computed with different height of building? How are the results?

P. HECQ

No, we didn't since the case studied is the real one.

J. L. WOODWARD

How is particulate deposition treated? The exponential approach did not appear to be used (Atomic Energy Commission method).

P. HECQ

It is treated as a constant deposition velocity

M. L. WILLIAMS

What was the magnitude of the fugitive emission rates you calculated and how did they compare with the stack emission rates?

P. HECQ

The fugitive emission rates are of the order of 1 to 5 kb/h; they are five times greater than the stack emission rates.

A COMPARISON WITH EXPERIMENTAL DATA OF SEVERAL MODELS FOR DISPERSION OF HEAVY VAPOR CLOUDS

J.L. Woodward
Exxon Research and Engineering Co.
P.O. Box 153, Florham Park, NJ 07932

J.A. Havens
University of Arkansas
Fayetteville, Arkansas 72701

W.C. McBride
Energy Resources Co., Inc.
3344 N. Torrey Pines Ct.
LaJolla, Ca 92037

J.R. Taft
Deygon-Ra, Inc.
P.O. Box 3227, LaJolla, CA 92038

It has been recognized in recent years that models developed to describe the dispersion of neutrally buoyant plumes or clouds are inadequate to describe negatively buoyant heavy vapor clouds formed, for example, by accidental spills of volatile liquids. Subsequently, numerous models have been proposed to describe heavy vapor clouds, often with nonexistent or very limited comparison with experimental data, beyond those used to adjust or "calibrate" model parameters. We compare here several heavy vapor dispersion models on a nearly consistent basis against the same sets of experimental data.

MODELS AND DATA SETS CHOSEN FOR COMPARISON

The models selected for comparison represent two types, the K-theory (or eddy diffusivity) type and the "top hat" or uniform concentration cloud type. In the former class are the numerical models ZEPHYR and MARIAH. In the latter class are models by Germeles and Drake (1975) and Eidsvik (1980). The HEGADAS II model by Colenbrander 1980), also considered here, can be described as an "advanced top hat" model, which assumes a particular form of concentration profile, although it also utilizes an eddy diffusivity (K-theory) approach. The Germeles and Drake (GD) model typifies "first generation" models which incorporate air entrainment in the vertical via constant coefficients. The GD model requires that a transition be made to a neutrally buoyant Gaussian model for the far field solution. The Eidsvik model typifies "second generation" top hat models which incorporate horizontal (cloud edge) and vertical air entrainment via nonconstant coefficients which are dependent on the Richardson number. The Eidsvik model does not require transition to a Gaussian model. The ZEPHYR model of Energy Resources Co., and the MARIAH model of Deygon-Ra, Inc. are three-dimensional numerical solutions of the partial differential equations of mass, energy, and momentum transfer. Both ZEPHYR and MARIAH are similar to the SIGMET model documented by Havens (1979). Although these models solve the same set of equations, there are significant differences in the numerical solution methods. ZEPHYR uses a particle-in-cell technique coupled with and explicit finite difference approach. MARIAH uses an implicit finite difference method which allows larger step sizes thereby reducing computer costs.

The above models were compared with experimental data from the following programs : the British Health and Safety Executive (HSE) releases of 40 m^3 of freon/air mixtures at Porton Down, England (Picknett, 1978), specifically HSE Trials 6, 8, 20; the Esso/API spills of LNG onto water at Matagorda Bay, Texas (Feldbauer, et. al., 1972), specifically, Esso Trials 11, 16, 17; and the Dutch Ministry of Social Affairs release of freon (Buschmann, 1975; Van Ulden, 1974). After reviewing nearly all available data, we concluded that the above experiments provide the best data available to date for comparison with models. The conditions for the selected tests are summarized in Table 1. They include a mixture of instantaneous releases and finite-time releases; spills on land and over water; with low, medium and calm wind speeds.

The comparison made here shows that different models are applicable over different ranges of conditions. That is, models complement one another. For example, K-theory models are needed in cases where terrain effects and obstacles are important (as in decisions related to dike design). Yet, K-theory models may not be well suited for studying the dispersion of toxic substances which must be analyzed to very low concentrations often occuring at large distances. Numerical errors can become significant in such cases whereas top hat models are less susceptible to such errors.

APPROACH AND OBJECTIVES

Previous model comparison studies have concentrated on the ability of models to predict the effects of very large (25,000 m^3

Table 1

Experimental data utilized in the comparison study

Data Set	General Description	Spill Size (m^3)	Spilling Time(s)	Atmospheric Stability[a] Reported	Atmospheric Stability[a] Used for MARIAH	Wind Speed (m/s)	Amb. Temp. (°C)	Rel. Hum. %
HSE 6	Freon 12/Air (Specific gravity 1.7) onto land.	40 (Vapor)	"Instantaneous"	C-D	A	2.8[b]	8	87
HSE 8	Ditto, SG = 2.0	40 (Vapor)	"Instantaneous"	F-G	F	Calm	Unavail.	Unavail.
HSE 20	Ditto, SG = 2.1	40 (Vapor)	"Instantaneous"	C	A,C	5.3[b]	12	49
Esso 11	LNG onto water	10.2 (Liquid)	35	D	D,F	8[c]	27	78
Esso 16	LNG onto water	7.57 (Liquid)	28	B	D	Calm	18	62
Esso 17	LNG onto water	8.37 (Liquid)	31	D	D	4[c]	18	85

(a) Pasquill Stability Class
(b) At 2.0 m height
(c) At 5.5 m height
(d) At 10 m height

of liquid) spills over water (Bowman et al., 1979; Havens, 1978; Taft and McBride, 1979). For this purpose, the available data base is clearly inadequate (since the required extrapolation for spill sizes is over four orders of magnitude). Yet the available data may be adequate for the more limited purpose of testing the adequacy of models to predict the effects of relatively small spills on land or water. In these cases, minor differences in model predictions can be important, and it is to these differences that the model comparison is addressed.

The top hat and HEGADAS II models were programmed from the literature descriptions. No adjustments were made to the model parameters originally reported. Discussions were held with the authors to clarify some points of detail not included in the original papers. For example, the precise height to use with wind speed calculations is often unspecified (we used half of the cloud height for cloud advection), as is the treatment of humidity and water condensation. A complete description of each of the models, how they were programmed, and how input data were obtained is left to a later paper (Havens, 1981).

MARIAH and ZEPHYR model calculations were performed by the model originators (Deygon-Ra, Inc. and Energy Resources Co., Inc. respectively). An effort was made to use the same wind speed profiles and vapor generation rates for a given case with all models. However, Deygon-Ra made an independent assessment of the atmospheric stability for each experiment based on reported vertical temperature gradients and used different values listed in Table 1. To evaluate the significance of this revision, the sensitivity of Deygon-Ra's model to atmospheric stability was determined for two cases. This sensitivity proved to be small. Reference was made to the original experimental data. The British Health and Safety Executive supplied unpublished information on the exact location of sensors for the HSE tests.

MODEL COMPARISON RESULTS - CLOUD DIMENSIONS

Model and Data Limitations Restricted the Range of Comparisons Possible

One of the first conclusions to emerge is that each model is limited in its range of applicability. In addition, the selected data sets are incomplete and of limited range as summarized in Table 2. Table 2 indicates there are no gas sensor data available for Esso Trial 16 or the Van Ulden experiment (a serious limitation). As indicated by a G in Table 2, the MARIAH and ZEPHYR models are unable to predict responses of the very low (5 and 10 cm) sensors used in HSE Trials 6, 8 and 20 without the use of a finer, high-resolution grid (at higher cost/run).

Table 2

Applicability of Models to Chosen Data Sets

Wind Speed (see Table 1)	Calm 0	Calm 0	Low 2.8 m/s	Med. 5.3 m/s	High 8 m/s	Low 4 m/s	Low 3.0 m/s
Data set / Model	HSE 8	ESSO 16	HSE 6	HSE 20	ESSO 11	ESSO 17	Van Ulden
MARIAH	G	S	✸,G	✸,G	✸	✸	S
ZEPHYR	G	S	✸,G	✸,G	✸	✸	S
Eidsvik	✸	S	✸	✸	✸	✸	S
HEGADAS	N	N	Q	Q	✸	✸	S
Germeles & Drake (GD)	✸	S	N	N	N	N	S

✸ = Applies (at least for some sensors)
N = Not applicable, model limitations (a)
G = Grid limitations to matching sensor data
Q = Questionable to apply model for instantaneous release
S = No sensor data available

(a) The HEGADAS model relies on "observers" which float with the wind; therefore zero wind cases cannot be modeled. The heavy gas portion of the GD model is not invoked for the indicated cases because of the GD "transition criterion".

The HEGADAS II model (hereafter referred to as HEGADAS) assumes a quasi-steady representation of the cloud and appears less applicable for instantaneous releases such as the HSE trials. It is marked in Table 2 as questionable (Q) for such cases. The HEGADAS model does not apply for calm wind conditions, and is marked with an N.

The Germeles and Drake model is shown in Table 2 as not applicable for four of the seven experiments. This is because the model consists of two submodels, a "gravity spreading" or heavy gas submodel and a neutrally buoyant (Gaussian) submodel. The required transition from the former to the latter is made (according to the originally published version of the model) when the cloud edge speed falls below the wind speed. This is a highly restrictive requirement since in some cases the cloud speed is always below the wind speed. In other cases, the transition occurs very early in the experiment, and the heavy gas portion of the model, which is of primary interest here, has little influence on GD model predictions.

Comparisons with Cloud Dimensional Data of Limited Value in Stressin Models

Figure 1 compares several model predictions with observed cloud half widths as a function of time for HSE Trial 8. This is a near zero wind speed case, in which an initial column of gas spread with radial symmetry upon release. The MARIAH and ZEPHYR models fit observed data very well, but with different assumptions as to what concentration represents the visible edge of the cloud (shown in parentheses on Figure 1). Since the cloud was made visible with a smoke grenade, and the initial ratio of smoke particle concentration to freon gas concentration was not determined, the "correct" visible edge concentration is unresolvable. In any event, the K-theory model predictions are insensitive to the assumption of what percent freon corresponds to a visible amount of marker smoke (between 0.2 and 1 % concentrations). That is, the models predict a sharp concentration gradient at the edge of the cloud since concentrations between 0.2 and 1 % fall within the same grid cell (1,5 m wide cells).

The top hat models match cloud width primarily by adjusting the parameter α_1, in the "gravity intrusion formula" :

$$\frac{dR}{dt} = \alpha_1 \left[g \left(\frac{\rho - \rho_a}{\rho} \right) H \right]^{1/2} \qquad (1)$$

where R is cloud radius, H is cloud height, ρ and ρ_a are cloud density and air density, and g is the gravitational acceleration. Figure 1 shows that for HSE Trial 8, $\alpha_1 = 1.3$, (used by Eidsvik) produces a better fit with observed data than does $\alpha_1 = 1.4$ (used by Germeles and Drake). However, the GD model also invokes the

Boussinesq approximation in Equation 1 (ρ_a can be substituted for ρ in the denominator).

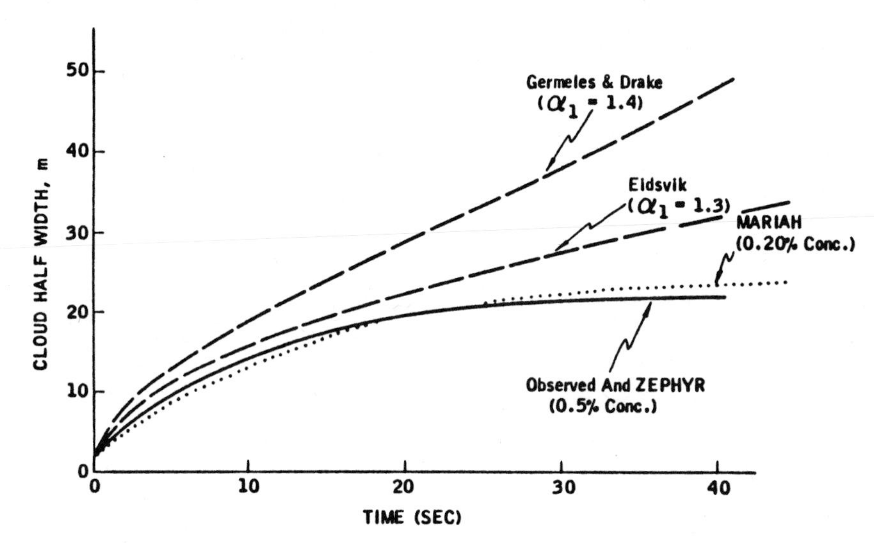

Figure 1 - Predicted and observed cloud radius for HSE trial 8

Unfortunately, a favorable comparison of model prediction against the cloud radius alone, as in Figure 1, is not an indication that the models will correctly predict concentrations in the cloud. This is because the relationship of cloud radius to time is insensitive to the amount of air entrained in the cloud. This was shown for isothermal clouds experimentally by Britter (1980) and theoretically by Picknett (1978), who rearranged Equation 1 in terms of the initial (constant) values of cloud height, radius and cloud density, H_o, R_o and ρ_o, to :

$$R \frac{dR}{dt} \cong \alpha_1 \left[g H_o \left(\frac{\rho_o - \rho_a}{\rho_a} \right) \right]^{1/2} \qquad R_o = \text{constant} \qquad (2)$$

or $R^2 \alpha\, t$

Thus, models in general and top-hat models in particular can predict cloud radius vs. time without reference to air entrainment terms (Air entrainment is directly related to cloud height in top-hat models). Comparison with sensor data is much more important than comparison with cloud dimensional data.

Model Predictions Very Sensitive to Wind Profile Assumptions

Figures 2 and 3 illustrate that, in general, model predictions are very sensitive to wind profile assumptions. Wind speeds, measured at four heights in HSE Trial 6 produced a range of values

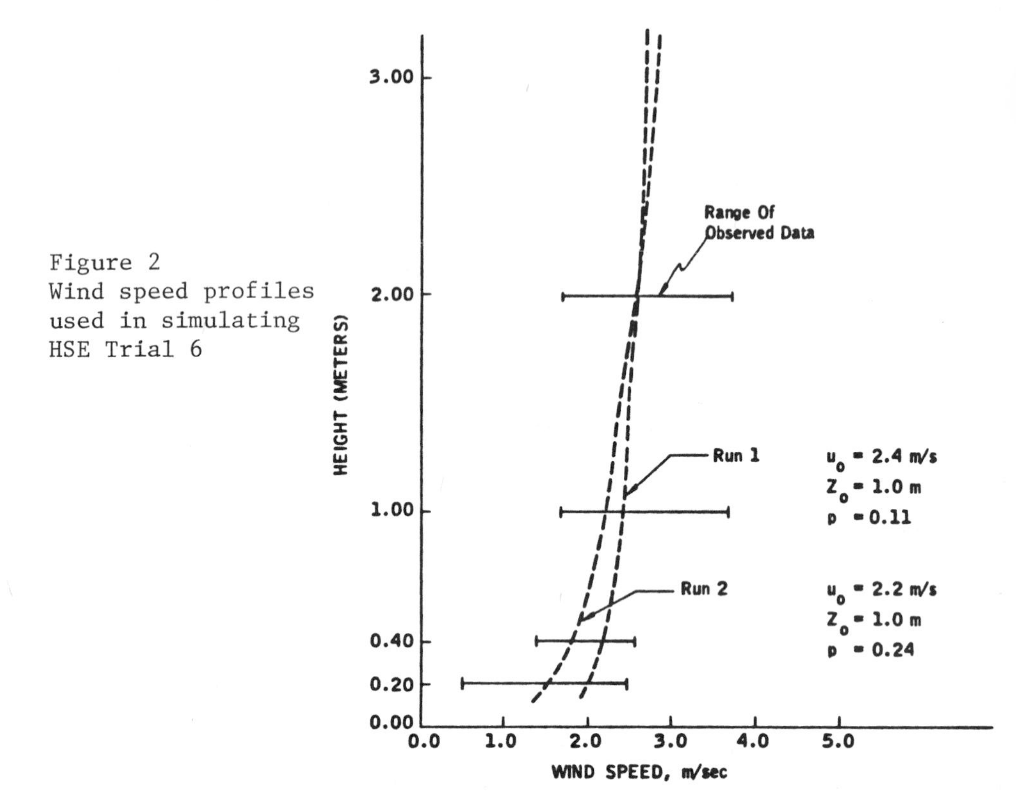

Figure 2
Wind speed profiles used in simulating HSE Trial 6

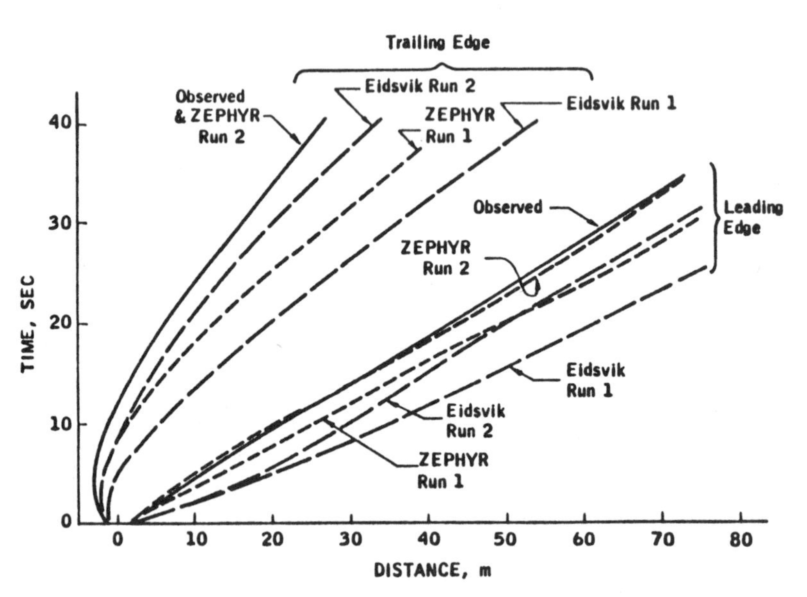

Figure 3 - Leading and training edge comparison for HSE Trial 6

as shown by the horizontal bars in Figure 2. Any number of "correct" wind speed profiles having the formula :

$$\frac{u}{u_o} = \left(\frac{z}{z_o}\right)^p \qquad (3)$$

(or a logarithmic formula) can be drawn through the data in Figure 2. Two such profiles are shown, both with the same reference height, $z_o = 1$ m, but differing in u_o and p. These profiles, labeled "Run 1" and "Run 2" were used with the ZEPHYR and Eidsvik models to produce the model predictions in Figure 3 labeled correspondingly. Model predictions change appreciably with the two wind profiles. Model treatment of wind profile is thus, very important.

Figure 3 shows that the Run 2 wind profile, which corresponds to a greater surface roughness, provides the better fit with data for both models. The ZEPHYR model using the assumed cloud edge concentration of 0.5 % matches data better than the Eidsvik model for both wind profile assumptions. In each case, model predictions tend to lead the data, for both the leading edge and the trailing edge. This would be expected for trailing edge data because heavy gas tends to move very slowly next to the ground where wind speeds approach zero. Models generally use a wind speed vertically averaged over some distance above the ground,thus erring toward high wind speed.

Model predictions for the leading edge are ahead of the observed data because cloud inertia and acceleration to wind speeds are crudely treated. The MARIAH and ZEPHYR models allow for inertia, both of the tent contents and of the air in the wake of the tent, and are capable of a perfect match with data. Top hat models assume instant acceleration to wind speed. This assumption could be improved by using acceleration due to the drag force on a cylinder, since drag coefficients could be estimated.

Noninstantaneous - LNG Spills over Water

Esso Run 16 was a noninstantaneous spill of 7.57 m^3 of LNG over water in which the wind died just as the test began. Consequently, the vapour cloud from the spill did not reach the line of sensors. However, an overhead videotape record of cloud dimensions has been analyzed recently by the Jet Propulsion Laboratories (JPL) using digital photographic processing techniques originally developed for space flight pictures. The resulting data are shown in Figure 4 as the shaded area for cloud half-width vs. time. With an irregular-shaped cloud, at least four different cloud diameters could be defined : along the axis of the release barge (D_1), perpendicular to this axis (D_2), and the two axes bisecting the first two (D_3 and D_4). Of these, $0.5D_2$ and $0.5D_4$ are plotted and 0.5 D_1 falls between those plotted.

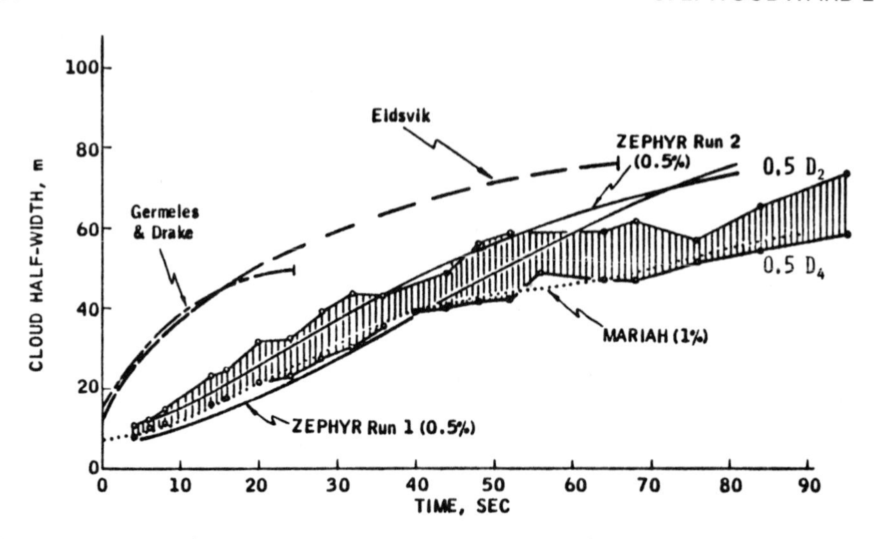

Figure 4 - Predicted and observed cloud radius for Esso Run 16

The Eidsvik and GD models significantly overpredict cloud diameter vs. time. The heavy gas portion of the GD model applies only for the first 25 seconds of the response.

The MARIAH and ZEPHYR models, which assume somewhat different cloud edge concentrations, fit the observed data very well. Two ZEPHYR runs were made by varying the assumed vaporization rate (assuming 0 and 75 % of the LNG vaporized in the air. The resulting source rate curves from Esso Run 16 are very nearly identical to that shown in Figure 10 for Esso Run 11. These predictions were before receiving the revised data from JPL.

Altogether, though, matching observed cloud dimensions alone is insufficient evidence for accepting model accuracy. Matching cloud composition data is more important, and this is discussed next.

MODEL COMPARISON RESULTS - SENSOR RESPONSES

Sensor Height Is Critical With HSE Trials

Very few sensor response were obtained during the HSE trials, and many of those available are from sensors placed very close to the ground (5 - 10 cm). K-theory models have difficulty matching such low sensors without resorting to a very fine grid which increases computing costs prohibitively. On the other hand, top-hat models fail to predict any response at all for several of the higher sensors (1 m and 2 m high), which K-theory models are able to match well.

These points are illustrated in Figures 5 and 6 for HSE Trial 6. Figure 5 shows that the Eidsvik and HEGADAS models strongly overpredict low-lying sensor responses in the early part of the response, but do better in the tail of the response. Part of the reason for the Eidsvik model overprediction is because it predicts an early arrival of the leading edge of the cloud (see Figure 3). Since concentration drops rapidly with time, especially close to the release point, an early response will tend to overpredict. In addition, the HEGADAS Model is not designed for an instantaneous release. Since it accounts only for wind entrainment and ignores entrainment caused by the spreading cloud, it would be expected to overpredict concentrations close to the source for an instantaneous release.

The MARIAH and ZEPHY models also overpredict the responses shown in Figure 5, in fact even more so than is portrayed in the figure. That is, the MARIAH and ZEPHYR predictions are for a height of 28 cm. Since concentration decreases vertically, a prediction for a 10 cm height would give even higher concentrations.

The ZEPHYR model predicts a bimodal response, since the heavy gas tends to concentrate in the edges of the cloud (forming a toroidal shape in calm winds). A bimodal response was observed only for Sensor 3, but photographic records show a generally toroidal-shaped cloud.

The K-theory model predictions at heigths of 1 and 2 m (Figure 6) agree very well with observations. Runs 1 and 2 by the ZEPHYR model refer to the alternate wind speed profiles in Figure 2. Clearly, concentration predictions are very sensitive to wind speed profile assumptions, just as cloud dimensions are.

Similar conclusion apply for HSE Trial 20 as illustrated in Figure 7. Again the K-theory models match observations very well. The Eidsvik model predicts a cloud height below one meter (thus no response for 1 and 2 m high sensors).

Figure 8 illustrates that the Eidsvik model matches observations for HSE Trial 8 better for far sensors than for near. In fact, for a sensor at 37.5 m from the source, the agreement with data is excellent. This seems to be a basic property of top hat models. If entrainment velocity parameters were to be adjusted to provide a better fit with data at the near sensors, then the fit at the far sensors may suffer.

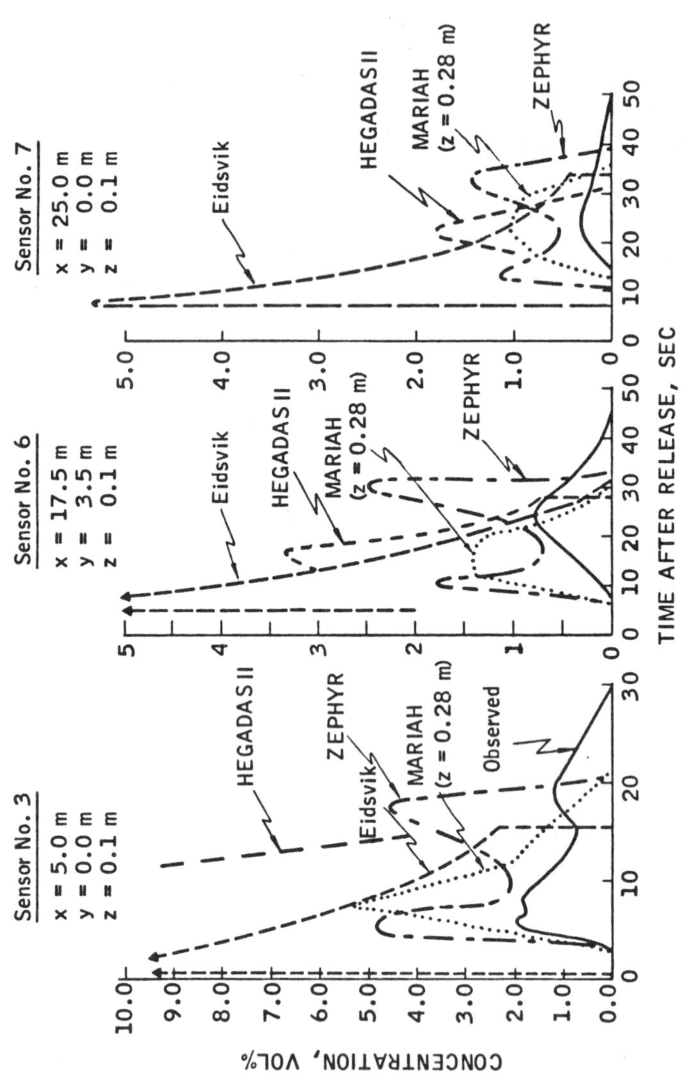

Figure 5 - Comparison of concentration responses for HSE trial 6 (Low sensors)

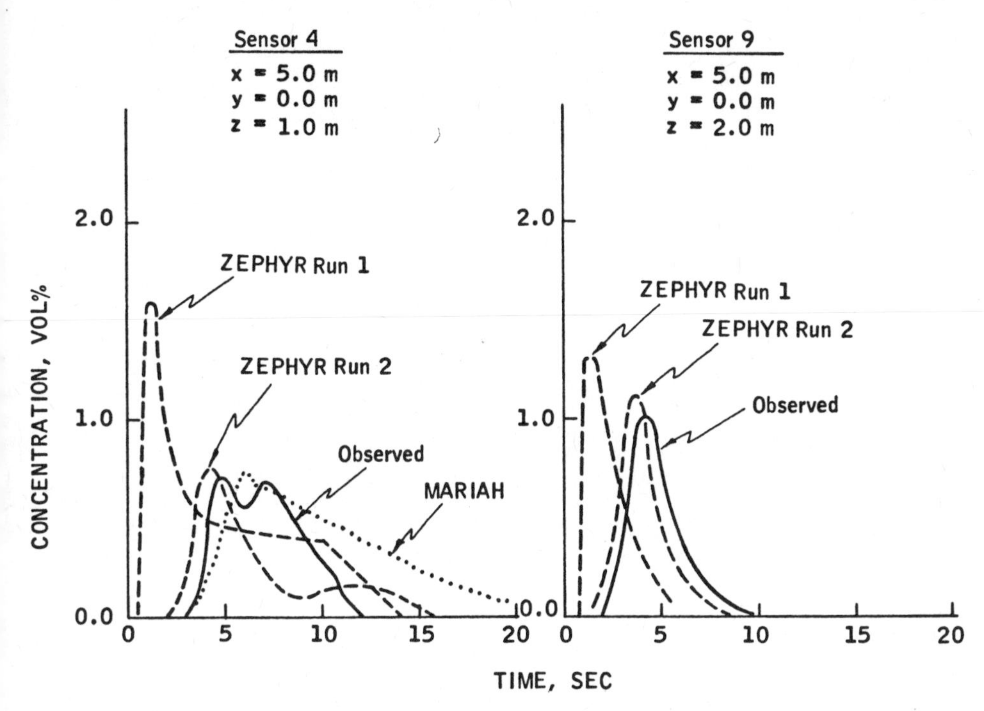

Figure 6 - Comparison of concentration responses for HSE trial 6

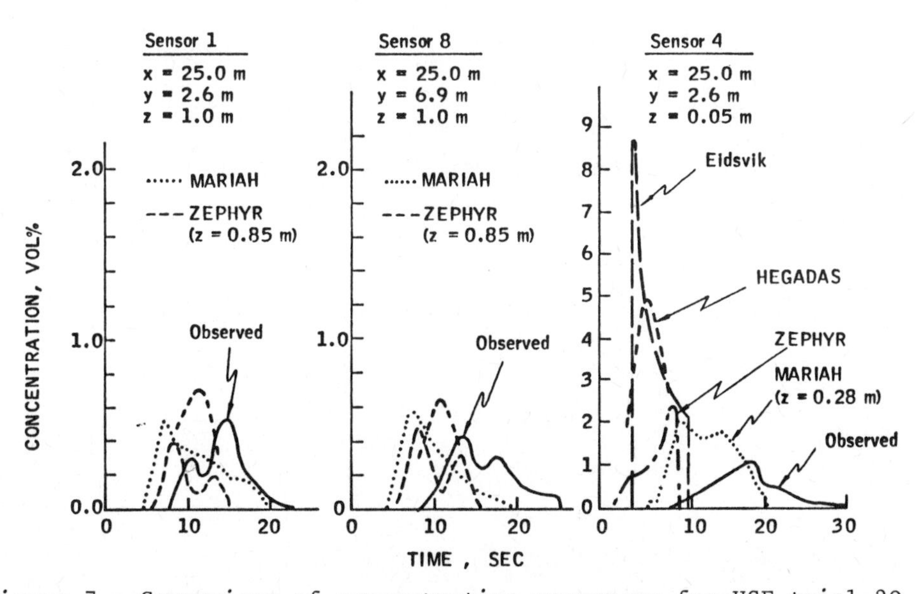

Figure 7 - Comparison of concentration responses for HSE trial 20

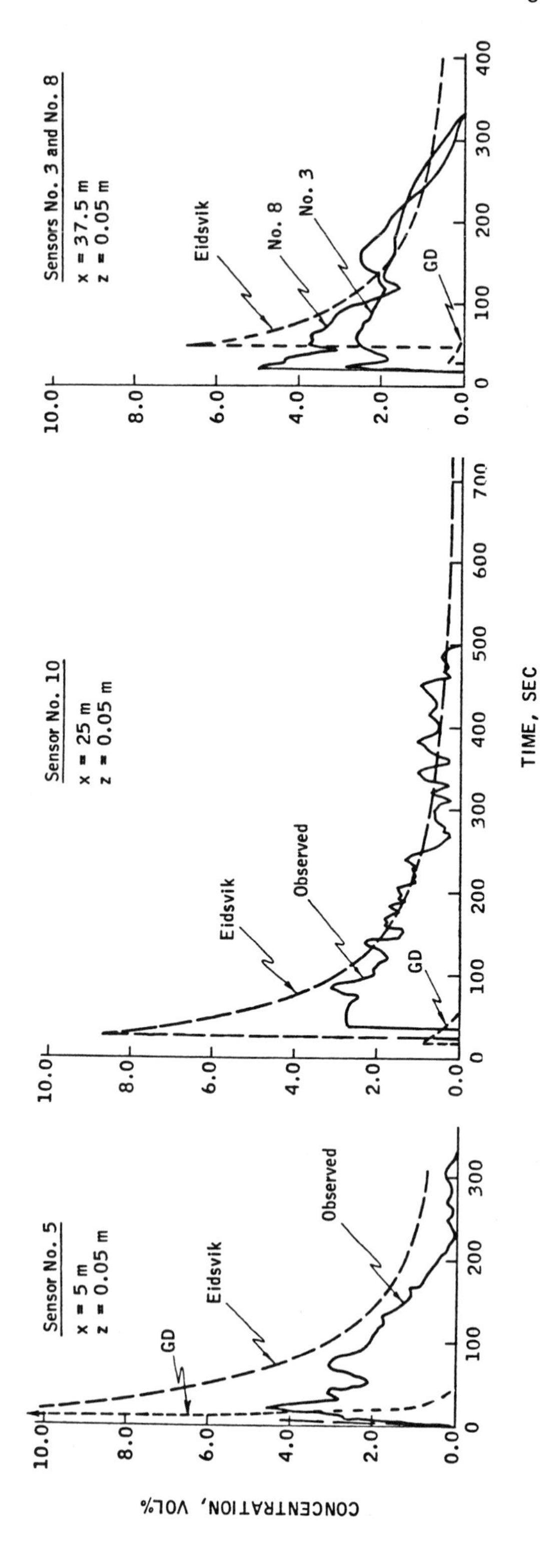

Figure 8 - Comparison of concentration responses for HSE trial 8 with Eidsvik model

The GD model predictions in Figure 8 are seriously in error, predicting much faster air entrainment and cloud height than the data indicated. (The GD model also overpredicts cloud radius as shown in Figure 1). Figure 9 contrasts the GD model prediction of cloud height with the very low (∿ 50 cm) prediction of the Eidsvik model. The observed cloud heights fall between the Eidsvik and GD predictions. Thus, the Eidsvik model overpredicts cloud radius (Figure 1) but underpredicts cloud height (Figure 9), and these compensating errors produce a good match for composition (Figure 8). Needless to say, compensating error is not a sound basis for model verification in general. In particular, it is not a good basis for extrapolation to other situations, since the Eidsvik model was calibrated against the HSE trials and would be expected to match these well.

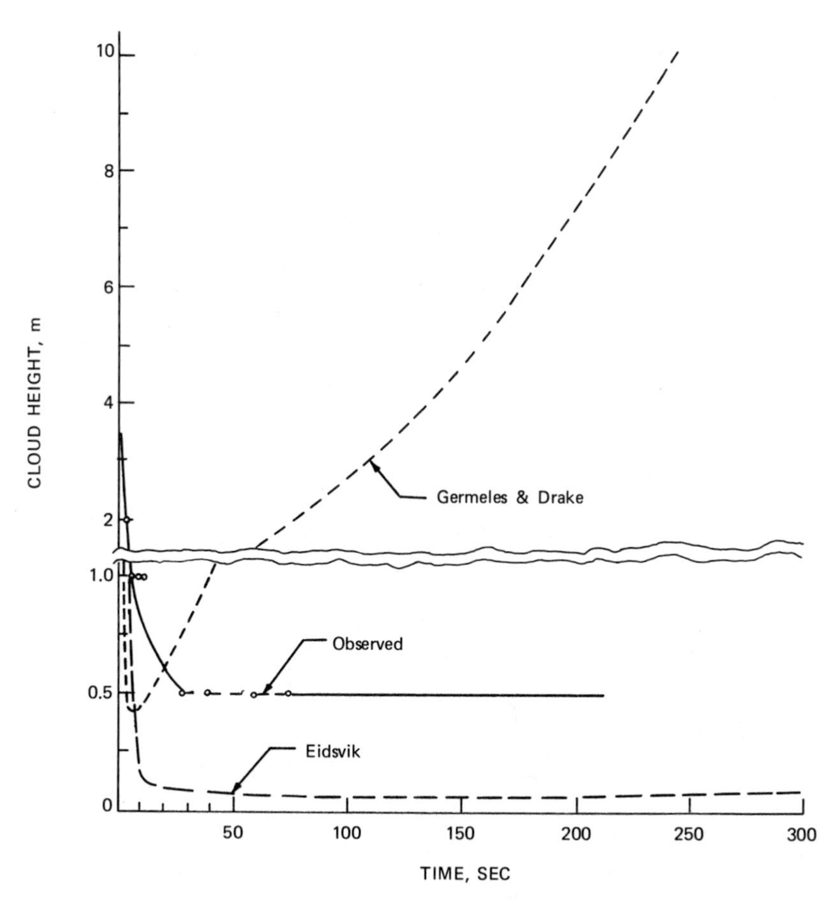

Figure 9 - Comparison of cloud height responses for HSE trial 8

Models Differ for Esso LNG Spills

In the Esso/API Matagorda Bay tests, LNG was forced out of a nozzle and traveled in an arched path through the air before reaching the water surface. The liquid pool diameter was visible in the stiff wind (and measured to be 29.3 m) and this fact has been used to establish the LNG vaporization rate over water. This vaporization rate is a point of controversy, though, because of uncertainty as to how much LNG vaporized in the air before hitting the water. Estimates range from 17 % to 75 % vaporized (May, 1980; Colenbrander, 1980).

The effect of these various assumptions on the vapor source rate for Esso Run 11 is illustrated in Figure 10. If no LNG is assumed to vaporize in the air, the peak-shaped curve results, and the evaporation flux is 0.195 kg/m^{2}s. If 75 % of the LNG evaporates in the air, the evaporation flux is 25 % of 0.195 or 0.049 kg/m^{2}s. Therefore, the source rate is high during LNG discharge (for 35 seconds) followed by an abrupt dropoff to a lower value. Fortunately, both source rates have nearly the same shape and peak values, so model predictions are insensitive to the assumptions affecting vaporization rate. This was shown for Esso Run 16 in Figure 4 (ZEPHYR Run 1 assumes no vaporization in the air, Run 2 assumes 75 % vaporization).

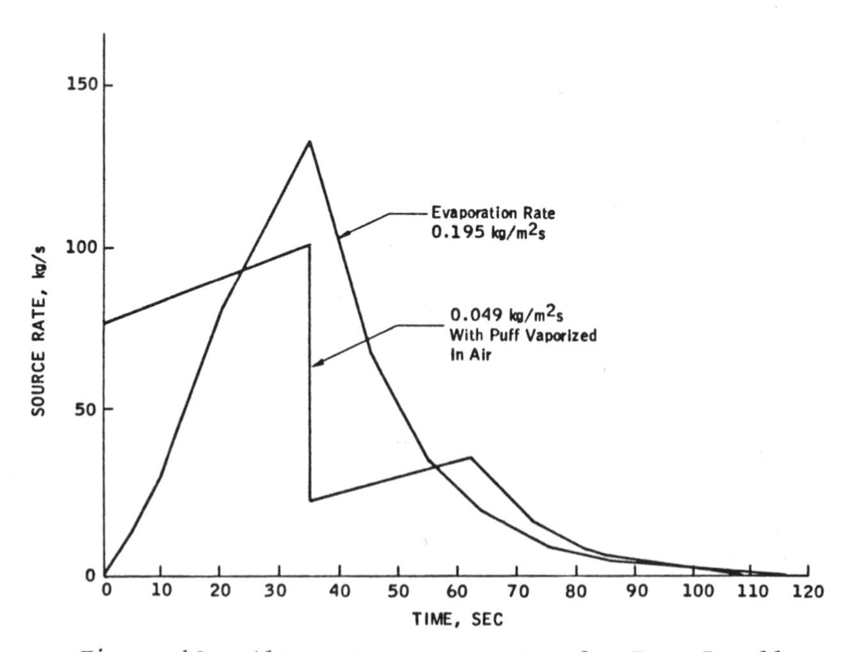

Figure 10 - Alternate source rates for Esso Run 11

Figures 11 and 12 compare predictions with observed data for Esso Runs 11 and 17 and Sensors 1A and 9. In both runs, these sensors gave the highest peak values and, although at the end of the line of sensors, are assumed to be in the cloud centerline. The data for Esso Run 17 are somewhat suspect since very little dilution seems to occur as the cloud travels between the two lines of sensors. In addition, wind speed changes may have occured which extended the response duration. In light of such uncertainties, all of the models compared in Figure 12 can be considered to be adequate. For Esso Run 11 (Figure 11) the Eidsvik moel, which applies the assumption of instantaneous release to a distinctly finite release time situation, is unsatisfactory. It predicts far too narrow a cloud (and short duration response) and overpredicts peak concentrations. For Figure 11, the MARIAH and HEGADAS predictions of peak concentration are within a factor of two of observed. ZEPHYR and EIDSVIK overpredict by more than a factor of two which is unsatisfactory, though conservative.

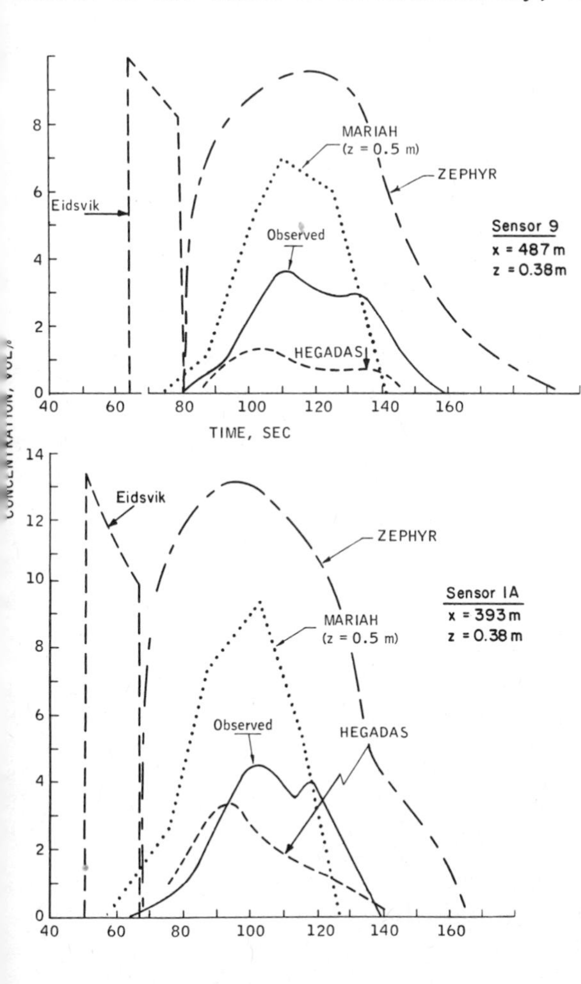

Figure 11
Comparison of composition responses for Esso Run 11

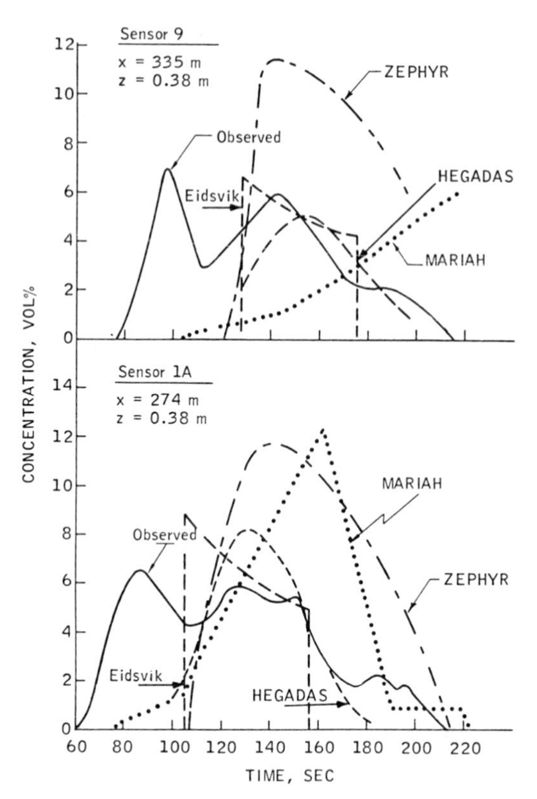

Figure 12 - Comparison of composition responses for Esso Run 17

Plan View LFL Contours are Useful

Figures 13 and 14 show model predictions for Esso Runs 11 and 17 plotted as lower flammable limit (LFL) contours at several times. Such a plot reveals the true power of the HEGADAS and K-theory models. These produce realistic-looking contours as opposed to top-hat models where the LFL contour is a single circle (the Eidsvik model prediction in Figures 13 and 14). HEGADAS, MARIAH and ZEPHYR (which was omitted in Figure 13 for clarity) predict long-narrow LFL contours in a high wind (Figure 13) and wider, rounder contours which are advected less distance in a low wind (Figure 14). ZEPHYR produces the most conservative predictions, consistent with Figures 11 and 12. Unfortunately, the comparison is exaggerated, since the contours for ZEPHYR and MARIAH are plotted for the 38 cm sensor height, whereas for HEGADAS and Eidsvik the contours are for 50 cm high.

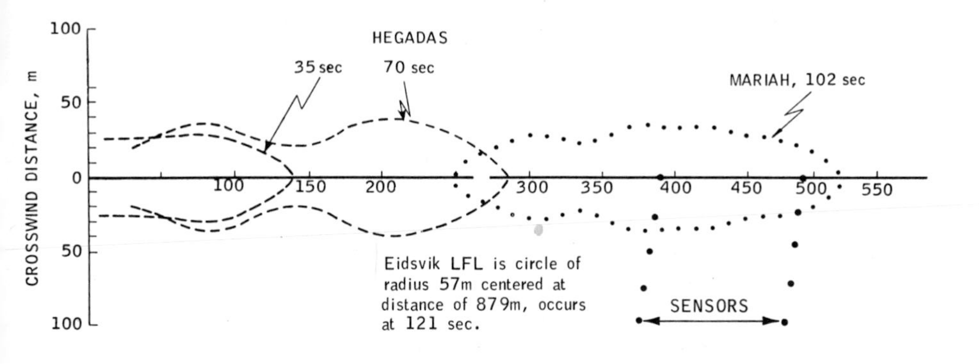

Figure 13 - Plan view LFL contours for Esso Run 11

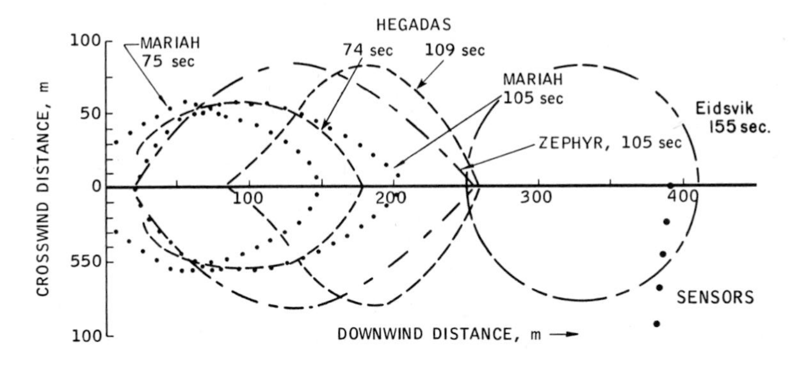

Figure 14 - Plan view LFL contours, for Esso Run 17

Van Ulden Experiment

Space limitations preclude a complete description of the model comparisons made in this study except to summarize some important observations. First the cloud height relationship reported for the Van Ulden experiment (Figure 4 in Van Ulden, 1974) may very possibly be anomolousand misleading. In any event, the experiment is of little value in assessing model performance. A vapor cloud of freon-12 was generated by rapidly boiling liquid freon over a bath of warm water. The resulting cloud was rendered visible for approximately 40 seconds by suspended water droplets which condensed in the cold (-29.8 °C boiling point) freon gas. No concentration sensor data are available, only rough measurements of initial cloud dimensions. These measurements and calculations say the initial cloud would be 5.4 m high with a 12 m radius, giving a cloud specific gravity of 1.25 (which requires cloud temperatures very near ambient, 20 °C, and dilution by air by a factor of 14.5). The actual cloud density could be greater than this if part of the freon formed aerosol droplets, if water droplets were present in significant mass, or if the actual cloud temperature were well below ambient temperature (and they were at least below the dew point). It is likely that rainout could occur in a cold dense cloud contributing to an apparent rapid drop in cloud height.

The MARIAH model predictions show cloud height remaining nearly constant at 5.4 m for 40 sec rather than rapidly dropping to below one meter after 15 seconds as was reported. The data are insufficiently substantiated to call this prediction incorrect.

CONCLUSIONS

The models which best match each experiment evaluated here are :

Eidsvik - HSE Trial 8 Compositions (far field)
HEGADAS - Esso Runs 11 and 17
MARIAH/ZEPHYR - HSE Trials 6 and 20, Esso Run 16, HSE Trial 8 cloud dimensions and near field concentrations

Incidentally, since the HEGADAS model neglects gravity spreading in the downwind direction and heat transfer from the water surface, and yet does well in matching Esso Runs 11 and 17, these effects may not be important.

The K-theory models match well all of the various types of experiments considered here. They are more versatile than top-hat models, but have been more costly in computer time. This cost disadvantage is rapidly decreasing, however,and with costs under $100-200 per run, MARIAH and ZEPHYR must now be considered practical. Top-hat models still cost less to run, and are, thus, useful to extend the range of conditions studied when used in conjunction with K-theory models.

The Eidsvik and HEGADAS Models seem to be considerably better than first generation top-hat models. Both match experimental data in the far field better than close to the source. Both match the low lying sensors in the HSE trials. The Eidsvik model does not match the higher (1 - 2 m) sensors. The K-theory models, on the other hand, do well against higher sensors, and reasonably well against low sensors in spite of grid size limitations. K-theory models match data equally well in the near and far field.

The Eidsvik and HEGADAS models are complimentary, since HEGADAS does not apply for calm winds or instantaneous releases, yet Eidsvik does. However, Eidsvik does not do well with non-instantaneous releases. These points are summarized in Table 3.

Table 3

Recommended Range for Choosing Between Eidsvik and HEGADAS Models

Wind Speed / Release Rate	Calm & Low <2 m/s	Medium to High >2 m/s
Instantaneous	Eidsvik	Eidsvik
Finite (> 20 sec)	Neither*	HEGADAS

* K-Theory models are also not validated for this case except by comparisons of cloud dimensions.

The K-theory and HEGADAS models allow plotting of realistic LFL contours. It is not yet clear which of these models produces the more conservative LFL contour predictions. In the low wind speed cases considered, HEGADAS and K-theory models gave similar predictions.

The GD model strongly overpredicts air entrainment for HSE Trial 8. It also predicts there will be no important gravity spread phase with high wind speeds. The other models studied, including the Eidsvik top hat model, predict otherwise. The GD model is not applicable to four of the seven experiments we analyzed. The Eidsvik model is superior to the GD model in providing a smooth transition to neutral buoyancy and in providing for heat transfer and both top and edge entrainment.

Model responses should be compared against compositional data and not only cloud dimensional data. The comparisons made here app only for relatively small spills. Inasmuch as adequately described physical principles are involved, they can predict behaviour beyond the original data base for which they are calibrated. However, the limitations to such extrapolation are not yet clear.

REFERENCES

Britter, R., (1980) "The Ground Level Extent of a Negatively Buoyan Plume in a Turbulent Boundary Layer", Atmos. Envir. 14, pp. 779-785.

Bowman, B. R., Sutton, S. B., and Comfort W. J., (1979) "The Impact of LNG Spills on the Environment : A Comparison of Dispersion Models and Experimental Data", 25th Annual Technical Meeting, Institute of Environmental Sciences, Seattle, Washington.

Buschmann, C. H. (1975) "Experiments on the Dispersion of Heavy Gas and Abatement of Chlorine Clouds", Proc. Symp. on Transport of Hazardous Cargoes by Sea, Jacksonville, Florida.

Colenbrander, G. W. (Sept. 1980) "A Mathematical Model for the Transient Behavior of Dense Vapor Clouds", 3rd Loss Prevention Symp., Basel, Switzerland.

Eidsvik, K. J., (1980) "A Model for Heavy Gas Dispersion in the Atmosphere", Atm. Env. 14, 769.

England, W. G., Teuscher, L. H., Hauser, L. E. and Freeman, B. (June 1978) "Atmospheric Dispersion of Liquefied Natural Gas Vapor Clouds Using SIGMET, a Three-Dimensional, Time-Dependent Hydrodynamic Computer Mode", Heat Transfer and Fluid Mechanics Institute, Washington St. U.

Feldbauer, G. F., May, W. G., Heigl, J. J., McQueen, W. and Whipp, R. H. (May 1972) "Spills of LNG on Water--Vaporization and Downwind Drift of Combustible Mixtures", Esso Research & Engineering Report EE.61E.72, Florham Park, N.J.

Germeles, A. E., and Drake, E. M., (1975) "Gravity Spreading and Atmospheric Dispersion of LNG Vapor Clouds", Fourth Int. Symp. on Transport of Hazardous Cargoes by Sea and Inland Waterways, Jacksonville, Florida.

Havens, J. A., (April 1978) "An Assessment of Predictability of LNG Vapor Dispersion from Catastrophic Spills into Water", Paper C2, Proc. 5th Int. Symp. on the Transport of Dangerous Goods by Sea and Inland Waterways, Hamburg. A fuller report was issued by the USCG, Cargo & Hazardous Materials Division, Washington, DC.

Havens, J. A., (Feb. 1979) "A Description and Assessment of the SIGMET LNG Vapor Dispersion Model", U.S. Coast Guard Report CG-M-3-79.

Havens, J. A. (1981) Submitted to J. Hazardous Materials.

May, W. G. (1980) Personal Communication.

Picknett, R. G., (Dec. 1978) "Field Experiments on the Behavior of Dense Clouds", Chemical Defense Establishment, Porton Down, Salisburg, Wilts, England, Report No. IL 1154/78.

Taft, J. R., and McBride W. M., (1979) "Validation Studies on Dense Gases Using the SIGMET Simulation Model", from L. E. Bell, Western LNG Terminal Associates, 705 Flower St., Los Angeles, California, 90017.

van Ulden, A. P. (1974) "On the Spreading of a Heavy Gas Released Near the Ground", Proc. 1st Int. Symp. on Loss Prevention and Safety Promotion in the Process Industries, Delft, Netherlands, pp. 221-226.

DISCUSSION

R. L. LEE — In what way do the models described in the paper handle variable terrain?

J. L. WOODWARD — Terrain and obstacles are handling by blocking out grid cells. The velocities at the blocked-out cells are set to zero.

A. P. VAN ULDEN — The authors devote a quite remarkable section to the Dutch Freon experiment. They state that no concentration sensor data are available. This is not correct. In fact sensor data are available as is clearly stated in the original paper (van Ulden, 1974). The sensor data have been used in the analysis and found to be consistent with the visual observations. Disregarding the sensor data and finding their model predictions to disagree with the observed cloud height, the authors claim that the latter "may very possibly be anomalous and misleading". This is a quite remarkable and misleading statement. Would you like to comment on this.

J. L. WOODWARD — Integrated or dosage response data are of little value in discriminating between model predictions. We prefer to make comparisons with time-dependent sensor responses. We thank Dr. van Ulden for calling our attention to the availability of one unpublished sensor response. However, with only one sensor we would still maintain that the experiment is not useful for model validation. Our work showed that many sensor responses are needed. Nevertheless, we would like to obtain this response, since it has bearing on his second question which has to do with his reported cloud height as a function of time. Photographs show a very irregular cloud shape at the arbitrarily chosen time which represents the designated beginning of the "gravity slumping" phase. Information is not available to correlate concentration with the observed visible edges of the cloud. The uncertainties in estimating the "initial" cloud height and diameter could produce substantial errors in the "initial" cloud composition which was calculated and not measured. Such uncertainties make any subsequent comparison of model responses with observations inconclusive.

A. BERGER

What makes ZEPHYR biomodal on Figure 5?

J. L. WOODWARD

The cloud formed in the HSE test had a distinctly toroidal shape. The passing of the front and near ridge in this toroid produces a biomodal response experimentally observed (see Figures 5, 6 and 7). The ZEPHYR model also predicts a definite high concentration ridge which passes sensor n° 3 with both the leading and trailing edges of the cloud.

APPLICATION OF THE SULFUR TRANSPORT EULERIAN MODEL (STEM) TO A SURE DATA SET

Gregory R. Carmichael and Leonard K. Peters

Department of Chemical Engineering, University of Iowa
Iowa City, Iowa and University of Kentucky
Lexington, Kentucky (USA)

INTRODUCTION

Regional scale models are important tools in the study of the atmospheric cycles of trace gases. For example, it is apparent that only after the regional cycle of a pollutant is understood can an efficient and cost-effective control strategy be developed. However, the observed distribution of a pollutant in the atmosphere results from complex interactions between the source distribution, the transport by the mean winds (both horizontally and vertically), the mixing by turbulent diffusion, the generation or depletion by chemical interaction with other trace species, and the removal by physical interaction with surfaces (dry deposition) and by encounter with a dispersed liquid phase (wet removal). Regional scale models provide a means of studying these complex processes.

Regional scale models are also important in the study of global cycles of trace gases since in many cases the transformation and removal processes occur with time and space scales such that only a small fraction of the material emitted into the lower troposphere may be transported to the free troposphere. Sulfur dioxide is an important example.

Thus, in order to understand regional and global cycles of pollutants and other trace gases in the atmosphere, it is necessary to characterize and model those processes that occur on a regional scale. Through the use of regional scale models it is possible to trace sources and to test theories of transport, transformation, and removal by comparing model predictions with measurements from monitoring networks and field studies.

In this paper, a regional scale combined transport/chemistry model (referred to as STEM, Sulfur Transport Eulerian Model) is applied to SO_x transport in the eastern United States. The model solves the governing coupled, three-dimensional advection-diffusion equations for SO_2 and sulfate by use of a Galerkin finite-element method. The model also considers spatial variations in topography and spatial and temporal variations in mixing layer heights, the wind field, and the dry deposition velocities. A photochemical SO_2 oxidation mechanism is incorporated into the transport model with the rate parameterized using diurnally varying radical species concentrations and a fixed heterogeneous rate constant.

MODEL DESCRIPTION

The regional transport of SO_2 and sulfate is modeled within an Eulerian framework. A block description of the model is presented in Figure 1. The modeling region with the grid system shown in Figure 2 was adopted so that grids contained within (1,15) to (26,15) to (26,32) to (1,32) are equivalent to the grid system used in the SURE experiment (Hidy et al., 1976). This was done so that the SO_2 emission inventory from that study could be used.

The mathematical analysis is based on the coupled, three-dimensional advection-diffusion equation for SO_2 and sulfate:

$$\frac{\partial c_\ell}{\partial t} + u_j \frac{\partial c_\ell}{\partial x_j} = \frac{\partial}{\partial x_j}\left[K_{jj} \frac{\partial c_\ell}{\partial x_j}\right] + R_\ell + S_\ell \tag{1}$$

where c_ℓ is the concentration of species ℓ, u_j is the velocity vector, K_{jj} is the eddy diffusivity tensor ($K_{ij} = 0$ for $i \neq j$ has been assumed), R_ℓ is the rate of formation or loss by chemical reaction, and S_ℓ is the emission rate.

The model has the capability of handling variable topography and variable vertical modeling regions. At present, the irregular vertical region is mapped into a dimensionless rectangular region with the top set at 3000 meters. The vertical structure of the model is shown schematically in Figure 3. There are eleven vertical layers with higher resolution between the surface and 450 meters. The vertical height is chosen so that it is well above the maximum mixing layer height. This permits simulation of transport in stable layers aloft and its subsequent re-entrainment into the mixing layer.

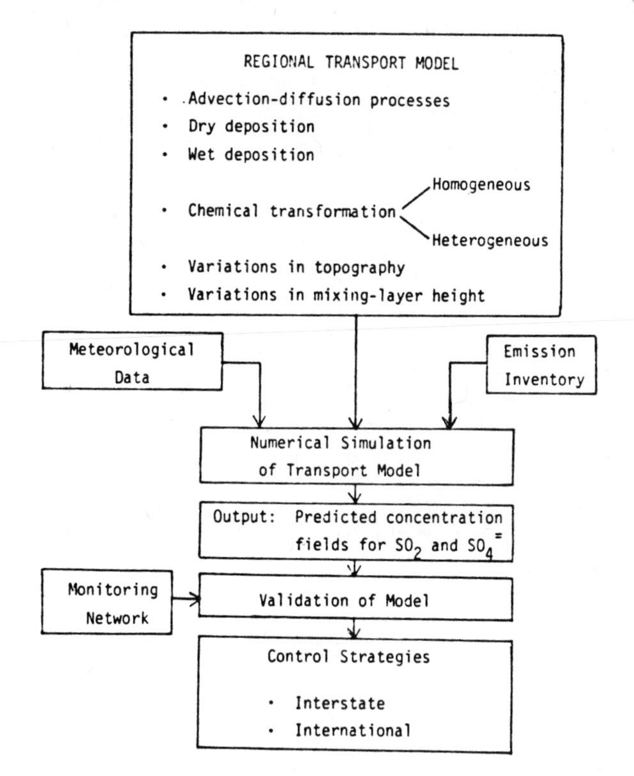

Figure 1. Block diagram of STEM.

The boundary conditions are as follows:

$$u\Delta H c_\ell - K_H \Delta H \frac{\partial c_\ell}{\partial x} = F_X \ (@\ x=0) \text{ or } G_X(@\ x=L_X) \tag{2}$$

$$v\Delta H c_\ell - K_H \Delta H \frac{\partial c_\ell}{\partial y} = F_y \ (@\ y=0) \text{ or } G_y(@\ y=L_y) \tag{3}$$

$$\frac{K_v}{\Delta H} \frac{\partial c_\ell}{\partial \rho} = V_d^\ell c_\ell \quad @\ \rho=0.01 \tag{4}$$

$$Wc_\ell - \frac{K_v}{\Delta H} \frac{\partial c_\ell}{\partial \rho} = G_z \quad @\ \rho=0.95 \tag{5}$$

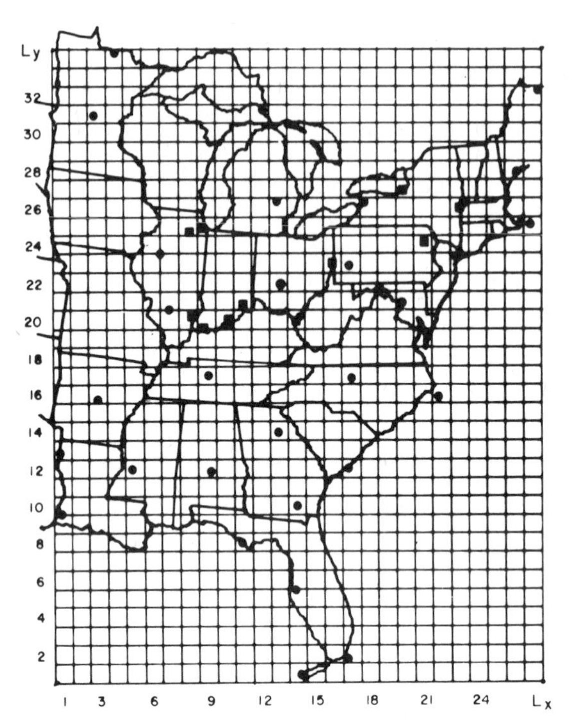

Figure 2. The modeling region and grid system used in STEM. The squares represent sulfate monitoring locations and the circles represent rawinsonde locations.

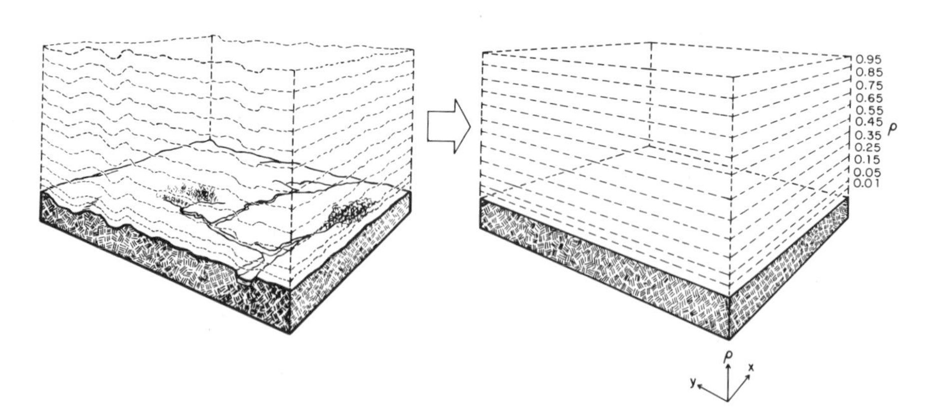

Figure 3. STEM handles a variable domain such as that shown above by transforming the irregular vertical region into a dimensionless rectangular region.

where ρ is the dimensionless vertical height and $\rho \epsilon$ [0,1], V_d^{ℓ} is the deposition velocity, ΔH is the vertical thickness of the modeling region, and W is the effective vertical velocity. Condition (4) represents dry deposition to the surface, and the other conditions are the prescribed total flux (advective and diffusive flux) at the boundaries.

It should be noted that the deposition condition is applied at $\rho = 0.01$ and not at the surface ($\rho = 0$). The diffusion of pollutant near the surface is dominated by small scale components of turbulence and must be simulated with a small grid spacing. Choosing the bottom grid to be near the top of the surface layer (~ 30m) enables the concentration at the surface and the concentration profile within the surface layer to be determined analytically using the flux-gradient relationships for the surface layer (Sheih, 1978).

The dynamic model uses a Galerkin method for the numerical solution of the partial differential equations. This method was selected based on results from a series of numerical experiments in one- and two-dimensional advection of initial wave forms (Carmichael et al., 1980). The numerical method was adapted to Eq. (1) using a time-splitting technique based on the work of Yanenko (1971). The Crank-Nicolson Galerkin approximation was applied to the resulting system of equations. (See Fairweather (1970) for a development and analysis of the method.)

METEOROLOGICAL DATA

The transport model requires input data derived from meteorological data. This model requires the mean wind components, eddy diffusivities, temperature, dry deposition velocities, and water vapor concentration at each grid point. The mean wind components and eddy diffusivities enter into the atmospheric diffusion equation as coefficients for the convection and diffusion terms, respectively, and the temperature and water vapor profiles are used to estimate the reaction rate constants. Unfortunately, not all of the necessary meteorological variables are routinely measured, and those available are only obtained at ground level or at discrete times. However, it is necessary to use the limited data available to provide estimates within the entire modeling region.

Figure 4 shows schematically how model inputs for STEM are derived from available meteorological data. The surface data and upper air data are obtained from the thirty-two National Weather Service stations within the model region measuring meteorological data aloft. The function modules then process this raw data to provide estimates of the necessary inputs.

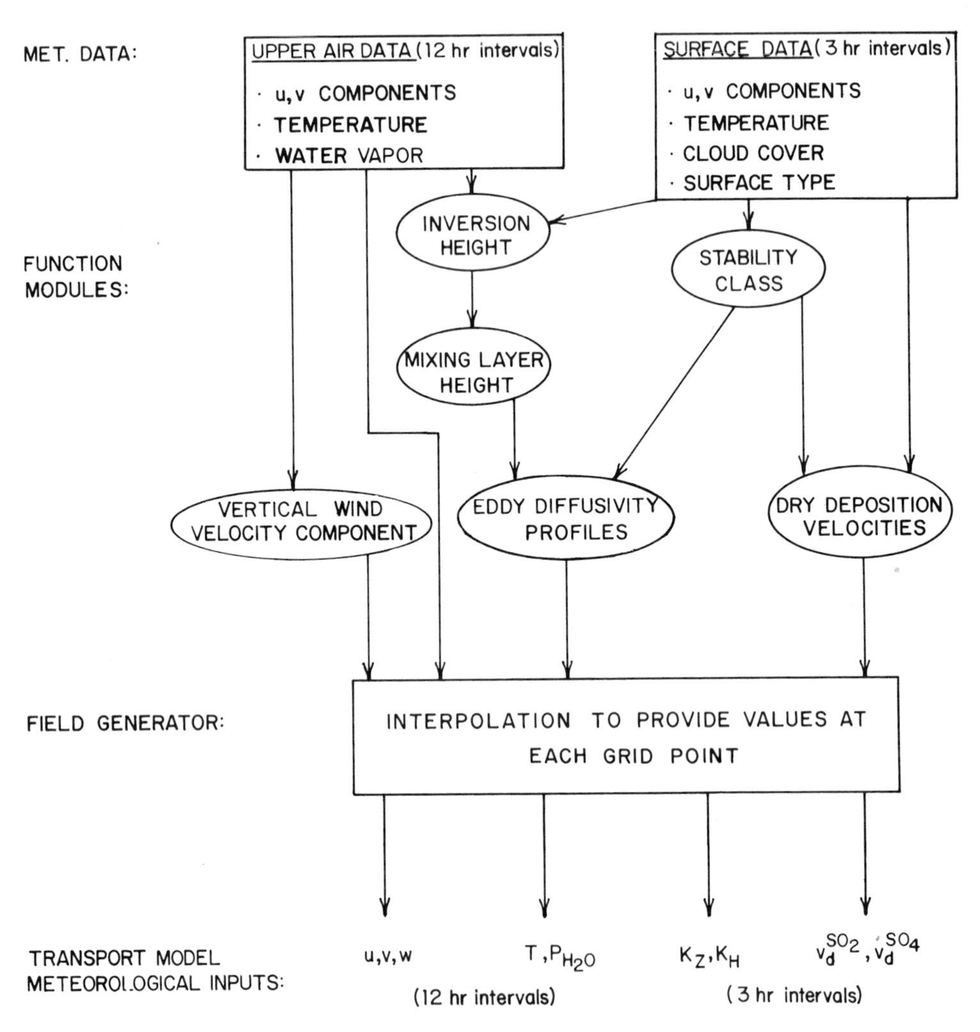

Figure 4. Schematic diagram of how inputs for STEM are derived from available meteorological data.

The vertical component of the mean winds is derived from the discrete two-dimensional wind data in the vertical velocity module. The method adopted yields a divergence-free wind field at the required altitudes. At any given level, the $\bar{u}$ and $\bar{v}$ velocity components are found by vertically interpolating between measured values, and the horizontal wind field at each grid point is estimated by horizontal interpolation using a weighting factor of $1/r_i^2$. The vertical velocity component is found by numerically solving the continuity equation using an explicit second-order finite difference procedure with the constraint that $\bar{w}=0$ at $\rho=0.01$. The problem of interpolating the wind field between measured times is not addressed, and the wind field is held constant for the twelve-hour period centered around the time of measurement.

The eddy diffusivity module based on the work of Myrup and Ranzieri (1976) calculates values consistent with theory and observation. This module estimates the Monin-Obukhov length scale, L, based on stability class, aerodynamic roughness length, z_0, and the evaporation rate at the surface. Input parameters to the vertical eddy diffusivity model are the wind speed at a known height, z_0, L, the depth of the mixing layer, stability class, and the inversion height. Within the mixing layer, the eddy diffusivity values are multiplied by a roll-off function to agree with eddy diffusivity data within the mixing layer which show a broad maximum that falls to near zero at the base of the inversion. Furthermore, under neutral or stable conditions the magnitude of the eddy diffusivities are bounded, consistent with the limiting expressions previously suggested by Yordanov (1972). The horizontal diffusivities are calculated as multiples of the vertical coefficients as recommended by Ragland et al., (1975).

Dry deposition velocities are calculated in that module from estimates of the aerodynamic and surface resistances based on surface windspeed, surface roughness, surface evaporation rate, and stability (Carmichael, 1978). Average values of the calculated deposition velocities (at 20-30m) throughout the modeling region for the period of July 4 -July 10, 1974 (average includes day and night values for all simulation days) are $V_d^{SO_2}=0.44$ cm s^{-1} and $V_d^{SO_4}=0.26$ cm s^{-1}. The most commonly used values for SO_2 and sulfate at ground level are 0.8 cm s^{-1} and 0.1 cm s^{-1}, respectively. Fowler (1978) has calculated an annual average SO_2 deposition velocity for the English countryside of 0.6 cm s^{-1} and concluded that a value of 0.6 to 0.8 cm s^{-1} provides a reasonable estimate of annual ground level deposition for countryside dominated by grass. The same may not be true of other areas, e.g., cropland, forest, urban area, etc. A smaller average deposition velocity for SO_2 was calculated in the present model since the countryside of the eastern United States is principally cropland and forest. Furthermore, when calculating deposition velocities for $z > 1$ m, there is additional resistance due to the surface layer. Fisher (1978) has used the deposition velocity of 0.5 cm s^{-1} at the top of the surface layer in his long range transport model. The present average value of $V_d^{SO_4}$=0.26 cm s^{-1} is in keeping with the results of Hicks and Wesely (1978).

In the other modules the stability class is estimated by a procedure given by Turner (1961) and the inversion heights are estimated using the method described by Holzworth (1964). The mixing layer heights are described by a diurnally varying function.

The necessary input fields are estimated by horizontal interpolation at each vertical level using a wèighting factor of $1/r_i^2$.

CHEMISTRY OF SO_2 AND SULFATE

The SO_2 oxidation to sulfate is broken into a homogeneous and a heterogeneous component.

Homogeneous Chemistry

The homogeneous reaction mechanism used in STEM is shown schematically in Figure 5 and the resulting SO_2 rate expression is given by

$$\frac{d[SO_2]}{dt} = -(k_8[OH] + k_9[HO_2])\,[SO_2] \tag{6}$$

The chemical mechanism for estimating the [OH] and [HO_2] concentrations is shown schematically in Figure 6 and the chemical reactions and rate constants are presented in Table 1. This reaction subset was selected by evaluating the rates of individual reactions obtained from a detailed (72 reactions) kinetic modeling of NO_x-air-SO_2-H_2O system chamber runs (Kocmond and Yang, 1976; Bradstreet, 1973). The details of these studies are presented elsewhere (Carmichael, 1978; Carmichael and Peters, 1979). However, the detailed mechanism showed good agreement between the measured and predicted SO_2 reaction rates. (A correlation coefficient of 0.90 was obtained from the comparison of predicted and Kocmond and Yang's measured SO_2 concentrations.) In addition, the trends within the data were maintained. The predicted and observed rates increased as the initial NO concentration, the relative humidity, and the initial NO_2 concentration increased.

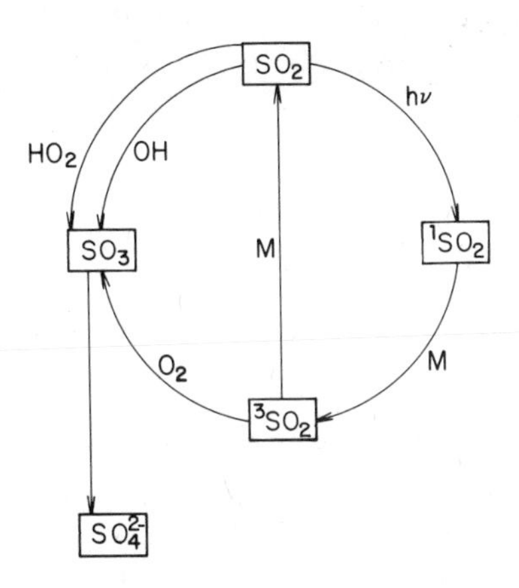

Figure 5. Schematic diagram of the homogeneous gas phase chemistry treated in STEM.

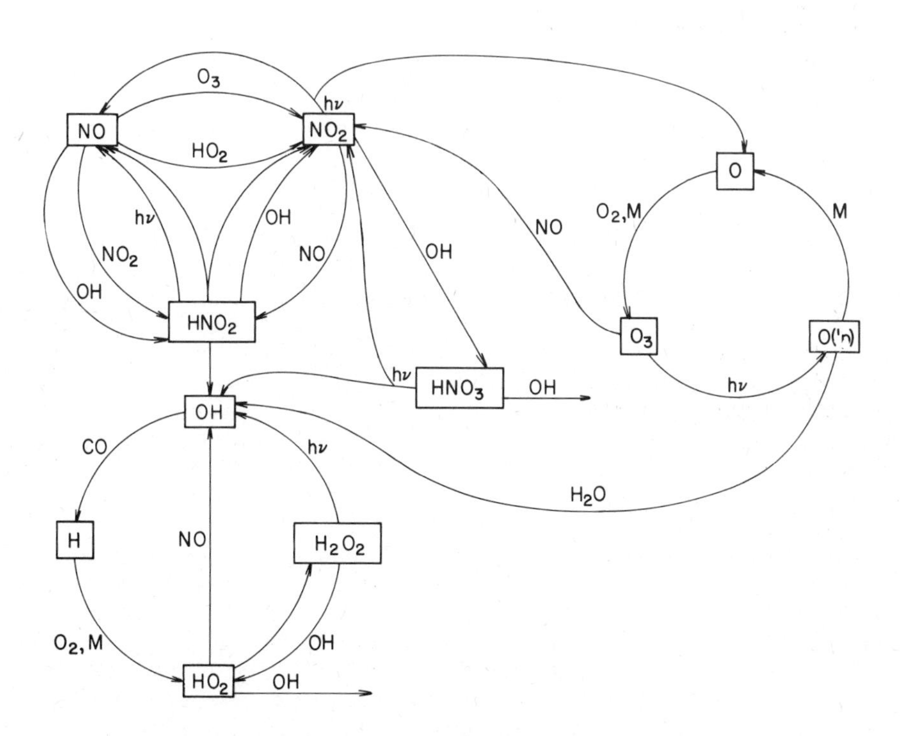

Figure 6. Schematic diagram of the non-SO_2 homogeneous gas phase chemistry treated in STEM.

Table 1. The Simplified Homogeneous Sulfur chemistry Model

Reaction	Rate Constant	Reaction	Rate Constant
$O_3 + h\nu \xrightarrow{1} O(^1D) + O_2$	$1.01 \times 10^{-2} \exp(-A/\cos\sigma)$, $A = 3.2 - \sqrt{4.173 - 0.64\,(z/10{,}000 - 2.0)^2}$	$CO + OH \xrightarrow{14} CO_2 + H$	4.41×10^{2}
$O(^1D) + H_2O \xrightarrow{2} 2\,OH$	7.35×10^{5}	$HO_2 + HO_2 \xrightarrow{15} H_2O_2 + O_2$	$2.81 \times 10^{4} \exp(-500/T)$
$O(^1D) + M \xrightarrow{3} O + M$	$8.40 \times 10^{4} \exp(107/T)$	$2\,HNO_2 \xrightarrow{16} NO + NO_2 + H_2O$	1.00×10^{-1}
$O + O_2 + M \xrightarrow{4} O_3 + M$	$3.01 \times 10^{-6} \exp(510/T)$	$NO + NO_2 + H_2O \xrightarrow{17} 2\,HNO_2$	2.00×10^{-6}
$SO_2 + h\nu \xrightarrow{5} {}^1SO_2$	$1.00 \times 10^{-2} \cos\sigma$	$H + O_2 + M \xrightarrow{18} HO_2 + M$	6.35×10^{-4}
${}^1SO_2 + M \xrightarrow{6} {}^3SO_2 + M$	2.00×10^{4}	$HO_2 + OH \xrightarrow{19} H_2O + O_2$	3.00×10^{5}
${}^3SO_2 + M \xrightarrow{7} SO_2 + M$	2.00×10^{2}	$NO + O_3 \xrightarrow{20} NO_2 + O_2$	$3.09 \times 10^{3} \exp(-1450/T)$
$SO_2 + OH \xrightarrow{8} HSO_3$	1.00×10^{3}	$NO_2 + OH \xrightarrow{21} HNO_3$	1.50×10^{4}
$SO_2 + HO_2 \xrightarrow{9} OH + SO_3$	1.2	$OH + HNO_2 \xrightarrow{22} NO_2 + H_2O$	1.00×10^{4}
$HO_2 + NO \xrightarrow{10} OH + NO_2$	$6.45 \times 10^{5} \exp(-1200/T)$	${}^3SO_2 + O_2 \xrightarrow{23} SO_3 + O$	1.50×10^{3}
$OH + NO \xrightarrow{11} HNO_2$	3.00×10^{3}	$H_2O_2 + h\nu \xrightarrow{24} OH + OH$	$(5.74 \times 10^{-4} + 1.88 \times 10^{-7}z - 1.5 \times 10^{-11}z^2) \cos\sigma$
$HNO_2 + h\nu \xrightarrow{12} OH + NO$	$(1.65 \times 10^{-1} + 4.95 \times 10^{-5}z - 4.50 \times 10^{-9}z^2) \cos\sigma$	$HNO_3 + h\nu \xrightarrow{25} NO_2 + OH$	$(5.93 \times 10^{-5} + 2.26 \times 10^{-8}z - 1.70 \times 10^{-12}z^2) \cos\sigma$
$NO_2 + h\nu \xrightarrow{13} NO + O$	$B \exp(-A/\cos\sigma)$, $A = 1.2 - \sqrt{1.25 - 0.16\,(z/10{,}000 - 2.15)^2}$, $B = 60\left[-0.122 + \sqrt{0.02102 - 1.6 \times 10^{-3}(z/10{,}000 - 1.15)^2}\right]$	$OH + HNO_3 \xrightarrow{26} NO_3 + H_2O$	2.30×10^{2}
		$OH + H_2O_2 \xrightarrow{27} HO_2 + H_2$	$1.79 \times 10^{4} \exp(-750/T)$

Units for the rate constants are in ppm and min; z is the altitude in meters; T is the temperature in °K; and σ is the solar zenith angle.

The model to estimate the homogeneous gas phase SO_2 oxidation rate based on the reaction subset described in Table 1 requires the concentrations of H_2O, M, CO, NO and NO_2 as inputs. Once these are specified, the concentrations of HNO_2, HNO_3, H_2O_2, HO_2, O_3, and OH are calculated by use of the pseudo-steady state relationship. Finally, the SO_2 oxidation rate can be determined. The rate expression for SO_2 is uncoupled from the expressions for OH and HO_2 by using the concentration of SO_2 at the previous time step in the calculation of OH and HO_2. Since the rate of SO_2 oxidation in the atmosphere is on the order of 1-2% hr^{-1}, the error introduced by this simplification is small. The algebraic expressions for [OH], [HO_2], [HNO_2], [HNO_3], and [H_2O_2] are coupled and require an iterative solution, but only one linear differential equation, that for SO_2 needs to be solved. A block diagram of this procedure is presented in Figure 7. It should be noted that in addition to the reaction rate, the concentrations of OH, HO_2, O_3, HNO_3, HNO_2 and H_2O_2 are estimated at each grid point at each time step. This is important since the concentrations of O_3, HNO_3 and H_2O_2 are necessary in the estimate of heterogeneous oxidation and removal mechanisms.

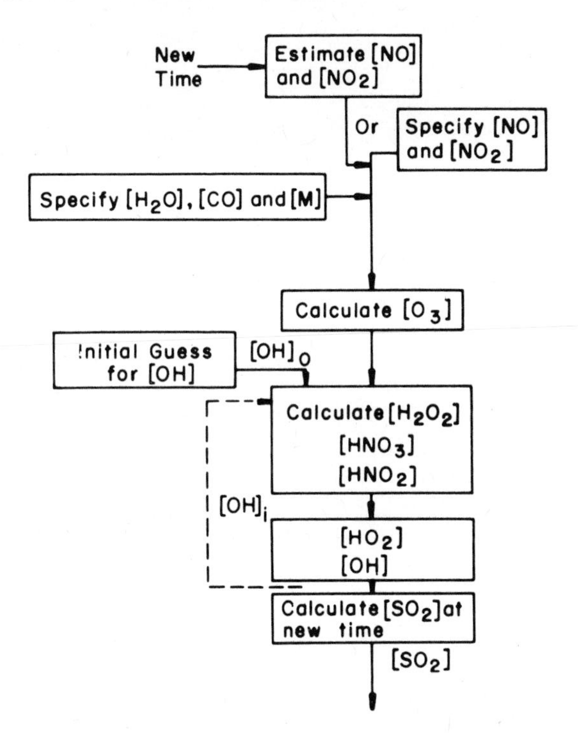

Figure 7. Block diagram of the calculation procedure used in STEM for the homogeneous chemistry.

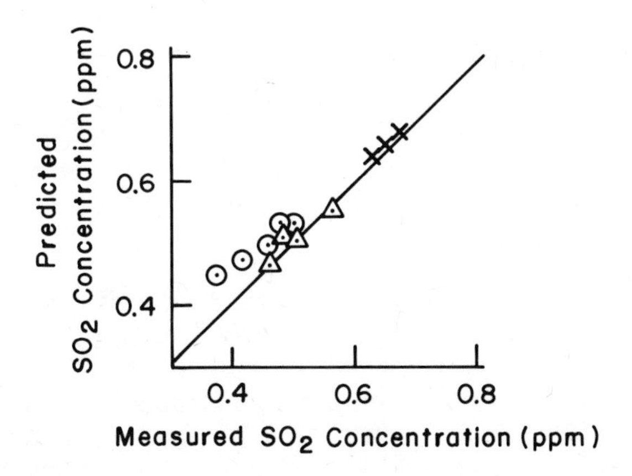

Figure 8. Comparison of the simplified chemistry model predictions with the data of Kocmond and Yand (1976).

Three of Kocmod and Yang's runs were simulated using the simplified kinetic model. Their reported NO and NO_2 concentration-time curves were input as data for each run. The results of these simulations are presented in Figure 8. This model agrees well with the experimental data (correlation coefficient of the data presented in Figure 8 is 0.99) and compares very well with those values predicted by the detailed model. The computation times for the detailed model and the simplified model are in the ratio of approximately 40:1.

Species concentrations generated by this simplified kinetic model are compared to those predicted by detailed urban and natural tropospheric chemistry models in Tables 2 and 3, respectively. The most important observation from these comparisons is that this simplified model predicts OH concentrations in the range of 3×10^{-8} - 3×10^{-7} ppm. These values appear to be consistent with those values reported in the literature. An indirect determination of [OH] by Calvert (1976) made by combining laboratory OH reaction rate constants with measured diurnal patterns of several hydrocarbon species found the OH concentration in Los Angeles to be $0.1 - 8 \times 10^{-7}$ ppm. Wang et al., (1975) used the technique of laser-induced fluorescence to measure the OH radical concentration in the ambient

Table 2. Urban-Troposphere Concentrations Generated by the Simplified Homogeneous Sulfur Chemistry Model

Predicted Concentrations (in ppm)

Species	SO_2 Parameterization Model A	B	C	Observed	Predicted by Graedel et al. (1976)
$[O_3]$max	4.5×10^{-2}	1.0×10^{-1}	5.3×10^{-2}	$4\text{-}6 \times 10^{-1}$	4.5×10^{-2}
[OH]max	1.0×10^{-7}	1.6×10^{-7}	1.1×10^{-7}	$(.1 - 3) \times 10^{-6}$	6×10^{-7}
$[HO_2]$max	8.1×10^{-6}	1.7×10^{-5}	7.0×10^{-6}		2×10^{-5}
$[HNO_2]$max	1.1×10^{-2}	7.2×10^{-3}	1.0×10^{-2}	3.2×10^{-3}	3.4×10^{-4}
daytime min	4.8×10^{-4}	4.0×10^{-4}	7.7×10^{-4}		
$[H_2O_2]$max	6.5×10^{-4}	1.6×10^{-3}	2.7×10^{-4}	4×10^{-2}	6.1×10^{-2}
$[HNO_3]$max	1.1×10^{-1}	1.1×10^{-1}	1.1×10^{-1}	6×10^{-3}	3.4×10^{-3}

CASE A SO_2 parameterization using NO and NO_2 conc. traces
B SO_2 parameterization using NO and NO_2 const. with time, NO = 0.005, NO_2 = 0.03
C SO_2 parameterization using NO and NO_2 const. with time, NO = 0.01, NO_2 = 0.03

air of Dearborn, Michigan, in August and found the peak values to vary from 8×10^{-7} ppm to 2×10^{-6} ppm. Davis et al., (1976) used an airborne mounted UV tunable laser system to measure the OH concentrations at an altitude of 7 km during the month of October. They report values from 4×10^{-8} ppm to 4×10^{-7} ppm, while surface measurements by Perner et al. (1976) via long path laser absorption spectroscopy during the period from August to November found OH concentrations to generally be less than 1.5×10^{-7} ppm.

Finally, it should be emphasized that many of the reaction rate constants are temperature dependent and that the photochemical rate constants vary with solar zenith angle and altitude.

Heterogeneous Chemistry

In the atmosphere, the heterogeneous oxidation of SO_2 occurs via oxidation by O_2 in the absence of catalyst, catalytic oxidation by O_2, and oxidation by O_3. The reported rate constants for the uncatalyzed oxidation by dissolved O_2 vary by two orders of magnitude, with most reports on the low end indicating that this reaction is not extremely important in the atmosphere. The catalyzed oxidation of SO_2 is most important in urban areas and stack plumes under conditions of high humidity and high catalyst concentrations; it is unlikely to be significant in rural areas. Recent measurements of the rate of oxidation of SO_2 by dissolved O_3 also differ by two orders of magnitude. If the higher rate constant is correct, then the oxidation by O_3 is important even at atmospheric background ozone concentrations.

Table 3. Natural-Troposphere Concentrations Generated by the simplified Homogeneous Sulfur Chemistry Model

Species	Kummler's Model	*SO_2 Parameterization Model			
		a	b	c	d
NO	4.1×10^{-4}	4.1×10^{-4}	4.1×10^{-4}	4.1×10^{-4}	4.1×10^{-4}
NO_2	1.7×10^{-4}	1.7×10^{-4}	1.7×10^{-4}	1.7×10^{-4}	1.7×10^{-4}
O_3	1.2×10^{-2}	7.4×10^{-3}	7.4×10^{-3}	7.4×10^{-3}	1.3×10^{-2}
HNO_3	1.6×10^{-3}	3.5×10^{-4}	2.3×10^{-4}	5.6×10^{-4}	4.8×10^{-4}
HNO_2	5.7×10^{-7}	2.0×10^{-6}	1.2×10^{-6}	4.5×10^{-6}	1.5×10^{-6}
H_2O_2	7.7×10^{-3}	2.6×10^{-3}	1.4×10^{-3}	6.2×10^{-3}	2.5×10^{-3}
OH	1.1×10^{-7}	4.9×10^{-8}	2.9×10^{-8}	1.1×10^{-7}	2.7×10^{-7}
HO_2	2.3×10^{-5}	2.1×10^{-5}	1.5×10^{-5}	3.3×10^{-5}	2.3×10^{-5}

*a $[H_2O]$ used in Kummler's Model, $[H_2O] = 4.1 \times 10^3$ ppm

b $[H_2O] = 1.72 \times 10^3$ ppm

c $[H_2O] = 1.72 \times 10^4$ ppm

all runs $[SO_2]_0 = 0.0024$ ppm.

d $[H_2O] = 1.72 \times 10^4$ ppm, new rate constants (see Table 1), rate constants evaluated at 300^oK and 1000 m.

Assuming oxidation by dissolved O_3, O_2, and metal catalysts occur simultaneously, the combined heterogeneous rate expression can be written as

$$\frac{d[SO_4^{2-}]}{dt} = -1.5 \frac{d[SO_2]}{dt} = \frac{60\ K_1 H v^*}{[H+]} \left(k_{O_3} K_{O_3} P_{O_3} + \frac{k_{O_2} K_2}{[H+]} + k_{cat} K_2 [M+]\right)[SO_2] \qquad (7)$$

where $d[SO_4^{2-}]/dt$ is in ppm min^{-1}, $[SO_2]$ in ppm, H is Henry's law constant, v^* is the aqueous volume per volume of air, $k_{O_3} = 1 \times 10^5 - 3.3 \times 10^5\ \ell\ mol^{-1} s^{-1}$; $K_{O_3} = 2.2 \times 10^{-2} mol\ell^{-1} atm^{-1}$; P_{O_3} is the O_3 partial pressure, $k_{O_2} = 6 \times 10^{-3} s^{-1}$ for $3 < pH < 5$, $k_{cat} = 2 \times 10^8 \ell^2 mol^{-2} s^{-1}$, $[M^+]$ is the concentration of the metal catalyst, $[H^+]$ is the hydrogen ion concentration, and $K_2 = 2.4 \times 10^{-10}$ exp (1671/T). By choosing typical rural values of $[O_3]$ = 50 ppb, a pH of 4.6 for moderately polluted continental clouds, $[M^+] = 5 \times 10^{-7} mol\ \ell^{-1}$, $k_{O_3} = 2 \times 10^5 \ell\ mol^{-1} s^{-1}$, T = 300°K, and $v^* = 3 \times 10^{-7}$, Eq. (7) reduces to

$$\frac{d[SO_2]}{dt} = -3.69 \times 10^{-5} [SO_2] \qquad (8)$$

The contribution by the ozone reaction is 92%, 5% by the uncatalyzed reaction, and 3% by the metal catalyzed oxidation.

Due to the uncertainties in the heterogeneous reaction mechanisms the fixed reaction rate given by Eq. (8) is used in the combined transport/chemistry model. However, Eq. (7) and the inclusion of the H_2O_2 reaction (oxidation by dissolved H_2O_2 appears fast and could be important in the atmosphere at H_2O_2 concentrations on the order of 1 ppb) does provide a method of linking the homogeneous and heterogeneous chemistry of SO_2. This is currently being investigated.

RESULTS AND DISCUSSION

Several important aspects of STEM need to be emphasized. The output from STEM consists of hourly and 24-hour averaged SO_2 and sulfate values at each grid point, hourly estimates of the SO_2 reaction rate and OH, H_2O_2, HO_2, O_3, HNO_3, and HNO_2 concentrations at each grid point, accumulated deposited amounts of SO_2 and sulfate at each surface cell, and a regional daily mass balance indicating the amount of SO_2 emitted, transformed to sulfate, deposited on the surface, and transported in and out of the region boundaries.

Also included in STEM are features which allow the interaction of elevated plumes with the dynamics of the mixing layer and the overall vertical mixing. These features have been studied using a subset of STEM. Detailed results of these studies have been presented by Carmichael et al. (1980) and Lin (1980). Those results demonstrate that the model is capable of simulating nighttime storage of pollutants aloft and trapping of pollutants near the surface, daytime re-entrainment and fumigation, and other features of pollutant pumping caused by diurnal variations in the mixing layer height and vertical mixing.

Results obtained during a 24 hour simulation using the SURE emission data, a constant wind field ($\overline{u} = \overline{v} = 5$ m s $^{-1}$), a representative eddy diffusivity profile (constant mixing layer height of 1500 m), and temperature, water vapor, and dry deposition velocity fields derived from meteorological data for July 4, 1974 are presented in Figures 9 to 14. Results using these simplistic but representative wind and eddy diffusivity fields illustrate the general capabilities of STEM.

The initial conditions for this simulation were $[SO_2]_0 = 40$ μg m^{-3}, $[SO_4]_0 = 5$ μg m^{-3}, $[SO_2]_b = 20$ μg m^{-3} and $[SO_4]_b = 2$ μg m^{-3}, where the subscript b denotes background (outside model region).

Presented in Figures 9 and 10 are contours of the calculated 24-hour average SO_2 and sulfate values, respectively. (The grid numbering system is the same as that used in Figure 2.) The bulk transport follows the $\overline{u} = \overline{v}$ wind vector as shown by the elongation of the contours and the concentration gradients along the south and west boundaries. High SO_2 values occur in regions of large emissions along the Ohio River Valley and the New York-Boston corridor. SO_2 values in the southeast are dominated by an erroneous entry in the NEDS summary.

Vertical SO_2 profiles after simulation times of 12 and 24 hours are presented in Figure 11. The profiles show the positive vertical concentration gradients at the surface resulting from dry deposition. Sometime after 12-hours, material emitted downwind is transported to this location. This is shown by the large increase in the concentrations in the 24-hour profile. The profile at 24-hours is typical of the distribution of material within the mixing layer.

Contours of the estimated pseudo-first order reaction rate constant at noon are presented in Figures 12 and 13. Values at ground level range from 1 - 1.7% hr^{-1} while values at level 6 (~ 800m) vary from 1.7 - 3% hr^{-1}. The lower values at ground level are due to the higher NO_x concentrations and the lower solar actinic flux near the surface.

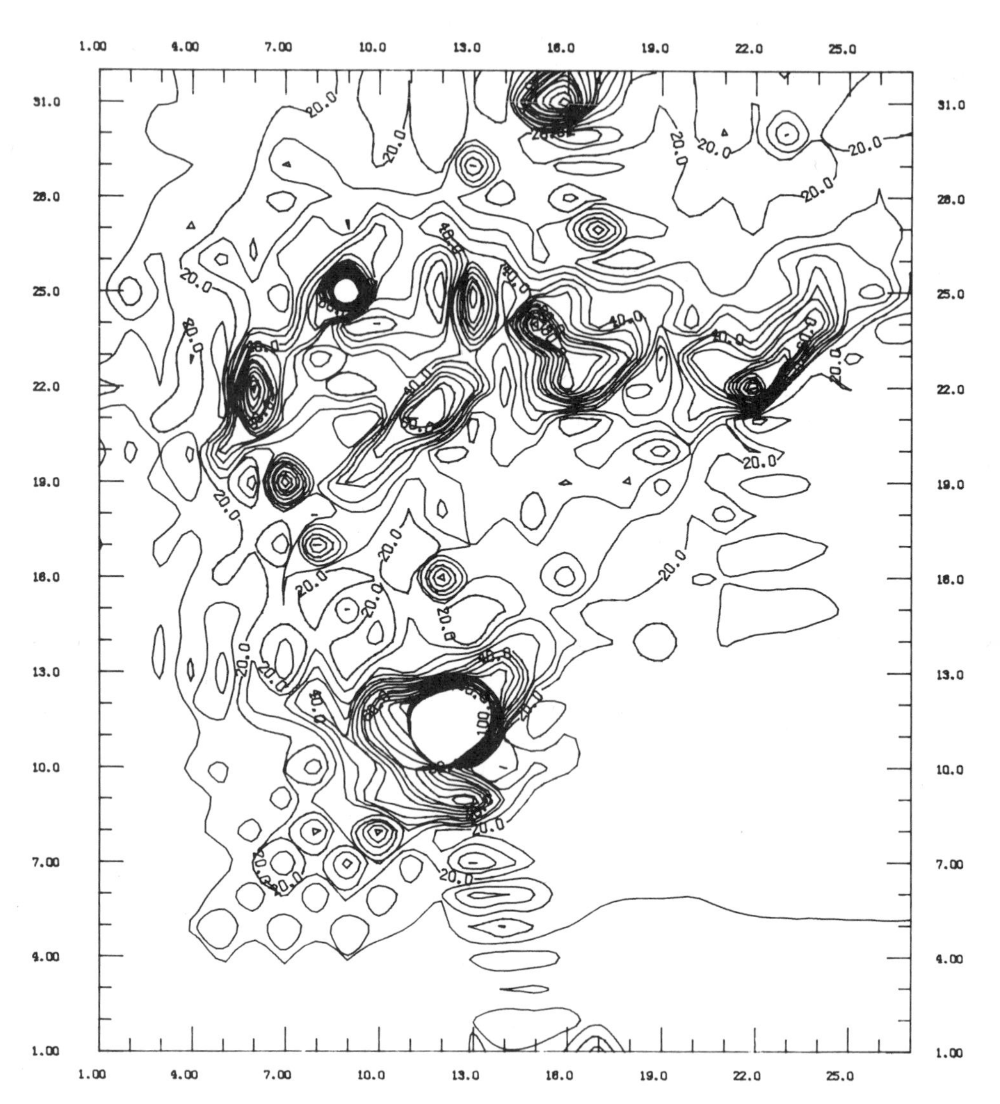

Figure 9. 24-hr averaged SO_2 concentrations in $\mu g\ m^{-3}$ at level 1.

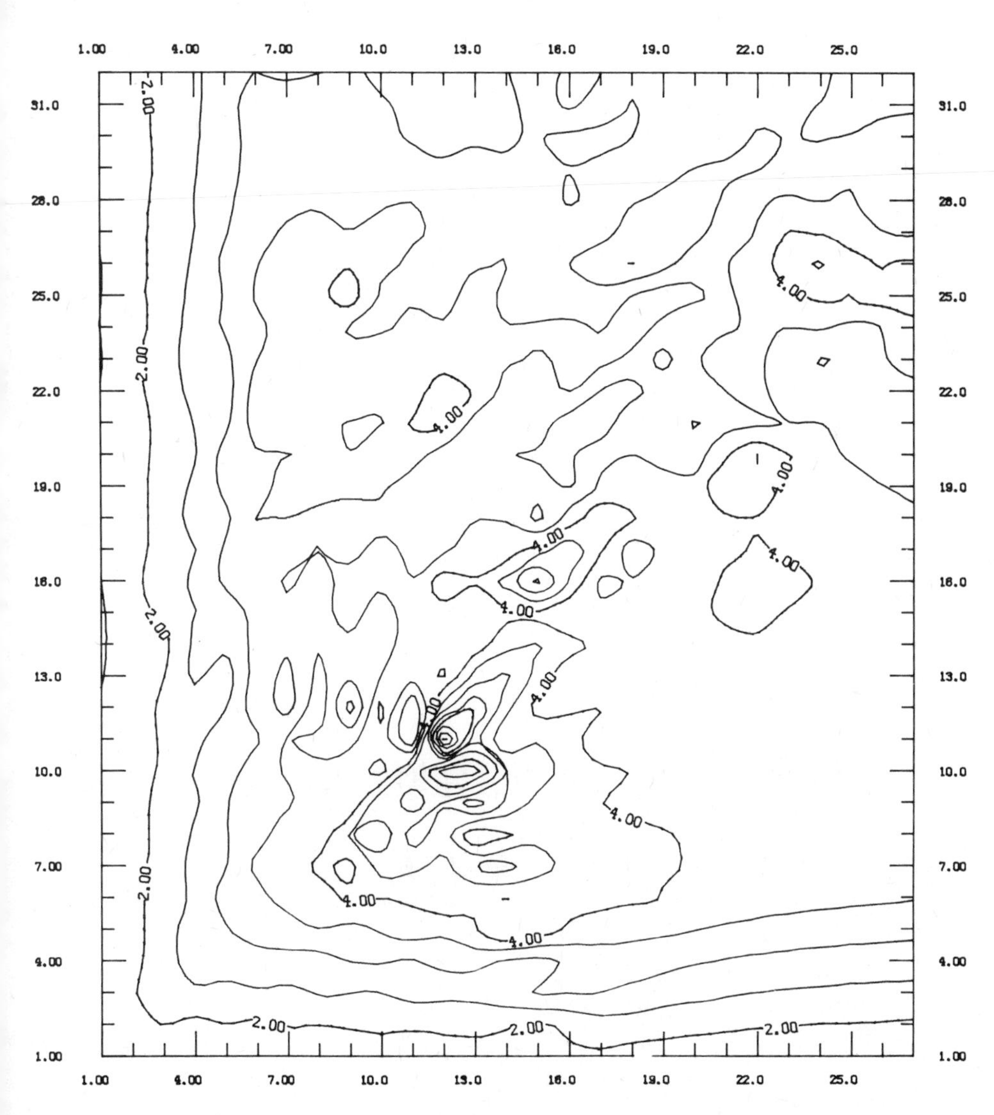

Figure 10. 24-hr averaged sulfate concentrations in $\mu g\ m^{-3}$ at level 1.

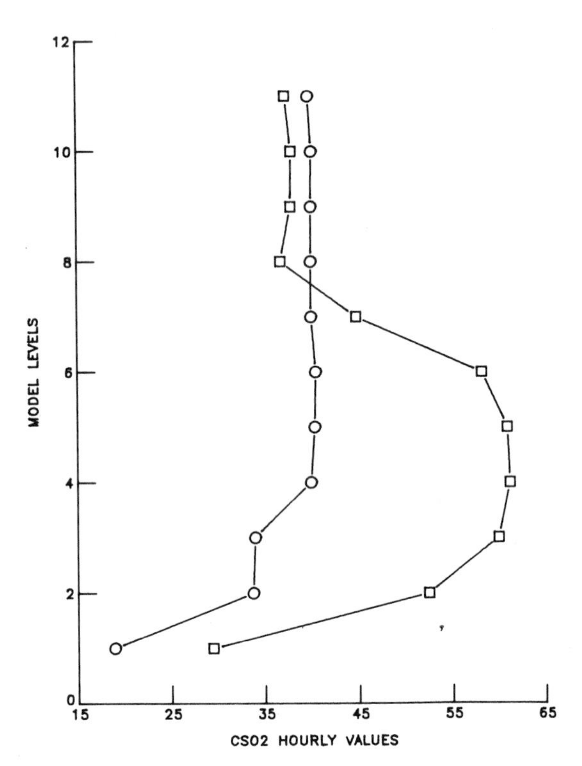

Figure 11. Vertical SO_2 concentration profiles in $\mu g\ m^{-3}$ at grid point (21, 26) at 12 hours (circles) and 24-hours (squares).

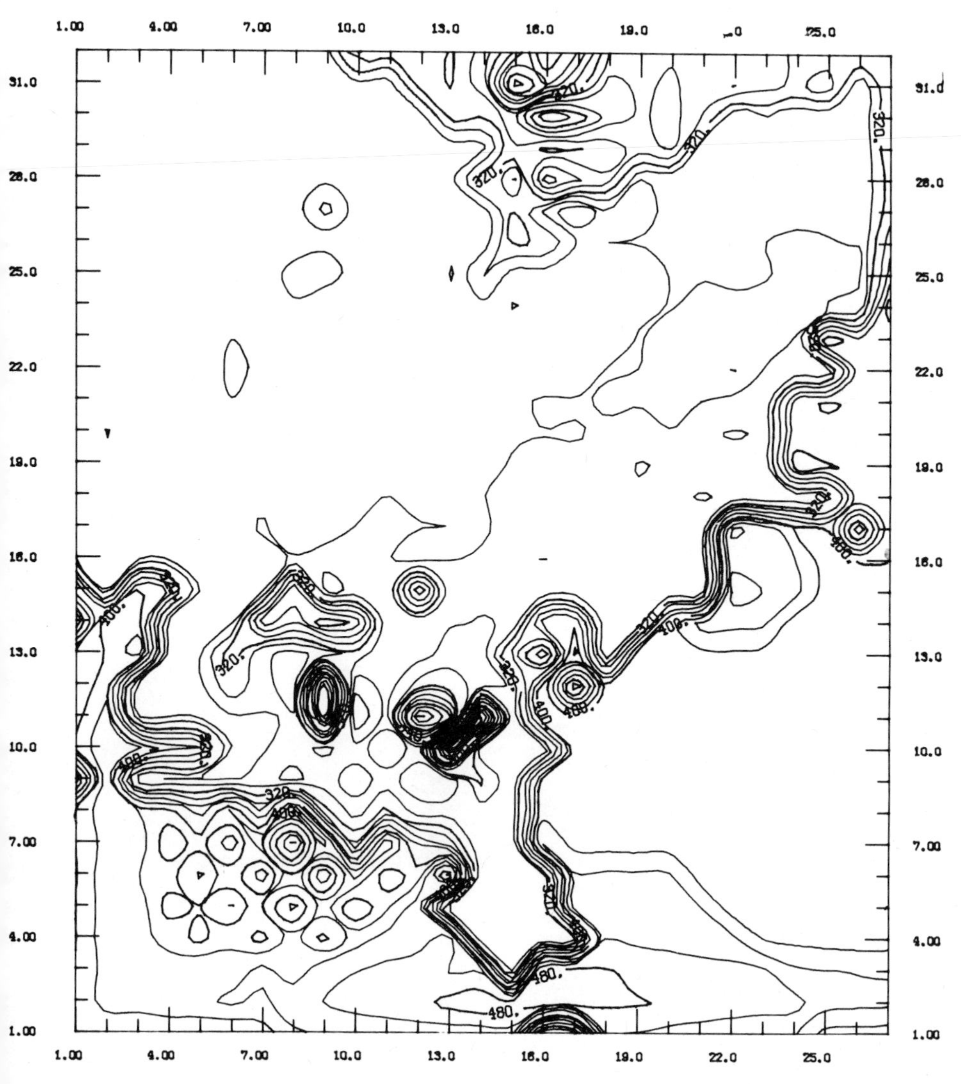

Figure 12. Pseudo-first order reaction rate constants (x 108) in s-1 for noon at level-1.

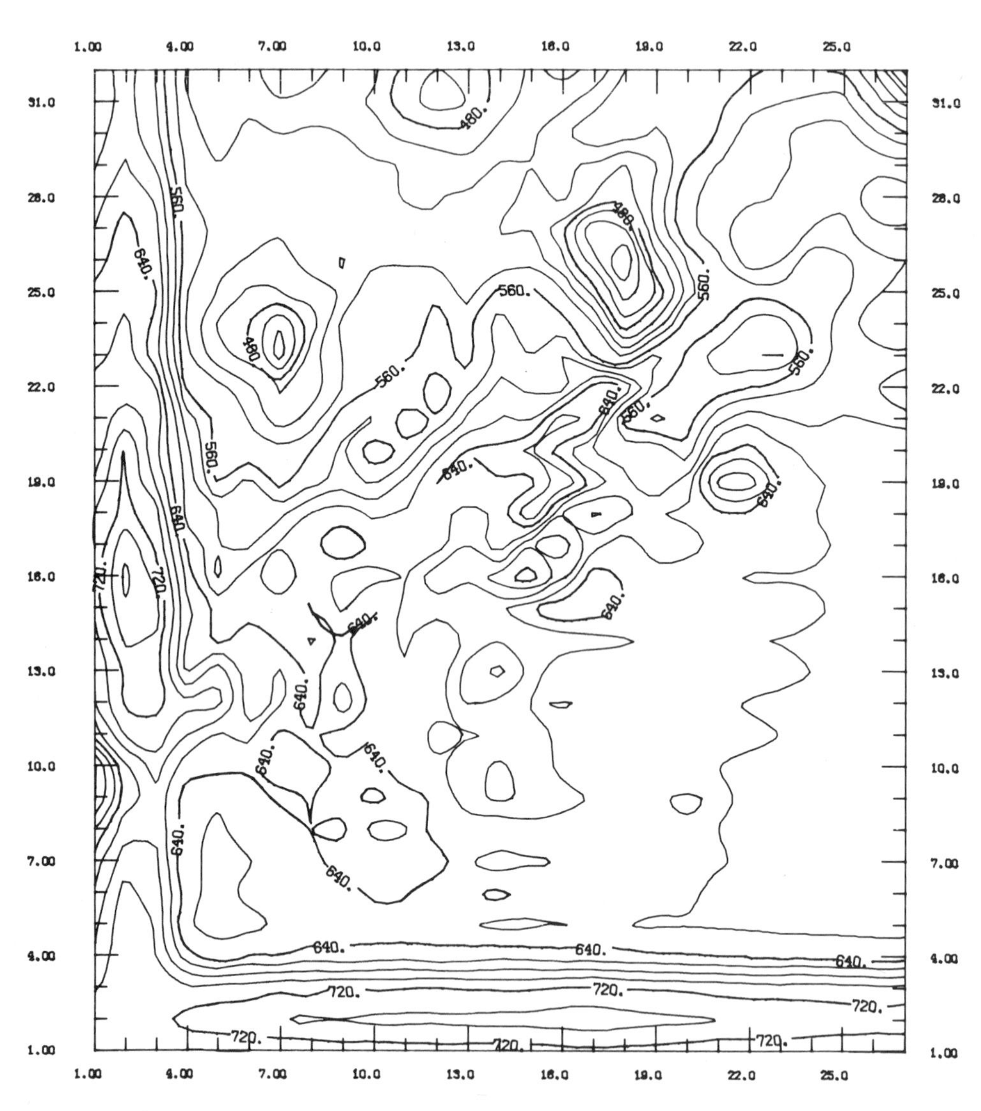

Figure 13. Pseudo-first order reaction rate constants (x108) in s-1 for noon at level-6.

The dry deposition of sulfate is presented graphically in the form of contours in Figure 14. The 160-contour corresponds to a 24-hour average deposition flux of 24 μg m-2 hr-1. The general structure of the deposition amounts is similar to the sulfate concentration contours, but differences due to the non-uniform deposition velocity field do arise.

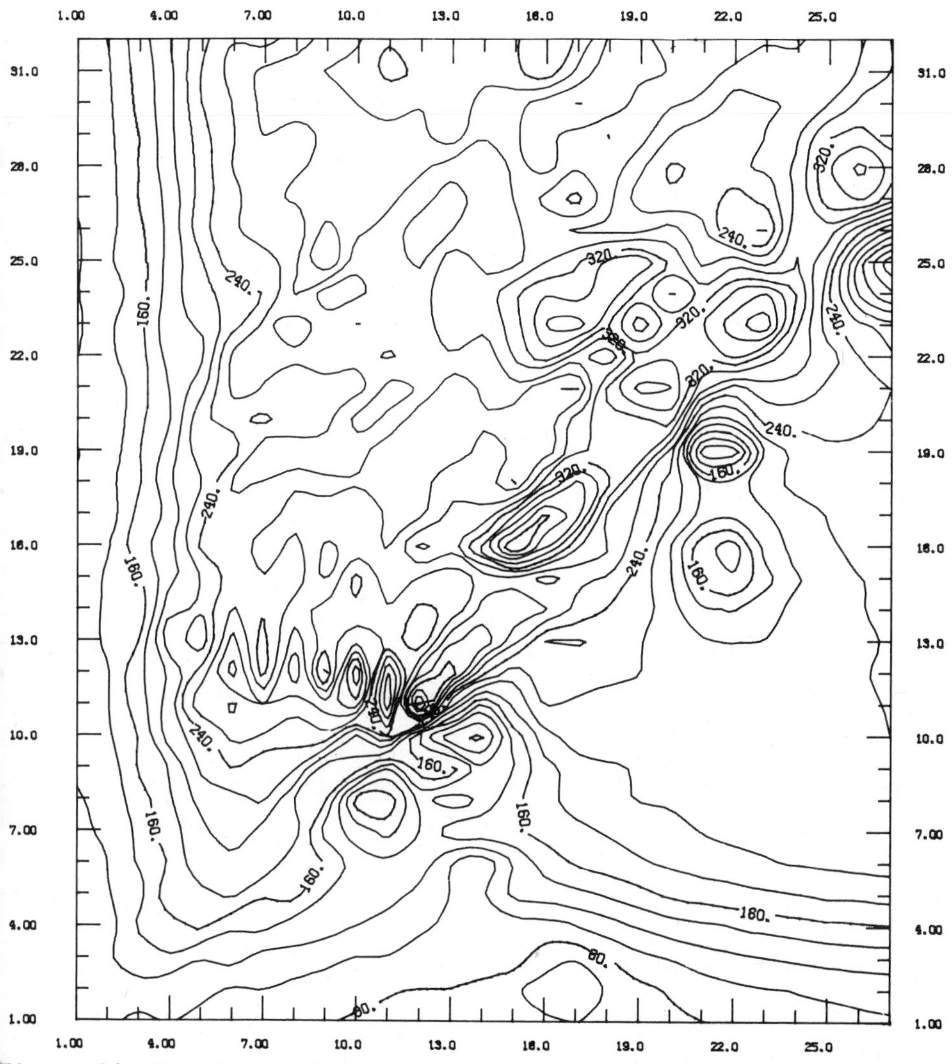

Figure 14. Dry deposition amounts (x102) of sulfate over 24-hour period in μg m^{-2} s-1.

Finally, the SO_2 and sulfate mass balances for the 24-hour period are summarized in Table 4. The sulfate mass balance is very good. The total sulfate either remaining in the region or transported out of the region is estimated from the source and productio terms to be ~1.2×10^{17} μg and that actually remaining in the regio is calculated to be 1.1×10^{17} μg. This leaves ~ 1×10^{16} μg to be transported out of the region. This number is consistent with the inflow term. The SO_2 balance shows an accumulation. However, reca that this simulation was the 24-hour period immediately following t initiation of the sources. Thus, this simulation presents the wors (most stringent) test of the mass balance. For runs on consecutive days the impulse due to the initiation of the sources will not be present.

As shown in the table, the amount of SO_2 deposited exceeds tha of sulfate by a factor of ten. This is due to the combination of higher average deposition velocities and concentrations of SO_2. In addition, for SO_2 the amount deposited exceeds the amount reacted b an order of magnitude, whereas for sulfate, the amount formed by reaction is approximately equal to the amount deposited.

Table 4. 24-hour Model Region Sulfur Balances

	SO_2 μg (as SO_2)	SULFATE μg (as SO_4)
emitted/day	1.3×10^{17}	0
in region at t = 0	6.69×10^{17}	8.62×10^{16}
transported into region/day	1.27×10^{17}	1.34×10^{16}
reacted/day	-5.03×10^{16}	$+ 7.55 \times 10^{1}$
deposited/day	6.82×10^{17}	5.40×10^{16}
in region at t = 24 hrs.	3.58×10^{17}	1.12×10^{17}

ACKNOWLEDGEMENTS

This research was supported in part by the National Aeronautic and Space Administration under Research Grant NAG 1-36 and in part Battelle Pacific NW Labs through the USEPA MAP3S/RAINE program.

Special thanks go to Pamela Rarig for her work on the computer program and computer graphics, to Kay Chambers for making the line drawings, and to Jane Frank for typing the manuscript.

REFERENCES

Bradstreet, J.W., 1973, 66th Annual Meeting of the Air Pollution Control Assoc., Chicago, IL.

Calvert, J.G., 1976, Hydrocarbon involvement in photochemical smog formation in Los Angeles Atmosphere, ES & T, 10:256.

Carmichael, G.R., 1978: "The Regional Transport of SO_2 and Sulfate in the Eastern United States," Ph.D. Thesis, Department of Chemical Engineering, University of Kentucky.

Carmichael, G.R., Kitada, T., and Peters, L.K., 1980, Application of a galerkin finite element method to atmospheric transport problems, Computers and Fluids, 8:155.

Carmichael, G.R., and Peters, L.K., 1979, Numerical simulation of the regional transport of SO_2 and sulfate in the eastern United States, in: "Proceedings of the Fourth Symposium on Turbulent Diffusion and Air Quality," Reno, NV, American Meteor. Soc.

Carmichael, G.R., Yang, D-K, and Lin, C., 1980, A numerical technique for the investigation of the transport and dry deposition of chemically reactive pollutants, Atm. Envirn., 14:1433.

Davis, D.D., Heaps, W., and McGee, T., 1976, Direct measurements of natural tropospheric levels of OH via an aircraft borne turnable dye laser, Geoph. Res. Letters, 3:331.

Fairweather, G., 1978, "Finite Element Galerkin Methods for Differential Equations, Lecture Notes in Pure and Applied Mathematics, Vol. 34," Marcel Dekker, Inc., New York, NY.

Fisher, B.E.A., 1978, The calculation of long term sulfur deposition in Europe, Atm. Envirn., 12:489.

Fowler, D., 1978, Dry deposition of SO_2 on agriculture crops, Atm. Envirn., 12:369.

Graedel, T.E., Farrow, L.A., and Weber, T.A., 1976, Kinetic studies of the photochemistry of the urban tropophere, Atm. Envirn., 10:1095.

Hicks, B.B., and Wesely, M.L., 1978, MAP3S Newsletter, U.S. Dept. of Energy, April/June.

Hidy, G.M., Tong, E.Y., Mueller, P.K., Rao, S., Thomson, E., Berlandi, F., Muldoon, D., McNaughton, D., and Majahad, A., 1976, "Design of the Sulfate Regional Experiment," PB-251 701.

Holzworth, G.C., 1964, Estimates of mean maximum mixing depths in the contiguous United States, Mon, Wea. Rev., 92:235.

Kocmond, W.C., and Yang, J.Y., 1976, "Sulfer Dioxide Photooxidation Rates and Aerosol Formation Mechanisms: A Smog Chamber Study," Calspan Corp., PB-260 910.

Kummler, R.H., and Baurer, T., 1973, A temporal model of tropospheric carbon-hydrogen chemistry, JGR, 24:5306.

Lin, C., 1980, "Regional Scale Pollutant Transport Modeling,: M.S. Thesis, Chemical and Materials Engineering Program, University of Iowa.

Myrup, L.O., and Ranzieri, A.J., 1976: "A Consistant Scheme for Estimating Diffusivities to be Used in Air Quality Models," PB-272 484.

Perner, D., Ehhalt, D.H., Patz, H.W., Platt, E.P., Roth, E.P., and Volz, A., 1976, OH-radicals in the lower troposphere, Geophys. Res. Lett., 3:466.

Rayland, K.W., Dennis, R.L., and Wilkening, K.E., 1975: "Boundary-Layer Model for Transport of Urban Air Pollutants," Presented at the Nat'l Meeting of AIChE, March.

Sheih, C.M., 1978, A puff-on-cell model for computing pollutant transport and diffusion, J. Appl. Meteor., 17:140.

Turner, D.B., 1961, Relationships between 24-hour mean air quality measurements and meteorological factors in Nashville, TN, JAPCA 11:483.

Wang, C.C., Davis, L.I., Wu, C.H., Japer, S., Niki, H., and Weinstock, B., 1975, Hydroxyl radical concentrations measured in ambient air, Science, 189:797.

Yanenko, N.N., 1971: "The Method of Fractional Steps," Springer, Berlin.

Yordanov, D., 1972, Simple formula for determining the concentration distribution of high sources, Atm. Envirn., 6:389.

DISCUSSION

A. VENKATRAM

How does your model compare with the SURE data set?

G. R. CARMICHAEL

The model is currently simulating July 1979 and July 1978 episodes. Model comparison with these data will be available in two to six months.

H. N. LEE

Since the finite element technique is used, is any reason why the irregular vertical region is transformed to the rectangular region?

G. R. CARMICHAEL

We have elected to transform the region to facilitate the implementation of the L.O.D. numerical method. This allows changes in the grid system to be handled in an easy manner.

PARTICLE SIMULATION OF INHOMOGENEOUS TURBULENT DIFFUSION

Lutz Janicke

Dornier System GmbH

D-7990 Friedrichshafen

INTRODUCTION

The dispersion of aerosols by turbulent diffusion can be modelled on a computer by following the paths of a group of particles, whose actual turbulent velocities are simulated by random movements. This particle simulation is a useful model of turbulent diffusion - as far as statistical averages are considered -, if the most important properties of the actual physical process are reproduced on the computer.

Mass conservation and positivity of mass density are obviously guaranteed. Being interested in predicting the long time behaviour of the concentration field we demand that the equilibrium state is correctly reproduced:

- the particle density has to be constant in space,
- mean value and variance of the particle velocities must be the same as in the ambient air.

For inhomogeneous conditions it is not obvious how to meet these requirements. In the following we shall describe a method to analyse the properties of a class of particle simulation models analytically and for some simple cases, relations will be given for the correct form of the model parameters.

THE SIMULATION MODEL

If we denote the position and the velocity of a particle by $\underset{\sim}{r}(t)$ and $\underset{\sim}{v}(t)$ respectively*, a class of simulation models uses the following algorithm to advance in time from t_n to $t_{n+1} = t_n + \Delta t$:

$$\underset{\sim}{r}_{n+1} := \underset{\sim}{r}(t_{n+1}) = \underset{\sim}{r}_n + \Delta t \, \underset{\sim}{v}_n, \tag{1a}$$

$$\underset{\sim}{v}_{n+1} := \underset{\sim}{v}(t_{n+1}) = \hat{\underset{\sim}{u}}(\underset{\sim}{r}_{n+1}) + \hat{\underset{\sim}{v}}(\underset{\sim}{r}_{n+1}, t_{n+1}) . \tag{1b}$$

$\hat{\underset{\sim}{u}}(t)$ is an adjustable velocity field, $\hat{\underset{\sim}{v}}$ is generated by a Markov process:

$$\hat{\underset{\sim}{v}}_{n+1}(\underset{\sim}{r}) := \hat{\underset{\sim}{v}}(\underset{\sim}{r}, t_{n+1}) = \underset{\sim}{\alpha}(\underset{\sim}{r}) \cdot \hat{\underset{\sim}{v}}_n(\underset{\sim}{r}) + \underset{\sim}{w}_{n+1}(\underset{\sim}{r}) . \tag{2}$$

The $\underset{\sim}{w}_n(\underset{\sim}{r})$ are random numbers with probability density $p(\underset{\sim}{w}, \underset{\sim}{r})$, statistically independent for different values of n. The tensor $\underset{\sim}{\alpha}(\underset{\sim}{r})$ is diagonal and related to the correlation times $T_i(\underset{\sim}{r})$ by

$$\alpha_{ij} = \delta_{ij} - \Delta t \, \tau_{ij}, \quad \tau_{ij} = \delta_{ij}/T_i . \tag{3}$$

All particles are moved independently.

The scheme used by Lamb et al.[1] to simulate road emission is of this type. Prescribing the mean velocity of the air, $\underset{\sim}{u}(\underset{\sim}{r})$, and the variance of the turbulent velocities, $\sigma_i(\underset{\sim}{r})$, they set

$$\hat{\underset{\sim}{u}}(\underset{\sim}{r}) = \underset{\sim}{u}(\underset{\sim}{r}) \tag{4a}$$

and choose for $p(\underset{\sim}{w}, \underset{\sim}{r})$ a gaussian distribution with first and second moment

$$\overline{\underset{\sim}{w}} = o, \tag{4b}$$

$$(\overline{\underset{\sim}{w}\underset{\sim}{w}})_{ij} = \delta_{ij} \left[1 - \alpha_{ii}^2 (\underset{\sim}{r})\right] \sigma_i^2 (\underset{\sim}{r}). \tag{4c}$$

---

*Vectors are represented by wavy underlined symbols, the subscripts x,y,z denoting the components are sometimes replaced by i = 1,2,3.

This choice is correct under homogeneous conditions. However, if σ_i depends on $\underset{\sim}{r}$, considerable errors in the computed particle density occur, as can be seen in the following example.

We take a 1-dimensional periodic system with $0 \leq x \leq L_1 = 16$, $u = 0$, $T(x) = 5$, $\sigma(x) = .15\ [3-\cos\ (\pi x/L_1)]$ and start the simulation with 5000 particles uniformly distributed in x, the velocity distribution is a Maxwellian of a width $\sigma(x)$. The size of the time step is $\Delta t = 1$ and the particles are reflected at the boundaries. Fig. 1 shows that the model approaches an equilibrium state with a highly varying particle density. Apparently the region with small σ acts like a trap by freezing the particle movement, thereby increasing the particle density. However, this error can be avoided by adjusting the model parameters $\underset{\sim}{\overline{w}}$ and $\underset{\approx}{\overline{ww}}$.

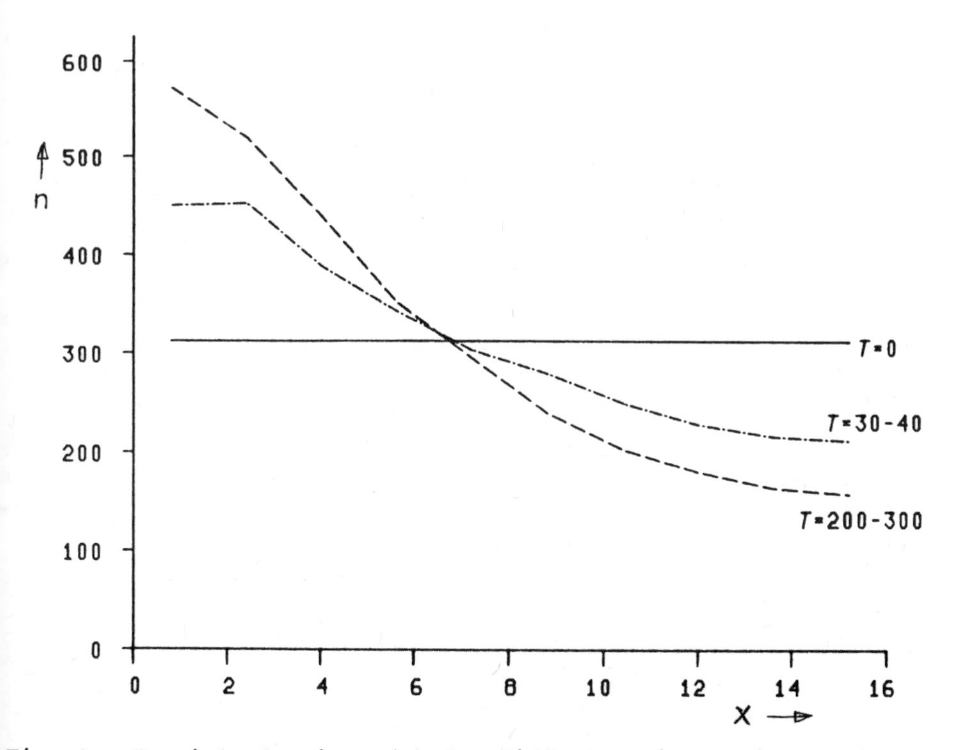

Fig. 1. Particle density n(x) for different times t in a model with $\overline{w}$ = o. Time averages are taken to smooth statistical fluctuations.

THE KINETIC EQUATION OF THE MODEL

Let $f(\underset{\sim}{r},\underset{\sim}{v},t)$ be the distribution density in phase space for the particles of our simulation model. Then the particle density n and the mean velocity $\underset{\sim}{V}$ are given by

$$n(\underset{\sim}{r},t) = \int f(\underset{\sim}{r},\underset{\sim}{v},t)\, d^3\underset{\sim}{v}, \tag{5a}$$

$$\underset{\sim}{V}(\underset{\sim}{r},t) = \frac{1}{n} \int \underset{\sim}{v} f(\underset{\sim}{r},\underset{\sim}{v},t)\, d^3\underset{\sim}{v}. \tag{5b}$$

In order to do approximations in a systematic way we introduce

L_o = typical length,
σ_o = typical velocity,
T_o = typical correlation time = $O\,(L_o/\sigma_o)$,

and consider small time steps,

$$\varepsilon = \Delta t/T_o \ll 1. \tag{6}$$

We assume $|\underset{\sim}{w}| = O\,(\sqrt{\varepsilon}\,\sigma_o)$ and $|\overline{\underset{\sim}{w}}| = O\,(\varepsilon\,\sigma_o)$. Using $\varepsilon \ll 1$ and Equations (1) and (2) for a particle with velocity $\underset{\sim}{v}'$ and position $\underset{\sim}{r}'$ at time t, we get after one time step with random velocity increment $\underset{\sim}{w}$ the new position $\underset{\sim}{r}$ and velocity $\underset{\sim}{v}$:

$$\underset{\sim}{r} = \underset{\sim}{r}' + \Delta t\, \underset{\sim}{v}' \tag{7a}$$

$$\underset{\sim}{v} = \underset{\sim}{v}' \cdot \left[\alpha(\underset{\sim}{r}') + \Delta t\, (\nabla\hat{\underset{\sim}{u}}(\underset{\sim}{r}')) \cdot \alpha\, (\underset{\sim}{r}')\right] + \Delta t\;\; \hat{\underset{\sim}{u}}(\underset{\sim}{r}') \cdot \tau(\underset{\sim}{r}') + \underset{\sim}{w}(\underset{\sim}{r}')\,. \tag{7b}$$

Therefore, the distribution function at time t + Δt is given by

$$f(\underset{\sim}{r},\underset{\sim}{v},t+\Delta t) = \iiint f(\underset{\sim}{r}',\underset{\sim}{v}'t)\; \delta(\underset{\sim}{r}-\underset{\sim}{r}' - \Delta t\underset{\sim}{v}')\; \delta(\underset{\sim}{v}-\underset{\sim}{v}' \cdot \left[\alpha + \Delta t\, (\nabla\hat{\underset{\sim}{u}}) \cdot \alpha\right] - \Delta t\hat{\underset{\sim}{u}}\cdot\tau - \underset{\sim}{w})\; p\,(\underset{\sim}{w}, \underset{\sim}{r}')\, d^3\underset{\sim}{w}\, d^3\underset{\sim}{r}'\;\, d^3\underset{\sim}{v}'\;\,. \tag{8}$$

Using $\varepsilon \ll 1$ again we get finally

$$\frac{\partial f}{\partial t} + \underset{\sim}{v} \cdot \frac{\partial f}{\partial \underset{\sim}{r}} = (\tau_I - \nabla\cdot\underset{\sim}{\hat{u}})f + \Big[\tau \cdot (\underset{\sim}{v}-\underset{\sim}{\hat{u}}) - \underset{\sim}{v} \cdot \nabla\underset{\sim}{\hat{u}}$$
$$-\underset{\sim}{\hat{w}}\Big] \cdot \frac{\partial f}{\partial \underset{\sim}{v}} + \frac{1}{2}\,\omega\cdot\cdot\frac{\partial^2 f}{\partial \underset{\sim}{v}\partial \underset{\sim}{v}} \tag{9}$$

with $\tau_I = \mathrm{tr}(\tau) = \Sigma 1/T_i$, $\underset{\sim}{\hat{w}} = \overline{\underset{\sim}{w}}/\Delta t$, $\omega = \overline{\underset{\sim}{w}\underset{\sim}{w}}/\Delta t$. This equation describes all properties of our simulation model in the limit of small time steps. It should be noted, that only the first and second moment of the probability density p enter, its detailed structure has no influence.

THE CORRECT CHOICE OF THE MODEL PARAMETERS

In most applications the meteorological parameters depend on height z only. Our goal is to find the conditions giving the correct equilibrium state described in the introduction. We therefore specialize the case where all model parameters and the solution f itself are independent of x and y and $u_z = w_x = w_y = 0$. Then with

$$\begin{aligned} T(z) &\equiv T_z \\ v &\equiv v_z, \\ g(z) &= T\hat{w}_z, \\ \Omega^2(z) &= \frac{1}{2}\,T\omega_{zz} \end{aligned}$$

we get for the reduced particle distribution function F,

$$F(z,v,t) = \int f(\underset{\sim}{r},\underset{\sim}{v},t)\, dv_x\, dv_y,$$

by integrating Equation (9) with respect to v_x and v_y:

$$\frac{\partial F}{\partial t} + v\,\frac{\partial F}{\partial z} = \frac{1}{T}\left[F + (v-g)\,\frac{\partial F}{\partial v} + \Omega^2\,\frac{\partial^2 F}{\partial v^2}\right] . \tag{10}$$

Equilibrium is defined by $\frac{\partial}{\partial t} \equiv 0$. Introducing a new space coordinate s by

$$s = \int_0^z \frac{dz'}{T(z')} , \tag{11}$$

we have to study

$$v\,\frac{\partial F}{\partial s} = F + (v-g)\,\frac{\partial F}{\partial v} + \Omega^2\,\frac{\partial^2 F}{\partial v^2} . \tag{12}$$

Taking the velocity momenta of this equation we get with

$$q_n(s) = \int_{-\infty}^{+\infty} v^n \; F(s,v)dv$$

for the first four equations of the infinite hierarchy:

$$q_1' = 0, \quad (13a)$$

$$q_2' = gq_o - q_1, \quad (13b)$$

$$q_3' = 2gq_1 - 2q_2 + 2\Omega^2 q_o, \quad (13c)$$

$$q_4' = 3gq_2 - 3q_3 + 6\Omega^2 q_1. \quad (13d)$$

For the correct equilibrium state we have

$$q_o = \text{const},$$

$$q_1 = 0,$$

$$q_2 = q_o \sigma^2.$$

From Equation (13b) it follows immediately that

$$g = \frac{d}{ds}\sigma^2 \quad \text{or} \quad (14)$$

$$\overline{w}_z(z) = \Delta t \frac{d}{dz}\sigma_z^2.$$

The tendency of the model to trap particles in regions with low σ must be compensated by additional momenta given to the particles in each time step. With $g = 0$ an equilibrium state with $q_2' = 0$ or $q_o \sim \sigma^{-2}$ would result as shown in Fig. 2. The improvement by the described momentum correction can clearly be seen.

If the correlation time T is short, we can neglect the left hand side of Equation (13c) and get the exspected result $\Omega = \sigma$. Using to lowest order $q_4 = 3\sigma^2 q_2$, valid for a Maxwellian of a width σ, we get from (13d)

$$q_3 = -\frac{1}{2}\frac{d}{ds}\sigma^4. \quad (15)$$

The skewness of the velocity distribution function is characterized by

$$\gamma_1 = \frac{q_3}{q_o \sigma^3} = -2\, T_z \frac{d\sigma_z}{dz}. \quad (16)$$

Now the first correction of Ω^2 can be calculated:

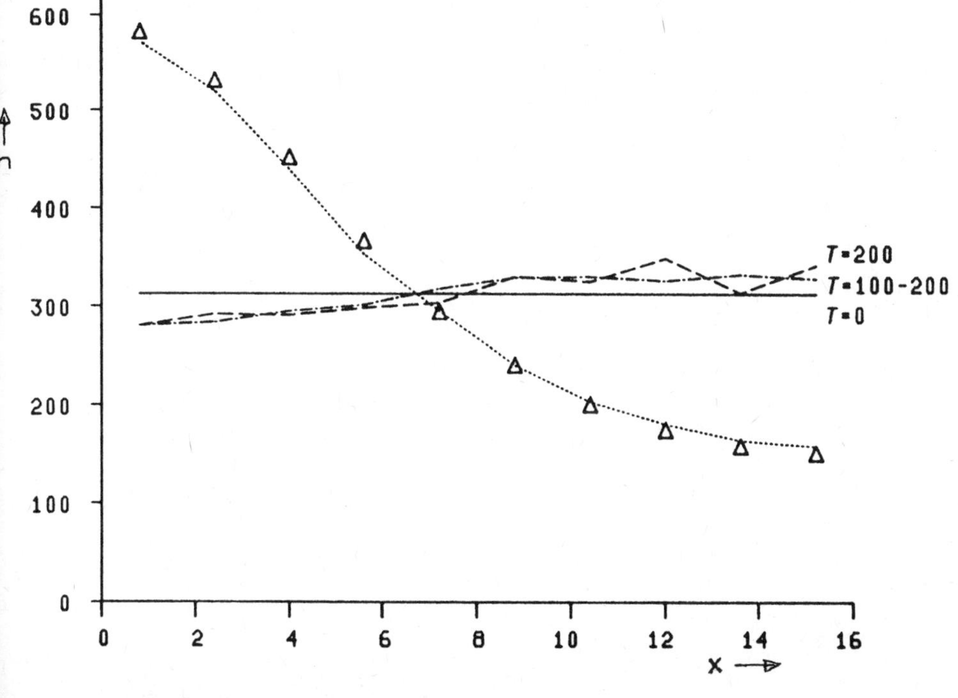

Fig. 2. Particle density n(x) for different times t in a model with momentum correction (Equation(14)). The dotted line is the final state in the uncorrected model agreeing well with the theoretical prediction $n \sim \sigma^{-2}$ (symbol Δ)

$$\Omega^2 = \sigma^2 - \frac{1}{4}\frac{d^2}{ds^2}\sigma^4 \qquad \text{or} \tag{17}$$

$$\overline{w_z^2} = 2\frac{\Delta t}{T_z}\left[\sigma_z^{\ 2} - \frac{1}{4} T_z \frac{d}{dz} T_z \frac{d}{dz}\sigma_z^{\ 4}\right].$$

This relation points to a difficulty in simulating turbulent diffusion in an unstable atmosphere. Using

$$T_z = \text{const}, \ \sigma_z = 1.3\ u^* \ (1-3z/L)^{1/3}$$

with u^* = friction velocity and L = Monin-Obukhov length, the correction term is small only if

$$(Tu^*/L)^2 \ll 0.5.$$

With T_z = 30 sec (Lamb et al. [1]) only weakly unstable stratifications can be treated. The situation can be improved by using a functional form of T(z) where $T \rightarrow 0$ with $z \rightarrow 0$ (Hanna [2]). If the correction term in Equation (17) is not small, a direct solution of Equation (12) is necessary to find the correct form of Ω^2.

The parameters w_x^2, w_y^2 and u cannot be determined from the 1-dimensional model. However, for small correlation times we can set again

$$\overline{w_x^2} \approx 2\frac{\Delta t}{T_x}\sigma_x^{\ 2}, \tag{18a}$$

$$\overline{w_y^2} \approx 2\frac{\Delta t}{T_y}\sigma_y^{\ 2}, \tag{18b}$$

$$\hat{\underset{\sim}{u}} \approx \underset{\sim}{u}.$$

CONCLUSION

The particle simulation model of inhomogeneous turbulent diffusion, where the particle velocities result from a Markov process, can be described by a kind of kinetic equation. This equation can be used to determine the model parameters by applying the condition that the equilibrium state is correctly reproduced.

In case, the meteorological parameters depend on height only, the result is that the vertical component of the random velocities must have a nonzero mean and a variance that is corrected for long correlation times. Difficulties in the case of an unstable stratification are predicted, where restrictions on the spatial dependence of the vertical correlation time exist.

REFERENCES

1. R.G. Lamb, H. Hogo and L.E. Reid, A Lagrangian Approach to Modeling Air Pollutand Dispersion, Report EPA-600/4-79-023 (1979).
2. S.R. Hanna, Effects of Release Height on σ_y and σ_z in Daytime Conditions, Proc. 11th NATO-CCMS Int. Techn. Meeting on Air Pollution Modeling and its Application, Part I, 198:215 (1980).

DISCUSSION

H. VAN DOP

If you would disperse your particles in a volume, which is strongly turbulent in, say the upper half, and very weakly turbulent in the lower half of the volume, then you would expect that once a "marked" particle enters the low turbulent regime, it will not be able to emerge from it. Accumulation will occur in the lower volume, which would contradict the entropy law. How do you resolve this problem?

L. JANICKE

The problem is not primarily associated with the simulation model but rather with your picture. The model "knows" where the turbulence becomes weaker and prevents accumulation by pushing the particles out of this region by means of the additional momentum given by Equation (14).

The problem with your picture can be resolved by observing that the turbulent flow can be regarded as incompressible. If each particle is identified with a certain volume element the deposition in the lower half is possible only by exchanging volume elements. With constant particle density in the whole volume the probability for being trapped in the lower half is the same as that for coming free into the upper half. Therefore, no accumulation occurs.

THE APT COMPUTER PROGRAM FOR THE NUMERICAL SOLUTION OF PROBLEMS IN ATMOSPHERIC DISPERSION

A. Ghobadian, A. J. H. Goddard and A. D. Gosman

Mechanical Engineering Department
Imperial College
Exhibition Road
London, ENGLAND

INTRODUCTION

The analysis of the dispersion of pollutants issuing from sources near the ground into the atmospheric boundary layer has traditionally been accomplished via so-called Gaussian plume models. It is assumed that the flow field properties are invariant in the lateral and down-wind directions which, along with other simplifying assumptions, allows the concentration field to be described by a Gaussian form; simplicity being one of the models' main attractions. Such models may be modified to take account of the duration of release and the frictional drag of the underlying surface (Clarke 1979). Dispersion may be treated by 'K' profile models (for example Maul (1977)) but it is difficult to specify the diffusivity profile from routine meteo-rological measurements.

In practise the atmospheric boundary layer often suffers perturbations caused by changes in surface roughness and/or heat flux producing internal layers, which destroy horizontal homogeneity and alter the vertical characteristics. These features in turn may have profound effects on the dispersion processes as discussed, for example, by Raynor et al (1980), such that simple models are rendered inapplicable. The present paper describes a more fundamentally-based method for simulating the effects of changes in surface conditions of the kind described above through numerical solution of the ensemble-averaged partial differential transport equations governing the flow, heat transfer and mass transfer processes, in which the turbulent fluxes are represented by a second-order turbulence model. The method is embodied in a computer code named APT, standing for Atmospheric Pollutant Transport.

APT was specifically devised for circumstances in which (i) the flow and dispersal processes are amenable to the usual ensemble averaging employed in turbulence theory and (ii) the flow is of the two- dimensional boundary-layer kind, with spatial variations confin to the vertical and windward directions. Moreover the plume is assumed to possess negligible momentum and buoyancy, and the topography is taken to be flat. Within this framework, the method i quite general and, in particular, admits wide windward variations in surface roughness and heat flux as well as atmospheric conditions.

The accuracy of the method has been assessed in a preliminary way by reference to wind-tunnel investigations of point and line source releases into neutral boundary layers on smooth and rough surfaces; an account of this work is given by El Tahry et al (1981) who show that there is good accord between APT predictions and experimental data. Further applications of the method to the prediction of hypothetical atmospheric dispersion problems have been presented by Harter et al (1980) and Goddard et al (1980); unfortunately however the quantitative assessment of the method for such situations has been impeded by the usual difficulties of identifying detailed atmospheric data for well-defined circumstances.

The purposes of the present paper are twofold. Firstly, the ability of APT to simulate the effects of windward changes in surfac conditions is demonstrated and assessed. Assessment is by compariso with wind profile measurements at a coastal site where the transitio from sea to land conditions gives rise to formation of an internal boundary layer, while the demonstrations consist of calculations showing the effects of different levels of surface heat flux and roughness on the growth of such layers and their influence on the dispersion process.

The second purpose is to describe work in progress to extend the APT methodology to allow simulation of Coriolis effects. Inclusion of these effects into the flow model is outlined and comparisons are made with the detailed predictions of Deardorff (1972) for neutral conditions.

OUTLINE OF PREDICTION METHOD

In this section the mathematical model embodied in the original APT method and the numerical method of solution are firstly outlined; a more detailed account is available in the thesis of El Tahry (1979). Then the extension to include Coriolis effects is described.

Mathematical model

The assumption of stationary, two-dimensional, fully-turbulent boundary layer behaviour, which excludes consideration of strongly-unstable conditions and Coriolis effects along with the other constraints mentioned earlier, gives rise to the following equations for the fluid dynamics in Cartesian tensor notation:

$$\frac{\partial(\rho U_1)}{\partial x_1} + \frac{\partial(\rho U_3)}{\partial x_3} = 0 \qquad (1)$$

$$\frac{\partial(\rho U_1^2)}{\partial x_1} + \frac{\partial(\rho U_1 U_3)}{\partial x_3} = -\frac{\partial P}{\partial x_1} - \frac{\rho\partial\overline{u_1 u_3}}{\partial x_3} \qquad (2)$$

where U and u represent the ensemble-average and fluctuating velocities respectively, subscripts 1 and 3 refer to the longitudinal and vertical directions, ρ and P are respectively density and pressure and the overbar denotes ensemble averaging. The distribution of temperature T in the boundary layer, which is also assumed to be two-dimensional, is governed by:

$$\frac{\partial(\rho U_1 T)}{\partial x_1} + \frac{\partial(\rho U_3 T)}{\partial x_3} = -\frac{\rho\partial\overline{u_3 T'}}{\partial x_3} \qquad (3)$$

where T' is the fluctuating component. Finally, the concentration C of plume material is governed by the three dimensional concentration equation:

$$\frac{\partial(\rho U_1 C)}{\partial x_1} + \frac{\partial(\rho U_3 C)}{\partial x_3} = -\frac{\partial}{\partial x_2}\left(\rho\overline{u_2 c}\right) - \frac{\partial}{\partial x_3}\left(\rho\overline{u_3 c}\right) \qquad (4)$$

where c is the fluctuating component.

The turbulent fluxes in the above equations, which have the general form $\rho\overline{u_i\phi}$, where ϕ may be u_j, T' or c, are obtained from second-order transport models having their origins in the work of Launder et al (1975, 1976) and Gibson and Launder (1978). The details are too lengthy to be presented here but may be found in El Tahry (1979). Briefly, each flux component has its own transport equation of the kind:

$D(\rho\overline{u_i\phi})/Dt$ = Diffusion + (Production by Mean Gradients) + (Production/Destruction by Buoyancy) + Redistribution - (Destruction by Molecular Action)

where D/Dt is the substantive derivative and the labels on the right hand side denote groups of terms representing the physical processes in question. These differential equations are reduced to algebraic ones on the assumption that the transport of $\rho\overline{u_i\phi}$ by convection and diffusion is proportional to the transport of turbulence energy $k \equiv u_i u_i/2$ by the same process. It is therefore only necessary to solve a differential transport equation for k and an additional one for the turbulence dissipation rate ε, which effectively provides th length scale for the transport processes: both equations are, withir the boundary layer context, of the general form of (3), but with additional source and sink terms. The solution of these equations f the major part of the boundary layer is matched to a simplified representation of the flow near the ground based on one-dimensional equilibrium flow theory, so as to avoid problems of applicability ar numerical resolution.

Numerical solution procedure

The foregoing equations are solved by an implicit, non-iterativ forward-marching finite-difference procedure which is conventional in many respects, the most noteworthy features being the use of computational grids which are: (i) different for the plume and boundary layer, in order to allow for the usually disparate scales c the two; and (ii) self-adjusting in a way which confines the calcul- ations to regions of significant variations in the dependent variabl These features, along with the other effort-saving facets of the modelling described earlier, give rise to a particularly economical prediction method, as will be demonstrated.

Inclusion of Coriolis Forces

For the purpose of initial development and assessment of a vers of the method incorporating Coriolis effects a non-stationary horizontally-homogeneous flow model has been developed, whose govern equations are:

$$\frac{\partial(\rho U_1)}{\partial t} = - \frac{\partial}{\partial x_3}\left(\rho\overline{u_1 u_3}\right) + f(U_2 - U_{2,g}) \tag{6}$$

$$\frac{\partial(\rho U_2)}{\partial t} = - \frac{\partial}{\partial x_3}\left(\rho\overline{u_2 u_3}\right) + f(U_{1,g} - U_1) \tag{7}$$

where x_1 and x_2 are in the easterly and northerly directions respectively and $f \equiv 2\Omega\sin\theta$ is the component of the earth's rotation vector in the x_3 direction. The quantities $U_{1,g}$ and $U_{2,g}$ are the components of the geostrophic wind which prevails at the outer edge of the boundary layer and is governed, in the absence of turbulence, by the following reduced forms of the equations of motion:

$$0 = -\frac{\partial P}{\partial x_1} + fU_{2,g} \quad (8)$$

$$0 = -\frac{\partial P}{\partial x_2} - fU_{1,g} \quad (9)$$

The temperature distribution within the boundary layer is obtained from

$$\frac{\partial(\rho T)}{\partial t} = -\frac{\partial}{\partial x_3}\left(\overline{\rho u_3 T'}\right) \quad (10)$$

The turbulent flux terms in these equations are obtained from the same type of closure model employed in the earlier method, but in this case the Coriolis effects are also allowed for in the development of the stress equations: full details will be provided in a later publication. The equations are solved by the same numerical procedure, now employed in a time-marching sense. At this stage the plume dispersion process has not yet been included in the analysis.

ILLUSTRATIVE APPLICATIONS OF APT

Three examples have been chosen to illustrate the ability of the APT method to model downwind changes in surface roughness and sensible heat flux. The first example simulates published experimental work where the vertical windspeed profile downwind of a shore-line has been measured in order to examine the growth of the internal boundary layer. In the second example a hypothetical shore-line situation has been simulated and a limited parametric study carried out to examine both the influence of roughness and sensible heat flux upon growth of the internal boundary layer and upon dispersion from an elevated stack. The third example simulates fumigation due to substantial changes in sensible heat downwind such as might be encountered in a plume above suburban and urban regions.

Internal Boundary Layer Depth at a Beach Site

Echols and Wagner (1972) have made wind profile measurements at a site near the upper Texas coast and, in addition to deriving local land and sea roughness lengths, have detected an internal boundary layer at a tower 90 metres inland. The terrain between shore and tower was essentially flat sand and clay marshland. Echols and Wagner reported a small Richardson number and hence they ignored buoyancy influences on the velocity profile; they interpreted their daytime wind profile above and within the internal boundary layer to yield a roughness length of approximately 0.0002 cm over water and approximately 3.5 cm on land. They attributed the low value over water to possible smoothing effects of oil and other pollutants. The fact that they discounted heating at the ground suggests that the value of roughness length that was reported may represent an upper limit.

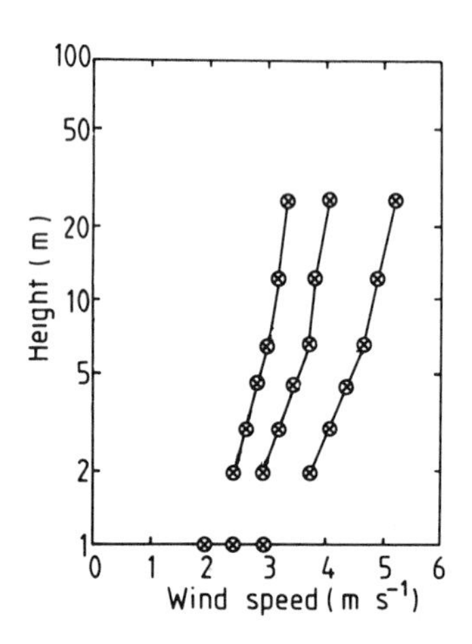

Figure 1. Wind speed as a function of height, for three cases studied by Echols and Wagner, whose experimental points are shown, together with APT predictions (continuous line).

APT calculations of wind profile have been performed using as input data a geostrophic windspeed chosen to match the 10 metre wind-speed inferred from Echols and Wagner's data and sea and inland roughness lengths as determined by these authors. A preliminary fetch over water under neutral conditions was allowed in the APT calculations. With the boundary conditions assumed inclusion of a plausible sensible heat flux over land did not have a significant effect upon wind speed profiles. Figure 1 shows APT calculated wind-speed profiles compared with the Echols and Wagner measured profiles. The agreement is good, with the APT internal boundary layer height decreasing with increasing windspeed as suggested by Echols and Wagner. While not a completely independent test, such a comparison does give confidence in the ability of the APT model to simulate the development of the internal boundary.

Parametric Study of the Effect of a Shoreline Roughness and Sensible Heat Flux Change upon Dispersion

The growth of an internal boundary layer at a shore-line is an important consideration in the siting of meteorological instruments and in the assessment of dispersion from a stack located at the shoreline. This question has been extensively reviewed from the point of view of available experimental data and simple models by Raynor et al (1980) and the reader is refered to this paper for a summary of the work of many other authors. In the restricted study reported here with the APT model, a neutral flow over the sea with an initial fetch of 20 kilometres and a roughness length of 5×10^{-4} metres has been assumed. With a zero sensible heat flux over land, calculations have been performed with three roughness lengths, 2, 10 and 40 cm, approximately simulating open grass land, root crops and park land

Table 1. Internal boundary layer heights (m) from parametric study for various roughness lenghts (z_o) and heat fluxes (H).

Case	z_o (cm)	H ($W\ m^{-2}$)	Distance from change in terrain (m)			
			500	1000	2000	3000
1	2	0	24	40	67	99
2	10	0	35	54	95	148
3	40	0	45	74	140	210
4	10	30	45	116	250	530
5	10	55	60	170	460	630*
6	10	124	114	280	620*	-

* Internal and main boundary layer heights coincide.

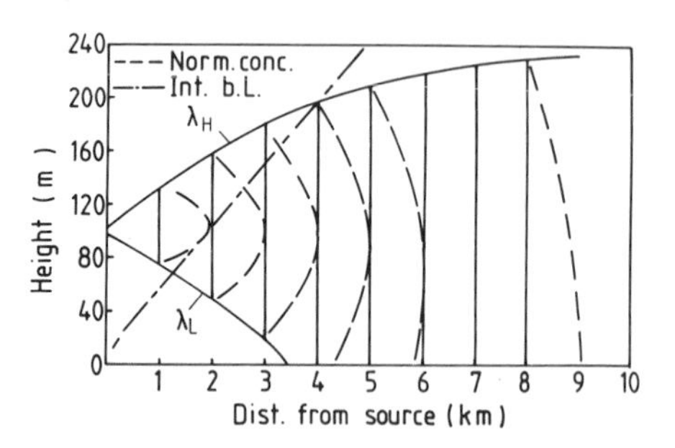

Figure 2. Case 2 of parametric study. Spread of plume and internal boundary layer growth plotted for neutral flow over sea and 10 cm roughness length and zero heat flux inland.

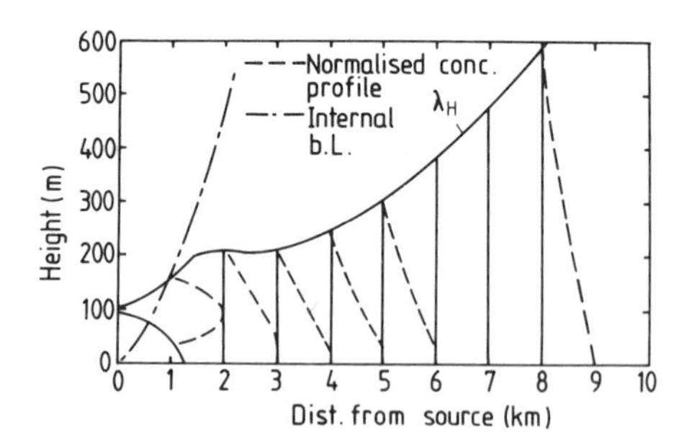

Figure 3. Case 5 of parametric study. As for Figure 2 with 10 cm roughness length and sensible heat flux 55 W m^{-2} inland.

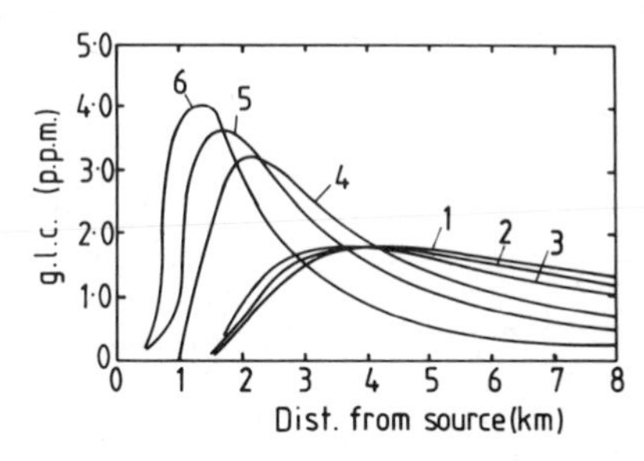

Figure 4. On-axis ground level concentrations for the six cases of the parametric study, where cases 1-3 represent increasing roughness inland with zero heat flux and cases 4-6 increasing heat flux with 10 cm roughness length.

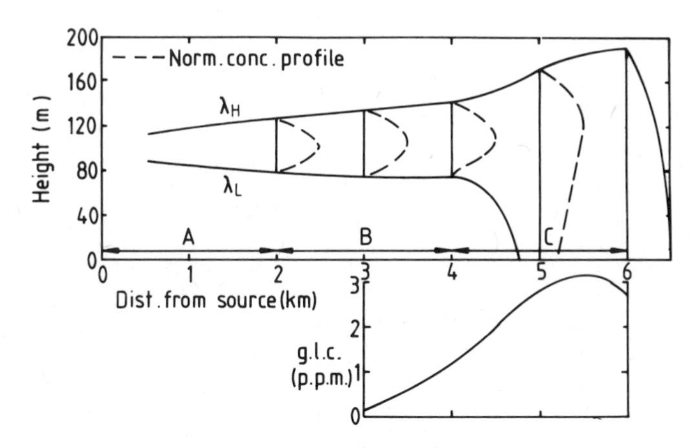

Figure 5. Plume dispersal in moderately stable flow (region A), passing over simulated suburban (B) and urban (C), with sensible heat fluxes -10, 30 and 100 W m^{-2} respectively.

respectively. In a second series of three calculations a roughness length of 10 cm has been retained and three alternative sensible hea fluxes of 30, 55 and 124 W m^{-2} have been assumed, corresponding approximately to stability categories C/D, C and B respectively. In all cases a geostrophic windspeed of 6 metres per second has been imposed. A 100 m stack releasing pollutant at an arbitrary rate of 0.3 kg s^{-1} has been assumed, sited at the shoreline.

Normalised plume profiles and 50% concentration boundaries are illustrated in Figures 2 and 3 as a function of downwind distance, fc two of the six cases, namely a roughness length of 10 cm with zero ar 55 W m^{-2} sensible heat flux respectively.

For each of the six cases the height of the internal boundary layer, defined where the fractional change in turbulent energy produc by the internal layer equals 0.1, has been determined as a function c downwind distance. Using this criteria, internal boundary layer development in each of the six cases is given in Table 1. The strong effect of heating changes on internal boundary layer growth may be se clearly. This dominant effect of heating changes can also be seen ir Figure 4 where calculated on-axis ground level concentration distributions are plotted.

A Simulation of Plume Fumigation due to Suburban and Urban Heating

The arrangement for calculation is illustrated in Figure 5, wher a pollutant is released, at 100 m and at the same rate as in the previous example, into a moderately stable flow established by an initial fetch of 8 km with a heat flux of -10 W m^{-2} (corresponding tc category E conditions) and a roughness length of 0.1 m. After a further 2 km under identical conditions the plume encounters a 2 km 'suburban' region with a positive heat flux of 30 W m^{-2} and this is followed by a further 2 km of 'urban' characteristics with a flux of 100 W m^{-2}. It should be noted that in this example the roughness changes are of substantially less significance than the heating chang In Figure 5, λ_H and λ_L denote the upper and lower levels of 50% of maximum concentration; normalised concentration profiles within these limits are also given at selected downwind distances.

Figure 6 shows momentum diffusivity plotted at the end of each of the 2 km regions. The growth of the internal boundary layer is very clear, having reached 96 m at 4 km downwind of the source and 250 m at 6 km downwind. The very large rise in diffusivity correspon to the rapid increase in dispersion shown in Figure 5 leading to the maximum ground level concentration estimated at 5.5 km downwind of the source. In Figure 7 are shown the potential temperature distributions as a function of height at the end of each region. At the end of the first region stable conditions clearly prevail while conditions are approximately neutral at the end of the second.

Unstable conditions prevail at the end of the fully urban region.

While the limitations of a two-dimensional calculation of suburban and urban effects must be recognised, this example does serve to illustrate the potential application of the APT model. The economy of the method is illustrated in that the calculation took 120 s CPU time on a CDC 6600 computer.

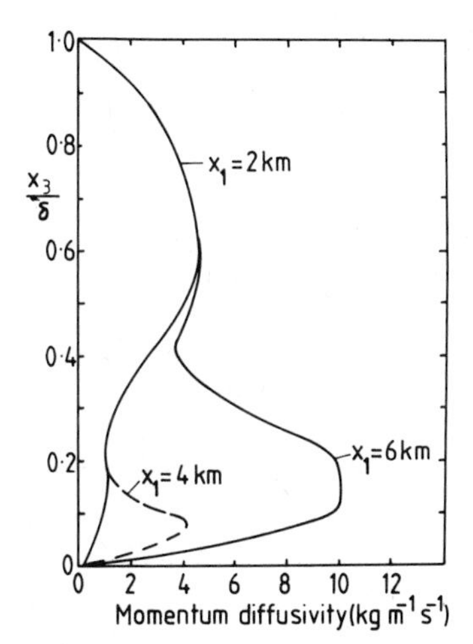

Figure 6. Momentum diffusivity plotted as a function of height (non-dimensionalised by boundary layer depth), at the end of regions A, B and C of Figure 5.

PREDICTION OF ATMOSPHERIC BOUNDARY LAYER WITH ALLOWANCE FOR CORIOLIS EFFECTS

In order to ensure the validity of modelling for this problem, the horizontally homogeneous time dependent version of the program has been applied to a relatively simple case, the stationary neutral boundary layer, which has been analysed in detail by earlier workers.

The results of Deardorff (1972) have been selected for comparison. This is the first step before experimental data from measurements in more complex atmospheric boundary layers are simulated.

A latitude of 45° North was assumed and the height of the calculation domain taken to be 1,100 m, where the velocity was taken to be geostrophic. The friction velocity u_* was calculated from the one-dimensional flow model for a surface roughness height of 2.4 cm. Solution was achieved by starting from an arbitrary condition and allowing the calculation to evolve to a steady state.

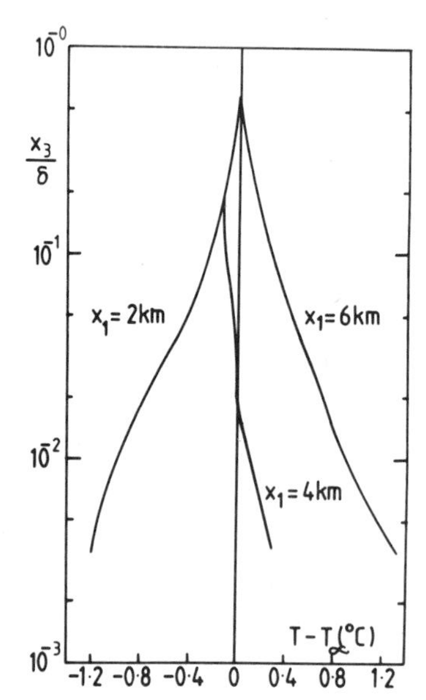

Figure 7. Potential temperature difference $T - T_\infty$, where T_∞ applies outside the boundary layer, plotted as a function of non-dimensional height, at the end of regions A, B and C of Figure 5.

Deardorff has applied 3-dimensional sub-grid modelling to predict this situation, thus directly simulating the larger scales of turbulence and modelling only the small scales. Wynggard et al (1974) have solved the ensemble-averaged equations, as in the present study, but using a version of the second order closure developed by Lumley and

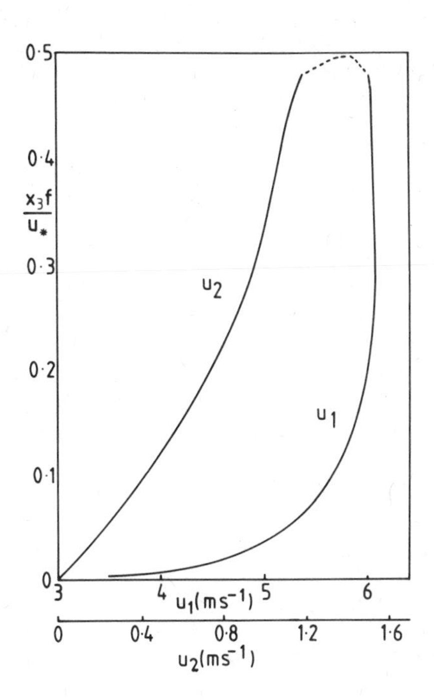

Figure 8. Neutral boundary layer. Velocities in surface direction (U_1) and in lateral direction (U_2), plotted as a function of height x_3 non-dimensionalised by u_*/f, f the Coriolis parameter.

Khajeh-Nouri (1974). Both Deardorff and Wynggard et al prescribed u_* and set the gradients of velocity and temperature to zero at the top of the calculation domain.

Figure 8 shows the variation of the velocity components in the direction of the surface wind and in the direction normal to this yielded by the present calculation, as a function of non-dimensional height. It can be seen that the windspeed U_1 reaches the geostrophic level at a non-dimensional height of about 0.2, slightly exceeds this above this height and then eventually returns to it. The lateral component U_2 increases steadily from zero to its geostrophic value. The rapid increase in velocity gradient near the top of the boundary layer shown by the dotted lines confims the conclusion of Wynggard et al that the height of the calculation domain should strictly be greater than the present value (which was chosen to correspond with Deardorff's calculations) for the boundary conditions to be applicable.

The component $\rho\overline{u_1u_3}$, non-dimensionalized by the surfa-

The stress component $\rho\overline{u_1u_3}$, non-dimensionalized by the surface stress, given by the present method is plotted as a fuction of non-dimensionalized height in Figure 9. The positive value at the top of the layer confirms the earlier conclusion that the boundary conditions should strictly be applied at a greater height. Also shown are the calculations of Deardorff, where a boundary condition of zero stress at a dimensionless height of 0.45 was assumed. Evidently the two results agree quite closely in the lower region. In Figure 10 is shown the variation of the stree component $\rho\overline{u_2u_3}$ with non-dimensional height, together with the results of Deardoff. Here too the agreement between the two calculations is very good.

Finally Figure 11 shows the variation of mean square vertical

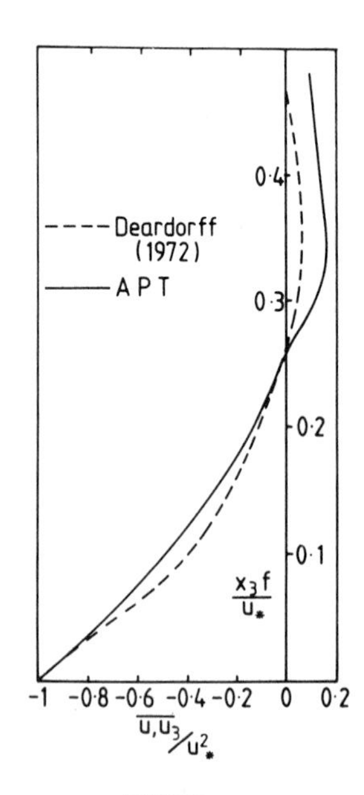

Figure 9. Turbulent stress $\rho\overline{u_1u_3}$, non-dimensionalized by surface stress, plotted as a function of non-dimensionalized heig

velocity fluctuations with height. Again generally good agreement is seen with the results of Deardorff ; the largest discrepancy is close to the ground where the present result is around 1.1 compared to Deardorff's level of about 1.3. The APT result ismuch smaller than that reported by Wynggard et al. (1974) (close to 1.75) ; the reason for this is that in the present model the redistribution term includes the surface contribution of the stress transport equation in the manner described by Gibson and Launder (1978), which serves to reduce the vertical fluctuations near the ground to levels close to those which exist in practice.

DISCUSSION

In this paper the modelling and numerical basis of the APT method have been outlined, potential applications have been indicated.

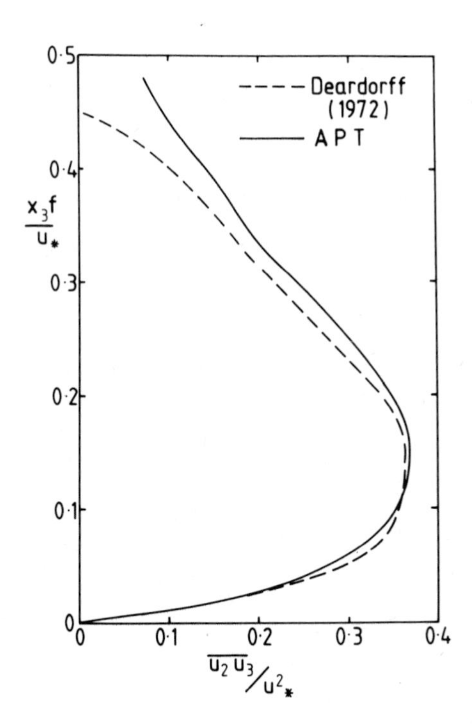

Figure 10. Turbulent stress $\rho\overline{u_2u_3}$, non-dimensionalized by surface stress, plotted as a function of non-dimensionalized height.

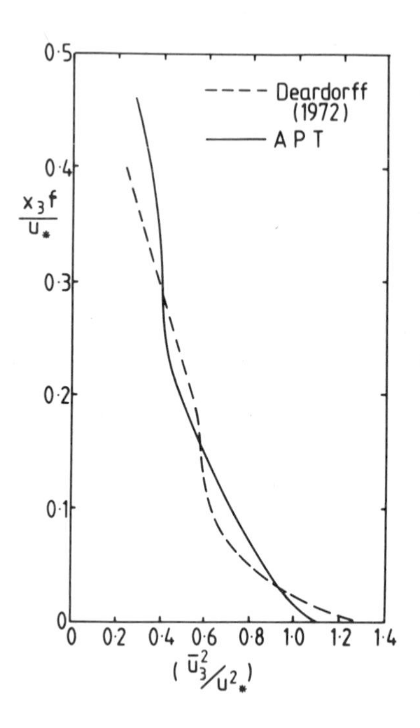

Figure 11. Turbulent stress $\rho\overline{u_3^2}$, non-dimensionalized by surface stre stess, plotted as a function of non-dimensionalized heigh

and the model has been assessed by comparison with both measurements and predictions of a more elaborate model.

Developments in the immediate future are intended to incorporate the Coriolis effect into the calculation of plume concentration, to consider the accuracy and range of application of the method under homogeneous flow conditions and to incorporate a treatment of plume meander. Plume meander must be treated separately because the mean quantities calculated by APT correspond to short time averages incorporating the micrometerological scale but not large scale effects. Because plume meander may be site dependent and because of the likely need to consider both pessimistic and average conditions it is intended to make treatment of plume meander a user option and to treat it simply by superposition.

ACKNOWLEDGEMENTS

The research reported here has been sponsored by the Safety and Reliability Directorate of the United Kingdom Atomic Energy Authority, but the views presented are solely those of the authors. Our thanks are due to Miss A N Ford who has prepared the manuscript.

REFERENCES

Clarke R.H., 1979, A model for short and medium range dispersion of radionuclides released to the atmosphere, Report NRPB-R91.

Deardorff, J.W., 1972, Numerical investigation of neutral and unstable planetary boundary layers, J. Atmos. Sci., 29:91

Echols, W. T. and Wagner, N. K., 1972, Surface roughness and material boundary layer near a coastline, J. Appl. Meteorology, 11:658.

El Tahry, S., 1979, Turbulent plume dispersal, PhD Thesis, University of London.

El Tahry, S., Gosman, A. D. and Launder, B. E., 1981, The two- and three-dimensional dispersal of a passive scalar in a turbulent boundary layer, Int. J. Heat Mass Transfer, 24:35.

Gibson, M. M. and Launder, B. E., 1978, Ground effects upon pressure fluctuations in the atmospheric boundary layer, J. Fluid Mech., 86:491.

Goddard, A. J. H., Ghobadian, A., Gosman, A. D., Harter, C. A. and Kaiser, G. D., 1980, APT - a computer program for the numerical solution of problems in atmospheric dispersion, in:- Proc. 14th IRCHA Coll. on Atmos. Pollution, Paris.

Harter, C. A., Kaiser, G. D., Goddard, A. J. H., Gosman, A. D., Ghobadian, A. and El Tahry, S., 1980, APT - a computer program for the numerical solution of problems in atmospheric dispersion and some applications to nuclear safety, in:- Proc. C.E.C. Seminar on radioactive releases and their dispersion following a hypothetical reactor accident, Risø.

Launder, B. E., 1976, Heat and mass transport, in:- Topics in Applied Physics (P. Bradshaw Ed.), Volume 12, Springer.

Lumley, J. L. and Khajeh-Nouri, B., 1974, Computational modelling o turbulent transport, Adv. Geophys., 18A:169.

Maul, P. R., 1977, The mathematical modelling of the meso-scale transport of gaseous pollutants, Atmos. Env., 11:1191.

Raynor, G. S., Michael, P. and SethuRamen, S., Meteorological measu ment methods and diffusion models for use at coastal nuclea reactor sites, Nuclear Safety, 21:6:749.

Wynggard, J. C., Cote, O. R. and Rao, K. S., 1974, Modelling the atmospheric boundary layer, Adv. Geophys., 18A:193.

DISCUSSION

W. GOODIN

How do you account for gradient of $\overline{U_2C}$ in lateral direction in a two dimensional flow field?

A. GHOBADIAN

The equation for $\overline{U_2C}$ looked at in detail has contributions from the lateral gradient of the mean concentration field, hence it is a function of lateral direction as well as other directions in a two dimensional flow field.

A THREE DIMENSIONAL, FINITE ELEMENT MODEL FOR SIMULATING HEAVIER-THAN-AIR GASEOUS RELEASES OVER VARIABLE TERRAIN

Robert L. Lee
Philip M. Gresho
Stevens T. Chan
Craig D. Upson

Lawrence Livermore National Laboratory
University of California
Livermore, CA 94550

INTRODUCTION

Due to an increase of activities over recent years with the transport and storage of liquefied gaseous fuels (e.g. liquefied natural gas), it has become necessary to make careful assessments of the environmental risks associated with such operations. For example, the collision of a liquefied natural gas (LNG) carrying tanker can conceivably release a large amount of potentially flammable methane vapor into the atmosphere. Verified numerical models capable of predicting the gravitational spread and atmospheric dispersion processes following an LNG spill can provide qualitative and quantitative estimates of the motion of such a combustible vapor cloud. Solutions from these models can also be used as initial conditions for other models which treat the deflagation and detonation processes if accidental ignition of the NG/air mixture occurs. LLNL is involved in experimental and model verification studies for the Department of Energy with programs which encompass each aspect of the potential scenarios. Some of the major unresolved questions regarding LNG research are described in Mott et al. (1981). In this paper, we focus attention on the modeling of the vapor dispersion phase only.

Over the past decade, a number of dispersion models have been proposed and utilized to predict the lower flamablility limit (LFL) of the LNG vapor cloud. Havens (1977) has evaluated seven models, several of which were widely cited in the literature, and has made independent calculations with each. In the calculations, comparable (if not identical) initial data were used to attempt to simulate essentially the same dispersion problem. The results indicated that for an instantaneous release of 25,000 m^3 of LNG onto water, order of magnitude differences in the LFL of the dispersing cloud are predicted by the various models. Of the models considered, Havens concluded that the relatively simple Germeles and Drake (1975) model and the more complex SIGMET model (England et al., 1978) are the most rational approaches for estimating the LFL limit of the vapor cloud.

The Germeles and Drake one-dimensionel model considers accidental spills over water and consists of three sequential but distinct calculational phases, namely : (i) vapor generation; (ii) gravitational spread of the cloud, and (iii) Gaussian dispersion. In this model, the spill results in a cylindrically-shaped cloud which is assumed to spread radially outwards until buoyancy effects become insignificant. After this time, Gaussian dispersion is invoked and the cloud spreads as a neutrally-buoyant trace pollutant. The limitations of such an idealized model are rather obvious; however, Havens has noted that reasonable results can be obtained with this model provided the spill volumes are small. More recently, improved versions based on similar simplified one-dimensional approaches (but employing better parameterizations for the more important physical processes) have been developed and are reported to compare well with specific field experiments (see for example, Cox and Carpenter, 1979; Eidsvik, 1980 and Colenbrander, 1980).

In contrast to the simple Gaussian-type models, the SIGMET model solves the three-dimensional (3D), time-dependent conservation equations of mass, momentum, energy and species concentration. The resulting motion is assumed to be hydrostatic with topography effects incorporated via a sigma co-ordinate system. With this more general formulation, the dense, cold vapor cloud may now interact with the local flow pattern and even influence the turbulence structure. However, the validity of the hydrostatic approximation for such vapor dispersion calculations is rather questionable (particularily over complex terrain) and is probably highly inaccurate in describing the motion of the cloud near the source area.

It is generally recognized that any numerical model developed to treat the most general LNG spill scenarios will be very complex and therefore computationally expensive. The search for appropriate simplifications to the complex model is therefore crucial. On the opposite extreme, an overly simplified version may be so limited in its applicability that the overall usefulness of the model would be severly restricted. With this in mind, we developed a 3D, nonhydrostatic, anelastic model using the finite element method for simulating LNG spills over variable terrain. Although the current model is still evolving (see for example Chan et al., 1981a, 1981b), we have been careful, from the onset, to introduce only those simplifying assumptions which expedite the nummerical procedures but do not seriously compromise the true physics. In the following sections we will attempt to justify our current non-Boussinesq and non-hydrostatic approach by comparing solutions between our model and variants in which the widely used Boussinesq and/or hydrostatic assumptions are invoked. Finally, a 3D simulation of an LNG spill over the variable terrain of the China Lake test site is presented.

MODEL EQUATIONS AND NUMERICAL PROCEDURE

In the present model, the spread and dispersion of LNG vapor is predicted by solving the 3D conservation equations as in the SIGMET approach. The form of our equations is different, however, and is based on the results of many numerical experiments in which various approximations were systematically tested but later discarded as being inappropriate for the problem. The governing equations currently used in our model, written for the mean (time-averaged) quantities in a turbulent flow field, are :

$$\frac{\partial(\rho\underline{u})}{\partial t} + \rho\underline{u}\cdot\nabla\underline{u} = -\nabla p + \nabla\cdot(\rho\underline{\underline{K}}^{m}\cdot\nabla\underline{u}) + (\rho-\rho_h)\underline{g} \qquad (1)$$

$$\nabla\cdot(\rho\underline{u}) = 0 \qquad (2)$$

$$\frac{\partial\theta}{\partial t} + \underline{u}\cdot\nabla\theta = \nabla\cdot(\underline{\underline{K}}^{\theta}\cdot\nabla\theta) + \frac{C_{P_N}-C_{P_A}}{C_P}(\underline{\underline{K}}^{\omega}\cdot\nabla\omega)\cdot\nabla\theta + S \qquad (3)$$

$$\frac{\partial\omega}{\partial t} + \underline{u}\cdot\nabla\omega = \nabla\cdot(\underline{\underline{K}}^{\omega}\cdot\nabla\omega) \qquad (4)$$

and

$$\rho = \frac{PM}{RT} = \frac{P}{RT\left(\frac{\omega}{M_N} + \frac{1-\omega}{M_A}\right)} \qquad (5)$$

where $\underline{u} = (u, v, w)$ is the velocity, ρ is the density of the mixture, p is the pressure deviation from an adiabatic atmosphere at rest, with corresponding density defined as ρ_h, $\underline{g}$ is the acceleration due to gravity, θ is the potential temperature deviation from an adiabatic atmosphere, S is the source term for temperature (e.g., latent heat), ω is the mass fraction of NG vapor, and K^m, K^θ and K^ω are the (diagonal) eddy diffusion tensors (which are to be parameterized using, at least initially, K-theory) for the momentum, energy, and NG vapor, respectively, and Cp_N, Cp_A, and $Cp = \omega Cp_N + (1-\omega)Cp_A$ are the specific heats for NG vapor, air, and the mixture, respectively. In the equation of state, P is the absolute pressure, R is the universal gas constant, T is the absolute temperature $(T/(\theta+\theta_o) = (P/P_o)^{R/\mathcal{M}C_p})$, and M_N, M_A, $\mathcal{M}$ are the molecular weights of NG air, and the mixture (suitably averaged), respectively. The above set of equations, together with appropriate initial and boundary conditions, are solved to yield velocity, pressure, potential temperature, mass fraction of NG vapor, and density of the mixture.

The appropriate initial conditions for the above set of equations are any "solenoidal" velocity field (satisfying $\nabla \cdot (\rho_o \underline{u}_o) = 0$) and a suitable initial temperature and concentration distribution for the problem to be simulated. For example, an instantaneous spill can be modeled by initially prescribing a spatial variation of temperature and concentration of the NG vapor over only a portion of the computational domain, with the remaining part unperturbed. Continuous or finite-duration spills can be simulated via appropriate mass injection boundary conditions with the initial fields being unperturbed.

Equations (1) through (4) are discretized spatially by the finite element method in conjunction with the method of weighted residuals (see for example, Gresho et al., 1980 and Chan et al., 1981b). The time-stepping procedure employs the explicit forward Euler method except for the pressure, which must be computed implicitly. In 3D, we use the simplest 8-node isoparametric "brick" element consisting of piecewise trilinear approximating functions for the velocity, temperature and concentration, and piecewise-constant approximmation for the pressure. Our code also contains a 2D option in which the analogous 4-node, bilinear element is employed.

A GENERALIZED ANELASTIC APPROACH

Both the Boussinesq and anelastic approximations have been widely used in the literature for modeling the flow of incompressible fluids. In the Boussinesq approximation, the density is treated as constant except in the gravitational buoyancy term. Under this approximation, the ideal gas equation of state is replaced by the following equation relating the perturbation quantities:

$$\frac{\delta\rho}{\rho_o} = -\frac{\delta\theta}{\theta_o}, \qquad (6)$$

where ρ_o is the reference density associated with the reference temperature θ_o. This expression can, in turn, be substituted into the buoyancy term to completely eliminate density as a variable from the equations. Whereas the Boussinesq approximation is known to be accurate for modeling convection within shallow fluid layers in which the fractional density change is small, it is less appropriate for problems associated with deep convection problems where mean density variations over a layer can be more significant. For such problems, Ogura and Philips (1962) showed that the anelastic approximation, obtained by replacing the (constant) density in the equations by a specified mean vertical density distribution ($\rho_h(z)$), is a better alternative.

The density close to the source area within a cold NG vapor cloud is often 60 % greater than that of the ambient air and varies significantly in both space and time. It is reasonable to expect that such large density variations would fall beyond the limit of applicability of either the Boussinesq or anelastic approximations. While compressibility effects are surely present in the case of interest here, we believe that they are of secondary importance because the Mach number is always very small ($\leq \sim .05$) and thus, acoustic waves and their effects are negligibly weak. On this premise, we have invoked a generalized anelastic approximation obtained by replacing (i) $\nabla\cdot(\rho \underline{u}) = -\partial\rho/\partial t$ by (ii) $\nabla\cdot(\rho \underline{u}) = 0$ (see eq. (2)). A linearized analysis for simple wave motion associated with the resulting (generalized anelastic) set of inviscid equations show that no wave solutions exist for an adiabatic atmosphere. For a stably stratified situation, only internal gravity waves (no acoustic waves) are (appropriately) possible. This new approximation results in two computationally attractive features; first, our time steps are not limited by acoustic effects since sound waves are filtered a priori and second, the solution algorithms developed originally for the Boussinesq version can be used to solve the current set of equations.

In the preceding discussion, it is important to note that the proper interpretation of (ii) is "acoustic density variations in time are of very small amplitude and occur so quickly that it is a good approximation to assume that density is always in equilibrium with the other thermodynamic variables." The variation of density with time is then determined implicitly by the time variation of temperature, pressure, and composition (again, neglecting sound waves) via the ideal gas law. It is not appropriate, in this context, to interpret (ii), via (i), as $\partial\rho/\partial t = 0$, since ρ does indeed vary with time.

This concept may be made less abstract by considering a simple "model problem" from ordinary differential equations given by

$$\dot{y}_1 = -y_1 + y_2 \quad ; \quad y_1(t=0)=1, \tag{7a}$$

$$\dot{y}_2 = y_1 - 100y_2 \quad ; \quad y_2(t=0)=\beta, \tag{7b}$$

where $\beta \leq O(1)$. The ratio of the "time constants" (approximately 100) has been chosen to mimic the time scale ratio in the fluid dynamics equations (i.e. the time scale associated with the speed of sound is about two orders of magnitude smaller than the other physically relevant time scales). A close approximation (3 to 4 digits) to the exact solution of this coupled system is :

$$y_1 = (1+.01\beta)e^{-.99t} - .01(\beta-.01)e^{-100t} \tag{8a}$$

$$y_2 = .01(.99+.01\beta)e^{-.99t} + (\beta-.01)e^{-100t} \, . \tag{8b}$$

Now consider the "anelastic" analog, obtained by invoking equilibrium in (7b); i.e. $y_2 = .01y_1$ for all t. The exact solutions to the anelastic model is then simply

$$y_1 = e^{-.99t} \quad (9a)$$

$$y_2 = .01e^{-.99t}. \quad (9b)$$

In the "nominal case", $\beta = 1$ and the solutions from (8) and (9) differ appreciably for very small time ($t \lesssim \sim .01$); e.g. $y_2(0) \simeq 1$ from (8b), but only .01 from (9b). At longer time, however, the rapidly-decaying component in (8) is negligible and we have (for $t \gtrsim \sim .05$) :

$$y_1 \simeq 1.01e^{-.99t}$$

$$y_2 \simeq .01e^{-.99t},$$

which is very close to the anelastic solution. On the other hand, for the "best case", $\beta = .01$ (y_2 is in equilibrium at $t = 0$), and the exact solution (to $\sim$ 4 digits) is, from (8),

$$y_1 = e^{-.99t}$$

$$y_2 = .01e^{-.99t};$$

i.e. it agrees with the anelastic solution uniformly in time. Final a "worst-case" result would be $\beta >> 1$, for which the solutions in (8) and (9) differ significantly for all time; i.e. the anelastic approximation is inappropriate.

We believe that the conditions associated with the release of heavier-than-air gases is close to the "best case" in most scenario On the other hand, an extreme situation, for which the anelastic model would generate erroneous results, might be the sudden burstin of a very high pressure tank, for which case the fluid dynamics would be compressible and the released gas would expand at high velocity behind the shock wave; this case corresponds to $\beta >> 1$ in the model problem.

Finally, from a computational viewpoint, the generalized anelas equations (1-5) differ from the corresponding Boussinesq version by effectively only several more algebraic manipulations involving the density. Therefore, the additional computational cost associated with this more general model is quite small in practice (our curren code, in fact, contains both versions as options). In the next sec- tion, results from our generalized anelastic model will be compared to those from the Boussinesq model for the case of a 2D instantaneo LNG spill.

TWO-DIMENSIONAL SIMULATIONS OF AN INSTANTANEOUS RELEASE OF LNG VAPOR IN A CALM ATMOSPHERE

In order to demonstrate the performance of the current anelastic model in simulating LNG spill scenarios, we will present results from a 2D calculation of an instantaneous release of LNG vapor in the absence of wind. This particular simulation is of great practical interest since, without advective transport due to a mean wind, experimental observations indicate that the vapor cloud typically presists close to the surface over a long period of time, thus creating a "worst-case" situation. These results are then compared with those from models which employ (i) the Boussinesq approximation, and (ii) the hydrostatic approximation in order to assess the utility of these simplications for modeling this class of flows. These two approximations have been routinely used in meteorological calculations and, if valid, could potentially enhance the computational efficiency of LNG vapor dispersion calculations.

We assume, for this simulation, that the vapor cloud initially exists as a thin "slab" over a flat surface. The temperature within the slab is -160°C but quickly increases (as a smooth Gaussian function) to the ambient temperature (20°C) along the top and at the edges of the slab. The (half) domain considered is 240 m x 24 m and the graded mesh contains 552, 4-node elements. Due to symmetry, the motion of only half of the vapor cloud (initially with dimensions 100 m x 4 m) is considered in the calculations. Insulated boundary conditions are employed for both temperature and NG concentration at all boundaries while no-slip and symmetry conditions apply at the bottom and left side, respectively. At the top, u and the natural boundary condition, $(-p + \rho K^m \partial v/\partial y)$, are zero; likewise, at the right (outflow) end, $(-p + \rho K^m \partial u/\partial x)$ and $\partial v/\partial x$ are set to zero. For these calculations, all turbulent eddy exchange coefficients (K's) for heat, momentum and mass transfer are constants equal to 0.1 m^2/sec, a value typically used for stable atmospheric conditions.

Results from the Generalized Anelastic Model

In Figure 1 we show the velocity and temperature fields at 10 and 30 seconds after the spill as predicted by the anelastic model. The dominating effect of gravitational spread is demonstrated as the density front (originally located at approximately 100 m) advances with an elevated head, toward the right and generates strong, vortical motion immediately behind it. The maximum velocities at the two times are approximately 3.6 and 2.6 m/sec respectively, suggesting that the front itself decelerates as the cloud gradually reduces in negative buoyancy due to mixing of the cold NG vapor with the warmer ambient air. The spatial oscillations displayed in the velocity field are caused by the inability of the coarse mesh to resolve the fairly sharp front (the turbulent

Reynolds number is several hundred in this simulation). The motion of this density front is also shown clearly by the temperature contours. We have carried the time integration for this problem beyon a time of 60 seconds with no apparent numerical difficulties, despi the presence of some wiggles in the velocity field.

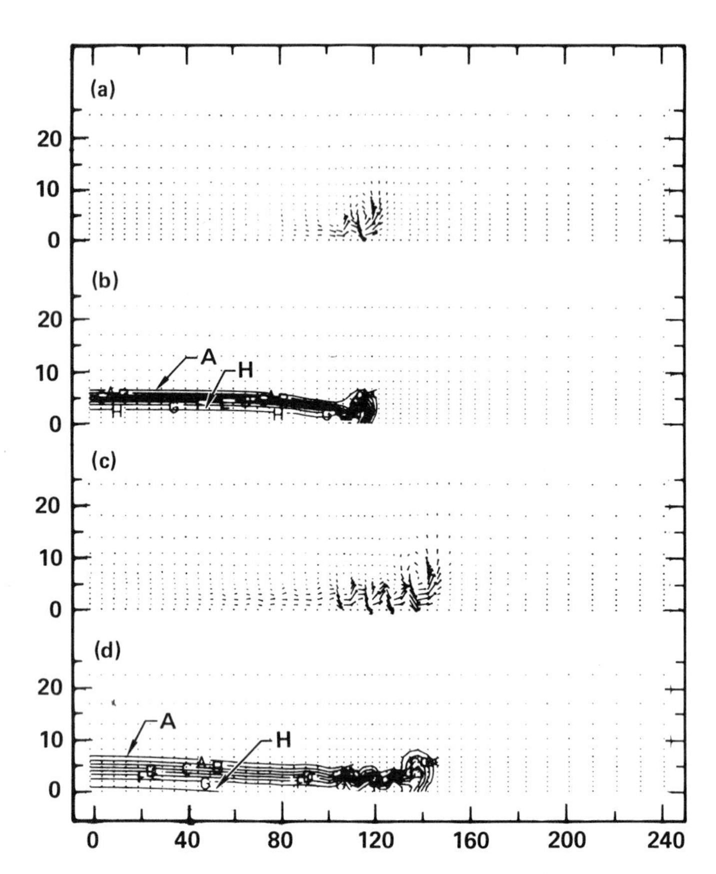

Fig. 1. Predicted fields with the generalized anelastic model. Results at: time = 10, (a) velocity (v_{max} = 3.57), (b) temperature; time = 30, (c) velocity (v_{max} = 2.57), (d) temperature. Temperatur contours vary in - 20° increments from A = -10° to H = -150°.

Results from the Boussinesq Model

We now present results based on the previous simulation of a 2D LNG spill but obtained via the Boussinesq option of our model. The following simplifications are introduced into eqs. (1) through (4) : (i) $\rho=\rho_o$; i.e. the constant density at the ambient temperatur and, (ii) $(\rho-\rho_h)/\rho_o \rightarrow - \theta/\theta_o$. Since the equation-of-state (5) is no longer required, the variation of NG concentration, ω, during

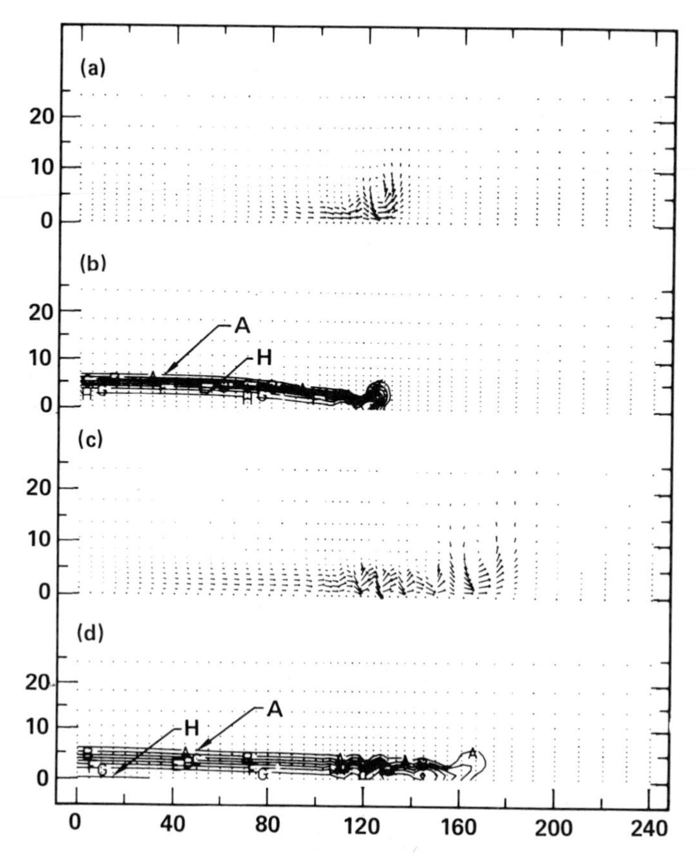

Fig. 2. Predicted fields with the Boussinesq model. Results at: time = 10, (a) velocity (v_{max} = 4.8), (b) temperature; time = 30, (c) velocity (v_{max} = 4.00), (d) temperature. Temperature contours vary as in Fig. 1.

the dispersion process does not alter the buoyancy force as in the anelastic case. The NG/air mixture is therefore considered, as far as the equation of motion is considered, as a single species gas with an initial temperature of -160°C. In this particular problem, the magnitude of the buoyancy acceleration, $(\rho-\rho_o)g/\rho_o$ for the mixture at this temperature is approximately the same for both the anelastic and Boussinesq cases, thus rendering a meaningful comparison in the sense that both simulations are subjected to approximately the same initial acceleration.

In Figure 2 we again show the velocity and temperature field at 10 and 30 seconds. The resulting fields appear qualitatively similar to those from the anelastic calculation (see Figure 1). However, the density front is now advected significantly farther downstream due to higher maximum velocities near the front. The differences in the frontal mean speed can be explained by noting that the constant density (ρ_o) employed in the Boussinesq model underestimates the actual NG cloud density and results in an effectively larger acceleration compared to the more accurate anelastic approximation. This behavior is consistent with the results of Daly and Pracht (1968) in their study of density-current surges, in which the density front velocity obtained from a single fluid, Boussinesq calculation is higher than that from a more exact two-fluid computation. This same study also suggests that the Boussinesq approximation is sufficiently accurate for density ratios below about 1.2, a result which we have independently verified.

There is some evidence, based on further calculations, that the Boussinesq and anelastic solutions agree much better under conditions in which gravitational spreading of the cloud is not the dominant factor (e.g., for cases when advective transport of the cloud by a pre-existing mean wind is also important). In our present formulation, however, the anelastic version is almost as computationally efficient as the Boussinesq version so that there is no point in selecting the latter, more restrictive, model.

Results from the Hydrostatic Model

A hydrostatic model based on the finite element method (Chan, et al. 1980) has been developed to assess the validity of this approximation for LNG vapor dispersion calculations. This model assumes that the vertical momentum equation is replaced by one in which the vertical pressure gradient completely balances the buoyancy force. For simplicity, the Boussinesq approximation is invoked although, in retrospect, a more general system (which includes the equation-of-state (5)) could be solved almost as easily. The hydrostatic assumption can make the pressure field much easier to compute since pressure is then no longer an implicit variable. As will be seen from the results, however, the range of applicability of such a model can be rather restrictive.

The hydrostatic code was used to simulate the 2D, instantaneous spill problem described in the previous section. The solutions shown in Figure 3 are at approximately the same times as the earlier calculations. In this case, the predicted fields are obviously unacceptable. The isotherms already display significant oscillations at the early time of 10 secs and the computed flow field is totally unphysical. After a time of 28.3 secs, the model even experienced difficulties in converging in spite of the small time-steps employed. The unphysical behavior of these results is believed

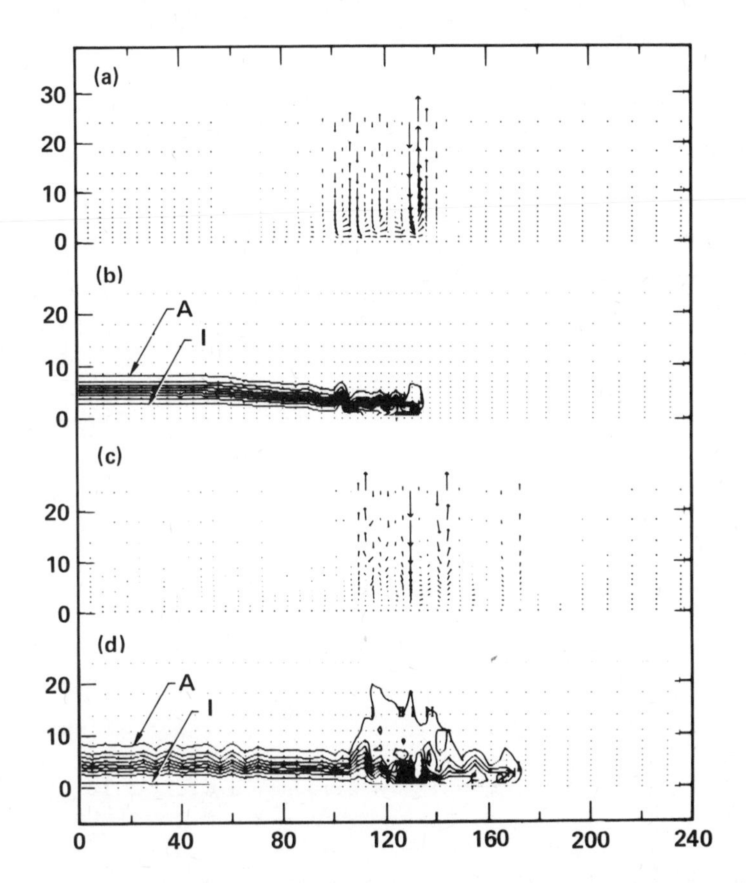

Fig. 3. Predicted fields with the hydrostatic model. Results at : time = 10, (a) velocity (v_{max} = 7.75), (b) temperature; time = 28.3, (c) velocity (v_{max} = 19.71), (d) temperature. Temperature contours vary in - 20° increments from A = +10° to I = -150°.

to be caused primarily by the appearance of a strong recirculation zone within the flow field which, apparently, cannot be handled by a hydrostatic model. (Vorticity generation via horizontal density gradients is not a hydrostatic process).

Additional numerical calculations (Chan and Gresho 1979 and Chan et al. 1980) suggest that good agreement between hydrostatic and Boussinesq (non-hydrostatic) models can be achieved only for large eddy diffusivities (∿10 m^2/sec) under either calm or windy conditions. For spills which occur under the "worst-case" situations, i.e. calm and stable conditions with little turbulent diffusion, a hydrostatic model is invalid as has been demonstrated. In more general LNG spill scenarios, it is expected that a wide range of eddy diffusivities will be required to adequately describe the dynamics in both the near and far field. It thus appears that only a non-hydrostatic approach can be used with confidence to treat the full spectrum of dynamical behavior.

THREE DIMENSIONAL SIMULATION OF AN LNG RELEASE OVER VARIABLE TERRAIN

An important aspect of the present model is its ability (via the finite element technique) to handle simulations over variable topography with no additional complexity to the user. The behavior of spills which occur over variable terrain can be quite different from those occurring over a flat surface. Field experiments conducted by LLNL at the China Lake test site (Koopman, et al. 1981) revealed that the motion of the vapor cloud can be significantly altered even for gentle terrain if winds are light and the atmosphere stable. In order to demonstrate and test the effectiveness of the code in tackling problems over topography, we have simulated one of the more interesting China Lake spill experiments. This particular field experiment was performed for a spill of 28.4 m^3 of LNG during the early evening when the winds were low ($\sim$ 3m/sec) and the lapse rate stable. For such conditions, the dominant transport mechanism is gravitational via differential densities and sloping terrain.

Figure 4 shows a perspective view of a portion of the 3D finite element mesh used to approximate the local terrain near the spill pond. The computational domain is contained within 500 m x 400 m in the horizontal and 20 m in the vertical and consists of 6400 (40 x 20 x 8) elements. Initial conditions for the simulation were obtained from an isothermal calculation carried out to approximately steady state conditions during which no NG vapor was released. Constant eddy diffusivities of 0.4 m^2/sec in the vertical and 2.0 m^2/sec in the horizontal were used for both the inital field and spill computations. In contrast to the instantaneous vapor release problem modeled in the 2D calculations, we now consider a finite-duration, continuous release of NG vapor. For this problem, LNG is assumed to "boil-off" as an area source over 12 of the 30 elements representing the spill pond surface. The source itself is modeled as a vertical injection of NG vapor occurring over a period of 108 secs (at which time the injection is terminated) with a velocity (w_o) of 0.1 m/sec, a temperature of -160°C (the LNG boiling temperature) and a normal mass flux balance given by $K^\omega\ \partial\omega/\partial n = w_o(1-\omega)$. Away from the source, we imposed $\underline{u} = 0$ and $\partial\theta/\partial n$ $0 = \partial\omega/\partial n$ at the ground. The remaining boundary conditions employed were : specified $\underline{u}$, T, ω at the inlet plane, natural boundary conditions at the outlet, and symmetry conditions at the top and along the two lateral surfaces.

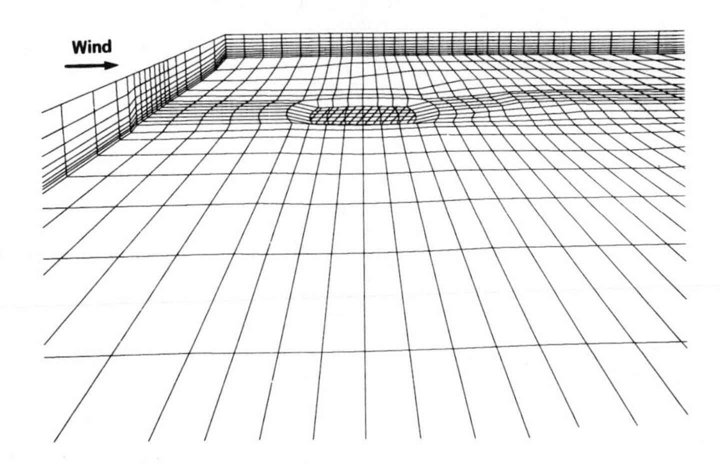

Fig. 4. View of a portion of the 3D finite element mesh with spill pond cross-hatched.

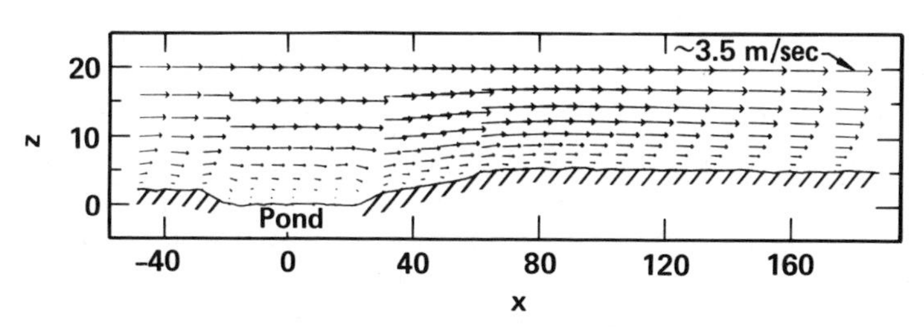

Fig. 5. Inital wind profile in a longitudinal plane through the pond centerline.

In Figure 5, we show the initial wind profile along a longitudinal cross-section through the pond centerline. Terrain effects on the initial flow field are noticeable but not dominant because of the fairly gentle slopes. The horizontal velocity vectors on a plane 1 m above the surface (with topographical contours superimposed) at 140 secs after the spill occurred ($\sim$ 32 secs after the vapor injection stopped) are depicted in Figure 6 (see also Chan et al. 1981 b). Both the gravitational spread (even upwind of the spill) and the influence of the sloping terrain on the overall motion is clearly evident. The surface concentration contours (in volume fraction) at t = 140 and 200 secs, shown as solid lines in Figure 7, further demonstrate the gravitational spreading of the cloud. Also apparent is the shifting of the cloud centerline ($\sim$ 23° at 200 secs) away from the mean wind direction as a result of the "slumping" downhill motion. This simulation predicts, interestingly, that the

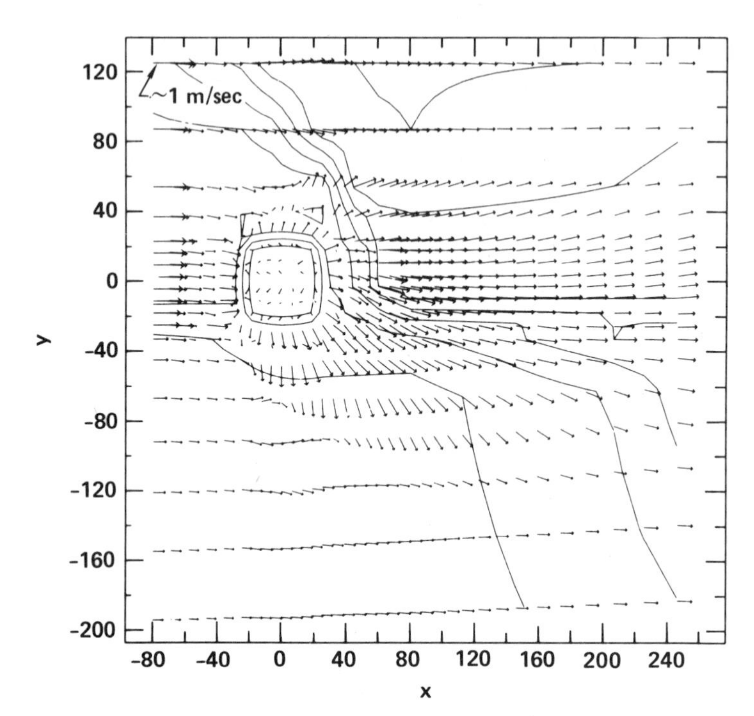

Fig. 6. Horizontal velocities 1 m above the ground at 140 seconds after the spill. Topography contours at 1 m increments are superimposed.

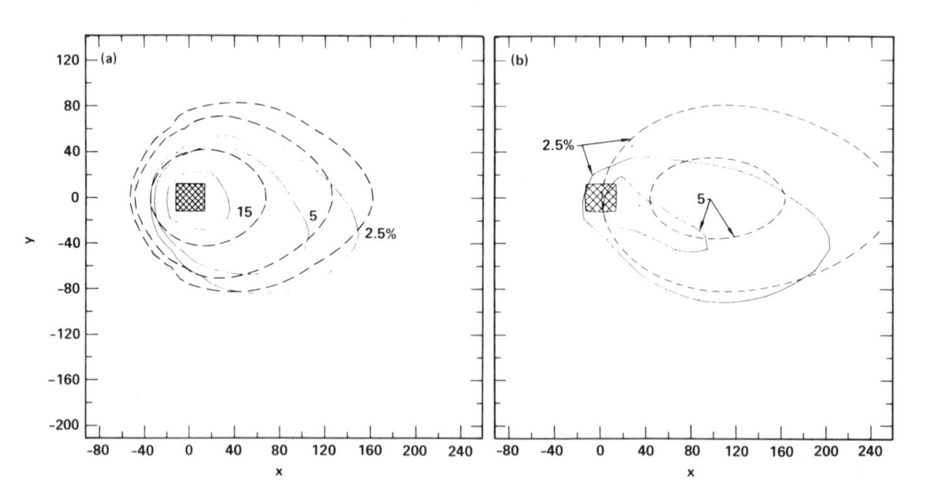

Fig. 7. Predicted ground NG concentrations (in volume fraction) at : (a) time = 140 secs.; (b) time = 200 secs., for variable (——) and flat (----) terrain. The spill area is cross-hatched.

flammable zone (5% to 15% volume fractions) appears to first grow but then retract at later times.

In order to emphasize the significant effect of even this relatively mild terrain upon the movement of the vapor cloud, we also compare these results from those of a similar calculation, but performed over flat terrain. The planar surface concentrations at 140 and 200 secs are superimposed (shown dashed) upon the variable terrain results in Figure 7. Here, in the absence of a basin-like topography near the source region to impede gravitational spreading in the cross-wind direction, the flammable zone is considerably wider. As expected, the cloud centerline now coincides with the mean wind direction.

The 3D variable terrain simulation agrees qualitatively with the field experiment. It is noteworthy that the predicted gravity-induced flow which occurs in all directions near the vapor source, and the downslope flow causing a shift of the cloud centerline, have been observed in the experiments. The measurements, however, also showed a thinner, more persistent cloud which can perhaps be modeled by employing a significantly smaller vertical diffusivity ($\sim$ .05 m^2/sec) requiring, concomittantly, a much finer mesh. In the simulation, the mean wind velocity near the ground is probably significantly lower than that of the field values thus causing errors in predicting advection velocities close to the ground. Such errors can probably be reduced by imposing "slip" rather than noslip conditions at the surface. This idea is currently being implemented and tested. Finally, with regard to the computational time for this 3D simulation which comprises of about 45,000 equations, the average cost per time step (0.4 second of real time) is approximately 0.3 second CPU and 1.3 seconds of I/O on the CRAY.

SUMMARY AND CONCLUSIONS

We have developed a 3D finite element model to predict the dispersion processes associated with heavier-than-air gaseous releases in the atmosphere. Both 2D and 3D examples were presented to demonstrate the flexibility of the model in simulating the motion of the vapor cloud generated after an LNG spill over flat or variable terrain. We have also presented results to justify our slightly more complex generalized anelastic model rather than one employing the Boussinesq and/or hydrostatic approximations which are routinely used for planetary boundary layer flows but can be inappropriate for this application.

Although the present version still contains some crude parameterizations of the real problem, more realistic physics involving variable eddy exchange coefficients, improved source emission modeling, and employing initial wind profiles more representative of the measured values, are all currently being introduced into the model. It is encouraging, however, that even the early version appears to exhibit many of the features which have been observed in the field experiments.

ACKNOWLEDGEMENTS

This work is performed under the auspices of the U.S. Department of Energy by the Lawrence Livermore National Laboratory under contract No. W-7405-Eng-48.

REFERENCES

CHAN, S.T. and P.M. GRESHO, 1979, A Comparsion of Hydrostatic and Nonhydrostatic Models as Applied to the Prediction of LNG Vapor Spread and Dispersion, Lawrence Livermore National Laboratory Report UCID-18097.

CHAN, S.T., P.M. GRESHO and R.L. LEE, 1980, Simulation of LNG Vapor Spread and Dispersion by Finite Element Methods, Appl. Num. Mod., 4:335.

CHAN, S.T., R.L. LEE, P.M. GRESHO and C.D. UPSON, 1981a, A Three Dimensional, Finite Element Model of Liquefied Natural Gas Releases in the Atmosphere, Proc. Fifth Symp. on Turb. and Diff., and Air Poll., Amer. Met. Soc., Atlanta, GA, March 9-13, 1981.

CHAN, S.T., P.M. GRESHO, R.L. LEE and C.D. UPSON, 1981b, Simulation of Three-Dimensional Time-Dependent Flows by a Finite Element Method, AIAA Fifth Comp. Fluid Dyn. Conf., Palo Alto, CA, June 22-23, 1981.

COX, R.A. and R.J. CARPENTER, 1979, Further Development of a Dense Vapor Cloud Dispersion Model for Hazard Analysis, Symp. 'Schwere Gase', Battelle-Inst., Frankfurt am Main, Sept. 3-4, 1979.

COLENBRANDER, G.W., 1980, A Mathematical Model for the Transient Behavior of Dense Vapor Clouds, Third Int. Loss Prevention Symp., Basel, Switzerland, Sept. 15-19, 1980.

DALY, B.J. and W.E. PRACHT, 1968, Numerical Study in Density-Current Surges, Phy. Fluids, 11:15.

EIDSVIK, K.J., 1980, A Model for Heavy Gas Dispersion in the Atmosphere, Atm. Environ., 14:769.

ENGLAND, W.G., L.H. TEUSCHER, L.E. HAUSER, and B.E. FREEMAN, 1978, Atmospheric Dispersion of Liquefied Natural Gas Vapor Clouds Using SIGMET, a Three-Dimensional Time-Dependent Hydrodynamic Computer Code, Proc. 1978 Heat Transf. and Fluid Mech. Inst., June 1978

GERMELES, A.E. and E.M. DRAKE, 1975, Gravity Spreading and Atmospheric Dispersion of LNG Vapor Clouds, Fourth Int. Symp. Transp. Hazardous Cargoes by Sea and Inland Waterways, Jacksonville, Florida, USA, 26-30, October 1975.

GRESHO, P.M., R.L. LEE and R.L. SANI, 1980, On the Time-Dependent Solution of the Incompressible Navier-Stokes Equations in Two and Three Dimensions, in "Recent Advances in Numerical Methods in Fluids, vol. 1," C. TAYLOR and K. MORGAN, ed., Pineridge Press, Swansea, U.K.

HAVENS, J.A., 1977, Predictability of LNG Vapor Dispersion from Catastrophic Spills into Water: An Assessment, DOT document CG-M-09-77, April, 1977.

KOOPMAN, R.P., L.M. KAMPPINEN, W.J. HOGAN and C.D. LUND, 1981, Burro Series Data Report, LLNL/NWC 1980 LNG Spill tests, Lawrence Livermore National Laboratory Report UCID-19075.

MOTT, W.E., M. GOTTLIEB, J.M. CECE and H.F. WATER, 1981, LNG Research: The Questions to be Answered, A.G.A. Transmission Conf., Atlanta, Ga., May 1981.

OGURA, Y and N. PHILLIPS, 1962, Scale Analysis of Deep and Shallow Convection in the Atmosphere, J. Atm. Sci., 19:173.

DISCUSSION

A.P. VAN ULDEN

Your continuity equation $\nabla.\rho u = 0$ seems to generate a divergent flow when density gradients are present. Since $\nabla.u = -\nabla\rho/\bar{\rho}$. This violates your assumption of incompressibility. Have you considered this drawback of your approach?

R.L. LEE

The assumption used in this case is not incompressibility in the conventional ($\nabla.\underline{u}=0$) sense but rather density variations occur so rapidly that the density field can be assumed to be in "quasi-static" equilibrium with the other thermodynamic variables. The idea is demonstrated in the paper in the form of approximate solutions to an ordinary differential equation system. So far, our numerical results using this generalized anelatic approximation have been quite reasonable and we have experienced no numerical difficulties pertaining to this assumption.

J.L. WOODWARD

What is the computing cost of your model?

What vaporization rate did you use for LNG vaporizing on water for the China Lake tests? (Do you have any new data to report on this)?

R.L. LEE

1. On a CRAY machine, the CPU time required for a typical 3D problem is about equal to real time. The associated I/D time required is, however, higher.

2. We assumed a steady-state vaporization rate of 10 cm/sec which is based on a liquid regression rate of 0.04 cm/sec. No new data are available on actual vaporization rates on water.

G.W. COLENBRANDER

You describe in your paper two-dimensional simulations of an instantaneous release of LNG vapor in a calm atmosphere. The results are interesting but may not be valid any more if features which are believed to be important for LNG vapour cloud dispersion are taken into account e.g. non-constant K-values and the way they are influenced by local vertical density gradients. Heat transfer from the earth's surface may also play an important role in this respect, while it is ignored in the numerical simulation.

R.L. LEE

The 2-D calculations are not intended to be used as faithful simulations of any particular spill scenario. Rather it is hoped that the simplified model (e.g. constant K values) with some parameters which are relevant to the real situation will provide a reasonable setting to evaluate the validity of the Boussinesq and hydrostatic assumptions. If the assumptions are inappropriate for this simple situation it is unlikely that the same assumptions could be valid for the more complex case of variable K and realistic surface boundary conditions.

G.W. COLENBRANDER

Since the model does not include the effect of ground heating on turbulence level, shouldn't these calculations be called numerical experiments rather than LNG simulations.

D. ERMAK

The purpose of the two simulations was to show the effect of terrain on the dispersion of an LNG spill under the terrain conditions at the China Lake test site and this goal was achieved. Currently, the FEM model does not include the effect of ground heating on the turbulence level and this is one of several aspects of the model which requires additional experimental and theoretical research. However, whether these simulations are to be called LNG simulations or numerical simulations appears to be just semantics. The model includes many of the physical phenomena which occur in an LNG spill and the properties of the released gas in the calculation were those of LNG vapor. In this sense, the calculations were LNG simulations.

NUMERICAL SOLUTION OF THREE-DIMENSIONAL AND TIME-DEPENDENT ADVECTION-DIFFUSION EQUATIONS BY "ORTHOGONAL COLLOCATION ON MULTIPLE ELEMENTS"

Hans Wengle

Institute for Fluid Dynamics, LRT
Hochschule der Bundeswehr München
8014 Neubiberg

1. INTRODUCTION

To predict the dynamic behavior of trace contaminants in given three-dimensional and time-dependent flow fields, a transport equation must be solved of the general form

$$\frac{\partial c}{\partial t} + u\frac{\partial c}{\partial x} + v\frac{\partial c}{\partial y} + w\frac{\partial c}{\partial z} = \frac{\partial}{\partial x}(K_x\frac{\partial c}{\partial x}) + \frac{\partial}{\partial y}(K_y\frac{\partial c}{\partial y}) + \frac{\partial}{\partial z}(K_z\frac{\partial c}{\partial z}) \quad (1)$$

where c is the mean concentration of a chemically inert contaminant, u,v, and w are the given components of the mean wind field, and K_x, K_y and K_z are given eddy diffusivities. The concentration field may be three-demensional and time-varying due to the three-demensional and time-dependent flow field and, in general, the eddy-diffusivities are also functions of position and time. In a wind field not varying in time, the concentration field may still be time-dependent due to unsteady source emission. In many cases, the turbulent transport in the main wind direction, say x-direction, may be neglected in comparison to the advective transport in the same direction, and it is sufficient to solve the less general form

$$\frac{\partial c}{\partial t} + u\frac{\partial c}{\partial x} + v\frac{\partial c}{\partial y} + w\frac{\partial c}{\partial z} = \frac{\partial}{\partial y}(K\frac{\partial c}{\partial y}) + \frac{\partial}{\partial z}(K_z\frac{\partial c}{\partial z}) \quad (2)$$

The x-direction is a so-called "marching"-direction, and significant savings of computing time are possible if (2) is solved instead of (1). Therefore, in this paper, we will consider the numerical solution of equation (2).

In recent times, a number of three-dimensional and time-depend wind fields are available from numerical solutions of so-called meso scale meteorological models. Fig. 1 gives an example of a three-dim sional and time-dependent wind field : the time-dependent atmospheri boundary-layer flow is disturbed by an urban area characterized by surface roughness and time-varying surface heat flux. An elevated point source with time-varying emission rate is located in the vicinity of the heated rough island. To determine, for example, the sur face maximum of the concentration field of a contaminant and its location for a number of possible positions of the point source, the atmospheric diffusion equation (2) must be solved many times. There is motivation, therefore, to develop efficient numerical methods for the numerical solution of the advection-diffusion equation (2).

The class of weighted residual methods offers a number of promising numerical solution methods, e.g. Galerkin and collocation methods. In particular, the so-called orthogonal collocation methods have experienced considerable success in applications to problems in chemical engineering. In a collocation method, the unknown solution is expanded in a series of basic functions including unknown coefficients. The expansion is substituted into the differential equation and the residual is set to zero at selected points, the so-called collocation points. For the solution of differential equations, collocation was systematically used for the first time by Frazer, Jones and Skan (1937). Later, the method was further developed and applied to problems in transport phenomena and chemical reactor theory by Villadsen and Stewart (1967). For theoretical

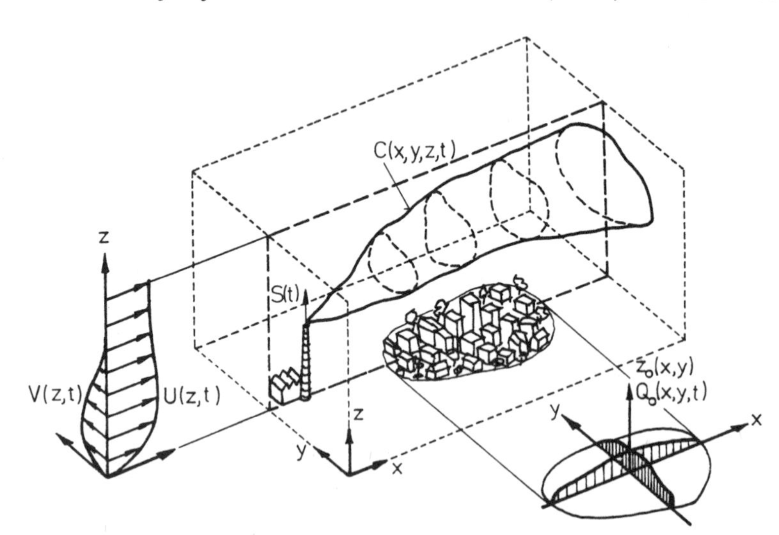

Fig. 1 Time-dependent concentration field of a point source plume in a 3D and time-varying flow field over a rough heated island.

aspects associated with orthogonal collocation methods the reader is referred to the book of Villadsen and Michelson (1978) and the reviews of Finlayson (1980, 1975). The efficiency and accuracy of the method "orthogonal collocation on multiple elements" for the solution of atmospheric diffusion problems was demonstrated by Wengle, Van den Bosch and Seinfeld (1978). A point source problem was solved by Fleischer and Worley (1978) using "global orthogonal collocation".

It is the purpose of this paper to demonstrate how the more flexible method "orthogonal collocation on multiple elements" can be used to solve the three-dimensional and time-dependent atmospheric diffusion equation (2) and to measure the accuracy of the numerical solutions against an analytical steady-state solution of a test-problem derived from (2).

2. DIMENSIONLESS FORM OF THE TRANSPORT EQUATION

To complete the mathematical description of the unsteady diffusion of an inert contaminant from an elevated point source, we define proper initial and boundary conditions. For example, we assume that the plume is confined between ground (z=0) and an elevated stable layer at z=H and that the point source is located at $(x_s=0, y_s, z_s)$ and begins emitting at time t=0 with constant source strenght S into an atmosphere without background concentration. If we further select a characteristic length scale H and a characteristic time scale H/u(H) of the problem, we may define the following dimensionless variables :

$$X=x/H,\ Y=y/H,\ Z=z/H,\ T=tu(H)/H \tag{3}$$

$$U=u/u(H),\ V=v/u(H),\ W=w/u(H)$$

$$K2=K_y/u(H)H,\ K3=K_z/u(H)H$$

$$C=cu(H)H^2/S$$

With these definitions we obtain a dimensionless formulation of the transport equation (2) :

$$\frac{\partial C}{\partial T} + U\frac{\partial C}{\partial X} + V\frac{\partial C}{\partial Y} + W\frac{\partial C}{\partial Z} = \frac{\partial}{\partial Y}(K2\frac{\partial C}{\partial Y}) + \frac{\partial}{\partial Z}(K3\frac{\partial C}{\partial Z}) \tag{4}$$

$$T=0:\quad C(X,Y,Z,0)=0 \tag{5}$$

$$Z=0:\quad (K3.\partial C/\partial Z)\ (X,Y,0,T)=0 \qquad Y=0:\quad C(X,0,Z,T)=0$$

$$Z=1:\quad (K3.\partial C/\partial Z)\ (X,Y,1,T)=0 \qquad Y=Y_{max}:\quad C(X,Y_{max},Z,T)=0$$

$$X=0:\quad C(0,Y,Z,T)=\delta(Y-Y_s)\,\delta(Z-Z_s)/U(0,Y_s,Z_s,T)$$

The constant source strength is no longer a determining parameter of the problem. The numerical representation of the Delta-function in the boundary condition at X=0 causes steep gradients close to the point source, and creates a difficult numerical problem for any numerical solution method.

3. THE SOLUTION PRINCIPLE

In this paper, orthogonal collocation is used to transform the continuous mathematical formulation (4), (5) into a discretized formulation. As already mentioned, the unknown solution is expanded into a finite series of basis functions and the expansions are substituted into the differential equation. The residual (defined as the difference between the evaluation of the continuous form and the discretized form of the differential equation) is required to be zero at certain selected points, the so-called collocation points. If we choose orthogonal polynomials as the basic functions and the zeroes of certain orthogonal polynomials as collocation points, the method is called "orthogonal collocation". A single expansion can be used to represent the dependent variable over the entire spatial domain in each direction ("global collocation") or we can first subdivide the computational domain in each coordinate direction into smaller subelements, and then use a separate expansion for each subelement ("multiple element approach" , "collocation on finite elements" or "global spline collocation").

For example, using the method "global collocation", the major steps in the numerical solution of the advection-diffusion equation (4) proceeds in the following way (see also Fig.2) :

a) Choice of basic functions : to expand the unknown solution separately in each direction, we select the Lagrange interpolation polynomials l_i. With that particular choice, the unknown coefficients of the expansions are precisely the values of the dependent variable at the collocation points.

b) Choice of collocation points : if the collocation points are selected to be the roots of so-called Jacobi polynomials, experience shows that we may expect high accuracy. For the calculations here, we always used the zeroes of Legendre polynomials (a special case of Jacobi polynomials).

c) Evaluation of partial derivatives : we obtain expressions for the partial derivatives in (4) by differentiating (term-by-term) the expansions in basic functions. The coefficients in the resulting expansions for the partial derivatives are the derivatives of the Lagrange polynomials at the collocation points. They need to be evaluated only once since the collocation points are fixed.

d) Solution of the resulting set of algeraic equations : by substitution of the expansions for the partial derivatives into the partial differential equation (4) and, in addition, into the boundary conditions (5), we obtain, after rearranging, a set of algebraic equations of the form A.C=b. The structure of the coefficient matrix A determines significantly the computing cost for the evaluation of the solution vector C. Solving directly this system of equations, and having applied "global collocation," the resulting coefficient matrix A is completely filled with non-zero entries (see Figure 3a). If we apply "orthogonal collocation on multiple elements," the matrix A will have a banded structure with a lot of zero-entries even within the band (fig. 3b). Solving the system of algebraic equations by an alternating-direction-implicit-(ADI-) approach, we will have to solve a series of small sets of equations each of which having a coefficient matrix with block-diagonal structure and with small blocks (Fig. 3c).

4. NUMERICAL SOLUTION BY "ORTHOGONAL COLLOCATION ON MULTIPLE ELEMENTS"

In contrast to "global collocation," we first subdivide the computational domain into smaller subregions (see Fig. 4). Then, we normalize the independent variables such that they vary from 0 to 1 within each of the subelements. To provide a finite set of albebraic equations we follow the steps (a) to (d) of section 3. Unfortunately, the number of unknowns will be larger than the number of equations resulting from solving the differential equations at all interior collocation points and from the boundary conditions. Therefore, additional equations must be provided by requiring continuous first derivatives normal to the boundaries between the subelements. For the partial derivative in time we take the most simple approach and use no internal collocation points at all; this will be equivalent to an implicit finite-difference Euler step. The x-direction also will be a so-called "marching"-direction, and we use at most one internal collocation point. It further follows that we may use the result at the end of a $\Delta X(ix)$-element as the initial condition for the calculations in the following $\Delta X(ix+1)$-element.

As a result of using the multiple element approach we are able (a) to use more subelements in those regions where steep gradients are to be expected (e.g. close to the point source) and (b) to evaluate the resulting set of algebraic equations faster than when global collocation is used. Now, the coefficient matrix has a banded structure with a large number of zero entries even within the band. However, the band width will be relatively large in comparison to the size of the coefficient matrix. For a time-dependent problem, the computing time will still be quite large if we solve this set of collocation equations directly.

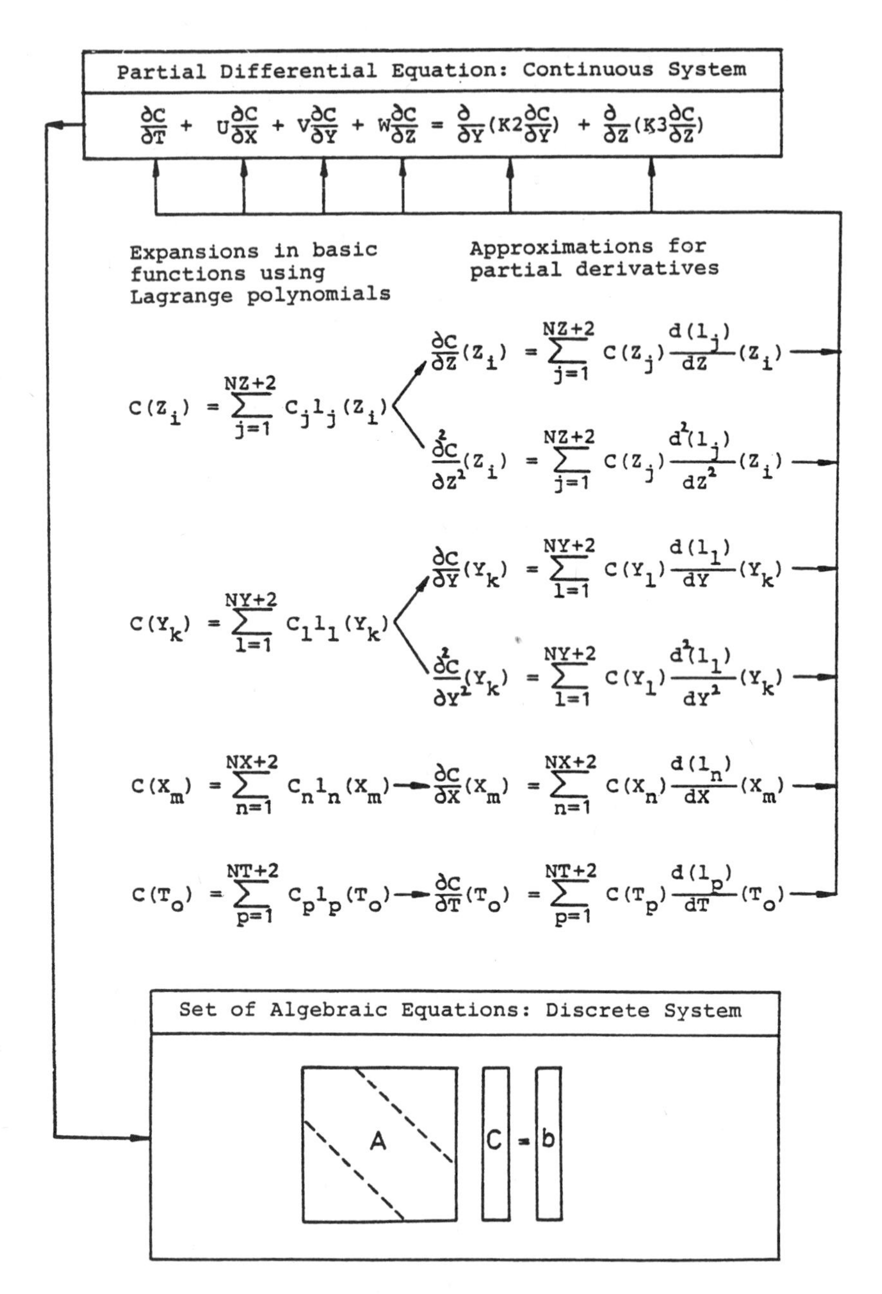

Fig. 2 Steps in the numerical solution of the atmospheric diffusion equation by "global orthogonal collocation" and by direct solution of the resulting collocation equation.

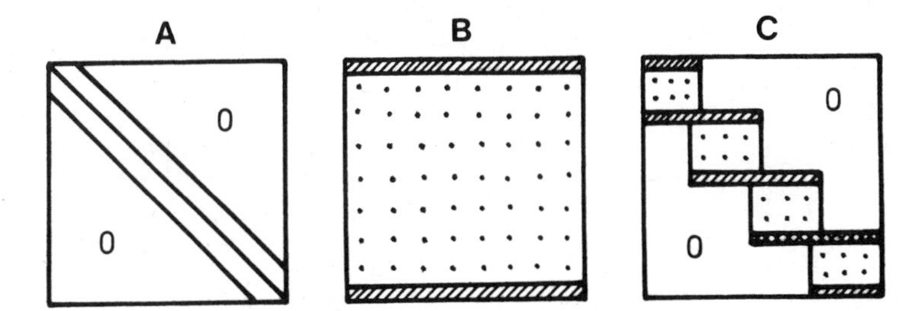

Fig. 3 Different structures of the coefficient matrix resulting from the collocation equations
(a) full matrix (one-dimensional problem)
(b) banded structure
(c) block-diagonal structure
First and last row from boundary conditions, intermediate hatched rows from condition of continuous first derivatives normal to the boundaries of the subelements.

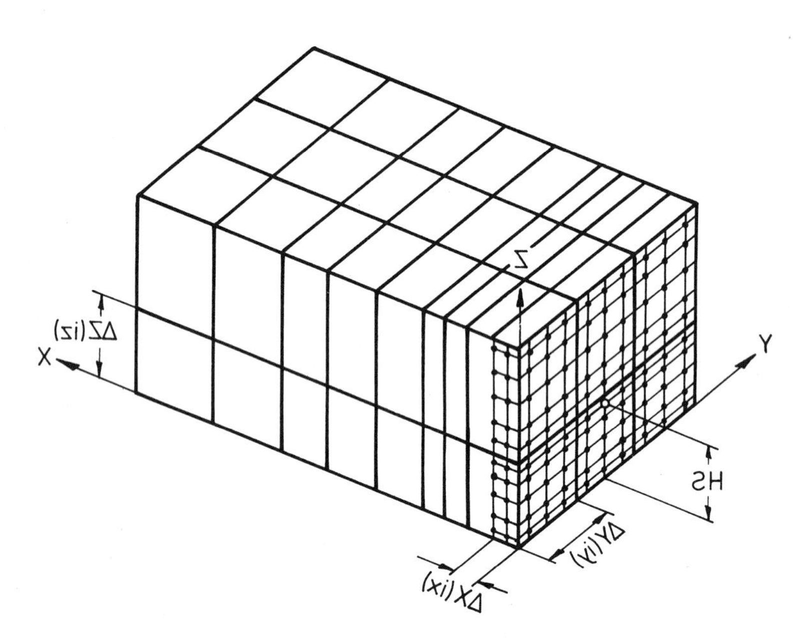

Fig. 4 Subdivision of computational domain into smaller subregion X(ix), Y(iy), Z (iz) with NZ = NY = 5 and NX = 1 internal collocation points

To improve the ratio of computing time to physical time for the solution of a three-dimensional and timedependent problem, we used an ADI(alternating-direction-implicit0-approach to solve the set of collocation equations. To advance the system (4) over one time-element of width ΔT, two intermediate values C^{*} and C^{**} are evaluated in applying the following scheme, given by Brian (1961):

$$\frac{C^{*}-C^{n}}{\Delta T(n)/2} = -U\frac{\partial C^{*}}{\partial X} - V\frac{\partial C^{n}}{\partial Y} - W\frac{\partial C^{n}}{\partial Z} + K2\frac{\partial^2 C^{n}}{\partial Y^2} + K3\frac{\partial^2 C^{n}}{\partial Z^2} + \frac{\partial K2}{\partial Y}\frac{\partial C^{n}}{\partial Y} + \frac{\partial K3}{\partial Z}\frac{\partial C^{n}}{\partial Z} \quad (6)$$

$$\frac{C^{**}-C^{n}}{\Delta T(n)/2} = -U\frac{\partial C^{*}}{\partial X} - V\frac{\partial C^{n}}{\partial Y} - W\frac{\partial C^{**}}{\partial Z} + K2\frac{\partial^2 C^{n}}{\partial Y^2} + K3\frac{\partial^2 C^{**}}{\partial Z^2} + \frac{\partial K2}{\partial Y}\frac{\partial C^{n}}{\partial Y} + \frac{\partial K3}{\partial Z}\frac{\partial C^{**}}{\partial Z} \quad (7)$$

$$\frac{C^{n+1}-C^{n}}{\Delta T(n)/2} = -U\frac{\partial C^{*}}{\partial X} - V\frac{\partial C^{n+1}}{\partial Y} - W\frac{\partial C^{**}}{\partial Z} + K2\frac{\partial^2 C^{n+1}}{\partial Y^2} + K3\frac{\partial^2 C^{**}}{\partial Z^2} + \frac{\partial K2}{\partial Y}\frac{\partial C^{n+1}}{\partial Y} + \frac{\partial K3}{\partial Z}\frac{\partial C^{**}}{\partial Z} \quad (8)$$

It follows from (6) to (8) that the system of collocation equations is solved in three steps: in each step the unknowns form partial derivatives with respect to one particular coordinate are treated implicitely while the unknowns from partial derivatives with respect to all other coordinates are taken from the last time step or from an intermediate step. From a physical point of view, we therefore advance the system first by advection in x-direction, then by advection and turbulent diffusion in z-direction and, finally, by advection and turbulent diffusion in y-direction. From a computational point of view, we solve in each of the three steps a series of sets of algebraic equations each of which has a small coefficient matrix with block-diagonal structure. The blocks are small and have all the same simple structure; the block size is determined by the number of internal collocation points on a line in a particular coordinate direction within a subelement.

5. NUMERICAL SOLUTION OF A TEST-PROBLEM

To test the numerical method and different versions of solution procedures, we selected a test-problem representing unsteady diffusi of a contaminant from an elevated point source in a constant flow fi with (U,V,W) = (U_0,0,0) and constant eddy diffusivities. Only the surface at X=0 represents a reflecting boundary. With these assumptions, the mathematical formulations (4) and (5) reduce to

$$\frac{\partial C}{\partial T} + U\frac{\partial C}{\partial X} = K2.\frac{\partial^2 C}{\partial Y^2} + K3.\frac{\partial^2 C}{\partial Z^2} \tag{9}$$

$$T=0 : \quad C(X,Y,Z,0)=0 \tag{10}$$

$$Z=0 : \quad (K3.\partial C/\partial Z)\ (X,Y,0,T)=0 \qquad Y=0 : \quad C(X,0,Z,T)=0$$

$$Z=1 : \quad (K3.\partial C/\partial Z)\ (X,Y,1,T)=0 \qquad Y=Y_{max} : \quad C(X,Y_{max},Z,T)=0$$

$$X=0 : \quad C(0,Y,Z,T)= \delta(Y-Y_s)\ \delta(Z-Z_s)\ /\ U(0,Y_s,Z_s)$$

This special case has an analytical solution for the steady-state ($t\rightarrow\infty$), which will be very useful in evaluating the corresponding numerical solution of (9), (10) once the concentration field has achieved steady-state :

$$C(X,Y,Z) = \frac{1}{4\pi X(K2.K3)^{1/2}}\, e^{-\frac{UY^2}{4K2.X}} \left\{ e^{-\frac{U(Z-Z_s)^2}{4K3.X}} + e^{-\frac{U(Z+Z_s)^2}{4K3.X}} \right\} \tag{11}$$

We obtained numerical solutions of test-problem (9) , (10)

(a) by direct computation of the steady-state and
(b) by computation of the steady-state via a series of time-dependent solutions.

Hereby, we evaluated the set of collocation equations

(a) by direct solution of the set of algebraic equations or
(b) by applying an alternating-direction-implicit approach.

In all the examples shown here, the point source was located at (x_s,y_s,z_s) = (0,750 m, 250 m) at the last internal collocation point in the first vertical subelement (Fig. 4). The computational domain with dimensions $(x_{max}, y_{max}, z_{max})$ = (1500 m, 1500 m, 1250 m) was subdivided into two vertical subelements (NEZ=2), three horizontal (cross-wind) elements (NEY=3), and 10 or 20 horizontal (along-wind) elements (NEX=10 or 20), see Fig. 4. The mean wind speed was taken as U_o=1m/sec and for the eddy diffusivities we have arbitrarily chosen values of K_z=K_y=20 m^2/sec.

In fig. 5, the numerical solution of the steady-state concentration field is compared to the analytical solution (11). This steady-state result has been obtained directly by direct solution of the collocation equations, using three internal collocation points in z- and y-direction (NZ=NY=3) within each subelement and using no internal collocation point in x-direction (NX=0). Fig. 5 also shows the significant improvement in accuracy by increasing the number of subelements in x-direction from NEX=10 to NEX=20. The locations and the widths of these subelements are indicated on the abscissa of Fig. 5a.

The concentration distributions in Fig. 6 have been obtained by solving the time-dependent problem and using the ADI-approach (6) to (8). To reach a steady-state result comparably accurate to the direct solution (Fig. 5), the ADI-approach required more internal collocation points (NZ=NY=5, NX=1). Fig. 7 shows the time-development of the point-source plume : cross-sections of the three-dimensional concentration field in wind direction (at $Y=Y_s$) are shown on the left-hand side and concentration distributions across the plume at dimensionless distances x/H=0.2 and 0.8 are shown on the right-hand side. The results at dimensionless times T=0.5, 1.0 and 3.0 correspond to the number of time steps it=10, 20 and 60. For the example of Fig.7, the surface concentration distributions $C(X,Y_s,0,T)$ are shown if Fig.6. As we assumed a constant wind speed $U_o=1.0$ and no turbulent diffusion in wind direction, a sharp concentration front moves away from the source. The position of that concentration front at dimensionless time T should be at $X=U_o.T$. In Fig. 6 this position in indicated by vertical dashed lines for T=0.25, 0.5, 0.75 and 1. The calculated concentration fronts are not very steep but smoothed out by numerical errors acting like a diffusive process and therefore, this effect is called "artificial" or "numerical" diffusion. Therefore, for the calculation of moving steep concentration fronts, more than one internal collocation point within each subelement in x-direction is required. Fig. 6 gives an indication of possible improvement of the result by increasing that number of collocation points from NX=0 to NX=1.

The approximate computing time for the steady-state solution evaluated by direct solution of the collocation equations (NZ=NY=3, NX=0, NEX=20, see Fig. 5) was 40 seconds on a Burroughs B7800. If time-dependent solutions were desired, this would be the computing time per time step. Using the ADI-approach, the same steady-state was reached after about 60 time steps (DT=0.05) and, for comparable accuracy, NZ=NY=5, NX=1 and NEX=10 was necessary. The computing time for one time step was about 10 seconds, leading to a ratio of computing time to physical time of about 1/6 which is equivalent to a ratio of about 1/12 on a CDC CYBER 175. For all these calculations, no attempts have been undertaken to optimize the FORTRAN programs with respect to computing time .

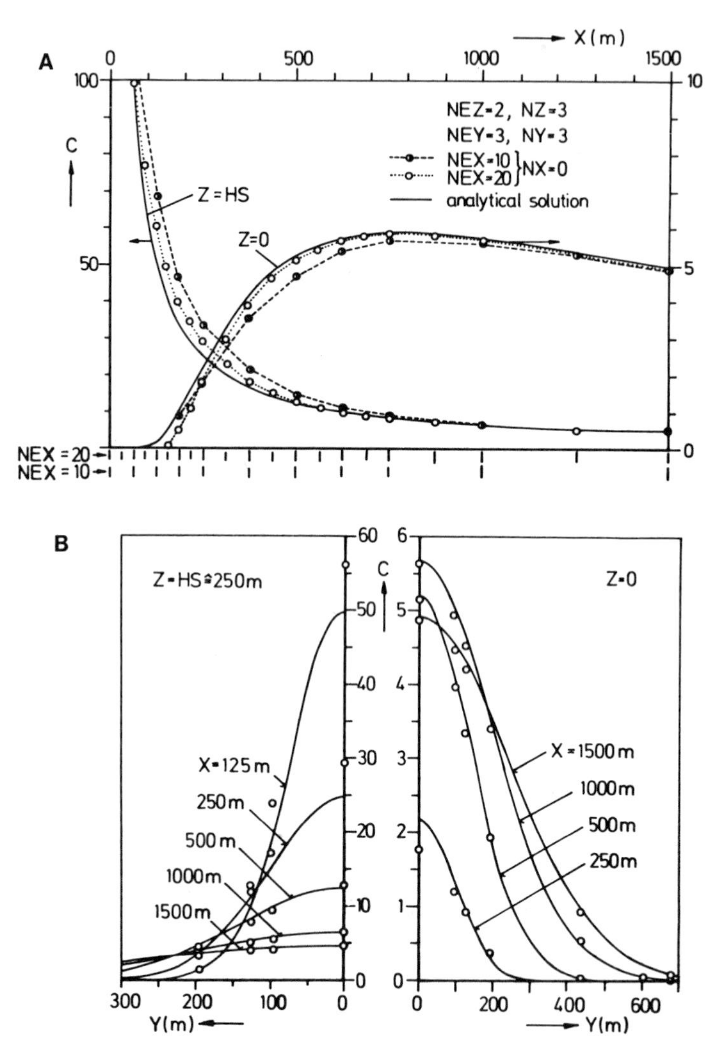

Fig. 5 Dimensionless steady-state concentration behind an elevated point source $C = cu(H)H^2/S$: (a) in wind direction at ground (Z = 0) and at source height (Z = HS) for $Y = Y_s$, (b) in crosswind direction at Z = 0 and Z = HS at different dimensionless distances o ⊙ : computed solution, ———— : analytical soln.

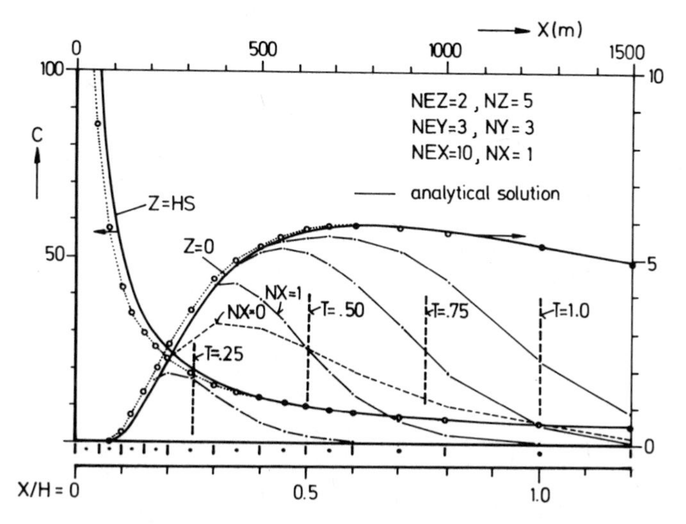

Fig. 6. Time-dependent concentration distribution behind an elevated point source, in wind direction at Z=0 for dimensionless times T=0.25,0.50,0.75,1.0 ------ computed with NX=0, —.— with NX=1 o computed steady-state, ——— analytical soln.

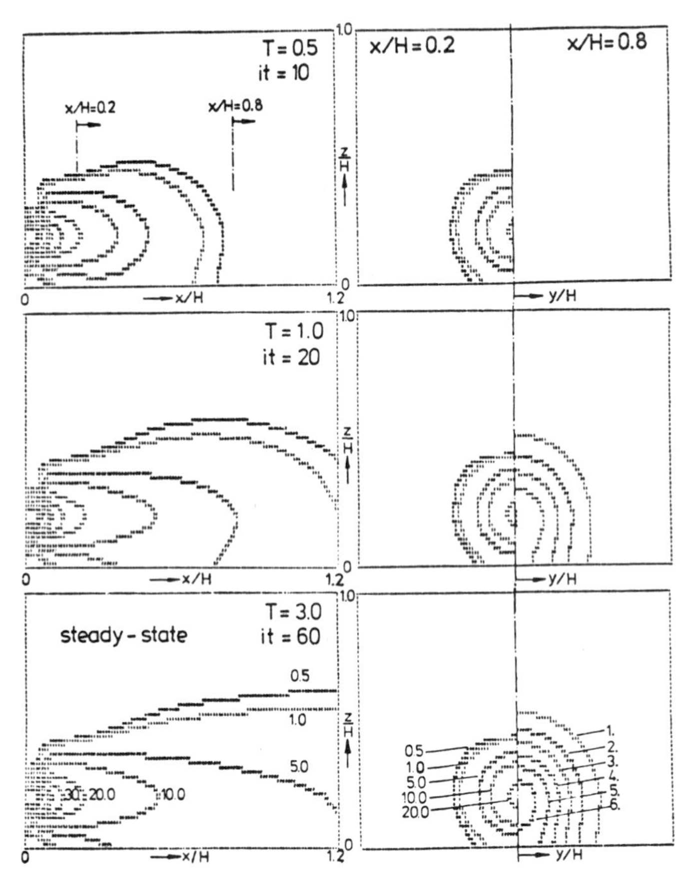

Fig. 7 Time-dependent development of a point source plume. Computed concentration distributions in wind direction (at Y = Ys) and across wind at x/H = 0.2 and 0.8 with NZ = NY = 5, NX = 1, NEZ = 2, NEY = 3, NEX = 10, DT = 0.05 and using an ADI-procedure.

6. CONCLUSIONS

For the numerical solution of three-dimensional and time-dependent diffusion behind point sources, the solution method "orthogonal collocation on multiple elements" is a flexible and efficient method, as long as the number of collocation points remaints relatively small. If the steady-state only is desired, about the same computational effort is required for the direct solution, and for the ADI-solution of the collocation equations. If transient states are desired, the ADI-approach appears to be superior. However, the ADI-procedure needs more collocation points to reach comparable accuracy. To improve the method with respect to computing time, one should concentrate on an efficient design of the ADI-procedure and on the method for solving sets of algebraic equations with block-diagonal structure of the coefficient matrix. With regard to accuracy, a careful control of the concentration field immediately behind the point source is advisable and, to predict with sufficient accuracy a moving steep concentration front, more than one internal collocation point should be used within the subelements in "marching"-direction.

REFERENCES

FRAZER, R.A., JONES, W.P. and SKAN, S.W., 1937, Approximations to functions and to the solution of differential equations, Great Britain Aero.Res.Conf., London, No. 1799, reprinted in: Great Britain Air Ministry Aero. Res.Comm.Tech.Tep. 1, 516-549.

VILLADSEN, J.V. and STEWART, W.E., 1967, Solution of boundary-value problems by orthogonal collocation, Chem.Engng.Sci. 21, 1483-1501.

VILLADSEN, J.V. and MICHELSEN, M.L., 1978, "Solution of Differential Equation Models by Polynomial Approximation" ; Prentice-Hall, Englewood Cliffs, N. Jersey.

FINLAYSON, B.A., 1975, Weighted residual methods and their relation to finite-element methods in flow problems in : Gallagher, R.H., Oden, J.T., Taylor, C. and Zienkiewicz, O.C., "Finite Elements in FLuids", Vol. 2, John WILEY and Sons, London.

FINLAYSON, B.A., 1980, Orthogonal collocation on finite elements - progress and potential, Mathematics and Computers in simulation 22, 11-17.

WENGLE, H., VAN DEN BOSCH, B. and SEINFELD, J.H., 1978, Solution of atmospheric diffusion problems by pseudo-spectral and orthogonal collocation methods, Atmospheric Environment 12, 1021-1032.

FLEISCHER, M.T. and WORLEY Jr., F.L., 1978, Orthogonal collocation - application to diffusion from point sources, Atmospheric Environment 12, 1349-1357.

BRIAN, P.L.T., 1961, A finite-difference method of higher order accuracy for the solution of 3-dim. transient heat conduction problems, A.I.Ch.E. J. 7, 367-370.

ANALYSIS OF A FINITE RELEASE OF A POLLUTANT FOLLOWED BY AN ABRUPT CHANGE OF WIND DIRECTION

Dov Skibin

Nuclear Research Centre - Negev
P.O. Box 9001, Beer-Sheva, Israel

INTRODUCTION

The releases of pollutants to the atmosphere may be divided into three groups according to the length of the release:
(i) Instantaneous release of a puff, (ii) Continuous (or quasi-continuous) release which creates an (quasi) infinite plume, (iii) The large group of intermediate cases of finite releases, which produce finite plumes. Many different methods exist, enabling the calculation of the concentration of the pollutant and the hazards resulting from its release. The methods range from simple gaussian models to very elaborated three dimensional numerical models. The major effort until the present time, concentrated in finding solutions to the first two groups of cases. The case of a steady wind direction and speed is fairly easy to handle, even with simple hand calculations, for the case of a puff as well as a (finite or infinite) plume, e.g. Slade (1968), Turner (1970).

Trajectory models or a trajectory-gaussian combination may be very helpful and relatively easy to use also under accident conditions, in the case of a puff release. If the release is not instantaneous, each puff (from which the cloud is comprised) may follow a different trajectory, according to the changing dynamics of the flow field. It follows, therefore, that general treatment of the composite real case of a finite release with a variable flow field, can hardly be expected without the use of a versatile computer program, provided the meteorological data is available. Under emergency conditions, the computer program has to be very efficient and the computer readily available, in order to give real time hazards evaluations of the released pollutant.

Shirvaikar et al. (1969) presented an elaborated method of computing the dose from a finite plume, taking into account changes in wind direction and speed. The method was used to compute the effect of wind persistence climatology on the downwind dose.

The present paper presents a model of the case of a finite release of a pollutant, followed by an abrupt change of wind direction. The model is simple enough to enable its field use as a hazards evaluation procedure under emergency situations, without the need of a computer. Perhaps more interesting than the model itself, is the dynamic picture of the hazards fields and the operational conclusions that may be drawn from it, influencing the procedures of environmental monitoring.

THE ACCIDENT MODEL

It is assumed that the pollutant released to the atmosphere is a γ emitting radioisotope. In addition to its relevance for accidents in nuclear facilities, this case is more general than the release of a chemical pollutant. In addition to the breathing hazards one faces also the problem of external radiation from the cloud and from the material deposited on the ground.

The release of the isotope starts at $t = 0$ and continues at a constant emission rate of $Q = 5$ Ci/s for one hour. Wind direction remains steady and uniform until $t = T_o \equiv 1.5$h, when a front passes the area and causes a sharp 90° veering of the wind. The resultant finite plume is then transported at right angles to its previous axis. For simplicity it is assumed that wind speed and stability are constants.

The main objective of this paper is to obtain a simple scheme that will give the general picture of the resulting hazards, the evolution, dynamics and time-table of the accident. We will therefore utilize the gaussian distribution to simulate the structure of the radioactive cloud. In the initial stage of the accident the concentration in the cloud is therefore given by

$$C = \frac{Q}{\pi\sigma_y\sigma_z\bar{u}} \exp\left(-\frac{y^2}{2\sigma_y^2} - \frac{z^2}{2\sigma_z^2}\right) \qquad (1)$$

assuming a ground level release; x and y are the downwind and crosswind distances, z the height above the ground and the sigma's are the standard deviations of the concentration in the cloud in the respective directions. $\bar{u}$ is the mean wind speed and the emission rate is $Q = 5$ Ci/s over most of the cloud's bulk except at its front and back edges. Fig. 1 (a,b) and Fig. 2 give the ground level concentration distribution of the cloud at t = 0,5, 1 and 1.5 h, respectively, for $\bar{u} = 5$ m/s and

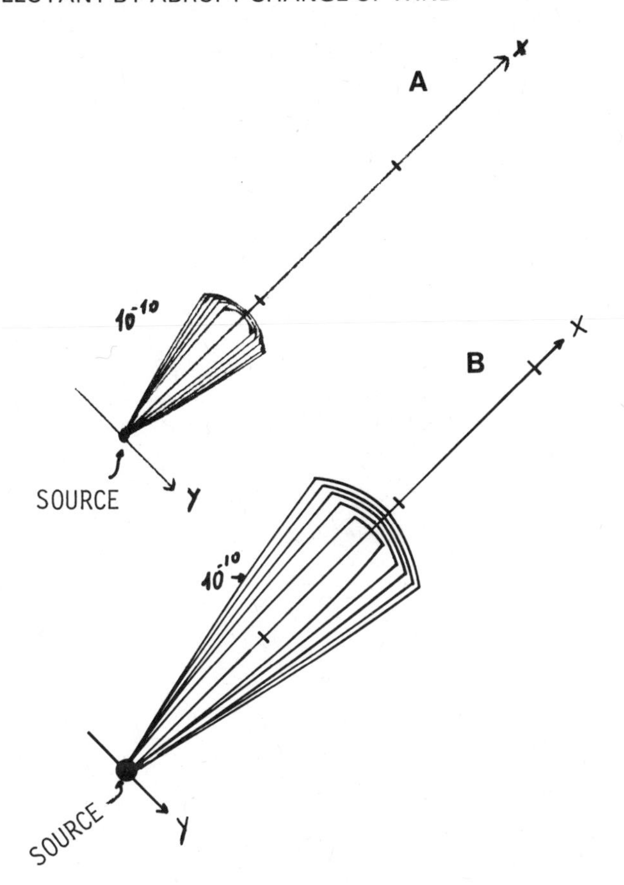

Fig. 1. Normalized concentration isopleths of the cloud at times t = 0.5 and 1 h (a and b, respectively). The external isopleth has a value of $(C\bar{u}/Q) = 10^{-10}\ m^{-2}$, the interval between consecutive isopleths is an order of magnitude. The isopleths also give the external radiation from the cloud in units 0.1 R/s.

Pasquill's stability D. Since we are not interested in the exact structure of the front and back ends of the cloud, the isopleths were roughly closed by hand drawing at these ends.

The semi-infinite homogeneous model is an approximation of the cloud's structure. Nevertheless it may be used as a good approximation to represent the field of direct radiation from the cloud, especially in cases of low radiation energy (E) (Slade, 1968). The proportion coefficient between the concentration and the external cloud radiation is 0.246E. If we take E = 0.4 MeV, the isopleths in Fig. 1 give also the external radiation from the clouds in units 0.1 R/s.

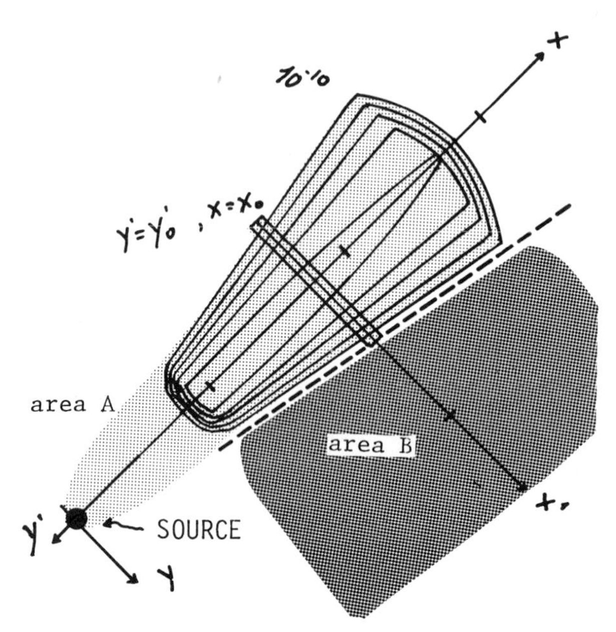

Fig. 2. Same as Fig. 1 but for time t = 1.5 h. Area A is the area over which the initial cloud, shown in this figure, passed. Area B is the area over which the cloud passed after the change of wind direction. The dashed line divides between area A (on the left hand side) and area B (on the right hand side of the line). A segment of the cloud at $x = x_o$ is also shown.

At $t = T_o \equiv 1.5$ h the wind direction changes to north-west. From this moment on we may rotate the coordinate axes, according to the usual convention, since the cloud is transported in the y direction, which now the new x' direction (Fig. 2). Naturally $y' \equiv x$ and $z' \equiv z$. The dispersion in the z' direction is a smoth continuation of the process before the wind direction change, so that $\sigma_z' = \sigma_z(x + x')$ for all the points of the cloud having the same x (or y'), $x' = \bar{u}\ (t - T_o)$. Consider a section of the cloud through $x = x_o$ ($y' = y_o'$, Fig. 2). The concentration in this section is normally distributed around the axis. The section and its axis are transported in the x' (y) direction at a speed $\bar{u}$. Longitudinal dispersion acts in the x' direction and continues to increase the width of this section in this direction. The information concerning longitudinal dispersion is relatively scarce. Nicola (1970) found a ratio of $\sigma_x/\sigma_y \approx 4$ for a krypton traces at $x = 200 \sim 800$ m, but the distribution was asymmetric. Since information on $\sigma_x(x)$ is insufficient and since the integrated concentration of the cloud is more interesting than its exact form, we will assume isotropy on the horizontal dispersion coefficient i.e. $\sigma y' = \sigma_x$. ($\sigma_y = \sigma_x$). The discussion in the appendix leads to the conclusion that at $t = T_o$ the variation with x (y') of the concentration in the cloud is relatively small, (at least down to $x' \sim x$). As a first approximation we may therefore treat the cload at $t > T_o$, a quasi homogeneous line source i.e. the dispersion in the y' direction is small compared to that in the x' and z' directions.

Utilizing this simple model, the position and shape of the cloud at $t > T_o$ may easily be drawn as follows. The cloud's axis is transported at a speed $\bar{u}$ in the x' direction. At the time t it is at $x' = \bar{u}(t - T_o)$. The shape of the cloud may be taken from Turner's isopleths by "cutting" a "sliding" segment the length of which is 18 km (1 h release at $\bar{u}$ = 5 m/s). The front edge of this segment is at $x_f = \bar{u}(t - T_o) + 27000$ and its back edge is at $x_b = \bar{u}(t - T_o) + 9000$. We then take this segment and copy it around the axis. Figs. 3 and 4 show the position of the cloud at t = 2 h and 3 h. The process is very easy and may be used even in field or emergency conditions.

The external radiation from the cloud at a certain (x,y) point, situated inside the cloud before it changes direction (area A in Fig. 2), increases relatively sharply at time $t = x/\bar{u}$,

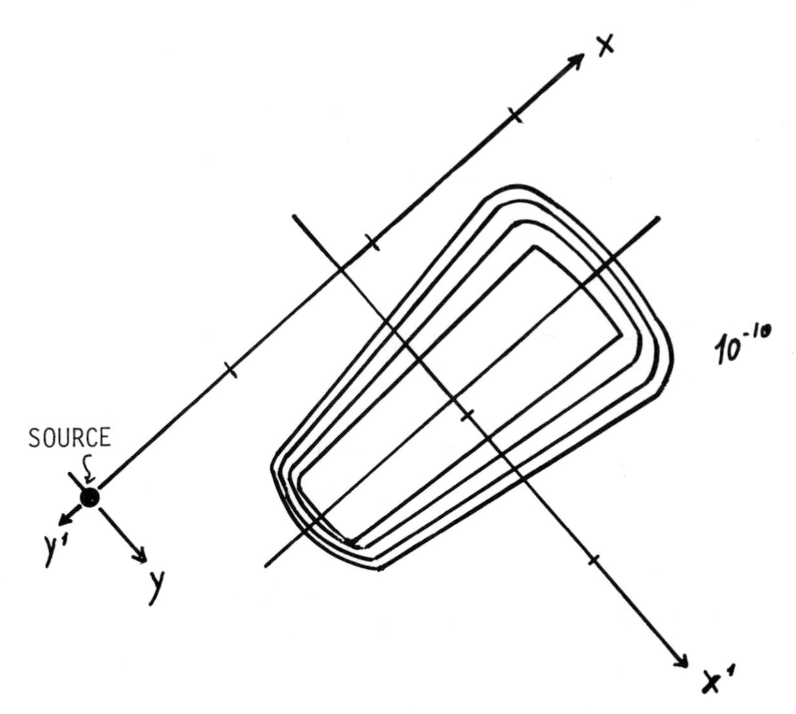

Fig. 3. Same as Fig. 1 but for t = 2 h. x' is in the new wind direction.

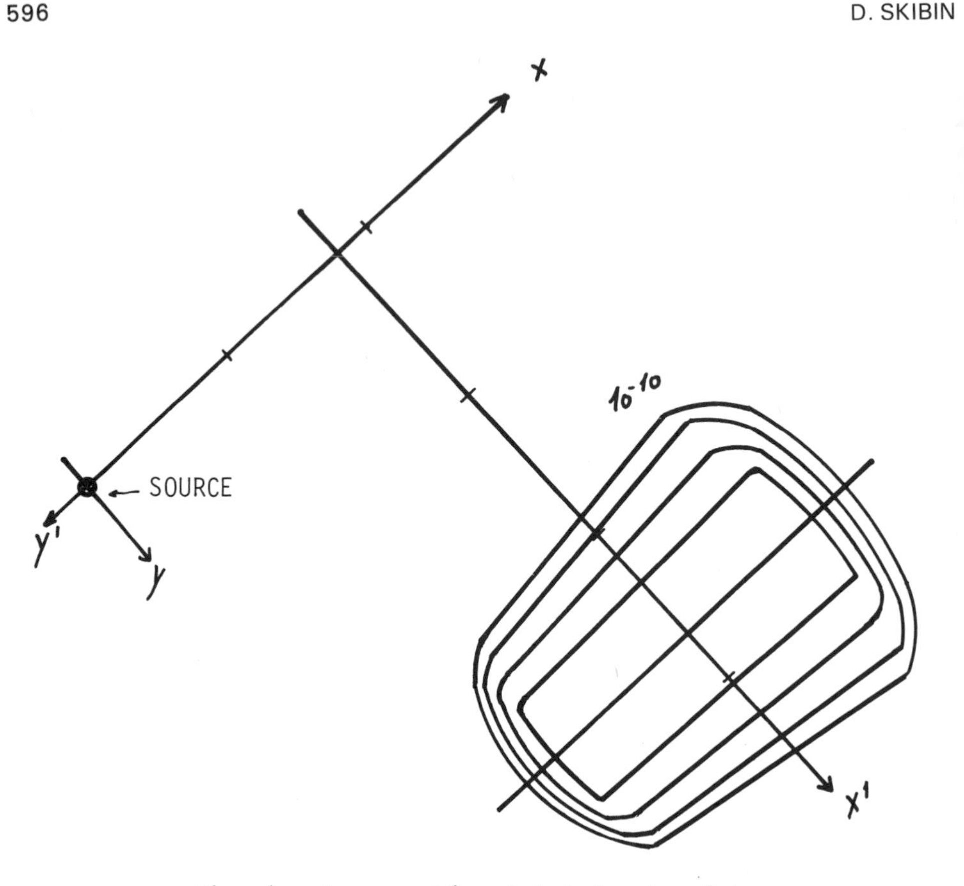

Fig. 4. Same as Fig. 1 but for t = 3 h.

to the level given by Figs. 1 or 2. It remains constant until $t = T_o$. Then it decreases (if $y < 0$), or increases and then decreases (if $y > 0$), according to the gaussian distribution. Similarly, the radiation at a distant point (x', y') in area B (Fig. 2) increases when the cloud approaches and then decreases as the cloud's axis passes overhead. The Total Integrated External Dose (TIED) from the cloud may be calculated accordingly.
In area A it is given by ($Q/\bar{u} = 1$, $0.246E = 0.1$)

$$\mathrm{TIED}_{(x,y)} = 0.1\left[\frac{C\bar{u}}{Q}\right]\left[T_o - \frac{x}{\bar{u}}\right] + \frac{0.1}{\pi\sigma_z}\int_{-\infty}^{y}\frac{1}{\sigma_y}\exp\left[-\frac{y^2}{2\sigma_y^{\,2}}\right]dy \qquad (2)$$

In area B the TIED equals the crosswind integrated radiation, and it is given by

$$\mathrm{TIED}_{(x',y')} = \frac{0.1}{\pi\sigma_z(x+x')}\int_{-\infty}^{\infty}\frac{1}{\sigma_y}\exp\left[\frac{y^2}{2\sigma_y^{\,2}}\right]dy = \frac{0.1(2/\pi)^{1/2}}{\sigma_z(x+x')} \qquad (3)$$

A map of the TIED is given in Fig. 5. In area B the isodose lines are actually lines of equal x + x', parallel to the bisector between the -x and x' axes. In area A the TIED is a combination of the isopleth field weighted by the time during which the cloud was overhead (a function of x, $\bar{u}$ and the accident time-table).

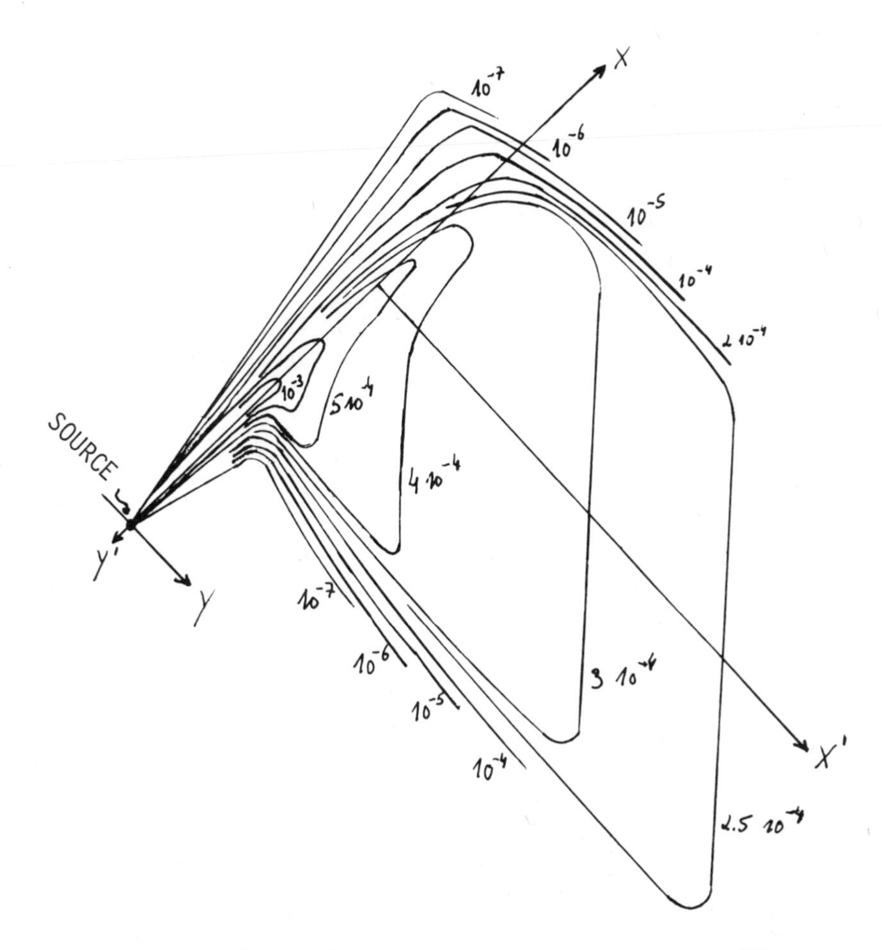

Fig. 5. The total integrated cloud dose (R).

The Total Integrated Concentration (TIC) is proportional to the TIED. If we assume a constant breathing rate during the cloud passage, the Total Integrated Breathing Dose (TIED) is also proportional to the TIED. Similarly, if we use the deposition velocity (V_d) concept the surface contamination field will also be proportional to the TIED field. Using the infinite quasihomogeneous approximation (appropriate for low energy radiation) the field of radiation from deposited activity is also proportional and has also the form of Fig. 5. The proportion coefficients are simple and may readily be computed so that one may simply copy Fig. 5 and change th numbers on the isolines in order to represent the desired field.

It should be noted that the resultant map of radiation from de-position prepared this way, is valid only after the cloud has passed over the area of interest. This map should therefore be covered, an then uncovered gradually according to the time-table of the cloud passage. The total radiation field at a certain point increases as the cloud comes overhead and as the surface deposition is accumulated. It then decreases as the cloud passes and we are left only with the final radiation from deposition.

DISCUSSION

Finite release is more likely to occur in real accidents, than instantaneous or continuous releases. After the accident, when the release starts, there is a tendency to identify and map the leading edge of the cloud and its contour. This may be done using mobile or airbone radiation detectors. However one does not always know the release rate, and moreover, one has to determine when the release ends. Mapping the cloud may occupy the monitors at a distance from the source so that we may not be able to answer those questions. It is therefore essential also to continuously monitor the radiation field close to the source. A set of permanent stations situated not too far around the potential source, may be very helpful for these purposes. Steady wind direction is the simplest case. Mapping the cloud in this case can easily be achieved for instance by saw-teeth penetrations into the cloud in the direction perpendicular to the wind. However, if the wind direction changes and we continue to monitor the cloud in the same manner and direction, it will escape us between consequent penetrations, if we are on its downwind side ($x',y < 0$). It will of course chase and contaminate us if we are on its upwind edge. In both cases, when we map the cloud this way it will appear tilted relative to its real axis, at an angle which depends on the monitoring speed relative to the wind speed and on the monitoring direction (in the plus or minus x direction). Sawteeth monitoring by penetrating in the y' (x) direction, perpendicular to the wind direction after it has changed, will be relatively inefficient, if not misleading. Two things may happen : one may loose the cloud completely if one monitors in its downwind edge, or one may not be able to identify its axis and edges, if one moves in the y' direction inside (below or above) the contour of the cloud.

Monitoring the radiation field from deposited activity should be done only after the cloud has passed, in order to be able to distinguish it from the direct radiation from the cloud.
The resultant map will have the form of Fig. 5, which has the "classical" form only downwind from the source in the x direction.

This field is relatively weak, "smeared" and changes relatively slowly in the x' direction (mainly due to vertical dispersion). Examples of such fields are usually not reported in the literature since people usually look for the steady cases which are typical and "classical" in form. An example in which the authors stated that wind direction changed, was measured during an HTO diffusion experiment by König et al. (1973).

If the direction change-ϕ does not equall 90°, one may still predict the resultant fields approximately by using Turner's (1970) Eq. (5.19) which includes $(\sin\phi)^{-1}$ and should not be used for $\phi < 45°$. A similar result is given by Shirvaikar et al. (1969), Eq. (13). The homogeneous line approximation may become worse as $\phi - 90°$ increases.

ACKNOWLEDGEMENT

I am indebted to Dr. E. DORON for useful discussions and to E. DOAR for his help.

REFERENCES

KONIG, L. A., NESTER, K., SCHUTTELKOPF, H. and M. WINTER, 1973, Experiments conducted at the Karlsruhe Nuclear Research Center to determine diffusion in the atmosphere by means of various tracers, Symp. on the Behaviour of Radioactive Contaminants in the atmosphere, IAEA-SM-181/4.

NICOLA, P. W., LUDWICK, J. D. and J. V. RAMSDELL Jr., 1970, An inert gas tracer system for monitoring the real-time history of a diffusing plume or puff, J. App. Meteorol. 9:621.

SHIRVAIKAR, V. V., RAMESH K. KAPOOR and L. N. SHARMA, 1969, A Finite plume model based on wind persistance for use in environmental dose evaluation, Atmos. Environ. 3:135.

SLADE, D. H. Ed., 1968, Meteorology and atomic energy 1968, TID-24190.

TADMOR, J. and Y. GUR, 1969, Analytical expressions for the vertical and lateral dispersion coefficients in atmospheric diffusion, Atmos. Environ., 3:688.

TURNER, D. B., 1970, Workbook of atmospheric dispersion estimates, U.S. Pub. Health Service, PB 191482.

APPENDIX

THE QUASI-HOMOGENEOUS LINE SOURCE

Consider the finite plume just before the wind direction changes (Fig. 2). After the direction change the plume becomes a line source. The wind is now perpendicular to the line. Fig. 6 shows schematically a segment of the line, moving downwind and dispersing. The line source may be considered homogeneous by an observer at (x', y') if the contribution from this segment to the concentration at this point is insignificant. Since the segment

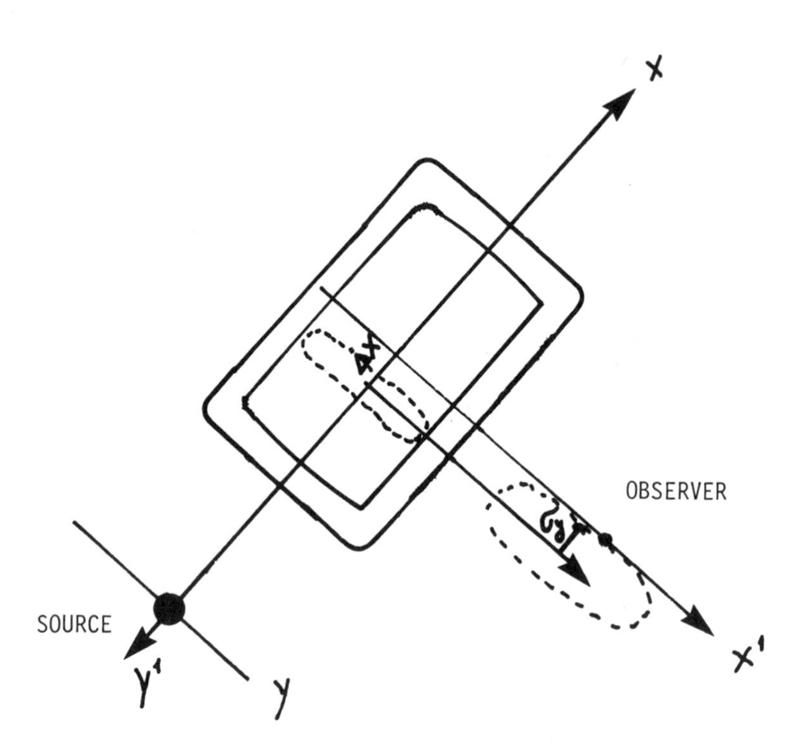

Fig. 6. Schematic presentation of a segment of the cloud, moving downwind, dispersing, growing and affecting the area within the range of σ_y form its center.

disperses and becomes larger as it moves, its contribution to the concentration at x' is limited to distances of the order of $\sigma_y(x')$ from its center. Therefore only segments closer than $\Delta x \sim \sigma_y(x')$ will contribute to the concentration at the observation point (x',y'), and the condition for homogeneity is $\Delta C/C \ll 1$, where $\Delta C = (dC/dx)\Delta x$ and $\Delta x \sim \sigma_y(x')$. Since $C(x) \alpha (\sigma_y \sigma_2)^{-1} \rightarrow dC/dx = - C d(\ell n \sigma_y + \ell n \sigma_z) dx$

For order of magnitude evaluation purpose let us assume that the two terms on the right hand side are of the same order and that $\sigma_y = ax^b$ (Tadmon and Gur. 1970), i.e.

$$(dC/dx)/C = (2/\sigma_y)(d\sigma_y/dx) = 2b/x$$

$$\Delta C/C\Big|_{\Delta x \simeq \sigma_y(x')} = (dC/dx)(\Delta x/C) = 2b\sigma_y(x')/x' = 2bax'^b/x$$

Taking $b \simeq 0.9$

$$\Delta C/C \leq 0.1 \rightarrow 1.8ax'^{0.9}/x \leq 0.1 \rightarrow x'^{0.9}/x \leq 0.056/a$$

For the neutral category $\bar{D}$, $a \simeq 0.07$ and the condition for homogeneity is $x' < x$. Further downwind the cloud experiences additional dispersion in the y' direction due to the relative nonhomogeneity of the finite cloud. This will appear as elongation and homogenization of the cloud in the y' direction. The isolines in Fig. 5 will have the same appearance but further downwind they will be more parallel to the x axis and will be more stretched in the $\pm$ y' directions.

DISCUSSION

B. NIEMANN

It seems to me that the situation you are modeling is more that of a plume traveling down a valley and encountering a crosswind from a side canyon, since a front is normally a very dynamic weather circulation and would probably cause more horizontally and vertical mixing of the plume than your modeling assumes. Would you agree ?

D. SKIBIN

The sudden change of 90° in wind direction over the area of the plume is indeed a simplification. On the other hand, when watching an arriving front with clouds you can see the wall of clouds moving and passing and sometimes the wind changes by 90° within a few minutes over a wide area. The simplification is intended to give an idea about the general form of the hazards fields and to enable its field real time use under emergency conditions without the need of a computer and a sophisticated program.

TWO-DIMENSIONAL UNSTEADY GRAVITY CURRENTS

A.P. van Ulden and B.J. de Haan

Royal Netherlands Meteorological Institute
P.O. Box 201
3730 AE De Bilt, The Netherlands

INTRODUCTION

Since long it has been recognized, that even a small density surplus can have a significant effect on the motion of a dense fluid. Probably the first to report experimental data on this subject was W. Schmidt (1911). This author describes a number of laboratory experiments with cold air. These experiments show, that a gravity current is formed, when an amount of cold air is released on the ground. This current attains a velocity relative to the ambient lighter fluid, that is of the order $\sqrt{g'H}$, where $g' = g\Delta\rho/\rho$ is the reduced gravity, $\Delta\rho$ the density difference, ρ the density and H the depth of the current. Schmidt related his experimental results to the outflow of cold air in the atmosphere.

Similar features have been observed for heavy gases. When an amount of heavy gas is released on a horizontal surface, an axisymmetric gravity current develops. The radial velocity of the leading edge also is of the order $\sqrt{g'H}$ (Van Ulden, 1974).

Another example of a gravity current is the lock exchange flow. When a lock, that separates two fluids of different density, is opened, the dense fluid will flow under the lighter fluid. Comprehensive series of experiments on lock exchange flow have been reported by Keulegan (1957) and by Huppert and Simpson (1980). These experiments show clearly that the velocity of a gravity current does not solely depend on $\sqrt{g'H}$. Huppert and Simpson show that the fractional depth H/D, where D is the total depth of the channel, is an important parameter. Furthermore, when the current becomes sufficiently thin, viscous forces will be important (Fay, 1969).

It is the purpose of this paper to analyse the various forces that determine the behavior of an unsteady gravity current. We describe inertial, viscous and bouyancy (pressure) forces. We results with experimental data by Keulegan (1957) and Huppert and Simpson (1980).

DESCRIPTION OF THE MODEL

We describe the unsteady gravity current, that is formed when an amount of dense fluid is suddenly released on the surface at the beginning of a long channel. Following Huppert and Simpson (1980), we represent the current by equal-area rectangles (Figure 1). We thus neglect mixing between the two fluids. The volume of the current is:

$$V_c = XH , \tag{1}$$

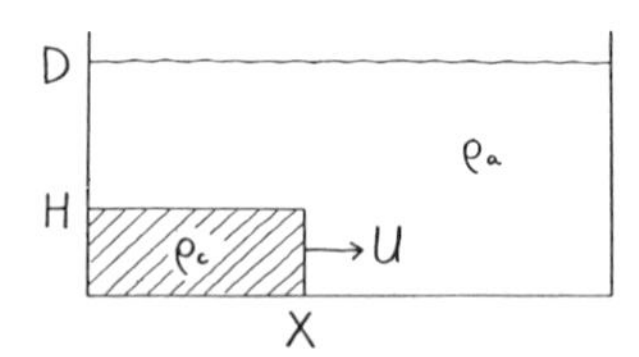

Figure 1. The flow described in this paper. The dashed area area represents the gravity current.

where X is the length of the current and H its depth. It follows from the continuity equation, that the layer averaged velocity in the current increases linearly with the distance x from the origin origin:

$$u(x) = (x/X)U . \tag{2}$$

Here u(x) is the layer averaged velocity in x and:

$$U = dX/dt, \tag{3}$$

is the velocity of the leading edge.

From (1) and (2) it follows that the total momentum of the current equals:

$$M = \tfrac{1}{2}\rho_c V_c U \ , \qquad (4)$$

where ρ_c is the density of the current. Since ρ_c and V_c are constant, the time derivative of M is:

$$dM/dt = \tfrac{1}{2}\rho_c V_c \ (dU/dt) \ . \qquad (5)$$

According to Newtons law, the time rate of change of total momentum equals the sum of all forces that act on the current:

$$dM/dt = \sum_i F_i \ , \qquad (6)$$

where F_i denote the various forces.

The driving force on the current is the buoyancy force. This force is obtained by integrating the static pressure jump $\Delta P(z)$ at the leading edge over the depth of the current. Since $\Delta P = g\Delta\rho(H-z)$, where $\Delta\rho$ is the difference between the densities of the two fluids, this yields:

$$F_b = \int_o^H g\Delta\rho(H-z)dz = \tfrac{1}{2}g\Delta\rho H^2 \ . \qquad (7)$$

Here the subscript b stands for buoyancy.

There are two retarding forces. The first force acts on the leading edge of the current. It is due to the presence of ambient fluid, that has to be accelerated by the advancing head. The resulting force can be written as:

$$F_a = \tfrac{1}{2}\alpha\rho_a \ HU^2 \ , \qquad (8)$$

where ρ_a is the density of the ambient fluid and where α is a drag-coefficient. In general α will depend on the shape of the head and on the configuration of the flow around the head. It will be determined later from experimental data. Here we wish to notice that in this parameterisation of the force on the head, we have absorbed all non-hydrostatic and dissipative features of the flow around the head.

The second retarding force acts on the horizontal boundaries of the current and is mainly due to the no-slip condition at the surface. This causes a viscous stress per unit length (Huppert and Simpson, 1980):

$$\tau(x) = \varepsilon\nu\rho_c \, u(x)/H \,, \tag{9}$$

where ε is an empirical constant of $O(1)$ and ν the kinematic viscosity. This parameterisation is only valid for a smooth surface. Upon integrating (9) over the length of the current, we obtain the total viscous force as:

$$F_v = \tfrac{1}{2}\varepsilon\nu\rho_c \, UX/H \,. \tag{10}$$

Our integral momentum equation now reads:

$$dM/dt = F_b - F_a - F_v \,, \tag{11}$$

or with (5) - (10):

$$\tfrac{1}{2}\rho_c V_c \frac{dU}{dt} = \tfrac{1}{2}g\Delta\rho_c H^2 - \tfrac{1}{2}\alpha\rho_a \, HU^2 - \tfrac{1}{2}\varepsilon\nu\rho \, \frac{UX}{H} \,, \tag{12}$$

This equation can be manipulated into a more convenient form. By dividing (12) by $\tfrac{1}{2}g\Delta\rho H^2$ and by using $g' = g\Delta\rho/\rho_a$ and $H = V_c/X$ we obtain:

$$\frac{\rho_c X^2}{\rho_a g' V_c} \frac{dU}{dt} = 1 - \frac{\alpha U^2 X}{g' V_c} - \frac{\varepsilon\nu\rho_c UX^4}{\rho_a g' V_c^3} \,. \tag{13}$$

We thus have normalized the momentum equation with the buoyant force. As we will see later (figure 6), this ensures, that important terms in the equation are of $O(1)$.

In (13) the parameters α and ε have to be evaluated empirically. This will be done in the subsequent sections. Also we will investigate the behaviour of the Froude number that is defined as:

$$Fr = U/\sqrt{g'H} \,. \tag{14}$$

As we will see, (13) is a powerful tool in the analysis of unsteady gravity currents.

THE VISCOUS-BYOYANCY PHASE

It can be seen from (13) that the viscous force will dominate over the inertial terms in the equation when X becomes sufficiently large. The current then has reached its viscous-buoyancy phase. The viscous force balances the buoyancy force and (13) reduces to:

$$\frac{\varepsilon \nu U X^4}{g' V_c^3} = 1 , \tag{15}$$

where we have assumed that $\rho_a/\rho_c \simeq 1$.
It follows from (1), (14) and (15) that the Froude number is proportional to $X^{-7/2}$ and vanishes rapidly when the length of the current increases.

Since $U = dX/dt$, (15) can easily be solved for X.
The result is:

$$X = (5g' V_c^3 t/\varepsilon\nu)^{1/5} . \tag{16}$$

Apart from the numerical factor $(5/\varepsilon)^{1/5}$ this relation has also been found by Huppert and Simpson (1980). The difference arises from the fact that H&S assumed a different shape for the current and took $\varepsilon = 1$. H&S further report pertinent data on the length of gravity currents in the viscous-buoyancy phase. From their figure 4 we estimate that

$$\varepsilon \simeq 1.5 . \tag{17}$$

With this value the numerical factor $(5/\varepsilon)^{1/5} \simeq 1.27$.
This is close to the "theoretical" constant 1.41 suggested by H&S. For our computations in following sections we assume that ε is constant and use the empirical value as given by (17). This enables us to tell under what conditions the viscous term in (13) can be neglected.

THE INERTIA-BUOYANCY PHASE

The inertia-buoyancy phase proper is a phase in which the flow is independent of the Reynolds number. This implies in the first place that the surface stress is small, i.e. that:

$$\varepsilon\nu UX^4/g'V_c^{\,3} \ll 1 \ . \tag{18}$$

The momentum equation (13) then reduces to:

$$\frac{X}{U^2}\,Fr^2\,\frac{dU}{dt} = 1 - \alpha Fr^2 \ , \tag{19}$$

where again we have taken $\rho_a/\rho_c \simeq 1$. In this section we shall deduce a relation for α depending on the fractional depth, in order to solve (19).

In equation (19) α may still be a function of the Reynolds number. There is some evidence (Keulegan, 1957; Simpson and Britter, 1979) that the shape of the gravity current head is more or less independent of the Reynolds number for Re > 7000. It thus is reasonable to assume that the drag coefficient α is Re-independent for Re > 7000. In the following analysis we assume that the gravity current is in the inertia-buoyancy phase if both Re > 7000 and $\varepsilon\nu UX^4/g'V_c^{\,3} < 0.05$.

It appears that many of the data by Keulegan (1957) meet these requirements. We will analyse these data first. In figure 2 we have plotted the measured Froude number against the fractional depth H/D. The data show a fairly small scatter and a significant variation of the Froude number with the fractional depth. In order to make further analysis manageble we have fitted a simple analytical curve to the data. The curve consists of three parts. The first fits the bulk of Keulegans data. It is given by:

$$Fr = 0.5(H/D)^{-0.4} \ , \tag{20}$$

and used for $0.155 < H/D < 0.67$. This curve is plotted in figure 2. From this curve we can compute α for the indicated range of fractional depths. This can be done because for a given experiment H/D varies with time. In fact initially H/D = 1 (for Keulegans experiments) and it decreases in time. So during a given experiment the Froude number increases.

For the range $0.155 < H/D < 0.67$ we can deduce a simple relation for α. After differentiating (1) and using (3) we find:

$$\frac{d}{dt}\,H = -\,\frac{UH}{X} \ , \tag{21}$$

and we find after equating (14) and (20) and differentiating the result:

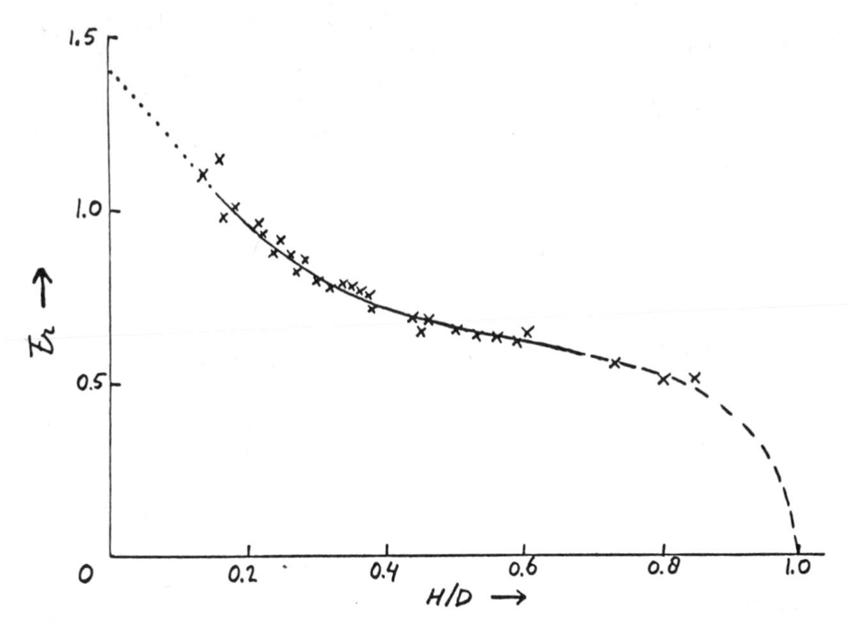

Figure 2. The Froude number as a function of the fractional depth for a gravity current with initial fractional depth H/D = 1. The data refer to the inertia-buoyancy phase. The curves are described in the text. The crosses are from four experiments by Keulegan (see also Huppert and Simpson, 1980, figure 5).

$$\frac{d}{dt} U = 0.1 \frac{U}{H} \frac{d}{dt} H .$$

Substituting the above results into (19) we obtain:

$$\alpha = \frac{1}{Fr^2} + 0.1 .$$

With (20) this yields:

$$\alpha = 4(H/D)^{0.8} + 0.1 \text{ for } 0.155 < H/D < 0.67 . \qquad (22)$$

This function is shown in figure 3. We see that the drag coefficient depends strongly on H/D. For the remaining intervals for H/D we propose to use:

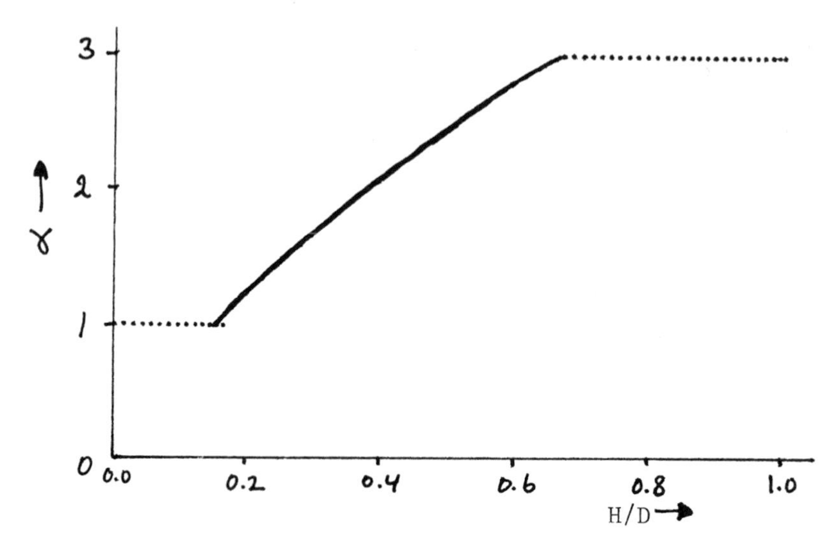

Figure 3. The drag coefficient as a function of fractional depth.

$$\alpha = 1.0 \quad \text{for} \quad 0 \leq H/D \leq 0.155 \,, \tag{23}$$

and

$$\alpha = 3.0 \quad \text{for} \quad 0.67 \leq H/D \leq 1 \,. \tag{24}$$

These values provide plausible extensions to the Froude number curve (20) in figure 2 as we shall see.

If α is a constant during a time interval $t_o \leq t \leq t_1$ we can find a relation for the Froude number depending on the current height H. We differentiate (14):

$$\frac{dU}{dt} = \sqrt{g'H}\,\frac{d}{dt}\,Fr + \tfrac{1}{2}\,\frac{g'Fr}{\sqrt{g'H}}\,\frac{d}{dt}\,H \,,$$

substituting (21) and writing $\frac{d}{dt}\,Fr = \frac{d}{dH}\,Fr\,\frac{d}{dt}\,H$, we find:

$$\frac{dU}{dt} = -\,\frac{U^2}{X}\,\frac{H}{Fr}\,\frac{dFr}{dH} - \tfrac{1}{2}\,\frac{U^2}{X} \,.$$

Equating the above with (19) we find the following differential equation:

$$-\frac{H}{2}\frac{d}{dH}Fr^2 = 1 - (\alpha-\tfrac{1}{2})Fr^2 .$$

The solution of this equation reads:

$$Fr^2 = \beta + (Fr_o^2-\beta)\left(\frac{H}{H_o}\right)^{2/\beta} , \qquad (25)$$

where

$$\beta = \frac{1}{\alpha-\frac{1}{2}}$$

$$Fr_o = Fr(t=t_o) , \qquad (26)$$

$$H_o = H(t=t_o) .$$

The extensions in figure 2 are obtained from (25). The curve for $\alpha = 3.0$ starts with the initial condition Fr = 0 for H/D = 1. It further passes through the few data for great fractional depths and is continuous in H/D = 0.67. The curve for $\alpha = 1$ starts with Fr = 1.05 for H/D = 0.155 and provides an almost linear extrapolation to Fr = 1.41 for H/D = 0. The latter extension is less certain, but we shall see later that it provides a good fit to the data by Huppert and Simpson (1980) for small fractional depths. We wish to make clear here that all the curves that we have constructed in this section are purely empirical.

Now we have completed our analysis of the Keulegan data on the inertia-buoyancy phase, we are left with the problem what occurs when both inertial and viscous forces are significant. We propose to use a simple interpolation between the description of the viscous-buoyancy phase and that of the inertia-buoyancy phase. Thus we use (13) with $\varepsilon = 1.5$ and α as in figure 3. We thus neglect a possible Reynolds number dependence of α.

The full equation (13) cannot be solved analytically, but it is easy to solve numerically. Numerical computations are compared with experimental data in the next section.

COMPARISON WITH OTHER EXPERIMENTAL DATA

Huppert and Simpson (1980) report a number of experiments on lock-exchange flow. These experiments have several interesting features. In the first place the initial fractional depth was varied in a fairly wide range. In the second place, as we saw in section 3, several experiments were continued long enough to include a viscous-buoyancy phase. This enables us to verify several important features of our model.

In figure 4 we show the comparison with the H&S experiments 1, 2 and 9. It is seen that the model performs quite well. Since most of the data relate to fractional depths less than 0.155 our choice to take $\alpha = 1$ for $H/D < 0.155$ seems adequate. Furthermore in these experiments the viscous force is about equal to the drag force for $X \simeq 5$ m. The data around $X = 5$ m thus refer to the mixed phase, in which both viscous and inertial forces are important. The upper part of the curves represent the beginning of the viscous-buoyancy phase. Thus the model also reproduces the proper viscous effects.

In figure 5 we show experiment no. 7. Again the model performs well. Let us analyse experiment 7 in some more detail. In figure 6 we give for this experiment the various forces that act on the gravity current. The forces are normalized with the buoyant force as in (13). Thus the normalized buoyant force equals unity by definition. Initially - for a very brief period - this force is balanced by the inertia of the gravity current itself, that is

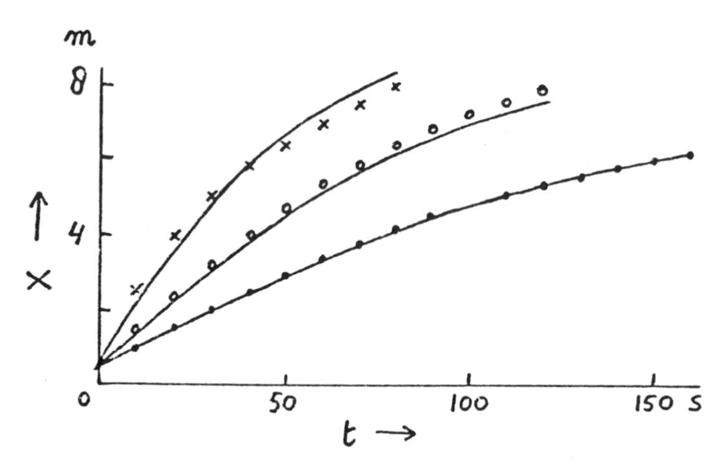

Figure 4. The length of the gravity current as a function of time. The curves are computed with (13). The data are from Huppert and Simpson (1980):
• $D = 0.149$ m, $H_o = 0.149$ m, $X_o = 0.390$ m, $g' = 0.091$ ms^{-2}
o $D = 0.149$ m, $H_o = 0.149$ m, $X_o = 0.396$ m, $g' = 0.287$ ms^{-2}
x $D = 0.449$ m, $H_o = 0.150$ m, $X_o = 0.392$ m, $g' = 0.648$ ms^{-2}

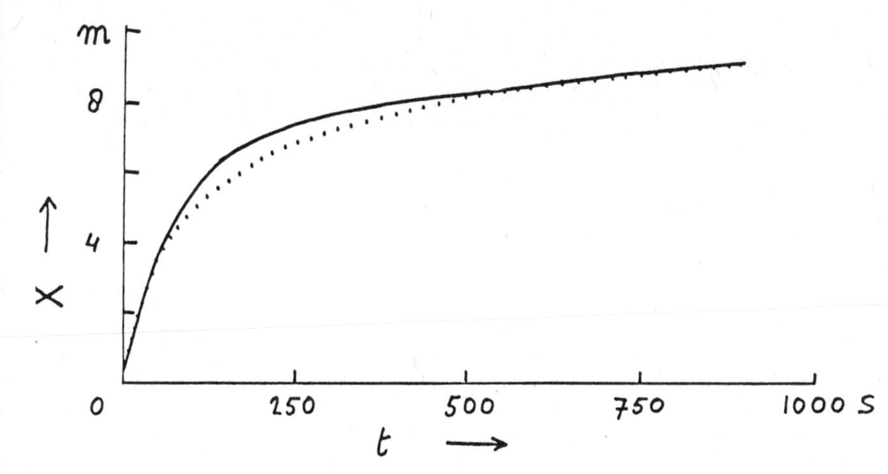

Fig. 5. As figure 4, The dotted line is obtained from experiment 7 : D = 0.44 m, H_o = 0,15 m, X_o = 0.391 m, g' = 0,094 ms^{-2}

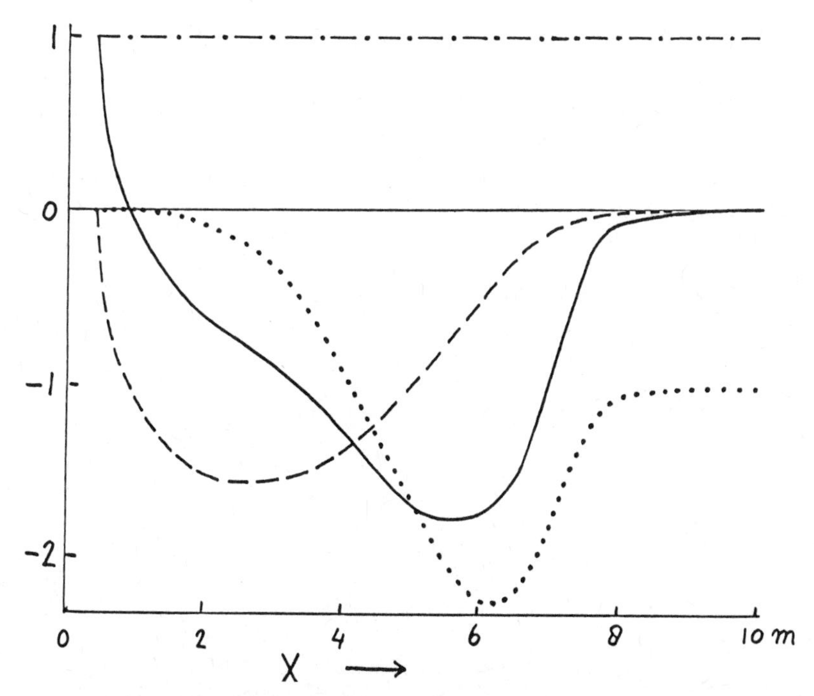

Fig. 6. Non-dimensional forces as a function of the length of the current. Computations for experiment 7. .-.-.- buoyant force, —— time rate of change of total momentum, ---- drag force on the head, ... viscous stress. See also (13)

current picks us speed. It vanishes as $X \simeq 7$ m, when the current has been slowed down by the viscous stress. We see that inertial forces become insignificant beyond $X = 8$ m. Then the buoyant force is balanced by the viscous stress and the current is in its viscous-buoyancy phase. The viscous stress is significant for $X > 2$ m. So there is an extended interval say from $X = 2$ m to $X = 8$ m in which the current is in a mixed phase. The corresponding time interval ranges from $t \simeq 25$ s to $t \simeq 400$ s. So most of the data from experiment 7 refer to the mixed phase. The same is generally true for the other experiments by Huppert and Simpson.

We conclude this section with a discussion on the Froude number. In figure 7 we have plotted again the analytical curve from figure 2. This curve we will call the limiting curve in the following. In the same figure we show our computed Froude numbers for the experiments 1, 7 and 14 by Huppert and Simpson. The main difference between these experiments is the difference in initial

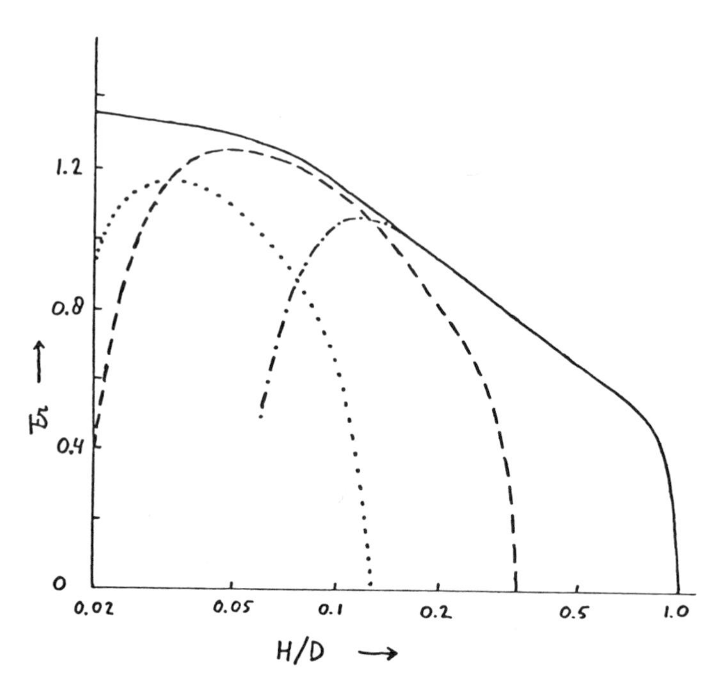

Figure 7. The Froude number as a function of fractional depth.
——— limiting curve from figure 2.
.-.-.- experiment 1: H /D = 1; computed with (13)
------ experiment 7: H /D = 0.34; computed with (13)
...... experiment 14: H /D = 0.13; computedwith (13).

fractional depth. Experiment no. 1, with an initial fractional depth H/D = 1 follows nicely the limiting curve. For H/D = 0.15 viscous effects begin to be important and the curve starts to deviate from the limiting curve. The second curve - that for experiment no. 7 - starts at H/D = 0.34 and approaches closely the limiting curve near H/D = 0.1. Then it falls off due to viscous effects. Experiment no. 14 starts with H/D = 0.13. This curve does not reach the limiting curve at all.The viscous stress begins to affect the current before it has had the time to approach the Froude number of the limiting curve. It now becomes clear why we call the upper curve in figure 7 the limiting curve. This curve gives the maximum Froude number an unsteady gravity current can reach for a given value of the fractional depth. Apparently there is no unique relation between the Froude number and the fractional depth. This relation depends on the initial fractional depth and on the magnitude of the viscous stress.

This analysis suggests that the Froude numbers for small fractional depths, that have been reported by Huppert and Simpson (1980) may have been affected by these effects. This concerns at least the data obtained from lock-exchange experiments. H&S reported that Fr = 1.19 for $H/D < 0.075$. The present analysis indicates that the Froude number can be as high as 1.4 when viscous effects are absent and when H/D is smaller than about 0.01. This is a relevant result for atmospheric gravity currents, because in the atmosphere very small fractional depth may easily occur.

CONCLUSIONS

We have developed a simple dynamic model for the unsteady gravity current. In the model buoyant, viscous and inertial forces are described by separate terms. The importance of these terms has been investigated. It has been shown that generally the inertia of the gravity current itself is important except in the viscous buoyancy phase. It has also been shown that the surface stress may readily be important for a long shallow current.

The Froude number dependence on fractional depth of the current has been studied. We have shown that it may vary widely, depending on the initial conditions. It appears that no unique relation between the Froude number and the fractional depth may be expected. The present analysis indicates that the Froude number can be as high as 1.4, when the fractional depth is vanishingly small and the viscous stress negligible.

The model reproduces well the observations on unsteady gravity currents by Keulegan (1957) and by Huppert and Simpson (1980). The model illustrates nicely the main features of the dynamics of unsteady gravity currents.

REFERENCES

Fay, J.A., 1969. The spread of oil slicks on a calm sea, in: "Oil on the Sea". D.P. Hoult, ed., Plenum.

Huppert, H.E. and Simpson, J.E., 1980. The slumping of gravity currents, J. Fluids Mech., 99: 785-799.

Keulegan, G.H., 1957. An experimental study of the motion of saline water from locks into fresh water channels, Nat. Bur. Stand. Rep. 5168.

Schmidt, W., 1911. Zur Mechanik der Böen, Meteorologische Zeitschrift, August 1911.

Simpson, J.E. and Britter, R.E., 1979. The dynamics of the head of a gravity current advancing over a horizontal surface. J. Fluid Mech.., 94: 477-495.

Van Ulden, A.P., 1974. On the spreading of a heavy gas released near the ground. Int. Loss Prevention Symp., The Netherlands. C.H. Buschman ed., Elsevier, Amsterdam: 221-226.

DISCUSSION

D. MORGAN

In the experiments how well is rectangular shape maintained and how is X defined?

A. P. VAN ULDEN

The experiments show that the rectangular shape is quite well maintained except for a brief period after the release (Huppert and Simpson, 1980). X is defined as the distance of the leading edge from the origin.

LAGRANGIAN MODELING OF AIR POLLUTANTS DISPERSION FROM A POINT SOURCE

E. Runca[*] G. Bonino[**] M. Posch[***]

[*] IIASA, Laxenburg, Austria; Istituto di Cosmogeofisica del CNR, C.so Fiume 4, Torino, Italy
[**] Istituto di Cosmogeofisica del CNR, C.so Fiume 4, Torino, Italy; Istituto di Fisica Generale della Universita, Torino, Italy
[***] IIASA, Laxenburg, Austria; Technische Universität Karlsplatz 13, A-1040 Wien, Austria

INTRODUCTION

The ensemble average concentration due to an Instantaneous Point Source of an inert pollutant ($<C_{IPS}>$) is given by the probability density of pollutant particle displacement $G(\underline{x},t|\underline{x}_s,0)$, where $\underline{x}_s$ is the source location and $\underline{x}$ is the particle location after a time t from the release. For a source of unitary strength we can then write :

$$< C_{IPS}(\underline{x},t) > = G(\underline{x},t|\underline{x}_s,0) \tag{1}$$

For a Continuous Point Source of strength S(t) the equivalent of (1) is :

$$< C_{CPS}(\underline{x},t) > = \int_0^t G(\underline{x},t|\underline{x}_s,t')S(t')dt' \tag{2}$$

The evaluation of G is the key problem in studies of atmospheric diffusion. Following the Eulerian approach (for the sake of simplicity, we drop the vector notation and consider the problem to be one-dimensional), the process of evaluation of G starts from the continuity equation, which, neglecting molecular diffusion takes the form :

$$\frac{\partial C_{IPS}}{\partial t} + \frac{\partial (uC_{IPS})}{\partial x} = 0 \tag{3}$$

In (3) u is the flow velocity. Taking the ensemble average of (3) we get :

$$\frac{\partial \langle C_{IPS} \rangle}{\partial t} + \frac{\partial \langle u C_{IPS} \rangle}{\partial x} = 0 \tag{4}$$

The process of ensemble averaging introduces the new unknown $\langle uC_{IPS} \rangle$. The definition of an equation for $\langle uC_{IPS} \rangle$ is the well-known closure problem (see e.g., Lewellen and Teske, 1976). An alternative approach is the transformation of (3) and (4) into an integro differential equation. Roberts (1961) derived the following integrodifferential equation for G :

$$\frac{\partial G(x,t)}{\partial t} = \frac{\partial^2}{\partial x^2} \int_0^t dt' \int_{-\infty}^{+\infty} Q(x-x', t-t') G(x', t') dx' \tag{5}$$

In (5), $Q(x-x',t-t')$ accounts for the correlation in both time and space of the pollutant particle velocities.

If Q takes the form (no correlation in both space and time)

$$Q = K\, \delta(x-x')\delta(t-t') \tag{6}$$

where $\delta(\cdot)$ is Dirac's function, then (5) reduces to the classic K-model :

$$\frac{\partial G(x,t)}{\partial t} = K \frac{\partial^2 G(x,t)}{\partial x^2} . \tag{7}$$

Assuming correlation only in time, (6) changes to :

$$Q = \delta(x-x') R(t-t') \tag{8}$$

where $R(t-t')$ is the time correlation function and (5) becomes :

$$\frac{\partial G(x,t)}{\partial t} = \frac{\partial^2}{\partial x^2} \int_0^t R(t-t') G(x,t') dt' \tag{9}$$

Equation (9), suggested by Bourret (1960), can be transformed to the telegraph equation if $R(t-t')$ has the form :

$$R(t-t') = \langle u^2 \rangle\, e^{-\frac{t-t'}{T}} \tag{10}$$

where T is the turbulence time scale.

With assumption (10), equation (9) can, after derivation in time, be reduced to the form :

$$\frac{\partial^2 G}{\partial t^2} = -\frac{1}{T} \frac{\partial G}{\partial t} + \langle u^2 \rangle \frac{\partial^2 G}{\partial x^2} \tag{11}$$

which is the well-known telegraph equation discussed amply by Goldstein (1951).

If the correlation is only in space, we have

$$Q = D(x-x')\,\delta\,(t-t') \tag{12}$$

where D is the space correlation function. Substitution of (12) in (5) gives :

$$\frac{\partial G(x,t)}{\partial t} = \frac{\partial^2}{\partial x^2}\int_{-\infty}^{+\infty} D(x-x')G(x',t)\,dx' \tag{13}$$

We note now that the right hand term of (13) can be put in the form :

$$\frac{\partial^2}{\partial x^2}\int_{-\infty}^{+\infty} D(x-x')G(x',\,t)\,dx' = -\frac{\partial}{\partial x}\int_{-\infty}^{+\infty} G(x',\,t)\frac{\partial}{\partial x'}D(x-x')\,dx' \tag{14}$$

since $\frac{\partial}{\partial x}D(x-x') = -\frac{\partial}{\partial x'}D(x-x')$.

Integration by parts of the integral of the right hand term of (14) shows that (13) can be written as follows :

$$\frac{\partial G(x,\,t)}{\partial t} = \frac{\partial}{\partial x}\int_{-\infty}^{+\infty} D(x-x')\,\frac{\partial G(x',\,t)}{\partial x'}\,dx' \tag{15}$$

Equation (15) is the pseudo-spectral model suggested by Berkowicz and Prahm (1979).

In contrast with the Eulerian, the Lagrangian approach attempts to compute G directly from its definition. This implies, unless a functional form of G (e.g., Gaussian) is assumed a priori, the generation of an ensemble of pollutant particle trajectories from which G can be computed. In order to generate such ensemble the turbulent flow velocity must be simulated. This can be done by means of laboratory models (Willis and Deardorff 1979, 1980) or numerical turbulence models (Deardorff 1974; Lamb 1978).

Another approach which is now receiving considerable attention is the generation of the turbulent flow velocity according to a Monte Carlo type algorithm which produces velocity fluctuations with the same statistics of the considered turbulent flow (Reid 1978; Lamb et al. 1979; Hanna 1980; Lamb 1980). This is discussed in the following for both homogeneous and nonhomogeneous turbulence.

HOMOGENEOUS TURBULENCE

First, let us observe that for sufficiently small Δx we can write :

$$G(x, t|x_s,0) = \frac{P(x,t|x_s,0)}{\Delta x} \tag{16}$$

where $P(x,t|x_s,0)$ is the probability that a particle will be in $(x \pm \frac{\Delta x}{2})$ after a time t. $P(x,t|x_s,0)$ can be estimated by the limi

$$P = \lim_{k\to\infty} \frac{1}{k}\sum_{j=1}^{k} \phi_j(x,t|x_s,0) \tag{17}$$

where

$\phi_j(x,t|x_s,0) = 1$ if the j-th particle is in $(x \pm \frac{\Delta x}{2})$ at time t

$\phi_j(x,t|x_s,0) = 0$ in other cases

To compute $\phi_j(x,t|x_s,0)$ the particle trajectories must be generated. The following algorithm can be applied :

$$n = 0:\ u_j^\ell(\Delta t,x_s) = u_j^e(x_s)$$

$$n > 1:\ u_j^\ell((n+1)\Delta t,x_s) = \alpha\, u_j^\ell(n\Delta t,x_s) + \rho_j((n+1)\Delta t) \tag{18a}$$

$$x_j((n+1)\Delta t,x_s) = x_j(n\Delta t,x_s) + u_j^\ell((n+1)\Delta t,x_s)\ .\ \Delta t \tag{18b}$$

where $u_j^\ell((n+1)\Delta t,x_s)$ is the velocity of the j-th particle (release at x_s) after (n+1) intervals of time Δt. This velocity is used to advance the particle from $x_j(n\Delta t,x_s)$ to $x_j((n+1)\Delta t,x_s)$. At the first time step the particle velocity coincides with the velocity recorded at the source location (suffixes ℓ and e indicate respectively Lagrangian and Eulerian quantities). α is a positive number less than one and $\rho_j((n+1)\Delta t)$ is a random contribution to the velocity of the j-th particle in the (n+1)th time step, which has to be generated by a Monte Carlo method.

In order to generate $\rho_j((n+1)\Delta t)$ we need to know the probabilit density of ρ, which we indicate by P_ρ.

Due to the hypothesis of homogeneous and stationary turbulence, the probability density of u^ℓ, which we indicate with P_u^ℓ, must coincide at all times with the probability density of the velocity recorded at any fixed point. This is indicated by P_u^e.

The probability densities of $u^\ell((n+1)\Delta t,x_s)$ and $u^\ell(n\Delta t,x_s)$ are therefore identical and if we furthermore assume P_u^e to be Gaussian, that is :

$$P_u^e \equiv N(<u>, <u^2>) \tag{19}$$

we deduce, since $P^{\ell}_{u} \equiv P^{e}_{u}$ is the convolution of $P^{\ell}_{\alpha u} \equiv P^{e}_{\alpha u} \equiv N(\alpha <u>, \alpha^2 <u^2>)$ and P_{ρ}, that

$$P_{\rho} \equiv N\left((1-\alpha) <u> ; (1-\alpha^2) <u^2>\right) \tag{20}$$

With the additional assumptions of

$$<\rho(n\Delta t)\,\rho(m\Delta t)> = 0 \qquad m \neq n$$
$$<u(n\Delta t)\,\rho(m\Delta t)> = 0 \qquad m > n$$

it can be shown that algorithm (18) generates trajectories with Lagrangian time correlation :

$$R(t) = \alpha^{t/\Delta t} \qquad t = n\Delta t, \tag{21}$$

Lagrangian integral time scale :

$$T = \frac{\Delta t}{1-\alpha}, \tag{22}$$

and produces particle displacements with :

$$\sigma^2(n\Delta t) = 2<u^2>\frac{\Delta t^2}{(1-\alpha)^2}\left[\frac{1-\alpha^2}{2}\,n - \alpha\,(1-\alpha^n)\right]. \tag{23}$$

Equation (23) must be compared with Taylor's result :

$$\sigma^2_{Ta}(n\Delta t) = 2<u^2>T^2\left[\frac{n\Delta t}{T} - (1-e^{-\frac{n\Delta t}{T}})\right] \tag{24}$$

obtained with

$R(t) = e^{-\frac{t}{T}}$ (Note that this is the limit of equation (21) for $\Delta t \to 0$ with α defined by (22)).

Comparison of (23) with (24) is done in Figure 1, which displays normalized values of both σ and σ_{Ta} for $\alpha = 0.5$ ($\Delta t = 0.5T$; see (22)). Figure 1 shows that for particle travel times not exceedingly larger than T, a time step equal to half of the Lagrangian time scale provides (in the case of homogeneous and stationary turbulence) a satisfactory approximation of the corresponding continuous process.

Of fundamental interest in the application of a Lagrangian Monte Carlo method is the number of marked particles which must be followed in order to approximate the limit (17). In order to perform this analysis we have considered the classic case of dispersion from a line source between an inversion layer and a ground assumed to be both impermeable to the diffusive matter. Taking the mean wind constant, which is directed along the x-axis (oriented perpendicularly to the line source), and assuming the horizontal diffusion negligible with respect to advection, the numerical modelling of dispersion from a Continuous Line Source reduces to the simulation of an instantaneous one-dimensional puff. This diffuses in the vertical axis between two impermeable boundaries and translates downwind at constant velocity.

For homogeneous stationary turbulence with $R(t) = e^{-t/T}$, the analytical solution (assuming the probability density of the velocity fluctuations to be Gaussian) is given by :

$$<C_{CLS}(x,z)> = \frac{S}{\sqrt{2\pi}\,\sigma_z\,U} \sum_{n=-\infty}^{+\infty} \left\{ \exp\left[-\frac{(z-h+2nz_i)^2}{2\sigma_z^2}\right] + \exp\left[-\frac{(z+h+2nz_i)^2}{2\sigma_z^2}\right] \right\} \quad (25)$$

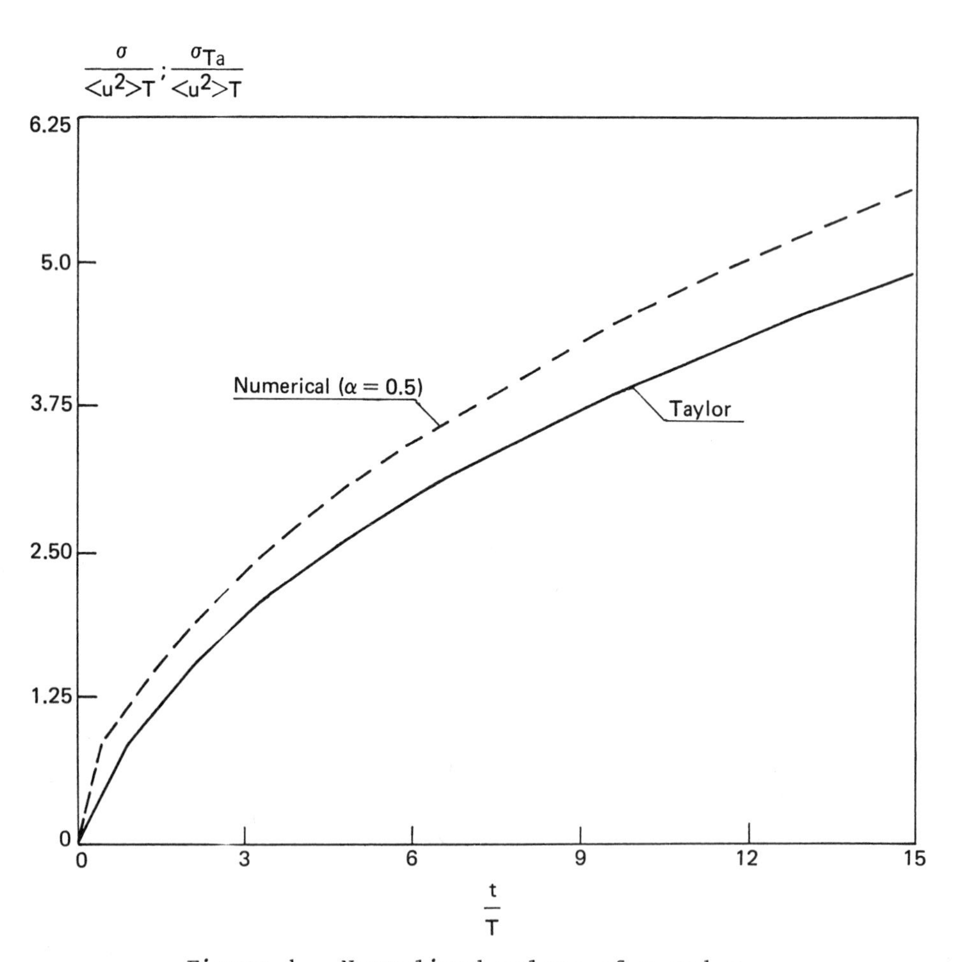

Figure 1. Normalized values of σ and σ_{Ta}.

where U is the mean wind speed, z_i is the inversion layer height, h is the source height, x = U . t and σ_z is given by (24) (with $< u^2 >$ replaced by $< w^2 >$).

Results achieved with algorithm (18) (for clarity, note that u must be replaced by w and x by z) are reported in Figures 2 and 3. The boundaries (ground and inversion layer) were taken into account by reflecting the position of those particles found at a given time step below the ground or above the inversion layer and reverting their respective velocity.

To generalize results of Figure 2 and 3, w, U, x, z, t, and C_{CLS} have been normalized to w_*, U, $\frac{z_i U}{w_*}$, z_i, $\frac{z_i}{w_*}$, and $\frac{S}{U z_i}$, respectively. w_* is a characteristic velocity of the turbulent vertical velocity fluctuations. It can be interpreted as the convective velocity scale (see Willis and Deardorff (1978) for the definition of w_*). Results are reported for h (source height) = $\frac{z_i}{4}$, $< w^2 > = w_*^2$, $T \frac{w_*}{z_i}$ = 0.2, and $\alpha = 0.5$. Note that with the adopted normalization, the dimensionless value of the uniform concentration profile which is obtained at a downwind location, sufficiently distant from the source, takes the value one.

Figure 2 compares analytical values given by equation (25) with the numerical ones for three sets of particles. The dimensions of the grid cell used to compute G (see (16)) were defined by $\frac{\Delta x}{U \Delta t} = 1$ and $\frac{\Delta z}{z_i} = 0.1$. Results show that the accuracy of the numerical results increases with the increasing number of particles. Due to the nature of the method, as expected, the convergence is not uniform. Even for a large number of particles the numerical solution displays a considerably erratic behaviour.

Figure 3 plots the root mean square of the deviation of the numerical from the analytical values for an increasing number of particles. The root mean square of the deviations has been computed over the whole integration region and also by taking into account only the values at ground level. Results indicate that, for the considered discretization, the error stabilizes after approximately 2000 particles.

NONHOMOGENEOUS TURBULENCE

In this section we discuss the extension of algorithm (18) to the case of nonhomogeneous turbulence.

With reference to the two-dimensional line source model described above, we note that for this ideal model far from the source, the particles should be uniformly distributed independent of the turbulence structure. As we have noted above, with the adopted normalization, the concentration far from the source should approximate at all heights the value of one. This is true for the homogeneous case (see Figure 2). However, if algorithm (18) is applied, with

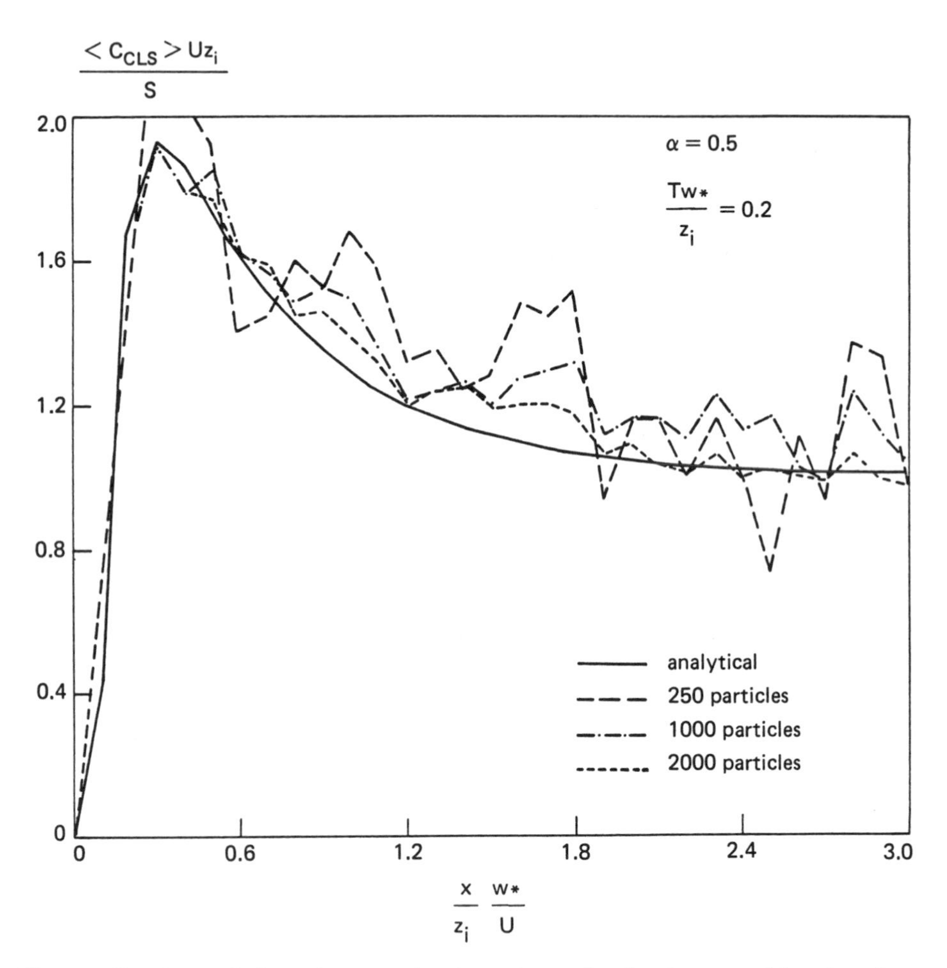

Figure 2. Ground level numerical and analytical normalized values of $<C_{CLS}>$ for $h = \frac{z_i}{4}$

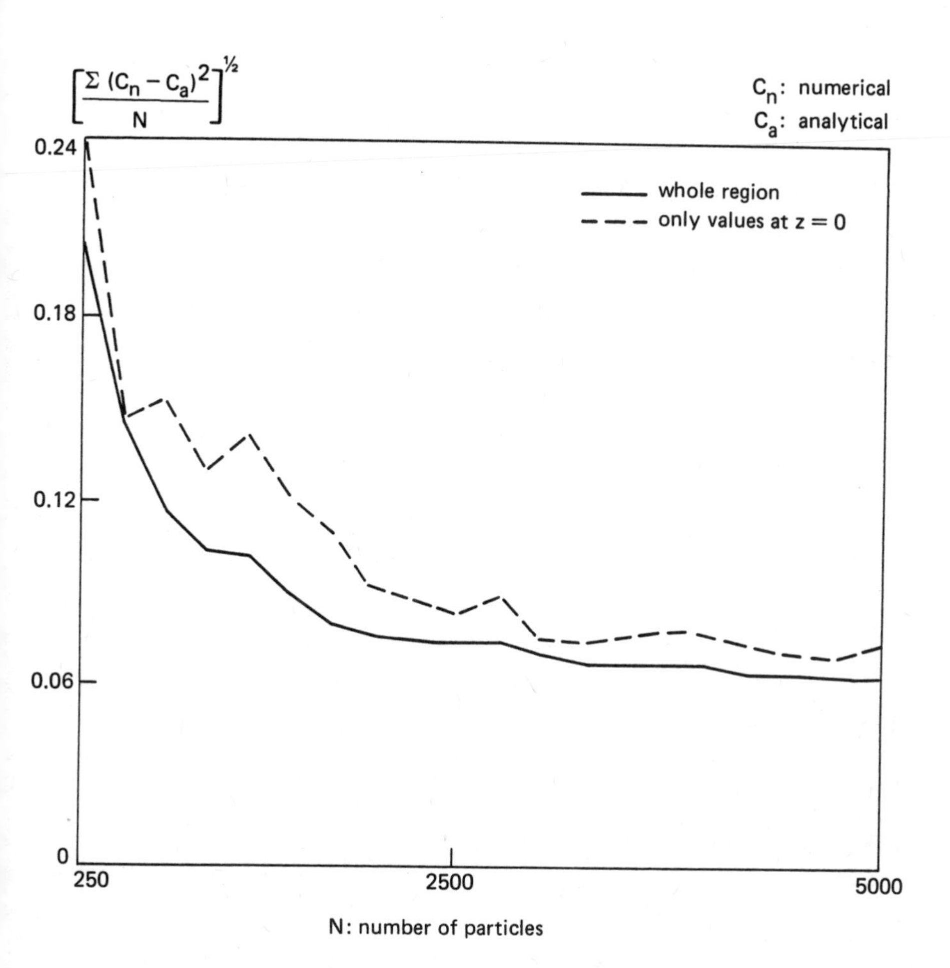

Figure 3. Root mean square of the normalized deviations of the numerical from the analytical values for increasing number of particles. Source height $h = \frac{z_i}{4}$; other parameters as in Figure 2.

assumption (20), to a case in which $< w^2 >$ and α are functions of z, the uniform profile cannot be achieved. We attempt below to identify what could be done in order to extend algorithm (18) to the nonhomogeneous case and we reserve ourselves to apply numerically some of the concepts discussed below in another study.

First, algorithm (18) is rewritten, taking into account that the parameters are now dependent on the particle location. By replacing u with w in order to refer directly to the geometry of the considered two-dimensional line source model, we obtain :

$$n = 0 \quad w_j^{\ell}(\Delta t, h) = w_j^{e}(h) \tag{26a}$$

$$n \geq 1 \quad w_j^{\ell}((n+1)\Delta t, h) = \alpha(z_j(n\Delta t)) \quad w_j^{\ell}((n\Delta t), h) + \rho_j \; ((n+1)\Delta t, z_j(n\Delta t))$$

$$z_j \; ((n+1)\Delta t, h) = z_j(n\Delta t, h) + w_j^{\ell} \; ((n+1)\Delta t, h) \; \Delta t \tag{26b}$$

As in the previous section use of (26) depends on the availability of P_ρ . Unfortunately, we cannot proceed to derive it as in the previous section. To understand which approach needs to be followed in order to apply (26) let us write (26b) for n = 1. Noting that for n = 1, $z_j \; (\Delta t) = h + w_j^{e}(h)\Delta t$ we get :

$$w_j^{\ell}(2\Delta t, h) = \alpha(h+w_j^{e}(h)\Delta t) \; w_j^{e}(h) + \rho_j(2\Delta t, h+w_j^{e}(h)\Delta t). \tag{27}$$

To proceed further, we define by $W_{\alpha w}^{\ell}$, the probability density of $\alpha(z_j(n\Delta t)) \; w_j^{\ell} \; (n\Delta t, h)$. From (26), we can formally write :

$$P_w^{\ell} = W_{\alpha w}^{\ell} \star P_\rho \tag{28}$$

where $\star$ indicates a convolution product.

Equation (27) shows that, for n = 1, $W_{\alpha w}^{\ell}$ can be derived, in principle, from the knowledge of P_w^{e}. If we could know P_w^{ℓ}, then P_ρ could be determined by inverting the convolution product (28). The process could then be repeated for n = 2 etc., until a full understanding of P_ρ is achieved. P_w^{ℓ} can be determined from trajectories simulated by means of a turbulence model as the one developed by Deardorff (1974). We are planning to perform such an experiment in the near future. However, this approach might prove too cumbersome. As an alternative, we have also planned the following procedure.

So far, we have focused on an ensemble of trajectories. Some interesting properties of P_ρ can be deduced if the focus is put on a single trajectory, and specifically on the distribution of the velocity of the j-th particle at time $(n+1)\Delta t$ once its velocity at time $(n\Delta t)$ has been established. Clearly, the probability density of w_j^ℓ $((n+1)\Delta t, h)$ coincides with the probability density of ρ at the location $z_j(n\Delta t)$, shifted by the quantity $(\alpha((n+1)\Delta t, z_j(n\Delta t))$. $w_j^\ell(n\Delta t, h))$. It can now be deduced that if Δt is very small, P_ρ must approach Dirac's function and α must tend to one. This is consistent with (20) and (22), derived in the homogeneous case. It is also reasonable to assume that if there is no specific force acting on the particle, the expected value of ρ must be zero and consequently

$$< w_j^\ell((n+1)\Delta t, h) > = \alpha((n+1)\Delta t, z_j(n\Delta t)) \, w_j^\ell(n\Delta t, h). \tag{29}$$

This is true in the homogeneous case, for which $< u > = 0$ (see (20)). For the nonhomogeneous case, we cannot assume $< \rho > = 0$. On the contrary, we expect $<\rho>$ to depend on the spatial variation of $< {w^e}^2 >$; in other words, the expected velocity of the j-th particle in the (n+1)th time step, in addition to the contribution from the velocity of the previous step (see (28)), must also receive a contribution dependent on $< {w^e}^2 >$. To find this contribution, we write Newton's law along the vertical coordinate, that is :

$$\frac{d\, w_j^\ell(t,h)}{dt} = F_z\,(z_j(t),\, x_j(t),\, t) \tag{30}$$

where $F_z(z_j(t), x_j(t), t)$ is the vertical force per unit of mass acting on the j-th particle at $\{z_j(t), x_j(t), t\}$. With

$$w_j^\ell(t, h) = w^e(z_j(t), x_j(t), t) \tag{31}$$

(30) can be put in the Eulerian form :

$$\frac{\partial w^e}{\partial t} + u^e \frac{\partial w^e}{\partial x} + w^e \frac{\partial w^e}{\partial z} = F_z \tag{32}$$

For an incompressible two-dimensional flow, the following relationship holds :

$$\frac{\partial u^e}{\partial x} + \frac{\partial w^e}{\partial z} = 0 \tag{33}$$

Making use of (33), (32) can be written as :

$$\frac{\partial w^e}{\partial t} + \frac{\partial u^e w^e}{\partial x} + \frac{\partial {w^e}^2}{\partial z} = F_z \,. \tag{34}$$

Taking the ensemble average of (34) under the assumption of stationary and horizontally homogeneous turbulence, we get :

$$\frac{\partial < {w^e}^2 >}{\partial z} = < F_z > . \tag{35}$$

Thus, for vertically nonhomogeneous turbulence, (29) must be replaced by

$$< w_j^{\ell}((n+1)\Delta t,\ h) > \ = \ \alpha((n+1)\Delta t,\ z_j(n\Delta t))\, w_j^{\ell}(n\Delta t,\ h)\ +$$

$$\Delta t \left[\frac{\partial < w^{e^2} >}{\partial z} \right]_{z_j(n\Delta t)} \tag{36}$$

or in other words

$$< \rho > = \Delta t\ \frac{\partial < w^{e^2} >}{\partial z} \ . \tag{37}$$

This result was also obtained through a different approach by Janicke in his paper included in this proceedings. *

To obtain additional information about the distribution of ρ, we recall (see above) that for $\Delta t \to 0$, P_ρ tends to Dirac's function. Thus, by making an analogy with the homogeneous case, we are led to write :

$$< \rho^2 > = [\, 1-\alpha^2(z_j(t))\,] \ . \ < w^{e^2}(z_j(t)) > \tag{38}$$

We consider (38) as a working hypothesis to be verified either by means of (28), that is, by checking that we produce the correct P_w^{ℓ}, or by comparing concentration values produced by algorithm (26) and equations (37)-(38), to those simulated by Lamb (1978). Clearly, the functional form fo P_ρ must be known in both cases. We plan to conduct two sets of experiments, one with P_ρ Gaussian an the other with P_ρ equal to the P_w^e recorded to the particle location. This way of proceeding, compared with the one proposed at the beginning of this section (see (28) and the paragraph following it) avoids the computation of $W_{\alpha w}^{\ell}$. However, we might not be able to verify our assumptions about P_ρ and $< \rho^2 >$; in this case P_ρ will have to be deduced by inverting (28).

CONCLUSIONS

Lagrangian Monte Carlo simulation of air pollutants dispersion downwind from a point source is an alternative approach to the Eulerian one, which has been made possible by the development of computer technology. However, its application depends on the proper definition of the probalility density of the "random" forces which act on the particle at any given time step. This has been deduced for homogeneous stationary turbulence. For the nonhomogeneous case the procedures which the authors intend to investigate in order to identify it have been presented.

* Our paper was written following the conference and we would like to acknowledge Dr. Janicke's work, which stimulated the above analysis.

ACKNOWLEDGEMENTS

This analysis could not have been done without the helpful assistance of Dr. Robert G. Lamb.

REFERENCES

BERKOWICZ, R., and PRAHM, L.P., 1979, Generalization of K-theory for turbulent diffusion. Part I. Spectral turbulent diffusivity concept, J. Appl. Met., 18:266-272.

BOURRET, R, 1960, An hypothesis concerning turbulent diffusion, Canad. J. Phys., 38:665-676.

DEARDORFF, J.W., 1974, Three-dimensional numerical study of the height and mean structure of a heated planetary boundary layer, Boundary-Layer Met., 7:81-106.

GOLDSTEIN, S., 1951, On diffusion by discontinuous movements and on the telegraph equation, Quart. J. Mech. Appl. Math., 4:129-156.

HANNA, S.R., 1980, Effects of release height on σ_y and σ_z in daytime conditions, Proc. 11-th NATO/CCS Int. Tech. Meeting on Air Pollution Modeling and its Application, Amsterdam.

LAMB, R.G., 1978, A numerical simulation of dispersion from an elevated point source in the convective planetary boundary layer, Atmospheric Environment, 12:1297-1304.

LAMB, R.G., HUGO, H., and REID, L.E., 1979, A Lagrangian approach to modeling air pollutant dispersion, EPA Report, EPA-600/4-79-023.

LAMB, R.G., 1981, A scheme for simulating particle pair motions in turbulent fluid, J. Comp. Physics, 39:329-346.

LEWELLEN, W.S., and TESKE, M.E., 1976, Second-order closure modelling of diffusion in the atmospheric boundary later, Boundary-Layer Met., 10:69-90.

REID, J.D., 1979, Markow chain simulations of vertical dispersion in the neutral surface layer for surface and elevated releases, Boundary-Layer Met., 16:3-22.

ROBERTS, P.H., 1961, Analytical theory of turbulent diffusion, J. Fluid Mech., 11:257-283.

WILLIS, G.E. and DEARDORFF, J.W., 1978, A laboratory study of dispersion from an elevated source within a modeled convective planetary boundary layer, Atmospheric Environment, 12:1305-1311.

WILLIS, G.E. and DEARDORFF, J.W., 1980, A laboratory study of dispersion from a source in the middle of the convectively mixed layer, Atmospheric Environment, 15:109-117.

DISCUSSION

A. VENKATRAM : Is your analytical solution consistent with the assumptions of the numerical model?

E. RUNCA : Yes, the probability density of ρ is Gaussian. This generates particle displacements which have a Gaussian distribution. The standard deviation of this distribution converges for $\Delta t \to 0$ to the one given by Taylor with Lagrangia time correlation $R = e^{-t/T}$

H. VAN DOP : There is experimental evidence that the veloci distribution in the mixed layer is not Gaussian Did you account for this in your model?

E. RUNCA : We are working on the extension of the algorit which generates turbulent flow velocities, to the case of nonhomogeneous turbulence. This accounts for non-Gaussian distribution of the velocity.

PHYSICAL AND MATHEMATICAL SIMULATION OF Pb DISPERSION IN AND AROUND A Pb INDUSTRY IN A URBANIZED AREA

P. Hecq, Cl. Demuth, A. Berger

Institut d'Astronomie et de Géophysique G. Lemaître
Université Catholique de Louvain
B-1348 Louvain-la-Neuve, Belgium

ABSTRACT

This study is related to the general simulation of gas and particles dispersion in and around a manufactory. The physical and the mathematical models used are described ; the results of both experiments are commented and qualitatively compared to the field data. Some technical solutions to tentatively reduce the pollution level due to this type of industry are also analysed.

INTRODUCTION

The existence of a Pb-manufactory closed to a small city has lead to critical pollution situations. Measurements inside the urban area show indeed concentration levels higher than normally allowed by the Belgian law. Figure 1 provides a description of the area which has been investigated.

The problem was, first, to simulate the dispersion of particles and gaz in and around the manufactory and, second, to analyze the efficiency of some technical solutions proposed to reduce the pollution level downwind in the urban area.

In such complex situations (urban and industrial buildings), the usual way to estimate the impact of several factors is through a numerical simulation or a physical experiment using a wind tunnel or a water tank.

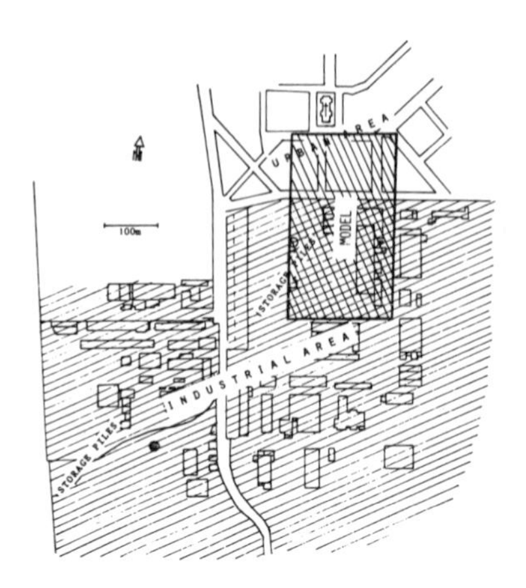

Fig. 1 Map of the lead manufactory surrounded by the urban area. The mean height of industrial buildings is between 10 and 30m. The houses are around 15m high.

In this particular study, a wind tunnel experiment and a numerical model have been applied.

CHARACTERISTICS OF THE WIND TUNNEL

The wind tunnel used is "open circuit" ; the range of wind speed is continuously adjustable from 0 to 8 m/s. The working section is circular with a working diameter of 1.2 m.

The gaz emitted in order to simulate the dispersion is a mixture of amonia and chloride acid. The materials used for the simulation of dust dispersion are talc and sand.

CHARACTERISTICS OF THE MATHEMATICAL MODEL

The computation of the wind field was made by a 2-D numerical model based on the stationnary solution of the continuity equation and of the horizontal equation of motion for the mean values of a 2-D turbulent flow within the Boussinesq approximation[1].

If x is oriented along the wind and z along the vertical, the equations are :

$$
(1)\left\{\begin{array}{l}
\dfrac{\partial \bar{u}}{\partial x} + \dfrac{\partial \bar{w}}{\partial z} = 0 \\[2ex]
\dfrac{\partial \bar{u}}{\partial t} + \dfrac{\partial}{\partial x} \overline{u}^2 + \dfrac{\partial}{\partial z} \overline{wu} = - \dfrac{\partial}{\partial x} \dfrac{\bar{p}}{\rho} + \nu \nabla^2 \bar{u} + \dfrac{\partial}{\partial x} K_{xx} \dfrac{\partial \bar{u}}{\partial x} + \dfrac{\partial}{\partial z} K_{xz} \dfrac{\partial \bar{u}}{\partial z} \\[2ex]
\dfrac{\partial \bar{w}}{\partial t} + \dfrac{\partial}{\partial x} \overline{uw} + \dfrac{\partial}{\partial z} \overline{w}^2 = -g - \dfrac{\partial}{\partial z} \dfrac{\bar{p}}{\rho} + \nu \nabla^2 \bar{w} + \dfrac{\partial}{\partial x} K_{zx} \dfrac{\partial \bar{w}}{\partial x} + \dfrac{\partial}{\partial z} K_{zz} \dfrac{\partial \bar{w}}{\partial z}
\end{array}\right.
$$

with : $\bar{u},\bar{w}$ wind mean components
$\bar{p}/\rho$ mean pressure divided by the density
g gravity
K_{ij} turbulent diffusivities
ν coefficient of molecular diffusion (1.5×10^{-5} m^2/s)

In these equations, K_{xz} and K_{zx} are neglected, K_{xx} and K_{zz} are constant and their values are respectively 0.1 m^2/s and 2 m^2/s. The initial pressure $\bar{P}$ is given by the hydrostatic equation :

$$\frac{\partial}{\partial z}\frac{\bar{P}}{\rho} = - g$$

The system (1) is solved by the method of the donnor cell.

DETERMINATION OF THE SCALES FOR THE PHYSICAL MODELLING

The scale of a physical model is not only given by the ratio between the width of the wind tunnel and the dimension of the investigated area, as it depends also of the type of problem which is analysed. This is why three successive scales have been retained:

1) a rather large scale (1/50) has first been choosen in order to simulate local phenomena, such as recirculations near buildings or walls. This has allowed to validate the numerical model, and to settle the size of the recirculation areas. The area covered was 15 x 10 m.

2) an intermediate scale (1/300) has been used as a base for the final model ; in this experiment, it is mainly the wind action on the storage piles and the particles transport that have been simulated. In this case the area covered was 350x200m.

3) finaly, a scale of 1/400 has been retained for the final model which cover an area of 700x520m ; here gaz dispersion and dust circulation have been simulated.

SIMILARITY CRITERIA AND SIMULATION OF THE ATMOSPHERIC BOUNDARY LAYER

Similarity criteria for physical modeling of atmospheric flows have been reviewed by many authors [2,3,4,5,6,7]. Since our main purpose in this study was to visualise the dispersion and to have qualitative informations about it, it is not necessary to enter into many details about all the criteria. Let us just mention that the following characteristics of the simulation have been retained:

- geometric similitude with caracteristic mean heights between 6 and 10 cm.
- wind speed in the tunnel was between 0.5 and 3.0 m/s.
- Reynolds number of the order of 6 10^5 (at this value, the flow is

completely rough and the friction coefficient is independant of Re).

- only neutral stability is simulated.
- thickness of the boundary layer was supposed to be 400.

As only the adiabatic case will be simulated, the wind profile can be given by a power law :

$$U(z) = U_1 . z^m$$

This has been simulated by a grid with horizontal bars (fig. 2) and for the case presented here, $U_1 \simeq 0.425$ and $m \simeq 0.3$.

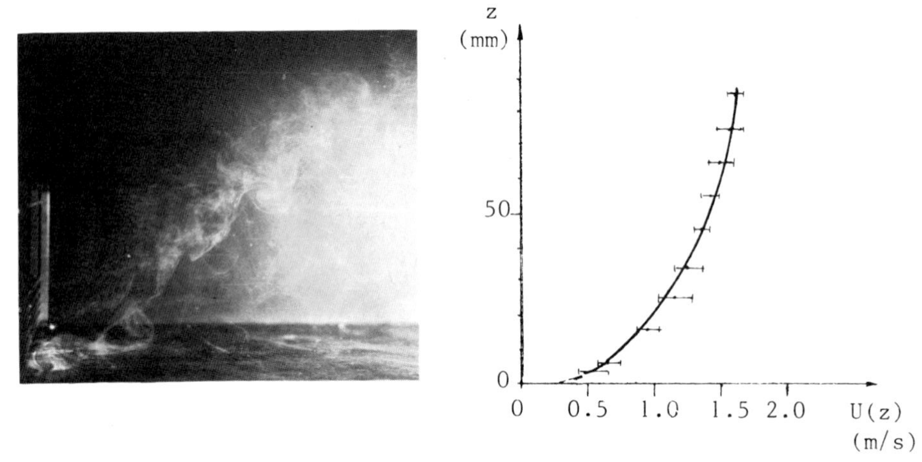

Fig. 2 Mean wind profiles observed and measured in the boundary layer.

SIMULATION OF A TWO-DIMENSIONAL LOCAL FLOW (1/50 SCALE) AROUND OBSTACLES

The purpose of the first simulation was to detect the local characteristics of the wind around a wall and a building. As shown on Figs 3 and 4, the flow comes from the right accross the street and towards the wall and the building. The wall and the building are 3m high and the street is 22m large. In the first case (fig. 3) the wall is at the same level as the building and in the second, its foot is at the roof level of the building (fig. 4). For each case, the numerical simulation is presented against the physical one.

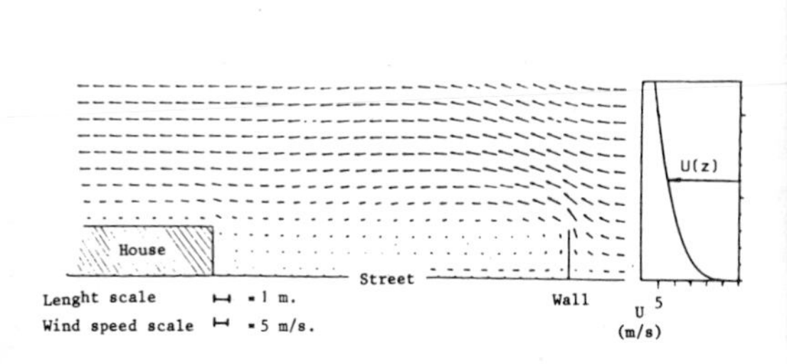

Fig. 3 Two dimensional physical and numerical simulation of the local flow around a wall and a house. This is the cross-section AB indicated on Fig. 1. The wall and the building are 3m high and the street is 22m large ; maximum wind speed in tunnel is 3 m/s.

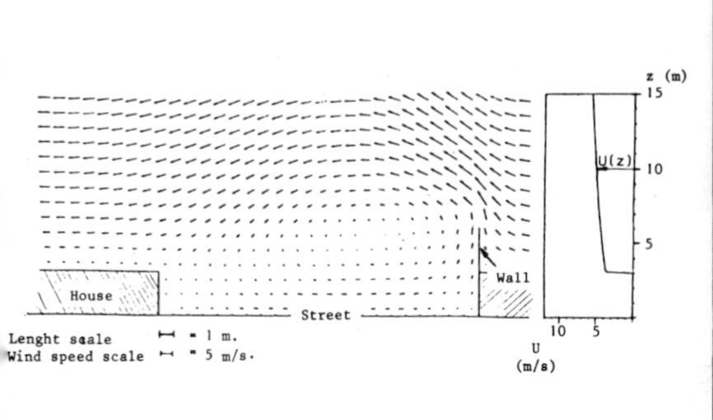

Fig. 4 Two-dimensional physical and numerical simulation of local flow around a wall and a house. This is the cross-section CD indicated on Fig. 1. Some caracteristics as in Fig. 3.

In the physical experiment, the smoke trails visualise the stream lines and the recirculation area ; in the numerical simulation, the wind field is directly obtained and from it, recirculation area can be deduced. The comparison of both results shows a good agreement.

The main interesting feature in these experiments is the recirculation area behind the wall where an important fall-out of particles can be expected. This has been observed on the field : this recirculation area extends as far as half the street in the first case (fig. 3) and in the second case, it covers all the street.

The next experiment was to simulate the flow when different type of walls are used. The results are summarized here under :

a) doubling the wall height (3 to 6 m) (fig. 5)

In the first case (wall at the same level as the building) the influence of a higher wall is not important ; the dead-area behind the wall is only a little bit larger. But on the second case, the effect of rising the wall is more evident : the eddy before the wall is still there but the dead-area behind is much larger and the building is less influenced by the flow field coming from the industry ; the fall-out will be localy less but spacially more extended.

Fig. 5 Two-dimensional physical simulation of the local flow around a wall and a building. The configuration is the same as in Fig. 3 and 4 but the wall is 6m high.

b) wall with holes

In this case, the influence of the permeability of the wall has been studied ; the presence of holes modify especially the structure of the turbulence[2] ; when an opaque wall is replaced by a permeable one, the recirculation area disappears and the turbulence scale is getting smaller. It can be deduced from this that the decreasing of the turbulence scale and of the horizontal wind speed gradients are favourable to sedimentation ; but the destruction of recirculation allows a transport of particles over larger distances.

c) oblique wall

The last solution which has been analysed in this simulation of the local wind field is related to an oblique wall. In this experiment, it has been possible to show (fig. 6) that the swirling area before the wall is greater and the flow behind the wall is not much changed.

Fig. 6 Two-dimensional physical simulation of the local wind around a wall and a building. Some configuration as in figs 3 and 4 but with on oblique wall.

SIMULATION OF THE WIND FIELD AND SEDIMENTATION AT AN INTERMEDIATE SCALE (1/300)

The area covered on this simulation (350x200m) is shown on Figure 7. The model has been oriented along the direction of the prevailing wind (S-SO) ; 3 heights for the wall have been retained: 3,6 and 15m. The resulting wind field and the dust transport are successively reported.

Figure 8 shows the wind field for these three wall heights. Only the wall of 15m influences the wind field a little : the flow is deflected by the wall and passes over the first houses.

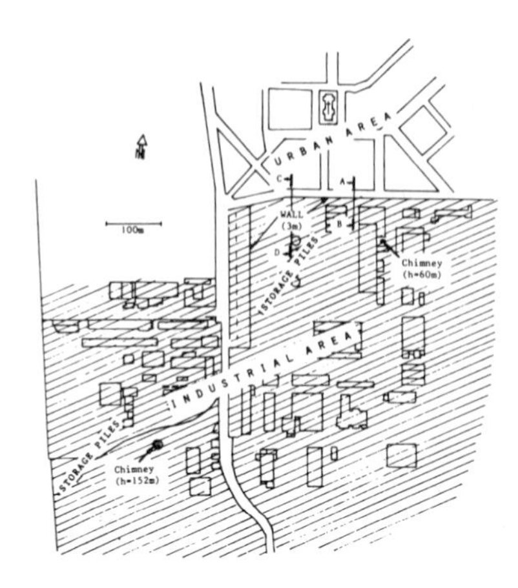

Fig. 7 Map of the manufactory and of the urban area with the area covered by the 1/300 model.

Fig. 8 Wind field above the urban area for three wall height (3,6 and 15m) (the wind blows from the right (S-SO) at ∿ 3 m/s.

The dust transport was investigated in order to study the action of the wind on dust and minerals stock piles. Piles are simulated by little piles of fine dust. Within the framework of this study, it was difficult to satisfy completely the criteria of similitude for sedimentation, mainly because of the large dispersion in the granulometry distribution. Nevertheless, the qualitative results obtained are significant. Fig. 9 shows the simulation with the actual wall size, the stockage area being at the right of the picture. Fig. 10 shows the same simulation for walls of 6 and 15m height.

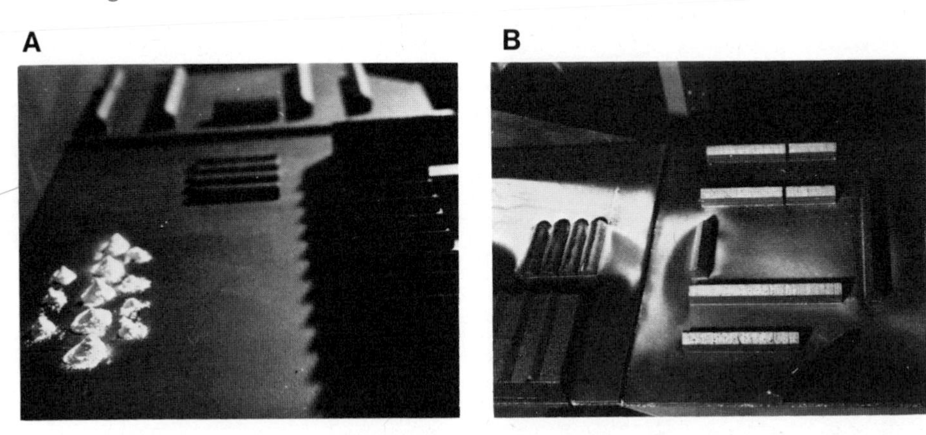

Fig. 9 Action if the wind on stock piles. The left picture 9A shows the situation before the simulation (piles simulates 6-10m high stock piles) ; the right picture 9B shows the results after 5 minutes of simulation with a wind speed in tunnel of 4.5 m/s ; direction is S-SO ; wall is 3m high.

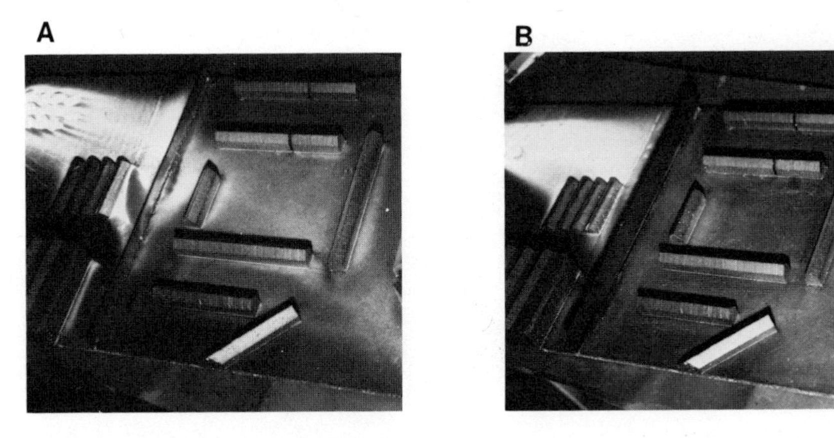

Fig. 10 Action of the wind on stock piles. Results of a 5 minutes simulation as in Fig. 9B but the wall is now respectively 6m high (left) and 15m (right).

With these results, it is possible to locate the accumulation areas (white areas on the picture) and to deduce the effect of the wall height. As already noted, with the wall of 6 meters high, the sedimentation decreases very little (the situation doesn't change very much) but with the wall of 15m high, the result is perfectly clear : there is no more deposit of particles beyond the wall.

SIMULATION OF THE DISPERSION AND SEDIMENTATION OVER THE WHOLE AREA

A simulation all-over the industrial and urban area is here presented. Fig. 11 shows the area covered by this particular study and fig. 12 presents 2 views of the model. The dimensions of the model are 1.35m by 1.75m. The dispersion of both gaseous pollutants and particulates from stock piles is investigated.

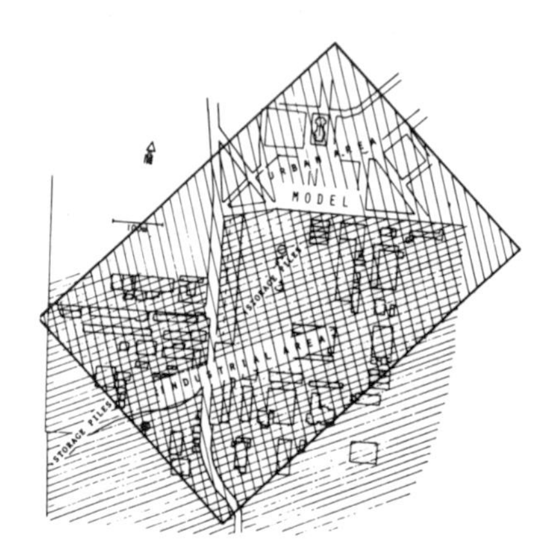

Fig. 11 Map of the manufactory and of the urban area with the area covered by the 1/400 model.

Fig. 12 Two views of the model used in the third simulation. The left picture is west oriented; the urban area is on the left and the industrial buildings the right; the right picture is east oriented; the two main chimneys are visible.

DISPERSION OF GASEOUS POLLUTANTS

The gaseous pollutants are supposed to originate from a stack of 60 or 152m ; three wind speed have been systematically prescribed. 0.5, 1.2 and 1.8 m/s. Two different rate of emission have been simulated : weak and strong. And finally, walls of three different heights have been considered. Fig. 13 shows some typical dispersion features for the chimney of 152m and fig. 14 shows plumes for the chimney of 60m.

Instantaneous plume (low wind, low pollutant flow rate)

Mean plume (low wind, low pollutant flow rate)

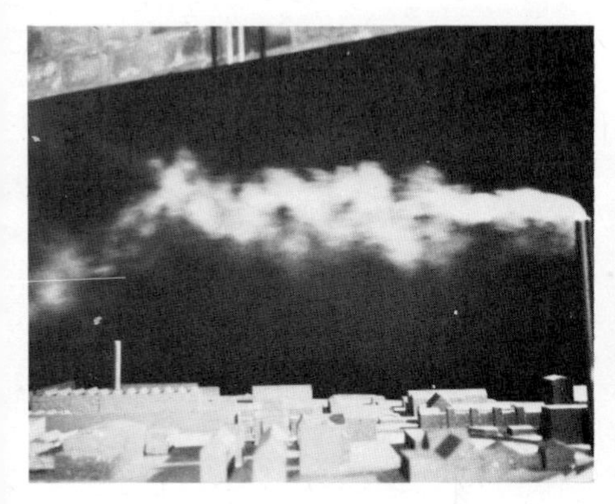

Mean plume (intermediate wind and pollutant flow rate)

Mean plume (strong wind, great pollutant flow rate)

Fig. 13 Different simulated plumes for the chimney of 152m. S-SO direction for the wind ; neutral stability ; wall 6m high.

Instantaneous plume (low wind, low polluant flow rate)

Instantaneous plume (intermediate wind and pollutant flow rate)

Mean plume (intermediate wind and pollutant flow rate)

Fig. 14 Simulated plumes for the chimney of 60m. Wind from S-SO; neutral stability ; wall 15m high.

It is clear from these experiments that the height of the wall has no influence on gas dispersion from such stacks.

DISPERSION OF DUST FROM STOCK PILES

As before, the aim of this simulation was, first, to find the accumulation areas all-over the region and to compare them with the real situation on the field. The improvement due to a rising of the wall was then analysed. A wind speed of 4.5 m/s in the tunnel was supposed to blow steadily and the pictures were taken after 7 minutes ; these "forced" experimental conditions are far away from reality, but they allow to quickly provide visible results.

Fig. 15 shows the different views of the simulation with a wall of 3m.

Situation before the simulation

General view

Accumulation areas

Results of the simulation

Fig. 15 Simulation of the dispersion from stock piles in tunnel The results are obtained after a 5 minutes simulation with a wind speed of 4.5 m/s ; direction is S-SO ; neutral stability ; wall of 3m .

Note the accumulation areas (arrows) in the street, behind the wall, inside the block of houses and behind the building at the left of the picture (see fig. 11 for the location of these points).

Fig. 16 gives values of the deposit ($mg.m^{-2}\ d^{-1}$) at some places in the urban area[8,9] ; the qualitative repartition of the different values is in agreement with the experimental ground-true measurements. Figs 17 and 18 show the result of the simulation for a wall of 6 and 15m.

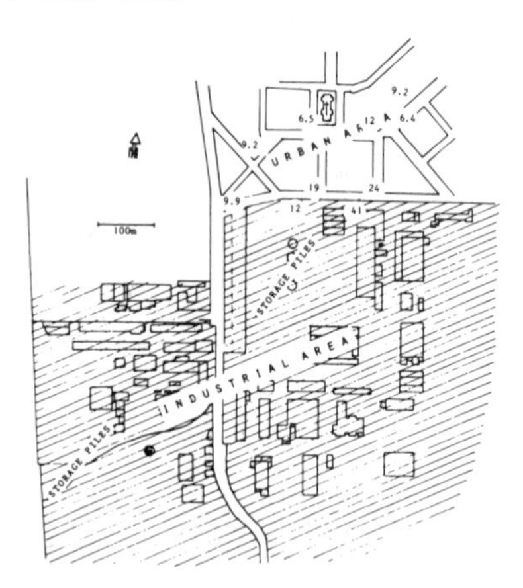

Fig. 16 Map of the area investigated with measured values of the deposit in the urban area (unit : $mg.m^{-2}.d^{-1}$).

General view

Accumulation areas

Fig. 17 Simulation of dispersion from stock piles. The results are obtained after 5 minutes simulation with a wind speed in tunnel of 4.5 m/s ; direction is S-SO ; neutral stability ; wall of 6m.

General view

Accumulation areas

Fig. 18 Simulation of dispersion from stock piles. Same characteristics as for fig. 17 by the height of the wall is here 15m.

These different pictures show clearly the existence, in the urban area, of many places of dust accumulation. Moreover, the simulations have confirmed the weak influence of a wall of 6 meters but, as already noted in the previous experiments, with a wall of 15 meters, it is possible to reduce effectively the deposits.

CONCLUSION

The purpose of this research was to try to better understand the local phenomena governing dispersion, entrainment and sedimentation of pollutants around obstacles as buildings and walls.

The different simulations have enabled to study the problem at different scales and to distinguish their most important characteristics. The impact of different technical solutions have also been tested and among them the construction of a wall of 15m seems to be the best one.

The objective was not to give a quantitative accurate estimation but rather to provide qualitative information as realistic as possible ; in this sense, the results obtained are significantive.

ACKNOWLEDGMENT

This research was financed by the Prime Minister Office for Science Policy within the framework of the National Programme "R&D - Environment-Air" and by the Ministry of Public Health.

We wish to acknowledge these two organizations for their support during this project.

REFERENCES

1. Institut d'Astronomie et de Géophysique G. Lemaître UCL Etude de la pollution par des particules sédimentables aux abords de l'usine Métallurgie-Hoboken - Etude du champ du vent autour d'une paroi. Scientific Report 1978/6 - Université Catholique de Louvain, Chemin du Cyclotron 2, 1348 Louvain-la-Neuve.
2. J. Armit and J. Counihan, Simulation of the atmospheric boundary layer in a wind tunnel, J. Fluid Mech. 28 : 48-71 (1967).
3. T.A. Hewett, J.A. Fay and D.P. Hoult, Laboratory experiments of smokestach plumes in a stable atmosphere, Atmospheric Environment 5 : 767-789 (1971).
4. P. Mecy, Reproduction et similitude de la diffusion dans la couche limite atmosphérique - Communication présentée au Comité Technique de la Société Hydrotechnique de France le 26 novembre 1968, La Houille Blanche n°4 : 327-344 (1969).
5. J.P. Schon and P. Mery, Méthodes de simulation de la couche limite atmosphérique neutre dans une soufflerie aérodynamique, E.D.F. - Bulletin de la direction des études et recherches, Série A n°1, 65-80 (1971).
6. L. Vadot, Technique de laboratoire pour l'étude des problèmes de pollution de l'air - Proceedings of the third International Clean Air Congress, Düsseldorf, B90-B91 (1973).
7. J.M. Snyder, Similarity criteria for the Application of Fluid Models to the study of Air Pollution Meteorology, Boundary Layer Meteorology 3 : 113-135 (1972).
8. Werkgroep Stofreductie van de Commissie Milieuhygiène en Industrie : "Reductie van stofenrissie door gebruik van winaschermen" ; Ministerie van Volksgezondheid en Milieuhygiene, Dokter-Reijersstraat 12, Leidschendam (1978).
9. Programme National de Recherche et de Développement dans le domaine de l'Environnement Air : "Bindrage van het Nationaal R-D Programme leefmilieu-lucht betreffende de samering van de toestand te Hoboken - Eindverslag - Rapport Scientifique-Programmation de la Politique Scientifique (Services du Premier Ministre) rue de la Science 8, 1040 Bruxelles.
10. J.E. Cermak, Laboratory simulation of the Atmospheric Boundary-Layer, AIAA Journal ç n°9 (1971).
11. J.E. Cermak, Applications of Fluid Mechanics to wind engineering. A Freeman Scholar lecture. Reprinted from Journal of Fluids Engineering (1974).
12. R. Dymen, A model of pollution control - Compressed air magazine oktober (1978).
13. G. Fumarola, J. Testino, G. Ferraiolo and Mancinir, Wind erosion of storage piles and dust dispersion in scale model wind tunnel experiments - The Fourth International Clear Air

Congress, Tokyo, May 16-20 (1977).
14. J. Halitsky, Wake and dispersion models for the EBR-II building complex, Atmospheric Environment 11 : 577-596 (1977).
15. J. Halitsky, Wind tunnel modeling of heated stack effluents in disturbed flows - Atmospheric Environment 13 : 449-452 (1978).
16. I.M. Kennedy and J.H. Kent, Wind tunnel modelling of carbon monoxyde dispersal in city streets, Atmospheric Environment 11 : 541-547 (1976).
17. A.G. Robins and J.P. Castro, A wind tunnel investigation of plume dispersion in the vicinity of a surface mounted cube, Atmospheric Environment 11 : 291-311 (1977).
18. L. Vadot, Etude de la diffusion des panaches de fumée dans l'atmosphère, CITEPA, Paris (1965).
19. J.M. Vincent, Model experiments on the nature of air pollution transport near buildings, Atmospheric Environment 11 : 765-774 (1977).

DISCUSSION

H.VAN DOP — Did you make an analysis of your particle size distribution ?

P. HECQ — No.

Ch. BLONDIN — Did you respect the effect induced by height scale ratio on density forces in choosing your particles ? (In other words, what is the difference between the Froude Number $F^2 = U^2/g'H$ in the atmosphere for Pb particles and in your wind tunnel for the particles cloud ?)

P. HECQ — It is not possible in a wind tunnel simulation to satisfy the similitude criteria for the Froude Number; but the fact that the accumulation areas found by the physical simulation are located at the same places as in the real situation allows to think that this factor has not a great influence.

J.W.S. YOUNG — Did you conduct any experiments by putting a 15 m high wall around the piles? It would seem to be a cheaper solution.

P. HECQ — Yes sure; but due to stock piles manipulations it is not a good practical solution; moreover, the stock piles are not the only sources.

B. WEISMAN

Did you test the effect of a partially porous wall in the second or third set of physical simulations?

P. HECQ

No, we didn't because it wasn't a good solution in regards of the posed problem.

4: EVALUATION OF MODEL PERFORMANCE IN PRACTICAL APPLICATIONS

Chairmen: M. Williams
J. Irwin
A. Berger

Rapporteurs: P. K. Misra
R. Y. Yamartino
B. Vanderborght

OBSERVATIONS AND SIMULATIONS OF CARBON MONOXIDE CONCENTRATIONS IN THE WAKE OF AN URBAN GARAGE

Ulrike Pechinger[+], Helga Kolb[++], Hans Mohnl[+] and Richard Werner[++]

[+] Zentralanstalt für Meteorologie und Geodynamik
[++] Institut für Meteorologie und Geophysik
both : Hohe Warte 38, A-1190 Vienna, Austria

ABSTRACT

A full scale dispersion experiment was carried out in Vienna, Austria, from June to November 1979 to investigate the effects of emissions from a 25 m urban parking garage on ambient carbon monoxide concentrations. A comparison of concentrations measured in the building cavity with those yielded by standard formulas from the literature is made.

INTRODUCTION

The entrainment of plumes emitted from short stacks into the wakes of buildings can result in maximum ground-level concentrations that are significantly greater than those found for similar sources in the absence of buildings (Huber, 1980). Non uniform flow fields and increased turbulent intensity result in a limited applicability of the Gaussian plume model. To investigate the building wake effect on dispersion of pollutants numerous wind tunnel studies were conducted. A review on flow and dispersion near buildings including several formulas for calculating concentration fields is given by Hosker (1980).

As agreement between concentrations calculated by different formulas is very poor a full scale research experiment of carbon monoxide dispersion from an artificially ventilated 25 m urban parking garage within a built-up area was carried out. A comparison of calculated concentrations, derived from several standard formulas with those, measured in the experiment, is made and results are discussed.

DESCRIPTION OF THE EXPERIMENT

A small garage in a densly populated area with only light traffic in the surrounding neighborhood was chosen as the emission source to ensure low background levels of carbon monoxide concentration. During the day, automobile traffic in the garage and consequently emissions from the 1 m stack (cross sectional area : 0.4 m^2) at roof level were low. Carbon monoxide was therefore artificially introduced into the mechanical ventilation system of the garage to enhance emissions and raise concentrations to levels readily measureable by an UNOR 4N monitor. The mechanical ventilation system was layed out for an air flow rate of 10.400 m^3 per hour, and CO-concentration at stack was maintained at 2000 ppm. This level is about 40 times the maximum concentration level actually recorded in any garage effluent measured in Vienna. For six runs, a special movable source was used, consisting of a fan enclosed in a casing fitted with an air intake and a vertically oriented air outlet on which tubes up to 2 m height (cross sectional area : 0.04 m^2) can be mounted. The emission of this movable source was generally adjusted to 1580 m^3/h with 10 000 ppm CO. The object of developing and testing this movable source was to become independent of built-in ventilation systems in future experiments.

The building used in the study was 25 m tall, with a parking area in the basement. Surrounding buildings were about the same height and formed a square courtyard to the SE of the study building.

During 13 measurment periods, CO-concentrations were recorded at a total of six sites at roof level and in the yard-side wake of the garage (Figure 1). Each sensor recorded 5 second average concentrations at 2 minute intervalls. These data will subsequently be referred to as the "2 minute-concentrations". They were then used to construct longer period average concentrations. The duration of each observational period varied between 30 and 90 minutes.

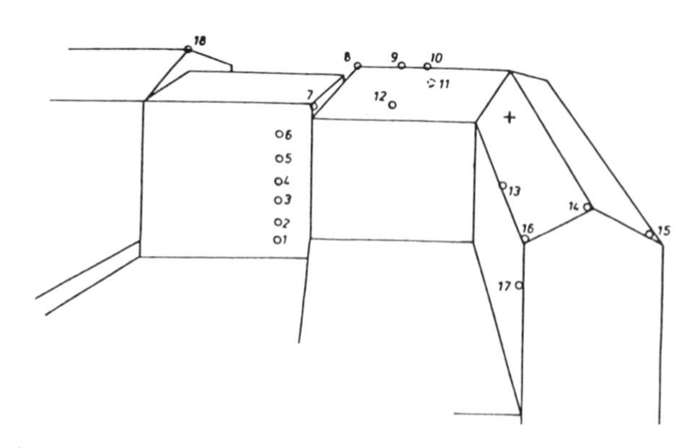

Figure 1 : Location of source and receptors

Only periods with suitable wind directions were chosen, i.e., when the effluent was transported towards the yard.

Preliminary investigations of the wind field at roof level showed surprisingly uniform spatial velocity conditions. Wind measurements from one site were therefore taken as representative of the entire roof area. The plume was made visible by addition of coloured smoke and oil fog to the effluent. Plume rise was evaluated visually from observations of the elevation angle on two sides of the source.

RESULTS

Measured concentrations

Figure 2 shows mean concentrations and standard deviations for several sites within the building wake for each measurement period. Numbers in the figure indicate height of the sensors above ground. Although there were noticable differences of mean concentrations between the various runs, within each run the values at the different sensors show good agreement. This indicates a well mixed cavity with no height dependence of concentrations. The large standard deviations are caused by strong fluctuations of the 2-minute concentrations due to small scale turbulence.

Although mean concentrations were expected to depend on the carbon monoxide source used, thermal stratification and wind conditions, no dependence on source, thermal stratification and wind speed was observed. As an example figure 3 shows the running half hour means of concentrations versus wind speed. Mean wind speeds during the experiment varied between 1 and 4 m/s. Observations of the spread of the coloured smoke showed that the plume was transported into the cavity during most of the time. Due to greater plume rise the downward transport occured to a lesser extent during light wind conditions (Kolb et al., 1980). As increasing wind speed enhanced downward transport and no wind speed dependence of concentrations was found, higher wind speeds must have enhanced dilution.

Of the parameters mentioned as possible causes for changes in mean concentrations only wind direction was not measured with the necessary accuracy. Visual observations showed that the plume was much more sensitive to directional fluctuations of the wind than the wind vane used. A certain homogeneity was attempted by evaluating only those cases when the plume was transported towards the yard. However this still includes all wind directions between west and north. Thus the differences of concentrations between the several runs are primarily attributed to changes of wind directions.

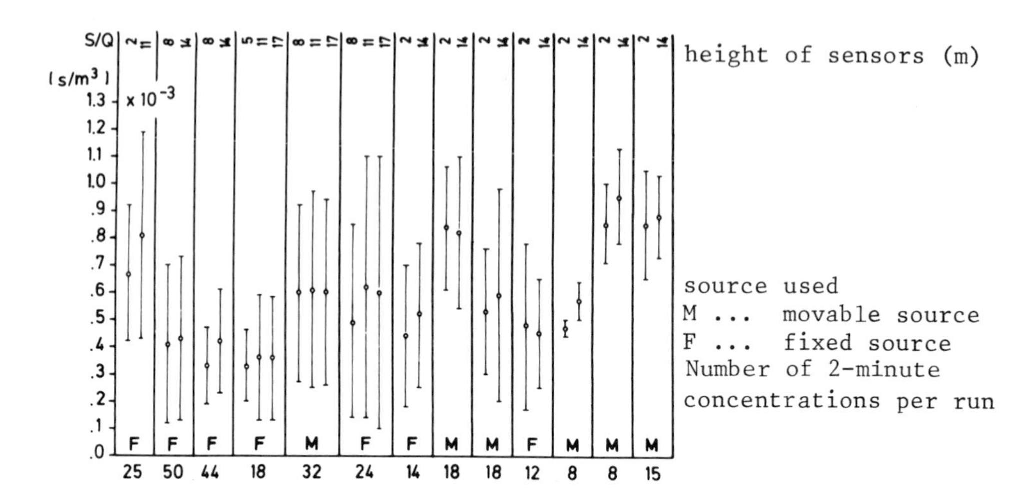

Fig. 2. Normalized mean concentration S/Q and standard deviation measured during each run at different sites in the wake

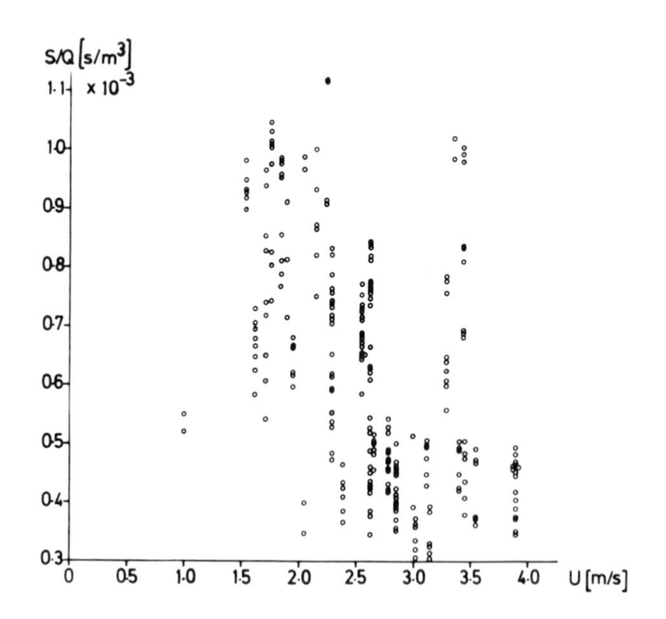

Fig. 3. Running half hour means of concentrations S/Q in the wake versus wind speed U

Comparison of measured and calculated concentrations

Measured concentrations are compared with those calculated by several standard formulas. The results of the formulas discussed in this study are represented in figures 4 and 5. Figure 4 gives comparisons of measured and calculated concentrations for each run ; the overall means and standard deviations are plotted in figure 5.

The formula by Scorer and Barret (1962)

$$S = \frac{2\,Q}{L^2} \quad (1)$$

developed for estimating typical maximum short-term concentrations close to a building with a low chimney shows no dependence on wind speed. This agrees with the results of the experiment. The formula is based on the assumption that worst conditions occur in moderate winds of about 5 m/s. Scorer and Barret assumed, that the wake is not developed in lighter winds and therefore concentrations are small. This however was not confirmed by the observations. Concentrations measured within the building cavity did not reveal a dependence on distance from source as assumed in Scorer's formula. Calculated concentrations are about a factor 2 to 10 larger than measured mean concentrations. They are even higher than most 2-minute concentrations.

A commonly used equation to estimate concentrations within a building wake is :

$$S = K\,\frac{Q}{A\,U} \quad (2)$$

where K is a coefficient, which ranges from :

0.5 to 20 (Vincent, 1977)
0.2 to 2.0 (Hosker, 1980)
0.5 to 2.0 (Gifford, 1968)
1.5 to 2.0 (Barry as cited at Gifford, 1968)
0.0 to 5.0 with 1.0 as a rough average (Halitsky, 1968)

Evaluation of K from the experiment yielded values between 0.3 and 4.2. This implies, that the upper value of K = 2.0, given by several authors can lead to an underestimation of concentrations.

Halitsky derived a formula for estimating concentrations in the vicinity of a building from wind tunnel tests for a clinical center :

$$S = \frac{P}{2.22\,M\left(3.16 + 0.1\,\frac{L}{\sqrt{A_e}}\right)^2}\,\frac{U_e}{U} \quad (3)$$

M = 4 for exhaust on roof and intake in the cavity

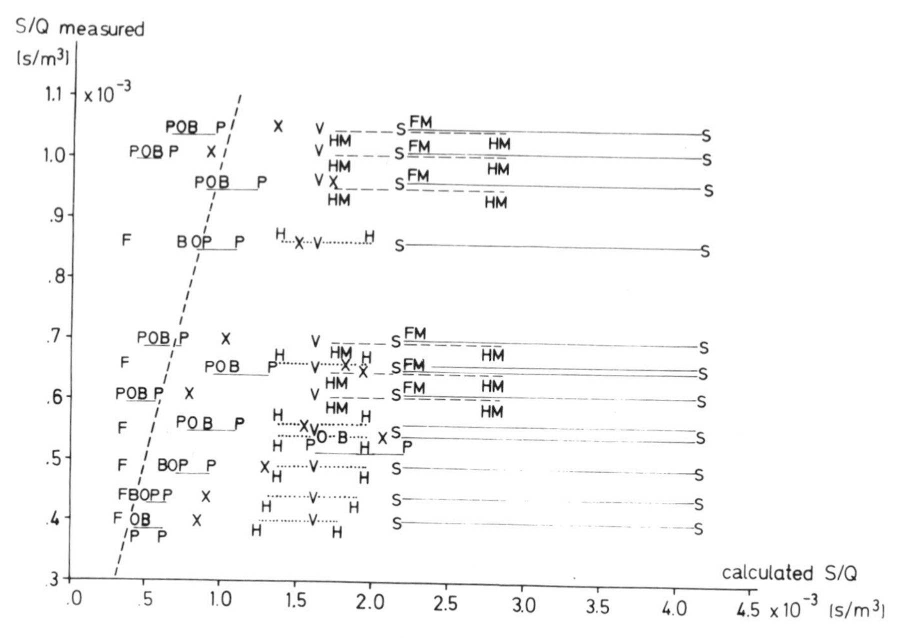

Fig. 4. Measured mean concentrations versus calculated concentrations for each measurement period

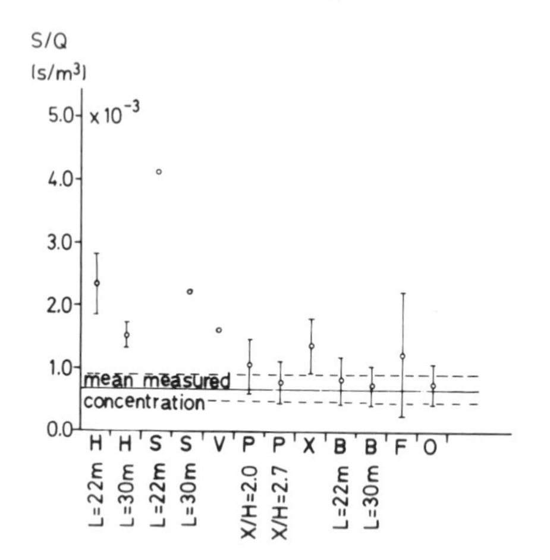

Fig. 5. Mean concentrations for all measurement periods as calculated by equations (1) to (8)

B ... equation (6)
F ... equation (8), fixed source
FM... equation (8), movable source
H ... equation (3), fixed source
HM... equation (3), movable source

O ... equation (2), K = 1
P ... equation (7)
S ... equation (1)
V ... equation (4)
X ... equation (5)

The average emission-velocity ratio was $U_e/U = 2.22$ with wind speed $U = 3.7$ m/s. According to Halitsky the factor U_e/U provides a correction for greater dilution at higher wind velocities. Concentration is independent of wind speed for emission-velocity ratios greater than 2.22. This agrees well with the results in the experiment as $U_e/U \geq 2.22$ was observed during most of the runs.

For the two sources used in the experiment widely different values served as input for equation (3). Concentrations calculated for the transportable source were however less than a factor 1.5 higher than those for the fixed source. For both sources calculated concentrations resulted in values that are a factor 1.5 to 5 higher than those measured. These results are better than those derived from equation (1), yet they still represent a noticable overestimation of concentrations. This might be due to lacking similarity as no information on A_e and entrainment of the plume into the cavity was available. Halitsky implies that because of wind gustiness in the atmosphere wind tunnel data can overestimate time-mean values in adiabatic conditions by a factor up to possibly 10.

Based on a theoretical model Vincent (1977) derived a simple formula for estimating mean concentrations in the cavity

$$S = \frac{Q}{A \frac{X}{T_d}} \qquad (4)$$

He assumed, that material is constantly emitted within the cavity and removed in a characteristic time T_d because of turbulent transfer. In the experiment the estimated mean residence time T_d varied between 120 and 240 s. For values of X/H between 2/0 and 2.7 (Pernpeintner, 1980), calculated concentrations are about a factor 1.5 to 4 greater than those measured for $T_d = 120$ s and 3.0 to 8.0 for $T_d = 240$ s. Considering the fact, that the plume was not completely downwashed into the cavity, a certain overestimation by Vincent's formula is to be expected.

According to a procedure, developed by Briggs (1973), concentration in the cavity can be calculated with the formula :

$$S = K \frac{Q}{U L_B^2} \qquad (5)$$

where $K \leq 3.0$ and typically 1.5. Concentrations estimated with this formula are about a factor 1.3 to 3.8 larger than those measured ($K = 1.5$). Differences become greater with decreasing wind speed. In average calculated concentrations are about two times the measured concentrations.

Downwind of the cavity region ground concentrations can be approximated by :

$$S = \frac{Q}{U\,(A_p + 2\,R_y R_z)} \qquad (6)$$

under the assumption, that the plume, downwashed into cavity, becomes a ground source with the initial cross sectional area A_p (Briggs, 1973). Concentrations calculated by equation (6) are lower than those by equation (5). Although the sites in the experiment were not located downwind of the cavity, measured mean concentrations for all runs, agree better with these lower values (figure Regarding each single measurement period calculated concentrations reach about 50 % to 300 % of measured values.

Like Vincent (1977) Pernpeintner (1980) assumes, that the pollution source, wholly contained in the cavity, is at equilibrium with the local pollutant flux normal to the cavity boundary. He derived a semi-empirical formula for calculating concentrations in the building cavity :

$$S = P\,E\left(\frac{U_e}{U}\right)\left(\frac{A_e}{H^2}\right)\left(\frac{\varepsilon}{\delta U}\right)^{-1}\left(\frac{A_c}{H^2}\right)^{-1} \qquad (7)$$

$$0 \leq E \leq 1$$

The terms $\varepsilon/\delta U$ and A_c/H^2 have to be determined by wind tunnel experiments. For his own set-up, a rectangular block, Pernpeintner gives the following expressions :

$$\frac{\varepsilon}{\delta U} = 0.07\left(\frac{L}{H}\right) + 0.05$$

and

$$\frac{A_c}{H^2} = 4.23\,\frac{X}{H}\sqrt{\frac{B}{H}}$$

with $2.0 \leq \frac{X}{H} \leq 2.7$

Assuming that the plume is continuously transported into the cavity (E = 1), in average of all runs calculated concentrations take about the same value as those measured for X/H = 2.7. For X/H = 2.0 they are about a factor 1.5 larger than measured. Concentrations estimated for each run are about 50 % to 400 % of those measured. As the entrainment factor was less than one during the experiment (an exact value cannot be derived), the formula does not overestimate concentrations like the other discussed formulas, in fact underestimations occur.

A very simple rule of thumb, given by Fett (1976) :

$$S = \frac{P}{1000} \quad (8)$$

for estimating concentrations in the cavity yields smaller concentrations than measured for the fixed source and larger concentrations for the movable source. The sole dependence of the ambient concentrations on the emission concentration is an oversimplification. The formula was however intended only to yield the order-of-magnitude of concentrations. It serves quite well for this purpose.

SUMMARY AND CONCLUSION

A comparison of calculated concentrations with those, measured in the cavity of a 25 m urban parking garage with a source at roof level (1 m stack) showed, that concentrations are generally overestimated. In the experiment the plume was not continuously transported downwards into the cavity due to gustiness of the wind. Wind tunnel experiments cannot fully reproduce these conditions. Briggs (equation (5)) assumes a continuous downwash of the plume and in the equations (4) and (7) the source is assumed to be located within the cavity. This is the main cause for overestimations.

There are also other discrepancies : In the widely used equation (2) and equations (5), (6) and (7) a wind speed dependence is assumed. Observations did not confirm this within the range of observed values ($\leq$ 4 m/s). Due to greater plume rise the downward transport occurs to a lesser extent during light wind conditions, while higher wind speeds enhance dilution. According to Halitsky this balance is not maintained for higher wind speeds. This wind range was not covered by measurements.

During all measurement periods a well mixed cavity with no height dependence of concentrations was observed. The dependence on transport distance of the effluent in equation (1) and (3) was therefore not confirmed.

In conclusion it is noted that calculated concentrations varied widely due to different and in general rather severe simplifications in wind tunnel studies as well as in the equations. The more sophisticated models require inpat data on flow structure that is not readily available. Therefore assumptions must be made, which influence results and the agreement with the observations. Coefficients (e.g. K, $\dot{E}$) derived from the experiment also are not of general value, as they depend on building configuration and emission situations in an as yet undefined way. In applying the models to practical questions the user should be aware of the uncertainty of the result.

LIST OF SYMBOLS

A ... cross sectional area of the building normal to the wind (L^2)
A_c ... surface area of cavity (L^2)
A_e ... area of effluent aperture (L^2)
A_p ... initial cross sectional plume area (L^2)
B ... building width (L)
E ... entrainment factor (dimensionless)
H ... building height (L)
K ... coefficient in equ. (2) (dimensionless)
L ... distance from source (L)
L_B ... minimum (building height, building width) (L)
P ... effluent concentration at the source (ML^{-3})
Q ... source strength (MT^{-1})
R_y ... plume half width (L)
R_z ... plume half depth (L)
S ... ambient concentration (ML^{-3})
T_d ... mean residence time of pollutant in the cavity (T)
U ... wind velocity (LT^{-1})
U_e ... velocity in the effluent aperture (LT^{-1})
X ... length of the cavity (L)
ε ... eddy diffusivity coefficient (L^2T^{-1})
δ ... length of the shear layer of the cavity (L)

ACKNOWLEDGEMENTS

This project was financed jointly by the Bundesministerium für Gesundheit und Umweltschutz and the Kuratorium für Umweltschutz, Austria.

REFERENCES

BRIGGS, G.A., 1973, Diffusion estimation for small emissions, ATDL contribution file No. 79
FETT, W., 1976, Kriterien und Empfehlungen zur Anlage von Abzügen für kontaminierende Gebäudeabluft, Institut f. Wasser, Boden- und Lufthygiene des Bundesgesundheitsamtes Berlin, Bericht 5/70
GIFFORD, F.A., 1968, An outline of theories of diffusion in the lower layers of the atmosphere, in : "Meteorology and Atomic Energy 1968", D.H. SLADE, ed., USAEC, Springfield
HALITSKY, J., 1968, Gas diffusion near buildings, in : "Meteorology and Atomic Energy 1968", D.H. SLADE, ed., USAEC,
HOSKER, Jr., R.P., 1980, Dispersion in the vicinity of buildings, sec. Joint Conf. on Appl. of Air Poll. Met. and Sec. Conf. on Indust. Met., New Orleans

HUBER, A.H., SNYDER, W.H., and LAWSON, R.E., 1980, The effects of a squat building on short stack effluents - a wind tunnel study, Environmental Protection Agency, EPA-600/4-80-055 Research Triangle Park, N.C.

KOLB, H., MOHNL, H., PECHINGER, U., REUTER, H., WERNER, R., 1980, Bericht über das Forschungsprojekt Emission von und Auswirkungen der Immission auf die Umgebung, Zentralanstalt für Meteorologie und Geodynamik, Vienna, Austria

PERNPEINTNER, A., 1980, Experimentelle Untersuchungen zur Ausbreitung von Schadstoffen in der Umgebung von Bauwerken, Fortschr.-Ber., VDI-2., Reihe 15, Nr. 15

SCORER, R.S., and Barret, C.E., 1962, Gaseous pollution from chimneys, Intern. J. Air. Wat. Poll., Vol. 6, 49-63

VINCENT, J.H., 1977, Model experiments on the nature of air pollution transport near buildings, Atm. Env., Vol. 11, 765 - 774.

INFLUENCE OF THE METEOROLOGICAL INPUT DATA ON THE COMPARISON BETWEEN CALCULATED AND MEASURED AEROSOL GROUND LEVEL CONCENTRATIONS AND DEPOSITIONS

I. Mertens, J. Kretzschmar and B. Vanderborght

Nuclear Energy Research Centre
B-2400 Mol
Belgium

INTRODUCTION

Within the framework of the Belgian National Research and Development Program Environment-Air of the Ministry of Science Policy a detailed investigation of the environmental behaviour of Sb-particulates emitted by a nonferrous metal industry situated in an open and flat region has been carried out over a period of approximately two years (1979-1980). The main purpose of this testcase was the development and verification of appropriate methodologies to deal with the different aspects of the impact of nonferrous metal industries upon the environment. The specific choice of Sb as testcase is due to the fact that the involved factory is the only Sb-emitter in the region, so that problems with background-levels were avoided, that the factory was willing to cooperate and that no direct health hazards existed.

The purpose of this paper is to present some of the results of the modelling undertaken within the context of the testcase. The required information on the emissions (three chimneys, many fugitive emissions), the local atmospheric conditions and the ground-level concentrations and depositions have been obtained by different teams of the Gent University (R.U.G.), the University of Antwerpen (U.I.A.), the Nuclear Energy Research Centre (S.C.K./C.E.N.) and the National Data Bank (I.H.E.). The lay-out of the monitoring network around the Sb-plant is given in Fig. 1. In each monitoring site 21 h-average ground-level concentrations and monthly deposition values were obtained. Meteorological data are measured half-hourly averages while the emission data are best estimates based on selective measurements, tracer releases and a systematic inventory and registration of the activities in the plant.

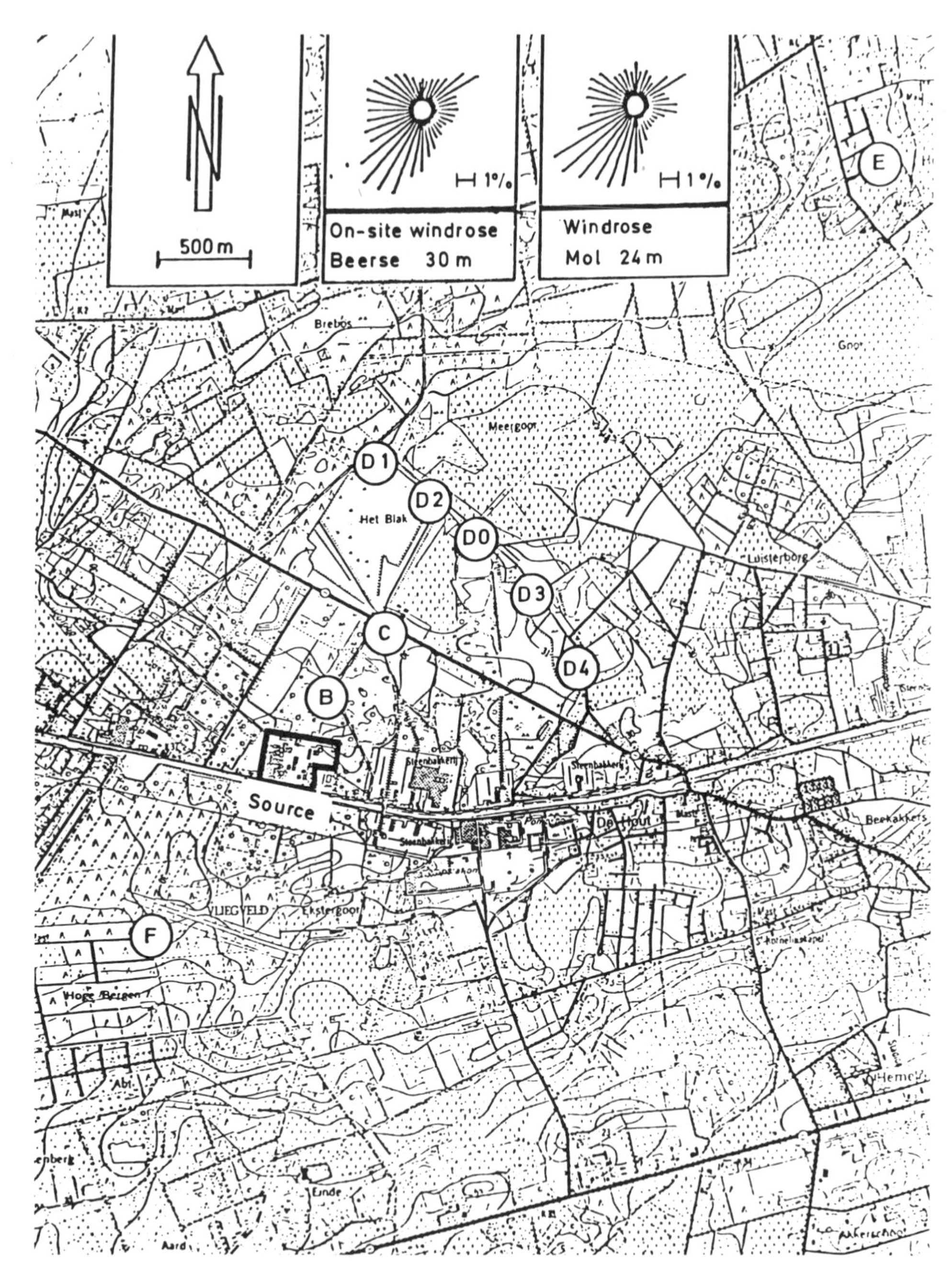

Fig. 1. Lay-out of the Sb-monitoring network (concentration and deposition)

The model calculations have been restricted to a period of 14 consecutive months during which the required input information was as complete as possible. Within a validation procedure, mainly based on the comparison between measured and calculated cumulative frequency distributions, the influence of the following model parameters and input data has been evaluated : on-site measured wind speed and wind direction against routinely obtained information at the nearest by meteorological station (Mol, 22 km), specific choice of the numerical value of the dry deposition velocity and finally the importance of the applied turbulence typing scheme with corresponding dispersion parameters.

DESCRIPTION OF THE MODEL

All calculations have been carried out by means of the Immission Frequency Distribution Model (IFDM), a bi-Gaussian dispersion code developed at the SCK/CEN[1,2] Mol, Belgium. The most important features of the model version actually used for this study are as follows :

- half-hourly averages as input data for the emissions and the meteorological parameters ;
- daily- and monthly averages obtained by taking the arithmetic average over consecutive half-hourly values ;
- turbulence typing scheme and corresponding dispersion parameters based on the ratio of the potential temperature gradient and the square of the average wind speed measured at 69 m above ground-level as described by Bultynck and Malet[3] ;
- plume rise formula of Stümke[4] for the chimney releases ;
- all fugitive emissions released at approximately 15 m above ground-level (no plume rise) ;
- transport speed in the Gaussian formula is the average wind speed at the effective release height obtained by means of the power law wind speed profile ($m = 0{,}53$ for very stable to $m = 0{,}10$ for very unstable conditions) ;
- all sources taken into account as point sources without any initial dispersion due to building effects ;
- deposition calculated by means of source depletion ;
- washout-rate equal to $2.10^{-4}\ s^{-1}$ and dry deposition velocity varying between 0,06 and 1,0 cm/s. No sedimentation was used as granulometric investigations indicated that almost all particulates were smaller than 1 µm.
- mixing height unlimited.

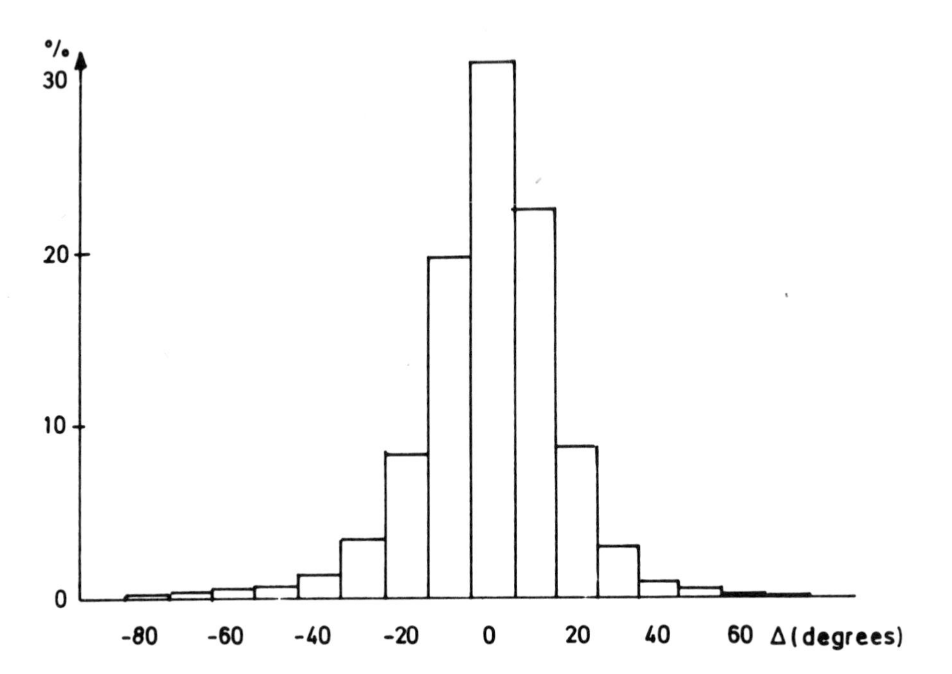

Fig. 2a. Histogram of the differences between the simultaneously measured wind direction (t_{av} = 30 min) in Mol and on-site

Fig. 2b. Histogram of the differences between the simultaneously measured wind speed (t_{av} = 30 min) in Mol and on-site

INFLUENCE OF WIND SPEED AND WIND DIRECTION

For impact studies it is common to use the available meteorological, climatological or synoptic data from the nearest by station. Especially for the tracer releases (part of the study of the fugitive emissions) it was nevertheless necessary to have on-site meteorological observations. Wind speed and wind direction at a height of 30 m above ground-level were therefore recorded continuously in monitoring station Do (Fig. 1). The obtained data set of half-hourly averages over the complete period of 14 months made it also possible to evaluate afterwards if on-site meteorological measurements over a sufficient period of time are also required for impact studies of the kind normally carried out by means of dispersion simulation models.

From the two wind roses given in Fig. 1 it is obvious that the statistics of the calculated concentration values in the different receptors (A to F) will not be significantly different if one or the other wind rose is used as input for the model. Note that the wind rose on the left was obtained locally in Do (h = 30 m) while the right one comes from the meteorological tower in Mol (24 m level) at a distance of some 22 km. The histogram of the wind direction differences between the simultaneously measured (half-hourly average) directions in Mol and on-site doesn't show any systematic bias as illustrated in Fig. 2a.

This is not true for the simultaneously measured (half-hourly average) wind speeds as shown in Fig. 2b. Wind speed is systematically higher on-site than in Mol. Despite the similarity of the landscapes, and the continuity between both areas in the flat northern part of Belgium, there seems to be a distinct difference in surface roughness possibly due to the fact that the Nuclear Energy Research Centre of Mol, where the meteorological tower is situated, is surrounded by pine trees in almost all directions, while this is not the case for the site where the experiments were carried out. The direct consequence of this phenomenon is that calculations carried out with the Mol wind speed data would contain a systematic error (wind speed too small, concentrations too high) which would not be detectable in a validation study, except when on-site wind speed measurements are available. This stresses the importance of on-site meteorological observations even if at a first glance the nearest by observation post seems to be representative.

INFLUENCE OF THE DRY DEPOSITION VELOCITY

As mentioned before wet and dry deposition were calculated by means of the source depletion approach. For wet deposition a constant washout-rate of 2.10^{-4} s^{-1} was used while for the dry deposition different numerical values for the dry deposition velocity v_d were tried out. In a recent review paper by Sehmel[5] values of 0,06 cm/s and 0,4 cm/s are given for the dry deposition velocity of Sb-particulates while v_d = 1 cm/s is a quite frequently used value for particulates in general. With each of the three v_d-value concentrations and depositions were calculated in each of monitoring sites A to F where simultaneously measured values were available (LIB high volume samplers with NAA for the concentrations, NILU deposit gauges with AAS for the depositions).
A typical result of the sensitivity of the model calculations for the numerical value of v_d is given in Fig. 3a, where for point Do (Fig. 1) the cumulative frequency distribution of the measured daily Sb-concentrations is compared with the corresponding cumulative frequency distributions of the calculated values obtained respectively with v_d = 0, 0.06, 0.4 and 1 cm/s.
The latter value for v_d gives the best agreement between calculated and measured values and the same holds for all the points on the D-line (Fig. 1). The influence of v_d is less pronounced in the points B and C, as those points are closer to the source, while for the most distant point E the model already underestimates the concentrations for v_d = 0 so that increasing v_d also increases this underestimation.

The influence of v_d will naturally be more pronounced on the deposition calculations as illustrated in Fig. 3b for site Do. From these results it's obvious that v_d = 1 cm/s gives the best agreement between the calculated and the measured monthly means of the dustfall (wet and dry) around the nonferrous plant. Similar calculations for the other monitoring sites proved that v_d = 1 cm/s always gives the best agreement so that it can be concluded that v_d = 1 cm/s is a reasonable choice for the testcase. This value is also supported by some experimental findings as the ratio between the measured average dustfall and the measured average concentration varies between a low 0,7 $\pm$ 0,3 for sites C and D2, and a high 2 $\pm$ 1,1 for D3.

Although some quite acceptable results were obtained here with a dry deposition velocity of 1 cm/s this doesn't proof at all that this is the magic number. A detailed analysis of all the available data from the different experiments carried out during the Sb-testcase showed that major uncertainties still exist as to the real physical meaning of dustfall measurements in general and the accuracy of actual dustfall measurements under field conditions.

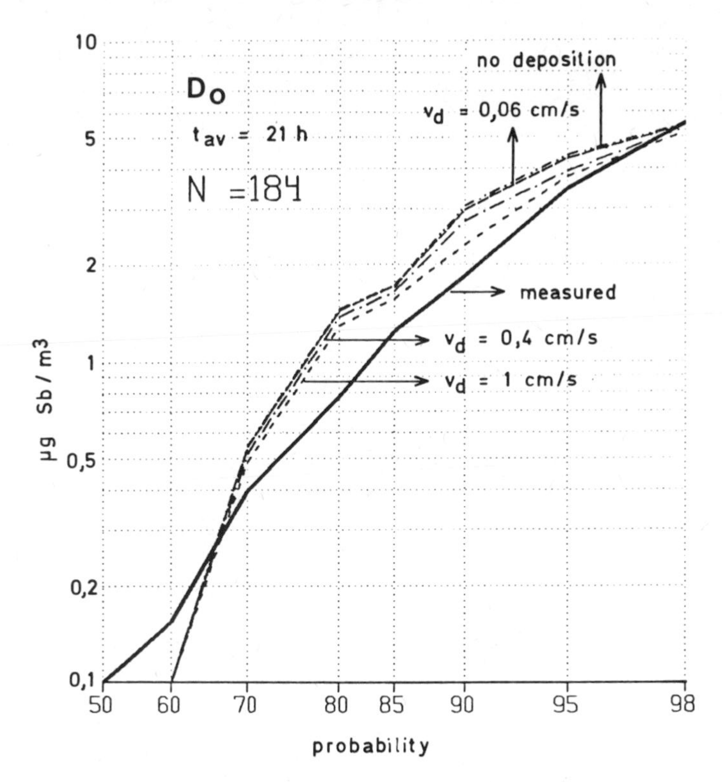

Fig. 3a. Influence of v_d upon the cumulative frequency distribution of the calculated daily averages in D_o and comparison with the measured CFD

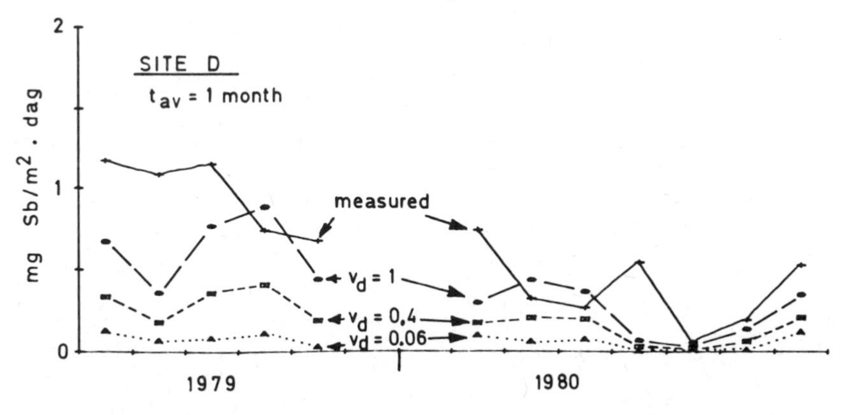

Fig. 3b. Influence of v_d upon the calculated monthly means for the total deposition (wet plus dry) in site D_o and comparison with the measured values

Somewhat related problems are the physics and the mathematical treatment and quantification of resuspended particulate matter and its influence upon the measured and calculated concentration and deposition values.

A final remark concerns the accuracy of the emission data used as input for the model calculations. By means of the model it has been shown that the contribution of the fugitive emissions in the concentrations and depositions in the different monitoring sites is much more important than the contribution of the chimney emissions. Although it is possible to obtain a certain degree of accuracy for the latter, it is also obvious that it is very difficult if not impossible, to quantify in an accurate way the many different fugitive emissions of a nonferrous plant and their variability as a function of time and space, especially if this exercise has to be continued over a long period of time. Those shortcomings of the model input data have also to be kept in mind whenever a model output is discussed.

INFLUENCE OF THE STABILITY CLASSIFICATION SCHEME AND THE CORRESPONDING DISPERSION PARAMETERS

The sensitivity of the bi-Gaussian model for the dispersion parameters $\sigma_y(x)$ and $\sigma_z(x)$ has been discussed by many different investigators. The purpose of the actual exercise is not to repeat well-known facts but to illustrate the previous findings by means of the calculated concentrations and depositions for the Sb-testcase. In order to do so the measured half-hourly average temperature gradient along the meteorological tower in Mol was used to determine the corresponding Pasquill stability class as described in the U.S. Nuclear Regulatory Guide[6]. All model calculations were redone using this data set and the corresponding dispersion parameters as given by Gifford[7]. The obtained concentration and deposition values are compared with the measured ones and the ones calculated previously by using the SCK/CEN turbulence typing scheme and corresponding dispersion parameters[3]. Note that the same meteorological measurements on the tower in Mol were used in both schemes. Fig. 4a illustrates the results close to the source while Fig. 4b does the same for a more distant receptor point. Close to the source there isn't much difference while farther away the results based on the Pasquill input show larger deviations from the measured values. (The same holds in first approximation for the depositions).
This phenomenon is due to the fact that the Pasquill parameters for the dominant stable and neutral conditions give narrower plumes than the SCK/CEN dispersion parameters. The correspondence between the stability categories in respectively Pasquill's and SCK/CEN's system is illustrated in Table 1.

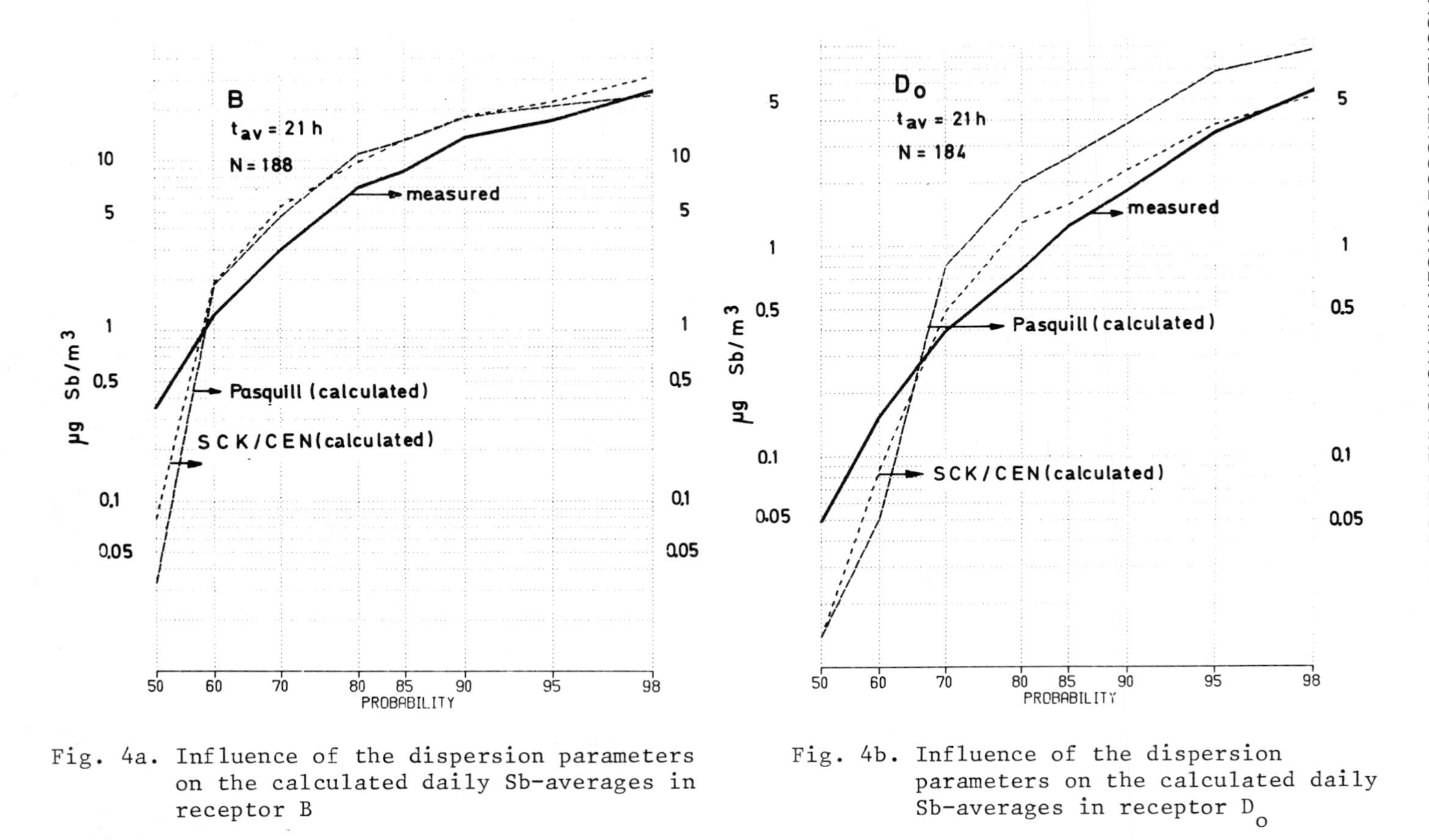

Fig. 4a. Influence of the dispersion parameters on the calculated daily Sb-averages in receptor B

Fig. 4b. Influence of the dispersion parameters on the calculated daily Sb-averages in receptor D_o

Table 1. Relative Frequency of Occurrence of Pasquill's and SCK/CEN's Stability Classes during the Test Period (14 m)

	E1	E2	E3	E4	E5	E6	E7	Σ
A	0	0	0	1,6	0,3	0	0	1,9
B	0	0	0,1	1,4	0,6	0	0	2,1
C	0	0	0,5	3,2	1,1	0	0	4,8
D	1,0	13,4	16,0	6,0	0,8	0	0	37,2
E	10,3	25,8	0,5	0	0	0	0,3	36,9
F	10,6	0,4	0	0	0	0	0,4	11,4
G	3,5	0	0	0	0	0	0	3,5
Σ	25,4	39,6	17,1	12,2	2,8	0	0,7	97,8

CONCLUSIONS

By means of the available data of a detailed study over fourteen months of the Sb-pollution around a nonferrous metal industry it was possible to demonstrate :

- that local onsite meteorological observations are recommended if atmospheric dispersion modelling has to be done in impact assessment studies. Even if the nearest by climatological, meteorological or synoptic station is not farther away than 25 km, significant differences in wind speed may exist due to the local environment ;
- that acceptable agreement between calculated and measured concentration and deposition values in monitoring sites at different distances from the source can be achieved by means of a simple bi-Gaussian dispersion model using the source depletion approach ;
- that major uncertainties still exist as to the physics and the real meaning of the measurement as well as the calculation of deposition and resuspension ;
- that the choice of a specific diffusion typing scheme, and the corresponding set of dispersion parameters, influences the model results significantly.

ACKNOWLEDGEMENT

Within the framework of the Sb-testcase of the National R & D Programme Environment-Air, under the direction of C. De Wispelaere (DPWB), it has been possible to collect the sufficient amount of reliable data for the proper application of the IFDM dispersion and deposition simulation model.

As modellists we greatly appreciate the efforts of the teams of R. Dams (RUG), F. Adams (UIA), A. Cottenie (RUG) and M. Legrand (IHE) as well as the support of our scientific, technical and administrative colleagues at the Nuclear Energy Research Centre in Mol.

REFERENCES

1. J.G. Kretzschmar, G. Cosemans, G. De Baere, I. Mertens and J. Vandervee, Some practical examples of the impact of individual sources upon the cumulative frequency distributions of the daily SO_2-concentrations in an urban and industrial area, in : "Proceedings of the Eight International Technical Meeting on Air Pollution Modeling and its Application", Nato CCMS (1977).
2. J.G. Kretzschmar, G. De Baere and J. Vandervee, The Immission Frequency Distribution Model of the SCK/CEN, Mol, in : "Modeling, Identification and Control in Environmental Systems", Vansteenkiste, ed., North-Holland Publ. Co. (1978).
3. H. Bultynck and L.M. Malet, Evaluation of the atmospheric dilution factors for effluents diffused from an elevated continuous point source, Tellus, 24 : 455 (1972).
4. H. Stümke, Vorschlag einer empirischen Formel für die Schornsteinüberhöhung, Staub, 23 : 549 (1963).
5. G.A. Sehmel, Particle and gas dry deposition : a review, Atm. Env., 14 : 983 (1980).
6. Anon., Regulatory Guide 1.23. Onsite meteorological programs, U.S. Nuclear Regulatory Commission (1972).
7. F.A. Gifford, Turbulent diffusion-typing schemes : a review, Nucl. Saf., 17 : 68 (1976).

DISCUSSION

M. SCHATZMANN

We saw that your model generally overestimated ground level concentrations in the vicinity of the plant. I wonder whether this might have been caused by an insufficient effective stack height calculation. Did you carry out a sensitivity analysis regarding the sensitivity of your results with respect to the source height ?

B. VANDERBORGHT

Yes we did, but the final result was not influenced by a change in plume rise for chimney emissions because the mass flow of the fugitive emissions at low altitude was about 5 to 10 times higher than for stack emissions and fugitive emissions contributed for more than 98 % to the ground level concentration within 1500 m from the factory.

M.L. WILLIAMS

Why did you not use a constant initial plume spread, σ_z , equal to the building height, for the fugitive sources ?

B. VANDERBORGHT

Initial plume spread is far from constant neither equal to the building height. We preferred having a good idea on the inaccuracy caused by using point source simulation for fugitive emissions, instead of improving the ratio of calculated to observed concentration by using an arbitrary chosen initial spread.

K.J. VOGT

The deposition velocity, which you have measured for 1-2 μm particles, is approximately one order of magnitude higher than the values resulting from the deposition experiments performed at the Jülich Nuclear Research Center. Could you comment on this difference.

B. VANDERBORGHT

There remains indeed an important uncertainty on the deposition velocity. Deposition velocities reported by Sehmel range over two orders of magnitude.
One of the conclusions of our study is that the determination of the deposition velocity is tremendously influenced by experimental conditions. Besides the apparatus used, also the surrounding of the sampling point is very important.

EPRI PLUME MODEL VALIDATION PROJECT

RESULTS FOR A PLAINS SITE POWER PLANT

Richard J. Londergan and
Herbert Borenstein

TRC-Environmental Consultants, Inc.
125 Silas Deane Highway
Wethersfield, CT 06109

INTRODUCTION

In recent years the Electric Power Research Institute (EPRI) has engaged in an extensive experimental project, the Plume Model Validation (PMV) project, designed to assess the ability of existing atmospheric dispersion models in predicting ground level impacts from tall, buoyant sources associated with electric power generating facilities. The PMV project has four primary objectives:

- Establish by statistically rigorous procedures the accuracy and uncertainty of ground-level concentrations predicted by available plume dispersion models.

- Assess model performance over a range of meteorological and source conditions at a given site, and determine the transferability of plume model performance from one site to another.

- Create and store extensive databases on observed power plant plume behavior and make these data available to the scientific community. Three terrain types are planned for evaluation: plains, moderately complex and complex.

- Develop and validate improved plume models.

The PMV project was initially designed through a workshop convened in 1978(1). Representatives of the utility industry, government agencies and laboratories, and the technical community met to formulate a master plan for field measurement programs and analytical studies to evaluate current models and develop improved predictive tools. The approach to be followed for evaluating model performance was carefully planned prior to the acquisition of field data. Specific features of the ground-level concentration pattern produced by the power plant plumes were identified as the basis for model evaluation. Statistical procedures were then formulated for comparing observed and predicted values of these features. The PMV methodology has been documented in two project reports (2,3) expounding the protocol and statistical methodology proposed in assessing the accuracy and reliability of the models. These documents together serve as a benchmark for performing endeavors of this type.

The EPRI PMV project is unusual in that several technical organizations have been contracted for the purpose of executing the program. Four primary contractors oversee the field program, perform the model evaluation, conduct quality assurance checks and provide technical management, respectively.

It is the aim of this paper to review the experimental design developed for the flat terrain site, and to present the results of model evaluations completed during 1980 using data obtained from March through July 1980. A resource document has been prepared that details the experimental program, data and results in depth(4).

EXPERIMENTAL DESIGN

The Plain's measurements program was performed in the vicinity of the Kincaid Generating Station of the Commonwealth Edison Company (CECo) located in central Illinois. The Kincaid facility is a baseload, coal-fired, mine-mouth plant having two 660-megawatt generators that are vented into a single 187-meter (600-foot) stack with an exit diameter of 9 meters. The plant is located in a flat, rural agricultural area and is isolated from other significant sources.

The model validation database consists of two types: routine and intensive. Routine measurements are taken continuously for multiple seasons on a fixed monitoring network, while intensive measurements are designed to yield spatially dense measurements of plume parameters over shorter periods of time, using a movable grid of group samplers plus remote sensing devices to characterize the plume aloft.

Routine Measurement Program

Routine measurements during 1980 were conducted continuously from March through August at the Plains site. The experimental design for the routine measurement program provided for continuous aerometric, meteorological and stack monitoring. Most of the data have been reduced and stored as 5-minute averages. A second routine measurement period began in March 1981 and ended in June 1981.

Air Quality Measurements. Air quality measurements (principally SO_2) were made at a total of 30 stations (Figure 1). The figure shows the location of the 10 existing CECo stations (1 through 10), and the 20 stations installed by the PMV project (A through T). The design of the PMV network was based on the results of an objective monitor-siting scheme which used historical meteorological data in identifying areas of expected maximum concentration measurements. Two of the 20 stations (S and T) were mobile units, occasionally deployed to areas closer to the stack when convectively unstable conditions were expected to produce significant concentrations in those regions.

Meteorological Measurements

Meteorological parameters were measured continuously at a selected site located approximately 1 kilometer east of the Kincaid plant. A 100-meter tower collected wind directions and speeds at four vertical levels; 10 m to 50 m and 10 m to 100 m temperature differences; wind component speeds and turbulence parameters at 100 m; and 100 m dew point. A nearby 10-meter tower was used to measure 2 m to 10 m temperature differences. Other meteorological measurements made at the site included continuous routine weather observations and vertical wind/temperature sounding (0000, 0600, 1200, 1500, 1800, 2100 GMT). A doppler acoustic sounder was operated at site S, measuring wind component speeds and mixing depths up to 600 m.

Source Measurements

Continous measurements of the stack-gas emissions and stack temperature were made at the 137-meter (450-foot) level of the stack. Backup data of SO_2 emission rate, exit velocity and temperature were calculated from plant operating information, daily fuel consumption data, hourly electrical load data and daily coal analyses.

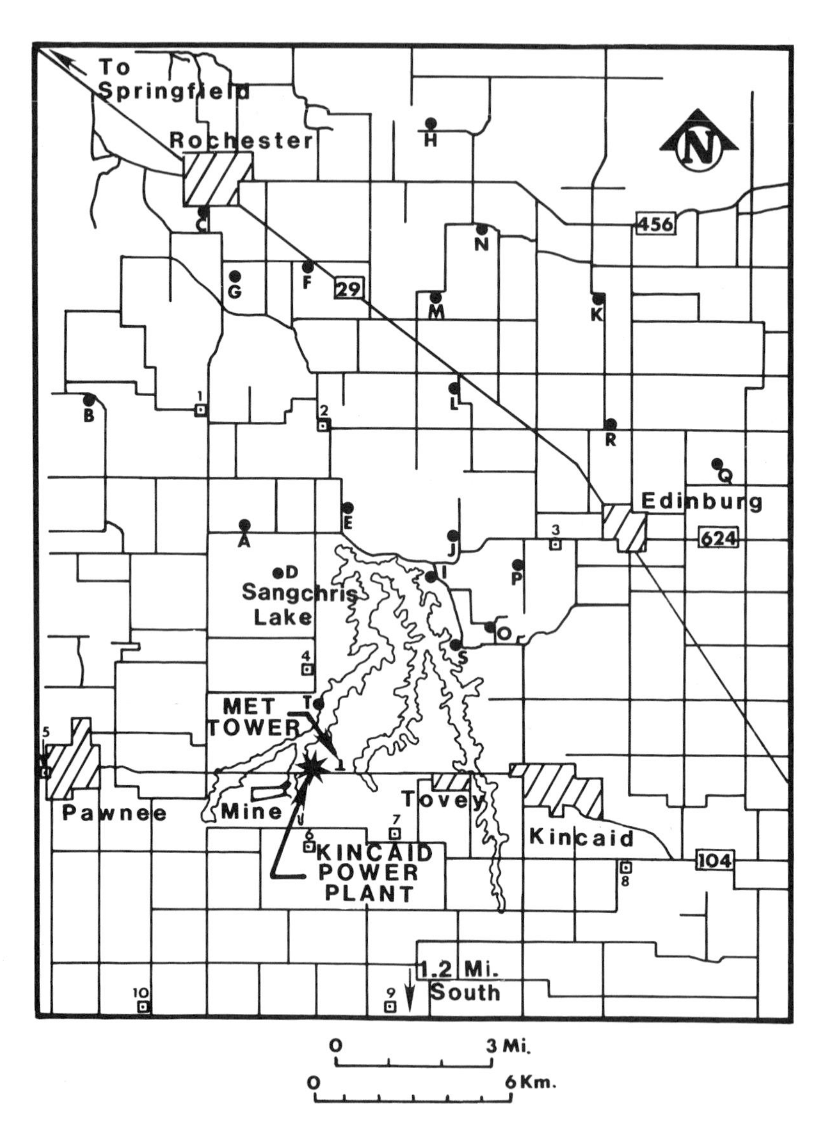

Figure 1. local area map. Lettered stations are PMV monitor locations. Numbered stations are existing CECo monitor locations.

Quality Assurance

A comprehensive quality assurance program was implemented for the PMV project. External audits performed by an independent third party on a quarterly basis were used to assess compliance of the routine field program with commonly practiced and accepted procedures. These procedures encompassed the ambient, meteorological and source monitoring activities of the routine measurement program.

INTENSIVE MEASUREMENT PROGRAM

Intensive measurements were conducted during a 3-week period in the spring (April 20 to May 10, 1980) and a 3-week period in the summer (July 9 to July 29, 1980) during which a complete range of atmospheric stability categories were covered (see Table I). A third intensive measurement period ran during a 3-week period in May 1981.

Tracer Measurements

Sulfur hexafluoride (SF_6) tracer gas was injected into the ductwork of the 187-meter stack continuously during test periods. For each tracer test, 200 sampler units were deployed on a series of five to seven arcs downwind from the source at stake

Table I. Summary of Tracer Sampling by Atmospheric Stability Condition

Turner Stability Class	Tracer-Test Hours		
	April-May 1980	July 1980	Total 1980
A	1	25	26
B	28	61	89
C	43	29	72
D	26	13	39
E	2	8	10
F	0	16	16
Morning Transition*	(13)	(20)	(33)
Evening Transition*	(6)	(12)	(18)
TOTAL	100	152	252

*Hours sampled during transitions are also included within the stability class totals.

locations and arc distances dependent on the stability category expected. The total grid network consisted of over 1,500 tracer sampling locations which were aligned on nearly concentric circles with radii of 0.5, 1, 2, 3, 4, 5, 10, 15, 20, 30, 40 and 50 km. Using the existing road network, the arcs were circumferentially staked at intervals ranging from 2° to 8° to mark all potential sampling locations. The sampling array selected for a particular test remained fixed for the duration of the test (6 to 9 hours).

Tracer sampling was performed with sequential samplers each designed to accrue 1-hour integrated samplers of SF_6 from 180 independent samples (consisting of a 2-second pulse every 20 seconds). A determination of the SF_6 samplers was carried out on electron-capture gas chromatographs which are sensitive to concentrations as low as 2-3 ppt.

Additional Plume Concentration Measurements

Remote Sensing. Two ground-based lidars were used for the remote sensing of plume height and geometry in the zone of transitional plume rise. A Differential Absorbtion Lidar (DIAL) employing prescribed wavelengths for the detection of SO_2 was located 2-3 km downwind of the stack, and a particle lidar was located an additional 2-3 km downwind. An airborne particle lidar was deployed to a greater distance downwind (about 10 km) for measurement of plume height and geometry beyond the area where final plume rise had been reached. Figure 2 depicts the orientation of the three lidar systems with respect to the plume and each other.

In-Situ Plume Measurements. In-situ plume chemistry data were measured for 33 hours for each intensive period with continuous analyzers for SO_2, NO/NO_x, O_3, and SF_6, which were housed in a Cessna 206 aircraft. An integrating nephelometer, temperature sensor and dew-point sensor were also used. The flight path adopted for making these measurements was a sawtooth pattern, illustrated in Figure 3. Each in-plume traverse was made as close to plume centerline height as possible for the entire 50 km net downwind distance. Each flight was terminated with a spiral descent intended for the measurement of the vertical mixing-layer structure.

Meteorological Measurements

Routine meteorological measurements continued through the intensive periods with some minor modifications. The Gill UVW

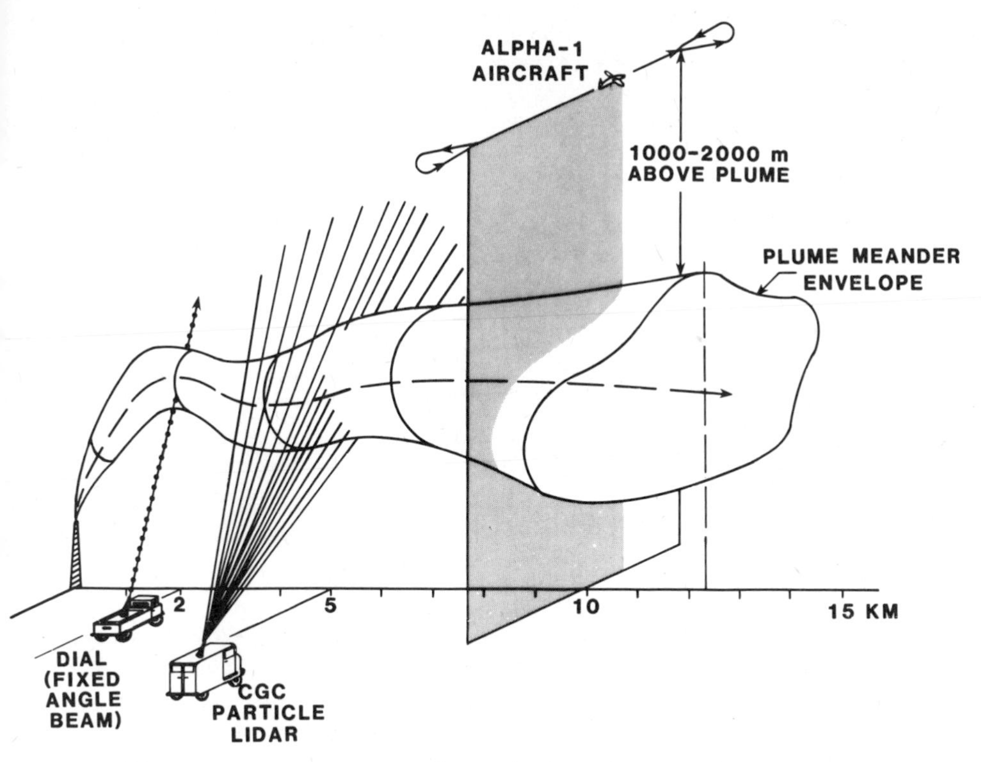

Figure 2. Remote sending of the plume with Lidars.

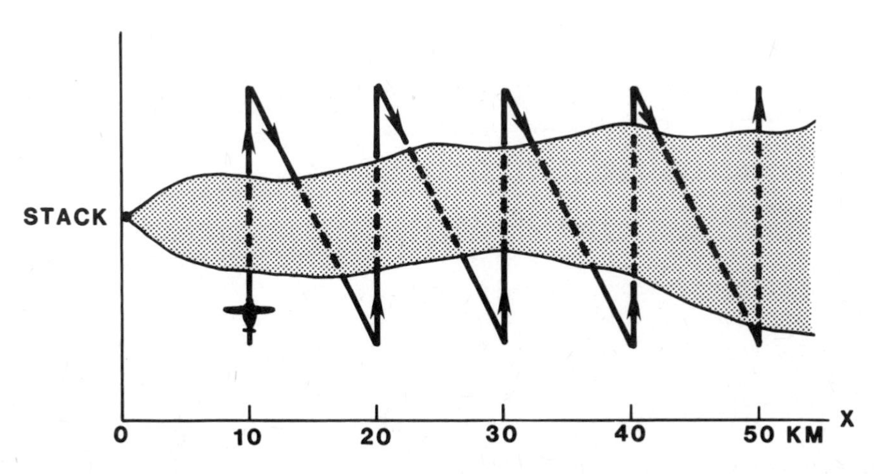

Figure 3. Flight pattern for in-plume sampling with the aerometric aircraft.

anemometer polled wind components at 5 hertz, which was increased from the routine 10-second scan rate. Additionally, T-sonde releases were made hourly during tracer release periods and a second doppler acoustic sounder was used at Station H, located about 20 km north of the plant (Figure 1).

Quality Assurance

With respect to the tracer facet of the intensive program, emissions, ambient sampling, and determination of SF_6 tracer concentrations were audited through third party quality assurance efforts. This was also true for in-plume measurement instrumentation and procedures. Although no universally accepted quality assurance procedure has yet been promulgated for remote sensing lidar equipment, the PMV project did institute third-party system audits of these instruments.

MODEL DESCRIPTIONS AND INPUT

For this presentation, results are reported for the three Gaussian models selected for evaluation:

- CRSTER - Selection was based on regulatory status as an approved model in EPA Guidelines. CRSTER is commonly used in the licensing applications for isolated power plants, even though it does not permit user flexibility in assignment of receptor locations.

- Multimax - Selected as a model that behaves similarly to CRSTER, but with the feature of allowing for user-specified receptor locations. Multimax has been modified [CRSTER Equivalent Model (CEQM)] to imitate the prediction of ground-level concentrations by CRSTER, but has retained the feature for permitting the assignment of receptors that are coincident with sites corresponding to actual field monitoring locations.

- TEM (Texas Episodic Model) - Characterized by a unique feature which provides for the adjustment of the horizontal dispersion parameter for various observation averaging times. TEM was included in the PMV project to investigate the effects of this feature on Gaussian model performance.

Input consist of source information and local meteorology. For source information, the weekly average and tracer duration average stack temperature and velocity were input to CRSTER/CEQM during routine and intensive measurement model executions, respectively, while TEM accepted hourly averages of these variables; hourly average SO_2 emission rates were used for all models. With respect to local meteorology, three separate sets of meteorological observations were used to operate the models. One was National Weather Service (NWS) data and the other two were acquired from on-site measurement data. Table II describes the three sets of meteorological data assembled for the PMV project.

MODEL EVALUATION METHODS

The model evaluation approach has been designed to provide an objective comparison of predicted and observed ground-level concentrations to assess the degree of uncertainty and bias in model predictions. The statistical methodology developed consists of a battery of parametric and nonparametric tests which provide estimated probabilities or confidence levels on model performance measures. A brief summary of these performance measures is presented below.

Tracer Experiments

Four performance measures were selected to characterize the spatially comprehensive observed and predicted concentration arrays:

- Magnitude of the maximum (highest) concentration by hour.

Table II. Summary of Meteorological Input

	National Weather Service	On Site 10 m	On Site 10 m ΔT
Air temperature	Springfield	10 m	100 m
Wind speed and direction	Springfield	10 m	100 m
Mixing depth	Peoria (CRSTER interpolation)	T-Sonde (CRSTER interpolation)	T-Sonde(00,06, 12,15,18,21, GMT, linear interpolation)
Stability Class	Turner Method(5)	Turner Method	10-50 m ΔT(6)

- Magnitude of the highest concentration by hour at a given arc distance.

- Location (distance and direction) of the highest concentration by hour.

- Plume width, that is, crosswind standard deviation of ground-level concentrations at a given distance from the source.

Routine SO_2 Measurements

The routine monitoring network was designed to provide a long-term record of power plant impacts with considerably less spatial density than the tracer experiments. The following four comparative measures were considered:

- Compare the 30 highest observed values with the 30 highest predicted values, regardless of location or time. Essentially, this entails a detailed comparison of the upper ends of the cumulative frequency distributions of observed and predicted concentrations.

- Compare the 30 highest observed values with the corresponding predicted value at the same station and time, and compare the 30 highest predicted values with the corresponding observed values.

- Compare the magnitude of the maximum (highest) concentration by hour.

- Compare the set of all observed values with the set of all predicted values for all hours and stations.

Recognizing the limits of detection of the SO_2 instrument, all data points for which the observed and predicted concentration values were below 15 ppb SO_2 were deleted from the analysis.

INTERIM RESULTS

The following results were derived from part of the data collected in 1980. Approximately 2,000 hourly events were available from SO_2 data and 140 events from tracer data. Any questionable or incomplete data blocks were excluded from the analysis, pending correction and/or completion. An augmented database will become available late in 1981 when the third and final phase of Plains site measurements is completed (that is, spring 1981 data).

Maximum Concentrations

Table III lists the maximum observed and predicted tracer and SO_2 concentrations for the two models and the three meteorological input sets.

The highest 1-hour concentrations of sulfur dioxide predicted by CEQM on the routine network were consistently greater than the highest observed value of 391 ppb. TEM predicted lower maximum 1-hour SO_2 concentrations than CEQM. For two of the three input sets, the highest value predicted by TEM was below 391 ppb. The highest predicted 1-hour concentrations of SF_6 on the tracer network were generally lower than the highest observed values for both models. When results for the highest

Table III. Comparison of Highest Observed and Highest Predicted 1-Hour Concentration Unmatched in Time and Space

	Sulfur Dioxide Data (ppb)		
	NWS Meteorology	On Site NWS Equivalent	On Site 100 m + ΔT
Observed High	391	391	391
CEQM Predicted High	475	397	558
TEM Predicted High	421	321	319

	Tracer Data (ppt)			
	First Intensive			Second Intensive
	NWS Meteorology	On Site NWS Equivalent	On Site 100 m + ΔT	On Site NWS Equivalent
Observed High	566	265	265	948
CEQM Predicted High	493	468	432	809
TEM Predicted High	127	117	176	671

concentration are compared for input sets, or for tracer and SO_2, no consistent pattern is evident for CEQM. TEM, however, is consistently underpredicted.

Cumulative frequency distributions of maximum hourly predicted and observed SO_2 concentrations in the routine network are presented in Figure 4 for CEQM, using NWS meteorology. Distinct differences between the observed and predicted distributions are evident. At lower concentration values (between 10 and 50 ppb), more events were observed than were predicted; by contrast, concentrations above 250 ppb were predicted more often than observed. The cumulative frequency distribution for maximum hourly predicted and observed SF_6 concentrations on the tracer grid are presented in Figure 5 for CEQM, using NWS meteorology for the spring 1980 tracer experiments. The resulting patterns differ from those in Figure 6 and indicate a persistent tendency toward underprediction.

Model skill for predicting the magnitude of plume impacts hour-by-hour was limited. Table IV summarizes the statistics calculated for the comparison of observed and predicted hourly values for the spring 1980 tracer measurement program. CEQM demonstrated predicted maximum concentration values within a factor of 2 of the observed hourly values at least 50 percent of the time. TEM tended toward frequent underprediction. Only 23 percent of the predicted maximum concentration values were within a factor of 2 of the observed values. Correlation coefficients for observed versus predicted maximum concentrations were 0.19 to 0.29, again indicating significant scatter. Analogous results for the routine network (Table V) indicate even less agreement between hour-by-hour observed and predicted values. This result is expected since the sparse routine network may not resolve either the maximum observed or maximum predicted concentration for a given hour.

The tracer measurements indicate the magnitude of concentrations resulting from power plant impact at distances ranging from 1 to 50 km from the source. Figure 7 illustrates the maximum normalized tracer concentration observed at each sampling arc distance, for the spring and summer 1980 tracer experiments. (Because the tracer emission rate varied from one test to another, measured SF_6 concentrations were normalized by the emission rate.) At distances of between 3 and 20 km from the source, observed maximum values were similar in magnitude. At distances closer than 3 km, or greater than 20 km, lower maximum values were obtained.

While the Kincaid plant is considered a baseload plant, substantial variations in the power plant load and the SO_2 emission rate were recorded during 1980 routine measurements. As

Table IV. Comparison of Predicted and Observed Highest Tracer Concentration for Three Models Using Data Measured in the Spring and NWS Meteorological Observations

		CEQM (N=58)		TEM (N=52)		MSDM (N=51)	
		Obs	Pred	Obs	Pred	Obs	Pred
Frequency Distribution Comparison	Maximum (ppt)	566	493	566	127	566	244
	Upper Quartile (ppt)	181	125	172	54	172	106
	Median (ppt)	132	98	127	37	127	94
Prediction Skill	Correlation Coefficient	0.29		0.19		0.20	
	% Predicted >2 x Observed	12		10		14	
	% Predicted within ±2 x Observed	50		23		57	
	% <1/2 x Observed	38		67		29	
Bias	Probability that distributions are the same	<.01		<.01		<.01	

Table V. Prediction Skill Indicators for Highest SO_2 Concentration Values

	Routine SO_2 Results*	
	CEQM*	TEM*
Correlation Coefficient	0.19	-.03
Percent Agreement within a Factor of 2	18	18

*Model predictions based on NWS meteorology.

the power plant load decreases, plume buoyancy is also expected to decrease. The effect of variations in plant load upon SO_2

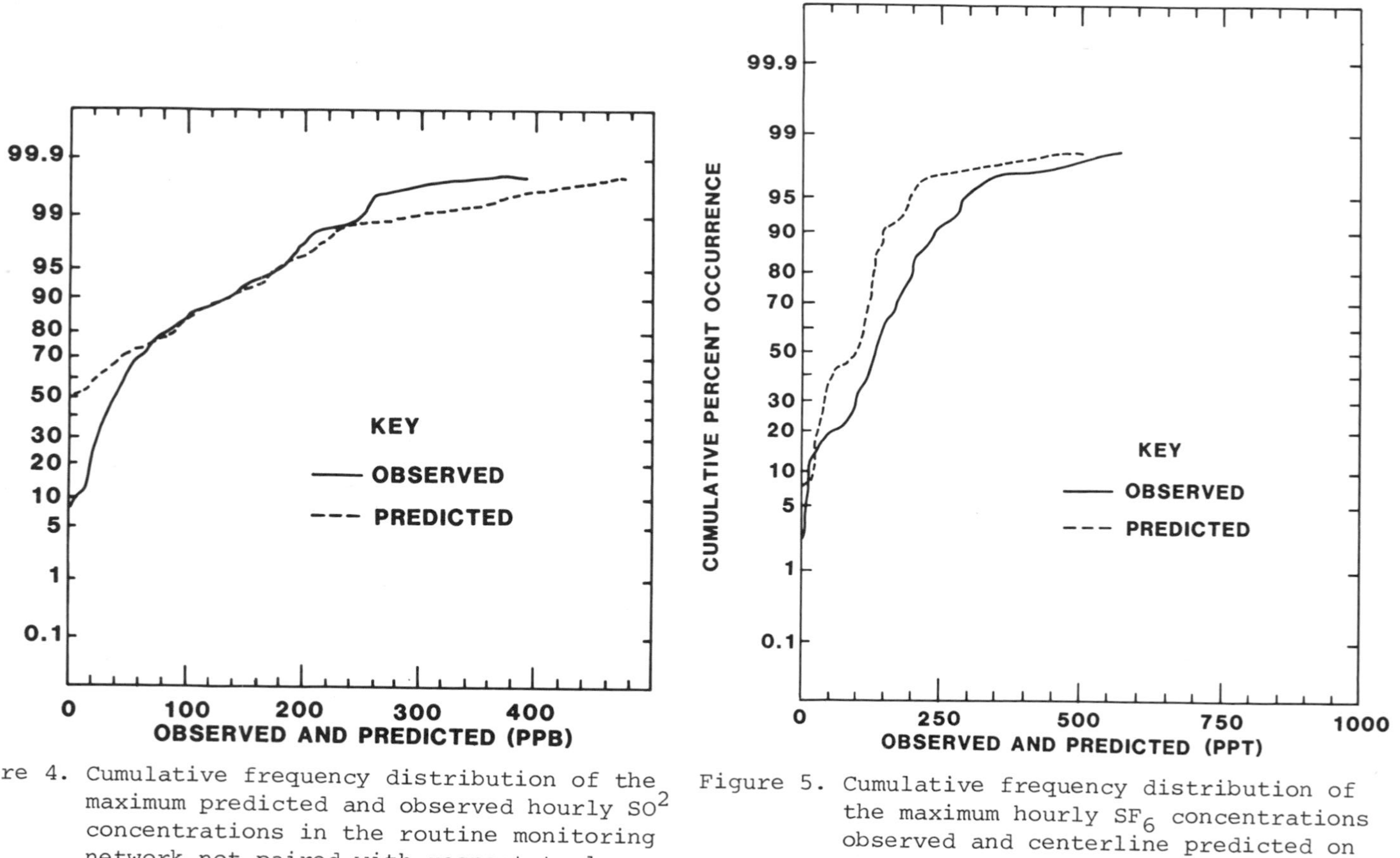

Figure 4. Cumulative frequency distribution of the maximum predicted and observed hourly SO^2 concentrations in the routine monitoring network not paired with respect to location. The data are for 482 hours during the period 4/21/80 to 7/28/80. The model used in CEQM with NWS meterology.

Figure 5. Cumulative frequency distribution of the maximum hourly SF_6 concentrations observed and centerline predicted on the tracer monitoring network. The data are for 58 hours during the period 4/24/80 to 5/10/80. The model used is CEQM with NWS meterology.

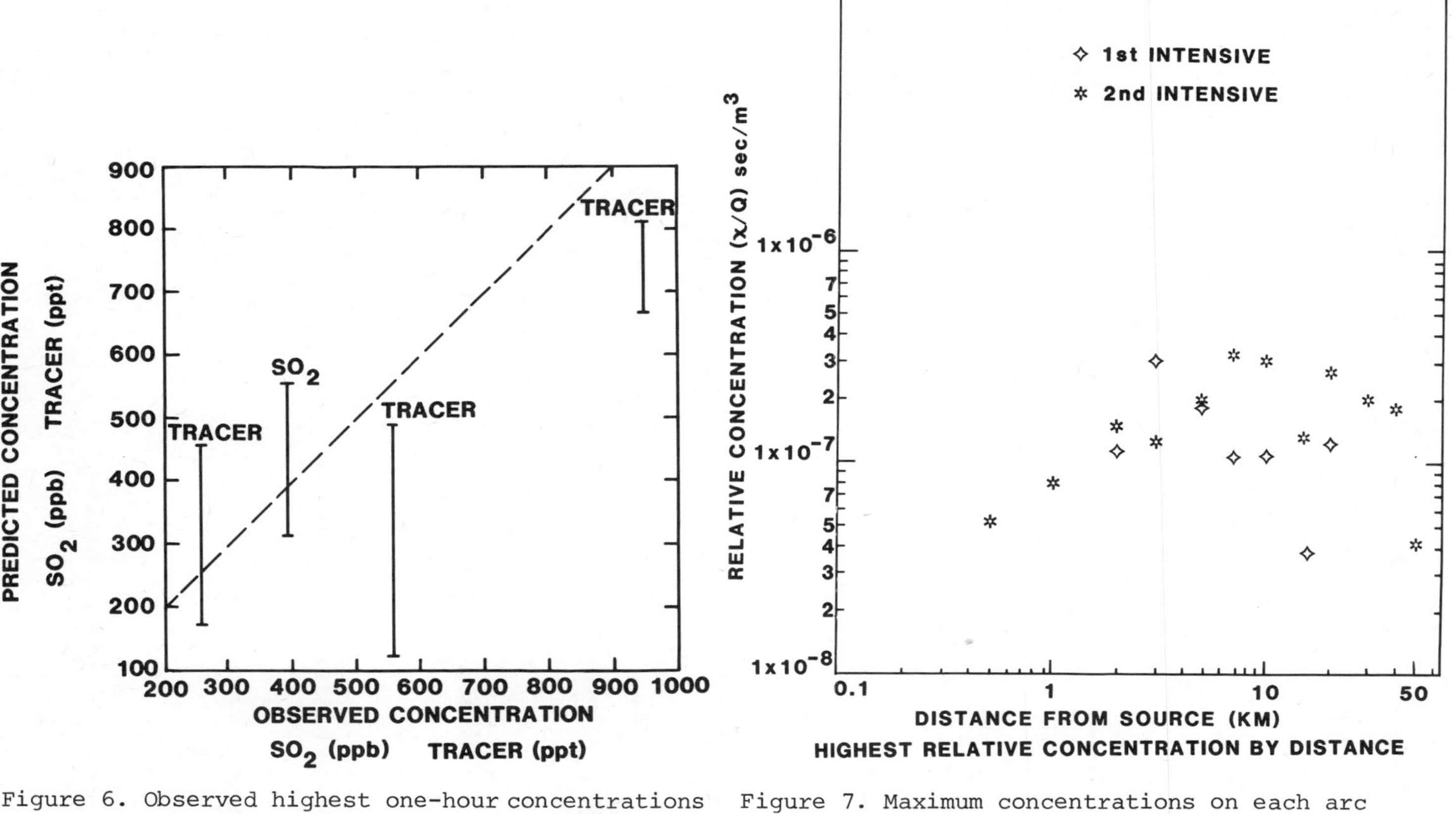

Figure 6. Observed highest one-hour concentrations with ranges of maximum predicted concentrations from CEQM and TEM for differing meteorological observing positions. (See Table 5-1).

Figure 7. Maximum concentrations on each arc normalized by emission rate for spring and summer intensive measurement periods during 1980.

impact is illustrated in Figure 8, with the highest normalized SO_2 concentrations plotted against the SO_2 emission rate. As expected, normalized concentration decreases with increasing emission rate (that is, increasing plume buoyancy). Interestingly, at Kincaid the magnitude of (unnormalized) observed SO_2 concentrations was relatively independent of the plant load. The effects of increased emissions and increased buoyancy tended to offset one another.

The influence of input measured wind direction on the error of predicted direction of plume travel is illustrated in Figures 9 and 10. Both bias and scatter were larger for the low level NWS wind direction than for the 100-meter wind direction measured at the site.

Observed plume widths were more than twice the plume width predicted by CEQM when the standard EPA specification of stability class was used.

CONCLUSIONS

This paper is an abridged synopsis of the first interim and partial results of EPRI's continuing PMV project, and is intended as a demonstration of our approach to defining the limits of error and uncertainty in the current operational uses of conventional plume models. The conclusions presented about plume model performance are tentative, pending final analysis with the complete measurement database from the Kincaid site. Diagnostic validation analyses are also contemplated using the remote sensing measurement data to compare observed and predicted plume behavior aloft.

The following quantitative evaluations of Gaussian model predictions have been observed:

- Any attempt to use plume models for estimates of short-term exposure at fixed receptors in a case-by-case basis may yield unreliable predictions. Correlation coefficients between maximum observed and predicted concentrations were below 0.3, indicating little skill.

- Model predictions of the cumulative frequency distributions of maximum 1-hour ground-level concentrations failed to reproduce the observed distributions according to statistical tests of bias and goodness-of-fit. In general, the models tended to underpredict the median, upper quartile and the highest of the maximum concentrations when tracer data were used.

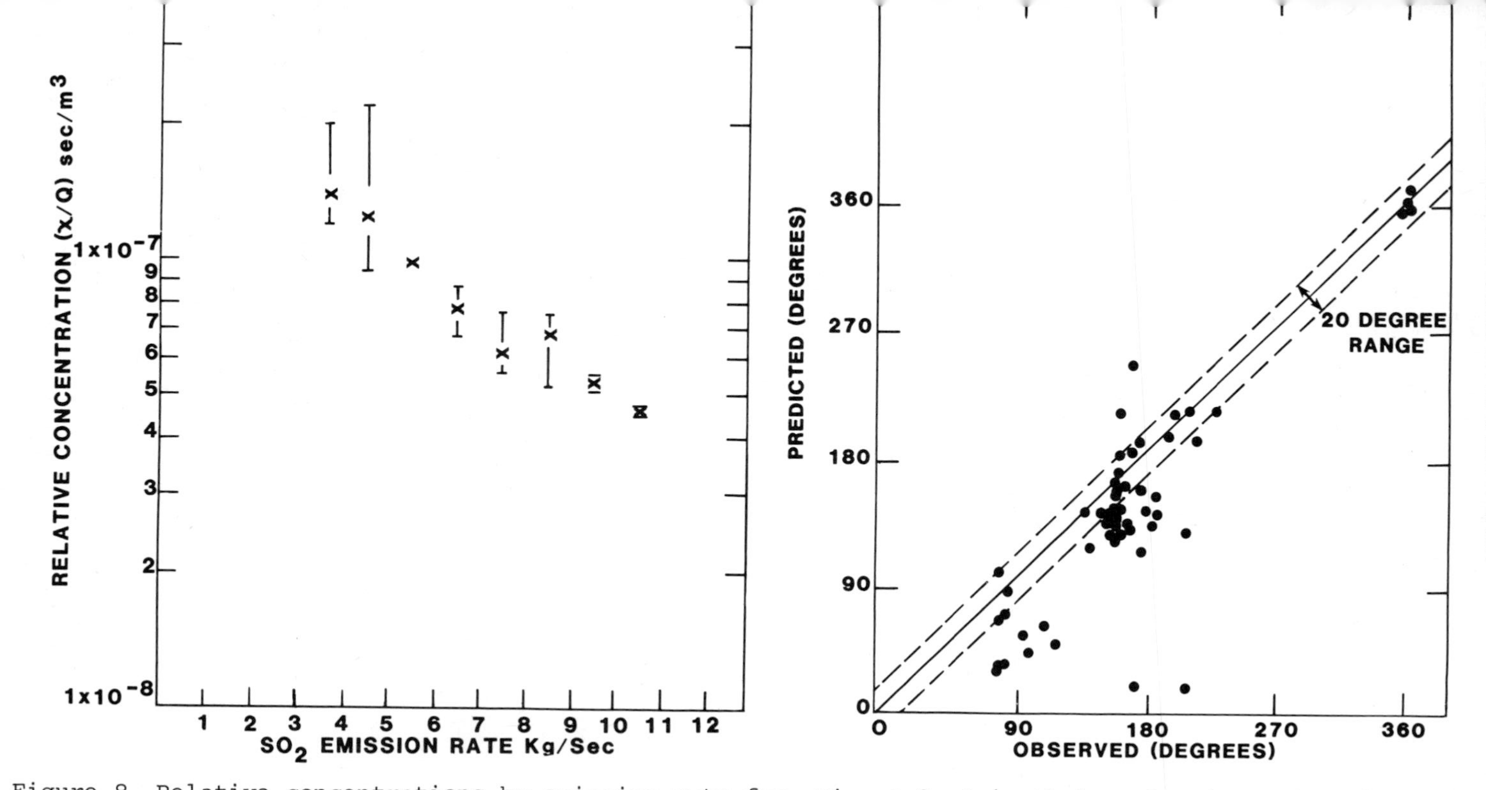

Figure 8. Relative concentrations by emission rate for highest thirty observed concentrations with total range indicated by vertical lines and arithmetic average by x. Routine SO_2 measurements at Kincaid site 4/21/80 - 7/27/80.

Figure 9. Azimuthal angles to maximum hourly observed and centerline predicted SF_6 concentrations on the tracer monitoring network. The data are for 58 hours during the period 4/24/80 to 5/10/80. The model used is CEQM with NWS meterology.

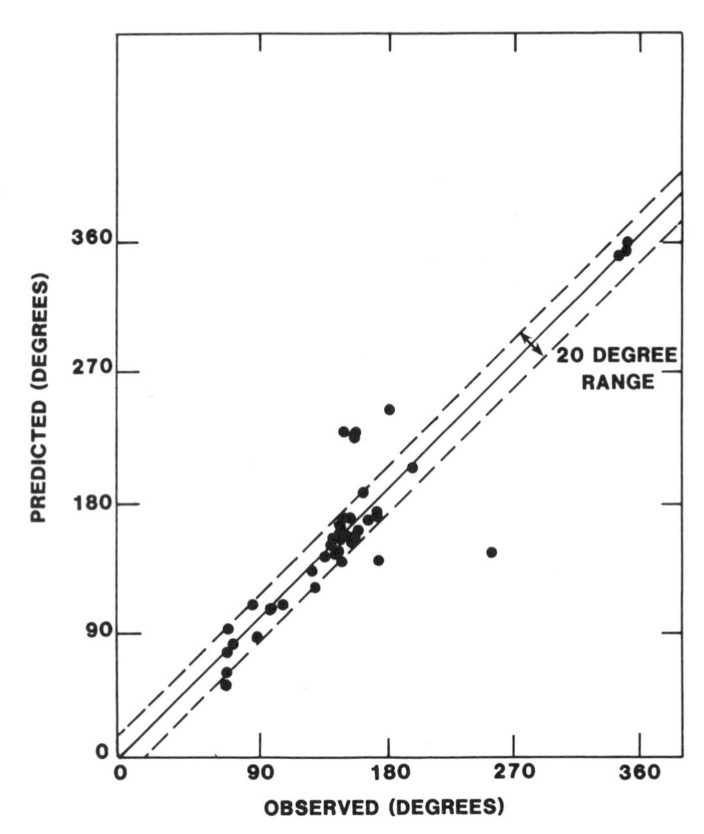

Figure 10. Azimuthal angles to maximum hourly observed and centerli predicted SF_6 concentrations on the tracer monitoring network. The data are for 48 hours during the period 4/24/80 to 5/10/80. The model used is CEQM with 100m wind data.

- The models failed to adequately replicate the plume footprint as characterized by the location (distance and direction to maximum) and width of the ground level plume. Plume axis location was often in error by more than 20 degrees of azimuth, and plume width was frequently underpredicted by a factor of 2 or more.

REFERENCES

1. "Plume Model Validation", Electric Power Research Institute, EA-917-SY, Palo Alto, California, (October 1978).

2. "Protocol for Plume Model Validation", Electric Power Research Institute, EA-1638, Palo Alto, California, (November 1980).

3. "Validation of Plume Models, Statistical Methods and Criteria", Electric Power Research Institute, EA-1673-SY, Palo Alto, California, (January 1981).

4. "Preliminary Results from the EPRI Plume Model Validation Project - Plains Site", Electric Power Research Institute, EA. 1788, Palo Alto, California, (January 1981).

5. D.B. Turner, "Workbook of Atmospheric Dispersion Estimates", Office of Air Programs, United States Environmental Protection Agency, No. AP-26, Research Triangle Park, North Carolina, (January 1974).

6. Proposed Revision 1 to Regulatory Guide 1.23, Meteorological Programs in Support of Nuclear Power Plants, Nuclear Regulatory Commission, Office of Standards Development, Washington, D.C., (September 1980).

DISCUSSION

G.L. POWELL

Figures shown in the presentation are different from these included in the preprint volume. Will both sets be included in the final published version of the paper?

R.J. LONDERGAN

Both are given in the report referenced in the preprint.

W.C. McBRIDE

Since the SF_6 and SO_2 data showed a reasonable correlation, why did the model overpredict for SO_2 and underpredict for SF_6 ? Was the model used for SO_2 and SF_6 data for the same periods ?

R.J. LONDERGAN The SO_2 measurements were taken continuously for a period of 17 weeks, but th tracer data used for validation analyses represen only 140 hours. Also, the monitoring networks do not coincide. Tracer measurements were made at distances where no SO_2 monitors were located. Either of these factors could lead to differences between SO_2 and tracer results, but we do not at this time know what produced the differences which were found.

P.G. WILLIAMS Were the Pasquill-Gifford sigma values corrected for averaging time before comparison with the measured plume width.

R.J. LONDERGAN For the CEQN model, no averaging time adjustment to the Pasquill-Gifford σ_y coefficients is made. Analysis of the TEM model, which does adjust the Pasquill-Gifford σ_y's, showed better agreement between observed and predicted plume widths.

COMPARISON OF SULFUR DIOXIDE ESTIMATES FROM THE MODEL RAM WITH ST. LOUIS RAPS MEASUREMENTS

D. Bruce Turner ★
John S. Irwin★
Meteorology and Assessment Division
U.S. Environmental Protection Agency
Research Triangle Park, NC 27711

INTRODUCTION

Since the passage of the Clean Air Act Amendments in August 1977, air quality simulation models have seen increased use for projecting air pollutant impacts of proposed sources for comparison with National Ambient Air Quality Standards (NAAQS) and Prevention of Significant Deterioration (PSD) increments. The U.S. Environmental Protection Agency has issued guidelines on air quality modeling (U.S. Environmental Protection Agency, 1978).

To properly utilize modeling in addressing air quality standards, the accuracy of modeling should be known. In order to accumulate a history of model performance, all available data bases suitable formodel evaluation should be used and the results of the evaluations made available to model users.

As part of the Regional Air Pollution Study (RAPS) conducted in St. Louis (Schiermeier, 1978; Strothmann and Schiermeier, 1979), ambient sulfur dioxide levels were routinely measured at 13 monitoring stations. Seven of these sites are within 13 km of downtown St. Louis, five are within 13 to 26 km, and the remaining site is about 50 km north of the downtown area. Also, as part of the RAPS, emission inventories were compiled including the pollutant SO_2. The sulfur dioxide emissions inventory allows calculation of hour-by-hour emissions. The inventory compiled for the calender year 1976 included 208 point sources and 1989 area sources.

★ On assignment from National Oceanic and Atmospheric Administration, U.S. Department of Commerce

To use the RAPS data in model evaluation, the RAM model (Novak and Turner, 1976; Turner and Novak, 1978 a and b) was modified to accomodate the large number of sources in the inventory and to utilize the emissions on an hourly basis. The model was executed for the calendar year 1976, producing estimated hourly sulfur dioxide concentrations at each of 13 receptors (corresponding to the position of each monitoring station) for the year. Data for each hour of the year-consisting of concentrations estimated from the model RAM for each of 13 locations, concentrations measured at each of the 13 locations, and additional information such as representative hourly meteorology and point and area emission totals - were furnished to SRI International for use in their evaluation of RAM for EPA. Their work has resulted in the contract report, Evaluation of the RAM using the RAPS Data Base (Ruff, 1981). Four volumes related to evaluation procedures had resulted from previous work under this contract (Ruff, 1980a; Javitz and Ruff, 1980; Ruff, 1980b; and Ruff, 1980c).

The information reported in this document is based upon the same data set furnished to SRI International and supplements the SRI evaluation. This paper examines extreme value statistics (highest and second-highest concentrations) from this data base.

The RAM model examined in this report has been the subject of other evaluations (Guldberg and Kern, 1978; Kummler, et al. 1979; Ellis and Liu, 1981; and Hodanbosi and Peters, 1981).

PRESENTATION OF DATA

In view of the way air quality simulation models are used within the regulatory framework, such as estimating the second-highest 3-hour concentration once a year or second-highest 24-hour concentration once a year, it seemed desirable to examine such "extreme event" concentrations for this 1-year period of record.

The highest, second-highest, fifth-highest and tenth-highest SO_2 concentrations, estimated and measured for each of the 13 locations are given in Tables 1 through 3. The principal averaging times considered are 1-hour, 3-hour, and 24-hour periods. In addition, Table 4 gives annual mean concentrations. Although the model yielded estimates of concentrations for each hour of the year, there were periods when the air sampling equipment did not yield valid measurements. The percent of valid data capture over the year ranged from 69% for station 101 to 86% for station116. Therefore in determining the annual average or selecting the highest hourly concentrations, the estimated concentration was discarded whenever there was no valid measurement at the corresponding time. In determining 3-hour and 24-hour averages, the average concentration was determined by averaging the concentrations for the hours of

Table 1. One-hour Sulfur Dioxide Concentrations (μgm^{-3})

STATION		HIGHEST	2nd HIGHEST	5th HIGHEST	10th HIGHEST
101	MEAS	487	482	476	385
	EST	1101	859	743	664
103	MEAS	1490	1095	605	442
	EST	622	519	407	393
104	MEAS	2533	2499	2208	2103
	EST	1292	1250	867	753
105	MEAS	487	479	400	369
	EST	842	781	737	571
106	MEAS	2025	1034	822	699
	EST	718	596	547	495
108	MEAS	487	434	377	332
	EST	856	580	466	427
113	MEAS	455	421	403	379
	EST	565	562	440	349
114	MEAS	1006	937	497	413
	EST	1217	869	750	585
115	MEAS	521	518	487	408
	EST	1675	1428	1084	1054
116	MEAS	673	584	542	413
	EST	548	546	411	377
120	MEAS	503	413	358	317
	EST	449	427	399	332
121	MEAS	560	523	314	303
	EST	518	512	448	366
122	MEAS	526	503	390	301
	EST	406	374	321	278

Table 2. Three-hour Sulfur Dioxide Concentrations (μgm^{-3})

STATION		HIGHEST	2nd HIGHEST	5th HIGHEST	10th HIGHEST
101	MEAS	330	326	265	230
	EST	586	523	469	447
103	MEAS	482	432	326	247
	EST	343	321	262	223
104	MEAS	2298	2145	1548	1421
	EST	691	609	522	378
105	MEAS	412	302	290	265
	EST	574	445	377	315
106	MEAS	1127	703	579	357
	EST	542	426	331	272
108	MEAS	316	294	252	224
	EST	330	325	249	219
113	MEAS	385	335	300	257
	EST	312	247	226	183
114	MEAS	619	572	303	258
	EST	425	425	376	280
115	MEAS	339	300	237	204
	EST	675	622	569	450
116	MEAS	559	484	345	231
	EST	284	264	187	163
120	MEAS	258	227	189	166
	EST	236	222	164	141
121	MEAS	288	259	201	178
	EST	288	223	194	176
122	MEAS	378	346	240	188
	EST	245	174	157	142

Table 3. 24-hour Sulfur Dioxide Concentrations (μgm^{-3})

STATION		HIGHEST	2nd HIGHEST	5th HIGHEST	10th HIGHEST
101	MEAS	160	155	121	101
	EST	336	299	236	202
103	MEAS	299	166	84	72
	EST	129	127	113	91
104	MEAS	1185	1179	755	557
	EST	218	213	158	138
105	MEAS	225	160	119	97
	EST	211	193	154	127
106	MEAS	443	414	165	116
	EST	171	170	130	105
108	MEAS	132	120	98	79
	EST	149	145	92	79
113	MEAS	139	124	106	84
	EST	121	83	74	61
114	MEAS	180	146	96	89
	EST	121	102	81	58
115	MEAS	122	103	89	68
	EST	242	238	140	91
116	MEAS	298	96	73	56
	EST	88	77	61	52
120	MEAS	91	85	74	58
	EST	62	56	49	43
121	MEAS	108	104	95	76
	EST	104	70	68	52
122	MEAS	128	127	85	74
	EST	80	62	61	53

Table 4. Annual Sulfur Dioxide Concentrations (μgm^{-3})

STATION	HOURS VALID DATA	UTM COORDINATES EAST	NORTH	MEAS	EST	EST/MEAS
INNER CITY[a]						
101	6088	744.18	4279.86	36.8	85.5	2.32
103	7369	747.59	4282.47	26.2	38.1	1.45
104	6688	747.31	4277.30	86.1	60.4	0.70
105	7206	743.71	4276.45	34.9	44.2	1.27
106	6433	738.66	4277.57	38.7	38.9	1.01
OUTER CITY[b]						
108	7284	748.41	4291.10	28.5	30.7	1.08
113	6780	737.74	4289.82	27.2	23.4	0.86
SUBURB[c]						
114	7083	744.32	4297.46	23.3	24.3	1.04
115	7165	757.11	4297.80	21.7	27.6	1.27
116	7571	762.78	4290.08	17.4	16.5	0.95
120	7292	723.08	4285.91	19.0	13.9	0.73
121	6913	732.41	4302.38	23.3	16.3	0.70
RURAL[d]						
122	6741	741.63	4329.22	18.9	12.6	0.67

[a] LESS THAN 6.5 km FROM STATION 101.

[b] FROM 6.5 TO 12.5 km FROM STATION 101.

[c] FROM 12.5 TO 25.5 km FROM STATION 101.

[d] GREATER THAN 25.5 km FROM STATION 101.

valid measurement. However, for the extremes reported in Tables 2 and 3, it was required that there be 3 hours and 22 hours (90 %) of valid data for the 3-hour and 24-hour periods, respectively. Where 8-hour concentrations are reported (Tables 5, 6 and 7), eight valid measurements were required.

DISCUSSION

Since regulatory use of models centers on estimates of second-highest concentrations, the measured and estimated second-highest concentrations from Tables 1 through 3 and similar information for the 8-hour averaging time were used fo form ratios of estimated to measured SO_2. These are given for each monitoring station in Table 5. At each station there is a tendency toward similar performance from one averaging time to another; that is, at locations wher concentrations tend to be overestimated, the tendency is for overestimation for most averaging times. Of the 13 sites compared, 4 overestimated and 5 underestimated for all 4 of the averaging periods presented. The extreme overestimate occurs at station 115 with a second-high 1-hour estimate 2.76 times the measured concen-

Table 5. Ratio of Estimates to Measurements for Second Highest

STATION	1-HR	3-HR	8-HR	24-HR
101	1.78	1.60	2.06	1.93
103	0.47	0.74	1.27	0.77
104	0.50	0.28	0.21	0.18
105	1.63	1.47	1.35	1.21
106	0.58	0.61	0.48	0.41
108	1.34	1.11	1.02	1.21
113	1.34	0.74	0.65	0.67
114	0.93	0.74	1.20	0.70
115	2.76	2.07	2.23	2.31
116	0.94	0.55	0.55	0.80
120	1.03	0.98	0.64	0.66
121	0.98	0.86	0.90	0.67
122	0.74	0.50	0.60	0.49

tration. The extreme underestimate occurs at station 104 with a second-high 24-hour estimate only 0.18 of the measured concentration; however, a sudden increase in measured concentrations at 104 during November and December 1976 was noted (Ruff, 1981), and it is suspected that a nearby source increased emissions during this period.

The spatial pattern was examined by plotting the ratios on maps by station locations. An example is shown for the 24-hour averaging time in Figure 1. The stations with over- or underestimates are not clustered in any specific part of the urban area. The cause of the over- or under-estimation is not kwown. It could be any or all of : the model does not well represent the physics of the atmosphere; the meteorological input is not appropriately representative of that producing the second-highest concentrations; there are inaccuracies in the estimated emissions.

Frequently, in using models to assist in regulatory matters, interest centers on the maximum concentrations. Table 6 provides network maxima for highest and second-highest concentrations for both measured and estimated for various averaging times. Because of the previously noted difficulty with station 104, the maxima for the network without station 104 is also given. Although comparing highest concentrations for monthly and quarterly averaging times would seem erroneous, the higher quarterly concentrations result from the relaxation of the criteria of percent valid data required from 90 % for monthly to only 80 % for quarters, and from the inclusion of several high concentration months in the quarterly data that are not included in the monthly data.

An estimate of air quality impacts from proposed sources to satisfy regulatory requirements would consider more than 13 receptor positions. Whether the experience reported here for the 13 receptors is representative of what might occur with a larger array of receptors is not known.

The ratios of these maximum second-highest measurements to estimates for the whole network and the network without station 104 are given in Table 7. Considering all stations, the model gives second-highest maxima about one-quarter the concentrations of the measurements for the 3-hour to 24-hour averaging times (primarily because of high concentrations at station 104). If station 104 is not considered, the ratios of second-highest estimates to second-highest measurements becomes on the order of 0.7 to 0.9 for the 3-hour to 24-hour averaging times. Note that maximum second-highest 1-hour concentrations are overestimated by 30 % and the annual maximum is overestimated by more than a factor of two. Also included in Table 7 is a summary of the number of stations with second-highest estimates within various categories of the measurements.

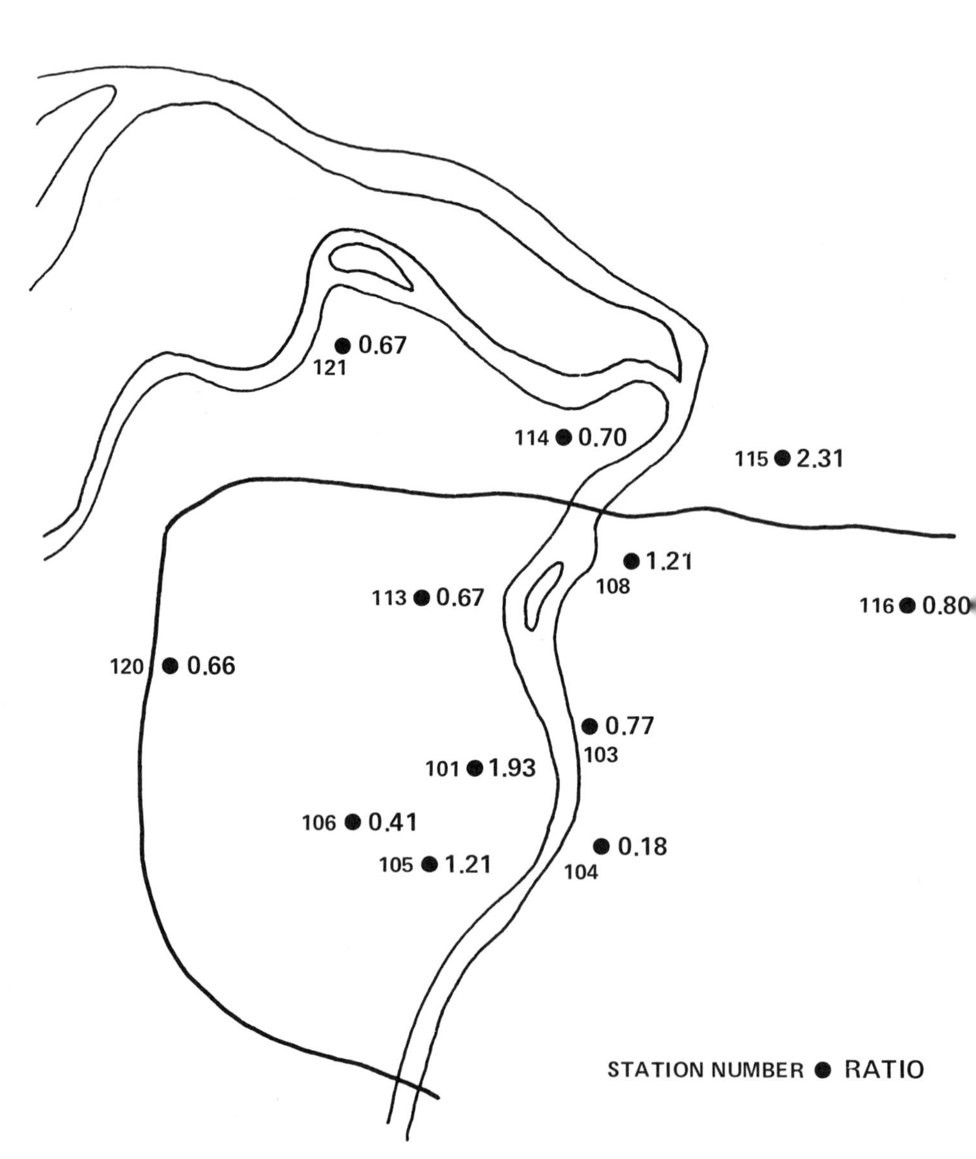

Figure 1. Ratio of estimated to measured second-highest 24-hr SO_2 concentrations

Table 6. Network Maximum Highest and Second Highest Concentrations (μgm^{-3})

NETWORK MAXIMA

	HIGHEST		2nd HIGHEST	
	MEAS (STATION)	EST (STATION)	MEAS (STATION)	EST (STATION)
1-HR	2533 (104)	1675 (115)	2499 (104)	1428 (115)
3-HR	2298 (104)	691 (104)	2145 (104)	622 (115)
8-HR	1799 (104)	620 (104)	1513 (104)	424 (101)
24-HR[a]	1185 (104)	336 (101)	1179 (104)	299 (101)
WEEKLY[b]	582 (104)	155 (101)	262 (104)	152 (101)
MONTHLY[c]	93.6 (106)	89.7 (101)	51.7 (103)	57.7 (105)
QUARTERLY[d]	224.2 (104)	122.1 (101)	43.4 (104)	67.2 (104)

NETWORK (WITHOUT 104) MAXIMA

	HIGHEST		2nd HIGHEST	
	MEAS (STATION)	EST (STATION)	MEAS (STATION)	EST (STATION)
1-HR	2025 (106)	1675 (115)	1095 (103)	1428 (115)
3-HR	1127 (106)	675 (115)	703 (106)	622 (115)
8-HR	756 (106)	486 (101,115)	604 (106)	424 (101)
24-HR[a]	443 (106)	336 (101)	414 (106)	299 (101)
WEEKLY[b]	186 (106)	155 (101)	94 (106)	152 (101)
MONTHLY[c]	93.6 (106)	89.7 (101)	51.7 (103)	57.7 (105)
QUARTERLY[d]	57.8 (106)	122.1 (101)	30.5 (105)	39.6 (103)

[a] 22 HOURS REQUIRED (90%)

[b] 152 HOURS REQUIRED (90%)

[c] 648 HOURS REQUIRED (90%)

[d] 1748 HOURS REQUIRED (80%)

Table 7. Summary

	1-HR	3-HR	8-HR	24-HR	ANNUAL[a]
ALL STATIONS:					
MAXIMUM OF SECOND HIGHEST CONCENTRATION					
MEAS	2499	2145	1513	1179	86.1
EST	1428	622	424	299	85.5
EST/MEAS	0.57	0.29	0.28	0.25	0.99
NETWORK WITHOUT 104:					
MAXIMUM OF SECOND HIGHEST CONCENTRATION					
MEAS	1095	703	604	414	38.7
EST	1428	622	424	299	85.5
EST/MEAS	1.30	0.89	0.70	0.72	2.21
NUMBER OF THE 13 STATIONS WITH SECOND HIGHEST CONCENTRATION ESTIMATE					
WITHIN A FACTOR OF 2 OF MEAS	10	11	9	9	12
OVERESTIMATED	6	4	6	4	7
UNDERESTIMATED	7	9	7	9	6
WITHIN ± 25% OF MEAS	4	3	3	4	5

[a] REFERS TO MEANS RATHER THAN SECOND HIGHEST.

The results presented here indicate overestimation of sulfur dioxide concentrations by the model at some locations but underestimation of concentrations at a larger number of receptors. Several of the previous studies (Guldberg and Kern, 1978; Hodanbosi and Peters, 1981) had indicated a dominant tendency for RAM to overestimate concentrations. Both of these evaluations used emissions for major sources for maximum operating conditions or for design capacity which would have a tendency to overestimate emissions some of the time. By contrast, the attempt was made with the St. Louis data to realistically estimate the emissions by applying data on emission variation. Since, for much of the simulated time, the procedure used with the RAPS data will result in lower emissions than will the procedure of using emissions for design capacity, this may be the principal reason that the results of this study differed from the tendency toward overestimation found in previous studies.

The reader should note that the model RAM uses single values each hour for each of the meteorological parameters. These are supposed to be representative of the urban area. For example, for this application the single hourly wind vector was derived from the determination of the flow using the wind information for 25 meteorological stations.The location and magnitude of maximum concentrations at specific receptors are extremely sensitive to the wind direction used for plume transport direction. The difficulty in estimating plume transport direction is one of the principal problems faced in doing hour-by-hour simulations to compare with measurements for the same time periods.

CONCLUSIONS

Extreme value statistics from measurements of sulfur dioxide at 13 monitoring sites in the St. Louis area from the RAPS data in 1976, and from simulated sulfur dioxide concentrations from the air quality simulation model RAM for the same sites and period have been compared. Greatest attention is given to second-highest concentrations for various averaging times for station-by-station comparisons and for the sampling network as a whole because of the frequent need to compare such rarely occurring extreme values to the NAAQS. For both second-highest 3-hour and 24-hour concentrations, the model underestimates concentrations at 9 of the 13 monitoring sites and also underestimates the network maximum. For annual concentrations (based on the hours of valid monitoring data at each monitor), the model overestimates the concentrations at 7 of the 13 monitoring sites and overestimates the network maximum by more than a factor of two (neglecting data for site 104).

ACKNOWLEGMENTS

The authors appreciate the assistance of Thomas E. Pierce in providing computational support to this project. We are especially grateful to Joan H. Novak and Ted Smith for the work with the emissions inventory, the modification to RAM to handle this large number of sources, for executing RAM for this 1-year period, and organizing the data files for delivery to the contractor. The assistance of Joan K. Emory is gratefully acknowledged.

REFERENCES

Ellis, Howard M., and Liu, Peter, C., 1981 : Review of the performance of the RAM model in predicted highest measured concentrations, J. Air Poll. Control Assoc., 31 (2) : 148-152.

Guldberg, Peter H., and Kern, Charles W., 1978 : A comparative validation of the RAM and PTMTP models for short-term SO_2 concentrations in two urban areas. J. Air Poll. Control. Assoc., 28 (9) : 907-910.

Hodanbosi, Robert F., and Peters, Leonard K., 1981: Evaluation of RAM model for Cleveland, Ohio. J. Air Poll. Control Assoc., 31 (3) : 253-255.

Javitz, Harold S., and Ruff, Ronald E., 1980 : Evaluation of the Real-Time Air Quality Model Using the RAPS Data Base, Volume 2. Statistical Procedures. EPA-600/4-80-013b. Environmental Sciences Research Laboratory, U.S. Environmental Protection Agency, Research Triangle Park, NC. 51 p.

Kummler, R.H., Cho, B., Roginski, G., Sinha, R., and Greenberg, A., 1979 : A comparative validation of the RAM and modified SAI models for short-term SO_2 concentrations in Detroit. J. Air Poll. Control Assoc., 29 (7) : 720 - 723.

Novak, Joan Hrenko, and Turner, D. Bruce, 1976 : An efficient Gaussian-plume multiple-source air quality algorithm. J. Air Poll. Control Assoc., 26 (6) : 570 - 575.

Ruff, Ronald E. 1980a : Evaluation of Real-Time Air Quality Model Using the RAPS Data Base, Volume 1. Overview. EPA-600/4-80-013a Environmental Sciences Research Laboratory, U.S. Environmental Protection Agency, Research Triangle Park, NC. 22 p.

Ruff,Ronald E., 1980b : Evaluation of the Real-Time Air Quality Model Using the RAPS Data Base, Volume 3. Program User's Guide. EPA-600/2-80-013c. Environmental Science Research Laboratory, U.S. Environmental Protection Agency, Research Triangle Park, NC. 130 p.

Ruff, Ronald E.,.1980c : Evaluation of the Real-Time Air Quality Model Using the RAPS Data Base, Volume 4. Evaluation Guide. EPA-600/4-80-013d. Environmental Sciences Research Laboratory, U.S. Environmental Protection Agency, Research Triangle Park, NC. 51 p.

Ruff, Ronald E., 1981 : Evaluation of the RAM Using the RAPS data Base. EPA-600/4-81-020. Environmental Sciences Research Laboratory, U.S. Environmental Protection Agency, Research Triangle Park, NC. 84 p.

Schiermeier, Francis A., 1978 : RAPS' field measurements are in. Environ. Sci. Technol., 12 (6) : 644 - 651.

Strothmann, Joseph A., and Schiermeier, Francis A., 1979 : Documentation of the Regional Air Pollution Study (RAPS) and Related Investigations in the St. Louis Air Quality Control Region. EPA-600/4-79-076. Environmental Sciences Research Laboratory, U.S. Environmental Protection Agency, Research Triangle Park, NC. 695 p.

Turner, D. Bruce and Novak, Joan Hrenko, 1978 : User's Guide for RAM. Vol. I. Algorithm Description and Use, Vol. II. Data Preparation and Listings. EPA-600/8-78-016 a and b. Environmental Sciences Research Laboratory, Research Triangle Park, NC. 60 and 222 pp.

U.S. Environmental Protection Agency, 1978 : Guideline on Air Quality Models. EPA-450/2-78-027, Office of Air Quality Planning and Standards, U.S. Environmental Protection Agency, Research Triangle Park, NC. 84 p.

DISCUSSION

B. VANDERBORGHT

When comparing measured and calculated highest values, do they have to occur simultaneously or do you compare highest concentrations of two different moments ?

J. S. IRWIN

The values presented in the paper are the highest values observed and estimated at the 13 sites. Hence, the comparisons are between ranked values (highest observed versus highest estimated, for instance) irrespective of when they occur.

A. BERGER

1) What is the model which was used to interpolate the observed second highest values at the grid points in your last slide ? What is your feeling about the reliability of this rule ?

2) Comment :

Research recently completed in the Institute of Astronomy and Geophysics, University of Louvain-la-Neuve, clearly demonstrates that the extreme and large concentration values do not follow a gaussian statistical distribution. The Fisher-Typett second asymptote gives the best results for values higher than the 95th percentile and the exponential-2 parameters law provides an excellent fit for the whole population of concentration data.

J. S. IRWIN

I used a $1/(\text{distance})^2$ interpolation. Such an interpolation is adequate given sufficient coverage of data values; however, it would appear that 13 values scatter in a 45 by 50 km area is insufficient. Hence, I used the results only to provide a visual display for a subjective evaluation. The scalar analyses and three dimensional depictions seem to help me to remember the differences in model performance.

LARGE SCALE VALIDATION OF A BI-GAUSSIAN DISPERSION MODEL

IN A MULTIPLE SOURCE URBAN AND INDUSTRIAL AREA

G. Cosemans, J. Kretzschmar, G. De Baere and J. Vandervee

Nuclear Energy Research Centre
B-2400 Mol
Belgium

INTRODUCTION

Gent, a moderate sized city of 300.000 habitants in the north of Belgium, is situated at the southern end of the sea-channel Gent-Terneuzen. On both sides of the channel, over a length of 15 km, an industrial zone has grown with a variety of activities such as electricity generation, oil refineries, metallurgical and chemical plants. The region around the city, world-famous for its flowers and plants, counts many greenhouses whereof the heating brings considerable quantities of SO_2 in the ambient air. Within the framework of the National R & D Programme Environment-Air, a detailed inventory of the emissions in this region has been made by several groups of researchers[1].

The automatic Monitoring Network of the Institute for Hygiene and Epidemiology[2] measures half-hourly ground-level SO_2-concentrations in this region in 12 monitoring sites. Meteorological data for as well the urban as the rural part of the region are obtained in the same network.

The Immission Frequency Distribution Model (IFDM)[3,4] has been applied to this region using the data for the year 1978. In the first part of this paper, measured concentrations are compared with calculated ones by means of pollution roses. Due attention will be given to the position of the SO_2-sources in this region relative to the SO_2-monitoring sites. In evaluating the calculated concentrations, problems such as SO_2-background and measured SO_2-concentrations influenced by other pollutants (CS_2) are to be dealt with. The second part of the paper investigates the performance of the model when applied with a randomly choosen set of 500 half-hours out of the year in order to evaluate the impact of a complex and time-varying source configuration.

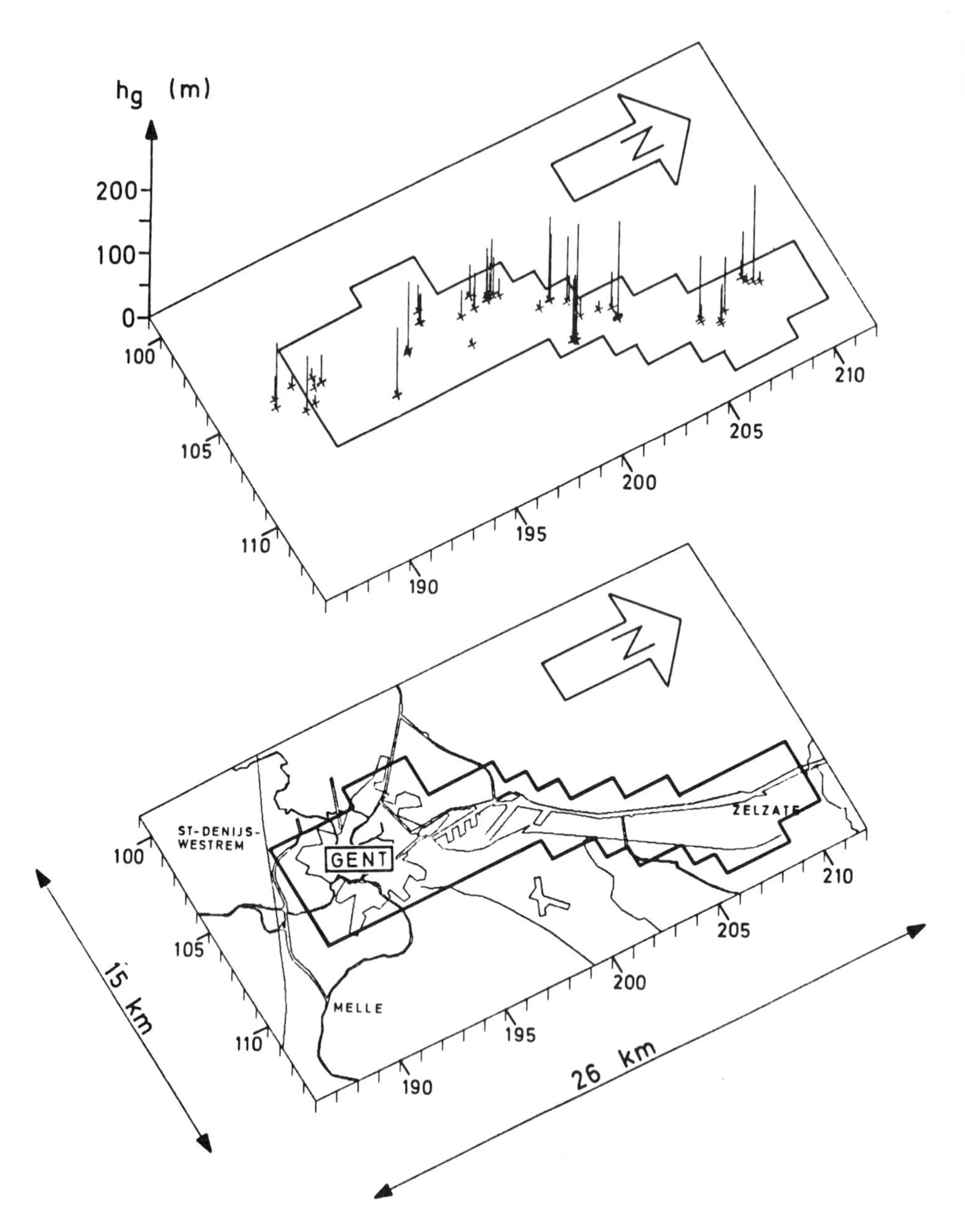

Fig. 1. Industrial SO_2-sources in the Gent area

THE IMMISSION FREQUENCY DISTRIBUTION MODEL (IFDM)

During the early seventies, the Belgian Nuclear Energy Centre developed the so-called IFDM in order to have a versatile tool for studying air pollution impact problems. The model has been extensively described in previous publications, as well as to its structure and possibilities[3,4] as to its validation on the region of Antwerpen[5]. The main properties of the model are : multiple-source bi-Gaussian approach, based on (half-)hourly observations of wind speed, wind direction and atmospheric stability, using yearly emission averages for the industrial point sources and hourly data for space heating emissions. The model is designed in such a way that whatever plume rise formula and stability classification scheme can be used ; the results of this paper are obtained by applying the second plume rise formula of Stümke[7] and the dispersion coefficients of the SCK/CEN as determined by Bultynck et al.[6]

Up to now, the model has been used in Belgium for many practical impact studies, as well in the nuclear as in the non-nuclear field. The model was also one of the participants in the NATO-CCMS pilot study on the Frankfurt common SO_2-data base[8].

SO_2-EMISSIONS IN THE GENT AREA

The position and the height of the individual industrial point sources is given in Fig. 1. The chimneys are mainly scattered along the Gent-Terneuzen sea-channel. A small spot of industrial sources is situated at the southern boundary of the city. The temporal variability of most of the industrial emissions is rather small : on a monthly base global emissions differ at most 20 to 25 % from the yearly average. A few sources however have a discontinuous operation.

The position and the importance of the area sources in this region are indicated in Fig. 2. House heating emissions are varying in time according to a degree-day system with a reference temperature of 18°C. For a day with extreme temperatures T_{min} and T_{max}, the number of degree-days is given by the expression :

$$MAX\ ((18 - (T_{min} + T_{max})/2),0)$$

The SO_2-emissions for a particular day are proportional to the number of degree-days calculated in this way. Within a particular day, the SO_2-emission is spread out over the different hours of the day according to a daily cycle. As a function of the type of building three different daily cycles have been determined by the emission inventory.

The SO_2-emissions originating from greenhouses are as important as the emissions due to house heating as shown in Fig. 2.

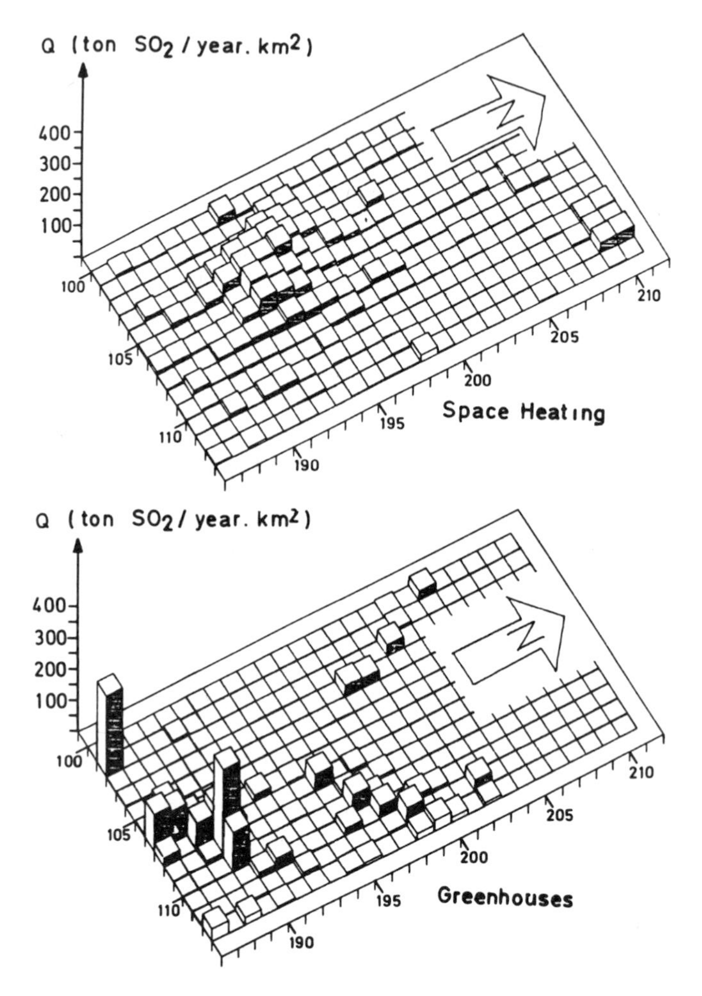

Fig. 2. SO_2-emissions originating from space heating and greenhouses in the Gent area during 1978

Depending on the plants grown in a greenhouse, the degree-day calculation for these emissions is based on a reference temperature of 6°C or 20°C.

A minor source of SO_2-emissions taken into account by the model is the Diesel-powered traffic. These emissions are only known for the central part of the region.

All the area sources are inventorised over squares of 1 km^2. More details concerning the SO_2-emissions in this region are given in reference 1.

METEOROLOGICAL DATA

Half-hourly averages of wind speed and wind direction at a height of 30 m are measured in as well the urban (M704) as the rural-industrial (M702) part of the region. The two sites are separated by a distance of 16 km (Fig. 3) in a completely flat region.

An analysis of the wind data (Fig. 4) was done in order to select the proper data for the calculation of the pollution roses and for the simulation with the model. Comparing the wind speed data from both sites, it is observed that the wind speed at the rural site is - most of the time - higher than at the urban site. The correlation between the wind speed in both places is however high ($r = 0.93$ for 11674 simultaneous observations). As both wind speed data-sets have missing data, the size of both sets was maximized by replacing missing data in one station by measured ones in the other station using regression equations calculated month by month.

The second scatter diagram of Fig. 4 shows that the simultaneously measured wind directions differ only a few degrees most of the time. Substituting missing data in one station by measured data from the other station has also been done for the wind direction. The resulting data-sets have grown in this way by more than 25 % compared to the sets of simultaneously measured data.

The substitution procedure to maximize the set of usable meteorological parameters is not without danger. Seen the preceeding analyses of the simultaneous measurements, the obtained data-set should only be used for the study of the global statistics of measured and calculated concentrations. If the individual values of measured and calculated concentrations are to be compared, only the simultaneously measured data may be used.

IMMISSION DATA

Half-hourly SO_2-concentrations are available from the IHE[2] for 12 monitoring sites identified on Fig. 3. The measured SO_2-concentrations have been used for the construction of pollution roses.

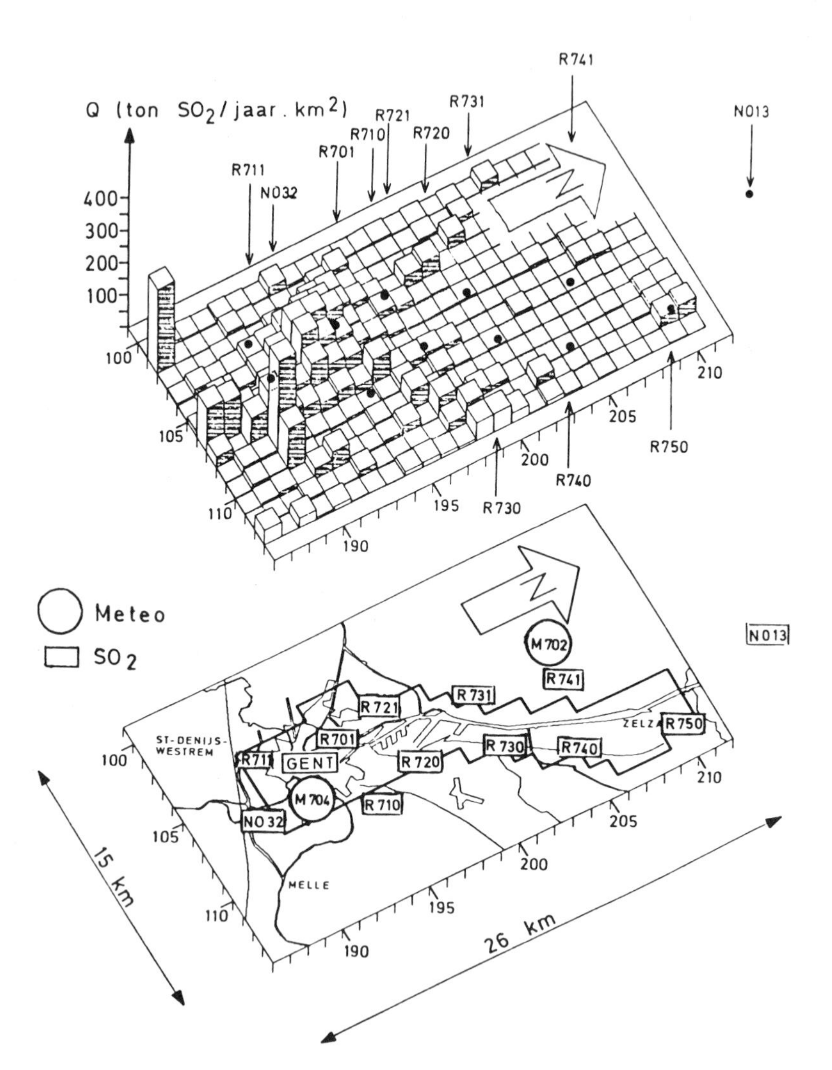

Fig. 3. Spatial repartition of all the area-sources (space heating + greenhouses) and position of the SO_2-monitoring (◯) and meteorological (□) stations of the automatic network in Gent (IHE) in 1978

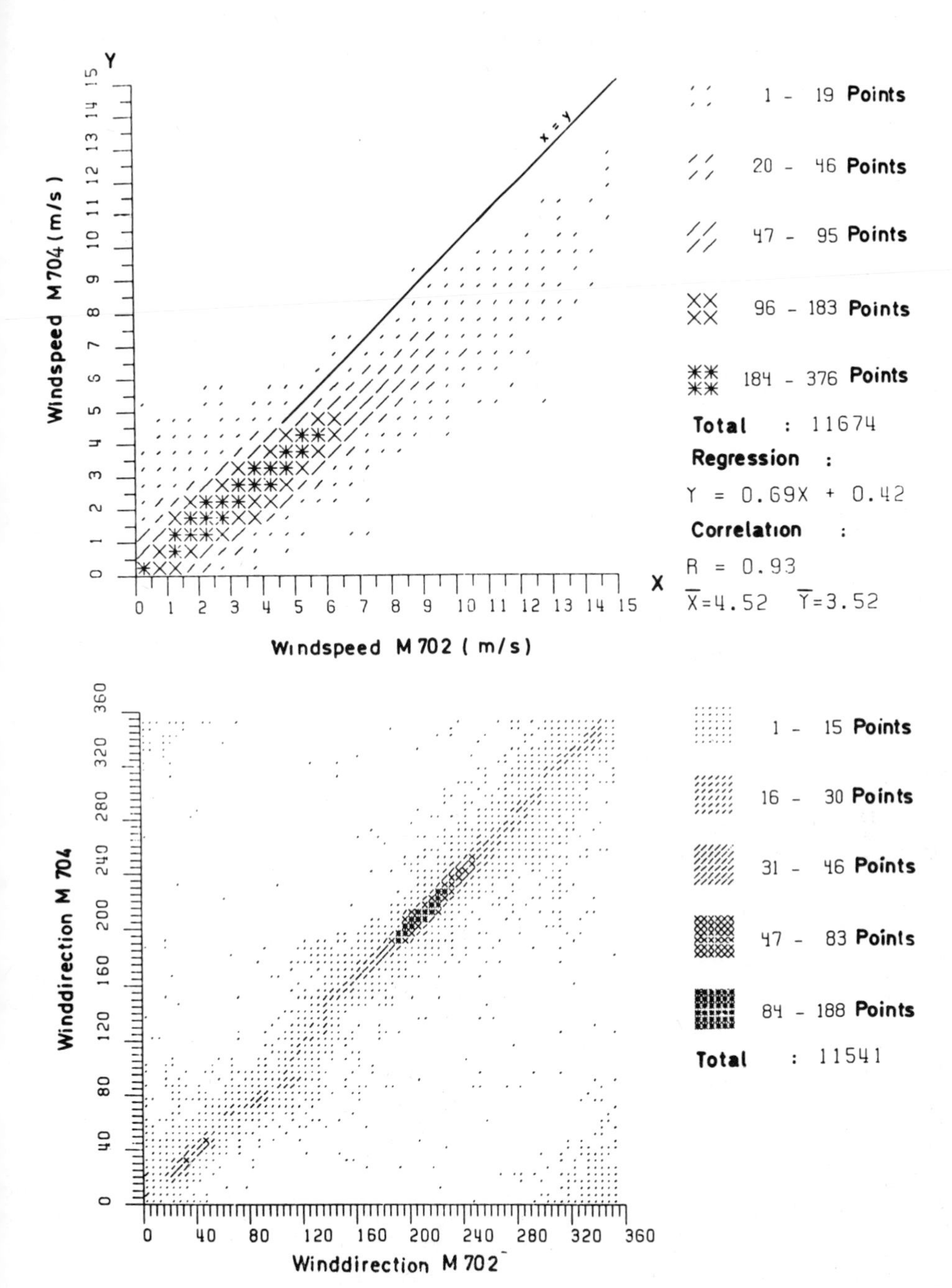

Fig. 4. Analysis of the wind data measured in the rural (M702) and urban (M704) meteorological station in 1978

The maximized set of wind directions from the urban meteorological station (M704) has been used for the six stations in the southern part of the region (R711, N032, R701, R710, R720, R721), the roses for the other stations have been constructed with the maximized data-set of M702. Only those half-hours having a wind speed of at least 1.5 m/s, and a wind direction not differing more than 20° from the wind direction measured during the previous half hour, have been used.

The pollution roses of the arithmetic means per wind sector of 10° ($\bar{X}$ min) are given in Fig. 5 (industrial sources are indicated by dots). In Fig. 10 the roses of the ninety-eight percentiles ($X_{.98,30\ min}$) are given. The influences of the most impotant industrial sources are clearly illustrated by the peaked form of the roses.

It must be noted however that some of the peaks in the 4 southerly stations (R711, N032, R701 and R710) are mainly due to the CS_2- and H_2S-emissions from a plant, whose position is indicated by the black square on the figures. These interferences are due to the fact that flame-photometric monitors are used in the automatic network. In a special study the IHE[9] determined the influence of these emissions upon the measured SO_2-concentrations and this quantitative information will be used in the validation of the model.

A common characteristic in all the measured pollution roses are the high SO_2-levels recorded by easterly winds. The origin of these immissions has to be situated outside the Gent region and they are obviously due to (unknown) sources not being present in the available emission input data for the model so that wind direction dependent SO_2-background levels are to be introduced in the validation procedure.

INFLUENCE OF THE SO_2-EMISSIONS OUTSIDE THE GENT REGION

In a first approximation the wind direction dependent SO_2-background in the Gent region has been evaluated in the following way. The numerical values of the pollution roses for a certain parameter (e.g. $\bar{X}_{30\ min}$) are placed in an array "wind direction - monitoring site". For each wind direction, the station with the minimum value is sought. Taking into account the position of this station with respect to the positions of the (known) SO_2-sources in the region it is decided if this minimum value is an acceptable estimate for the SO_2-background for that wind direction (Table 1). A further refinement in the method is obtained by repeating the analysis for respectively the summer and the winter period. The necessity to do so is illustrated by the following example. For the direction of 60° e.g. the SO_2-background over the entire year is $\sim$ 50 μg SO_2/m^3. During the summer months, this value is reduced to approximately 30 μg SO_2/m^3, but during the winter months it goes up to $\sim$ 70 μg SO_2/m^3.

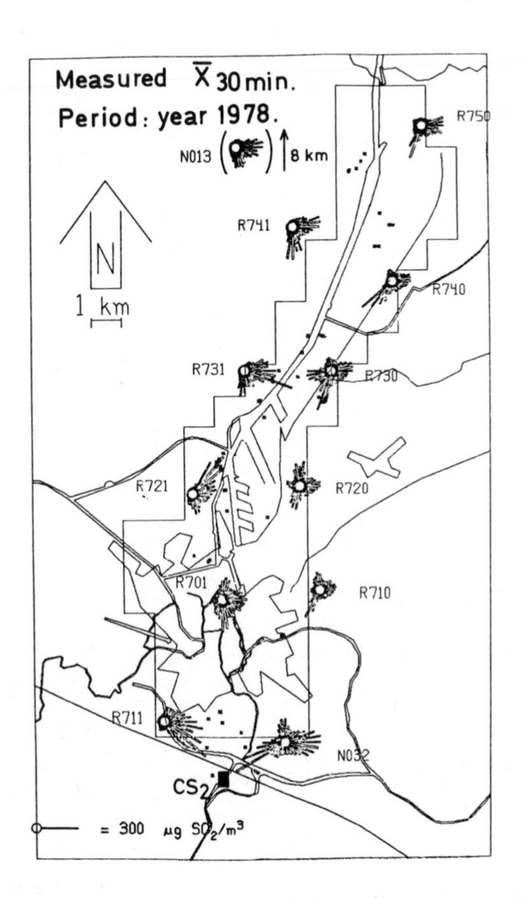

Fig. 5. Pollution roses of the arithmetic means of the measured concentrations in 1978

For the comparison of the pollution roses based on respectively measured and calculated concentrations, the average background level rose can be substracted from the pollution roses for the arithmetic averages of the measured concentrations which yields figure 6. This may not be done with the roses of the 98-th percentiles as percentiles are not additive. For the same reason the cumulative frequency distributions of the measured concentrations can not be corrected for the SO_2-background, except if the background level can be determined for each half hour and is substracted from the corresponding measured concentrations before the statistic is made. This procedure will be used in the evaluation of the random sample scheme.

Table 1. SO_2-background levels ($\mu g\ SO_2/m^3$) per wind direction determined from the pollution roses of the half-hourly SO_2-concentrations measured in Gent during 1978

dd	$\overline{X}_{30min}$	$X_{98,30min}$	dd	$\overline{X}_{30min}$	$X_{98,30min}$	dd	$\overline{X}_{30min}$	$X_{98,30min}$
10	5	20	130	40	130	250	5	30
20	20	80	140	40	120	260	5	30
30	20	130	150	30	120	270	5	20
40	20	130	160	30	120	280	5	20
50	30	180	170	20	80	290	5	20
60	50	200	180	20	100	300	5	30
70	80	200	190	30	120	310	5	30
80	90	200	200	30	150	320	5	30
90	70	200	210	30	110	330	5	30
100	60	200	220	20	80	340	5	20
110	50	140	230	10	60	350	5	20
120	50	140	240	5	50	360	5	20

COMPARISON BETWEEN CALCULATED AND MEASURED POLLUTION ROSES

For the subsequent simulation of the SO_2-situation in the Gent area, the IFDM was run with the following data :

- for the industrial sources, yearly emission averages and the maximized data-set of M702, the meteorological station in the north, were combined ;
- for space heating, hourly emissions in combination with the wind data from the urban meteorological station (M704) were used ;
- atmospheric stability information was obtained from the meteorological mast in Mol, some 80 km towards the north-east of Gent.

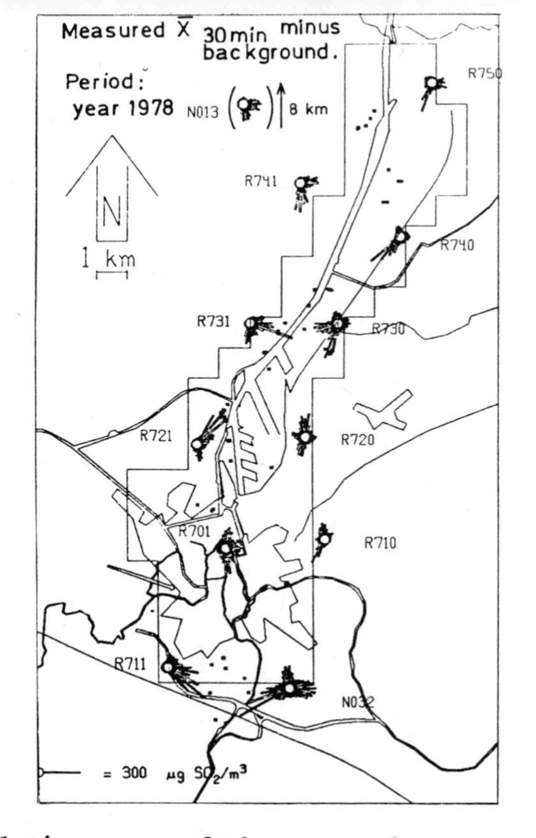

Fig. 6. Pollution roses of the measured average concentrations corrected for SO_2-background

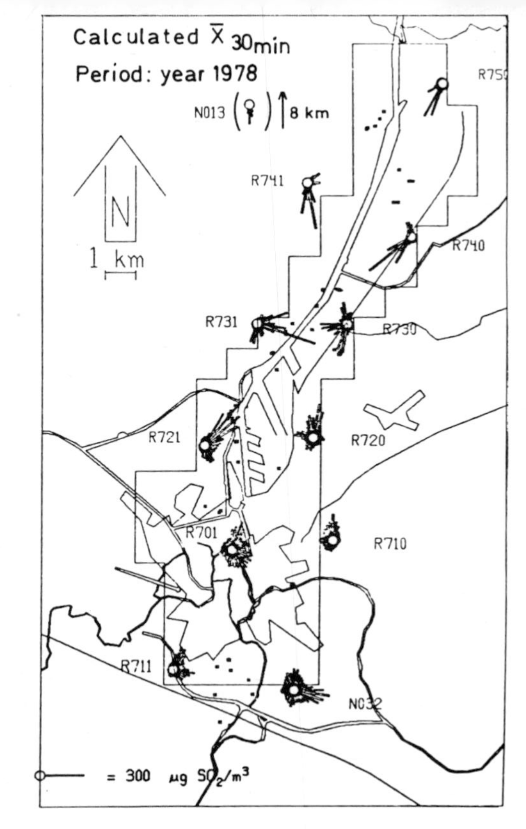

Fig. 7. Pollution roses of the calculated average concentration

Figure 7 gives the pollution roses ($\overline{X}_{30min}$) for the calculated concentrations. This figure is to be compared with figure 6 representing the measured roses corrected for the SO_2-background. The differences between both are given in Fig. 8 (calculated $<$ measured) and Fig. 9 (calculated $>$ measured). The following comments can be made. The peaks in the southerly stations R711, N032, R701, R710 and R720 (Fig. 8), all pointing to the indicated CS_2-source, illustrate the already mentioned positive interference caused by this source. On the same figure the northerly stations still show peaks pointing to the east. These peaks are largest in the N013-site this is not surprising as this station is situated outside the region for which the emissions have been inventorized, so that the emissions in its direct neighbourhood are unknown. For the other stations, (R750, R740 and R730), the remaining easterly peaks indicate that the SO_2-background, derived from the ensemble of all the receptors, was not sufficiently high to compensate completely the effects of the unknown SO_2-sources east of the Gent area. At least part of the easterly peaks in R741 may be attributed to the same phenomenon.

Apart of these almost trivial and easy to explain cases of undercalculation, there remain some peaks in R740 (SSW and SW), R730 (SSW, WSW) and R731 (ESE, SSE) that indicate an eventual underestimation of reality. These deviations however are not really surprising if one takes into account that half-hourly concentrations are used, so that the accuracy of the meteorological data becomes very important. Differences of only a few degrees in wind direction may alter the influence of a source upon a receptor in a very drastic way. In this context, it must be recalled that the missing wind data of the M702 meteorological station have been replaced by the wind directions measured in the urban station. A good example of the error that can be introduced by this is the difference peak directed to the south in station R741 on Fig. 8, suggesting a considerable underestimation of the reality. On Fig. 9 however it is seen that the concentrations calculated for the adjacent wind sector are too high. In reality both peaks must compensate so that the model gives a realistic simulation for this station. A similar situation is observed in station R740 for wind directions 220° to 240°.

Going down to the south, the roses in the stations R721, R720, R710, R701, N032 and R711 (Fig. 9) show peaks of overestimation for wind directions, that are adjacent to those wind directions for which high concentrations are measured as illustrated by the 98th-percentile roses of figures 10 and 11 for the stations R720 and R710. A detailed analysis of the involved data showed that this is due to the fact that the M702 wind data were used for the simulation of the impact of the industrial sources upon these urban receptors, while local wind data (M704) were used for the construction of the pollution roses for the measured SO_2-concentrations.

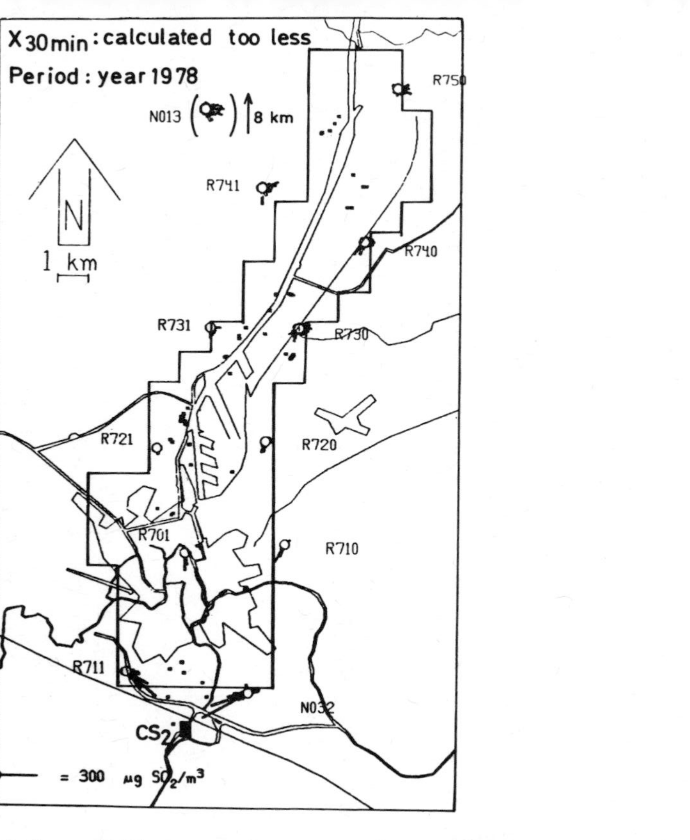

Fig. 8. Positive difference between the pollution roses of the measured concentration (corrected for background) and the situation calculated by the IFDM

Fig. 9. Negative difference between the pollution roses of the measured concentration (corrected for background) and the situation calculated by the IFDM

A straightforward comparison of the percentile roses (figure 10 and 11) is more difficult, as the previously used correction for the SO_2-background doesn't hold for percentiles. The figures illustrate however that the model gives a realistic picture of the actual situation. An analysis of the differences between the measured and calculated concentrations leads to conclusions similar to the ones obtained from the study of the $\overline{X}_{30min}$-roses.

COMPARISON OF FREQUENCY DISTRIBUTIONS

In the first two columns of Tables 2a and 2b, the measured and calculated $\overline{X}_{30min}$ and $\overline{X}_{.98,30min}$ are given for each station. SO_2-background and CS_2-interferences complicate the proper interpretation of these data.

In recent papers[10,11] some attention was given to the possibilities of random sampling for the analysis of ambient air quality data For a region as Gent, where the populations of the measured half-hourly concentrations have a geometric standard deviation between 2.5 and 4.5, theory[12,13] indicates that samples of 400 to 600 elements perform very good even for the 98th-percentile. A sample of this size - 450 half hours - taken at random out of the year 1978, has been selected. Tables 2a and 2b (3rd column) show that the 450 half hours are indeed very representative for the year in all the monitoring sites. As the sample has been taken from the half hours with simultaneous SO_2-measurements in a sufficient number of monitoring sites, the SO_2-background can be determined for each individual half hour, by using the minimum measured value as a reasonable estimate. Cumulative frequency distributions of the measured values, individually corrected for the estimated background level, are given in the fourth column of Table 2a and 2b.
The arithmetic averages can also be corrected for the CS_2-interferences as the IHE has experimentaly determined the influence of this source during the year 1978[9]. Applying this correction gives the fifth column of Table 2a.
For the 450 half hours in the sample, the model has been applied wit two different options for the industrial emissions

a) emissions averaged over the year ;
b) time varying emissions taking into account the weekly cycle in the activities and monthly averages of the emissions during working hours.

The differences in the results given by both approaches were very small : for the arithmetic average, the largest difference was 4 µg SO_2/m^3 in R720 and R730 ; for the 98-percentiles, the largest difference was observed for station R740 (457 µg SO_2/m^3 with constant emissions, 416 µg SO_2/m^3 with time varying emissions). In R750, a difference of 18 µg SO_2/m^3 was noted (290 respectively 308 µg SO_2/m^3). The parameters $\overline{X}$ and $X_{.98,30min}$ of the concentration calculated with the time varying emissions are given in the tables 2a and 2b under the heading "samp. calc.".

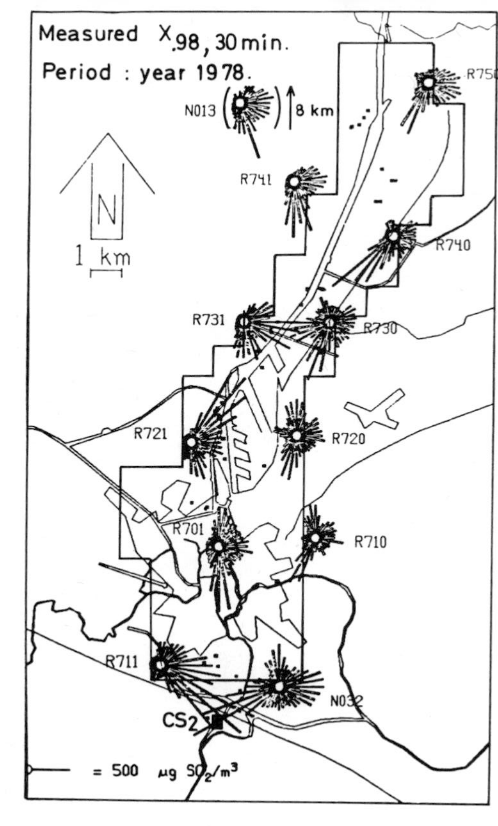

Fig. 10. 98th percentile of the measured half-hourly SO_2-concentrations per wind sector

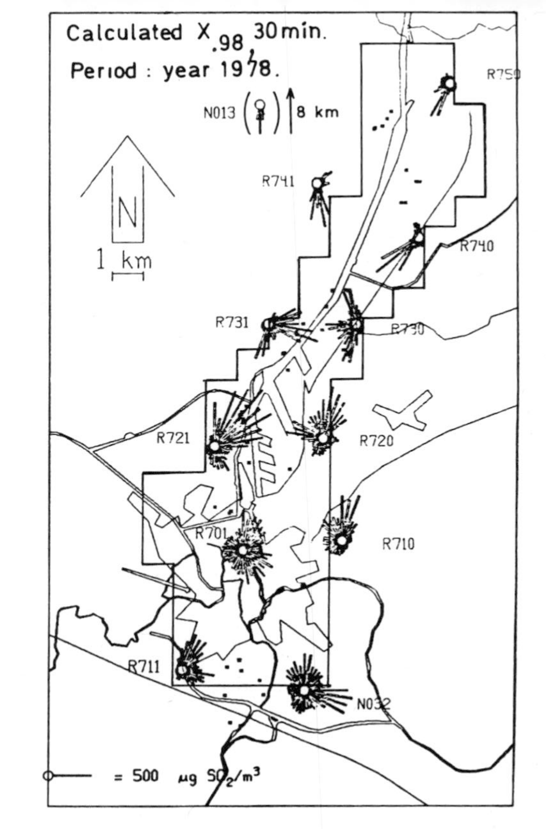

Fig. 11. 98th percentile of the calculated half-hourly SO_2-concentrations per wind sector

Table 2a. Comparison of the arithmetic averages of the measured and calculated concentrations in each of the SO_2-monitoring sites, using respectively all data and a random sample of size 450

station	arithmetic average ($\overline{X}_{30min}$)						ratio	ratio
	pop. meas.	pop. calc.	sample meas.	sample meas. -BG (A)	sample meas. $-BG-CS_2$ $(A)^★$	sample calc. (B)	(A)/(B)	$(A)^★/(B)$
R701	83	72	76	60	56	76	0,8	0,7
R710	68	46	63	48	35	46	1,0	0,8
N032	130	76	122	109	75	73	1,5	1,0
R711	62	34	57	44	37	35	1,3	1,0
R720	73	51	69	54		59	0,9	
R721	61	59	71	57		65	0,9	
R730	99	62	97	82		56	1,5	
R731	58	42	60	42		43	1,0	
R740	80	66	74	59		51	1,2	
R741	55	30	53	38		27	1,4	
R750	63	47	68	53		43	1,2	

with : pop. meas. : all the measured concentrations) 8000 to 12000 half hours per station
pop. calc. : all the calculated concentrations)
sample meas. : sample of the measured concentrations (n = 450)
sample meas. -BG : idem, corrected for SO_2-background
sample meas. $-BG-CS_2$: measured sample, corrected for SO_2-background and CS_2-interferences
sample calc. : calculated concentrations for the sample, using time varying industrial SO_2-emissions
(A), $(A)^★$, (B) : auxiliary symbols

Table 2b. Comparison of the ninety-eight percentile of the measured and calculated concentrations in each of the SO_2 -monitoring sites using respectively all data and a random sample of size 450

station	ninety-eight percentile ($X_{.98,30min}$)					ratio
	pop. meas.	pop. calc.	sample meas.	sample meas. -BG (A)	sample calc. (B)	(A)/(B)
R701*	345	294	336	255	364	0,7
R710*	351	194	339	302	236	1,3
N032*	589	297	562	562	281	2,0
R711*	438	217	376	275	233	1,2
R720	315	258	325	311	399	0,8
R721	441	546	576	535	590	0,9
R730	461	314	512	499	288	1,7
R731	375	361	426	346	343	1,0
R740	500	504	518	517	416	1,2
R741	324	301	309	219	225	1,0
R750	367	342	471	435	308	1,4

* stations with CS_2-interferences. Available information is not sufficient to compensate for the interferences at the 98 percentile level

Symbols : see Table 2a

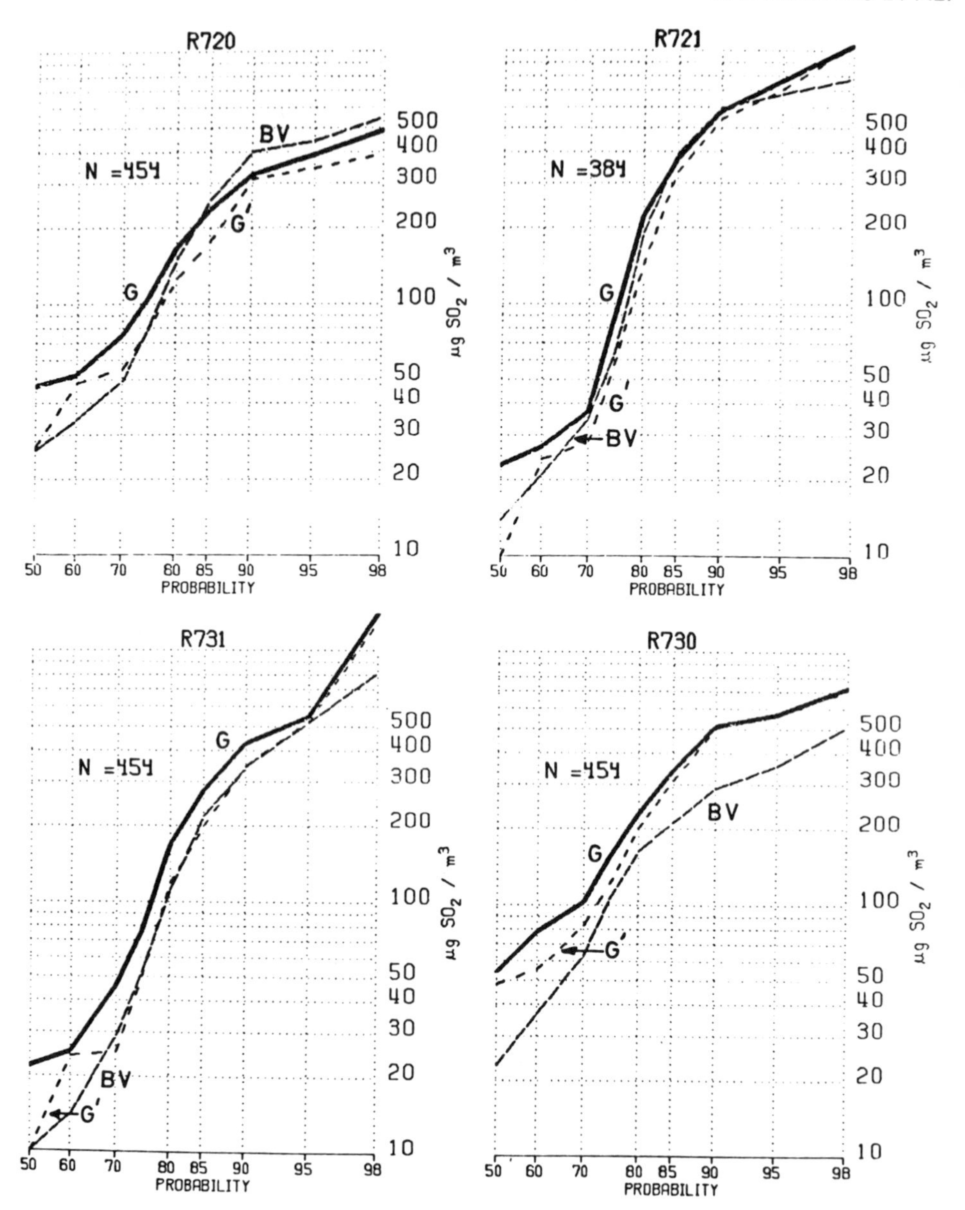

Fig. 12. Cumulative frequency distribution of the half-hourly SO_2-concentrations in Gent 1978.

G : sample of the measured concentration ;

G' : sample of the measured concentration corrected individually for SO_2-background ;

BV : sample of the calculated concentrations using time varying emissions (industry, space heating and traffic)

The last columns of these tables give the ratios between the measured parameters (corrected for background and where possible for CS_2) and the calculated ones for the sample. The overall similation turns out to be succesfully in all stations, except for the R730-site, where the model gave too low concentrations. Figure 12 gives the cumulative frequency distribution of the measured and the calculated (sample calc. in Tables 2a and 2b) half-hourly SO_2-averages.

CONCLUSIONS

The impact of the complex source configuration in the region of Gent has been calculated with the IFDM-model. The calculated concentrations have been compared with the measured concentrations by means of pollution roses and cumulative distribution frequencies, both based on the half-hourly data.

It has also been demonstrated that a sample of 450 half-hours chosen at random gives a sufficient data-set for the simulation of the ambient SO_2-situation in the region, and can be used for the evaluation and the application of an air quality model.

REFERENCES

1. "Emissieregistratie-Lucht Zone Gent-Zelzate", National R & D Programme Environment-Air, DPWB, Brussel (1980).
2. A. Lafontaine, J. Bouquiaux, G. Verduyn, M. Legrand and T. De Rijck, "Belgisch Automatisch Meetnet voor Luchtverontreiniging." Jaarrapport 1978, IHE, Brussel (1980).
3. Working Group Mathematical Models. "IFDM, The Immission Frequency Distribution Model", S.C.K./C.E.N., Mol (1977).
4. J.G. Kretzschmar, G. De Baere and J. Vandervee, The Immission Frequency Distribution Model of the S.C.K./C.E.N., Mol, Belgium, in : "Modeling, Identification and Control in Environmental Systems", Vansteenkiste, ed., North-Holland Publ.Co. (1978).
5. J.G. Kretzschmar, G. De Baere, J. Vandervee, Validation of the Immission Frequency Distribution Model in the Region of Antwerp, Belgium, in : "Proceedings of the seventh International Technical Meeting on Air Pollution Modeling and its Applications", Nato CCMS, (1976).
6. H. Bultynck and L. Malet, Evaluation of the atmospheric dilution factors for effluents diffused from an elevated continuous point source, Tellus, 24 : 455 (1972).
7. H. Stümke, Vorschlag einer empirischen Formel für die Schornsteinüberhöhung, Staub, 23 : 549 (1963).
8. Nato Modeling Panel, Practical demonstration of Urban Air Quality Simulation Models, part II, Nato CCMS (1978).
9. "Luchtverontreiniging door CS_2 te Gent", IHE, Dienst Lucht, Brussel (1980).

10. J.G. Kretzschmar and G. Cosemans, Random sampling against continuous monitoring for air quality monitoring systems, in : "Atmospheric Pollution", M. Benarie ed. Elsevier, Amsterdam (1980).

11. J.G. Kretzschmar and G. Cosemans, Random- and Minimax-Campaigns for the determination of the actual air pollution levels in an unknown region, Atmospheric Environment 15 : 1047 (1981)

12. A. Junker, Statistische Auswertung von Messwertkollektiven zur Ermitlung von Kenngrössen-Perzentil-Schätzung, Staub-Reinhalt. Luft 36 : 253 (1976).

13. J. Juda, Planung und Auswertung von Messungen der Veruntreinigungen in der Luft, Staub-Reinhalt.Luft 28 : 186 (1968).

DEVELOPMENT AND VALIDATION OF A MULTI-SOURCE PLUME DOWNWASH MODEL

Jawad S. Touma, CCM
Northwest Energy Services Company
P.O. Box 1090
Kirkland, WA 98033

Yi-Hui Huang
Consumers Power Co.
1945 W. Parnall Rd.
Jackson, MI 49201

Hua Wang, Ph. D. and John H. Christiansen
Dames & Moore
1550 Northwest Highway
Park Ridge, IL 60068

ABSTRACT

The objective of this study is to develop and validate a plume downwash model using site-specific plant operating data, meteorological data, and sulfur dioxide monitoring data at a multi-source power generating complex in Michigan. The base model selected in U.S. Environmental Protection Agency's Gaussian-Plume Multiple Source Air Quality Algorithm (RAM). This model is modified to include plume downwash algorithms developed by Briggs and Huber. During the course of the model validation study, the accuracy and reliability of the modified base model are assessed by alternating various modeling parameters and algorithms, such as (a) Pasquill-Gifford, Brookhaven and Huber dispersion coefficients; (b) Briggs trajectory and final plume rise formulae; (c) treatment of plume rise enhanced dispersion suggested by Pasquill and Irwin; (d) treatment of plume partial inversion penetration by Briggs; etc.

The evaluation of model performance focuses on assessing the validity and accuracy of model predictions in comparison with the field measurements. Emphases are placed on the adequacy of physical representation of various segments that comprise the model for assessing its validity, and various statistical comparisons between predicted and measured concentrations for assessing its accuracy. An improved plume downwash model has been developed as a result of this study.

INTRODUCTION

An atmospheric dispersion model should be validated to assure a measure of confidence in model prediction. Model validation involves a detailed investigation of input parameters (such as meteorological data, air quality data and emission data) and a comparison of modeling results with measured air quality data. The investigation often leads to alterations of modeling parameters and algorithms which will improve the accuracy and reliability of the model.

This paper describes the development and validation of a multi-source plume downwash model for a large power generating complex in Michigan. There are eight existing oil and coal-fired units at this complex, with a total combined net generating capacity exceeding 2000 MW. Several of the plant stacks are relatively short and suspected of experiencing building downwash.

A meteorological and air quality monitoring program, consisting of a 90-meter meteorological tower with three levels of measurements at 10, 60 and 90 meters, an accoustic sounder, and four sulfur dioxide (SO_2) monitors, have been established at the site. The hourly meteorological parameters collected from the tower include wind speed, wind direction, ambient temperature, and temperature difference at the different heights. The hourly mixing height data are derived from on-site accoustic sounder observations. Four continuous SO_2 monitors, located within a 2.5 km radius of the plant site, provide measured hourly ambient SO_2 concentration data. The hourly stack operating data and SO_2 emission rates are derived from actual hourly unit operating load and daily fuel analysis data. The model validation study is based on one year of hourly source operating, meteorological, and SO_2 monitoring data in 1978-79.

DESCRIPTION OF BASE MODEL

The base model chosen for this validation study is the rural version of the Gaussian-Plume Multiple-Source Air Quality Algorithm (RAM), developed by the U.S. Environmental Protection Agency (USEPA) for assessing the impact of multiple source complexes. RAM is designed primarily for determining ground-level concentrations of nonreactive pollutants from point and area sources over level or gently rolling terrain for averaging periods up to 24 hours. The model utilizes the Gaussian plume dispersion equations[1] and Briggs[2,3,4] distance-dependent buoyancy and momentum plume rise formulae for dealing with elevated point sources and Calder's[5] narrow plume hypothesis for dealing with area source emissions. The effects of limited mixing (or plume trapping) on plume dispersion are incorporated into RAM by the assumption that the plume is completely reflected at the top of the atmospheric mixing layer as well as at the ground, using the principle of multiple images developed by Bierly and Hewson[6]. There are two versions of RAM. The rural version, employing the Pasquill-Gifford dispersion coefficients[7,8] was used in this study.

Modification of the physical stack height to account for stack downwash effect is considered by RAM using the following equation recommended by Briggs[9] :

$$h' = h_s + 2D(V_s/U - 1.5); \quad \text{for } V_s < 1.5\ U$$

$$= h_s \quad ; \quad \text{for } V_s \geq 1.5\ U$$

where h' = modified stack height taking into account stack aerodynamic downwash,

h_s = physical stack height,

D = stack inside diameter,

V_s = stack gas exit velocity, and

U = wind speed at physical stack height.

PRINCIPAL MODIFICATIONS TO THE BASE MODEL

Various modeling algorithms and parameters are incorporated into the RAM as modeling options to assess the validity and accuracy of model predictions in comparison with field measurements. Principal modifications to the base model are described below.

Treatment of Building Downwash

Figure 1 presents the building downwash modeling algorithm used in this study to account for building downwash effects on lowering plume rise using an analytic scheme developed by Briggs[9]. This analytic scheme has been verified by Snyder and Lawson[10] in a series of wind tunnel studies.

A plume is considered to be within the region of building influence if the stack is located on or near the building or is within 3 l_b downwind and h' is less than $h_b + 1.5\ l_b$, where h_b is the building height and l_b is the lesser of building height (h_b) or building width (w_b). If the plume is within the region of building influence, there are several possibilities :

a) Part or all of the plume will be entrained into the building cavity-wake region if h' is less than $h_b + 0.5\ l_b$;

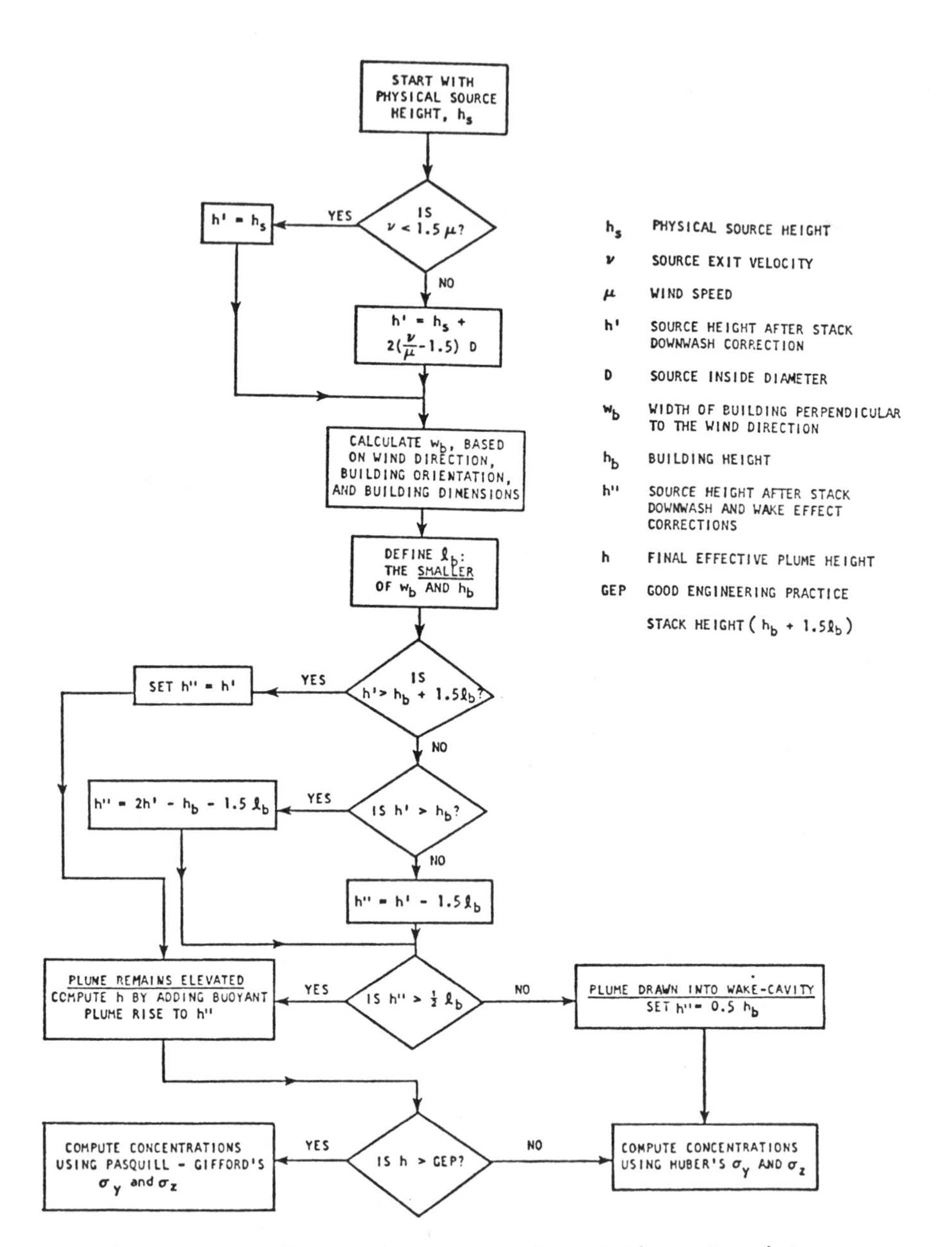

Figure 1 : Briggs-Huber downwash modeling algorithm

b) The plume will remain as an elevated source if h" is greater than or equal to $l_b/2$, where h" is the modified stack height taking into account both stack and building downwash. The following equations are used to compute h" :

$$h'' = 2h' - (h_b + 1.5\ l_b) \quad \text{for } h' \geq h_b$$

$$\text{or } h'' = h' - 1.5\ l_b \quad \text{for } h' < h_b$$

c) The plume is treated as a ground release if h" is less than $l_b/2$.

Choice of Different Dispersion Coefficients

Three different sets of atmospheric dispersion coefficients to characterize the rate of the crosswind and vertical spread of the plume are evaluated in this study. They are the Pasquill-Gifford dispersion coefficients[7,8], Brookhaven dispersion coefficients[11], and Huber's modified dispersion coefficients[12]. Huber's modified dispersion coefficients, however, are used only when the modified effective plume height, taking into account stack and building downwash effects, is less than the Good Engineering Practice (GEP) stack height, defined as :

$$h_e = h_b + 1.5\ l_b$$

The Pasquill-Gifford dispersion coefficients are expressed as a function of downwind distance by the following equations :

$$\sigma_y = \frac{x}{2.15} \tan\ (a-b \log x)$$

$$\sigma_z = cx^d$$

where

σ_y and σ_z = the horizontal and vertical dispersion coefficients, respectively,

x = downwind distance from the source,

a,b,c,d = empirical constants which vary with atmospheric stability class.

The Brookhaven dispersion coefficients are expressed as a function of downwind distance by the relations :

$$\sigma_y = ax^b$$

$$\sigma_z = cx^d$$

For elevated sources, the Brookhaven dispersion coefficients are considered as an appropriate choice for downwind dispersion calculations by Hanna, et al[13].

Huber's modified dispersion coefficients for enhanced plume dispersion in the building wake, presented below, are based on the concept that the scale of turbulent mixing in the immediate lee of a building is related to the length scale of the building :

$$\sigma_y = 0.7\, w_b/2 + 0.067\,(x - 3h_b) \quad ; \text{ for } 3\,h_b < x < 10\,h_b$$

$$= \sigma_y\,(x + S_y) \quad ; \text{ for } x \geq 10\,h_b$$

and

$$\sigma_z' = 0.7\,h_b + 0.067\,(x - 3h_b) \quad ; \text{ for } 3\,h_b < x < 10\,h_b$$

$$= \sigma_z\,(x + S_z) \quad ; \text{ for } x \geq 10\,h_b$$

where σ_y' and σ_z' = modified horizontal and vertical dispersion coefficients in the building wake, respectively,

σ_y and σ_z = horizontal and vertical dispersion coefficients in the absence of building influences, respectively,

w_b = building width,

h_b = building height,

x = downwind distance from the trailing edge of the building,

S_y = virtual source distance such that $\sigma_y'(10\,h_b) = 0.7\,w_b/2 + 0.5\,h_b$,

S_z = virtual source distance such that $\sigma_z'(10\,h_b) = 1.2\,h_b$.

Estimation of Plume Rise

Plume rise is calculated using Briggs' formulae[2,3,4] which can be either buoyancy-dominated or momentum-dominated depending on the temperature difference between stack gas and ambient air. The plume rise is calculated from the estimated plume trajectory which is distance-dependent. Modification to the base model includes the calculation of final plume rise due to buoyancy effect only, as used by the USEPA's Single Source (CRSTER) Model[14].

Treatment of Plume Rise Enhanced Dispersion

The horizontal and vertical spreads of a plume can be enhanced by entrainment of ambient air during plume rise, in addition to dispersion due to ambient turbulence. Pasquill[15] suggested that the induced enhancement of the vertical dispersion coefficient is on the order of $\Delta h/3.5$, where Δh is the plume rise. Irwin proposed that this plume rise enhanced dispersion applies to the horizontal dispersion coefficient as well. The effect of the plume rise enhanced dispersion can be represented by the following formulae :

$$\sigma_{y_e} = \sqrt{\sigma_y^2 + 0.1\ \Delta h^2}$$

$$\sigma_{z_e} = \sqrt{\sigma_z^2 + 0.1\ \Delta h^2}$$

where σ_{y_e} and σ_{z_e} are the plume rise enhanced dispersion coefficients, and σ_y and σ_z are the normal, unenhanced dispersion coefficients. The effect of the enhanced dispersion, due to plume rise, on ground-level concentration is evaluated in this study.

Treatment of Plume Penetration into Elevated Inversion Layer

Partial penetration of a plume into an elevated inversion layer occurs more frequently than does total penetration. The plume penetration ability depends not only on plume buoyancy, but also on the height and intensity of the elevated inversion layer. To determine the fraction of the plume rise that can penetrate into the inversion layer, Briggs[17] defines, "penetration factor" as follows :

$$P = 1.5 - Z_b/Z_{eq}$$

where Z_b = height of the elevated inversion base above the stack top,

and Z_{eq} = equilibrium (final) plume height above the stack top.

For a thermally buoyant, bent-over plume, Briggs has derived the following equation using a "rectangular plume" approximation for calculating Z_{eq} :

$$\frac{3}{2}\left[\frac{Z_{eq}}{Z_b}\right]^2 \left[\frac{Z_{eq}}{Z_b}\right]^{-1} + \frac{2}{9} = \frac{3F}{B'^2\ Us\ Z_b^3}$$

where F = plume buoyancy flux,

U = wind speed at stack top,

s = stability parameter,

B'= effective entrainment coefficient (B' 0.4).

The above equations indicate that : (1) no inversion penetration (P = 0) occurs if Z_{eq} is less than 2/3 Z_b; (2) 50 percent inversion penetration (P = 0.5) occurs if Z_{eq} is equal to Z_b; and

(3) full penetration (P = 1) occurs if Z_{eq} is greater than 2 Z_b. Furthermore, only 3.6 percent of the total plume buoyancy flux needed for full penetration would be required to achieve 50 percent plume penetration. The possibility of partial plume penetration into an elevated inversion layer is evaluated in this study.

METHODOLOGY FOR MODEL VALIDATION

The model validation study is based on a statistical comparison of the predicted and measured one-hour SO_2 concentrations in the upper percentiles of their respective frequency distributions at each of the four SO_2 monitors.

Prior to model validation, the measured hourly SO_2 concentration data are adjusted to eliminate the background concentration for that hour. The hourly background SO_2 concentration for each hour is calculated as the arithmetic mean of the measured hourly SO_2 concentrations from those monitors outside of a 120-degree downwind sector. This average background SO_2 concentration is subtracted from the measured concentrations at all monitors for that hour to yield the measured-minus-background concentrations.

Figure 2 presents a schematic diagram of the model validation methodology used in this study. Two different types of statistical analysis are employed to evaluate the accuracy of the multi-source plume downwash model, one based on the arithmetic means of the predicted and measured concentrations and the other based on cumulative frequency distributions. The arithmetic mean ratios of predicted and measured-minus-background concentrations are used to evaluate model accuracy. The log-probability plots of frequency distribution data are used to examine the degree of agreement between model predictions and measured values. These two model accuracy criteria are used to identify deficiences of the base model which, in turn, suggest necessary modifications to improve the predictive accuracy of the model.

RESULTS OF MODEL VALIDATION STUDY

Evaluation of Base Model

As shown in Table 1, the base model underpredicts the SO_2 concentrations at Monitor 1 through 3 but overpredicts at Monitor 4. The average ratio of predicted mean to measured-minus-background mean concentrations for all monitors is 0.55, indicating a substantial under-prediction by the base model. The log-probability plots of cumulative frequency distributions of predicted and measured-minus-background SO_2 concentrations for each monitor are shown in Figures 3 through 6. In these figures, both sets of cumulative frequency distributions are plotted side-by-side for easy comparison on a log-probability scale. In general, the base model showed sig-

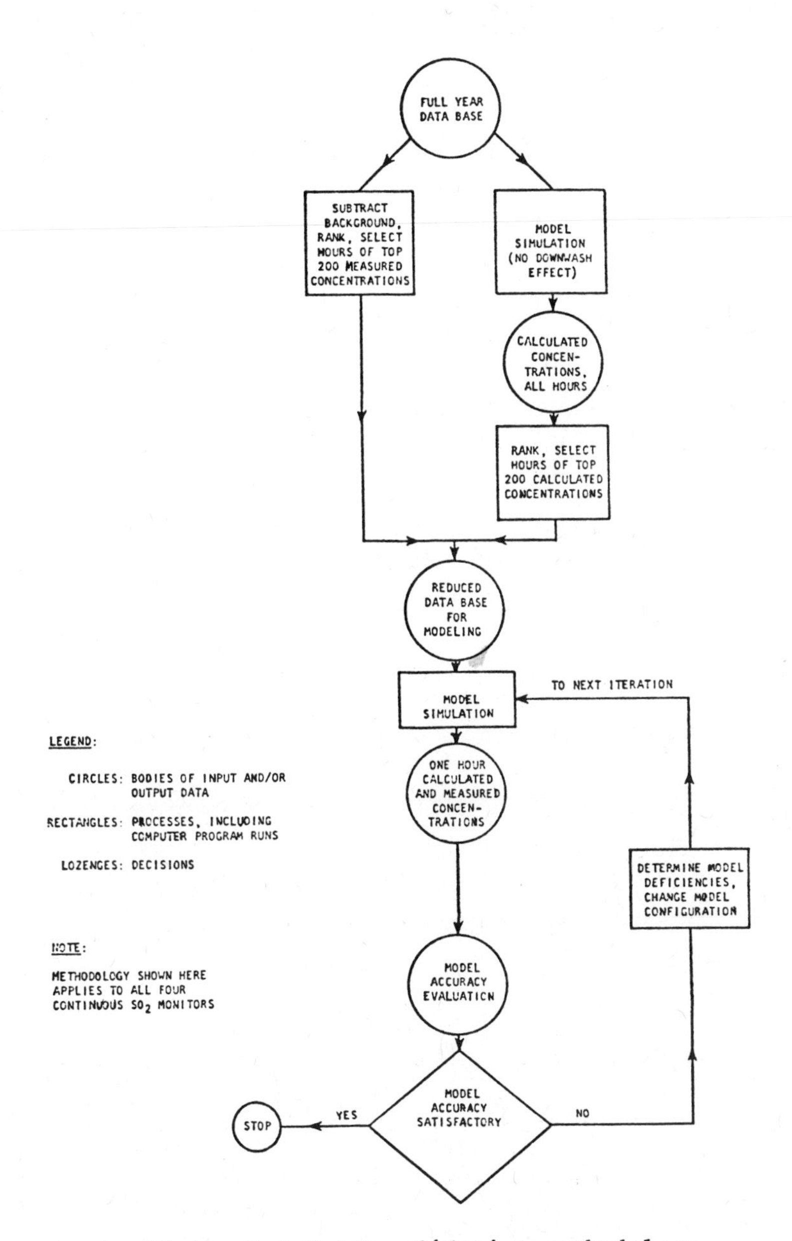

Figure 2 : Model validation methodology

Table 1

Model Accuracy Evaluation: Base Model
(No Building Downwash Effect)

	MONITOR 1	MONITOR 2	MONITOR 3	MONITOR 4	AVERAGE FOR ALL MONITORS
Calculated mean concentration for all hours modeled (g/m^3)	35.1	65.6	40.9	138.7	70.1
Measured mean concentration★ for all hours modeled (g/m^3)	87.8	186.1	134.6	121.2	132.4
Ratio of calculated mean to measured mean	0.400	0.352	0.304	1.144	0.529

★ With background subtracted

NOTE : The average ratio of calculated mean to measured mean concentration from four monitors is 0.55.

nificant under-predictions at low frequency ends and over-prediction at high frequency ends at all four monitors.

Configuration of Final, Validated Model

Through a series of model iterations employing various combinations of different modeling algorithms and parameters described previously, a final model configuration, which provides the best predictive accuracy of any of the modeling configurations evaluated, is obtained. This final, validated model configuration includes the following modifications to the base model :

1) Inclusion of Briggs-Huber building downwash effect,

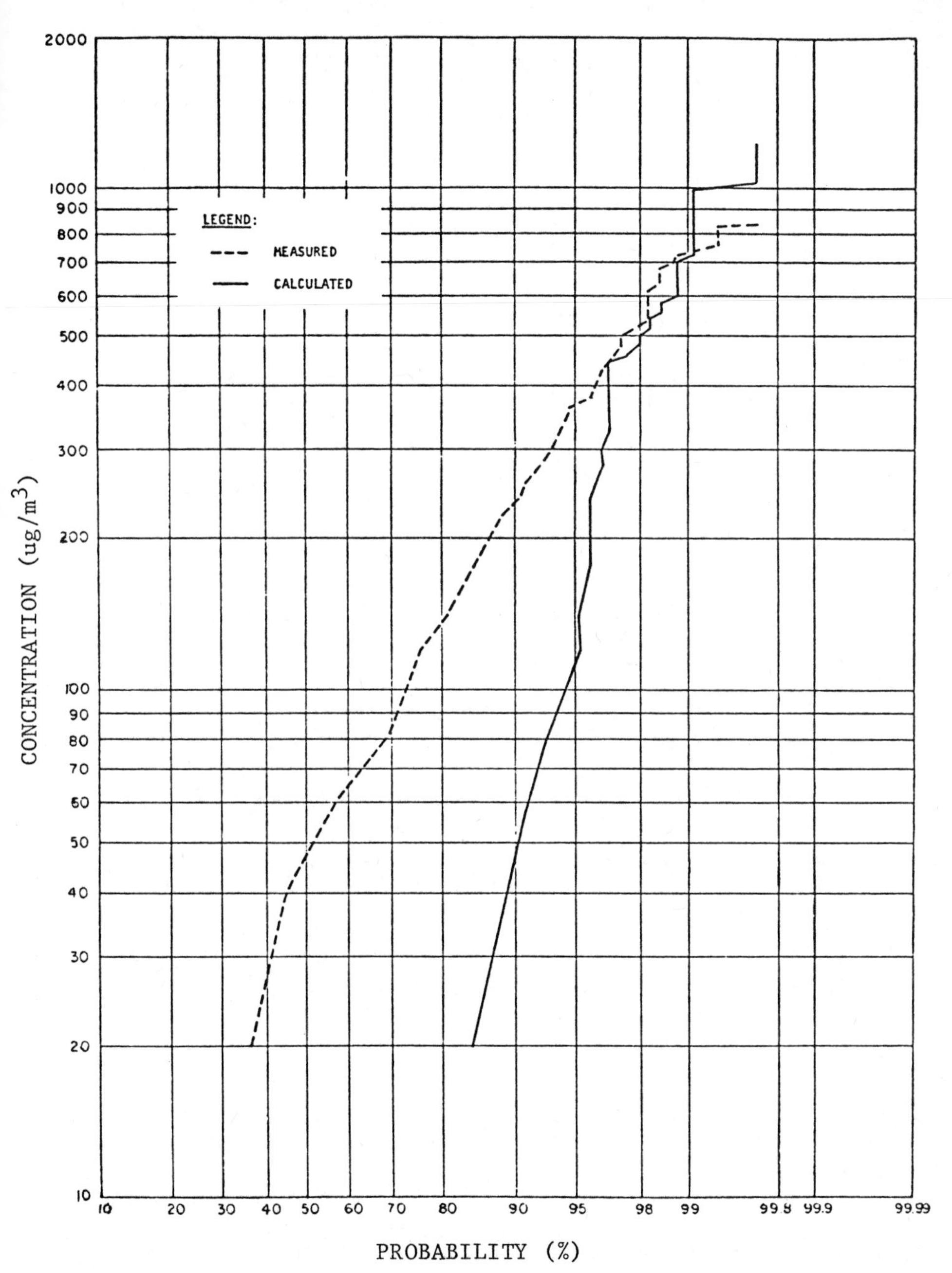

Figure 3 : Cumulative frequency distributions of one-hour average SO_2 concentrations at monitor 1 : unmodified model without wake effect

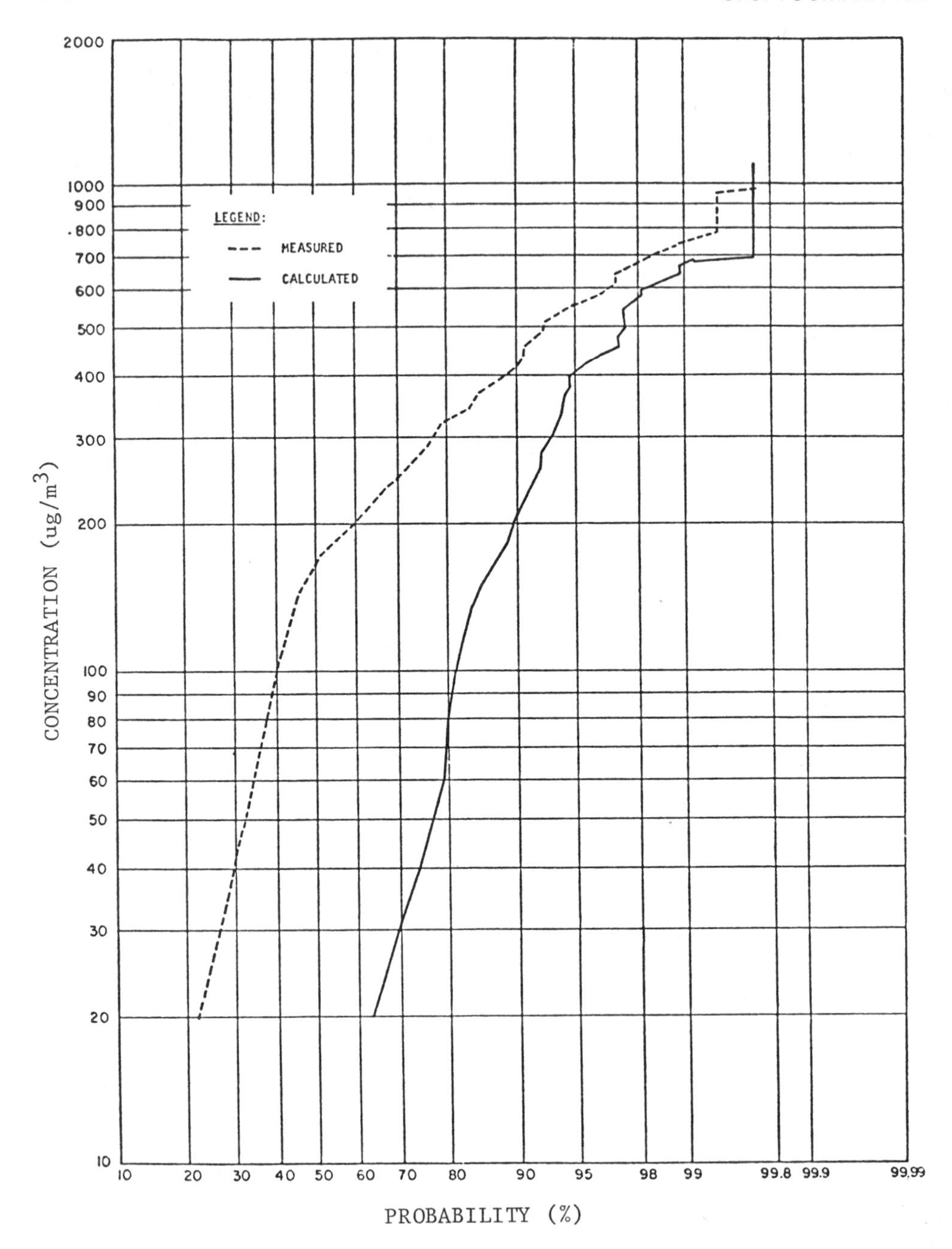

Figure 4 : Cumulative frequency distributions of one-hour average SO_2 concentrations at monitor 2 : unmodified model without wake effect

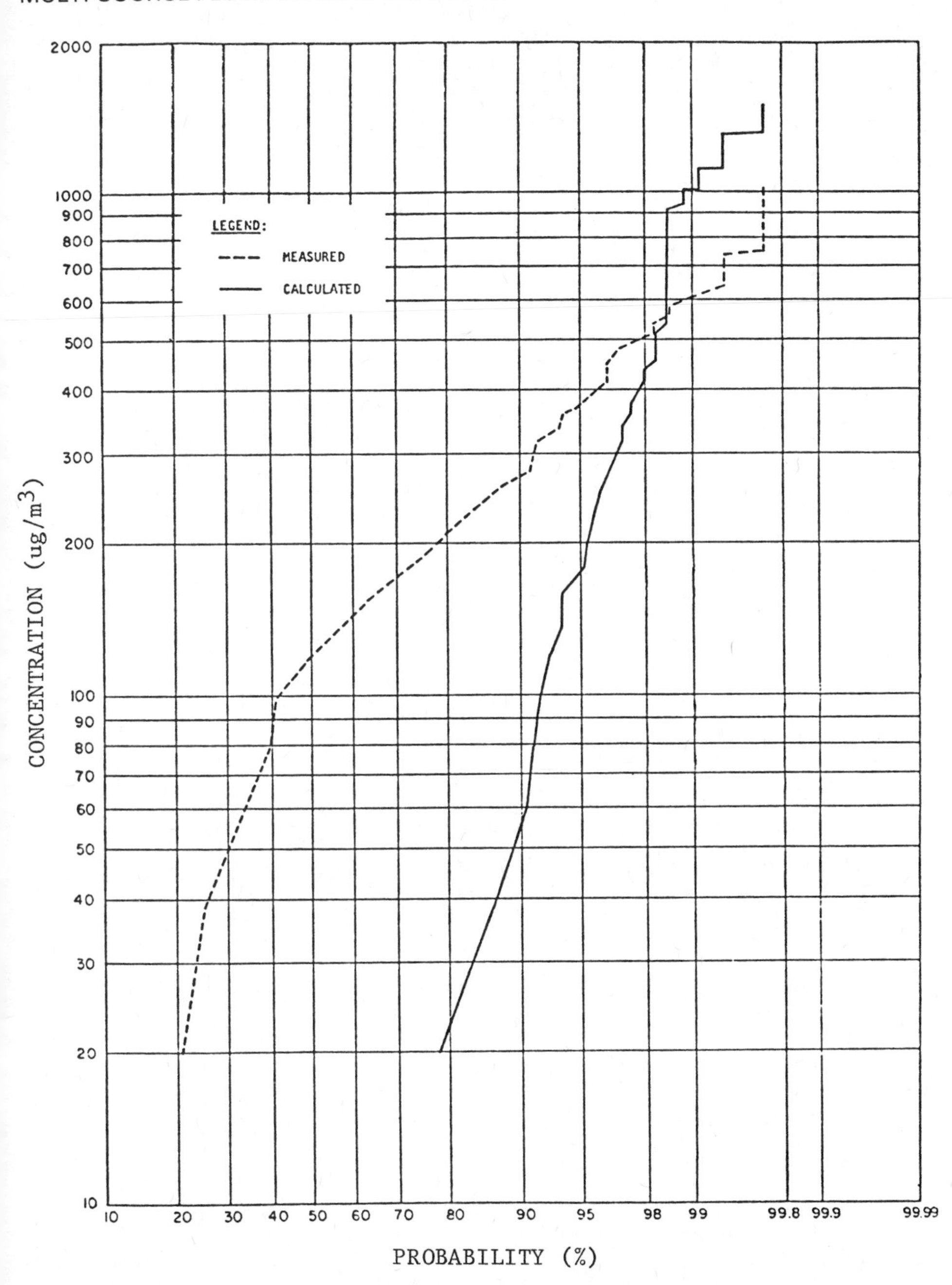

Figure 5 : Cumulative frequency distributions of one-hour average SO_2 concentrations at monitor 3 : unmodified model without wake effect

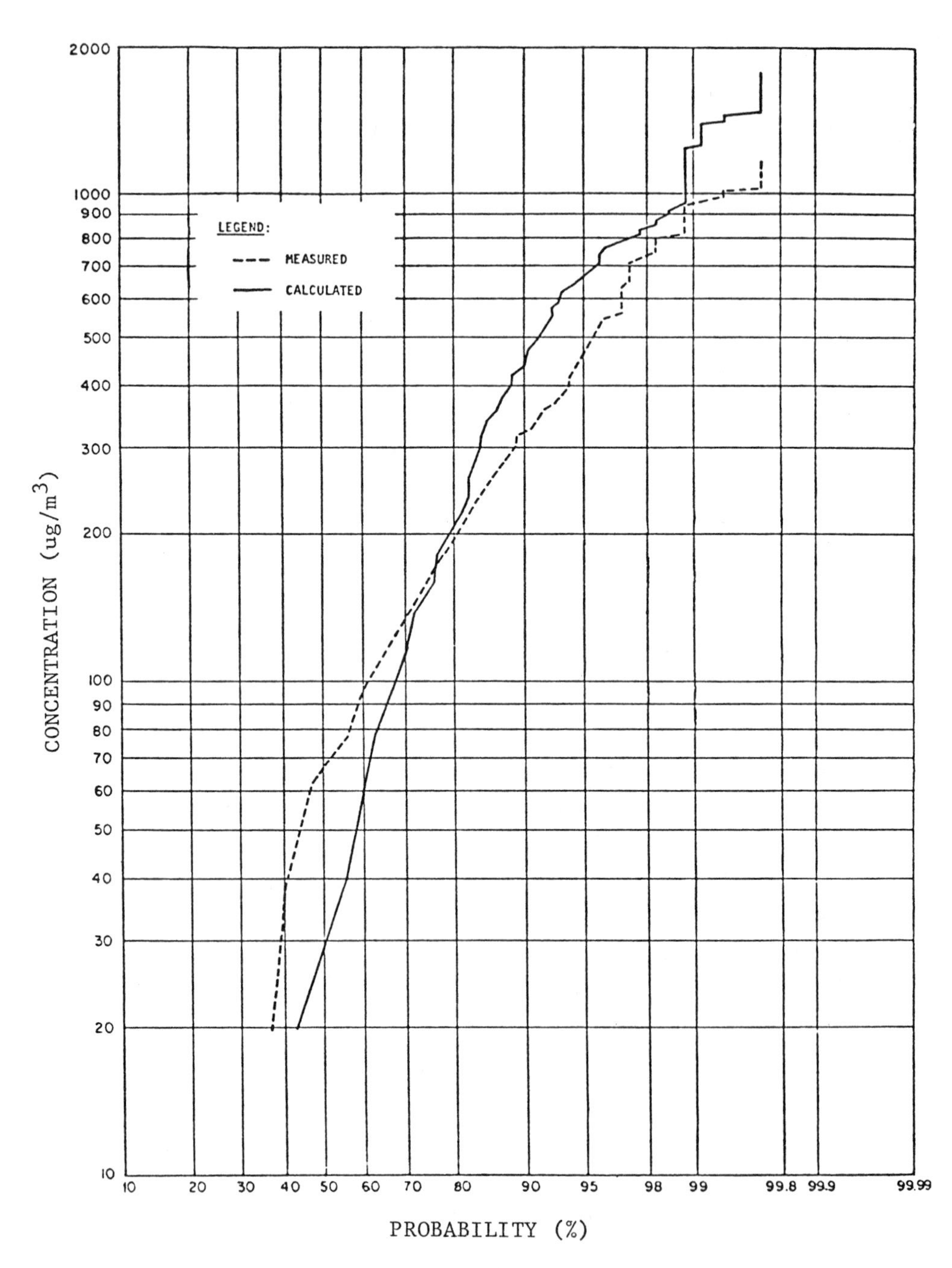

Figure 6 : Cumulative frequency distributions of one-hour average SO_2 concentrations at monitor 4 : unmodified model without wake effect

2) Selection of Brookhaven dispersion coefficients taking into account plume-rise enhanced dispersion, and

3) Adoption of Briggs' final plume rise formulae for determining effective stack heights.

For the near-water monitor (Monitor 4) only, the potential lake-induced stabilizing effect is represented by shifting all observed hourly unstable atmospheric conditions (Pasquill Classes A through C) by one class toward greater stability, prior to dispersion calculations at this monitor.

The model accuracy analysis results for the final model configuration, presented in Table 2, indicate a very substantial improvement in the degree of agreement between predicted and measured-minus-background SO_2 concentrations at all four monitors. This agreement is evident in the results for each monitor, as well as the average values over all monitors. The average ratio of the predicted mean to measured-minus-background mean concentrations for all monitors is 1.024, indicating an over-prediction by 2.4 percent.

Figures 7 through 10 present the cumulative frequency distributions of predicted and measured-minus-background SO_2 concentrations for all four monitors. Comparison of these figures with the corresponding frequency distributions for the initial iteration shown in Figures 3 through 6 reveals a distinct improvement in the degree of agreement between predicted and measured SO_2 concentrations in the final model configuration. The most significant improvement, in terms of the degree of agreement between model predictions and measured values, can be found in the upper percentile ranges of the cumulative frequency distribution curves.

CONCLUSIONS

The development and validation of a multi-source plume downwash model has been performed for a power generating complex. The data bases for the model validation consist of one year of hourly actual plant operating data, SO_2 emission data, on-site meteorological data, and measured SO_2 concentration data. Rural RAM is selected as the base model in which various alternative modeling algorithms and parameters are incorporated as modeling options. Based on the results of the model validation study, an improved plume downwash model is developed which provides the best agreement between the predicted and measured SO_2 concentrations in the upper percentile ranges of the respective cumulative frequency distributions. In terms of overall model accuracy, this final validated model configuration shows a slight over-prediction by approximately 2.4 percent.

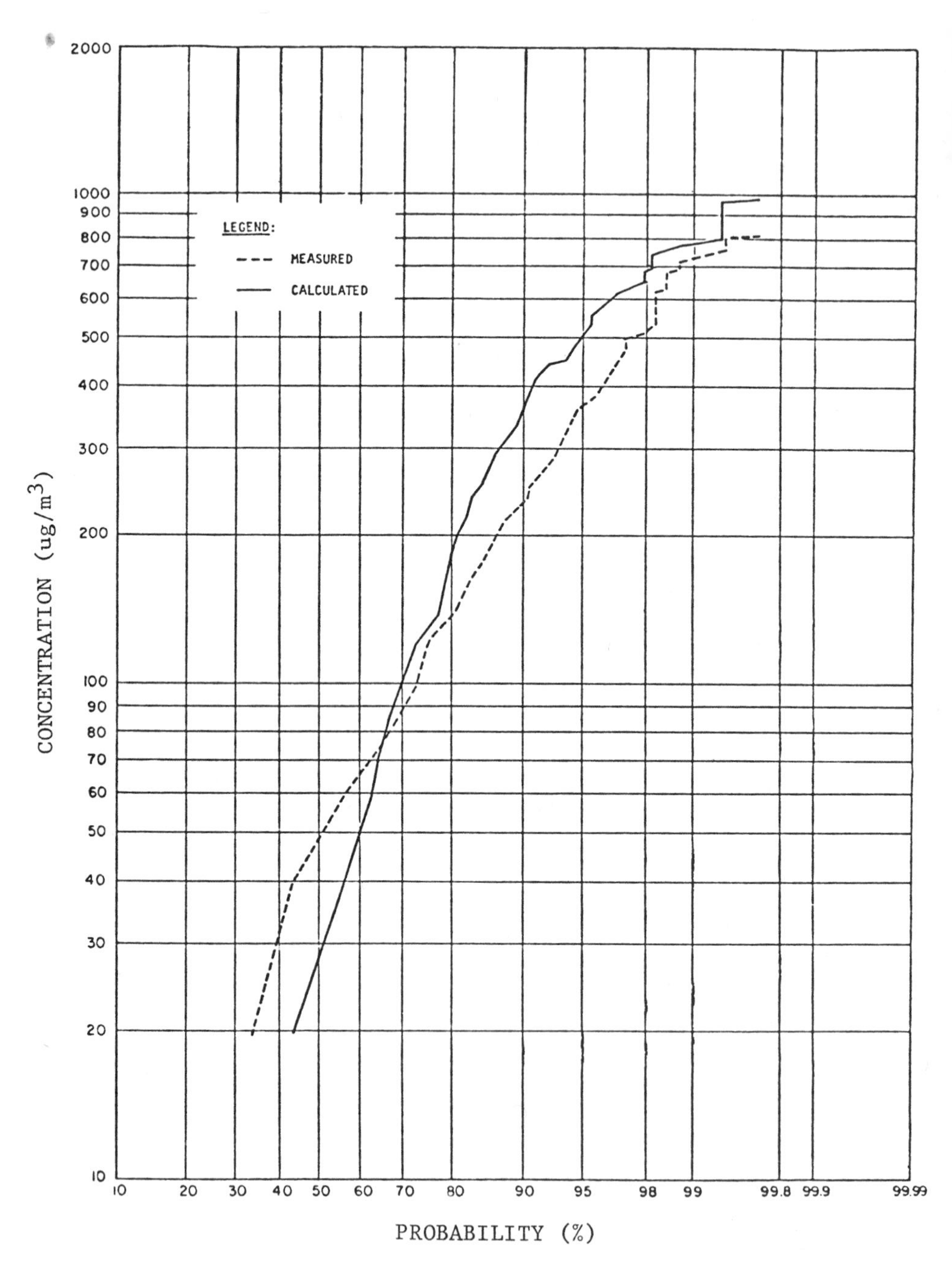

Figure 7 : Cumulative frequency distributions of one-hour average SO_2 concentrations at monitor 1 : final model configuration

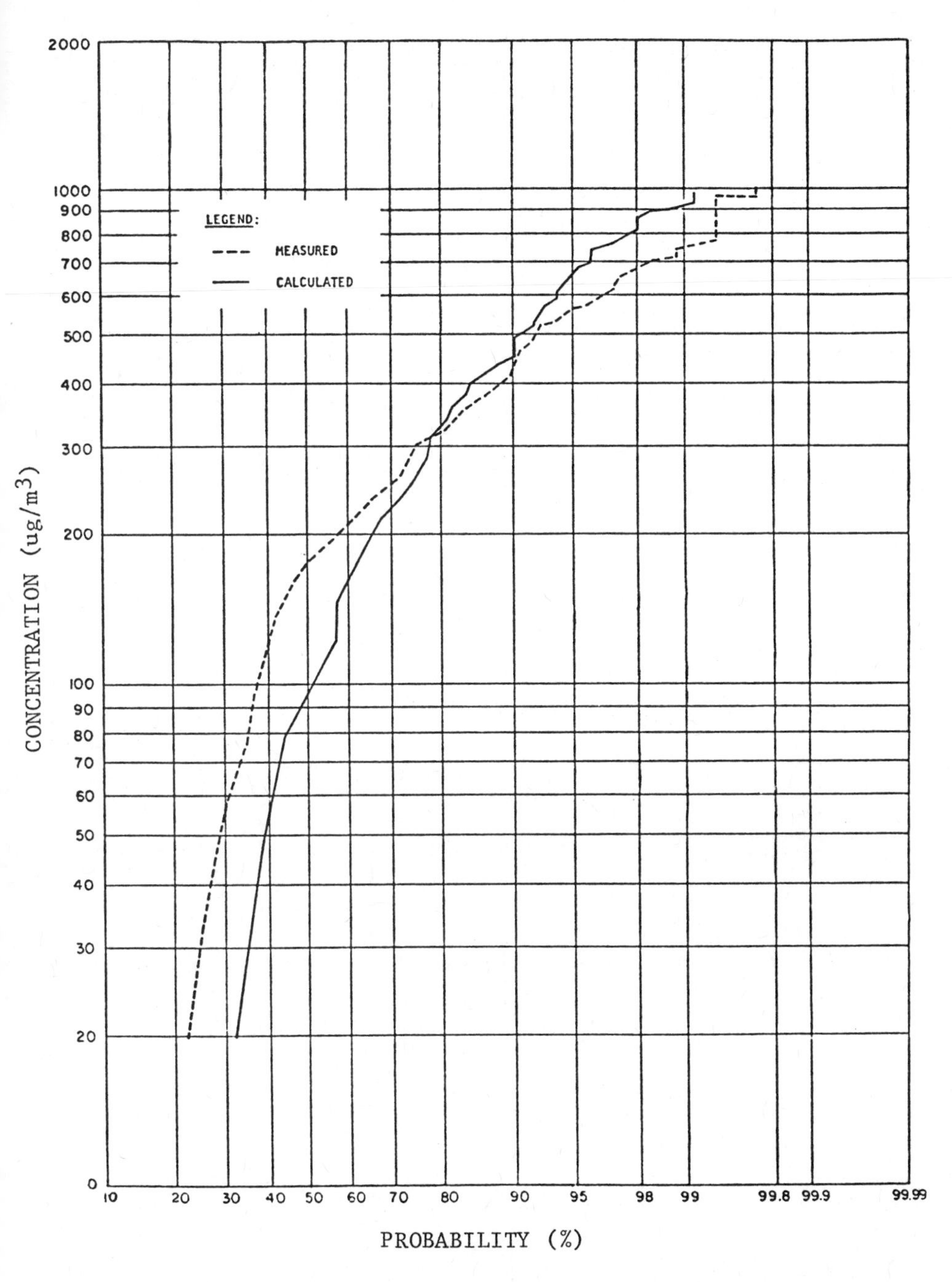

Figure 8 : Cumulative frequency distributions of one-hour average SO_2 concentrations at monitor 2 : final model configuration

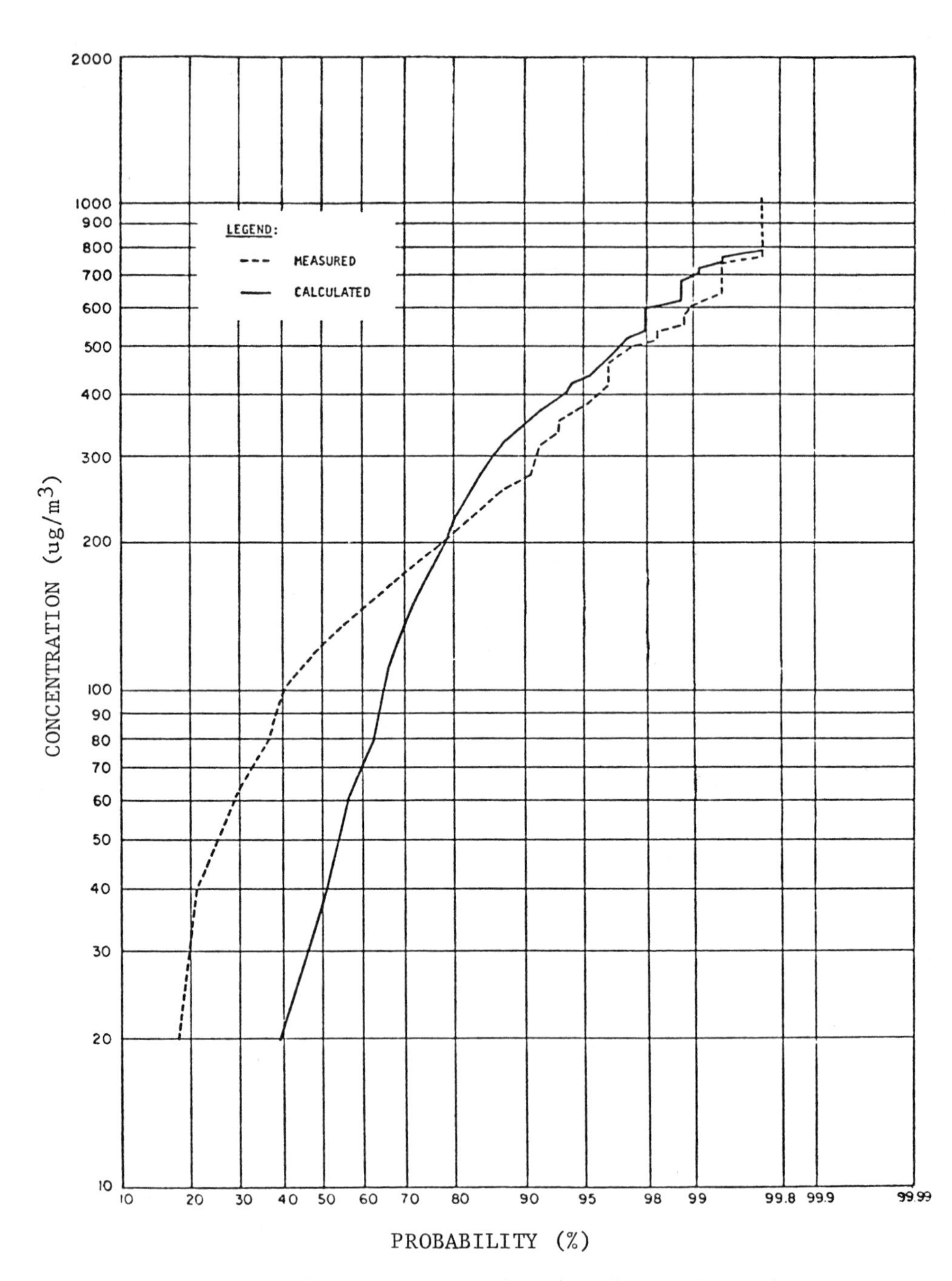

Figure 9 : Cumulative frequency distributions of one-hour average SO_2 concentrations at monitor 3 : final model configuration

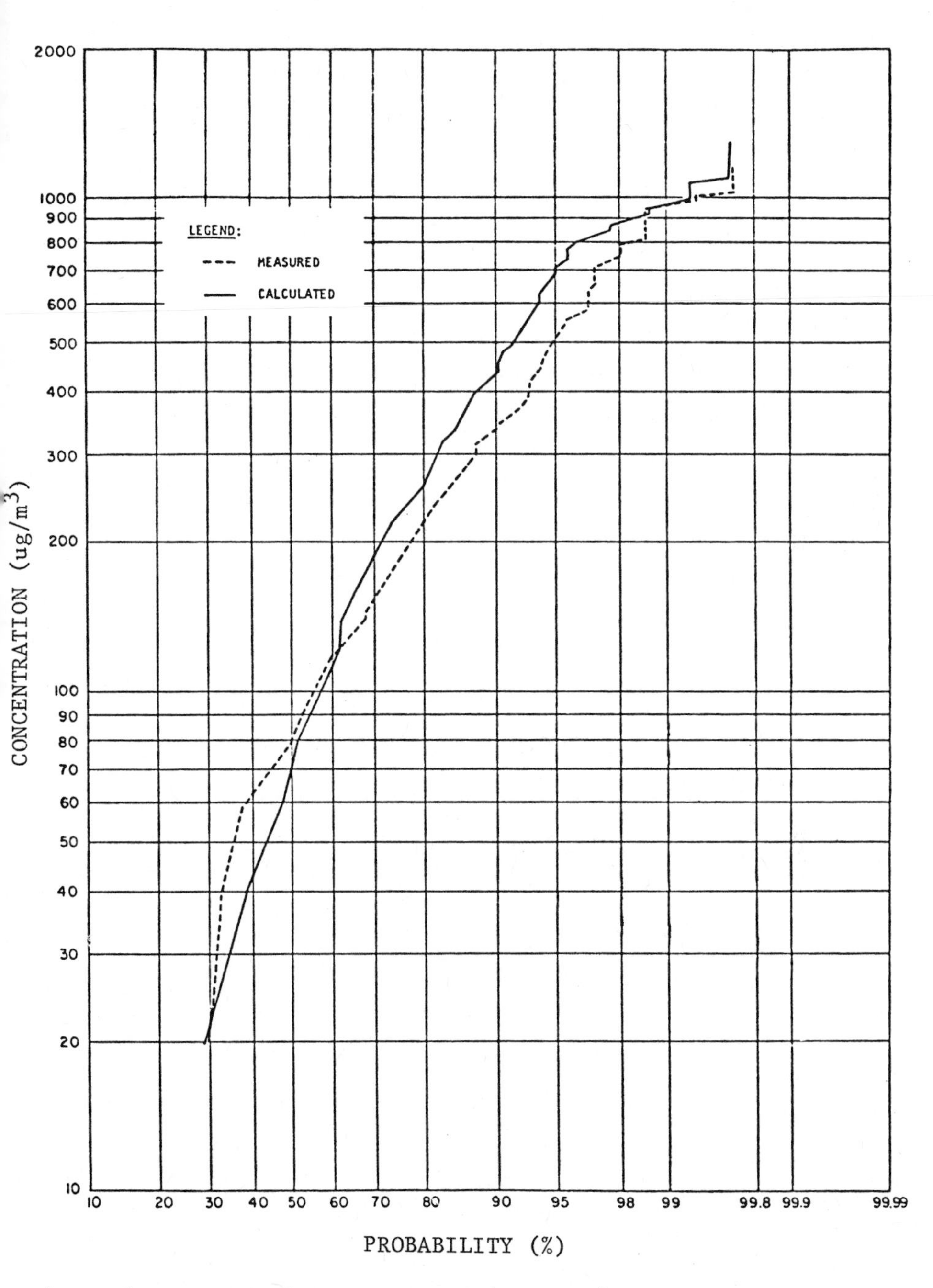

Figure 10 : Cumulative frequency distributions of one-hour average SO_2 concentrations at monitor 4 : final model configuration

Table 2

Model Accuracy Evaluation: Final Model Configuration

	MONITOR 1	MONITOR 2	MONITOR 3	MONITOR 4	AVERAGE FOR ALL MONITORS
Calculated mean concentration for all hours modeled (g/m^3)	111.9	170.5	113.5	165.0	140.2
Measured mean concentration ✶ for all hours modeled (g/m^3)	91.3	194.0	140.2	139.7	141.3
Ratio of calculated mean to measured mean	1.226	0.879	0.810	1.181	0.992

✶ With background subtracted.

NOTE : The average ratio of calculated mean to measured mean concentration from four monitors is 1.024

ACKNOWLEDGMENTS

The conscientious effort of Paul Slaughter and Frank Rochow who where responsible for the design and operation of the meteorological and air quality monitoring network for this study are appreciated.

REFERENCES

1. Turner, D. B. : Workbook of Atmospheric Dispersion Estimates, AP-26, Office of Air Programs, U.S. Environmental Protection Agency, Research Triangle Park, North Carolina, 1970.
2. Briggs, G. A. : Plume Rise, AEC Critical Review Series, TID-25075, U.S. Atomic Energy Commission, 1969.
3. Briggs, G. A. : "Some Recent Analyses fo Plume Rise Observations Proceedings of the Second International Clean Air Congress, Academic Press, pp. 1029-1032, 1971.

4. Briggs, G. A. : "Discussion of Chimney Plumes in Neutral and Stable Surroundings", Atmospheric Environment, Volume 16, pp. 507-510, 1972.
5. Calder, K. L. : "A Climatological Model for Multiple Urban Air Pollution", Proceedings of the First Meeting of the NATO/CCMS Panel on Air Pollution Modeling, 1971.
6. Bierly, E. W. and E. W. Hewson : "Some Restrictive Meteorological Conditions to be Considered in the Design of Stacks", Journal of Applied Meteoroloy, Volume 1, pp. 383-390, 1962.
7. Pasquill, F. : Atmospheric Diffusion, Second Edition, Ellis Horwood Ltd., Sussex, England, 1974.
8. Gifford, F. A. : "Use of Routine Meteorological Observations for Estimating Atmospheric Dispersion", Nuclear Safety, Volume 2, pp. 47-51, 1961.
9. Briggs, G. A. : "Diffusion Estimation for Small Emissions", ATDL contribution No. 79 (Draft), Air Resources, Atmospheric Turbulence and Diffusion Laboratory, NOAA, Oak Ridge, Tennessee, 1973.
10. Snyder, W. H. and R. E. Lawson : Determination of Heights for Stack Near Building - Wind Tunnel Study, EPA-600/4-76-001, Environmental Sciences Research Laboratory, U.S. Environmental Protection Agency, Research Triangle Park, North Carolina, February 1976.
11. Smith, M. E. : Recommended Guide for the Prediction of the Dispersion of Airborne Effluents, Second Edition, American Society of Mechanical Engineers, 1973.
12. Huber, A. H. : "Incorporating Building/Terrain Wake Effect on Stack Effluents", Preprints of the AMS-APCA Joint Conference on Applications on Air Pollution Meteorology, November 29 - December 2, 1977, Salt Lake City, Utah, pp. 353-356.
13. Hanna, S. R. et al. : "AMS Workshop on Stability Classification Schemes and Sigma Curves - Summary of Recommendations", Bulletin of the American Meteorological Society, Volume 58, pp. 1305-1309, December 1977.
14. U.S. Environmental Protection Agency, User's Manual for Single Source (CRSTER) Model, EPA-450/2-77-013, Office of Air Quality Planning and Standards, Research Triangle Park, North Carolina, July 1977.
15. Pasquill, F. : Atmospheric Dispersion Parameters in Gaussian Plume Modeling, Part II, EPA-600/4-76-030b, Environmental Sciences Research Laboratory, U.S. Environmental Protection Agency, Research Triangle Park, North Carolina, June 1976.
16. Irwin, J. S. : "Estimating Plume Dispersion - A Recommended Generalized Scheme", Proceedings of the Fourth Symposium of Turbulence, Diffusion, and Air Pollution, American Meteorological Society, Reno, Nevada, 62-69, 1979.
17. Briggs, G. A. : "Plume Rise Prediction", Lectures on Air Pollution and Environmental Impact Analysis, (D. A. Haugen, ed.), American Meteorological Society, 1975, pp. 59-111.

DISCUSSION

E. RUNCA

Is your model extendible to configurations different from the one for which it has been validated?

J. S. TOUMA

No.

/

DISPERSION MODELLING STUDIES IN SOME MAJOR URBAN AREAS OF THE UK

M.L. Williams

Warren Spring Laboratory
Stevenage, UK

1. SUMMARY

This paper presents an overview of some recent dispersion modelling work carried out at Warren Spring Laboratory (WSL), Department of Industry, and describes the more important features and results rather than giving a detailed description of each of the applications. The use of long and short period Gaussian plume models to London and Glasgow is discussed and the results are compared with measurements carried out in these areas. The use of acoustic sounder (SODAR) data to provide input data of particular use in dispersion modelling calculations of hourly average concentrations is also described.

2. LONG PERIOD MODELLING

In this section, rather than give a detailed discussion of the methods used in long period average dispersion modelling, two of the more important aspects of the results of some studies will be discussed. An assessment is given of the accuracy of the results obtained in some applications at WSL of the calculation of long period (in these cases annual) average concentrations by comparison with observation. This is followed by an investigation of the sensitivity of the results to the adopted vertical dispersion parameters σ_z. In the first part of this section a brief description of two recent WSL studies in Glasgow and London is given. More detailed accounts of these studies will be published elsewhere[1] so that the descriptions here are necessarily brief.

2.1 Glasgow

During the period Nov. 1977 to Sept. 1979 WSL carried out a monitoring/modelling study in and around Glasgow, Scotland, sponsored by the Scottish Development Department. Measurements of black smoke and SO_2 were made at roughly 25 sites in the city area, and also at outlying locations, using 24-hour samplers employing the H_2O_2 bubbler method for SO_2 and the smoke stain reflectance method for smoke based on the UK calibration curve.

At three sites in the city area measurements providing hourly average concentrations were obtained with total sulphur monitors using the flame photometric detector principle. Routine meteorological data (wind speed, direction and cloud cover) were obtained from Abbotsinch airport to the west of the city and also from the WSL site at Daldowie on the eastern edge of the city where, as well as measuring wind speed and direction, solar radiation and temperature, an acoustic sounder (SODAR) was operated.

An emission inventory was compiled for the study area with the help of the local authorities, particularly the Glasgow District Council, and other bodies. Domestic and road traffic sources were treated as 1 km^2 area sources; commercial and industrial sources were treated as individual point emitters in the city area and as 1 km^2 area sources outside the city.

The model used to calculate annual average smoke and SO_2 concentrations is a climatological Gaussian plume model developed at WSL and has been described previously[2]. The basic structure of the model will not be described in detail here, but the parameterization employed to simulate the dispersion over the urban area of Glasgow will be briefly discussed. This involved primarily the enhanced dispersion due to the increased thermal and mechanical turbulence in urban boundary layers.

2.1.1 Stability Categories. Values of the Smith stability parameter P[3] were obtained using the measured values of surface wind speed and solar radiation at the WSL meteorological site using the scheme developed by Smith[3]. During night hours, when no solar radiation data were available, cloud cover and wind speed data from Abbotsinch airport were used to estimate the value of P. These 'rural' stabilities were then corrected to allow for the effect of the urban heat load in Glasgow[4] by subtracting 1 from the stable rural categories ($P > 4.0$). In the modelling calculations wind roses (using harmonic mean wind speeds) for 8 stability categories were used with values of P such that $0 < P \leq 1$; $1 < P \leq 2$, ... $6 < P \leq 7$ and 'calms' (those hours when the wind speed was ≤ 1 m s^{-1}). For the calm category a value of P of 5.5 was calculated from the rural data and was corrected to 4.5; an isotropic wind rose with a wind speed of 0.75 m s^{-1} was used to simulate dispersion in the 'calm category.

2.1.2 Dispersion Coefficients and Surface Roughness. The dispersion coefficients, σ_z, of Pasquill-Gifford[5] were used. The effect of surface roughness was taken into account using the scheme proposed by Smith[3] who expressed the variation of σ_z with z_o for different downwind distances in nomogram form. The equation (2.1) provides a fit to these curves to within ∿ 10%:-

$$\sigma_z(x,\ P;\ z_o) = F(x,\ P)F^*(x,\ z_o)\ \sigma_z\ (x,\ 3.6;\ 0.1m) \qquad (2.1)$$

where $F^*(x,z_o) = 1.86-0.17\ \log_{10}x + (0.86-0.17\ \log_{10}x)\ \log_{10}\ z_o$.

The function F (x, P) was given the values proposed by Pasquill-Gifford, and is appropriate to a value of z_o = 0.1 m. A value of 0.75 m was used for z_o in Glasgow.

2.2 London

Although WSL has carried out dispersion modelling studies on the London area in the past, for example in relation to a period of elevated pollutant concentrations in 1975[6] the emission inventory was relatively crude, particularly in the spatial resolution which was 10 km x 10 km. Recently the Greater London Council (GLC) has compiled a detailed emission inventory of the GLC area for the year 1975/6 with a spatial resolution for area sources of 1 km. Additionally, data for some 350 point sources were also obtained. For the work described here, this inventory of SO_2 emissions was made available to WSL by the GLC. Dispersion modelling work using the inventory has also been carried out by the GLC with the CDM model, a widely used Gaussian plume model developed by the US EPA[12]. It is anticipated that both the GLC and the WSL work will be reported in the near future, so that only a brief description of the WSL work will be given here.

Meteorological data for the modelling work were obtained from the Meteorological Office site at the London Weather Centre in Holborn, a central city location on the north side of the river. The data obtained were wind speed, direction, solar radiation and cloud cover and these were used to produce stability/wind roses as described in section 2.1.1. The heat island was also simulated by the same procedure as described in section 2.1.1., but to incorporate the effect of the generally taller buildings in central London compared with Glasgow, the surface roughness length z_o was assumed to be 1.0 m. The effect of surface roughness was incorporated using equation 2.1.

The results of the dispersion modelling calculations for London have been assessed by comparison with the observed concentrations of SO_2 obtained as part of the UK National Survey of Smoke and SO_2 which uses the H_2O_2 bubbler method to provide 24 hour average concentrations of SO_2.

2.3 Assessment of Results

2.3.1 Accuracy of the Modelling Results. Annual average concentrations of smoke and SO_2 were calculated at 20 sites in Glasgow for the year 1977/8 and of SO_2 concentrations in London at 17 sites for the year 1975/6. There are several ways of assessing the performance of dispersion models using, for example, the correlation coefficient between observed and calculated values or the root-mean square error[8,9]. However, the results obtained here are compared with the observations using a technique which has not been used extensively, yet concisely presents the frequency distribution of the percentage error of the modelling calculation for each site or receptor[9]. Here percentage error is defined as 100 x (calculated - observed)/observed. A summary of the results for Glasgow and London is presented in this way in Figure 1, where included for comparison are the results of the earlier work carried out by WSL in the Forth Valley area of Scotland. This work was fundamental in the development of dispersion modelling at WSL and has been reported previously[2]. The plots in Fig. 1 display the fraction of the number of receptors for each application within a particular percentage error as defined above. For example, in the case of the annual average SO_2 calculations for London, at 75% of the 20 receptors the dispersion model yielded a concentration within ±30% of that observed. It should be stressed that, in presenting results in this way, a very important factor is the number of points used in the comparison. This, for obvious reasons, should not be small and in this type of application the minimum number should probably be greater than, say, 10.

Turning to the accuracy of the results themselves, this is probably of the magnitude which is to be expected of such practical applications of dispersion models, bearing in mind the comments of Pasquill[3] and of the American Meteorological Society Workshop[10]. In both these instances however, the accuracy to be expected of dispersion models has been expressed simply in terms of one number ("±x%" or "factor of x accuracy"). Clearly a more precise estimate of accuracy should involve two numbers, such as, in the case already cited for SO_2 in Glasgow, "75% probability of being within 20%". In absolute terms, the most accurate of the applications discussed here is the modelling of SO_2 in Glasgow followed by the calculation of smoke concentrations for that city where the 75th percentile of receptors corresponds to errors of 20% and 26% respectively. The behaviour of the Forth Valley SO_2 curve at higher percentiles illustrates the effect of a relatively small number of points for which the percentage error is large, so that in this method of performance assessment, the number of points used should always be borne.

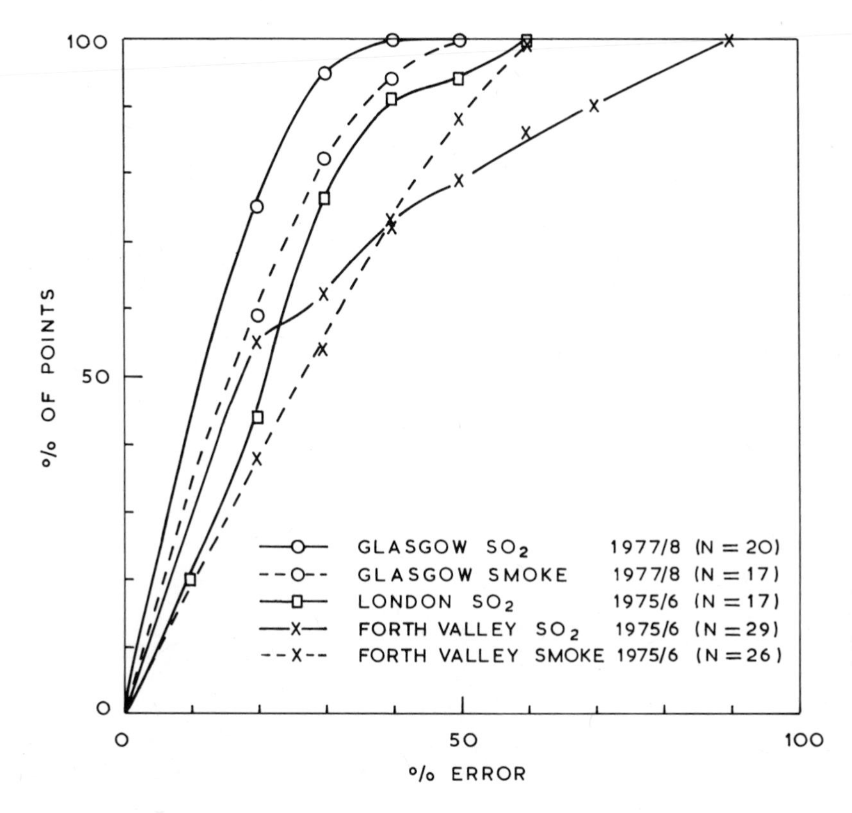

Figure 1. Plots of Percentage of Receptor Points within Percentages of Observed Concentrations for Annual Average SO_2 Concentrations.

2.3.2 Influence of the choice of vertical dispersion parameter σ_z by comparison with rural areas. There have been relatively few experimental studies of dispersion or turbulence structure over urban areas. Consequently, although the situation is by no means always straightforward, the choice of vertical dispersion parameters σ_z in applications to rural terrain is often aided by the existence of experimental data obtained from an area similar to that under study. For urban areas the most widely used experimental σ_z data are those obtained in the St. Louis experiment of McElroy and Pooler[11]. Other approaches to the problem of urban dispersion have used Pasquill-Gifford σ_z's with the appropriate surface roughness and heat island corrections where applicable (see for example the CDM model[12]). As part of our studies at WSL we have used both these approaches, with a third, namely the use of the F B Smith σ_z values with the surface roughness and heat island corrections discussed in section 2.1.1. No such corrections to the McElroy-Pooler values were made, the σ_z's in this case being evaluated using the forms given by Briggs[13]. The results of the application of each of these schemes to the Glasgow SO_2 calculations are shown in Fig. 2, all other aspects of the calculations being the same and as described in section 2.1. It is clear from the data that the Pasquill-Gifford and Smith σ_z's give similar results while the St. Louis/Briggs' scheme leads to a marked underprediction of the concentrations. Although the relevant results are not presented here, similar qualitative features were found in the calculations on London. It is worth pointing out here that the form of presentation used in Fig. 1 may not reveal any systematic error as is the case in Fig. 2(c).

3. SHORT PERIOD MODELLING

The model was applied to hourly average SO_2 concentrations on January 30th 1978, the day when concentrations were highest during a period of several days of elevated concentrations. Conditions were such that a shallow ground based stable layer formed in the early hours ($\sim$0200), persisted until $\sim$1200 hours when no elevated layer was discernible. The low level layer re-established itself around 1500 hrs. Wind speeds were relatively high between 0000-0400 hrs at 4-5 m s^{-1}, dropping to 1-2 m s^{-1} for the rest of the 24 hour period. Temperatures were also low throughout the 24 hour period, the maximum being 2.6°C, and the minimum -5.6°C.

3.2 Features of the Short Period Model

The basis of the model is the well-known Gaussian plume equation for a point source, incorporating perfect reflections from the ground and an elevated barrier at height L

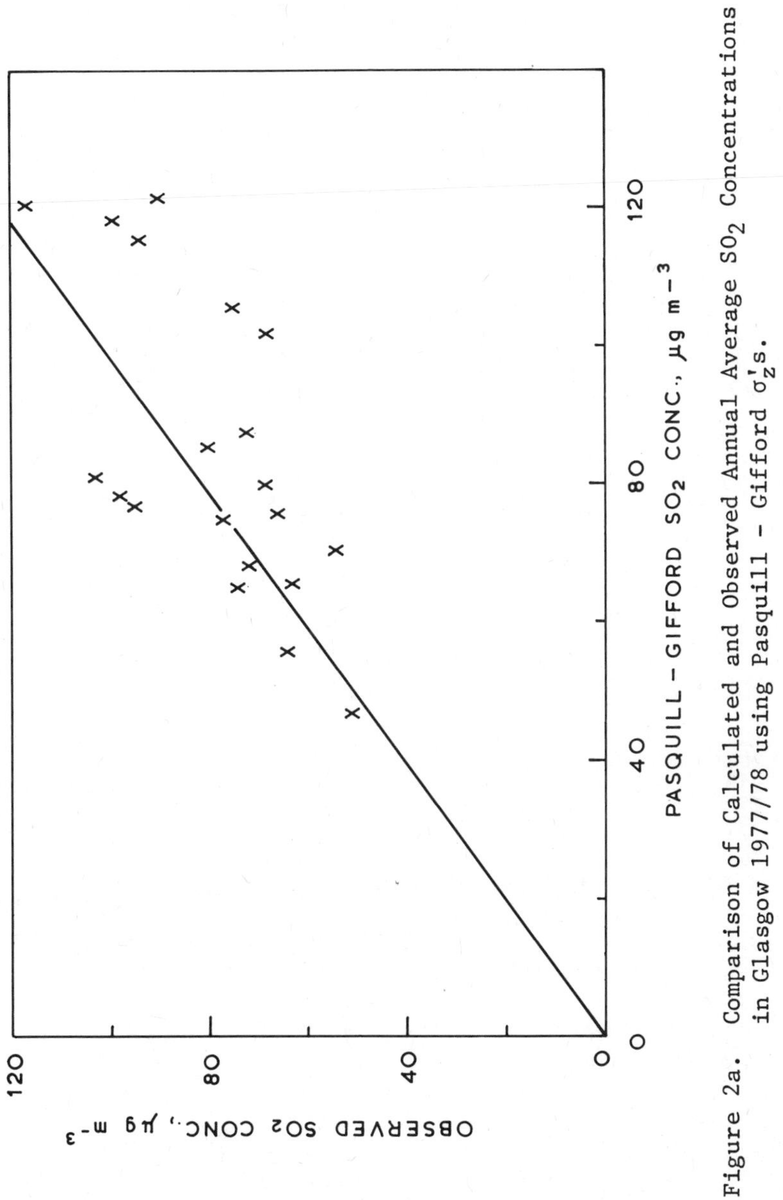

Figure 2a. Comparison of Calculated and Observed Annual Average SO_2 Concentrations in Glasgow 1977/78 using Pasquill - Gifford σ_z's.

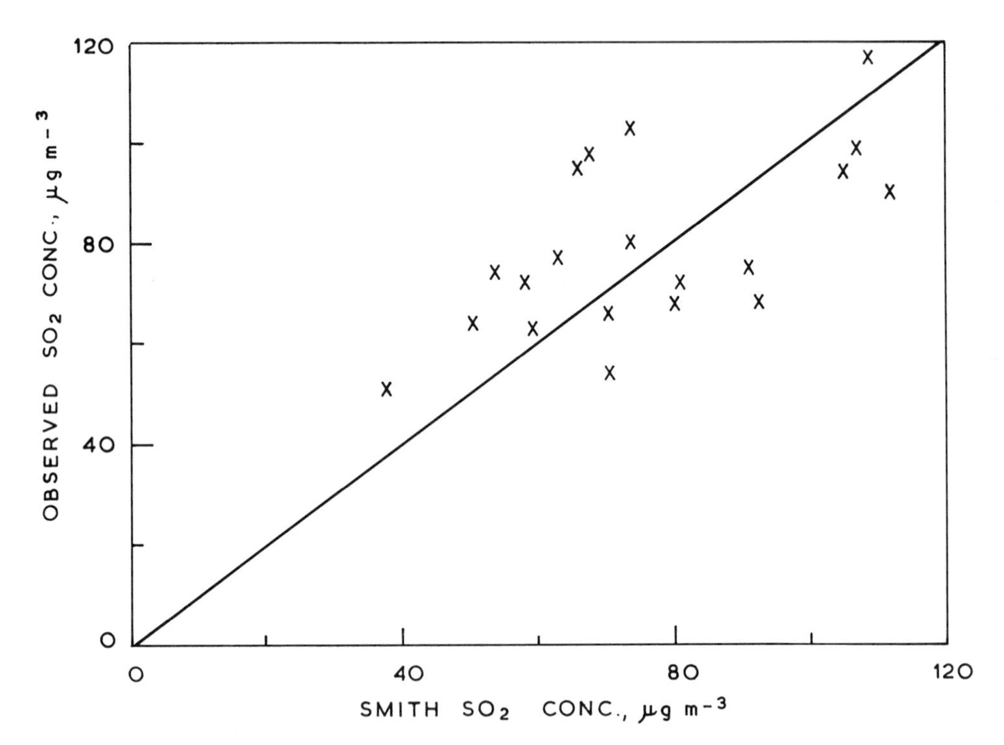

Figure 2b. Comparison of Calculated and Observed Annual Average SO_2 Concentrations in Glasgow 1977/78 using Smith σ_z.

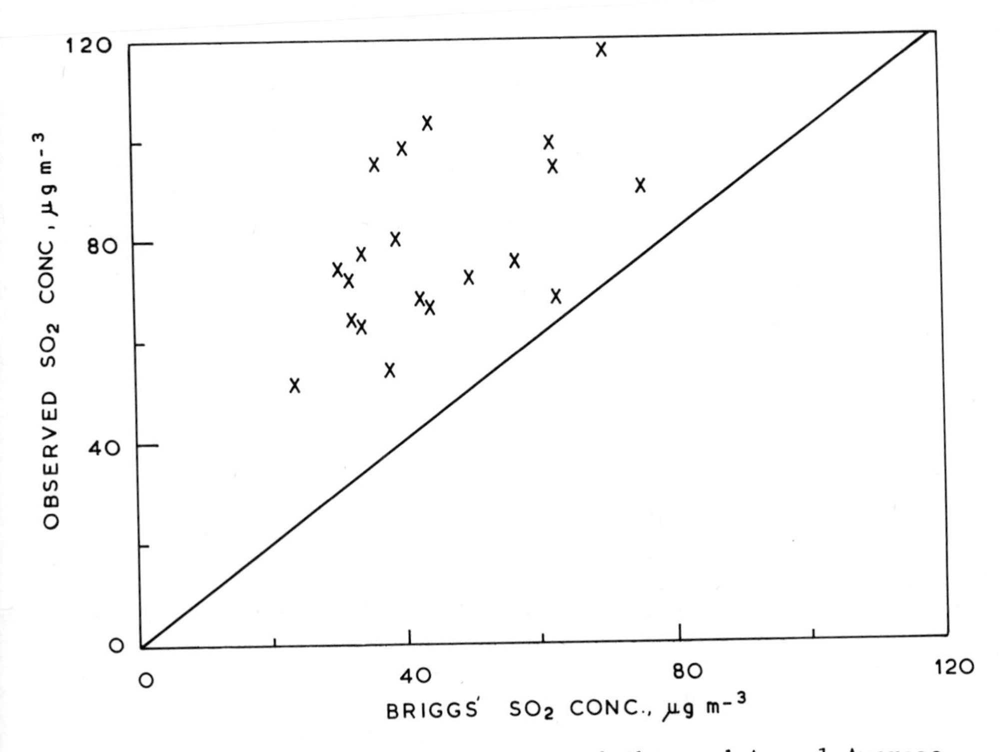

Figure 2c. Comparison of Calculated and Observed Annual Average SO_2 Concentrations in Glasgow 1977/78 using Briggs' Urban σ_z.

$$\chi = \frac{Q}{2\pi\, U\, \sigma_y\, \sigma_z} \exp\left(\frac{-y^2}{2\sigma_y^2}\right) \sum_{n=-\infty}^{+\infty} \left\{ \exp \frac{-(z-h+2nL)^2}{2\sigma_z^2} + \exp \frac{-(z+h+2nL)^2}{2\sigma_z^2} \right\} \quad (3.1)$$

in the usual notation, the infinite sum being evaluated until an additional term only made a difference of ~1%.

The formulation used for σ_y follows the suggestion of Moore[14] for incorporating the so-called 'microscale' turbulence and the lower frequency component of the horizontal spread due primarily to wind direction fluctuations. This is achieved by writing

$$\sigma_y^2 = \sigma_{yt}^2 + \frac{b^2\, T}{U_{10}} x^2 \quad (3.2)$$

where σ_{yt}^2 is the microscale term and is approximated by the Pasquill-Gifford 3 minute σ_y's. In the second term $b^2 = 0.0296$, T is the averaging time in hours, and x is the downwind distance in metres. Although this scheme was originally developed from an analysis of power station data, it has been recommended by the UK Dispersion Modelling Working Group[15] for use when no better or more relevant data are available.

One of the many difficulties in applying the steady state Gaussian plume equations to successive hours (or shorter periods) is that, particularly over urban distance scales in relatively low winds, the travel time of pollutant from source to receptor can exceed the averaging period of the calculated concentrations so that meteorological conditions, for example, which naturally have the same averaging time could change over time taken for the pollutant to travel from source to receptor. While it is of course questionable to use the Gaussian model in very low wind speeds and recognising that to account properly for this effect, time dependent equations should be used, we have attempted to allow for this phenomenon by writing for the concentration during averaging period t (here 1 hour)

$$\chi_t = \alpha\, \chi_{t-1} + (1-\alpha)\, \chi_t \quad (3.3)$$

where α is the proportion of χ_t arising from sources with travel time >1 hour and χ_t, χ_{t-1} are the steady-state Guassian concentrations at hours t and t-1 respectively.

Another feature of the model, as applied here, which is of interest in multiple source urban applications is the calculation of an average pollutant length scale defined as

$$\langle x \rangle_R = \sum_i \chi_i x_i / \Sigma \chi_i$$

where χ_i and x_i are the contribution of source i, distant x_i from receptor R. For the central city receptors in the Glasgow calculations, $\langle x \rangle$ is $\sim$2 km in neutral conditions and $\sim$10 km in F-type conditions.

The emissions from space heating sources were assumed to depend on temperature (or strictly, on degree days) as follows

$$E(T) = E_o [a + b (T_d - T)] \quad f(t)$$

where T is the temperature, E_o is the annual average emission rate, a = 0.33, b = 0.11 and T_d is the so-called datum temperature taken here to be 14.5°C. The diurnal variation factor f(t) was assumed to be 0.7 for $0 < t < 0600$ hrs, 1.15 for $0800 < t < 2000$ and 1.0 for $2000 < t < 2400$, from a consideration of typical diurnal power demand curves.

3.3 Results

The results of the calculations of hourly averages for January 30th 1978, the day with the highest concentrations, are compared in Figs 3 and 4 with those observed at two of the continuous total sulphur monitors described in section 2.1. The Waterloo Street site is in the centre of the city while the Blairbeth Road site is in a suburban area near the southern perimeter of the city. The error bars in the figures represent the values obtained with the higher and lower estimates of the mixing height derived from the SODAR, which were 25 m and 100 m respectively. The points representing calculated concentrations with no error bars correspond to hours with very little uncertainty as to the SODAR-derived mixing height. The shape of the calculated variation is much the same as that observed, which is perhaps not surprising since the relative variation of concentration depends primarily on wind speed, mixing height and emission rate (temperature), the diurnal variation of which should be predicted quite accurately. The absolute agreement is again probably of the order which may be expected for the application of short period models in multiple source areas; the concentrations are generally within a factor of 2. It is interesting to note the markedly poorer agreement at Blairbeth Road for the hours from 1100 onwards. This probably arises from the use of inaccurate wind direction data arising from uncertainties in the measurements at the low wind speeds at this time. This is clearly more important for a site such as Blairbeth Road on the edge of the city and also the emission inventory area, (so that sources outside the emission inventory may contribute) and illustrates the need for good quality meteorological as well as emission data.

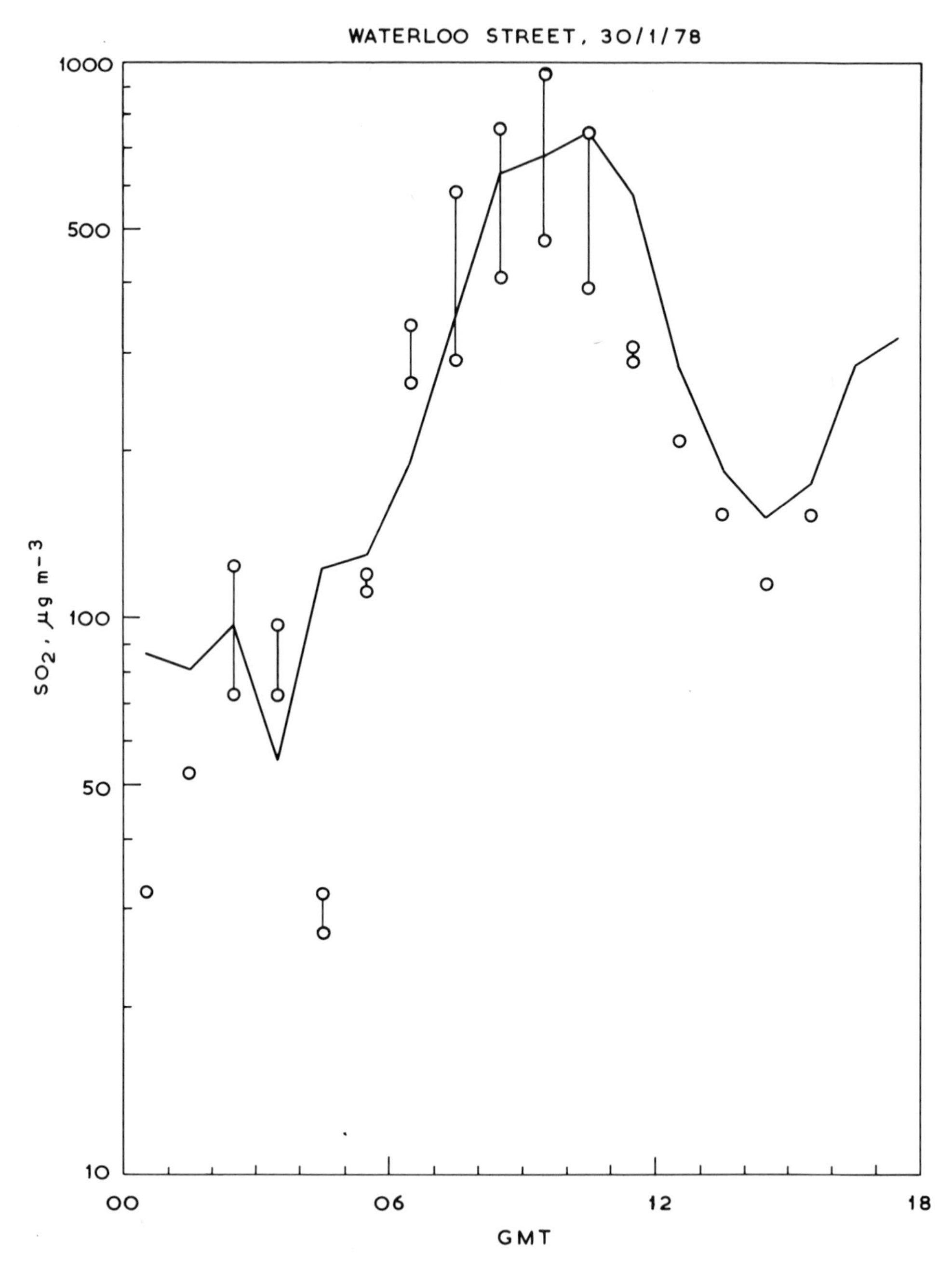

Figure 3a. Comparison of Calculated (-o-) and Observed (—) Hourly Average SO_2 Concentrations in Glasgow 30/1/78, City Centre Site.

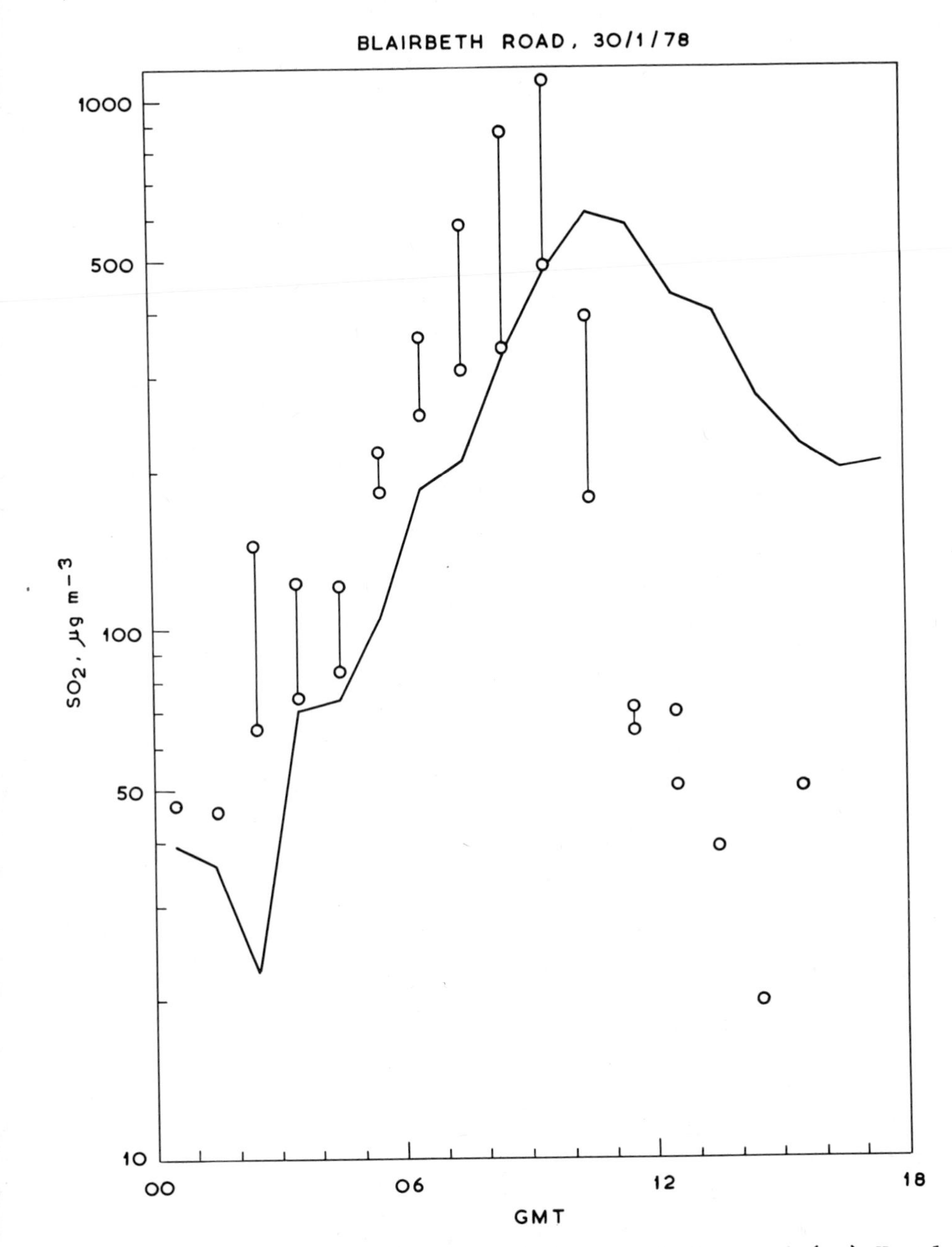

Figure 3b. Comparison of Calculated (-o-) and Observed (—) Hourly Average SO_2 Concentrations in Glasgow, 30/1/78, Suburban site.

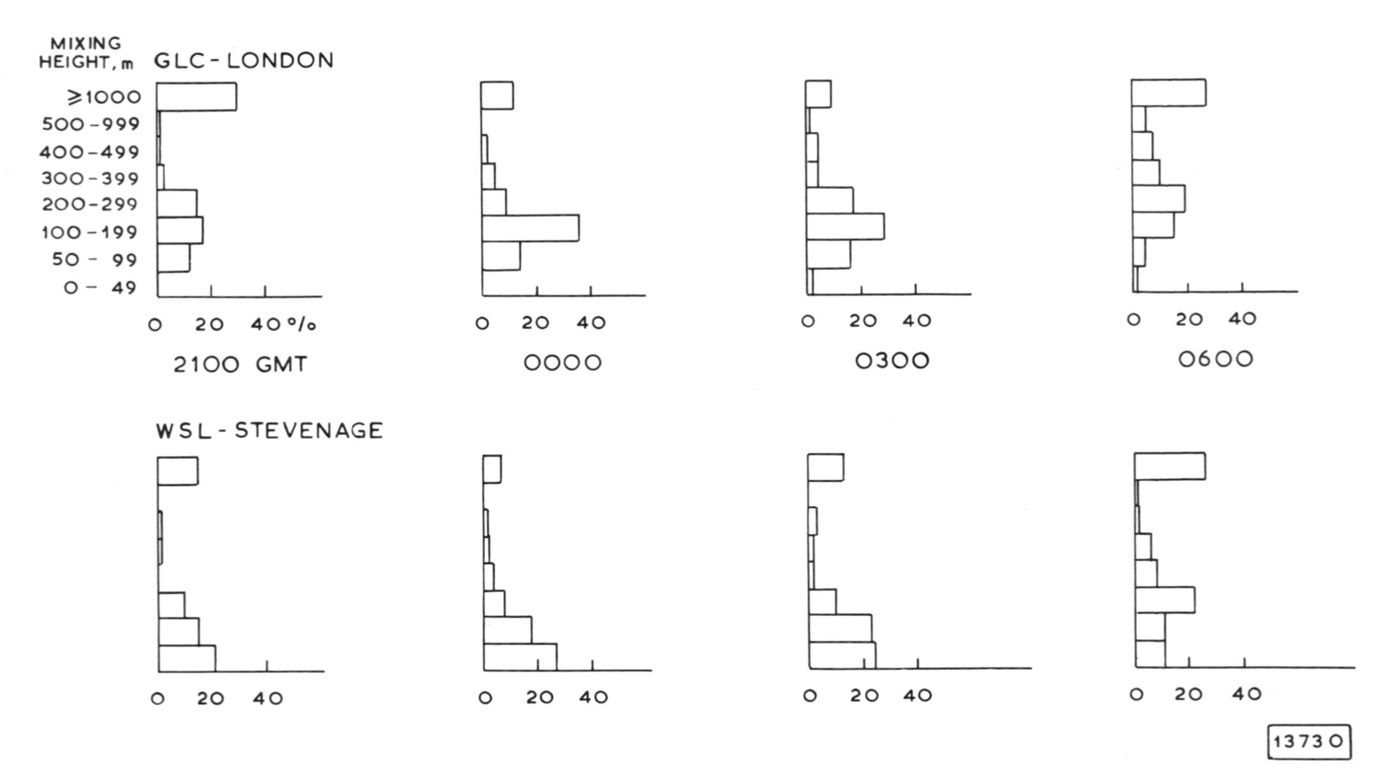

Figure 4. Distribution of Mixing Heights by Hour of Day (Night Hours) at the Central London and WSL Sites.

4. THE USE OF SODAR AS AN AID TO DISPERSION MODELLING

In section 3 the sensitivity of calculated short period concentrations to the precise value of the mixing height or elevated barrier to dispersion was illustrated. This is clearly of most importance when such heights are of the order of 100's of metres and it is in such cases that SODAR, with operating ranges of 500-1000 m on commercial instruments, can provide much useful information, even in the basic monostatic configuration. Some applications of SODAR to air pollution studies carried out by WSL have already been reported[16,17]. In this section some preliminary results of furhter work are described.

The foregoing sections have briefly discussed the limitations of experimental data on dispersion in urban areas. Accordingly in 1981, WSL, with the cooperation of the GLC installed a monostatic SODAR on the roof of County Hall, the GLC headquarters. The site is ∿30 m above street level, is unaffected by nearby reflections or sources other than aircraft and is in the centre of the city. The instrument is being operated continuously, together with an identical instrument at WSL, Stevenage, in a semi-rural (in metoerological terms) situation some 50 km due north of central London. Although the full programme of work has not been completed some preliminary results are presented here. Mixing heights have been obtained for each hour using the coding scheme and methods previously described[16,17]. To allow for the height of the GLC site, 30 m was added to the observed mixing heights for that site. The operating range for the instruments in this application is 1000 m so that if no echo is detectable the mixing height is coded as >1000 m. As an illustration of the type of analysis which can be performed, the data have been presented as frequency distributions of mixing heights for some representative night hours in Fig. 4. In this diagram the data for central London are compared with those obtained at WSL, Stevenage, for the period April-July 1981. Whereas at the semi-rural Stevenage location the majority of hours show some low-level stable layers generally of heights <200 m, there is a virtual absence of such layers below ∿50 m in the central city location where the maximum in the frequency distribution occurs at 100-200 m.

SODAR can thus provide invaluable data for use in short and long period dispersion modelling and this together with simultaneous continuous concentration measurements can prove invaluable in understanding the mechanisms giving rise to elevated pollutant concentrations in urban areas.

ACKNOWLEDGEMENT

The author wishes to thank his colleagues at WSL (G H Roberts, R A Maughan, A M Spanton and R T Giles) for carrying out much of the work described above. The cooperation of the Greater London Council, particularly Dr. D. J. Ball and Amorgie, in providing both the London emission inventory and a site for a SODAR instrument is gratefully acknowledged. The work was sponsored by the UK Department of the Environment and the Scottish Development Department.

REFERENCES

1. M. L. Williams et al', "The Measurement and Prediction of Smoke and Sulphur Dioxide in the East Strathclyde Region : Final Report", WSL Report, to be published, 1982).

2. A. W. C. Keddie et al., "The Measurement Assessment and Prediction of Air Pollution in the Forth Valley of Scotland - Final Report", Stevenage, WSL Report No. LR 279 (AP) (1978).

3. F. Pasquill, "Atmospheric Diffusion", Ellis Horwood, (1974).

4. Department of Energy, "Heat Loads in British Cities", Energy Paper No. 34, HMSO, London (1979).

5. D. Turner, "Workbook of Atmospheric Dispersion Estimates", US Dept. of Health and Welfare, (1969).

6. A. J. Apling et al, "The High Pollution Episode in London, December 1975", Stevenage, WSL Report No. LR 263 (AP), (1977).

7. D. J. Ball and S. W. Radcliffe, "An Inventory of SO_2 Emissions to London's Air", GLC Research Report No. 23, London, (1979)

8. M. Benarie, "Urban Air Pollution Modelling", McMillan, London (1980).

9. D. M. Rote, in "Atmospheric Planetary Boundary Layer Physics, ed., A. Longhetto), Elsevier, Amsterdam, (1980).

10. AMS Committee on Atmospheric Turbulence and Diffusion, Bull. Am. Meteor. Soc., 59 (8), 1025-1026, (1978).

11. J. L. McElroy and F. Pooler, "St. Louis Dispersion Study Vols I and II", US Dept. of Health, Eduction and Welfare, (1968).

12. A. D. Busse and J. R. Zimmerman, "User's Guide for the Climatological Dispersion Model", EPA Report No. EPA-RA-73-024 USA, (1973).

13. G. A. Briggs in "Lectures on Air Pollution and Environmental Impact Analysis", American Meteor. Soc., (1975).

14. D. J. Moore, "Calculation of ground level concentration for different sampling periods and source locations", in Atmospheric Pollution, Elsevier, Amsterdam, (1976).

15. R. H. Clarke (Chairman), "A Model for Short and Medium Range Dispersion for Radionuclides Released to the Atmosphere, National Radiological Protection Board Report NRPB-R91, (1979).

16. R. A. Maughan, "Frequency of potential contributions by major sources to ground level concentrations of SO_2 in the Forth Valley, Scotland", Atm. Environ., 13, 1697-1706, (1979).

17. R. A. Maughan, A. M. Spanton and M. L. Williams, "An Analysis of the frequency distribution of SODAR derived mixing heights classified by atmospheric stability". To be published in Atmos. Environ., (1982).

DISCUSSION

E. RUNCA

Did you analyze, in the validation study of "Long Period Average", the variation of the correlation coefficient between observed and simulated data as a function of the different sets of dispersion coefficient you have used?

M. L. WILLIAMS

Yes we did. The correlation coefficients for Pasquill Gifford and Smith σ_z's were rather similar and better than Briggs. However although one should calculate the correlation coefficient in such cases, I don't believe that by itself it is a very useful parameter. It can be misleading in that one can obtain good correlation coefficients but still have a very wide scatter about a line of perfect agreement.

A. BERGER

Have you used different σ_z' formulas for different meteorological conditions and/or different types of sources?

M. L. Williams

Yes. Although the Pasquill-Gifford curves were developed for low level emissions in rural areas we have used them in urban areas, as have many other workers, by shifting the stable categories towards neutral to simulate the urban effects. The Smith curves, although based to some extent on rural data, are intended to be more generally applicable (given the appropriate surface roughness). The Briggs or McElroy-Pooler curves were, of course, developed for an urban area, although in our experience they do not give as good results as the other schemes used.

RESULTS FROM ELEVATED-SOURCE URBAN AREA DISPERSION EXPERIMENTS

COMPARED TO MODEL CALCULATIONS

Sven-Erik Gryning

Meteorology Section
Risø National Laboratory
DK-4000 Roskilde, Denmark

Erik Lyck

Danish National Agency of Environmental Protection
Air Pollution Laboratory
Risø National Laboratory
DK-4000 Roskilde, Denmark

ABSTRACT. Atmospheric dispersion experiments are carried out in Copenhagen under neutral and unstable conditions in order to study the atmospheric dispersion process in a built-up area. The tracer sulphurhexafluoride is released without buoyancy from a meterological instrumented tower at a height of 115 m, and then collected at ground-level positions in up to three crosswind series of tracer-sampling units, positioned 2-6 km from the point of release. In addition to standard meteorological profile measurements along the tower, the three-dimensional wind fluctuations at the height of release are measured. Characteristic dispersion parameters are estimated from the measured tracer concentrations (averaging time 1 hour), and compared to the dispersion parameters that can be calculated from the atmospheric parameters by various methods. Some of these methods (based on a stability-index) have been in common use for a long time, other (based on wind-fluctuation measurements) reflect recent research. The wind-fluctuation based methods turned out to compare most favourably with the results from the experiments. Based on these experiments a half-empirical model is devised for the prediction of the lateral and vertical dispersion parameters, σ_y and σ_z, for elevated point sources in an urban area under neutral to unstable conditions.

INTRODUCTION

Air quality simulation models have become important tools in air quality planning. However, models are virtually useless until it can be shown that they provide a realistic simulation and meet tolerable accuracy requirements. Model evaluation is best accomplished using detailed data from special observational programs specifically designed for this purpose. The ongoing project described here investigates experimentally the ability of the atmosphere to disperse nonbuoyant effluents released from elevated point sources in an urban environment. Results from these experiments can be used for model evaluation.

DESCRIPTION

The experimental site is a mainly residental area in the northern part of Copenhagen. The roughness length for the area has been estimated to 0.6 m from measurements of vertical wind variance as well as from mean wind profiles. The site, the experimental procedure, the tracer analysis techniques and the tracer sampling-units are described in Gryning et al. (1978), Gryning et al. (1980) and Gryning and Lyck (1980b) and Gryning (1982).

The tracer, sulphurhexafluoride, was released at a height of 115 m from the TV-tower in Gladsaxe, Fig. 1. In addition to the routine measurements of wind speed (10, 60, 120, 200 m; 10 minutes' average), direction (10, 120, 200 m;10 minutes' average) and temperature (2, 40, 80, 120, 160, 200; 10 seconds every 10th minute), along the tower at Gladsaxe, the turbulent wind velocity components were measured during the experiments. These measurements at the height of release were performed with a combination of a lightweight cup anemometer, vane sensor and vertical propeller, Gryning and Thomson (1979).

In the experiments about 20 tracer sampling-units were used in each series. The sampling-units take three consecutive air samples, each collected over 20 min., so that the total averaging time of the three samples was 60 min. Tracer samples were analysed for content of SF_6 by use of electron capture detector gas chromatography. The uncertainty of the absolute tracer concentrations is about 20% and the short-term reproducibility is about 2%. The calibration results show that SF_6-concentrations could be detected down to about 2.10^{-12} ppp (parts SF_6 per parts of air) with a signal-to-noise ratio of 4.

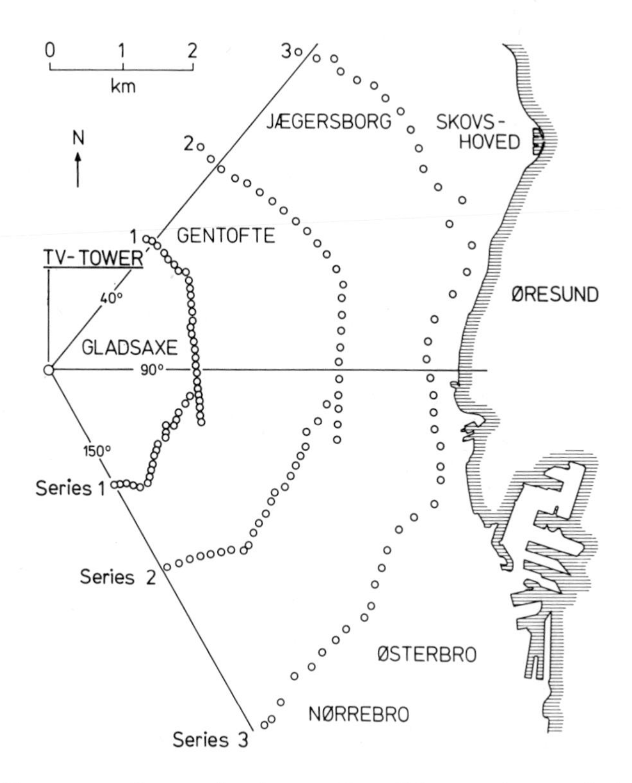

Fig. 1. The experimental site. Sampling-unit positions used at the experiments are indicated by o.

ANALYSIS

Analysis of tracer concentrations

During the experiments the tracer sampling-units were placed near ground-level at preselected positions along roads. Therefore, the measured tracer concentrations primarily allowed for an estimate of the lateral standard deviation of the concentration distribution,

$(\sigma_y)_{mea}$, and for an estimate of the crosswind-integrated concentration distribution at ground-level, $(\chi_{CWI})_{mea}$. As the vertical tracer concentration distribution was not measured, the vertical standard deviation, $(\sigma_z)_{est}$ was indirectly estimated from continuity consideration by assuming that the vertical tracer concentration distribution is of the Gaussian type. Then the crosswind-integrated tracer concentration at ground-level can be expressed in terms of the release height, h, the transport velocity of the plume, u, (here taken as the mean wind speed at the height h), the source strength of the tracer, Q, and the inversion height z_i, as

$$(\chi_{CWI})_{mea} = \sqrt{\frac{2}{\pi}}\,\frac{Q}{u.(\sigma_z)_{est}} \left\{ \exp\left\{-\frac{1}{2}\left(\frac{h}{(\sigma_z)_{est}}\right)^2\right\} + \exp\left\{-\frac{1}{2}\left(\frac{h}{(\sigma_z)_{est}}\right)^2\left(1-\frac{2z_i}{h}\right)^2\right\} + \exp\left\{-\frac{1}{2}\left(\frac{h}{(\sigma_z)_{est}}\right)^2\left(1+\frac{2z_i}{h}\right)^2\right\}\right\} \quad (1)$$

This expression implies specular reflection of the tracer at the ground and at the inversion. Knowing $(\chi_{CWI})_{mea}$, u, Q, h, and z_i, this equation was used to estimate $(\sigma_z)_{est}$.

Methods based on stability classifications

Methods to determine σ_y and σ_z, based on an indirect description of the atmospheric stability, have been put forward in the workbook by Turner (1970) and in ASME's Recommended Guide for the Prediction of Airborne Effluents, edited by Smith (1973). The dispersion parameters by Turner (1970) will be called $(\sigma_y)_{PGT}$ and $(\sigma_z)_{PGT}$ and those by Smith (1973) $(\sigma_y)_{Sm}$ and $(\sigma_z)_{Sm}$. Both sets of parameters are given as function of downwind distance as curves, separated according to a stability classification. In Turner (1970), the classification is based on the mean wind speed at 10 m and insolation, basically according to Pasquill (1974). No classification scheme has been recommended for the use of the Smith (1973) curves, we choose to put Smith's unstable class equal to B and C, and the neutral class equal to D in the Pasquill classification. The dispersion parameters that can be estimated from Turner (1970) are characteristic for rather low level sources, a rather smooth area, (grass, $z_o \sim 0.03$ m), and an averaging time of 3-10 min. The values that can be derived from Smith (1973) are for elevated sources, rough terrain ($z_o \sim 1$ m) and 1 hour averaging time.

Methods based on wind variances

These methods assume knowledge of the turbulent wind fluctuations in the crosswind and vertical directions, σ_θ and σ_ϕ, as well as the wind velocity u, all measured at the height of release.

Taylor's formula for the second moment of the displacement of a continuously released tracer from a point source in stationary, homogeneous turbulence can be written, Pasquill (1974), as

$$\sigma_j^2 = \sigma_i^2 . T^2 \int_0^\infty F_{L,i}(n) \left(\frac{\sin(\pi n T)}{\pi n T} \right)^2 dn; (i,j)=(v,y), (w,z) \qquad (2)$$

where σ_v^2 ($\sigma_v \simeq \sigma_\theta \cdot u$) and σ_w^2 ($\sigma_w \simeq \sigma_\phi \cdot u$) are the variances of the turbulent wind velocity components in the crosswind and vertical direction. $F_{L,i}$ are the normalized Lagrangian variance spectrum functions, T the travel time, and n the frequency.

Hay and Pasquill (1959) proposed to set $R_{L,i}(\beta_j t)=R_{E,i}(t)$, (i,j)= (v,y), (w,z) where $R_{L,i}$ is the Lagrangian and $R_{E,i}$ the Eulerian autocorrelation functions; β_j is a purely empirial, dimensionless Lagrangian-Eulerian time scale ratio. With this assumption connecting the Lagrangian and Eulerian flow properties, Equation (2) can be rewritten as

$$(\sigma_j)_\beta = T \cdot (\sigma_i)_{\tau, T/\beta_j} \qquad (i,j) = (v,y), (w,z)$$

where the double subscript $\tau, T/\beta_j$ indicates that the measured wind velocity record over the period τ, has been smoothed over the time interval T/β_j prior to the calculation of the square root variance. The subscript β indicates that σ_j is derived from this working approximation,which we call the β-method. A value of β_j is derived using the formula of Wandel and Kofoed-Hansen (1962)

$$\beta_j = 0.44 \frac{u}{\sigma_i} \qquad (i,j) = (v,y), (w,z)$$

The travel time, T, is derived by Taylor's hypothesis T=x/u, where x is the downwind distance from the point of release.

Under a number of rather weak assumptions, (Gryning, 1982), Equation (2) generally can be written

$$\sigma_j = \sigma_i \cdot T \cdot f_j(T/t_{L,j}) \qquad (i,j) = (v,y), (w,z) \qquad (3)$$

This formulation of Equation (2) is the starting point in the methods proposed by Pasquill (1976) and Draxler (1976), for estimating σ_y and σ_z from the variance of the wind fluctuations.

Pasquill (1976) argues that Equation (3) for the y-component within fairly close limits follows the relation

$$\sigma_y = \sigma_\theta \cdot x \cdot f_y(x) \tag{4}$$

For the determination of σ_z, Pasquill recommends use of the workbook curves (Turner, 1970). For dispersion over a city Pasquill (1976) suggests that the heat island effect can be taken into account by changing the estimated stability class one half class towards more unstable conditions. The roughness at the area is taken into account directly by enhancement of the σ_z-values by a factor which can be estimated from Smith (1972).

In contrast to Pasquill (1976), Draxler (1976) assumes that the f_j-function depends on the travel time T, i.e. Equation (3) is rearranged in the form

$$\sigma_y = \sigma_\theta \cdot x \cdot f_y(T/T_y)$$

$$\sigma_z = \sigma_\phi \cdot x \cdot f_z(T/T_z)$$

where the travel time, T, is normalized with T_j, being proportional to the Lagrangian time scales. Values of T_j are given by Draxler as a function of the stratification.

In the ASME guide, Smith (1973), formulas for σ_y and σ_z based on σ_θ and σ_ϕ, are suggested. These are not based on Taylor's formula, but on empirical power functions of distance and wind variances. For neutral and unstable conditions, the formulas read

$$\sigma_y = 0.045 . \sigma_\theta . x^{0.86}$$

$$\sigma_z = 0.045 . \sigma_\phi . x^{0.86}$$

The values of σ_y and σ_z that can be derived from Pasquill (1976) will be called $(\sigma_y)_{EPA}$ and $(\sigma_z)_{EPA}$; when derived from Draxler (1976) $(\sigma_y)_{Dr}$ and $(\sigma_z)_{Dr}$, and from Smith (1973) $(\sigma_y)_{S\theta}$ and $(\sigma_z)_{S\theta}$.

COMPARISON BETWEEN CALCULATION METHODS AND EXPERIMENTAL RESULTS

Ten out of a total number of 13 tracer experiments, carried out in the Copenhagen area in the period September 12, 1978 to July 19, 1979, were found suitable for analyses. Because of wind direction shifts the others were discarded. All the dispersion experiments were carried out in neutral and unstable conditions. In Table 1 important parameters for the 10 experiments are shown. Three experiments are performed under near neutral conditions (Pasquill stability class D), 5 experiments under slightly unstable conditions

Table 1. Meteorological conditions during the experiments and results from the analyses of the measured tracer concentrations. All parameters are one hour averages except for the experiment of April 30 in which 40 min. averages have been used. The symbol "-" indicates that the parameter was impossible to determine.

Experiment 1978-1979	Time of experiment	Assigned distance (km)	$(\sigma_y)_{mea}$ (m)	$\frac{u(\chi_{CWI})_{mea}}{Q}$ $(10^{-9}\cdot m^{-1})$	Mean wind speed 115 m (m/s)	Mean wind speed 10 m (m/s)	Wind fluctuations 115 m height lateral $\sigma_\theta(^o)$	Wind fluctuations 115 m height vert. $\sigma_\phi(^o)$	$\Delta T/\Delta z$ (2-40m) $(^oC/100m)$	Monin-Obukhov length (m)	Pasquill stability class
Sep. 14	1523-1623	3.9	375	4.4	8.9	-	7.5	4.5	-0.97	-	D
Sep. 20	1317-1417	1.9 3.7	254 444	2.2 0.8	3.4	2.1	16.0	14.7	-3.33	-46	C
Sep. 26	1140-1240	2.1 4.2	239 438	5.7 3.1	10.6	4.9	7.7	6.2	-3.07	-384	C
Oct. 19	1213-1313	1.9 3.7 5.4	184 284 404	4.1 3.1 2.2	5.0	2.4	9.6	8.4	-2.52	-108	C
Nov. 3	1320-1420	4.0	301	5.4	4.6	2.5	6.0	6.3	-2.06	-173	C
Nov. 9	1330-1430	2.1 4.2 6.1	185 279 376	4.5 3.9 3.3	6.7	3.1	6.9	6.4	-1.54	-577	C
Apr. 30	1302-1342	2.0 4.2 5.9		5.7 3.2 2.7	13.2	7.2	9.8	6.6	-3.99	-569	D
Jun. 27	1245-1345	2.0 4.1 5.3	290 595 786	5.1 2.5 1.7	7.6	4.1	11.8	7.0	-4.65	-136	B-C
Jul. 6	1250-1350	1.9 3.6 5.3	190 402 580	3.9 1.9 1.4	9.4	4.2	8.8	4.7	-7.63	-72	B-C
Jul. 19	1215-1318	2.1 4.2 6.0	236 460 623	4.8 3.3 2.7	10.5	5.1	9.7	5.8	-3.25	-382	D

(Pasquill C), and 2 under slightly unstable to unstable conditions (Pasquill B-C).

Lateral dispersion parameter

Available for the analysis of the lateral dispersion parameter are 21 values of $(\sigma_y)_{mea}$, (60 min. averaged tracer concentrations). The values of $(\sigma_y)_{mea}$ were compared to the estimates of the lateral spread parameter, calculated by the various methods, $(\sigma_y)_{..}$

The deviations between $(\sigma_y)_{..}$ and $(\sigma_y)_{mea}$ are evaluated by the fractional error, defined in the case of σ_y as

$$\text{fractional error} = \frac{(\sigma_y)_{..} - (\sigma_y)_{mea}}{0.5\left[(\sigma_y)_{..} + (\sigma_y)_{mea}\right]}$$

This quantity is used because it is logarithmically unbiased, i.e. a predicted value that is 1/n the measured value produces the same numerical fractional error as one that is n times the measured value. Fig. 2 shows, without taking notice of the differences in downwind distance, the fractional error between $(\sigma_y)_{mea}$ and σ_y, derived from the computational methods. The β - and Dr-methods (based on wind variances) as well as the PGT- and Sm-methods (based on stability classification) are seen on the average to estimate the lateral dispersion parameter well, whereas the EPA-method produces estimates of σ_y that are 37% too low on the average. In the mean the Sθ-method overpredicts $(\sigma_y)_{mea}$ by 27%. It should be noted that $(\sigma_y)_{PGT}$ is actually characteristic of a sampling time of 3-10 min. and therefore can be expected to be smaller than $(\sigma_y)_{mea}$. This is well in accordance with the average findings.

The root-mean-square (r.m.s.) fractional error clearly shows that the methods based on wind variances have the smaller scatter, a r.m.s. fractional error of about 20%, whereas the PGT- and Sm-methods based on stability classification produce results with r.m.s. fractional error of ≃ 30% (PGT) and ≃ 40% (Sm). It is noted that the β- and Dr-methods, which have the smallest r.m.s. fractional error of all, are both based on Taylor's diffusion formula.

Those methods based on an atmospheric stability classification have, as expected, large scatter compared to the wind variance methods. From the comparison it is evident that the methods based on wind variances produce estimates with the smaller r.m.s. fractional error. This therefore, might serve as a promising basis for further attempts to refine the methods to estimate σ_y from relatively simple measurements. It is equally evident that the use of certain of the wind variance-based methods in the Copenhagen area reveal biased estimates of σ_y.

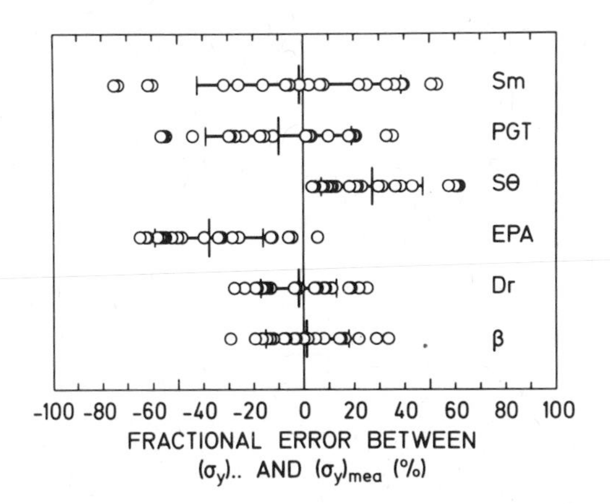

Fig. 2. Comparison of lateral dispersion parameters. The fractional error between the individual values of σ_y, computed from the various methods, and $(\sigma_y)_{mea}$ are indicated by o. Symbols refer to the computational methods, explained in the text. The center bar at each method indicates the mean fractional error. The spread in the distribution is evaluated by the root-mean-square fractional error. The outer bars indicate the mean fractional error plus or minus the roor-mean-square fractional error.

To investigate Pasquill's (1976) and Draxler's (1976) methods in greater detail, the normalized lateral spread, S_y, has been calculated. S_y is defined as

$$S_y \equiv \frac{\sigma_y}{\sigma_v T} \simeq \frac{\sigma_y}{\sigma_\theta x}$$

which, following Taylor's theory, for an infinitely long sampling time can be argued to be a function of T and the Lagrangian time scale $t_{L,y}$

$$S_y = f_y(T/t_{L,y})$$

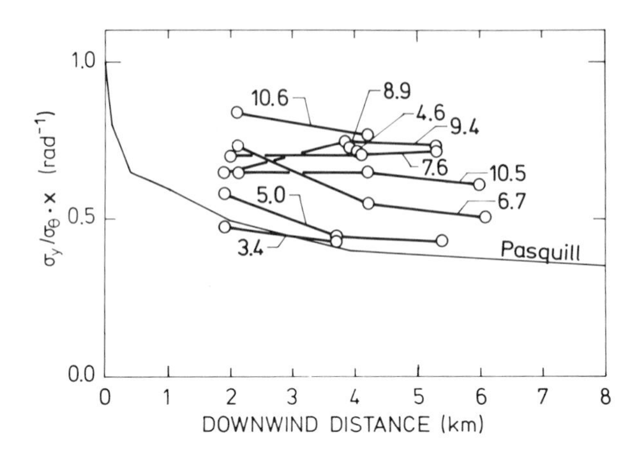

Fig. 3. The normalized lateral spread, $(\sigma_y)_{mea}/\sigma_\theta \cdot x$, as function of downwind distance. The mean wind velocity in units of m/s is indicated. Also shown is the f_y-function proposed by Pasquill (1976).

Pasquill (1976) argues that S_y mainly is a function of downwind dis tance. In contrast, Draxler (1976) assumes that the f_y-function takes the dependence T/T_y where T_y is proportional to the Lagrangia time scale. Figure 3 shows $\sigma_y/\sigma_\theta x$ as a function of x, which const tutes the dependence suggested by Pasquill (1976). The points display a rather large scatter. It is characteristic that all but one point lies above the f_y-function suggested by Pasquill (1976). The proposed x-dependence with no other parameters is seen to be a rather poor approximation to the actually measured normalized sprea At a downwind distance of approximately 2 km, Pasquill suggests f_y(2 km) = 0.5. The value of f_y in these experiments is seen to var in the range 0.48-0.84. For a downwind distance of approximately 4 km Pasquill suggests f_y(4 km) = 0.4. The experimental values of f_y covers the range 0.43-0.77, and for a distance of 6 km, Pasquill suggests f_y(6 km) = 0.37. The experimental values lie in the range 0.43-0.72.

It can be argued that the f_y-function in addition to the x-dependence also depends on the wind velocity; this dependence originates from the finite sampling time. Small values of f_y are associ ed with small wind velocities and vice versa. The measured f_y-valu

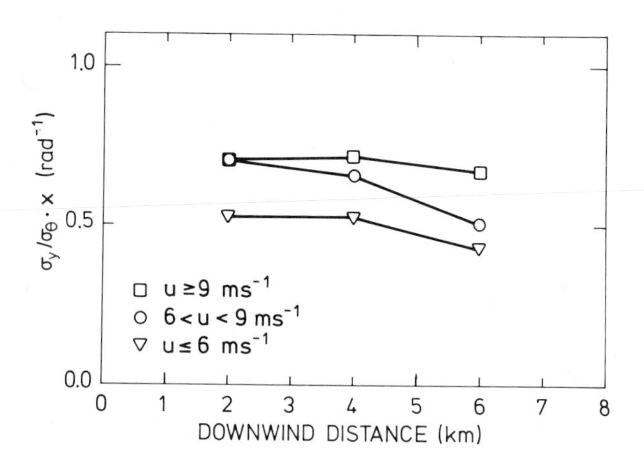

Fig. 4. Average values of the normalized lateral spread as function of downwind distance (∿2 km, ∿4 km, and ∿ 6 km). The values are divided into 3 wind velocity classes, characti-rized by $u \geq 9$ m/s, $6 < u < 9$ m/s and $u \leq 6$ m/s.

shown in Fig. 3, indeed support the existence of this relationship, although the points do not vary with wind speed in a completely systematic manner; this suggests that more than downwind distance and wind speed are required in order to describe the behaviour of the f_y-function. To throw the wind velocity dependence into relief, points characterized by wind velocities $u \geq 9$ m/s, $6 < u < 9$ m/s and $u \leq 6$ m/s has been averaged and the results are illustrated in Fig. 4. The f_y-function suggested by Pasquill (1976), when used for one-hour-averaged concentrations, corresponds to situation that are characterized by very low wind velocities. This finding agrees with the results from the pilot dispersion experiments at Risø, reported by Gryning and Lyck (1980b).

Draxler (1976) suggested that the f_y-function depends primarily on the time of travel, $T = x/u$, and introduces in that way the wind velocity in the f_y-function. This dimensionless spread is plotted in Fig. 5 as function of time of travel. The points display substantial scatter, although obviously not as large as in Fig. 3. The points do not seem to move systematically with wind velocity. The f_y-function for unstable conditions and elevated sources as

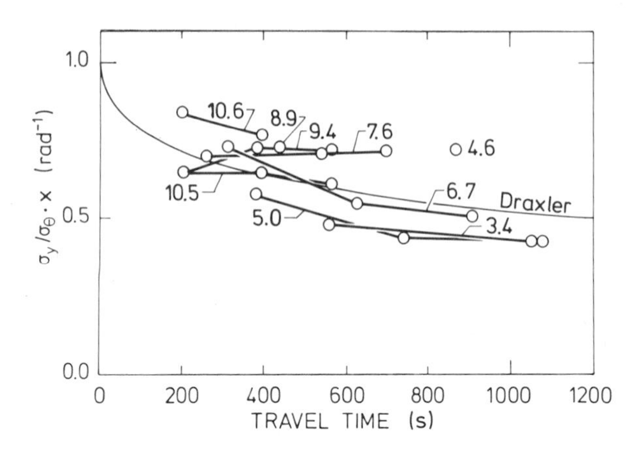

Fig. 5. The normalized lateral spread as function of the travel time, T = x/u. The mean wind velocity in units of m/s is indicated. Also shown is the f_y-function proposed by Draxler (1976).

suggested by Draxler, are also shown, the overall fit is seen to be rather fair.

Vertical dispersion parameter

Twenty-four values of $(\sigma_z)_{est}$, of which three are based on 40-min. averages and the rest are one-hour averages, are available for the analysis of vertical dispersion. We recall that $(\sigma_z)_{est}$ is bas on Equation (1). To estimate the uncertainty on $(\sigma_z)_{est}$ originatin from the uncertainty in the absolute calibration of the gas chromatograph, values of $(\sigma_z)_{est}$ are derived by increasing and decreasing the measured tracer concentrations by 20%, $(\sigma_z)_{est}$ is seen to be ve sensitive to variations in the absolute tracer concentrations. Owi to this substantial uncertainty in $(\sigma_z)_{est}$ it is not found reasonab to carry out a strict comparison between $(\sigma_z)_{..}$ and $(\sigma_z)_{est}$ along lines similar to the σ_y comparisons.

Due to the larger roughness of the urban area as compared to a rural area, larger values of σ_z are expected over a city than over a rural area. In an attempt to isolate this effect, Fig. 6 shows

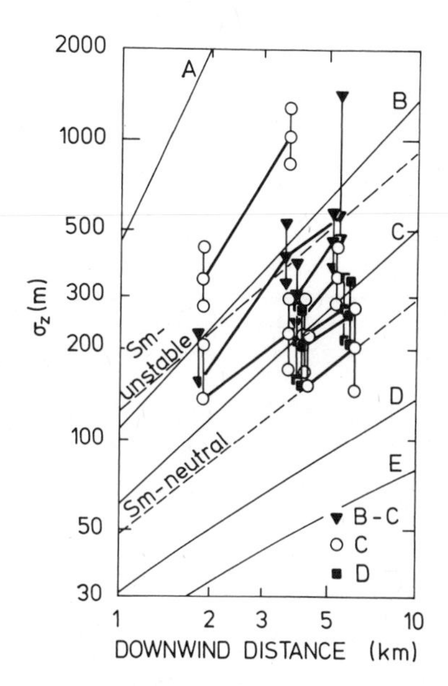

Fig. 6. Values of $(\sigma_z)_{est}$, separated according to the Pasquill stability classification, versus downwind distance. The effect on $(\sigma_z)_{est}$ from the uncertainty in the calibration of the gas chromatograph, is shown as bars. It was derived by increasing/decreasing the measured concentrations by 20%. For comparison, the PGT-curves (Turner, 1970) for the Pasquill stability classes A-E are shown. Also shown are the Sm-curves, (Smith, 1973) for neutral and unstable conditions.

the values of $(\sigma_z)_{est}$, plotted together with the curves for $(\sigma_z)_{PGT}$ and $(\sigma_z)_{Sm}$. From the figure it can be seen that for the experiments assigned Pasquill stability class D, all values of $(\sigma_z)_{est}$ are considerably larger than $(\sigma_z)_{PGT}$ even when taking into account the $\pm 20\%$

calibration uncertainty in the absolute tracer concentrations, and also larger, but to a lesser extent, than $(\sigma_z)_{Sm}$. For the Pasquill stability class C experiments, $(\sigma_z)_{est}$ is seen to scatter around $(\sigma_z)_{PGT}$ with the main part of the values being smaller than $(\sigma_z)_{Sm}$. Concerning $(\sigma_z)_{est}$ for the experiments assigned Pasquill stability class B-C, the points are seen to scatter around $(\sigma_z)_{PGT}$ as well as $(\sigma_z)_{Sm}$.

Maximum concentrations

From an air pollution point of view, absolute concentrations are of particular interest. In this chapter we investigate how well it is possible to calculate the maximum ground-level concentrations at the experimental distances by applying σ_y and σ_z derived from the aforementioned methods. To do this the normalized, ground-level concentration at the centreline, $u\chi_{max}/Q$ were calculated. This quantity was compared to the normalized measured tracer concentrations, $(u\chi_{max}/Q)_{mea}$, where for χ_{max} the maximum measured tracer concentration in the various series have been inserted. Figure 7 shows the values of $(u\chi_{max}/Q)_{..}/(u\chi_{max}/Q)_{mea}$. In the ratio $(u\chi_{max}/Q)_{..}/(u\chi_{max}/Q)_{mea}$ the numerator is based purely on the individual metho and meteorological data, whereas the denominator is based solely on measured tracer concentrations. If prefect agreement were obtained the ratio would turn out to be 1. However, $(u\chi_{max}/Q)_{mea}$ does not actually represent the real ground-level maximum concentration, because the measurements were made at discrete positions. The difference is believed to be minor, however. The concentrations derived with the use of the PGT-method have been corrected for the difference in averaging time. The correction is found following a formula in Turner (1970), $\chi_{60\ min.} = \chi_{10min.} \cdot (10/60)^p$ where p is in the range of 0.17-0.20. Here p = 0.17 was used. For the β-, Dr- and EPA-methods, which are all based on wind variances, the mean fractional error is seen to be less than the 20% that constitutes the uncertainty in the measured, absolute tracer concentrations. Thus none of these three methods can be judged to be superior in their ability to predict the mean of χ_{max}. The wind variance-based S -method has a mean fractional error of -58%, a result that is not surprising remembering the tendency of this method to overpredict both σ_y and, reported in Gryning and Lyck (1980a), σ_z. The stability-based methods, i.e. PGT and Sm, are seen to have a mean fractional error only slightly smaller than the lower limit of the uncertainty on the measured tracer concentrations.

The r.m.s. fractional error describes the ability of the metho to predict the variation of χ_{max}. Here, the wind variance-based methods, i.e. β, EPA, Dr, and Sθ, all come out with r.m.s. fractiona errors of approximately 35%, whereas the stability-based methods have substantially larger r.m.s. fractional errors, 64% for the Sm-method, and 53% for the PGT-method. It is interesting to note how well the EPA-method works, keeping in mind that this method on the average underpredicts σ_y with a mean fractional error of -37% and also underpredicts the crosswind-integrated concentrations by -34% (Gryning and Lyck, 1980a). As the maximum concentration is inversely proportional to σ_y, these effects to a certain extent cancel each other in calculations of the centreline concentrations, rendering the final result as good as the results from the β- and Dr-method.

For these experiments, $(u\chi_{max}/Q)_{..}/(u\chi_{max}/Q)_{mea}$ is seen to vary in the range 0.6-2.8 for the β-method, in the range 0.5-2.3. for the EPA-method, 0.6-2.8 for the Dr-method, 0.3-1.5 for the Sθ-method, and 0.2-2.7 for the Sm-method. The ratio for the widely used PGT-method is seen to vary in the range 0.3-4.0, or more than an order of magnitude.

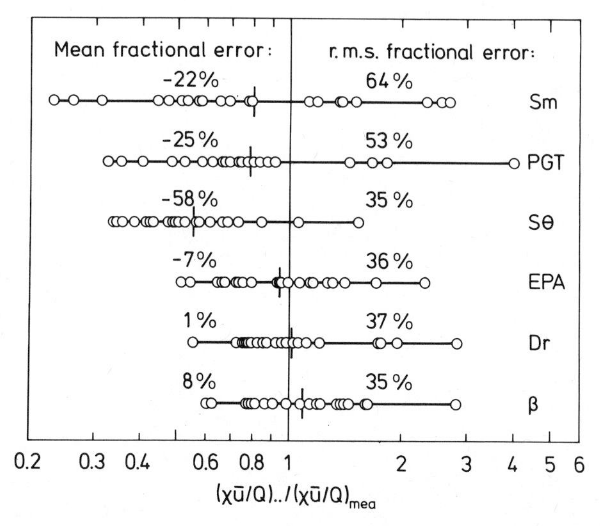

Fig. 7. The ratio of the computed maximum ground-level concentrations, $(u\chi_{max}/Q)_{..}/(u\chi_{max}/Q)_{mea}$, for all experimental distances. Symbols refer to the computational methods, explained in the text. The vertical bar at each method indicates the mean fractional error.

PROPOSED METHOD FOR PREDICTION OF σ_y AND σ_z FOR ELEVATED POINT SOURCES IN AN URBAN AREA IN NEUTRAL AND UNSTABLE CONDITIONS

The results from the tracer experiments in the Copenhagen area will constitute the basis for this attempt to device improvements to the methods of predicting σ_y and σ_z.

Lateral dispersion parameter

An inspection of Fig. 2 reveals that for predictions of the variation of σ_y, methods based on the measured wind variances clearly are superiors compared with stability index-based methods. This judgement is based on the r.m.s. fractional error. In this context the mean fractional error is of no interest; biased values can simply be scaled to give a correct mean value.

For practical calculations of σ_y, it is suggested to take into account the wind velocity dependence of the f_y-function, which is demonstrated in Fig. 3. Values of this revised f_y-function is shown in Table 2. The fractional error between $(\sigma_y)_{mea}$ and σ_y calculated by use of this revised f_y-function was calculated. The mean fractional error turns out to be -3%, and the r.m.s. fractional error 15%.

Table 2. Values of $f_y(x,u)$ to be used in the proposed procedure for evaluating $\sigma_y = \sigma_\theta \cdot x \cdot f_y(x,u)$ with σ_θ in radians

$x(m \cdot 10^3)$	0	1	2	4	6
$u \geq 9$ m/s	1	0.85	0.75	0.70	0.65
$6 \leq u < 9$ m/s	1	0.80	0.70	0.60	0.55
$u \leq 6$ m/s	1	0.70	0.55	0.45	0.40

Vertical dispersion parameter

Altough the wind variance-based methods were substantially better for predicting σ_y than were the stability-based ones, this is not the case for the prediction of the vertical dispersion parameter, judged from the ability to predict the measured crosswind-intergrated ground-level tracer concentration. From Fig. 4 in Gryning and Lyck (1980a) it can be seen that the r.m.s. fractional error for the individual methods for all practical purposes may be judged to be equivalent, although the wind variance based methods are seen to work a little better.

Concerning the stability-based methods, an increase of σ_z compared with $(\sigma_z)_{PGT}$ is expected due to the difference in surface roughness between the experimental site and the characteristic surface roughness for the PGT-curves. This effect is expected to be most pronounced under neutral conditions, and insignificant under fully convective conditions. This is fully reflected in Fig. 6. It is suggested that predictions of σ_z should be based on the PGT-curves, using the C-curve for experiments assigned Pasquill stability class D as well as C, while the A and B curves should be employed in the usual way. This proposal is limited to neutral and unstable meteorological conditions and release heights and surface roughness as existing in these experiments.

The mean fractional error between the value of the cross-wind-integrated concentrations that can be calculated by this new method, and the actually measured values is -3% and the r.m.s. fractional error turns out to be 35%.

ACKNOWLEDGEMENT

The authors appreciate the continuing support of H. Flyger, N.E. Busch and N.O. Jensen. E. Lund Thomsen carried out the calibration of the gas chromatograph and performed the tracer analysis. U. Torp furnished routine meteorological measurements. H.S. Buch of the Danish Meteorological Institute provided forecasts for the experiments. Finally K. Andersen devoted evenings and much efforts, carefully converting manuscript into typescript, and then cheerfully retyped the typescripts. The technical staff at the Air Pollution Laboratory and Meteorological Section provided practical support. The routine meteorological measurements at the TV-tower are performed by the Danish Meteorological Institute. The TV-tower belongs to the Danish Department of Post and Telecommunication. The study was sponsored by the Danish National Agency of Environmental Protection and Risø National Laboratory.

REFERENCES

DRAXLER, R.R., 1976, Determination of atmospheric diffusion parameters. Atmos. Environ., 10:99.

GRYNING, S.E., LYCK, E., and HEDEGAARD, K., 1978, Short-range diffusion experiments in unstable conditions over inhomogeneous terrain. Tellus, 30:392.

GRYNING, S.E., and THOMSON, D.W., 1979, A tall tower instrument system for mean and fluctuating velocity, fluctuating temperature and sensible heat flux measurements. J. Appl. Meteorol., 18:1674.

GRYNING, S.E., PETERSEN, E.L., and LYCK, E., 1980, Elevated source SF_6-tracer dispersion experiments in the Copenhagen area. Preliminary results I. Proceedings of the tenth international technical meeting on air pollution modelling and its application. NATO/CCMS Air Pollution Pilot Study, Rome, Italy, October 22-26, 1979, 119.

GRYNING, S.E., and LYCK, E., 1980a, Elevated source SF_6-tracer dispersion experiments in the Copenhagen area. Preliminary results II. In: European seminar on radioactive releases and their dispersion in the atmosphere following a hypothetical reactor accident. Vol. 2. Risø, April 22-25, 1980. (Commission of the European Communities, Luxemburg), 905.

GRYNING, S.E., and LYCK, E., 1980b, Medium-range dispersion experiments downwind from a shoreline in near neutral conditions. Atmos. Environ., 14:923.

GRYNING, S.E., 1982, Elevated source SF_6-tracer dispersion experiments in the Copenhagen area. Risø-R-446.

HAY, J.S., and PASQUILL, F., 1959, Diffusion from a continuous source in relation to the spectrum and scale of turbulence. In: Atmospheric diffusion and air pollution. Proceedings of a symposium held at Oxford, August 24-29, 1958. Ed. by F.N. Frenkiel and P.A. Sheppard (Advances in Geophysics, 6) Academic Press, London, 345.

PASQUILL, F., 1974, Atmospheric diffusion. The dispersion of windborne materials from industrial and other sources. 2. edition. Wiley, London.

PASQUILL, F., 1976, Atmospheric dispersion parameters in Gaussian plume modelling. Part II. Possible requirements for change in the Turner workbook values. EPA-600/4-76-030B. PB-258036:3 GA.

SMITH, F.B., 1972, A scheme for estimating the vertical dispersion of a plume from a source near ground-level. Proceedings of the third meeting of the expert panel on air pollution modelling. A report of the air pollution pilot study, Nato Comittee on the Challenges of Modern Society, Paris, France, XVII-2 to 14.

SMITH, M.E., (Ed.) 1973, Recommended guide for the prediction of the dispersion of airborne effluents. 2. American Society of Mechanical Engineers, New York.

TURNER, D.B., 1970, Workbook of atmospheric dispersion estimates. HPS Pub. 999-AP-26.

WANDEL, C.F., and KOFOED-HANSEN, O., 1962, On the Eulerian-Lagrangian transform in the statistical theory of turbulence. J. Geophys. Res., 67:3089.

DISCUSSION

A. VENKATRAM

Have you related σ_y and σ_z to the convective velocity scale w_x? Have you analysed your results within the framework of new theories on dispersion in the convective boundary layer?

S.E. GRYNING

Yes, I have compared the results from these experiments with the results from Willis and Deardorff's tank experiments, that simulate dispersion under convecting conditions. The general behaviour as reported by Willes and Deardorff of the ground-level crosswind-integrated concentrations for near-surface releases is supported by the results from the experiments in Copenhagen, but it is characteristic that, even considering the substantial scatter, they all are larger than the values reported by Willis and Deardorff for near-surface releases. This means that the value of the actually measured crosswind-integrated concentrations for the experiments in full scale are larger than those reported from the tank experiments. A tentative explanation for this difference might be found in the circumstances that a substantial part of the dispersion takes place in the lower region of the boundary layer, where mechanical turbulence is not negligible. Details of this comparison can be found in Gryning (1982).

CH. BLONDIN

Do you think that, according to your concluding remark which shows that the σ_y is not only a function of x, but also of mean velocity, the relevant parameter to estimate σ_y is the travel time?

S. E. GRYNING

Rather crude theoretical arguments, that can be found in Gryning (1982), suggest that the normalized lateral spread is a function of x and u (distance and wind-velocity). The x-dependence originates from the travel time filter in Taylors dispersion formula and the u-dependence originates from the filtering of the spectrum due to the finite sampling time. I do not presently think that x and u should be managed as $t = x/u$, as this is not suggested by the theoretical arguments referred to above, but I think that more parameters than x and u have to be included in order to describe more accurately σ_y. Using the x dependence only is surely insufficient in calculations of σ_y.

INFLUENCES OF DATA BASES ON MODEL PERFORMANCE

Ronald E. Ruff

SRI International
Menlo Park, California 94025

INTRODUCTION

Until recently, the ability to provide objective evaluations of air quality models has suffered from :

- The scarcity of adequate test data bases.
- The lack of a standardized statistical protocol for model evaluation.

These problems have been somewhat alleviated in recent years, because a number of governmental and industry groups have sponsored large field projects to collect adequate data bases for evaluating models, and because the American Meteorological Society (AMS), under sponsorship of the U.S. Environmental Protection Agency (EPA) has taken the initiative in standardizing a model-validation protocol along with specific performance measures (Fox, 1981). However, much of this work is still in an initial phase, with results not yet in the literature.

One recent study (Ruff, 1980) applied some of the AMS measures to evaluate the performance of the RAM air quality model using sulfur dioxide (SO_2) data from the St. Louis-based Regional Air Pollution Study (RAPS). In addition, this study used certain statistical tests to help diagnose the causes of poor performance at certain RAPS monitoring locations. These test results indicated that some of the poor performance could be attributed to inconsistencies in the data base.

This paper describes results from the diagnostic tests mentioned above. We conclude that a thorough statistical evaluation of the data base should be conducted as part of the model evaluation process.

APPLICATION OF THE RAM USING RAPS DATA

The RAM (Turner and Novak, 1978) was developed by EPA as a tool for estimating hourly SO_2 surface concentrations in relatively flat urban areas; these hourly estimates can be used to estimate SO_2 concentrations for three-hour, 24-hour, monthly, seasonal, and annual averaging times. This Gaussian-plume, multiple-source (point and area) model requires as input hourly emissions and meteorological data. Hence, the model output consists of SO_2 concentration estimates for each hour.

The RAPS was conducted in St. Louis because that geographical area already had a fairly good historic air quality data base and was one of the least complicated urban areas to model. (St. Louis has fairly level terrain and is free of pollutant intrusion from other large urban areas.) The heart of the RAPS is the 25-station air-monitoring network that collects data on the criteria pollutants (including SO_2) and meteorological parameters (wind, ambient temperature, dew point temperature, and temperature gradient). The geographical layout of the network highlighting the 13 stations used in our analysis is shown in Figure 1. The data base is rich not only in the SO_2 measurements needed to evaluate model performance, but also in the emission and meteorological measurements needed to understand the limitations of the RAM input parameters. Other documents (e.g., Schiermeier, 1978) give comprehensive documentation on RAPS.

The RAM was executed for each hour of the year 1976 using wind speed and direction interpolated from measurements at the 25 monitoring stations. The model-evaluation data base for each hour includes calendar date, time, wind speed, wind direction, atmospheric stability class, mixing height, total point-source emissions, total area-source emissions, and 13 pairs of observed and estimated SO_2 concentrations.

SUMMARY OF RAM PERFORMANCE

A host of performance measures and their confidence intervals were computed for RAM. These are not presented here because they are not particularly relevant to the theme of this paper. Instead, we summarize RAM performance in terms of visual displays. In this paper we are concerned with "bias" between observed and predicted concentrations. Bias exists when the model inherently overpredicts

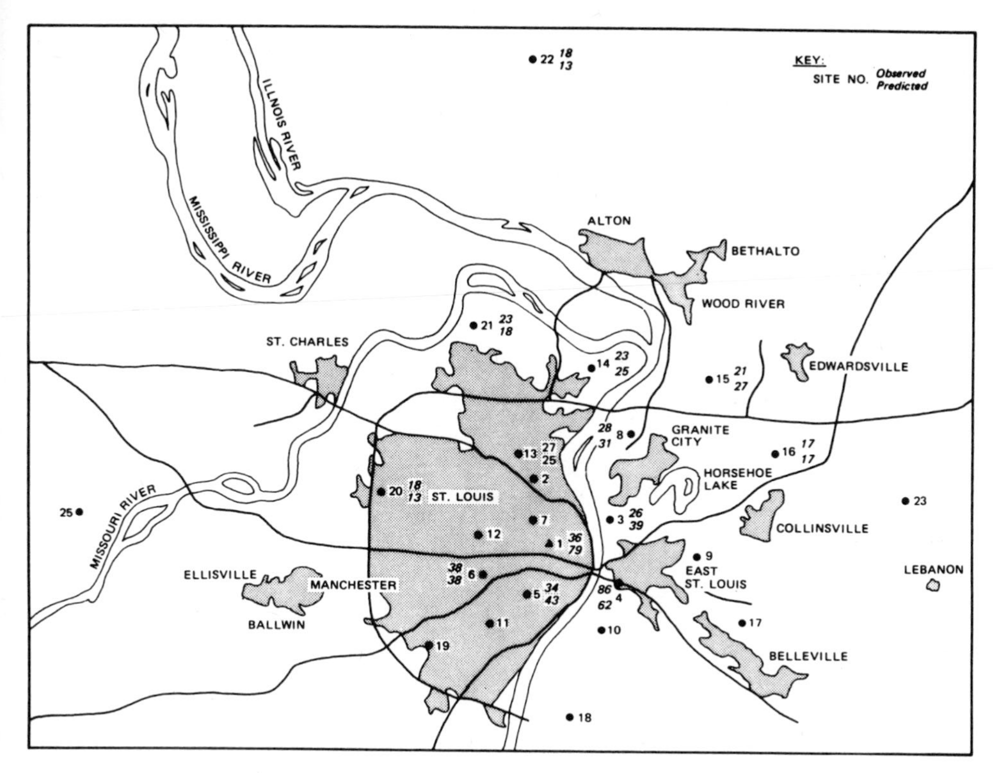

Fig. 1. RAPS SO_2 monitoring-stations showing annual averaged observed and RAMS-predicted concentrations (μg/m^3) (Add 100 to site designators to get standard RAPS site designator)

or underpredicts. (This contrasts to purely random behavior, in which case the model sometimes overpredicts and sometimes underpredicts but shows no preference for either.)

One very quick way of observing overall bias is to display observed and predicted annual concentrations on a geographical map as shown in Figure 1. Note that the agreement is good for some sites (e.g., 106 and 116) and not so good for others (e.g., 101 and 104): Site 101 overpredicts by a factor of two, while Site 104 underpredicts by about 30 percent. This analysis gives us a quick identification of bias for each site. The next step is to examine the frequency distributions for each site.

In Figure 2, cumulative frequency distributions of observed and predicted concentrations are displayed for Sites 101, 104, and 122. Distributions for other sites tend to fall within the extremes of agreement represented by the plots in Figure 2.

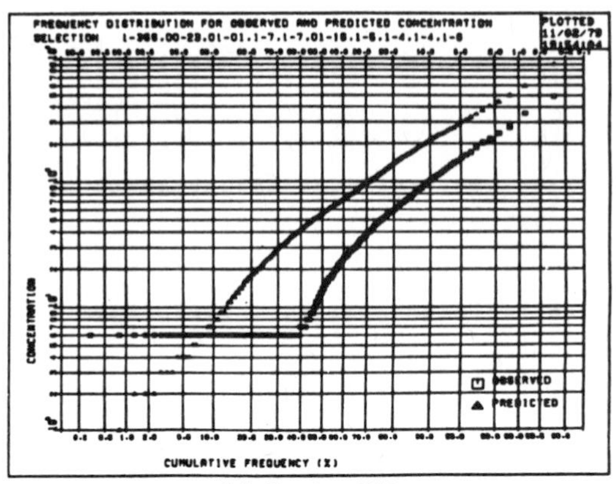

(a) Station 101

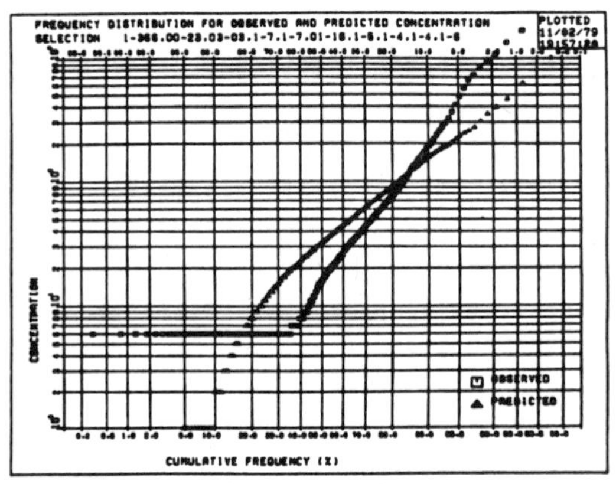

(b) Station 104

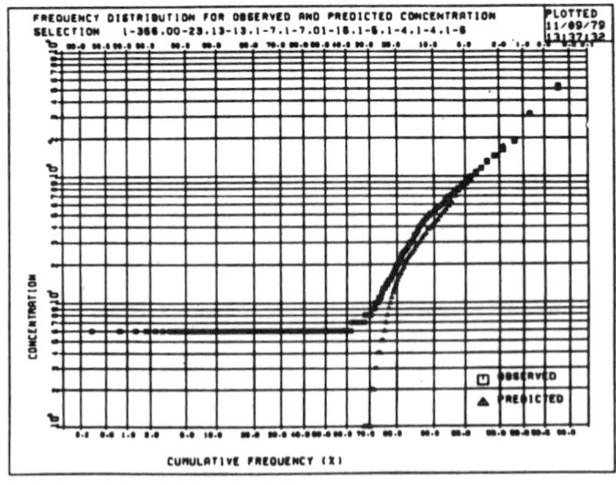

(c) Station 122

Fig. 2. SO_2 frequency distributions for 1976 hourly averaged observed and predicted concentrations ($\mu g/m^3$) for selected RAPS site. The minimum concentration in the RAPS data base is $6 \mu g/m^3$.

For Site 101, the observed and predicted concentrations differ (overpredict) almost uniformly throughout the distribution, but the crossover for Site 104 indicates the RAM tends to overpredict at the low concentrations and underpredict at high concentrations. For Site 122, we demonstrate a site of "good" agreement between observed and predicted concentrations. Distributions for Sites 103, 105, and 106 (not shown here) also exhibit a bias toward overprediction, although the magnitude of the bias is not as great as for Site 101. Bias was negligible for the other sites.

DIAGNOSIS OF PERFORMANCE AT SELECTED SITES

Based on the results presented above, it can be stated that an overprediction bias exists at several of the sites closest to the center of the city. One exception is Site 104, where the bias has two characteristics--overprediction for measured concentrations of 100 $\mu g/m^3$ or less and underprediction for values greater than 100 $\mu g/m^3$.

RELATING MISPREDICTIONS AMONG MONITORING SITES

The first step in our diagnostic procedure was to determine if the errors among sites were interrated. This was done by comparing the linear correlation coefficients of the residual concentrations among the sites for the same hourly time periods. (Residual concentrations are simply the difference between the observed and predicted concentrations). This procedure, called the interstation error-correlation test, resulted in the matrix of correlation coefficients given in Figure 3.

In reviewing the interstation error correlation coefficients presented in Figure 3, we find that errors at certain sites near Site 101 are moderately correlated to errors at Site 101. (The correlation coefficients between errors at Site 101 and Sites 103, 105, and 106 are 0.51, 0.54, and 0.43, respectively.) We also note that the errors for Site 104 do not correlate well with these other sites. The only other moderate correlation for sites not mentioned exists between Sites 115 and 116 (a correlation coefficient of 0.47). As stated previously, examination of the frequency distribution for Sites 115 and 116 did not reveal much bias.

Analysis of the interstation error correlation test in concert with other results leads to the following conclusions:

- Bias exists in the vicinity of Sites 101, 103, 105, and 106. Isolating the cause of errors at one of these sites may have benefit at the other sites as well.

INTER-STATION ERROR CORRELATION TESTS.

ENTER DATA ELEMENT ADDRESS OR 11 FOR FILE
ENTER OPTION 1 OR 2
SELECTION 98-998,00-23,01-13,1-7,1-7,01-16,1-5,1-4,1-4,1-6

NO. OF DATA: 598 NO. OF SITES: 13 IOPT: 2 SITE1: 0 SITE2: 0

CORRELATION MATRIX ...

SITE	101	103	104	105	106	108	113	114	115	116	120	121	122
101	1.00	.51	.23	.54	.43	.33	.28	.18	.21	.12	.08	.20	.11
103	.51	1.00	.27	.36	.27	.41	.26	.15	.08	.08	.10	.23	.10
104	.23	.27	1.00	.33	.24	.14	.10	.21	.18	.06	.17	.20	.15
105	.54	.36	.33	1.00	.44	.33	.23	.15	.16	.05	.12	.21	.12
106	.43	.27	.24	.44	1.00	.29	.50	.34	.06	-.03	.19	.30	.22
108	.33	.41	.14	.33	.29	1.00	.25	.20	.07	.07	.04	.16	.27
113	.28	.26	.10	.23	.50	.25	1.00	.29	.04	.02	.32	.30	.30
114	.18	.15	.21	.15	.34	.20	.29	1.00	.11	.11	.35	.11	.15
115	.21	.08	.18	.16	.06	.07	.04	.11	1.00	.47	.11	.01	.03
116	.12	.08	.06	.05	-.03	.07	.02	.11	.47	1.00	.08	-.02	.08
120	.08	.10	.17	.12	.19	.04	.32	.35	.11	.08	1.00	.22	.23
121	.20	.23	.20	.21	.30	.16	.30	.11	.01	-.02	.22	1.00	.28
122	.11	.10	.15	.12	.22	.27	.30	.15	.03	.08	.23	.28	1.00

95-PERCENT CONFIDENCE LIMITS MATRIX ...
45:>

Fig. 3. Interstation error correlation matrix.

- Results for Site 104 do not follow the spatial trend. Some of the causes for rather large errors at this site are not related to causes for errors at other sites (i.e., a highly localized problem).

DIAGNOSING THE CAUSE OF POOR PERFORMANCE AT SITE 104

We systematically ran further statistical tests to diagnose the causes of poor performance at sites close to the city center. We concluded that the overpredictions for Sites 101, 103, 105, and 106 were not related to inconsistencies in the data base. Instead, it appeared that the errors were related to the manner by which many small commercial sources were combined and represented as one large area source.

As it happens, the area source emissions are highest in the vicinity of Site 101. In fact, the westward adjacent grid has the highest emission in the inventory at 400,000 kg of SO_2 per year. Using techniques outlined by Turner (1970), we estimated that, for a neutral stability, the ratio of concentration to wind speed at Site 101 is 67 $(\mu g/s)/m^3$. This concentration, together with contributions from other nearby area sources, could yield a concentration consistent with the annual overprediction of 43 $\mu g/m^3$ at

Site 101. Other sites (103, 105, and 106) were assumed to overpredict for similar reasons (because the errors are statistically related).

As shown in the results from the interstation error-correlation test discussed earlier, Site 104 is somewhat of an anomaly; its residuals do not appear to be related to those from other nearby sites. Hence, rectifying the bias at Site 101 might also rectify some bias at Sites 103, 105, 106 and others, but perhaps not as much at Site 104. Therefore, we attempted to diagnose the causes of poor performance at Site 104 independently.

The most useful statistical test here was a discriminant analysis (Nie et al., 1975). The discriminant analysis allows one to analyze meteorological and emission conditions that correspond to some user-defined criteria, such as large overpredictions or large underpredictions.

Several discriminant functions were examined. All entailed separation of the residuals into three classes:

- Group 1: Agreement between the observed and predicted concentrations within preset limits (i.e., a symmetric threshold about the agreement limits of 600 $\mu g/m^3$).
- Group 2: Underprediction by more than the preset threshold of 600 $\mu g/m^3$.
- Group 3: Overprediction by more than the preset threshold of 600 $\mu g/m^3$.

The analysis revealed that of the 2342 cases examined, 2304 fell into Group 1 (i.e., agreed within the 600 $\mu g/m^3$ limit); 30 cases resulted in underprediction (Group 2); 8 cases resulted in overprediction (Group 3).

Examining the computer output tables of parameter means and standard deviations led to the following conclusions:

- Underpredictions occur most frequently during slightly stable conditions, low wind speeds, low mixing heights, very low temperatures (averaging 0.1333°C), and high point-source emissions.
- Overpredictions occur most frequently (in comparison to underpredictions) during more stable, lower wind speeds, and lower mixing heights. Estimated point-source emissions are much lower than average, and mean temperatures are higher than normal.
- Wind directions do not favor over- or underpredictions.

- The amount of underprediction is considerably higher than the amount of overprediction for the 30 cases in Group 2. The average underprediction was 1058 μg/m³.
- Because the standard deviations of model input values for each group are fairly high compared to the means, there are many exceptions to the general conclusions listed above.

The above results did not help us isolate the cause for the large mispredictions. The next step was to examine specific cases. Because the number of cases of underprediction was not large, it was a fairly simple matter to examine the 30 cases of underprediction on a case-by-case basis. This process revealed that most of the large underpredictions occurred during December, favoring no particular time of day nor wind direction and occurring quite often over extended periods of time (e.g., several hours). Figure 4 illustrates one particular extended period.

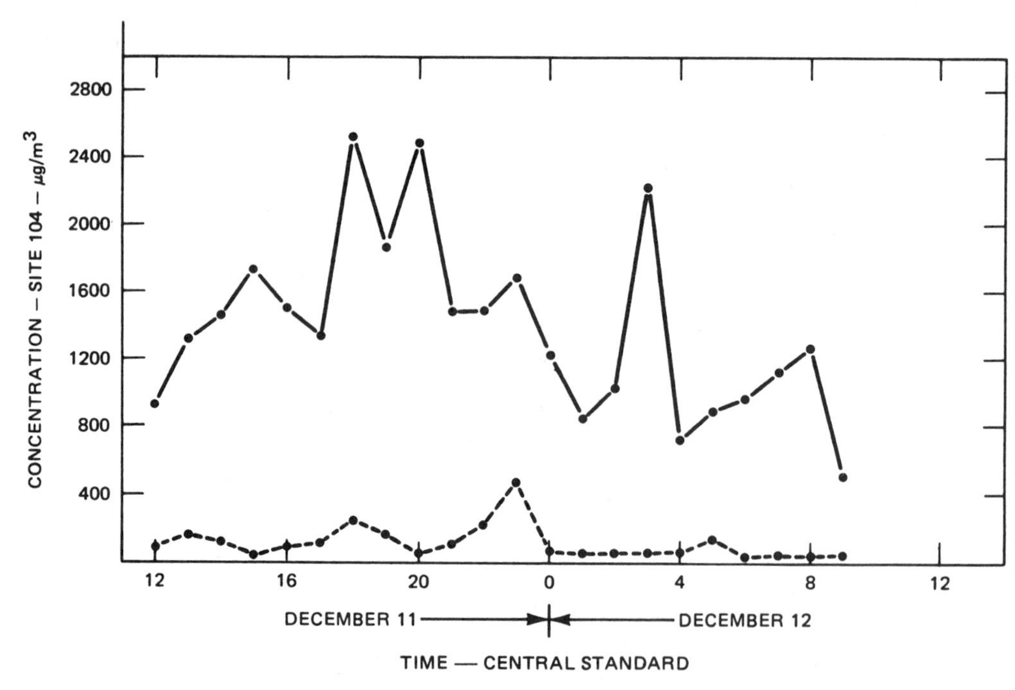

Fig. 4. Time sequence of high observed concentrations (solid line) at Site 104 during December 1976. (Predicted concentrations are shown by the dashed line.)

Table 1 contains a breakdown of the cases by 20° wind-direction sectors for the 22 (out of 30) cases for which wind speeds were 2 m/s or greater. The results in Table 1 show that large underpredictions occur primarily when the winds are out of the south to west sectors.

After much subjective analysis, the results indicated that there are at least two factors influencing the RAM performance at Site 104. Referring to Figure 2, it can be observed that for the lower concentrations (which occur most frequently), the model overpredicts, and for the high concentrations (which occur infrequently), the model severely underpredicts. The two extremes of the frequency plot could be a manifestation of these two controlling factors:

- The overprediction at lower concentrations bears many similarities to the overall performance at Site 101. For Site 101, it was postulated that overpredictions occurred because the emission inventory represented too many small, individual emission sources as one large area source. Furthermore, the interstation error-correlation test demonstrated moderate

Table 1. Site 104 Underprediction of 600 $\mu g/m^3$ and Greater by Wind Direction

Wind direction (degrees)	December 1976	Total 1976
0-20	0	0
20-40	1	1
40-60	0	0
60-80	0	0
80-100	0	0
100-120	0	0
120-140	1	1
140-160	2	2
160-180	0	0
180-200	2	2
200-220	0	1
220-240	3	4
240-260	4	4
260-280	4	4
280-300	0	0
300-320	0	0
320-340	0	0
340-360	3	3
Total	20	22

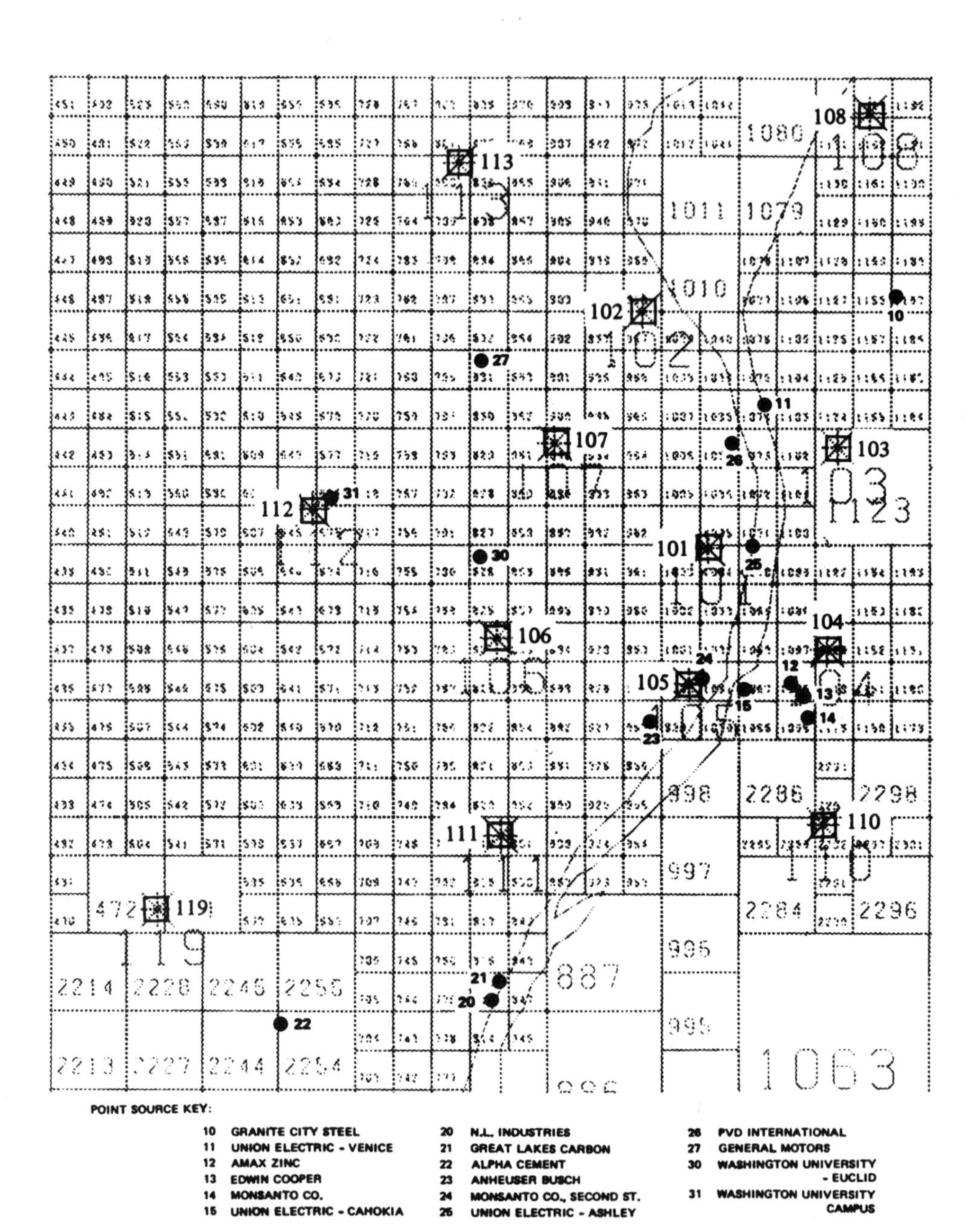

Fig. 5. Point source locations (•) in relation to RAPS monitoring stations with a background of area source grids.

correlations between Site 101 and other sites in the vicinity. We believe that the lower-concentration performance at Site 104 would improve if our remedy for Site 101 proved effective.

- The large underpredictions do not appear to be correlated among sites and hence are an anomaly at Site 104. The most plausible cause is that the large observed concentrations result from an uncategorized or poorly quantified emission source located in the vicinity of the site (most likely in the south to west sector). As seen in Figure 5, several substantial point sources are nearby. Because the underpredictions occurred over a brief period (mostly in December), we speculate that the error could have resulted from poorly quantified (or neglected) emissions from one of these nearby sources.

One of these sources of error is a problem in the handling of emission data by the model ; the other appears to be related to an error (by omission) in the emission data base.

CONCLUSIONS

Our results suggest that, despite extensive efforts to enchance the thoroughness and accuracy of the RAPS emissions inventory, a few large errors exist in the inventory of SO_2 emissions. Although their number is small, the large magnitude of these errors dominates the measures of RAM performance at one site. It would appear that better techniques are needed to screen the data bases prior to their use for model evaluation.

ACKNOWLEDGMENT

The research described here was supported under U.S. Environmental Protection Agency Contract 68-02-2770. The content of the publication does not necessarily reflect the views or policies of the U.S. Environmental Protection Agency, nor does mention of trade names, commercial products, or organizations imply endorsement by the U.S. Government. Mr. John S. Irwin, the EPA Project Officer, made substantial contributions to the technical effort.

REFERENCES

Fox, D. G., 1981, Judging Air Quality Model Performance-Review of the Woods Hole Workshop, in "Extended Abstracts, 5th Symp. Turbulence, Diffusion, and Air Pollution," American Meteorological Society, Boston, Massachusetts.

Nie, N. H., Hull, C. H., Jenkins, J. G., Steinbrenner, K., Bent, D. H., 1975, "Statistical Package for the Social Sciences," McGraw-Hill Book Co., New York, New York.

Ruff, R. E., 1980, "Evaluation of the RAM Using the RAPS Data Base ; Part 2 : Results," Final Report, EPA Contract 68-02-2770, SRI International, Menlo Park, California.

Schiermeier, F. A., 1978, Air monitoring milestones : RAPS field measurements are in, Env. Sci. & Tech., 12 : 6

Turner, D. B., 1970, "Workbook of Atmospheric Dispersion Estimates," Publication AP-26, Office of Air Programs, U.S. Environmental Protection Agency, Research Triangle Park, North Carolina.

Turner, D. B., and Novak, J. H., 1978, "Users Guide for RAM," EPA-600/8-78-016 a and b, U.S. Environmental Protection Agency (2 volumes).

DISCUSSION

M.L. WILLIAMS — The discrepancy between calculated and observed SO_2 concentrations at site 104 would seem to imply a very large source of SO_2 missing from the emission inventory ?

R.E. RUFF — That is our suspicion. There are several large sources to the south of site 104. We suspect that there were fugitive emissions from one of these sources during December which are not accounted for in the model input data base.

5: PARTICULAR STUDIES IN THE FIELD OF AIR POLLUTION MODELING

Chairman: A. Berger

Rapporteur: B. Vanderborght

HYDRAULIC MODELING OF THE ATMOSPHERIC BOUNDARY LAYER AT LARGE LENGTH SCALE RATIOS : CAPABILITIES AND LIMITATIONS

Pierre Bessemoulin

Etablissement d'Etudes et de Recherches Météorologiques
Direction de la Météorologie
73-77, rue de Sèvres - 92100 BOULOGNE

INTRODUCTION

Because many practical problems involve air flow patterns and turbulence characteristics over complicated topographies, it is most interesting to be able to explore the capabilities offered by physical modeling of the atmospheric boundary layer, either in a wind tunnel or in a water tank.

Recently, some authors (see MERONEY (1980) for instance) have shown that it is quite possible to reproduce successfully atmospheric flows in the laboratory at reduction scales as low as 1/5000.

The body of information presented here has been obtained over a topographic model in a water tank.

THE BASIS OF HYDRAULIC MODELING

The isothermal water flow we use is equivalent to an adiabatic atmosphere. As far as near neutral atmosphere is concerned, the only pertinent parameter is the Reynolds number, provided that :

- the topographical features of the model are undistorted
- the influence of Coriolis forces can be neglected. This means that we only deal with the lower part of the atmospheric boundary layer (surface layer), or we consider complicated and sharp topographies. In the latter case, the forcing induced by the ground features may be regarded much more important than Coriolis effects.
- the mean velocity and turbulence profiles of the approach flow are correctly modelled.

In other words, similarity is actually based on :

a - the use of high enough Reynolds numbers, so as to satisfy the following relation :

$$U_* Z_0 / \nu > 3$$

Where U_* is the friction velocity

Z_0 is the roughness parameter

ν is the kinematic vicosity of the fluid

This criterion (NIKURADSE) insures that the model is aerodynamically rough.

b - the vertical distribution of mean velocity and turbulence intensities of the approach flow.

Velocity distributions may be charcaterised either by an exponent α_p (power law), or by a friction velocity U_* and a roughness parameter Z_0 (logarithmic profiles).

Depending on the roughness of the upwind fetch, desired approach flows can be obtained with vortex generators, roughnesses...

With such devices, possible flow parameters range as follows in our water tank (7 m long, 1.2 m wide, 0.5 m deep) :

δ (boundary layer height) $\lesssim$ 15 cm
$0.12 \lesssim \alpha_p \lesssim 0.55$
$0.15 \lesssim I_{ground} \lesssim 0.30$
$0.03 \lesssim I_{\delta} \lesssim 0.06$

c - the height of the boundary layer, based on formulas similar with BLACKADAR and TENNEKES' (1968) one :

$$\delta \simeq \frac{U_*}{4f}$$

where f is the Coriolis parameter.

EXPERIMENTAL CONDITIONS

The present measurements have been obtained over a scale model of a part of the Rhône Valley (France), built at the scale 1/5000, without vertical distorsion.

The results concern :

- a mean velocity profile and a longitudinal turbulent intensity profile measured 20 cm (1 km "in-situ") downwind of a scaled hill which is in the field about 200 m high (Fig. 1).

- a data processing of a record of the instantaneous longitudinal velocity, measured at 8 mm over the model (40 m "in-situ"), slightly downwind of the preceding point. The local mean velocity was 11.5 cm/s.

Other experimental conditions were :

U_∞ = 18.5 cm/s

$\delta \simeq$ 10 cm (500 m in the field)

RESULTS

Mean wind speed and turbulence intensity profiles are characterised by the following features.

The power law exponent is 0.19, corresponding to a roughness parameter of some tenths meters, after COUNIHAN (1975), over flat terrain.

The friction velocity and the roughness length have been estimated from a logarithmic fitting of the velocity profile :

$$U_* = 1.273 \text{ cm/s} \; ; \; Z_0 = 3.\ 10^{-2} \text{ cm}$$

Taking into account a length scale ratio of 1/5000, this would correspond to a field value : 1.5 m.

This is consistent with other results over complex terrain (NAPPO (1977), THOMPSON (1978)).

The turbulence intensity normalised by the friction velocity is presented (Fig. 2 and 3) and compared with recent results obtained by SERRES (1978) (wind tunnel) and CRABOL (1978) (water channel).

It can be seen that the curve found in this study systematically lies under those from the authors quoted above. The reason is that the roughness of their model was lower, as it appears from the values of the ratios Z_o / δ :

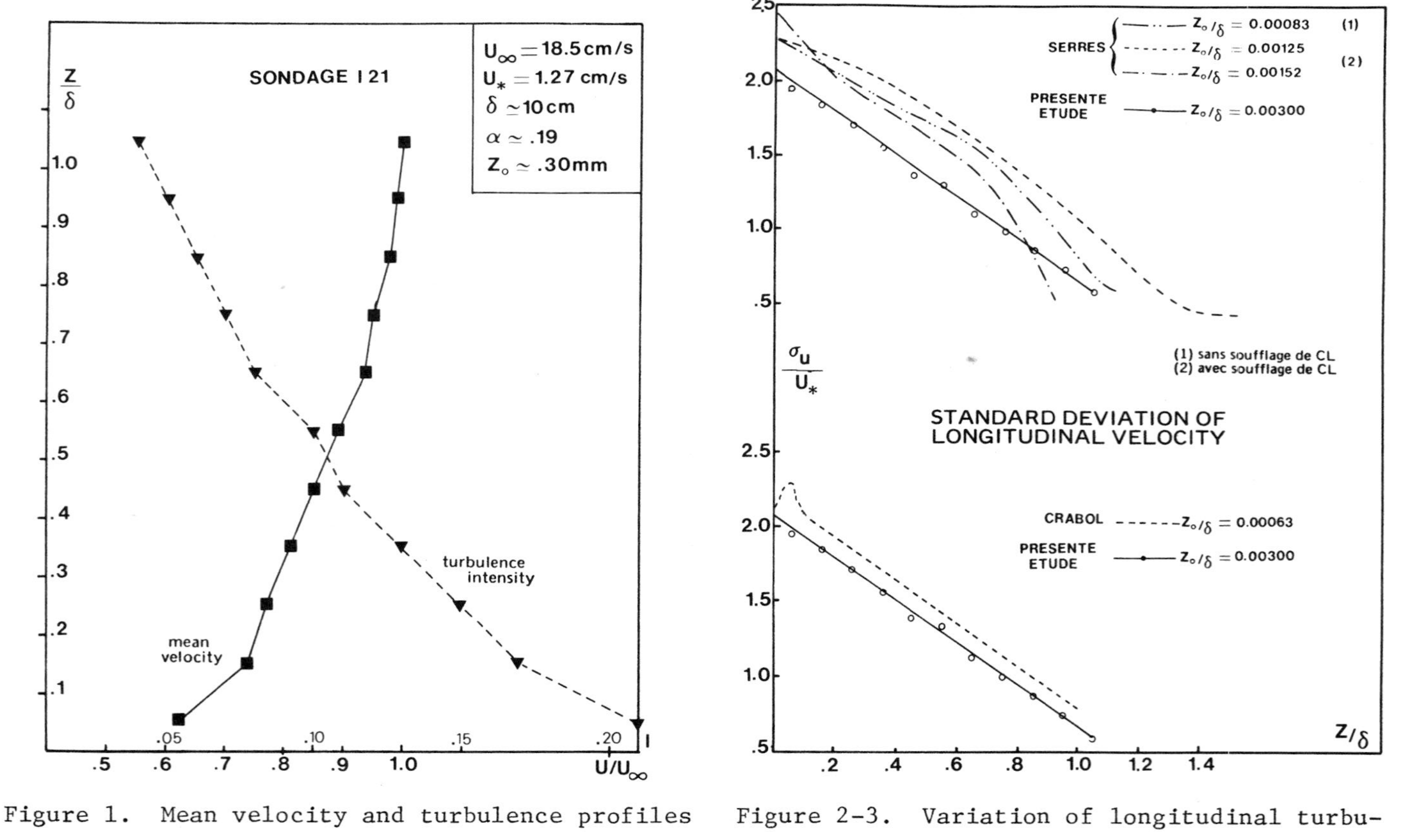

Figure 1. Mean velocity and turbulence profiles over the model.

Figure 2-3. Variation of longitudinal turbulence intensity compared with other results

SERRES : $8.5\ 10^{-4}$

CRABOL : $6.3\ 10^{-4}$

present study : $3.\ 10^{-3}$

The normalised velocity probability-density distribution is presented Fig. 4 as well as the Gaussian distribution having the same standard deviation.

The skewness is $4.\ 10^{-4}$ and flatness 2.783 (respectively 0 and 3 for a Gauss curve).

This is consistent with SERRES observations : in the lower part of the boundary layer, skewness is nearly zero. It takes negative values increasing in absolute value as soon as intermittency of the turbulence occurs ($Z \gtrsim 0.5\delta$).

The normalised turbulent spectrum (divided by the mean square value of the longitudinal component of turbulent energy σ_u^2 so that $\int F(n)/\sigma_u^2 . dn = 1$, is shown in Fig. 5, as a function of the reduced frequency $\frac{nZ}{U}$.

It can be seen that for a very limited area lying between $\frac{nZ}{U}$ values ranging from 10^{-1} to 3.10^{-1}, the slope of the spectrum is very close to -2/3, as predicted by KOLMOGOROV theory.

It must be remembered that this value for the slope is a necessary but not sufficient condition for the existence of an inertial sub-range (see for instance TENNEKES and LUMLEY (1973)).

The Taylor microscale λ_x , computed with a formula proposed by DRYDEN :

$$\frac{1}{\lambda_x^2} = \frac{4\pi^2}{\bar{U}^2} \int_o^\infty n^2\ F(n)\ dn$$

is 3.3 mm ; this is in good agreement with RAICHLEN (1967) results obtained in similar conditions.

Very often, Von KARMAN formula is used to represent analytically energy spectra. We have fitted the beginning of our spectrum ($\frac{n\ Z}{U} \leq 0.3$) with this expression :

$$\frac{n\ F(n)}{\sigma_u^2} = \frac{4n\ \frac{L_x}{\bar{U}}}{\left(1+70.8\ \frac{n^2 L_x^2}{\bar{U}^2}\right)^{5/6}}$$

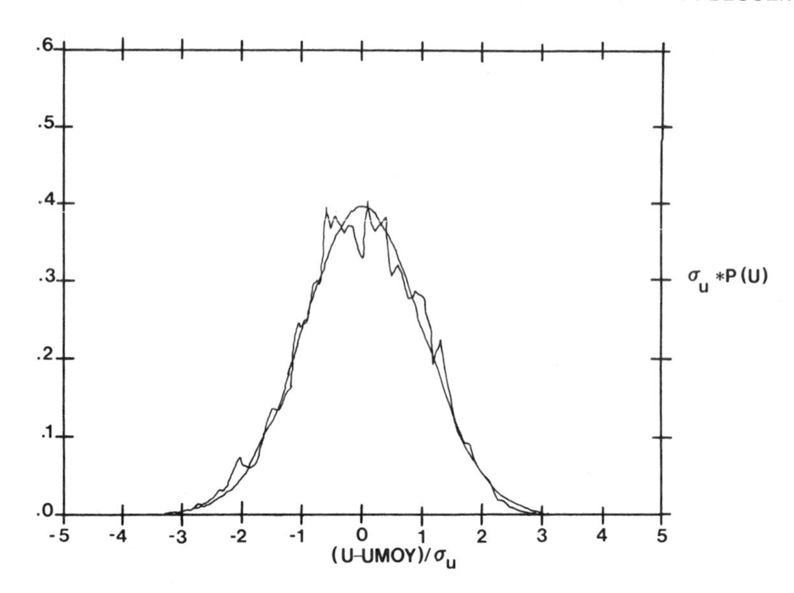

Figure 4. Normalised velocity probability density distribution in the surface layer.

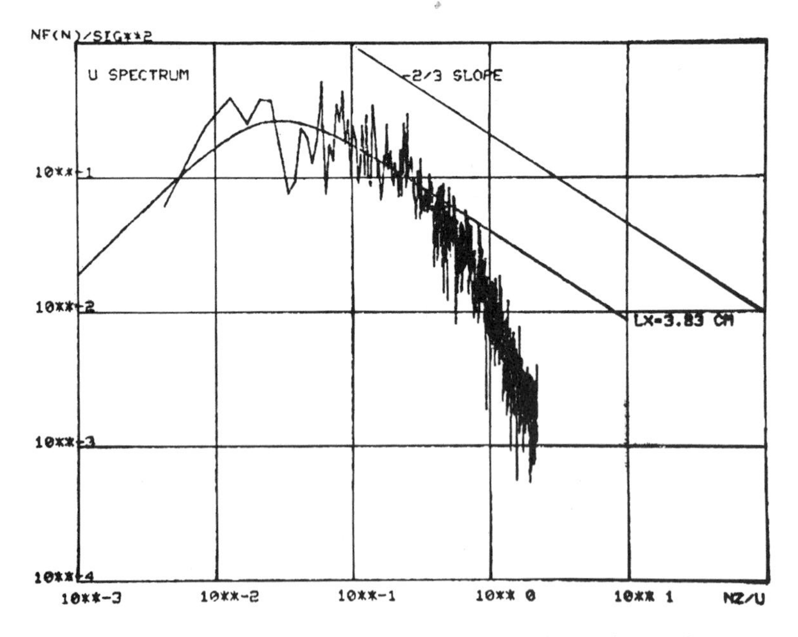

Figure 5. Turbulence spectrum at large length scale ratios.

The computed value for L_x (longitudinal integral length scale) is 3.83 cm, which for a length scale ratio of 1/5000 gives a full scale value : 191 m.

For Z_0 = 1.5 m, COUNIHAN formula for Z = 40 m lead to L_x = 110 m.

If the integral scale is computed from the autocorrelation curve R (t) (FFT of the power spectrum), by integration from t = 0 to the first zero value of the autocorrelation, one get 116 m, which is in good agreement with COUNIHAN.

In a recent paper, TEUNISSEN (1980) quoted that the estimates of integral scales from autocorrelation curves and spectra could differ significantly.

POSSIBILITIES FOR DIFFUSION STUDIES

Starting from Taylor's expression for the mean square fluid particle displacement in a stationnary homogenous turbulence :

$$\overline{y^2(t)} = \overline{v^2} \; t^2 \int_o^\infty F_L(n) \left(\frac{\sin \pi \; n \; t}{\pi \; x \; t} \right)^2 dx$$

it can be shown that only the largest eddies present in the flow contribute to the spread of a plume and that the ones smaller than the plume width have very little effects on the dispersion of the plume.

From the shape of the actual spectrum obtained in our water tank, it is obvious that for $\frac{n\,Z}{U} > 0.7$ (n > 10 Hz), eddies are not simulated with the appropriate spectral power. So, turbulent structures which mean longitudinal dimension are lower than 5000. 1/10. 11.5 $\simeq$ 60 m will not be correctly modelled. Thus, only large emissions may be simulated (area sources, cooling towers) provided that emission and stack Reynolds numbers are high enough.

CONCLUSIONS

These few results confirm that physical modeling can reproduce successfully wind and turbulence patterns over complex terrain, even at large scale ratios, as emphasized by MERONEY in some recent papers.

So, trajectories over such topographies can be built with confidence if velocity measurements and flow visualisation techniques are available.

Spectral studies suggest that, in general, diffusion studies cannot be undertaken at such large scale ratios, because small eddies are not will simulated.

REFERENCES

BACKADAR A.K-TENNEKES H. (1968) : "Asymptotic similarity in neutral barotropic atmospheric boundary layer", Journal of Atm. Science Vol 25, pp 1015-1020.

COUNIHAN J. (1975) : "Adiabatic atmospheric boundary layer : a review and analysis of data from the period 1880-1972", Atm. Environment, Vol 9, n° 10, pp 871-905.

CRABOL B. (1978) : "Contribution à l'étude de la simulation en laboratoire des transferts de masse en atmosphère neutre", Thèse de 3ème Cycle, PARIS VI.

MERONEY R.N. (1980) : "Wind-tunnel simulation of the flow over hills and complex terrain", Journal of Ind. Aerodynamics, Vol 5, pp 297-321.

MERONEY R.N. (1980) : "Physical simulation of dispersion in complex terrain and valley drainage flow situations". 11th ITM on Air Pollution Modeling and its application . NATO/CCMS, Amsterdam.

NAPPO C.J. (1977) : "Mesoscale flow over complex terrain during the eastern Tennesse Trajectory Experiment (ETTEX)". Journal of Applied Meteo, Vol 16, pp 1186-1196.

RAICHLEN F. (1967) : "Some turbulence measurements in water", Journal of Engineering, Mechanics Division, pp 73-97.

SERRES E. (1978) : "Etude de la simulation en soufflerie des basses couches de l'atmosphère. Application à la prévision de l'impact d'un site industriel sur l'environnement". Thèse de Docteur Ingénieur-Université C. Bernard-LYON.

TENNEKES H.-LUMLEY J.L. (1973) : "A first course in turbulence". MIT Press.

TEUNISSEN H.W (1980) : "Structure of mean winds and turbulence in the planetary boundary layer over rural terrain". Boundary Layer Meteo., Vol 19, pp 187-221.

THOMPSON R.S. (1978) : "Note on the Aerodynamic Roughness length for complex terrain". Journal of Applied Meteo., Vol 17 , pp 1402-1403.

FORMATION OF CLOUD PARTICLES FROM LARGE HOT PLUME IN CONTROLLED EXPERIMENTS

Pham Van Dinh*, B. Bénech*, N. Bleuse** and J.P. Lacaux*

(*) I.O.P.G. du Puy de Dôme, Centre de Recherches Atmosphériques, Campistrous Cidex B 47, 65300 Lannemezan (France)

(**) Ecole Nationale de Météorologie, 78390 Bois d'Arcy (France)

INTRODUCTION

For electrical production from nuclear energy, the existing cooling systems release their heat into the atmosphere as a combination of latent and sensible heat, such devices need a very important water flow. And the increasing consumption of electrical power implies new elaborating of plants more powerful but not heavy in water because it becomes more and more difficult to find appropriate sites near river or sea. For this reason, the forthcoming cooling tower will have to emit only waste dry heat. Such released heat over a small area can be intense. Thus there is legitimate concern about the effects of such heat source on natural atmospheric processes and the possible initiation of atmospheric processes which otherwise would not occur.

To evaluate the potentiel environmental impacts of waste heat released by such dry cooling towers, the COCAGNE* Project elaborated by the Centre de Recherches Atmosphériques was started by the design of an oil burning system on the "Plateau of Lannemezan", southwest of France. The device, named "Météotron", consists of an array of 105 oil burners deployed in a three-armed spiral pattern within 15000 m^2, it enables to emit some 1000 MW as dry heat-for details of the combustion budget, see Pham Van Dinh et al., 1981.

---

*COnséquence de la Convection Artificielle, Génération de Nuages et autres Effets.

Table 1. Specific instrumentation aboard the B33 aircraft used in this study (compiled from Hobbs et al., 1976 and Radke et al., 1980).

Measured Parameter	Detection Technique	Range used	Manufacturer
Size spectrum of fine aerosol	Electrical mobility analysis (EAA)	0.01-1.0 μm in 8 size intervals	TSI. Inc.
Size spectrum of large aerosol	Single particle 90° light scattering (Royco 202)	0.3 - 10 μm in 14 size intervals	ROYCO Instr., modified
Concentration of cloud condensation nuclei (CCN)	Light scattering/ thermal diffusion chamber	0.2 to 1.5 % supersaturation (20-5000 CCN.cm^{-3})	Univ. of Washington
Size spectrum of small cloud drops	Single particle forward light scattering (ASSP)	2.6 - 68 μm in 15 size intervals	PMS Inc.
Size spectrum of large cloud drops	Single particle diode occultation (CDP)	20 - 300 μm in 15 size intervals	PMS Inc.
Size spectrum of precipitation drops	Single particle diode occultation (PSP)	370 - 4700 μm in 15 size intervals	PMS Inc.
Liquid water content (LWC)	Hot wire resistance (JW)	0 - 2 g•m^{-3}	Johnson Williar (JW)
Light-scattering coefficient of the air	Integrating nephelometer	0 - 10^{-2} m^{-1}	Meteor. Researc Inc., modified
Total gaseous sulfur	Flame photometry	0 - 1000 ppb in 4 ranges	Meloy, model 285
Morphology and elemental composition of smoke	Filter/scanning electron microscope + X-Ray dispersive energy	0.1 - 10 μm	In-House

Forty experiments were carried out in 1978 and 1979 with a great deal of instrumentation : the heat source was observed with a two-level array of wind and temperature sensors over the source area and with a ground-based array around the source, the plume was tracked by means of instrumented aircrafts, photogrammetry stations, teledetection devices (lidar, radars, infrared imagery), and radio-sounding balloons were launched periodically on the experiment day, for detailed summary of instrumentation and atmospheric events, see Bénech et al. (1980). This paper presents the more recent results dealing with the particulate structure of the plume and the ambient atmosphere, as well as with the microphysics of the natural and induced clouds that occured during the 1979 program, when 15 experiments were carried out. Each burn lasted some 30 minutes, while airborne measurements required over 3 hours centered around the burn time.

INSTRUMENTATION AND DATA ANALYSIS PROCEDURE

The airborne measurements were performed aboard the University of Washington's Douglas B23 aircraft. Aerosol spectra were measured with the Electrical Mobility analyzer (EAA) and the single particle optical counter (ROYCO 202) which, in overlapping steps, covered the size range from 0.01 μm to 10 μm. Aerosol particles were dried before sizing from a semi-automatic bag sampler in nearly isokinetic conditions ; a whole run took about 85 seconds. Cloud particles spectra were measured by three single hydrometer probes (ASSP, CDP and PSP) which covered the size range from 2.6 μm to 4700 μm ; the droplets were sampled continuously well away from the boundary layer of the aircraft and the measurement was processed and displayed every 6 seconds corresponding to a sampling path of 360 m. Cloud condensation nuclei were measured by light scattering technique using a diffusing chamber. Because of temperature regulation problem encountered during the experiment, few data were available. The plume was easily located during the flight by the smoke and its typical smell and later retrieved from the SO_2 content and the light-scattering coefficient that were measured respectively by flame photometry and integrating nephelometry. The liquid water content (LWC) was monitored continuously by a hot wire device (Johnson Williams, JW). For further details and for other instruments, see Hobbs et al., 1976.

There was a discrepancy between the JW-LWC and the LWC computed from the hydrometeor spectra, even in the case of small drops. Since the JW might be regarded as more reliable than the ASSP, especially when the clouds contained only liquid drops smaller than about 40 μm in diameter (Levine quoted by Ruskin, 1976), the JW-LWC was used in this occurence as a standard to adjust the ASSP-LWC. The ASSP-LWC was plotted versus the JW-LWC relating to drops smaller than 60 μm, 40 μm and 32 μm ; the best regression line was found for drops smaller than 32 μm in diameter and expressed as JW-LWC ($g \cdot m^{-3}$) = 0.202 (ASSP-LWC) + 0.055. This equation may be compared to the Baumgardner and Vali (1980)'s one : JW-LWC ($g \cdot m^{-3}$) = 0.43 (ASSP-LWC) + 0.06 (expression retrieved from their figure showing the scattegram of ASSP-LWC with JW-LWC). The two relationships point out the drifting of the zero baseline, consequently the cloud samples with too low LWC (< 0.057 $g \cdot m^{-3}$) will not be taken into account by us.

About aerosol measurement, the most serious problem for the EAA was the question of the detection limit. The minimum concentration numbers per cubic centimeter were 2000 for the first channel (0.01 - 0.018 μm) to 20 for the last channel (0.56 - 1.0 μm), and furthermore the EAA counted by discrete values as a multiple of the minimum. So when the air contained few particles, the concentration number scattered strongly to any way, particularly through channels relating to fine particles, because of the high values for the minimum detection limits. For the Royco, as the minimum detection limit was as low as 0.003 particle per cm^3 for any channel, the major problem encountered was in the opposite way. When sampling within the plume, the count was often saturated in spite of the use of a dilution bag. Therefore, for comparison between plume particles and ambient ones, it may be more legitimate to use only the EAA count and through smaller size range, e.g. 0.1 - 1.0 μm.

DRY PLUME NATURE AND MICROSTRUCTURE

The plume was featured by smoke particles resulting from incomplete combustion of fuel-oil. Particles were sampled on Nuclepore filters above the burner and within the whole plume. Sized and counted by means of scanning electron microscope, the smoke aerosol consists of agglomerates of smaller particles with a spherical aspec of about 0.1 μm ; many agglomerates exhibit a chain-like structure. The frequency distributions versus particle size (Feret's diameter), established from numerous samples involving several thousands particles, show a mode at 0.4 μm for samples at both ground and high levels (Pham Van Dinh and Bénech, 1980). The mass spectrography analysis reveals that smoke particles essentially consist of carbon (at 97 %), while other elements (such as Cl, Na, Ca, K ... that are able to form hygroscopic nuclei) are not more than traces (Pham Van Dinh et al., 1981). The analysis by X-ray dispersive energy on single particles, which is a non-sensitive method for carbon element, reveals in some cases, the presence of silicon and sulfur at 1 % rate. The other elements at too low content or with atomic number lower than 11 cannot be analyzed.

Figure 1 shows an example of the vertical profile concentration of the aerosol in the 0.1 - 1 μm range in the ambient air and within the plume during the burning phase. Natural concentrations (A) were higher in the boundary layer than in the stable layer.

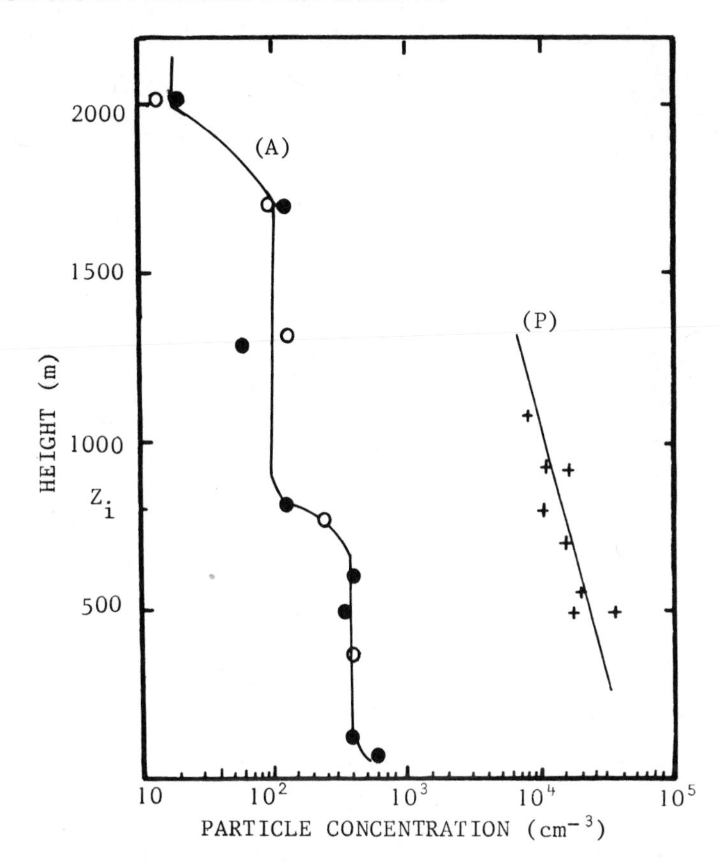

Figure 1 : An example of vertical profile of aerosol particles within the plume (P) and in the ambient air (A) through the (0.1 - 1 µm) diameter range (18 june 1979). Solid circles refer to samples prior to the burn, while open circles refer to samples after the burn.

Plume particles concentrations (P) reached up two orders of magnitude higher than the ambient concentrations, and slightly decreased with height.

The concentrations of cloud condensations nuclei (CCN) counted at 1 % of water sursaturation within the planetary boundary layer fluctuated strongly from day to day and in a lesser ratio within the same day. The whole variations range was from 200 to 4000 nuclei per cm^3. On June 7, at the same height, the variation factor reached up to 4, and on June 19, up to 5. The variability may be explained both by the heterogeneity and the origin of the airmass (continental and/or maritime). The Plateau of Lannemezan is some 30 km north of the Central Pyrenees and some 150 km away from the atlantic coast. The air mass, moving from the Ocean and then over cities, may be regarded as a modified maritime type. In the stable layers, the mixing was less strong, thus the air was less modified, and so the CCN concentrations were lower and less variable (50 to 500 nuclei per cm^3). On May 30, the CCN concentrations were low, 300 - 500 per cm^3 in the boundary layer and twice as high within the plume. This CCN increasing was too poor to be truly significant. Moreover, CCN measurements, on ground level by means of a diffusion chamber associated with a TV imaging device, provided much more nuclei than inside the plume, and so the heat column might be regarded rather as a nuclei carrier (from the ground) than as a true source. This result is in accordance with the hydrophobic nature of the aerosol freshly produced in the Meteotron plume (carbon particles) and with the very low rate of SO_2 oxidation. Dioxide sulfur content was high (e.g. up to 400 ppb) inside the plume while the ambient SO_2 content was lower than 10 ppb sulfate particles are efficient CCN indeed, however according to Roberts and Williams (1979) and Forest et al. (1979), the rate of oxidation of SO_2 to particulate sulfate is quite low ($\sim$ 0.25 % h^{-1}), so that gaseous sulfur is not likely to contribute significantly to CCN formation in our case.

MICROPHYSICS OF INDUCED AND NATURAL CLOUDS

Among fifteen burns made during the course of the 1979 program, twelve burns initiated cloud formation. Generally the induced clouds and the surroundings ones exhibited some similar characteristics, such as condensation level and radar reflectivity factor (radar wavelength : 8.6 mm), in some cases, the artificial cloud rose up over neighbouring clouds. Hereafter, the microphysical aspects of clouds, for five experiments, are investigated, such as droplet concentration, liquid water content (LWC) and mean diameter. The LWC is computed from drop size distribution after adjustment as above mentioned. No samples from artificial clouds are taken into account after fire extinction.

One datum point represents a 6 seconds-averaged measurement, the natural ones are plotted by a circle while the artificial ones by a cross (Fig. 2 through 4). The two populations are compared by the non-parametric statistical test of Wilcoxon at 95 % confidence level, and in addition the mean values were calculated for all three microphysical parameters.

<u>On May 30</u>, the natural clouds were changing : broken non-precipitating stratus on the beginning of the burn and later the overcast became more solid, therefore, for the comparative test, no samples were selected in the natural clouds after the fire extinction. Cumulus fractus were produced by the Meteotron. The base level was at 300 m (AGL) and the top at 700 m. All levels were investigated both for natural and artificial clouds, and the LWC, concentra-

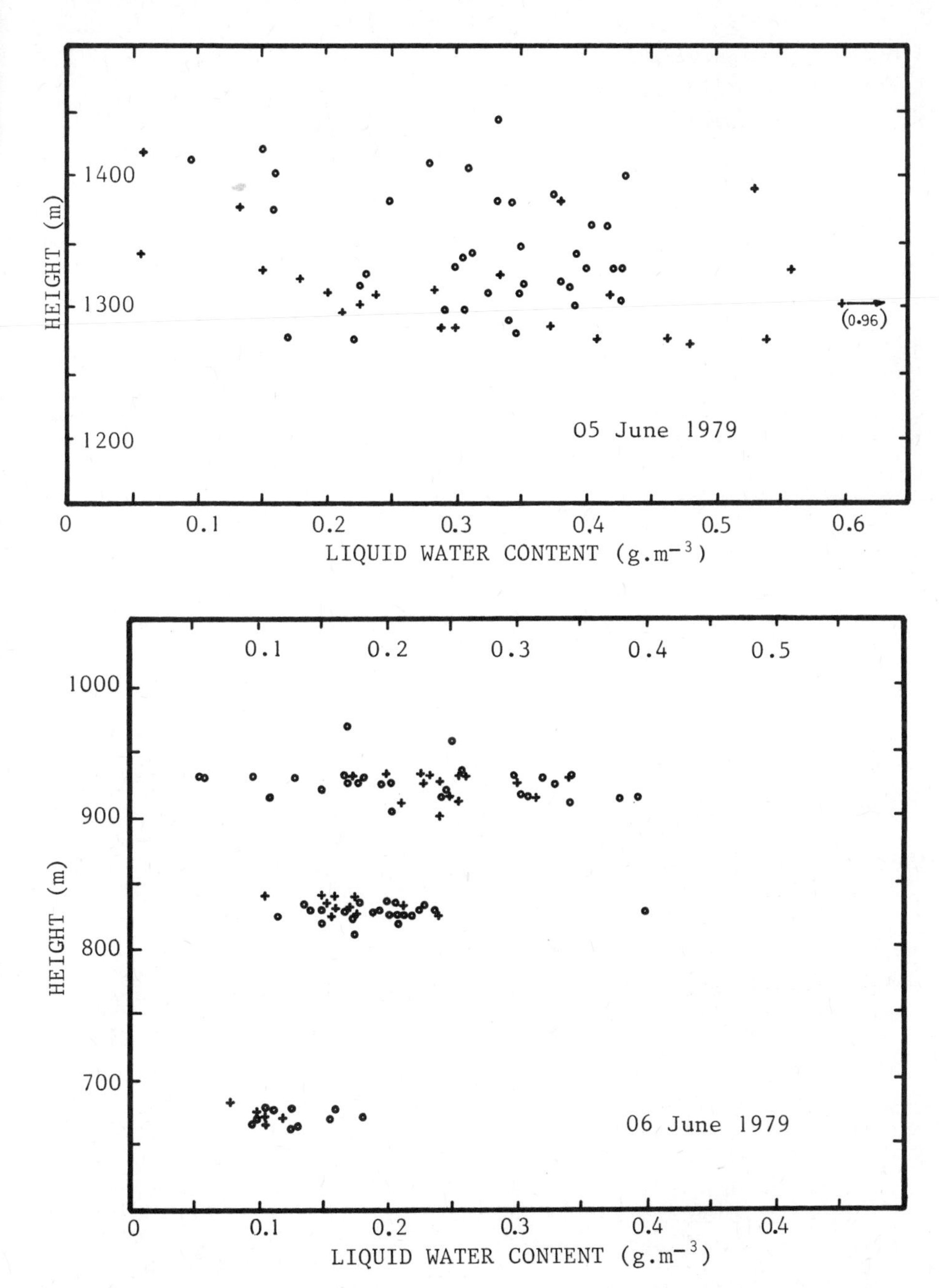

Figure 2 : Comparative liquid water contents sampled from natural stratocumulus (o) and Meteotron-affected stratocumulus (+) on June 5 and on June 6 1979. One datum point corresponds to a 6 seconds-averaged measurement (360 m flight path).

Table 2 . Means and Wilcoxon test for the microphysical parameters characterizing the natural and induced clouds (in brackets, the standard deviation).

	Natural clouds	Induced clouds	Wilcoxon test result at 95 % confidence level
Liquid water content ($g \cdot m^{-3}$)			
30 May 1979	0.09 (± 0.05)	0.08 (± 0.03)	not different
02 June 1979	0.10 (± 0.04)	0.09 (± 0.03)	not different
05 June 1979	0.31 (± 0.08)	0.33 (± 0.2)	different
06 June 1979	0.20 (± 0.08)	0.20 (± 0.07)	not different
11 June 1979	0.20 (± 0.15)	0.09 (± 0.03)	different
Drop concentration (cm^{-3})			
30 May 1979	210 (± 55)	230 (± 59)	not different
02 June 1979	130 (± 63)	185 (± 66)	different
05 June 1979	61 (± 18)	110 (± 51)	different
06 June 1979	81 (± 50)	100 (± 64)	not different
11 June 1979	85 (± 44)	130 (± 45)	different
Mean diameter (µm)			
30 May 1979	8.0 (± 1.7)	7.7 (± 1.0)	not different
02 June 1979	10.5 (± 1.0)	8.7 (± 1.0)	different
05 June 1979	17.9 (± 1.9)	13.2 (± 1.5)	different
06 June 1979	14.1 (± 3.9)	13.7 (± 3.9)	not different
11 June 1979	12.9 (± 1.0)	9.8 (± 1.1)	different

tions and mean diameters of which were not statistically different (table 2).

On June 2, shallow cumulus were present in the environment between 650 and 1150 m (AGL), the induced cumulus rose up to 1400 m, but they were not larger than the surrounding ones at the same heigh No samples were investigated prior to the burn nor when the Meteotro interacted with the natural clouds. No difference is noticed for the LWC, while the mean diameter is larger but the concentration lower for the natural clouds (table 2).

On June 5, a deep stratocumulus deck covered the site from near the ground to 1500 m, but only the upper levels (1250 - 1450 m) were penetrated by the aircraft because of the poor visibility. For this experiment and for the next, the interacting between the Meteo-

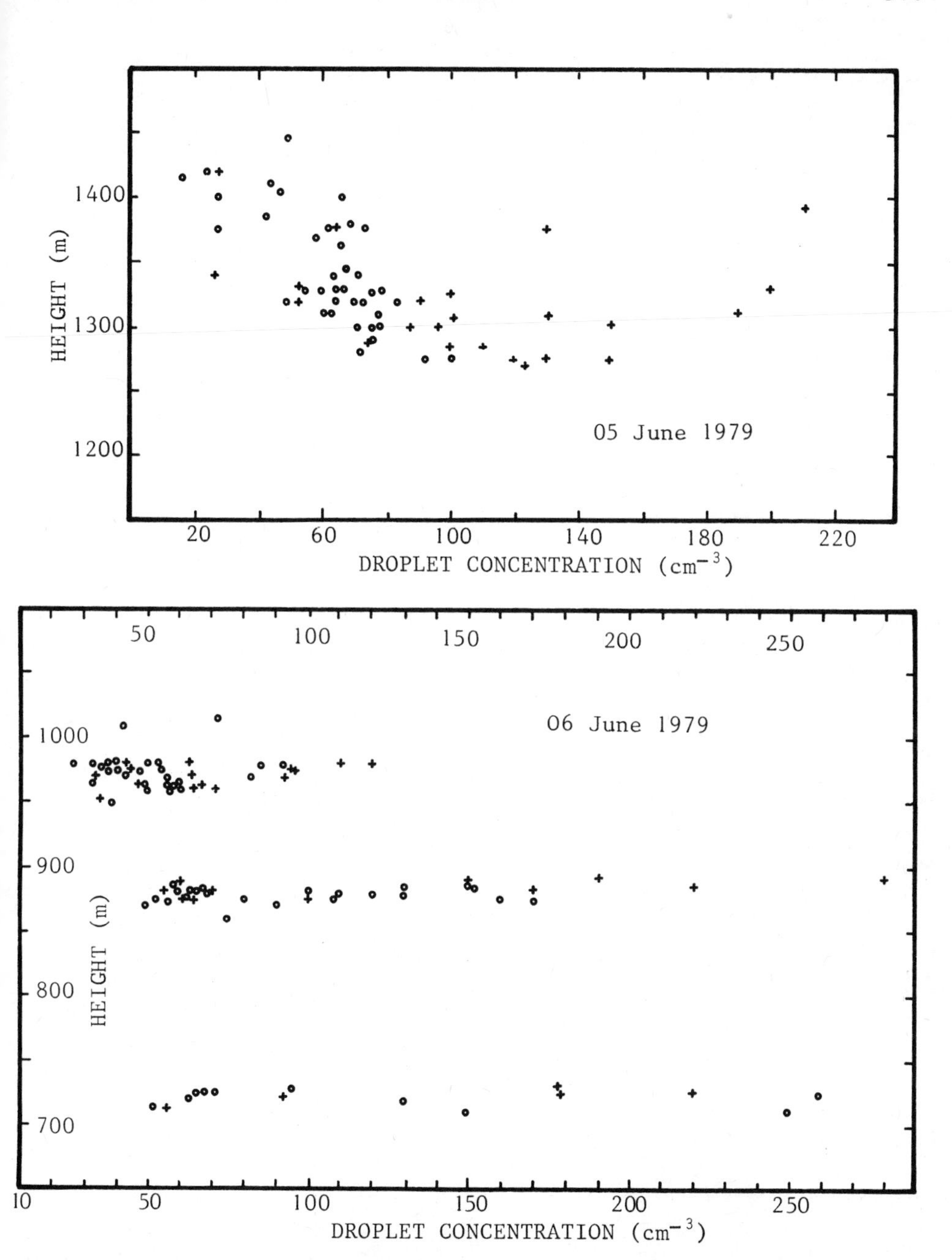

Figure 3 : Comparative droplet concentrations sampled from natural stratocumulus (o) and Meteotron-affected stratocumulus (+) on June 5 and on June 6 1979. One datum point corresponds to a 6 seconds-averaged measurement (360 m flight path).

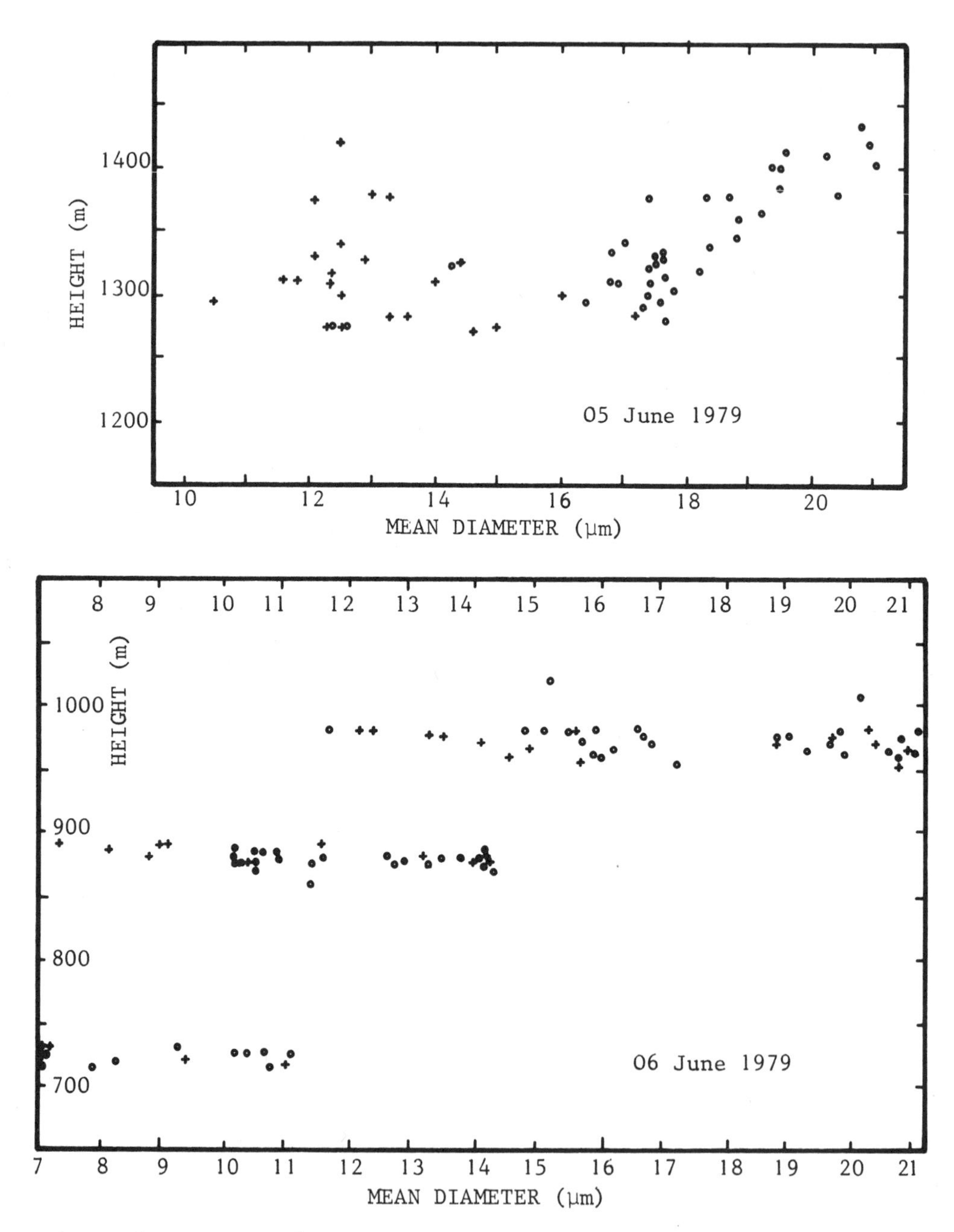

Figure 4 : Comparative droplet mean diameters computed from size spectra of natural stratocumulus (o) and Meteotron-affected stratocumulus (+) on June 5 and on June 6 1979.

tron and natural clouds were analyzed, therefore only affected clouds bordered by natural clouds of a similar length were examined. The clouds raised up by the Meteotron exhibited a slightly higher LWC, and more smaller droplets than the environmental ones (Fig. 2, 3, 4 ; table 2).

On June 6, the stratocumulus was less deep than on the preceding day (100 - 1100 m), and thus more cloud levels might be sampled (700 - 1050 m). Contrary to the preceding experiment, the Meteotron-affected clouds did not contain more liquid water, while their higher drop concentrations, as well as their slightly smaller mean diameters, did not present a significant difference from the natural ones (fig. 2, 3, 4 ; table 2).

The higher LWC in induced clouds on June 5 might be explained by the mixing ratio gradient. On June 5, the difference between the mixing ratio at ground level and upper layers was more pronounced as compared to June 6, and so the water vapour contribution from ground was more consistent : the Meteotron was an efficient water carrier (Fig. 2, 3, 4 : table 2).

On June 11, the Meteotron-induced cumulus had a size similar to the natural ones (at level 1000 - 1800 m), but their condensation level might be less lower ; their lower LWC and their smaller droplet mean diameters were quite different from the natural ones, as well as their higher particle concentrations (table 2).

CONCLUSION

The present state of data analysis about the 1979 COCAGNE program exhibits out some main features dealing with plume microstructure and microphysical characteristics of clouds.

In plume generated by the Meteotron, the aerosol consisted essentially of hydrophobic carbon particles which did not increase significantly concentrations of cloud condensation nuclei. Plume particles were about two orders of magnitude higher than surroundings ones.

Clouds were initiated by the Meteotron mainly when natural clouds existed. As compared to ambient clouds, their drop concentrations were higher, while the drop mean diameters were smaller, as well as liquid water contents. The differences are attested in most cases by the statistical Wilcoxon test. The liquid water content seemed more sensitive to water vapour content at ground level as emphasized on June 5. The microphysical characteristics of the Meteotron-induced clouds are indeed common to freshly-formed clouds.

ACKNOWLEDGMENTS

The airborne measurements were performed aboard the University of Washington's Douglas B23 aircraft. We would like to acknowledge particularly Prof. Radke, Eltgroth and Hobbs for their productive cooperation.

We are grateful to Dr Breugnot for his assistance for data processing.

This work was supported by a grant from Electricité de France, Division Etudes et Recherches.

REFERENCES

Baumgardner,D. and G. Vali, 1980 : Empirical evaluation of airborne liquid water measuring devices. VIIIth Int. Conf. on Cloud Physics, Clermont-Ferrand, France, 15-19th July 1980, Vol. II, 373-379.

Benech, B., J. Dessens, C. Charpentier, H. Sauvageot, A. Druilhet, M. Ribon, Pham Van Dinh and P. Mery, 1980 : Thermodynamics and microphysical impact of a 1000 MW heat-released source into the atmospheric environment. Third WMO Scientific Conf. on Weather Modification, 21-25 July 1980, Clermont-Ferrand, France Vol. I, 111-118.

Forest, J., R. Garber and L. Neuman, 1979 : Formation of sulfate amonium and nitrate in oil-fired power plant plume. Atmospheric Environment, 13, 1287-1297.

Hobbs, P.V., L.F. Radke and E.E. Hindman II, 1976 : An integrated airborne particle-measuring facility and its preliminary use in atmospheric aerosol studies. J. Aerosol Sci., 7, 195-211.

Pham Van Dinh and B. Benech, 1980 : Particulate pollution of the atmosphere due to liquid hydrocarbon fires. *In* Atmospheric Pollution 1980, Elsevier, Amsterdam, 249-254.

Pham Van Dinh, B. Benech et W. Diamant, 1981 : Evaluation des propriétés énergétiques et microphysiques d'une source de convection artificielle à partir d'une étude de combustion organisée de fuel-oil en milieu naturel. Accepted to be published in Atmospheric Environment.

Radke, L.F., B. Benech, J. Dessens, M.W. Elgroth, X. Henrion, P.V. Hobbs and M. Ribon, 1980 : Modifications of cloud microphysics by a 1000 MW of heat source and aerosol. Third WMO Scientific Conf. on Weather Modification, 21-25 July 1980, Clermont-Ferrand, France, Vol. I, 119-126.

Roberts, D.B. and D.J. Williams, 1979 : The kinetics of oxidation of sulfur dioxide within the plume from a sulphide smelter in a remote region. Atmospheric Environment, 13, 1485-1499.

Ruskin, E.R., 1976 : Liquid water content devices. Atmosph. Technology, 8, 38-42.

ACCOUNTING FOR MOISTURE EFFECTS IN THE PREDICTION OF BUOYANT PLUMES

Michael Schatzmann

Sonderforschungsbereich 80
University of Karlsruhe
Karlsruhe, W. Germany

Anthony J. Policastro

Div. Env. Impact Studies
Argonne National Laboratory
Argonne, Illinois, USA

ABSTRACT

In this paper, the theory of plume rise from stacks with scrubbers is critically evaluated. The significant moisture content of the scrubbed plume upon exit leads to important thermodynamic effects during plume rise which are unaccounted for in the usual dry plume rise theories. For example, under conditionally unstable atmospheres, a wet scrubbed plume treated as completely dry acts as if the atmosphere were stable whereas in reality, the scrubbed plume behaves instead as if the atmosphere were unstable. Even the use of moist plume models developed for application to cooling tower plume rise are not valid since these models employ (a) the Boussinesq approximation, (b) use a number of additional simplifying approximations which require small exit temperature differences between tower exit and ambient, and (c) are not calibrated to stack data.

Although these two theories are often used to predict plume rise from stacks with scrubbers, both theories contain unacceptable assumptions. This paper details the invalid approximations made in each theory. The direction and magnitude of the important errors are estimated for these models when applied to scrubber stack plumes.

INTRODUCTION

Sulfur-dioxide released mainly through tall chimneys of fossil-fired power plants or industrial steam generators can cause serious environmental problems. In highly industrialized areas, measures have to be taken to reduce ground-level SO_2-concentrations.

This can be done by either:

(a) increasing the height of the chimneys,
(b) desulfurization of the combustibles, or
(c) desulfurization of the flue gas.

Increasing the stack height is suitable only if the ground level concentration in the vicinity of the power plant needs to be reduced. Due to the increased height of rise of the plume, a greater rate of concentration decay is obtained before the lower edge of the plume touches the ground. A disadvantage to the tall stack, however, is that a larger area of deposition is affected and that the generation of unwanted bi-products like sulfuric acid and sulfate aerosols in the atomosphere are likely to be increased. Long range transport of SO_2 from west and central European power plants into Norway's clean air resorts has been reported with harmful effects on wood and fish populations in rivers caused by acid rain. Therefore, only the application of desulfurization techniques remains to provide a substantial contribution towards environmental protection. Since desulfurization of the combustibles (alternative (b)) seems to be by far the more costly alternative, flue gas desulfurization devices (alternative (c)) are being installed in a rapidly increasing number. Most of them employ a wet scrubbing technique by which the flue gas is washed with a $Ca(OH)_2$-solution. Sulfur-dioxide reacts with calcium producing $CaSO_4$ which can either be used in the fertilizer industry or can be dumped. In 1978 there were 49 scrubbers installed in the U.S., 562 in Japan, and 4 in West Germany. By 1985 these numbers will have been increased by at least a factor of 3[8].

In spite of the SO_2 scrubbing that is done, a considerable amount of SO_2 and other harmful contaminants remain in the flue gas because it is common for only a portion of the flue gas flux to be passed through the scrubber. As a result, the study of the dispersion of the contaminants in the environment is still of great importance.

ATMOSPHERIC DISPERSION OF SCRUBBED FLUE GAS PLUMES

In the calculation of downwind ground-level concentrations for "dry" plumes emitted by conventional industrial stacks, meteorologists generally employ a two-step procedure. Firstly, the height of rise of the plume Δh is determined by use of a simple plume rise formula. The effective stack height $h_{eff} = h_s + \Delta h$ is thereby obtained which is then input into a diffusion model. In this diffusion model, it is assumed that the waste gas is emitted free of excess momentum through an elevated continuous point source and dispersed by atmospheric turbulence (Fig. 1).

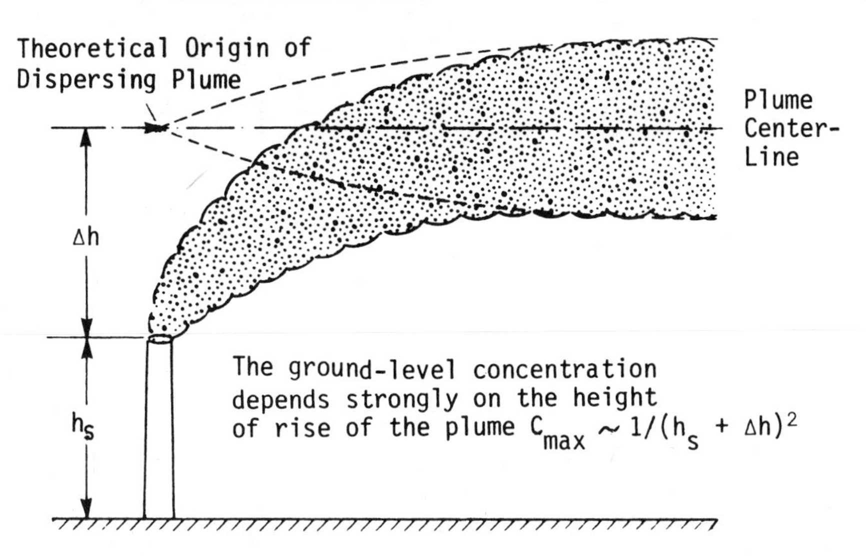

Figure 1: Far-field ground-level concentrations are commonly calculated by applying diffusion theory. In this approach, it is assumed that the plume is emitted from a virtual point source of zero buoyancy.

This procedure is also adopted for the calculation of scrubbed chimney plumes; in such cases we apply the same diffusion model as in the case of a dry plume. Since, in general, all concentrations are normalized by the stack exit value, the reduction of initial SO_2-concentration due to desulfurization is automatically taken into account. However, scrubbing of a plume reduces not only the SO_2-concentration, but it changes also other waste gas properties. Usually, the exit excess temperature is significantly reduced and the waste gas now contains a considerable amount of moisture. This additional moisture in a hot plume leads at some point in the dispersion to conditions of saturation and later condensation and evaporation of some of the water vapor. Physical reasoning and available numerical calculations show that the dynamic behavior and thus the height of rise of the wet scrubbed plume will be quite different in comparison to that of its otherwise equivalent dry counterpart, emitted under identical meteorological conditions. A substantial part of the waste energy contained in the flue gas appears in the latent form (water vapor). Whether the latent energy in the scrubbed plume is again converted into sensible heat (through condensation) depends in part on the saturation deficit and lapse rate of the atmosphere.

Since there is no proven method of determining Δh for scrubbed plumes, it has been common practice to assume that the contribution of latent heat to plume rise is completely negligible or, on the other hand, that all of it is ultimately converted into sensible heat and therefore contributes to initial buoyancy. Both extreme views prevail, although they cannot be supported, either from a physical point of view or from numerical or experimental studies. Without knowledge of the effective stack height, however, diffusion theory cannot be applied due to the lack of knowledge of the very model-sensitive input parameter h_{eff}. It is the purpose of this paper to elucidate these issues and to show how moisture effects should be accounted for in the prediction of buoyant chimney plumes.

SCRUBBED STACK PLUMES COMPARED WITH DRY STACK PLUMES AND COOLING TOWER PLUMES

Considerable work has been done by many researchers in predicting plume rise from conventional industrial stacks. These plumes are generally regarded as being dry although they contain some water vapor which came from the ambient air used during combustion and from the combustion process itself. Only the literature on plumes from cooling towers deals with the treatment of moisture effects on plume dispersion.

Comparing plumes resulting from the use of stack gas scrubbing techniques with dry plumes from chimneys and wet plumes from cooling towers, we note immediately that the exit parameters vary significantly (Fig. 2). Whereas in the case of wet cooling towers,

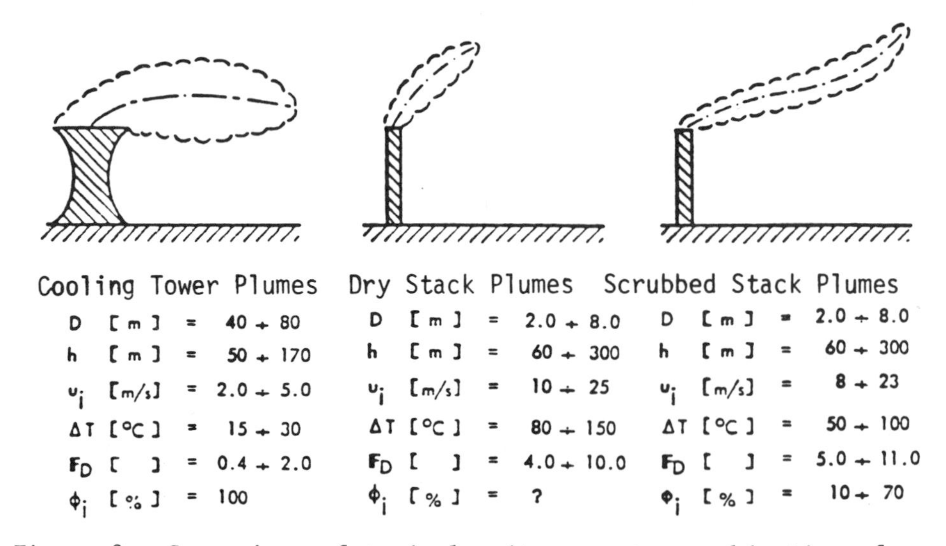

Figure 2: Comparison of typical exit parameter combinations for cooling tower plumes, scrubbed chimney plumes, and dry chimney plumes.

the exit excess temperature is in the range between 15°C and 30°C, we have 80°C to 150°C (or even more than 250°C [12]) for dry stack plumes. The excess temperatures of scrubbed plumes are typically about 60°C but may exceed 100°C in case of a bypass operation.

The exit relative humidities of scrubbed plumes vary over a wide range (10% to 70%) whereas cooling tower plumes are usually fully saturated at tower exit. Although the exit humidities of scrubbed stack plumes do not indicate saturation, a large amount of water vapor is still present in the plume at exit due to the high exit temperatures associated with those exit humidities. After a few diameters downwind, the plume generally becomes saturated due to the lower plume temperature resulting from mixing. The moisture content of dry plumes is, on the other hand, usually unknown but should be negligible with the exception of plumes coming from lignite- or peat-fired power plants.

Since the diameters and exit velocities of chimney and cooling tower plumes are very different, the densimetric Froude numbers can differ up to an order of magnitude. If k is defined as the ratio of ambient wind speed to effluent velocity, then large k values occur for cooling tower plumes and small k values for the chimney cases. For cooling tower plumes, large k values ($k>1.5$) imply rapidly bent-over plumes (with perhaps significant downwash) and a shorter near-field region. These large k values are, for cooling tower plumes, combined with low F_D values (0.4-2.0) indicating plumes which are rapidly bent-over and highly buoyant. From Fig. 2, cooling tower plumes have very different F_D and k values from chimney plumes, dry and moist. Among chimney plumes, the scrubbed and unscrubbed plumes have similar F_D and k values. The additional moisture in the scrubbed plumes leads to additional buoyancy which tends to partly counterbalance the effect of the lower chimney temperatures in the scrubbed plume case. It is therefore the thermodynamic effects which will really distinguish dry chimney and scrubber plume dispersion characteristics.

LITERATURE REVIEW ON THE EFFECTS OF MOISTURE ON PLUME RISE

As will be seen from the literature review below, none of the published models is appropriate to the simulation of the important physical phenomena (momentum transfer, buoyancy, thermodynamics, density differences) relating to the rise of scrubbed plumes.

The first major contribution to the theory of vapor plumes was by Morton [6] who considered only vertical plumes in a quiescent atmosphere. This integral approach developed by Morton was later extended to bent-over plumes by Csanady [2] and Wigley and Slawson [15]. These papers, however, concentrate mainly on the question of whether and where condensation can occur in a moist plume. Later, Wigley and Slawson [16] made a comparison of wet (i.e.

condensing) and dry plumes having the same exit excess temperature. In developing the energy equation for their moist plume model from the energy equation of a dry plume, Wigley and Slawson simply replaced the dry adiabatic lapse rate with the saturated adiabatic lapse rate; as a result, they were, with justification, strongly criticized by Briggs and Hanna [1]. The conclusion of the Briggs-Hanna remarks that all results of the Wigley and Slawson paper would be erroneous, was, however, premature.

A few years later, Weil [14] demonstrated that in the special case of a saturated plume in a saturated atmosphere this replacement was justified. The most important result of the Wigley and Slawson paper, however, is surely independent of the use of equations at all. They made clear that saturated plumes in a moist atmosphere show a significantly different stability behavior as compared to dry plumes in a dry atmosphere. Whereas the stability of a dry atmosphere is defined from the dry adiabatic lapse rate, $dT_\infty/dz = -\Gamma_{\infty d} = -0.00975\ ^oC/m$, the stability of the moist atmosphere depends on the saturation adiabatic lapse rate, $dT_\infty/dz = -\Gamma_{\infty s}$, or more precisely, on a slightly greater reference value, $\Gamma_{\infty m}$ when the total water content of the atmosphere changes with height [14]. Both, $\Gamma_{\infty s}$ and $\Gamma_{\infty m}$ are always smaller than $\Gamma_{\infty d}$. These latter values are not constant but depend on the local ambient pressure and temperature. For $T_\infty = 0^oC$ and $p_\infty = 1000$ mbars we have $\Gamma_{\infty s} = 0.00646\ ^oC/m$; for $T_\infty \cong 20^oC$ and the same pressure, $\Gamma_{\infty s} = 0.00428$ C/m is obtained. What is true for the stability of the atmosphere is also valid for the mode of rise of buoyant plumes (unstable, neutral or stable). For temperature gradients of the atmosphere between the dry and the saturated adiabatic lapse rate, $\Gamma_{\infty s} < \Gamma_\infty < \Gamma_{\infty d}$, the range of the so-called "conditionally unstable" atmosphere, the dry plume will behave in a stable manner, whereas the wet (i.e. condensing) plume shows an unstable mode of rise (Fig. 3). Thus, comparing a dry and wet plume with otherwise identical exit conditions in the same moist atmosphere, the wet plume will rise higher than its dry counterpart since it "sees" the atmosphere as always less stable. The reason for this is that the wet plume expands and cools as it rises often leading to condensation which releases latent heat. The wet plume in a moist atmosphere therefore cools but more slowly (at the saturated adiabatic lapse rate) than will a dry plume in a dry atmosphere (cooling at the dry adiabatic lapse rate). This argument follows providing that the relative humidities of the wet plume and atmosphere are sufficiently large that condensation occurs at all.

Comparing then (a) a saturated plume dispersing in a saturated atmosphere whose lapse rate is the local saturated adiabatic lapse rate with (b) an otherwise identical dry plume in a dry atmosphere with dry adiabatic lapse rate, Wigley and Slawson confirmed the former findings of Csanady [2] that the two plume paths differ by only a few percent since both plumes experience their environment

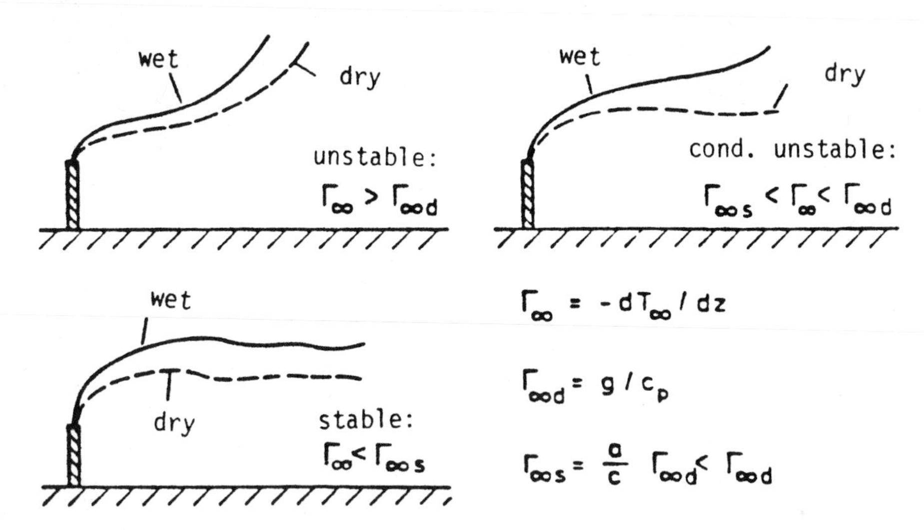

Figure 3 : Stability behavior of corresponding dry and wet buoyant plumes, according to [16].

to be neutral. This result, however, cannot be generalized and can only be true for plumes with very small exit excess temperatures (as was assumed in the Wigley-Slawson model calculations). Since water vapor is lighter than air, the intitial density defect of the moist plume can be significantly larger than that of a corresponding dry plume, provided that the initial excess temperature is large enough to saturate a substantial amount of water. For example, for an ambient pressure of 980 mbars, an ambient temperature of 0°C and an stack exit temperature of 65°C, the density defect of the dry plume would be $\Delta\rho / \rho_\infty = 0.192$ compared with 0.233 or 0.27 for moist plumes with 50% or 100% relative humidity, respectively.
All other initial data were assumed the same. Consequently, the saturated plume in a saturated atmosphere with $\Gamma_{\infty s}$ as the ambient lapse rate must rise higher than an equivalent dry plume in a dry atmosphere with $\Gamma_{\infty d}$ as the ambient lapse rate. The difference in the heights of rise can be significant depending on the values of the exit densimetric Froude numbers for the two cases which differ due to the presence of moisture effects on the density defect ratio. Taking into account the results presented in the papers by Hanna [3], Weil [14], and those published later by Wigley [19] and Wigley and Slawson [17], it can be concluded that their predictions tend to agree qualitatively. Quantitatively, however, they disagree remarkably and it is only partly clear to the reader where these differences come from.

All papers discussed here do not contain any verification of the predictions presented, either for dry plumes, for which sufficient detailed data are available [11] or for wet plumes, for which measurements are available at least for the parameter range of wet cooling towers. Furthermore, only dry and wet plumes with identical exit temperature excesses have been compared. The question of real interest, however, is to what extent the use of a scrubber changes the ultimate plume rise and the resulting downwind distribution of pollutants.

DISCUSSION AND EVALUATION OF LITERATURE APPLIED TO SCRUBBED PLUMES

The most pertinent work relating to plume rise from scrubber stacks are, in terms of theory, the papers by Wigley [19] and Schmitt [10], and in terms of experimental data, the papers by Sutherland and Spangler [33] and Meagher et al. [4,5]. Wigley in his numerical analysis of the effect of condensation on plume rise presents a number of predictions concerning the dependence of wet and dry maximum plume rise on atmospheric stability for both cooling tower plumes and scrubbed plumes. For scrubbed plumes he assumed an excess temperature of 55°C and fully saturated conditions for both the plume and the atmosphere. Although the initial density difference that such a plume exhibits is already significant, he used a Boussinesq-approximated model. Furthermore, the model used was not calibrated, either with dry or with wet data. Consequently, some doubts arise as to the reliability of the results. Since he presented in sufficient detail only results of cooling tower plume calculations, we repeated these calculations with a similar model but calibrated with a large number of cooling tower experimental data. As can be seen from the comparison in Figure 4, the results are qualitative similar but differ quantitatively by some one hundred percent.

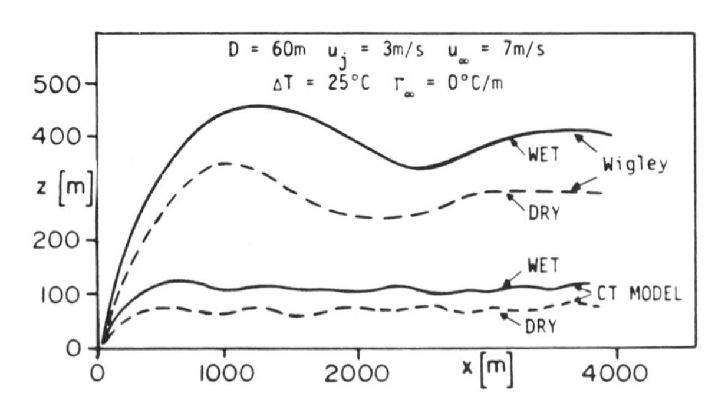

Fig. 4 :Comparison of wet (ϕ_j = 100%, ϕ_j = 100%) and dry (ϕ_j = 0%, ϕ_∞ =0%) cooling tower plumes with two similar models.

It seems to be very unlikely and is not supported by experimental evidence that a cooling tower plume in an isothermal (=stable) atmosphere with wind velocity more than twice as large as the plume exit velocity could rise up to 450 m above tower top level, as in Wigley's prediction.

Since the results of both models are qualitatively the same and the basic assumptions employed in both models for cooling tower application justified, the discrepancy must have been caused mainly by the different entrainment hypotheses used in the two models. A user required to make engineering calculations would get very different results, depending on which model he used. The conclusion here then is that quantitative results from numerical models can be relied upon only after accurate calibration of the model with experimental data appropriate to that application.

The need to have the models properly evaluated is further underscored in the discussion below of the work by Schmitt[10]. His (internal) report to the German Minister of Interior is the only theoretical approach we are aware of which addresses the question: to what degree can the effect of moisture in a scrubbed plume compensate for the loss of sensible heat of the plume (due to the scrubbing process) in plume rise.

Schmitt compared a scrubbed plume of 65°C exit temperature (T_j) and 49°C dew point temperature (T_{dp}) with a corresponding nearly dry unscrubbed plume of T_j=120°C and T_{dp}=38°C at exit. Schmitt, however, also uses a Boussinesq-approximated model of cooling tower origin; this model must also be regarded as being uncalibrated since he only tests it (for the dry plume applications) with one of Briggs formulas and not with measured data. The report does not provide enough details so that it remains unknown whether his model contains any further simplifications which are not justified for the parameter ranges of stack plumes under consideration.

The importance of some of the issues discussed earlier relating to plume thermodynamics is revealed in Fig. 5. In this Figure, we compare Schmitt predictions with predictions of our own model for identical cases. We ran the same cases with a similar Boussinesq-approximated wet cooling tower model but using an entrainment hypothesis calibrated with chimney data. Both models employ the equilibrium theory of thermodynamics which states that water vapor and liquid water must remain in equilibrium whenever liquid water is present.

An inspection of Fig. 5 reveals that significantly different predictions result between our model and that of Schmitt. The results do not even qualitatively agree. Since the lapse rate of the atmosphere is slightly above $\Gamma_{\infty d}$ and both plumes are moist enough to condense, our predictions result, as one should expect,

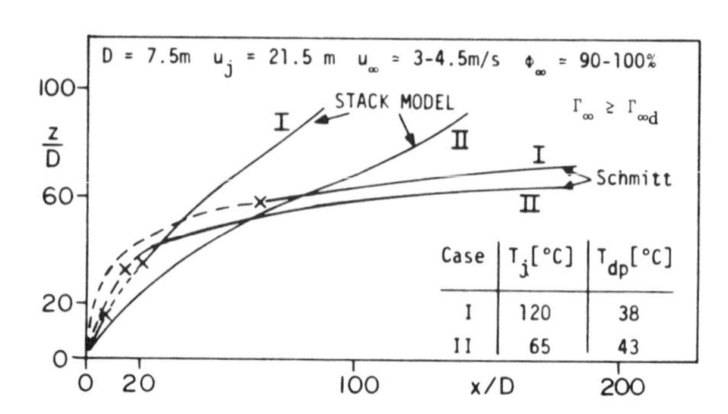

Figure 5: Comparison of predictions of nearly dry (I) and moist (II) plumes from two similarly simplified models.

in unstable plumes whereas Schmitt's plumes seem to be neutral. As can be also seen from Fig. 5, the invisible parts of the plumes are differently predicted (dotted lines). It seems very unlikely to us that even the scrubbed plume(II) with its small (T_j-T_{dp})-difference needs more than 30 diameters before it starts condensing in a nearly saturated atmosphere.

The conclusion here is again that models must be evaluated both qualitatively and quantitatively. They must be tested to determine whether they show the correct systematic behavior in their predictions. Finally they must be compared with experimental data.

As far as experimental data are concerned, there is a tremendous amount of cooling tower plume data[7] available; in addition for dry chimney plumes some data have been published which are suitable for model evaluation purposes[11]. As far as scrubbed plumes are concerned, the situation is much worse. There have been measurements carried out at TVA[4,5], but with emphasis on chemical reactions in such plumes. During the TVA data aquisition campaigns, the plume path in the near field and detailed exit and ambient conditions have not been measured.

The only plume rise measurements from scrubbed and unscrubbed buoyant stack plumes we are aware of are those published recently by Sutherland and Spangler[13]. These data were taken at the same plant during two subsequent winter periods. Since the purpose of their measurements was to test the performance of dry plume rise formulas as applied to scrubber plumes, no humidity measurements were taken during the plume study. For model evaluation applications, they are therefore only of limited value. The data (Fig. 6) confirm, however, the theoretical finding that there is a significant con-

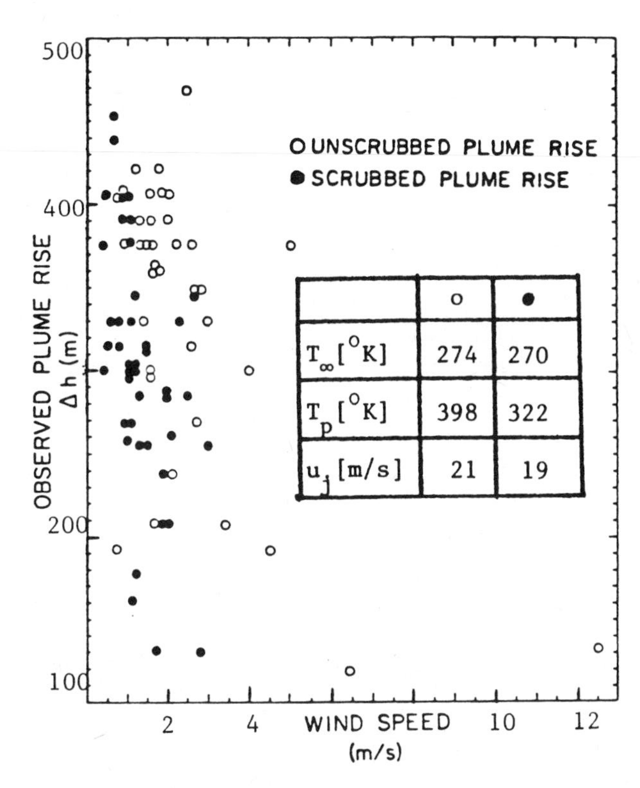

	o	●
T_∞ [°K]	274	270
T_p [°K]	398	322
u_j [m/s]	21	19

Figure 6: Comparison of observed plume rise heights for scrubbed and non-scrubbed stack plumes (from [13]).

tribution of thermodynamic effects to plume rise. The measurements were carried out in both cases at the same time of the year with, on average, very similar weather conditions and nearly similar exit conditions during each measurement campaign.

Although the exit temperature excess of the unscrubbed plumes was more than twice the temperature excess of the scrubbed plumes, the loss in height of rise of the moist plumes was obviously rather small. What is of much more striking evidence from this figure is the large scatter of the data points; to be noted is the scatter from the data which were sampled under similar stability conditions. The work of Sutherland and Spangler thus indicates once again that simple plume rise formulas are already of questionable reliability in the case of dry plumes. If moisture effects have to be taken into account as well, the problem becomes more acute due to the potential feedback of thermodynamic effects on plume dynamics for scrubber plumes and the different stability behavior that wet plumes show from dry plumes. The complexity of this problem requires that a more advanced model be used. Such a model must be sufficiently detailed that all important physical processes dominating plume rise is to be taken into account. On the other hand, the model should be simple enough that the effort in terms of computer time and storage demand is manageable. We believe that mathematical models of the integral type satisfy both requirements. Yet there are a number of additional requirements to such a model which result from our discussion here : Since the initial temperature excess and the density defect of scrubbed chimney plumes are large (above all in cases of bypass operation) such a model must avoid simplifications and approximations usually contained in cooling tower models which are now no longer justified. Such simplifications are typically :

A linearized equation of state ($\Delta\rho/\rho_\infty = \beta.\Delta T + \gamma.\Delta q + \delta.\Delta$ which is valid only for small excess temperatures, ΔT, small excess specific humidities, Δq, and small excess liquid water contents, $\Delta\sigma$,

First order approximations, in the calculation of the local saturation deficit, which can result in a significant error as soon as temperature differences become large, and

The Boussinesq-approximation, which has among one of its requirements that the density defect ratio be very small ($\Delta\rho/\rho_\infty \ll 1$).

As an example, the consequences resulting from a possible violation of the Boussinesq-approximation are subsequently quantified : As our calculations with a calibrated non-Boussinesq-approximated dry chimney plume model [9] show (Fig. 7), there might be more than a 20 % difference in the final height of rise when we introduce the Boussinesq approximation into the model and set the appropriate terms to zero. (The case presented in Fig. 7 represents a dry chimney plume case without scrubbing). The required model should also avoid the use of other assumptions common in cooling tower models which are not well justified in chimney plume models as, e.g., bentover plume assumption (initial momentum too large in case of chimney plumes), a different spreading rate for temperature than for moisture, etc.

Entrainment or turbulence models which have been proven to be successful in cooling tower plume calculations cannot be expected to give the same good results in scrubbed plume calculations due to the order of magnitude difference in densimetric Froude numbers. In the initial zone of cooling tower plumes, buoyancy forces can exceed inertial forces ($F_D < 1$) whereas in case of scrubbed chimney

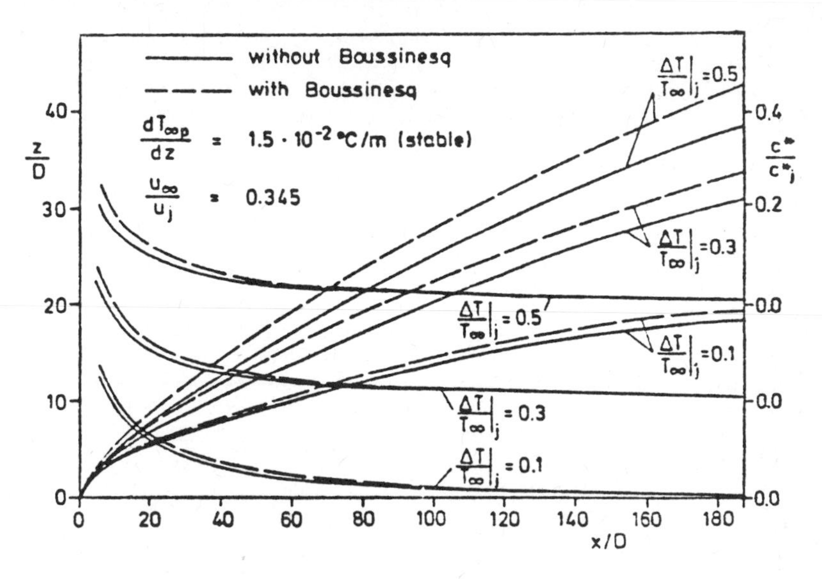

Figure 7: Trajectories and centerline dilutions for plumes in stable atmosphere calculated with a Boussinesq- and a non-Boussinesq approximated model.

plumes ($F_D > 5$) inertial forces always dominate (Fig. 2). Consequently, self-generated turbulence within the plume which governs near-field dispersion is expected to be mainly shear turbulence in case of chimney plumes whereas for cooling tower plumes a substantial contribution of buoyancy towards turbulence generation is expected.

CONCLUSIONS

As a result of the above discussion, we can conclude that existing formulas and models for dry plume rise from chimney stacks and wet plume rise from cooling towers are inappropriate for the prediction of plume rise from stacks using scrubbers. A reliable integral model is needed which employs an entrainment function or turbulence hypothesis calibrated with chimney data covering the whole range of parameters given in Figure 2. Such a model should neither contain (a) the Boussinesq-approximation nor (b) linearizations of the equation of state, nor (c) simplifications in the calculation of the moisture excesses (plume minus ambient). The latter moisture excess simplifications are often applied in the energy equation and for the determination of the visible (i.e., condensing) part of the plume. Each of these three commonly used simplifying approximations necessarily restricts the applicability of the model to low excess temperature cases. As soon as such an integral model has been developed for scrubber plume rise and proven to be reliable, tables or nomograms can be developed which can simplify the prediction of scrubber plume rise.

REFERENCES

[1] Briggs, G.A. and Hanna, S.R.: Comments on "A Comparison of Wet and Dry Bent-Over Plumes". Journ. Appl. Met., Vol. 11, pp. 1386-1387, 1972.
[2] Csanady, G.T.: Bent-Over Vapor Plumes. Journ. Appl. Met., Vol. 10, pp. 36-42, 1971.
[3] Hanna, S.R.: Rise and Condensation of Large Cooling Tower Plumes. Journ. Appl. Met., Vol. 11, pp. 793-799, 1972.
[4] Meagher, J.F., Stockburger, L., Bonanno, R.J., and Lurica, M.: Atmospheric Oxidation of Flue Gases from Partially SO_2-Scrubbed Power Plant. Tennessee Valley Authority, Report TVA/ARP-I 80/35, 1980.
[5] Meagher, J.F., Stockburger, L., and Bonanno, R.J.: Chemical Interactions in Scrubber Plumes, Data Supplement I. Tennessee Valley Authority, Report TVA/I-AQ-78-9, 1978.
[6] Morton, B.R.: Buoyant Plumes in a Moist Atmosphere. Journ. Fluid Mech., Vol. 2, pp. 127-144, 1957.
[7] Policastro, A.J., Carhart, R.A., Ziemer, S.E., and Haake, K.: Evaluation of Mathematical Models for Characterizing Plume Behavior from Cooling Towers, Vol. 1: Dispersion from Single and Multiple Source Natural Draft Cooling Towers. U.S. Nuclear Regulatory Commission, Report No. NUREG/CR-1581, 1980.
[8] Schaerer, B. (Hrsg.): Luftverschmutzung durch Schwefeldioxid-Ursachen, Wirkung, Minderung. Texte des Umweltbundesamtes, Berlin, 1980 (in German).
[9] Schatzmann, M. and Policastro, A.J.: Effects of the Boussinesq-approximation on the Results of Chimney Plume Calculations. Proceedings, 2nd Intern. Conf. on Num. Meth. in Therm. Probl., Venice, 1981.
[10] Schmitt, R.: Abschlussbericht zum FE-Vorhaben "Ueberhoehungsformeln für gereinigte, nasse Abgase" (?), Bericht an das Bundesministerium des Innern, 1980 (internal report, in German).
[11] Slawson, P.R. and Csanady, G.T.: The Effect of Atmospheric Conditions on Plume Rise. Journ. Fluid Mech., Vol. 47, pp. 39-49, 1971.
[12] Slawson, P.R., Davidson, G.A. and Maddukuri, C.S.: Dispersion Modeling of a Plume in the Tar Sands Area. Syncrude Env. Res. Rep. No. 1/1980.
[13] Sutherland, V.C. and Spangler, T.C.: Comparison of Calculated and Observed Plume Rise Heights for Scrubbed and Non-Scrubbed Buoyant Plumes. 2nd Joint Conf. on Appl. of Air Poll. Met., pp. 129-132, New Orleans, 1980.

[14] Weil, J.C.: The Rise of Moist, Buoyant Plumes. Journ. Appl. Met., Vol. 13, pp. 435-443. 1974.
[15] Wigley, T.M.L. and Slawson, P.R.: On the Condensation of Buoyant, Moist, Bent-Over Plumes. Journ. Appl. Met., Vol. 10, pp. 253-259, 1971.
[16] Wigley, T.M.L. and Slawson, P.R.: A Comparison of Wet and Dry Bent-Over Plumes. Journ. Appl. Met., Vol. 11, pp. 335-340, 1972.
[17] Wigley, T.M.L. and Slawson, P.R.: The Effect of Atmospheric Conditions on the Length of Visible Cooling Tower Plumes. Atmosph. Environment, Vol. 9, pp. 437-445, 1975.
[18] Wigley, T.M.L.: Condensation in Jets, Industrial Plumes and Cooling Tower Plumes. Journ. Appl. Met., Vol. 14, pp. 78-86, 1975.
[19] Wigley, T.M.L.: A Numerical Analysis of the Effect of Condensation on Plume Rise. Journ. Appl. Met., Vol. 14, pp. 1105-1109, 1975.

DISCUSSION

D. EPPEL

Does the mention of advanced mathematical models mean, that the full coupled momentum-concentration equations have to be solved ?

M. SCHATZMANN

Since the initial zone (e.g. the first few hundred meters) of a buoyant plume is clearly dominated by the action of forces (buoyant forces, inertial forces, pressure forces) we have no other choice than using the Navies-Stokes-equation in advanced mathematical models. The only alternative would be to use simple plume rise formulas which have already been proven to be of questionable reliability in case of dry plumes. As I tried to show during the talk in case of wet plumes the situation is even more complex. Plume rise formulas must lead to an even larger error.

PLUME BEHAVIOR IN STABLE AIR

F. L. Ludwig and J. M. Livingston

SRI International

Menlo Park, California 94025

INTRODUCTION

Because stable plumes tend to have less impact at ground level, the behavior of plumes from elevated sources in stable air has not attracted the same interest that plume behavior in unstable air has. However, some aspects of stable-plume behavior hold great interest. For example, concentrations in the plume remain high for long periods of time, which increases the importance of chemical reactions taking place among plume constituents or between constituents and the ambient air. The nature of the stable plume also has its effect during periods of fumigation, when the high concentrations within the plume are mixed to ground level by the onset of unstable conditions.

The behavior of a stable plume is particularly amenable to study with laser radar (lidar) instruments. Numerous studies (e.g., Mc Elory et al., 1981 ; Uthe et al., 1980a ; Uthe et al., 1980b) have demonstrated that the shape of plumes from elevated point sources in stable air can be clearly defined at distances of tens of kilometers from the source by mobile (either ground vehicle or airborne) lidars.

This paper discusses two measures of plume behavior in stable air : the near-instantaneous plume dimensions and temporal changes in plume position. The descriptions of plume dimensions are derived directly from lidar observations and are well supported by the data. The descriptions of variability of plume position, however, are almost completely hypothetical and are included at this time primarily to provide a basis for future data collection and analysis.

THE NATURE OF THE LIDAR DATA

The Mark IX lidar developed by SRI International (SRI) was usec to study Sulfur Transport And Transformation in the Environment (STATE) downwind from the Tennessee Valley Authority's (TVA's) Cumberland, Tennessee power plant. The lidar provided many depictions of vertical cross sections through the plume and the boundary layer at various distances downwind from the two 305-m stacks of this 2600 MW plant. Some examples of such cross sections have beer published by Uthe et al. (1980a). The use of the mobile lidar syste for studying plumes has been described by Uthe and Wilson (1979). Briefly, the instrument is driven back and forth under the plume anc fired vertically every 80 m. The received return, which is a function of particulate backscatter, provides an indication of particle concentration. Thus, it provides a portrait of plume dimension and a qualitative depiction of the density of the material within the plume as it is driven back and forth. Under stable conditions, such as those discussed here, plume boundaries have generally been sharp and unambiguous.

In addition to the lidar data, special meteorological measurements were made by TVA (Clarke et al., 1978, 1980; Crawford et al. 1979). Radiosondes and aircraft were used to obtain temperature profiles in the vicinity of the nearby Johnsonville and Paradise Steam Plants. The Johnsonville plant is about 36 km south-southwest of Cumberland; the Paradise plant is about 100 km northeast. The pilot balloon measurements of winds aloft were made at the Cumberlar location. These wind observations are very good descriptors for the downwind plume dimensions.

The coordinates of points on the plume outline were digitized to provide the basic data set from which the analyses and interpreta tions discussed in the next section were derived. Plume width, thickness, and cross-sectional area can be derived quickly when the coordinates are available in digital form. Figure 1 shows schematically how these measures were defined. Inasmuch as the lidar van was constrained to follow available roadways, a traverse normal to the plume axis was not always possible; the apparent width of the plume depends on the angle between the plume axis and the plane of the cross section. Measured widths were multiplied by the sine of the angle between the plane of the cross section and the line connecting the plume center to the source to correct for this effect Plume widths, thicknesses, and cross-sectional areas were tabulated and analyzed. The data were stratified into two categories: night (defined as the period from 1730 to 1000), when the air at plume height tended to be relatively stable; and day (from about 1100 to 1730), when conditions tend to be more turbulent. This paper deals only with the nighttime cases.

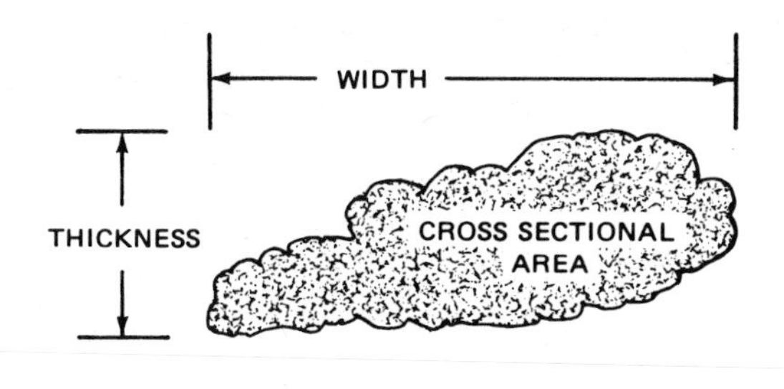

Fig. 1. Schematic representation of plume outline.

As noted above, width and thickness of the plume were the measured parameters. However, the most frequently used measures of plume dimension are σ_y and σ_z, i.e., the horizontal and vertical standard deviations of a Gaussian concentration distribution. Ching et al. (1979) suggested several ways to obtain σ_y and σ_z from lidar measurements. We chose the simplest, which assumes that the boundary of the plume in the lidar cross section occurs at a fixed fraction of the maximum backscatter values (at the center of the plume). With this assumption, it is possible to estimate σ from the measured dimensions. We have assumed that the visual boundary of a plume occurs at the point where the concentration falls to ten percent of that at the center. In the Gaussian distribution, this occurs at a point that is 2.15 σ from the center, so that

$$\sigma_y = (y_L - y_R)/4.3 \quad ,$$

where y_L and y_R are the coordinates of the left and right edges of the plume. An analogous relationship can be used to determine σ_z from the top and the bottom of the plume outline.

NEAR-INSTANTANEOUS PLUME DIMENSIONS IN STABLE AIR

Vertical Dimensions

Figure 2 shows the values of σ_z at different downwind distances when the air was generally neutral or stable. The figure also shows the line of best fit that relates σ_z to downwind distance. A functional form where σ_z is proportional to a power of downwind distance has been assumed. Figure 2 also shows the vertical dispersion coefficient curves (approximated by straight lines) given by Turner (1979) for neutral and stable cases. Although the fit is not

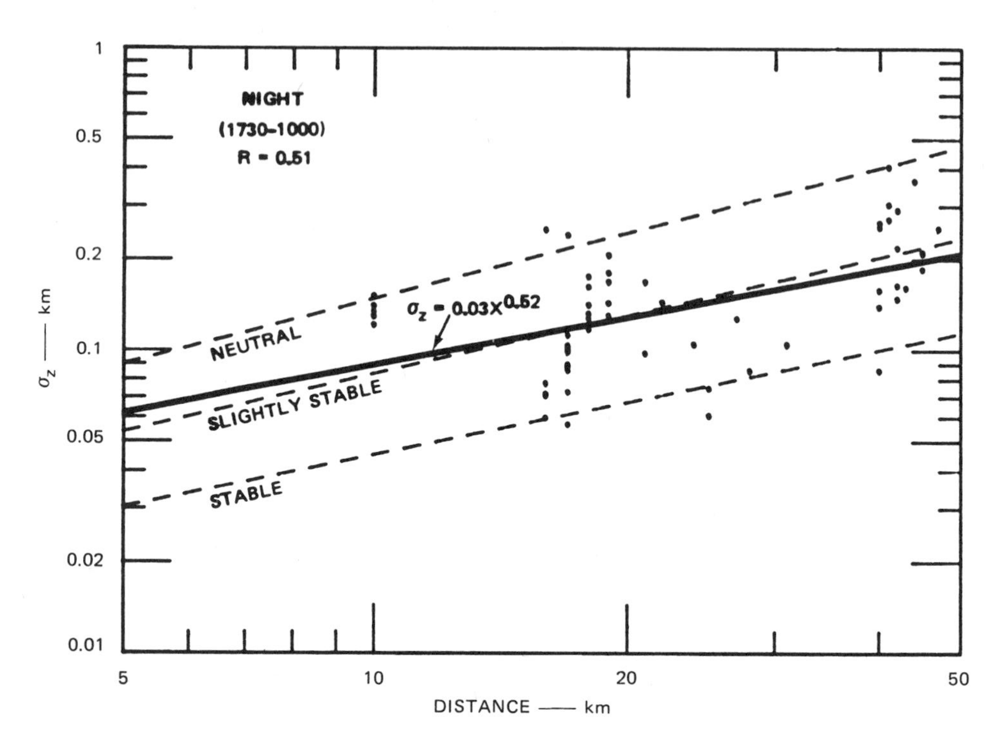

Fig. 2. Vertical standard deviation measured at night by the lidar at different distances downwind of the Cumberland Plant

particularly good (the correlation is 0.51), the observed values of σz do fall in the appropriate range of Turner values. During this same program, McElroy et al. (1981) obtained similar results with an airborne lidar. Their observations showed nearly constant values of σ_z--from about 50 to 100 m--over downwind distances of 1 km to 50 km. Their results and those shown in Figure 2 suggest only a weak dependence of plume thickness on downwind distance under stable conditions.

Horizontal Plume Dimensions

Figure 3 shows lateral standard deviation (σ_y) measured under generally neutral or stable conditions as a function of downwind distance. The line of best fit (with a correlation of 0.7) and Turner's (1969) curves are also shown. The observed nighttime lateral standard deviations are consistently larger than those shown by the Turner curves; more than 60 percent of the nighttime observations showed a lateral spread that was greater than what would be expected at the same downwind distance for neutral conditions.

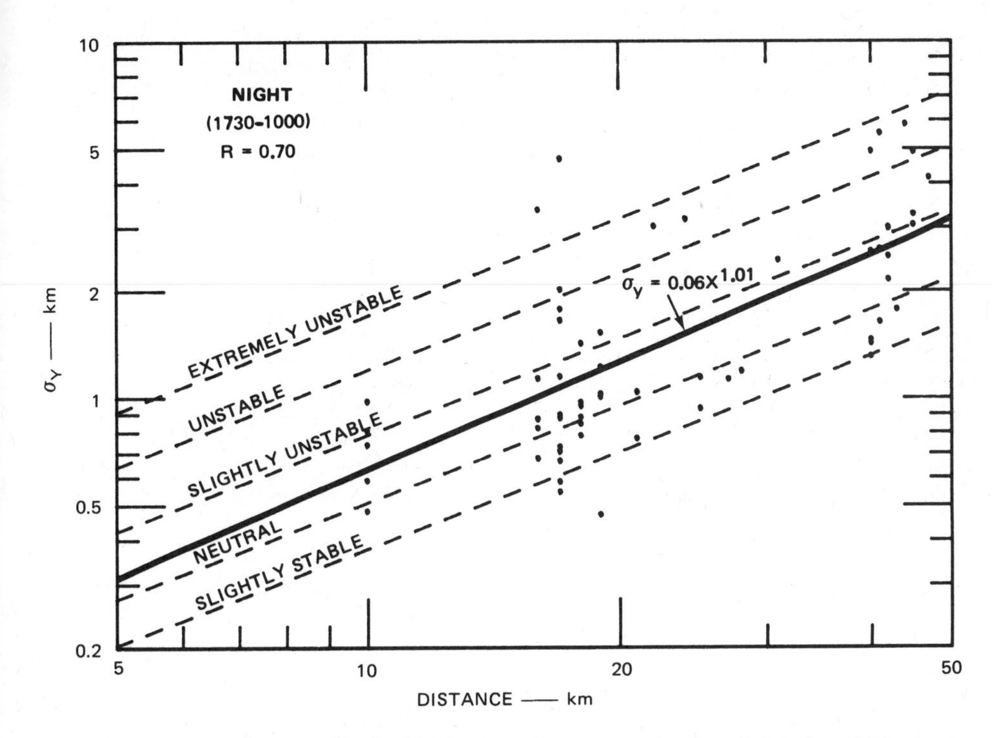

Fig. 3. Lateral standard deviations measured at night by lidar at different distances downwind of the Cumberland Plant

It should also be noted that the line of best fit indicates that σ_y is very nearly proportional to downwind distance.

Pooler and Niemeyer (1970) suggested that plume spread in stable air might be governed by a shear in the wind direction with height. Uthe et al. (1980a) presented striking examples of such behavior taken from the data set being discussed here. They found that plume dimensions corresponded closely to the downwind distance times the angular separation between wind vectors at the top and bottom of the plume, i.e.

$$W_s = X \sin(\Delta\Theta) \quad ,$$

where X is the distance downwind of the source and $\Delta\Theta$ is the absolute angular difference between the wind direction at the lowest and highest parts of the plume.

Data from compact, well-defined plumes were used to test the above relationship. Sometimes, wind shear reversed with height,

-ing in ">" shaped plumes (see Uthe et al., 1980a). In such -, the two plume segments were treated separately. Figure 4 -s the width measured by lidar (W_L) as a function of that esti- -ed on the basis of wind shear and downwind distance. The figure -so presents the best-fit line and lines that differ from it by -actors of 1.5 and 2.

It is apparent from Figure 4 that the estimate of the plume width based on wind shear is generally good, but tends to be less than the actual width for smaller values (in particular, for actual plume widths less than 12 km). Pooler and Neimeyer (1970) speculated that the rise and stabilization of the plume would account for an initial spread that would lead to underprediction on the basis of wind shear alone close to the source, but which would be of decreasing importance at greater downwind distances so that the wind shear effect would appear to dominate this one moved farther from the source. The data appear to confirm their speculation. Pooler and Neimeyer (1970) also noted that plume spread averages about 70 percent of that which is to be expected from wind shear alone. The slope of the best-fit line in Figure 4, 0.75, is certainly consistent with those findings.

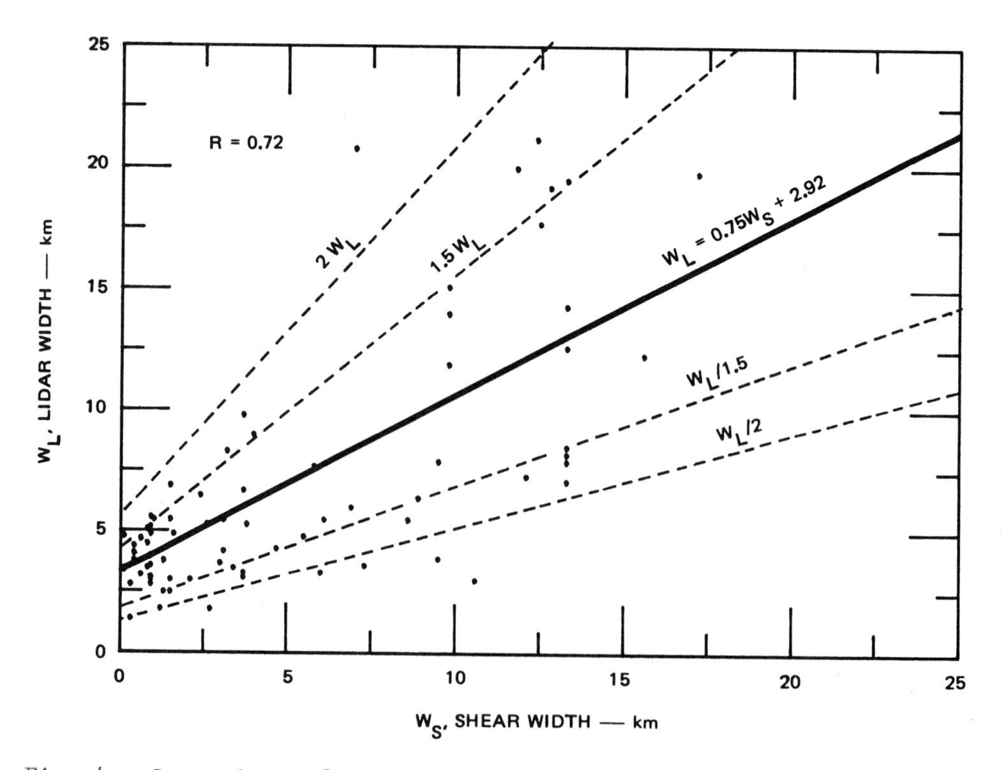

Fig. 4. Comparison of observed width of compact plumes with that estimated from wind shear

The implication in these findings and those of Pooler and Niemeyer (1970) is that there is some process (such as turbulent diffusion or initial plume expansion) that augments the widening by the wind shear. We have attempted to examine this effect by looking at those stable plumes where wind shear was not a major factor, either because there was little or no shear with height or because there was so little difference between the height at the bottom and top of the plume that the wind directions were essentially the same at both levels. Figure 5 shows the widths of such plumes plotted as a function of downwind distance. The best-fit line is also shown:

$$W = 0.71\ X^{0.5} \quad .$$

If we take the derivative with respect to X and make the necessary substitutions to express that derivative as a function of width, W, we get

$$dW/dX = 0.25/W \quad .$$

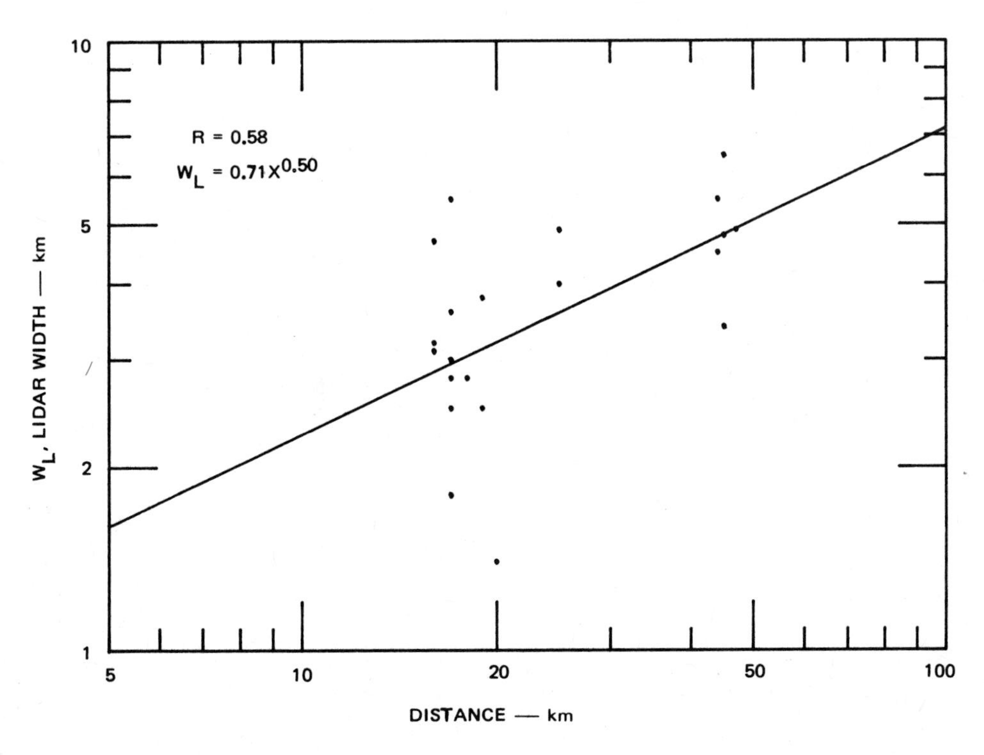

Fig. 5. Widths of compact plumes that were not affected by wind shear

The above relationship suggests that the growth due to non-shear effects is rapid at first, but slows when the plume grows larger. The rate at which shear effects cause the plume to grow is independent of plume width or downwind distance, however, so that a width derived solely from wind-shear effects will underestimate the actual width when that width is small because the non-shear effects will be causing significantly rapid growth. The shear-derived estimates will be more accurate as the plume grows larger.

A HYPOTHESIS REGARDING LONGER TERM PLUME BEHAVIOR IN STABLE AIR

The following description of plume behavior in stable air is an hypothesis that has not been fully tested. It is based on the fluctuating plume concept (cf. Gifford, 1959 and 1970), which breaks the description of plume behavior into two parts: the "instantaneous" dimensions of the plume and the variations in the location of the plume center. A fluctuating plume model is analogous to treating the problem with an ensemble of still photographs, each one showing the instantaneous appearance of the plume at a different time. The data supplied by the lidar cross sections for compact, stable plumes are generally consistent with this type of plume description. The fluctuating plume description is also advantageous when one is looking for a model that will facilitate the treatment of chemical reactions, because such reactions depend on the concentrations of the reactants and their products within the plume at any given time, rather than average of concentrations. The fact that a plume passes over one location at one time and another location at another time is of little consequence in determining the chemical reactions, although it does affect the distribution of longer-term average concentrations. The major item of importance for plume chemistry is the rate at which the "instantaneous" cross section of the plume changes as the materials move downwind.

It is usually assumed that the probability of finding the plume center at some particular location is statistically determined (cf. Gifford, 1970). We are hypothesizing that plume behavior under stable conditions is more orderly than that, because the physical processes that determine plume location are continuous and change rather slowly. Figure 6 shows the idealized motion of such a plume with time. We have hypothesized that the width of the plume (measured normal to the average center line) remains the same (W in the figure) and that the position moves sinusoidally from one side of center to the other with an amplitude B and a period T. For our purposes, it has been assumed that the two sides of the plume move in phase and have the same amplitude of motion. We have followed the example of Gifford (1970) and assumed that the instantaneous distribution is of the "top hat" variety, where concentrations within the plume are equal to Q/Au and are zero outside it. The emission rate is given by Q; A is the instantaneous cross-sectional area of the plume, and u is the wind speed.

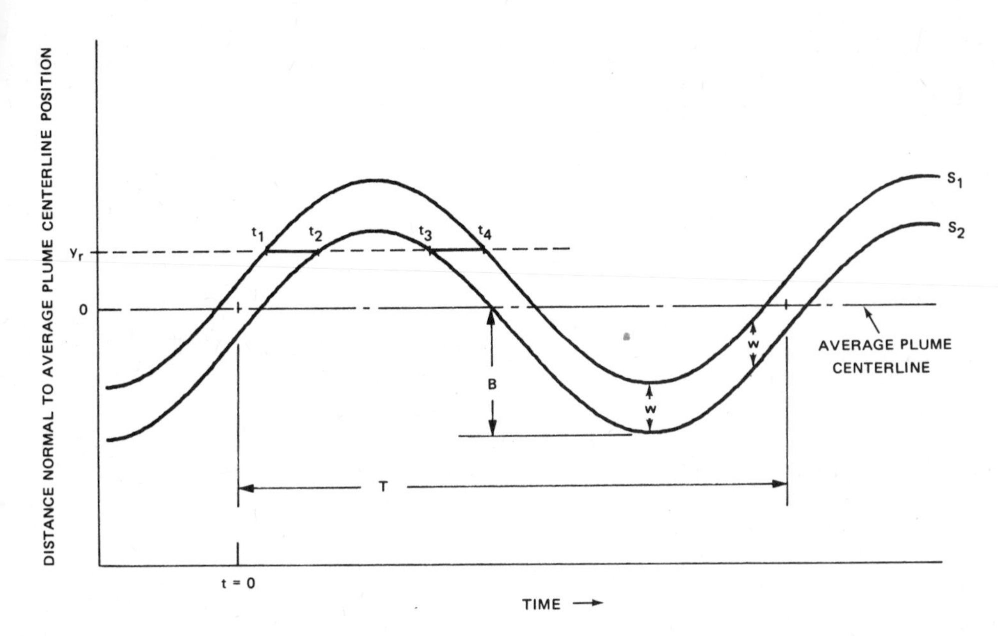

Fig. 6. Time variations of plume location in fluctuating-plume model

A point at a distance y_r from plume center, as shown in Figure 6, will sometimes be within the plume and sometimes outside it. The times when the plume is over the receptor are marked by the heavy solid lines in Figure 6 that extend from t_1 to t_2 and from t_3 to t_4. The symmetry of the sinusoidal motion means that $(t_4 - t_3) = (t_2 - t_1)$. During the period when the plume is over the point y_r, the concentration would be Q/Au; the concentration is zero at other times, so the average concentration over an integral number of periods would be given by:

$$C = 2(t_2 - t_1)\ Q/TAu \quad .$$

Thus, it is possible to estimate the longer-term average concentration at a point y_r if we know the width of the plume W and the amplitude of its fluctuation, B. The relevant time differences are given by

$$t_2 - t_1 = T\ [\text{arcsine}\ (Y_2) - \text{arcsine}\ (Y_1)]/2\pi \quad ,$$

where

$$Y_1 = (y_r - W/2)/B$$

$$Y_2 = (y_r + W/2)/B \quad .$$

In the above equations, y_r is measured from the average position of the plume center and the term W/2 accounts for the fact that the sine curves describing the edges of the plume are displaced from the plume center by a distance equal to half the plume on either side of the center. The amplitude of the sine curve is given by B. The terms Y_1 and Y_2 are equal to the value of the sine where curves S_1 and S_2, respectively, intersect the line $y = y_r$. The above relationships hold if the Y_1 is set equal to -1 whenever $(y_r - W/2)/B < -1$ and Y_2 is set equal to +1 whenever $(y_r + W/2)/B > +1$. The concentration averaged over an integral number of cycles of plume meander will then be given by

$$C = Q\,[\text{arcsine}\,(Y_2) - \text{arcsine}\,(Y_1)]/\pi Au \quad .$$

Figure 7 shows the long-term average concentration changes with distance from the plume center for different ratios of amplitude to instantaneous plume width. The graphs in the figure show only half the plume ; average concentrations in the other half are the mirror image of those shown. Figure 7 shows that this model leads to a bimodal average concentration distribution across the plume when the width of the plume is equal to, or smaller than, the amplitude of the plume meander. Under such conditions the peak concentrations are at the outermost point reached by the inner edge of the plume ; this point remains within the plume for the longest period of time during a cycle.

Figure 7 shows that the two modes get closer together as the amplitude of the plume meander gets smaller relative to the instantaneous width of the plume. When the amplitude of the meander is half the width of the plume, the two modes merge so that the peak in the average concentration distribution occurs at the central location of the plume meander. For meanderings that are less than half the width of the plume, there will be a relatively broad central peak.

Although we have assumed sinusoidal motion, qualitatively similar results would be obtained for any periodic motion that did not involve a discontinuous reversal of direction at the edges. For any periodically moving plume that slows down before reversing direction, the longest time will be spent at the extremes, and, hence, the highest long-term average concentrations will also be found there.

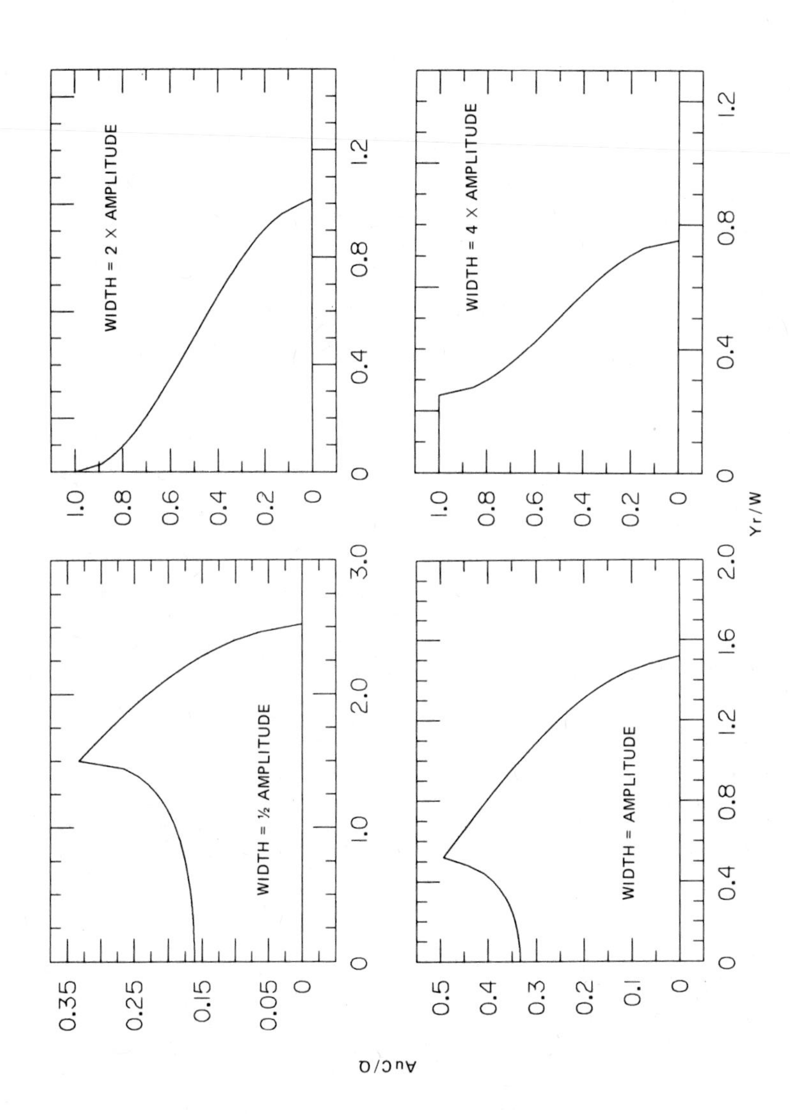

Fig. 7. Relative average concentration for points at different distances (measured in instantaneous plume widths) from the centerline of a sinusoidally fluctuating plume

If the meander of a plume is determined largely by the periodic fluctuations in the wind direction at the time the materials are emitted, then the amplitude will tend to remain relatively constant with downwind distance. However, the instantaneous width of the plume tends to increase with downwind distance, as was shown earlier. Therefore, one might expect to find bimodal long-term average concentrations near a source, but unimodal distributions at greater downwind distances.

As noted earlier, the preceding description of plume behavior has not been verified. One case has been identified in the literature that exhibits the hypothesized average concentration pattern near a source. Figure 8 (from Thuillier and Mancuso, 1980) shows hour-averaged lidar observations at a point 300-m downwind of a smoke release in stable air. The backscatter data have been averaged in the vertical to provide total backscatter along a vertical line at different crosswind distances. The figure shows a bimodal distribution, such as that which would be expected if the preceding description of plume behavior were accurate. Pooler (1981) has pointed out that this example is hardly an adequate confirmation, inasmuch as the bimodal character could have been caused by building configuration or wind-direction variations. Unfortunately, only two time periods with inversions were included in the tests reported by Thuillier and Mancuso (1980); the other was in the late afternoon and did not show the strong bimodal concentration distribution.

CONCLUSIONS

Lidar provides very useful descriptions of plume behavior and dimensions, especially under the stable conditions discussed here. These data and those collected by others (e.g., McElroy et al. 1981)

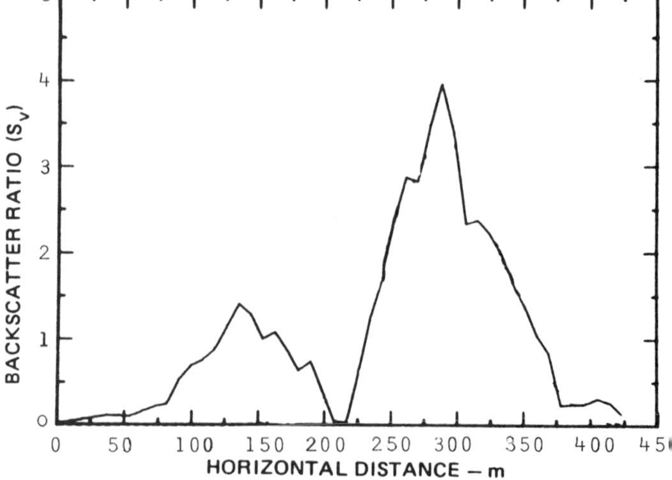

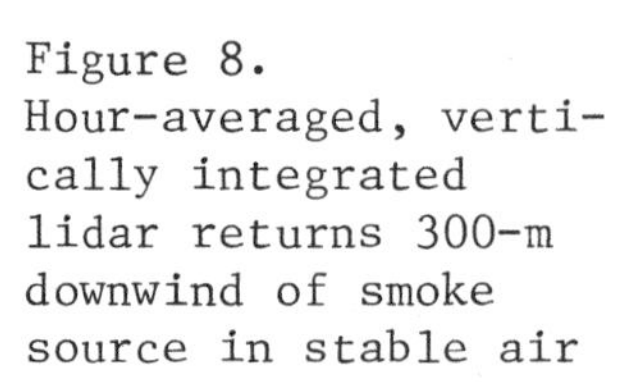
Figure 8. Hour-averaged, vertically integrated lidar returns 300-m downwind of smoke source in stable air

indicate that plumes from large elevated sources tend to have relatively constant thickness for long downwind travel distances when they are released into stable air. The plumes tend to be 50- to 100-m thick to distances of 50 km.

The width of such plumes generally increases with downwind distance, following the pattern described by Pooler and Neimeyer (1970). When wind direction changes with height, the plume tends to spread at a rate that is proportional to the difference between the wind direction at the highest and lowest points of the plume, i.e. the width of the plume is proportional to the product of the downwind distance and the wind direction difference. When little wind shear is present, the width tends to be proportional to the square root of the downwind distance.

The model of meandering plume behavior that has been presented should be amenable to testing. In fact, it may well be possible to test the hypothesis that has been given using only historical data bases. However, to do so would require considerable analysis and manipulation of the primary data sets.

ACKNOWLEDGMENTS

The research described has been funded at least in part with Federal funds from the U.S. Environmental Protection Agency under Contract 68-02-2970. The content of this publication does not necessarily reflect the views or policies of the U.S. Environmental Protection Agency, nor does mention of trade names, commercial products, or organizations imply endorsement by the U.S. Government. We gratefully acknowledge the helpful comments provided by Jason Ching and Francis Pooler of the EPA and by our SRI co-worker Edward Uthe.

REFERENCES

Ching, J., Clarke, J. and Pooler, F., 1979, Methodology to estimate sigmas from STATE data, memo to STATE Tennessee Plume Study participants, EPA Meteorology Division, Research Triangle Park, North Carolina (February).

Clark, R.E., Gillen, M.E., and Tibi, R.M., 1978, "Project STATE--The Tennessee Plume Study, VI. TVA Meteorological Analysis--August 2-18, 1978," Report AQB-I80-9, Tennessee Valley Authority, Division of Natural Resource Services.

Clark, R.E., Gillen, M.E., and Tibi, R.M., 1980, "Project STATE--The Tennessee Plume Study, VII. TVA Meteorological Analysis--August 20-28, 1978," Report AQB-I80/8, Tennessee Valley Authority, Division of Natural Resources Services.

Crawford, T.L., Reisinger, L.M., and Coleman, J.H., 1979, "Project STATE--The Tennessee Plume Study, V. TVA Wind Profile Data," Report I-AQ-78-16, Tennessee Valley Authority, Division of Environmental Planning.

Gifford, F.A., 1959, Statistical properties of a fluctuating plume dispersion model, Advances in Geophysics, 6:117-138.

Gifford, F.A., 1970, "Peak-to-Mean Concentration Ratios According to a 'Top-Hat' Fluctuating Plume Model," Contribution No. 45, Atmospheric Turbulence and Diffusion Laboratory, National Oceanic and Atmospheric Administration, Oak Ridge, Tennessee.

McElroy, J.L., Eckert, J.A., and Hager, C.J., 1981, Airborne down-looking lidar measurements during STATE 78, Atmos. Environ., 15:2223-2230.

Pooler, F., 1981: Personal communication.

Pooler, F. and Niemeyer, L.E., 1970, Dispersion from Tall Stacks: An Evaluation, Proceedings of the 2nd International Clean Air Congress, 1049-1056 (Academic Press, New York).

Thuillier, R.H. and Mancuso, R.L., 1980, "Buildings Effects on Effluent Dispersion from Roof Vents on Nuclear Power Plants," Final Report, Project 1973-1, Electric Power Research Institute.

Turner, D.B., 1969, "Workbook of Atmospheric Dispersion Estimates," Public Health Service Pulication 999-AP-26.

Uthe, E.E., Ludwig, F.L., and Pooler, F., Jr., 1980a, Lidar Observations of the Diurnal Behavior of the Cumberland Power Plant Plume, J. Air Poll. Cont. Assoc., 30:888-893.

Uthe, E.E., Nielsen, N.B., and Jimison, W.L., 1980b, Airborne Lidar Plume and Haze Analyzer (ALPHA-1), Bull. Am. Meteorol. Soc., 61:1035-1043.

Uthe, E.E. and Wilson, W.E., 1979, Lidar Observations of the Density and Behavior of the Labadie Power Plant Plume, Atmos. Environ., 13:1395-1412.

DISCUSSION

E. RUNCA

Did I understand correctly that there is some possibility that close to the plume the long term average concentration is bi-modal with maximum at the edges.

I just want to add that this shape of the concentration profile is reproduced by the telegraph equation.

F. L. LUDWIG

Yes, but this assumes that the amplitude of the meander does not change much with downwind distance while the width of the "instantaneous" plume spreads with the travel.

N.D. VAN EGMOND Did you try to relate the Lidar results with vertically integrated gasburdens, e.g. obtained by Cospecs, in order to arrive at a more quantitative interpretation.

F.L. LUDWIG No, but those data were collected by another organization (EMI) and such comparisons would probably prove interesting. In general the Cospec data were not collected during periods when stable plumes were present because the Cospec is limited to making measurements during daylight hours. According to Dr. E. E. Uthe, some comparisons of daytime cospec and lidar have been made.

B. YAMARTINO You use the often used factor 4.3 to relate plume width to the standard deviation. Has this been examined ? For example, does the lidar defined width correspond to the 10 % concentration boundaries ?

F.L. LUDWIG The choice of the factor 4.3 is arbitrary, but seems to provide values that are consistent with other estimates of σ's when applied to photographs. Otherwise its use seems to be justified largely by convenience and tradition. The lidar representations of plumes in stable air generally have sharp, well defined boundaries that suggest that the Gaussian distribution may not be appropriate so the use of σ's is artificial, but provides a basis for comparison with other studies. I would prefer to use the width and thickness as the primary measures of plume dimensions in stable air.

B. WEISMAN In trying to define the shear width of a plume from vertical wind profiles, are you not able to use plume rise prediction to define the maximum height of the plume ?

F.L. LUDWIG We did not attempt this, but J.Mc Elroy of the US EPA found reasonably good agreement with his airbone lidar observations of plumes from this same plant and the estimates of plume rise obtained from the standard Briggs equations. The suggestion that we use the plume rise formula to define the layer over which shear should be measured is a good one that should be tested.

PARTICIPANTS

List of participants at the 12th International Technical Meeting on Air Pollution Modeling and its Application. Palo Alto, August 25-28, 1981, Menlo Park, California, U.S.A.

AUSTRALIA

Williams P.G.	State Elec. Com of Victoria Herman Research Lab (SECV) Howard Street, Richmond 3121 Victoria

AUSTRIA

Pechinger U.	Zentralanstalt für Met. und Geodynamik Hohe Warte 38 A-1160 Wien
Runca E.	IIASA A-2361 Laxenburg

BELGIUM

Berger A.	Institut d'Astronomie 2, Chemin du Cyclotron 1348 Louvain-la-Neuve
Hecq P.	Institut d'Astronomie et de Géophysique Georges Lemaître Université Catholique de Louvain B-1348 Louvain-la-Neuve
Keyzer P. M., De	State University, Ghent Grote Steenweg Noord, 12 B-9710 Zwijnaarde
Vanderborght B.	Nuclear Energy Research Center SCK/CEN Dept. GKD B-2400 Mol
Wispelaere C., De	Prime Minister's Office for Science Policy Programming Wetenschapsstraat 8 1040 Brussels

CANADA

Atkinson G. B.	Gov't of Canada, Dept. of Envir. Atmospheric Environment Service Room 1000, 266 Graham Avenue Winnipeg, Manitoba 3RC 3V4
Chartier D. R.	Ontario Hydro K-D140 800 Kipling Avenue Toronto, Ontario M8Z554
Cheng L.	Alberta Research Council 700 Terrace Plaza 4445-Calgary Trail South Edmonton, Alberta, T6H 5R7
Christie A. D.	Atmospheric Environment Service 4905 Dufferin Street Downsview, M3H 5T4, Ontario
Lapozak S.	Atmospheric Environment Service 25 St. Clair E., Room 322 Toronto, Ontario, M4T 1M2
Ley B.	Ontario Ministry of Environment 880 Bay Street - 4th Floor Toronto, Ontario
Misra P. K.	Ontario Hydro Research 800 Kipling Avenue, KD140 Toronto, Ontario, M8Z 5S4
Olson M. P.	Atmospheric Environment Service 25 St. Clair E Toronto, Ontario, M4T 1M2
Schaedlich K.	Ontario Hydro 700 University Ave. H4J19 Toronto, Ontario
Shaw R.	Environment Canada 5th Floor, Queens Square 45 Alderney Drive Dartmouth, Nova Scotia B2Y2N6
Thomson R. B.	Alberta Environment 15th Floor, Oxbridge Place 9820-106 Street Edmonton, Alberta, 5TK 2J6

Turner H. E.	Atmospheric Environment Service 25 St. Clair E Toronto, Ontario, M4T 1M2
Weisman B.	MEP Co. 850 Magnetic Drive Downsview, Ontario, M3J 2C4
Wilson H.J.	Transport Canada Tower B. Place de Ville Ottawa, KIA ON8
Young J. W. S.	Atmospheric Enviroment Service 4905 Dufferin Street Donwsview, Ontario, M3H 5T4
DENMARK	
Conradsen K.	Bldg. 349 Tech. Univ. of Denmark DK-2800 Lyngby
Gryning S.-E.	Riso National Lab. Meteorology Section DK-4000 Roskilde
Lyck E.	Danish National Agency of Environmental Protection Air Pollution Lab & Riso Lab DK-4000 Roskilde
Nielsen L. B.	Millostyrelsens Luftforurenings Lab Bygning 306, OEH 2800 Lyngby
FEDERAL REPUBLIC OF GERMANY	
Eppel D.	Institut für Physik GKSS Forschungszent. Geest GmbH 2054 Geesthacht
Frank J.	Rhenisch-Westfalisher TUV Stenbenstrasse 53 4300 Essen 1
Janicke L.	Dornier-System GmbH Postfach 1360 7990 Friedrichshafen 1
Klug W.	Institut für Meteorologie Tech. Hochschule Darmstadt D61 Darmstadt

Morawa Ch. — Umweltbundesamt
Bismarckplatz 1
D-1000 Berlin 33

Rudolf B. — Deutscher Wetterdienst Zentralamt
Postfach 185
6050 Offenbach am Main

Schatzmann M. — Sonderforschungsbereich 80
University of Karlsruhe, SFB 80
Kaiserstrasse 12
D-7500 Karlsruhe

Schultz H. — Universität Hannover, Z.f.S.
Appelstr. 9A, D-3000 Hannover 1

Stern R. — Institut f. Geophysikalische
Wissenschaften
Thielallee 50, 1000 Berlin 33

Vogt K. J. — Kernforschungsanlage Julich GmbH
Postfach 1913
D-5170 Julich

Wengle H. — Institute for Fluid Dynamics
Hochschule der Bundeswehr
8014 Neubiberg

FRANCE

Blondin Ch. — Direction de la Météorologie
FERM/GMA
Rue de Sèvres 73-77
92100 Boulogne-Billancourt

Dinh P., Van — Institut de Physique
Centre de Recherches Atmospheric
Campistrous, 65300 Lannemezan

Mery P. — Electricité de France
6, Quai Watier
78400 CHATOU

Oppenau J. Cl. — Ministère de l'Environnement
14 Bd du Gal Leclerc
92521 Neuilly s/Seine

ISRAEL

Asculai E. — Nuclear Research Center Negev
P.O. Box 9001
Beer-Sheva 84190

Dinar N. — Israel Inst. for Biological Res.
P.O. Box 19
Ness-Ziona

Skibin D. — Nuclear Research Center Negev
P.O. Box 9001
Beer-Sheva

THE NETHERLANDS

Builtjes P. J. H. — MT.N.O.
Postbus 342
7300 AH Apeldoorn

Colenbrander G. W. — Koninklijke/Shell Laboratorium
EE Department
Badhuisweg 3
1031 CM Amsterdam-Noord

Dop H., Van — Royal Netherlands Meteor. Inst.
P.O. Box 201
3720 AE De Bilt

Egmond N. D., Van — Dutch National Inst. of Public Health
P.O. Box 1
Bilthoven

Haan B. J., De — Royal Netherlands Met. Institute
P.O. Box 201
3730 AE De Bilt

Ham J., Van — Study & Info. Centre TNO
P.O. Box 186
2600 AD Delft

Schneider T. — National Inst. of Public Health
P.O. Box 1
3720 BA Bilthoven

Ulden A. P., Van — Royal Netherlands Meteor. Inst.
P.O. Box 201
3730 AE De Bilt

Zwerver S.	Dokter Reyersstraat 12 Leidschendam/Den Haag PB439 - 2260 AK Leidschendam

TURKEY

Ulug S. E.	Chairman, Env. Eng. Dept. Middle East Technical University Ankara

UNITED KINGDOM

Cocks A. T.	CERL Kelvin Avenue Leatherhead, Surrey, KT22 7SE
Ghobadian A.	Mechanical Engineering Depart. Imperial College of Sci. & Tech. Exhibition Road, London SW7 2BX
Goddard A. J. H.	Mechanical Engineering Dept. Imperial Coll. Sci. & Tech. Exhibition Road, London SW7 2BX
Williams M. L.	Warren Spring Laboratory Gunnels Wood Road Stevenage, Herts

UNITED STATES

Bhumralkar C.	SRI International 333 Ravenswood Avenue Menlo Park, California 94025
Bornstein D.	San Jose State University Department of Meteorology San Jose, CA 95192
Carmichael G.	Chemical and Materials Eng. Pro 129 Chemical Bldg University of Iowa Iow City, IA 52248
Davis W. E.	Battelle N.W. 622-R Bldg. - 200 W Box 999 Richland, WA 99352

Dennis R. L. — Nat. Center for Atmos. Research
P.O. Box 3000
Boulder, Colorado 80307

Endlich R. — Atmospheric Science Center
SRI International
333 Ravenswood Avenue
Menlo Park, California 94025

Ermak D. — Lawrence Livermore Laboratory
P.O. Box 808
Livermore, CA 94550

Goodin W. — Dames & Moore
1100 Glendon Avenue
Los Angeles, Ca. 90024

Hakkarinen C. — EPRI
P.O. Box 10412
Palo Alto, Ca. 94303

Havens J. A. — Dept. of Chemical Engineering
University of Arkansas
Fayetteville, Arkansas 72701

Heffner D. — Chevron Research
576 Standard Avenue
Richmond, California 94802

Irwin J. S. — U.S. Environmental Protection Agency
Mail Drop 80
Research Triangle Park, N.C. 27711

Johnson W. B. — Atmospheric Science Center
SRI International
333 Ravenswood Avenue
Menlo Park, CA 94025

Larson R. — Chevron Research
576 Standard Avenue
Richmond, CA. 94802

Lee H. N. — Department of Meteorolgy
The University of Utah
819 Wm. C. Browning Bldg.
Salt Lake City, Utah 84112

Lee R.	L-262, P.O. Box 808 Lawrence Livermore National Lab Livermore, CA 94550
Liu M. K.	Systems Applications, Inc. 101 Lucas Valley Road San Rafael, CA 94903
Livingston J.	Atmospheric Science Center SRI International 333 Ravenswood Avenue Menlo Park, CA 94025
Londergan R. J.	TRC - Environmental Consultants 125 Silas Deane Highway Wethersfield, CT 06109
Ludwig F. L.	Atmospheric Science Center SRI International 333 Ravenswood Avenue Menlo Park, CA 94025
Martinez J. R.	Atmospheric Science Center SRI International 333 Ravenswood Avenue Menlo Park, CA 94025
Maxwelle C.	Atmospheric Science Center SRI International 333 Ravenswood Avenue Menlo Park, CA 94025
Mayerhofer P.	Atmospheric Science Center SRI International 333 Ravenswood Avenue Menlo Park, CA 94025
McBride W. C.	Energy Resources Co., Inc. 3344 N. Torrey Pines Court La Jolla, CA 92037
Michael P.	Brookhaven National Laboratory Upton, New Yrok 11973
Middleton P.	National Center for Atmospheric Research (NCAR) P.O. Box 3000 Boulder, Colorado 80307

Morgan D.	P.O. Box 808 Lawrence Livermore National Lab. Livermore, CA 94550
Murphy B. D.	Oak Ridge National Lab. Computer Sciences Division Union Carbide Corp. Nuclear Div. Oak Ridge, Tennessee 37830
Niemann B.	U.S. Environmental Protection Agency 609 Compton Road Raleigh, NC 27609
Perardi T. E.	Bay Area Air Quality Mgmt. Dist. 939 Ellis Street San Francisco, CA 94109
Pfenning D.	Energy Analysts 2001 Priestley Avenue Norman, Oklahoma 73069
Powell G. L.	Environmental Services Dept. Salt River Project P.O. Box 1980 Phoenix, Arizona 85001
Robinson L. H.	Bay Area Air Quality Mgmt. Dist. 939 Ellis Street San Francisco, CA 94109
Roth Ph.	Systems Applications, Inc. 950 Northgate Drive San Rafael, CA 94903
Ruff R. E.	Atmospheric Science Center SRI International 333 Ravenswood Avenue Menlo Park, CA 94025
Sabnis J.	Energy Resources Co., Inc. 3344 N. Torrey Pines Court La Jolla, CA 92037
Samson P. J.	Dept. of Atmos. & Oceanic Sc. University of Michigan 2218 Space Research Bldg. Ann Arbor, MI 48109

Shannon J.	RER, Bldg. 181 Argonne National Laboratory Argonne, Illinois 60439
Shelar G.	Pacific Gas & Electric Compangy 245 Market Street San Francisco, California 94106
Taft J. R.	Deygon-Ra, Inc. P.O. Box 3227 La Jolla, CA 92038
Touma J. S.	Northwest Energy Services Co. P.O. Box 1090 Kirkland, WA 98033
Venkatram A.	ERT Inc. 3 Militia Dirve Lexington, MA 02173
Vogel D.	Pacific Gas & Electric Co. 245 Market Street San Francisco, CA 94106
Wang H.	Dames & Moore 1550 Northwest Highway Park Ridge, Illinois 60068
Warner T. T.	The Pennsylvania State Univ. 503 Walker Bldg. Dept. of Meteo. University Park, PA 16802
Westbrook J. K.	Department of Soil Sci. & Biometeorology Utah State University UMC48 Logan, Utha 84322
Whitlatch W. F.	Pacific Gas & Electric 245 Market Street San Francisco, CA 94106
Whitten G. Z.	Systems Applications, Inc. 950 Northgate Drive San Rafael, CA 94903
Woodward J. L.	Exxon Res. & Eng. Company P.O. Box 153 Forham Park, New Yersey 07932

Yamartino B.
ERT Inc.
3 Militia Drive
Lexington, Mass. 02173

AUTHORS INDEX

SUBJECT INDEX